Multinary Alloys Based on IV-VI and IV-VI_2 Semiconductors

IV–VI and IV–VI_2 semiconductors have attracted considerable attention due to their applications in the fabrication of electronic and optoelectronic devices as infrared lasers and detectors. The electrical properties of these semiconductors can also be tuned by adding impurity atoms. Because of their wide application in various devices, the search for new semiconductor materials and the improvement of existing materials is an important field of study. Doping with impurities is a common method of modifying and diversifying the physical and chemical properties of semiconductors. This book covers all known information about the phase relations in multinary systems based on IV–VI and IV–VI_2 semiconductors, providing the first systematic account of phase equilibria in multinary systems based on IV–VI and IV–VI_2 semiconductors and making research originally published in Ukrainian and Russian accessible to the wider scientific community. This book will be of interest to undergraduate and graduate students studying materials science, solid state chemistry, and engineering. It will also be relevant for researchers at industrial and national laboratories, in addition to researchers of phase equilibria, inorganic chemists, and solid state physicists.

Key Features:

- Provides up-to-date experimental and theoretical information.
- Allows readers to synthesize semiconducting materials with predetermined properties.
- Delivers a critical evaluation of many industrially important systems presented in the form of two-dimensional sections for the condensed phases.

Vasyl Tomashyk is a leading researcher at the V.Ye. Lashkaryov Institute of Semiconductor Physics of the National Academy of Sciences of Ukraine. He obtained a master's degree in chemistry from Chernivtsi State University in 1972. He completed his doctorate in chemical sciences (1992) and is a professor (1999) and author of about 700 publications in scientific journals and conference proceedings and 13 books (eight of them were published by CRC Press), which are devoted to the physical-chemical analysis, the chemistry of semiconductors, and chemical treatment of semiconductor surfaces.

Multinary Alloys Based on IV-VI and IV-VI_2 Semiconductors

Vasyl Tomashyk

V.Ye. Lashkaryov ŁInstitute of Semiconductor Physics of the National Academy of Sciences of Ukraine

CRC Press is an imprint of the
Taylor & Francis Group, an **informa** business

Front cover image: Vasyl Tomashyk

First edition published 2025
by CRC Press
2385 NW Executive Center Drive, Suite 320, Boca Raton FL 33431

and by CRC Press
4 Park Square, Milton Park, Abingdon, Oxon, OX14 4RN

CRC Press is an imprint of Taylor & Francis Group, LLC

ISBN: 978-0-367-63927-3 (hbk)
ISBN: 978-0-367-64217-4 (pbk)
ISBN: 978-1-003-12346-0 (ebk)

DOI: 10.1201/9781003123460

Typeset in Times
by Newgen Publishing UK

This book is dedicated to the 150th anniversary of my alma mater
Yuriy Fedkovych Chernivtsi National University, Ukraine (founded in 1875)

Contents

Preface

A significant volume of semiconductor devices and circuits employs IV–VI and IV–VI_2 semiconductors, the most commonly used crystal material for integrated circuits. The IV–VI semiconductors are among the most interesting materials in solid state physics. Many of them crystallize as a rock-salt structure, and structural transitions are common in them. The most widely studied compounds in this group are PbTe, PbSe, PbS, SnTe, and GeTe. These materials have small gaps, which are usually less than 0.5 eV; hence they are good candidates for devices like infrared lasers and detectors. Despite their simple crystal structure, some of these compounds exhibit ferroelectric, paraelectric, and superconducting behavior. In addition, the temperature dependence of the energy gaps, the energy positions of impurity levels, the high doping levels found, the static dielectric constants, and the electronic structure of some of the alloys of these compounds appear to be anomalous compared to the "conventional behavior" of the diamond and zinc-blend semiconductors. These semiconductors either have already found use or are promising materials for infrared sensors and sources, thermoelectric elements, solar cells, memory elements, among others. The basic characteristics of these compounds, namely, narrow band gap, high permittivity, relatively high radiation resistance, high mobility of charge carriers, and high bond ionicity, are unique among semiconductor substances.

Though ternary and quaternary phase equilibria based on Si, Ge, Sn and Pb chalcogenides have been published as handbooks by V. Tomashyk – *Ternary Alloys Based on IV-VI and IV-VI_2 Semiconductors* (Taylor & Francis. 2022) and *Quaternary Alloys Based on IV-VI and IV-VI_2 Semiconductors* (Taylor & Francis, 2023) – data pertaining to phase equilibria in multinary systems based on these semiconductors appear dispersed in scientific literature. This reference book is intended to describe and illustrate the up-to-date experimental and theoretical information about phase relations based on IV–VI and IV–VI_2 semiconductors systems with five and more components. The book critically evaluates many industrially significant systems presented in the form of a two-dimensional section for the condensed phases.

In most cases, the properties of IV–VI and IV–VI_2 semiconductors can be modified by doping them with isovalent or heterovalent foreign impurities or by interaction with other binary compounds, which can lead to the formation of a solid solution or the formation of multinay compounds. Such materials have expanded and improved the properties of semiconductors that can be used to produce work items to create new electronic devices.

The book contains also practically all multinary compounds, including more than 400 minerals, which contain Si, Ge, Sn, or Pb and S, Se, or Te, as these materials could be used as precursors for obtaining the nanosized IV–VI and IV–VI_2 semiconductors.

The present book is aimed to collect and systematize all available data on the IV–E_1–...–E_n–VI multinary systems. It contains the results for about 663 multinary systems based on IV–VI and IV–VI_2 semiconductors illustrated with 38 figures. The information is divided into 12 chapters according to the number of possible combinations of Si, Ge, Sn, or Pb with S, Se, or Te. The chapters are structured first in order of increasing ordinal number of elements of group IV in the Periodic Table, that is, from Si to Pb compounds, and then in order of increasing ordinal number of chalcogen, that is, from sulfides to tellurides. The same principle is used to further describe the systems in each section, that is, in order of increasing ordinal numbers of the third and fourth components in the Periodic Table.

Missing literature and new literature data on ternary and quaternary systems based on IV–VI and IV–VI_2 semiconductors are listed in Appendices A and B, respectively. Appendix A contains new results on 95 ternary systems and 29 figures, while Appendix B contains new data on 225 quaternary systems and 43 figures.

Most of the figures are presented in their original form, although in some are, minor corrections have been made. If the published data varied essentially, several versions are presented in comparison. The content of system components is presented in mol% (this is not indicated in the figures). If the original phase diagram is given with mass%, this is indicated in the figure. The book includes the literature data from 1351 papers.

The book is meant for researchers at industrial and national laboratories and for university and graduate students who study materials science, phase equilibria in the systems based on the IV–VI and IV–VI_2 semiconductors, solid state chemistry, semiconductor chemistry, and engineering. It is also suitable for researchers in the field of phase relations, inorganic chemists, and solid state physicists.

About the Author

Vasyl Tomashyk is a leading researcher at the V.E. Lashkaryov Institute of Semiconductor Physics of the National Academy of Sciences of Ukraine. He obtained a master's degree in chemistry from Chernivtsi State University in 1972. He completed his doctorate in chemical sciences (1992) and is a professor (1999) and author of about 700 publications in scientific journals and conference proceedings and 13 books (eight of them were published by CRC Press), which are devoted to the physical-chemical analysis, the chemistry of semiconductors and chemical treatment of semiconductor surfaces.

Tomashyk is a specialist in the field of solid state and semiconductor chemistry, including physical-chemical analysis and technology of semiconductor materials, at the international level. He was head of research topics within the International program "Copernicus". He is a member of Materials Science International Team (Stuttgart, Germany, since 1999), which prepares a series of prestigious reference books under the title *Ternary Alloys* and published 14 chapters in this series and 35 chapters in Landolt–Börnstein New Series. Tomashyk is actively working with young researchers and graduate students and under his supervision 22 PhD theses have been already completed. For many years, he has been a professor in the Ivan Franko Zhytomyr State University in Ukraine.

Symbols and Acronyms

DFT	density functional theory
DMF	dimethylformamide
DSC	differential scanning calorimetry
DTA	differential thermal analysis
EDX	energy dispersive X-ray spectroscopy
EMF	electromotive force
EPMA	electron probe microanalysis
FTIR	Fourier-transform infrared spectroscopy
M	mol
MeOH	methanol
mM	mmol
ppm	parts per million
rpm	revolutions per minute
SEM	scanning electron microscopy
THF	tetrahydrofuran
(X)	solid solution based on X
XRD	X-ray diffraction

1

Systems Based on Silicon Sulfides

1.1 Silicon–Hydrogen–Sodium–Potassium–Calcium–Magnesium–Zinc–Aluminum–Oxygen–Sulfur

In the system containing these elements, the $(Na,K)_4Ca_2(Mg,Zn)_3Al_8(SO_4)_{10}(SiO_4)_2{\cdot}40H_2O$ multinary compound (mineral chessexite), which crystallizes as an orthorhombic structure with the lattice parameters a = 1370, b = 2796, c = 999 pm and a calculated density of 2.21 g·cm^{-3}, is formed (Sarp and Deferne 1982; Dunn et al. 1984a).

1.2 Silicon–Hydrogen–Sodium–Potassium–Calcium–Barium–Aluminum–Oxygen–Sulfur

In the system containing these elements, the multinary compound $(Ba,K)_4(Ca,Na)_6[(Si,Al)_{20}O_{39}(OH)_2](SO_4)_3{\cdot}0.5H_2O$ (mineral wenkite), which crystallizes as a hexagonal structure with the lattice parameters a = 1351.5, c = 746.5 pm, and the calculated and experimental densities of 3.16 and 3.10 g·cm^{-3}, respectively (Wenk 1973; Lee 1974) [a = 1352.8 ± 0.3, c = 747.1 ± 0.2 pm, and the calculated and experimental densities of 3.195 and 3.19 ± 0.01 g·cm^{-3}, respectively (Papageorgakis 1962; Fleischer 1963); a = 1351.17 ± 0.13 and 1351.46 ± 0.19, and c = 746.15 ± 0.11 and 746.49 ± 0.17 pm for two different crystals (Wenk 1966); a = 1337 ± 2, c = 770 ± 2 pm, and the calculated and experimental densities of 3.24 and 3.195 g·cm^{-3}, respectively (Yanulova et al. 1971); the calculated density is 3.21 g·cm^{-3} (Merlino 1974)], is formed.

1.3 Silicon–Hydrogen–Sodium–Potassium–Calcium–Aluminum–Carbon–Oxygen–Sulfur

In the system containing these elements, the multinary compound $K(Ca,Na)_6(Si,Al)_{10}O_{22}[SO_4,CO_3,(OH)_2]{\cdot}H_2O$ (mineral tuscanite), which crystallizes as a monoclinic structure with the lattice parameters a = 2403.6 ± 1.4, b = 511.0 ± 0.3, c = 1088.8 ± 0.8 pm, β = 106.95 ± 0.03°, and the calculated and experimental densities of 2.77 and 2.83 g·cm^{-3}, respectively (Orlandi et al. 1977) [a = 2403 ± 2, b = 511 ± 1, c = 1088 ± 2 pm, and β = 106.94 ± 0.09° (Mellini et al. 1977)], is formed.

1.4 Silicon–Hydrogen–Sodium–Potassium–Calcium–Aluminum–Oxygen–Sulfur

In the system containing these elements, some multinary compounds are formed. $Na_5K_{1.5}Ca(Al_6Si_6O_{24})(SO_4)(OH)_{0.5}{\cdot}H_2O$ (mineral alloriite) crystallizes as a tetragonal structure with the lattice parameters

DOI: 10.1201/9781003123460-1

$a = 1289.2 \pm 0.3$, $c = 2134.0 \pm 0.5$ pm, and the calculated and experimental densities of 2.358 and 2.35 g·cm^{-3}, respectively (Chuкanov et al. 2007; Rastsvetaeva et al. 2007; Poirier et al. 2009).

$Na_5K_2Ca[Al_6Si_6O_{24}](S_5)(SH)$ (mineral sulfhydrylbystrite) crystallizes as a trigonal structure with the lattice parameters $a = 1295.67 \pm 0.06$, $c = 1077.11 \pm 0.05$ pm, and the calculated and experimental densities of 2.368 and 2.391 ± 0.001 g·cm^{-3}, respectively (Sapozhnikov et al. 2015, 2017; Belakovskiy et al. 2017b).

$(Na,K)_6Ca_2(Al_6Si_6O_{24})(SO_4)_2 \cdot 0.5H_2O$ (mineral franzinite) crystallizes as a trigonal structure with the lattice parameters $a = 1291.6 \pm 0.1$ and $c = 2654.3 \pm 0.3$ pm (Ballirano et al. 2000; Jambor and Roberts 2001) [$a = 1288.4 \pm 0.9$, $c = 2658.0 \pm 2.1$ pm, and the calculated and experimental densities of 2.52 and 2.49 g·cm^{-3}, respectively (Fleischer et al. 1977b); $a = 1286.1 \pm 0.5$, 1290.6 ± 0.5, $c = 2662 \pm 3$, 2645 ± 2 pm, and the calculated and experimental densities of 2.57, 2.55 and 2.46 ± 0.02, 2.52 ± 0.02 g·cm^{-3}, respectively, for two different samples (Leoni et al. 1979)].

$(Na,Ca,K)_8(Al_6Si_6O_{24})(SO_4)_2(OH)_{0.5} \cdot H_2O$ (mineral biachellaite) crystallizes as a trigonal structure with the lattice parameters $a = 1291.3 \pm 0.1$, $c = 7960.5 \pm 0.5$ pm, and the calculated and experimental densities of 2.52 and 2.51 ± 0.01 g·cm^{-3}, respectively (Chuкanov et al. 2008, 2009; Rastsvetaeva and Chukanov 2008).

$(Na,Ca,K)_{56}(Al_6Si_6O_{24})_7(SO_4)_{12} \cdot 6H_2O$ (mineral farneseite) crystallizes as a hexagonal structure with the lattice parameters $a = 1287.84 \pm 0.02$, $c = 3700.78 \pm 0.12$ pm, and a calculated density of 2.425 g·cm^{-3} (Cámara et al. 2005; Piilonen et al. 2006).

$[Na_{82.5}Ca_{33}K_{16.5}](Si_{99}Al_{99}O_{396})(SO_4)_{33} \cdot 6H_2O$ (mineral fantappièite) crystallizes as a trigonal structure with the lattice parameters $a = 1287.42 \pm 0.06$, $c = 8721.5 \pm 0.3$ pm, and a calculated density of 2.471 g·cm^{-3} at 100 K (Cámara et al. 2010).

$(Na_{90}Ca_{36}K_{18})(Si_{108}Al_{108}O_{432})(SO_4)_{36} \cdot 6H_2O$ (mineral kircherite) also crystallizes as a trigonal structure with the lattice parameters $a = 1287.70 \pm 0.07$, $c = 9524.4 \pm 0.6$ pm, and a calculated density of 2.383 g·cm^{-3} (Cámara et al. 2012).

1.5 Silicon–Hydrogen–Sodium–Potassium–Calcium–Aluminum–Oxygen–Fluorine–Sulfur

In the system containing these elements, the multinary compound $Na_3(K_{17}Ca_7)Ca_4(Al_{24}Si_{24}O_{96})(SO_3)_6F_6 \cdot 4H_2O$ (mineral steudelite), which crystallizes as a hexagonal structure with the lattice parameters $a = 1289.53 \pm 0.01$ and $c = 2127.78 \pm 0.03$ pm, is formed (Chukanov et al. 2021d,e).

1.6 Silicon–Hydrogen–Sodium–Potassium–Calcium–Aluminum–Oxygen–Fluorine–Chlorine–Sulfur

Two multinary compounds are formed in this system. $(Na_{61}Ca_{32}K_{19})(Al_6Si_6O_{24})_{14}(SO_4)_{26}Cl_2F_6 \cdot 2H_2O$ (mineral sacrofanite) crystallizes as a hexagonal structure with the lattice parameters $a = 1289.857 \pm 0.014$ and $c = 7425.47 \pm 0.11$ pm at 30°C (Ballirano 2018) [$a = 1286.5$, $c = 7224.0$ pm, and the calculated and experimental densities of 2.446 and 2.423 g·cm^{-3}, respectively (Cabri et al. 1981); $a = 1289.45 \pm 0.04$ and $c = 7421.28 \pm 0.37$ pm (Ballirano et al. 1995); $a = 1290.3 \pm 0.2$ and 1288.57 ± 0.05, and $c = 7428.4 \pm 0.8$ and 7419.2 ± 0.3 pm for two different samples (Ballirano and Bonaccorsi 2005; Bonaccorsi et al. 2012]. The temperature dependence of the lattice parameters within the interval of 30°C–600°C are described by the equations: $a = (1291.7 \pm 0.5) - (2.3 \pm 0.3) \cdot 10^{-4}\,T + (7.0 \pm 0.5) \cdot 10^{-7}\,T^{-2} - (4.8 \pm 0.3) \cdot 10^{-10}\,T^3$ pm and $c = (7444 \pm 3) - (1.92 \pm 0.16) \cdot 10^{-3}\,T + (5.3 \pm 0.3) \cdot 10^{-6}\,T^2 - (3.37 \pm 0.16) \cdot 10^{-9}\,T^3$ pm (all temperatures are in K) (Ballirano 2018).

$(Na,K)_{10}Ca_5Al_6Si_{32}O_{80}(Cl_2,F_2,SO_4)_3 \cdot 18H_2O$ (mineral delhayelite) crystallizes as an orthorhombic structure with the lattice parameters $a = 652 \pm 1$, $b = 2483 \pm 6$, and $c = 707 \pm 1$ pm (Sokolova et al. 2005) [$a = 653 \pm 3$, $b = 2465 \pm 2$, $c = 704 \pm 3$ pm, and an experimental density of 2.60 ± 0.03 g·cm^{-3} (Fleischer 1959; Sahama and Hytönen 1959); $a = 2486 \pm 1$, $b = 707 \pm 2$, and $c = 653 \pm 1$ pm (Cannillo et al. 1970)].

1.7 Silicon–Hydrogen–Sodium–Potassium–Calcium–Aluminum–Oxygen–Chlorine–Sulfur

Three multinary compounds are formed in this system. $Na_{2.53}K_2Ca_{2.73}(Al_6Si_6O_{24})(SO_3)_{0.5}[(SO_3)_{0.47}](OH)_{0.99}Cl_{0.30}\cdot 0.85H_2O$ (natural analog of the mineral alloriite) crystallizes as a hexagonal structure with the lattice parameters $a = 1289.5 \pm 0.2$ and $c = 2127.6 \pm 0.4$ pm (Chukanov et al. 2021a).

$(Na,Ca,K)_8Al_6Si_6O_{24}(SO_4)_2Cl\cdot H_2O$ (mineral tounkite) crystallizes as a trigonal structure with the lattice parameters $a = 1275.7 \pm 0.3$, $c = 3221.1 \pm 0.5$ pm and a calculated density of 2.48 $g\cdot cm^{-3}$ (Rozenberg et al 2004) [in the hexagonal structure with the lattice parameters $a = 1284.3 \pm 0.3$, $c = 3223.9 \pm 0.8$ pm, and the calculated and experimental densities of 2.60 and 2.557 ± 0.004 $g\cdot cm^{-3}$, respectively (Ivanov et al. 1992; Jambor and Grew 1994a)].

$(Na,K)_{42}Ca_6(Al_6Si_6O_{24})_6(SO_4)_8Cl_2\cdot 3H_2O$ (mineral marinellite) crystallizes as a trigonal structure with the lattice parameters $a = 1288.0 \pm 0.2$, $c = 3176.1 \pm 0.6$ pm and the calculated and experimental densities of 2.40 and 2.405 ± 0.005 $g\cdot cm^{-3}$, respectively (Bonaccorsi and Orlandi 2003; Jambor and Roberts 2004).

1.8 Silicon–Hydrogen–Sodium–Potassium–Aluminum–Carbon–Oxygen–Sulfur

In the system containing these elements, the multinary compound $(Na,K)_8(Al_6Si_6O_{24})(SO_4,CO_3)\cdot 2H_2O$ (mineral vishnevite), which crystallizes as a hexagonal structure with the lattice parameters $a = 1272.28 \pm 0.03$ and $c = 519.80 \pm 0.03$ pm (Della Ventura et al. 2007) [$a = 1258.2 \pm 0.4$ and $c = 510.5 \pm 0.2$ pm (Semenov et al. 1984; Jambor et al. 1988); $a = 1278.9$ and $c = 523.6$ pm (Sosedko et al. 1989); $a = 1283.9 \pm 0.5$, $c = 527.2 \pm 0.1$ pm and a calculated density of 2.43 $g\cdot cm^{-3}$ for the mineral enriched by potassium (Pushcharovkiy et al. 1989)], is formed.

1.9 Silicon–Hydrogen–Sodium–Potassium–Aluminum–Oxygen–Sulfur

Two multinary compounds are formed in this system. $Na_6K_2(Al_6Si_6O_{24})(SO_4)\cdot 2H_2O$ (mineral pitiglianoite) crystallizes as a hexagonal structure with the lattice parameters $a = 1282.7 \pm 0.2$ and $c = 526.9 \pm 0.1$ pm at 25°C (Bonaccorsi et al. 2007) [$a = 2212.1 \pm 0.3$, $c = 522.1 \pm 0.1$ pm, and the calculated and experimental densities of 2.394 and 2.37 ± 0.04 $g\cdot cm^{-3}$, respectively (Merlino et al. 1991); $a = 2223.0 \pm 0.1$ and $c = 526.79 \pm 0.05$ pm (Bonaccorsi and Orlandi 1996)]. Upon heating, the cell parameters increased in the range 25°C–226°C, following paths that were best fitted by the following equations: $a = 12.821 + 4\cdot 10^{-4}\,T - 1\cdot 10^{-6}\,T^2$ A and $c = 5.264 + 2\cdot 10^{-4}\,T - 6\cdot 10^{-7}\,T^2$ A (all temperatures are in °C) (Bonaccorsi et al. 2007). In the range 226°C–403°C, there was a strong decrease in the values of both a and c, while for $T > 403$°C, there was a new expansion of the structure. The Fourier transform infrared (FTIR) data showed a major loss of H_2O in the range 200°C–400°C.

$Na_{15}K_9(Al_6Si_6O_{24})_3(OH)_2(SO_4)_2\cdot 7\text{–}8H_2O$ also crystallizes as a hexagonal structure with the lattice parameters $a = 2213.8$ and $c = 524.8$ pm (Klaska and Jarchow 1977). This compound was obtained in an alkaline solution at the hydrothermal conditions of 425°C to 525°C and at 70–90 MPa in addition to $Na_8(OH)_2(Al_6Si_6O_{24})\cdot 5H_2O$.

1.10 Silicon–Hydrogen–Sodium–Calcium–Aluminum–Carbon–Oxygen–Sulfur

In the system containing these elements, the multinary compound $Na_{28}Ca_4(Si_4Al_4O_{16})_6(SO_4)_6(S_6)_{1/3}(CO_2)\cdot 2H_2O$ (mineral slyudyankaite), which crystallizes as a triclinic structure with the lattice parameters $a = 905.23 \pm 0.04$, $b = 1288.06 \pm 0.06$, $c = 2568.1 \pm 0.1$ pm, $\alpha = 89.988 \pm 0.002°$, $\beta = 90.052 \pm 0.001°$, and $\gamma = 90.221 \pm 0.001°$ (Sapozhnikov et al. 2022a,b), is formed.

1.11 Silicon–Hydrogen–Sodium–Calcium–Aluminum–Oxygen–Chlorine–Sulfur

Two multinary compounds are formed in this system. $Na_6Ca_2(Al_6Si_6O_{24})(SO_4,S,Cl,OH)_2$ (mineral lazurite) has four polymorphic modifications (Kaneva et al. 2011). The first of them crystallizes as a triclinic structure with the lattice parameters $a = 906.7 \pm 0.3$, $b = 1289.6 \pm 0.3$, $c = 2570.8 \pm 0.6$ pm, $\alpha = 89.98 \pm 0.02°$, $\beta = 90.08 \pm 0.02°$, and $\gamma = 90.22 \pm 0.02°$ (Evsyunin et al. 1997) [$a = 909.1$, $b = 1285.7$, $c = 2571.9$ pm, $\alpha = \beta = \gamma = 90°$ (Sapozhnikov 1990; Jambor and Puziewicz 1991)].

The second modification crystallizes as an orthorhombic structure with the lattice parameters $a = 906.6$, $b = 1285.1$, and $c = 3854.9$ pm (Kaneva et al. 2011) [$a = 905.3 \pm 0.3$, $b = 1283.7 \pm 0.3$, $c = 3844.5 \pm 1.0$ pm, and a calculated density of 2.44 g·cm^{-3} (Evsyunin et al. 1998); $a = 907.2$, $b = 1283.0$, $c = 3849$ pm (Sapozhnikov 1992)].

The third modification of this mineral crystallizes as a cubic structure with the lattice parameter $a = 907.7 \pm 0.1$ pm and a calculated density of 2.306 g·cm^{-3} (Rastsvetaeva et al. 2002) [$a = 907$, 907.5 and 908.8 pm for three different samples (Hogarth and Griffin 1976); $a = 910.5 \pm 0.2$ and 905.4 ± 0.01 pm and a calculated density of 2.39 and 2.42 g·cm^{-3} for two different samples (Hassan et al. 1985); $a = 907.2$ pm (Sapozhnikov 1990; Jambor and Puziewicz 1991)].

The fourth modification of lazurite crystallizes as a monoclinic structure with the lattice parameters $a = 3636$, $b = c = 5140$ pm, and $\beta = 90°$ (Sapozhnikov 1990; Jambor and Puziewicz 1991).

The lazurite formula has now been redefined to $Na_7Ca(Al_6Si_6O_{24})(SO_4)(S_3)^{-}\cdot H_2O$ (Miyawaki et al. 2021a,b). It crystallizes as a cubic structure with the lattice parameter $a = 908.7 \pm 0.3$ pm (Sapozhnikov et al. 2021).

$Na_6Ca_2(Al_6Si_6O_{24})(SO_4,S_3,S_2,Cl)_2\cdot H_2O$ (mineral vladimirivanovite) crystallizes as an orthorhombic structure with the lattice parameters $a = 906.6 \pm 0.3$, $b = 1285.1 \pm 0.3$, $c = 3855.8 \pm 1.0$ pm, and the calculated and experimental densities of 2.436 and 2.48 ± 0.03 g·cm^{-3}, respectively (Sapozhnikov et al. 2011a,b, 2012; Belakovskiy et al. 2013).

1.12 Silicon–Hydrogen–Sodium–Calcium–Titanium–Niobium–Oxygen–Fluorine–Sulfur

In the system containing these elements, the multinary compound $(Na,Ca)_5Ca(Ti,Nb)_5(Si,S)_{12}O_{34}(OH,F)_8.5H_2O$ (mineral haineaultite), which crystallizes as an orthorhombic structure with the lattice parameters $a = 720.4 \pm 0.4$, $b = 2315.5 \pm 0.5$, $c = 695.3 \pm 0.2$ pm, and a calculated density of 2.28 g·cm^{-3}, is formed (McDonald and Chao 2004; Jambor and Roberts 2005).

1.13 Silicon–Hydrogen–Sodium–Aluminum–Carbon–Oxygen–Sulfur

In the system containing these elements, the multinary compound $Na_{7.88}(Si_6Al_6O_{24})(CO_3)_{0.73}(S_3)_{0.30}\cdot 1.32H_2O$ crystallizes as a cubic structure with the lattice parameter $a = 899.73 \pm 0.01$ pm and a calculated density of 2.249 g·cm^{-3}, is formed in this system (Climent-Pascual et al. 2009). It was synthesized as follows. A homogeneous mixture of nitrite sodalite (1 g), anhydrous sodium carbonate (0.8 g), sulfur precipitate (1 g), and activated charcoal (0.1 g) was heated gradually up to 700°C in a sealed alumina crucible and allowed to stand for 12 h. The obtained greenish-blue powder of the title compound was washed with distilled water and dried at 80°C for 6 h.

1.14 Silicon–Hydrogen–Sodium–Aluminum–Oxygen–Sulfur

Two multinary compounds are formed in this system. $Na_8(Al_6Si_6O_{24})(SO_4)\cdot H_2O$ (mineral nosean) crystallizes as a cubic structure with the lattice parameter $a = 908.4 \pm 0.2$ pm and a calculated density of 2.21 g·cm^{-3} (Hassan and Grundy 1989).

$Na_8(Al_6Si_6O_{24})(HS)_2$ (mineral sapozhnikovite) also crystallizes as a cubic structure with the lattice parameter a = 891.462 ± 0.007 pm and the calculated and experimental densities of 2.255 and 2.25 ± 0.01 g·cm^{-3}, respectively (Chukanov et al. 2021b,c, 2022). The synthetic analog of sapozhnikovite with the formula $Na_{7.27}Al_{6.06}Si_{5.94}O_{24}(HS)_{1.21}$ was synthesized using highly dispersed powder of metakaolin with $Na_2S \cdot 10H_2O$ (3 g) and 8 M NaOH (5 ml) aqueous solution (Shchipalkina et al. 2023). The components were loaded to the hydrothermal reactor with 25 ml polypropylene in a bottle and heated for 18 h at 230°C.

According to Shchipalkina et al. (2023), $Na_8(Al_6Si_6O_{24})(HS)_2$ undergoes a reversible phase transition at 950°C, drastic increase of linear expansion coefficient at 900°C, and is characterized by non-elastic behavior after cooling. The temperature dependencies of the unit-cell parameters were divided into two parts and described by quadratic a = (891.1 ± 1.2) – (0.0018 ± 0.0033)T + (0.0000138 ± 0.0000022)T^2 (from room temperature to 900°C) and linear a = (776.5 ± 1.3) + (0.1130 ± 0.0011)T (from 900°C to 1000°C) polynomial functions.

1.15 Silicon–Hydrogen–Potassium–Magnesium–Barium–Aluminum–Oxygen–Iron–Sulfur

In the system containing these elements, the multinary compound $(Ba,K)(Fe^{2+},Mg)_3((Si,Al,Fe)_4O_{10}(S,OH)_2$ (mineral anandite), which has two polytypes, is formed. The first polytypes (anandite-2O) crystallizes as an orthorhombic structure with the lattice parameters a = 543.9 ± 0.1, b = 950.9 ± 0.2, and c = 1987.8 ± 0.6 pm (Filut et al. 1985) [a = 546.8 ± 0.9, b = 948.9 ± 1.8, c = 1996.3 ± 1.1 pm, a calculated density of 4.04 g·cm^{-3} (Giuseppetti and Tadini 1972)]. Second polytypes (anandite-2M) crystallizes as a monoclinic structure with the lattice parameters a = 544.31 ± 0.03, b = 947.19 ± 0.06, c = 2004.2 ± 0.1 pm, and β = 95.046 ± 0.001° (Bujnowski et al. 2009) [a = 541.2 ± 0.5, b = 943.4 ± 0.5, c = 1995.3 ± 1.0 pm, β = 94°52' ± 10', and an experimental density of 3.94 g·cm^{-3} (Fleischer 1967b; Pattiaratchi et al. 1967)].

1.16 Silicon–Hydrogen–Copper–Zinc–Oxygen–Sulfur

In the system containing these elements, the multinary compound $CuZn_7(OH)_{13}[SiO(OH)_3SO_4]$ (mineral bechererite), which crystallizes as a trigonal structure with the lattice parameters a = 831.9 ± 0.2, c = 737.7 ± 0.3 pm, and the calculated and experimental densities of 3.51 and 3.45 ± 0.05 g·cm^{-3}, respectively, is formed (Giester and Rieck 1996; Hoffmann et al. 1997).

1.17 Silicon–Hydrogen–Copper–Aluminum–Lead–Oxygen–Chlorine–Sulfur

In the system containing these elements, the multinary compound $Pb_4(Al_3Cu)(Si_4O_{12})(Si_{0.5}S_{0.5}O_4)(OH)_7Cl \cdot 3H_2O$ (mineral bobmeyerite), which crystallizes as an orthorhombic structure with the lattice parameters a = 1396.9 ± 0.9, b = 1424.3 ± 1.0, c = 589.3 ± 0.4 pm, and a calculated density of 4.381 g·cm^{-3}, is formed (Kampf et al. 2012d, 2013g; Belakovskiy et al. 2015).

1.18 Silicon–Hydrogen–Copper–Aluminum–Oxygen–Arsenic–Sulfur

In the system containing these elements, the multinary compound $Cu_9Al(HSiO_4)_2[(SO_4)(HAsO_4)_{0.5}](OH)_{12} \cdot 8H_2O$ (mineral barrotite), which crystallizes as a trigonal structure with the lattice parameters a = 1065.0 ± 0.2, c = 2195.4 ± 0.7 pm, and the calculated and experimental densities of 3.00 and 2.90 g·cm^{-3}, respectively, is formed (Sarp and Cerny 2013; Sarp et al. 2014).

1.19 Silicon–Hydrogen–Copper–Aluminum–Oxygen–Sulfur

In the system containing these elements, the multinary compound $\{Cu_9Al[SiO_3(OH)]_2(OH)_{12}(H_2O)_6\}(SO_4)_{1.5}\cdot 10H_2O$ (mineral tiberiobardiite), which crystallizes as a trigonal structure with the lattice parameters $a = 1068.60 \pm 0.04$, $c = 2832.39 \pm 0.10$ pm, and a calculated density of 2.528 g·cm^{-3}, is formed (Biagioni et al. 2017b, 2018d).

1.20 Silicon–Hydrogen–Copper–Lead–Oxygen–Sulfur

In the system containing these elements, the multinary compound $Cu_2Pb_7(SO_4)_4(SiO_4)_2(OH)_2$ (mineral wherryite), which crystallizes as a monoclinic structure with the lattice parameters $a = 2078.9 \pm 0.4$, $b = 578.7 \pm 0.1$, $c = 914.2 \pm 0.3$ pm, and $\beta = 91.42 \pm 0.02°$ (Cooper and Hawthorne 1994; Jambor et al. 1995c) [$a = 2082 \pm 2$, $b = 579 \pm 1$, $c = 917 \pm 1$ pm, $\beta = 91.17 \pm 0.03°$ and a calculated density of 7.22 g·cm^{-3} (McLean 1970); an experimental density is 6.45 g·cm^{-3} (Fahey 1950)], is formed.

1.21 Silicon–Hydrogen–Calcium–Aluminum–Boron–Oxygen–Sulfur

In the system containing these elements, the multinary compound $Ca_6(Al,Si)_2(SO_4)_2[B(OH)_4](OH,O)_{12}\cdot 26H_2O$ (mineral charlesite), which crystallizes as a hexagonal structure with the lattice parameters $a = 1109.7 \pm 0.5$, $c = 2122 \pm 3$ pm, and the calculated and experimental densities of 1.74 and 1.84 g·cm^{-3}, respectively (Kusachi et al. 2008) [$a = 1116 \pm 1$, $c = 2121 \pm 2$ pm, and the calculated and experimental densities of 1.79 and 1.77 g·cm^{-3}, respectively (Dunn et al. 1983c)], is formed.

1.22 Silicon–Hydrogen–Calcium–Aluminum–Boron–Oxygen–Iron–Sulfur

In the system containing these elements, the multinary compound $Ca_3(Si,Fe,Al)(SO_4)[B(OH)_4]O(OH)_5\cdot 12H_2O$ (mineral buryatite), which crystallizes as a trigonal structure with the lattice parameters $a = 1114 \pm 1$, $c = 2099 \pm 5$ pm, and a calculated density of 1.895 ± 0.010 g·cm^{-3}, is formed (Malinko et al. 2001; Jambor et al. 2002).

1.23 Silicon–Hydrogen–Calcium–Aluminum–Yttrium–Thorium–Phosphorus–Oxygen–Sulfur

In the system containing these elements, the multinary phase $Ca(Y,Th)Al_5(SiO_4)_2(PO_4,SO_4)_2(OH)_7\cdot 6H_2O$ [mineral saryarkite-(Y)], which crystallizes as a tetragonal structure with the lattice parameters $a = 821.3 \pm 0.2$, $c = 655 \pm 1$ pm, and the calculated and experimental densities of 3.35 and 3.07–3.15 g cm^{-3}, respectively, is formed (Fleischer 1964; Krol' et al. 1964).

1.24 Silicon–Hydrogen–Calcium–Aluminum–Carbon–Oxygen–Sulfur

The phase diagram of the $CaO–Al_2O_3–SiO_2–CaSO_4–CaCO_3–H_2O$ subsystem as a part of the Si–H–Ca–Al–C–O–S multinary system has been resolved by calculation of the composition of the stable invariant points by Damidot et al. (2004). The mineral taumasite, $Ca_3Si(OH)_6(CO_3)(SO_4)\cdot 12H_2O$, was calculated to be stable over a wide range of aqueous phase compositions and is one of the major phases of this subsystem at 25°C. With respect to the mineral ettringite, $Ca_6Al_2(SO_4)_3(OH)_{12}\cdot 26H_2O$, the formation of thaumasite generally demands more sulfate ions in the aqueous phase, but thaumasite is also stable at pH as low as 8.5 compared to 9.5 for ettringite. In these low pH conditions, thaumasite is expected to form directly without need for the formation of precursor ettringite. At 25°C and 0.1 MPa pressure, the phase diagram of this subsystem is composed of 331 stable phase assemblages in equilibrium with the aqueous phase and contains 30 invariant points with 15 solids.

1.25 Silicon–Hydrogen–Calcium–Carbon–Oxygen–Sulfur

In the system containing these elements, the multinary compound $Ca_3Si(OH)_6(CO_3)(SO_4){\cdot}12H_2O$ (mineral thaumasite), which crystallizes as a hexagonal structure, is formed. The lattice parameters of this compound are given in Table 1.1. No phase transitions occur within 22.5 and 301 K (Gatta et al. 2012).

The structural changes during dehydration and subsequent decomposition in thaumasite were studied by in situ synchrotron powder diffraction between 30°C and 825°C (Martucci and Cruciani 2006). Between 27°C and 144°C, the thaumasite structure was observed to break down. Within this temperature range, the cell parameters increased as a function of temperature in a nearly linear fashion up to about 120°C, at which temperature, a slight change in the slope was observed. Above 127°C, the dehydration process proceeded very rapidly while the refined occupancy of water molecules dropped below a critical level, leading to instability in the thaumasite structure. At about 144°C, the crystal structure of thaumasite collapsed on losing the crystallization water and it turned amorphous. This result indicated that the dehydration/decomposition of thaumasite was induced by the loss of crystallization water. At about 677°C, anhydrite ($CaSO_4$) and cristobalite (SiO_2) crystallized from the thaumasite glass.

The high-pressure behavior of the thaumasite structure was investigated using synchrotron powder X-ray diffraction (XRD), up to 19.5 GPa (Ardit et al. 2014). It was shown that thaumasite retained the room-pressure hexagonal structure throughout the whole investigated pressure range while the pressure dependence of the refined unit-cell parameters can be cast into three different compression regimes, each

TABLE 1.1

Crystallographic data for mineral thaumasite, $Ca_3Si(OH)_6(CO_3)(SO_4){\cdot}12H_2O$

		$d_{calc.}$	$d_{meas.}$	
a (pm)	*c* (pm)	g·cm^{-3}		References
1099.2	1031.1	–	–	Font-Altaba 1960
1103	1040	–	–	Knill and Young 1960
1104 ± 2	1039 ± 2	–	–	Edge and Taylor 1971
1101.3 ± 0.2	1037.9 ± 0.5	–	–	Grubessi et al. 1986
1102.2 ± 0.2[a]	1037.4 ± 0.2[a]	1.894±0.001[a]	–	Jacobsen et al. 2003
1102.9 ± 0.2[b]	1038.3 ± 0.2[b]	1.890±0.001[b]	–	
1103.7 ± 0.2[c]	1039.3 ± 0.2[c]	1.885±0.001[c]	–	
1105.38 ± 0.06[d]	1041.11 ± 0.08[d]	1.876±0.001[d]	–	
1105.75 ± 0.01[e]	1041.63 ± 0.01[e]	–	–	Martucci and Cruciani 2006
1108.25 ± 0.02[f]	1044.47 ± 0.03[f]			
1103.60 ± 0.05[g]	1038.45 ± 0.06[g]	–	–	Gatta et al. 2012
1105.45 ± 0.05[h]	1041.31 ± 0.06[h]			
1105.67 ± 0.07[i]	1041.40 ± 0.07[i]	1.784[i]	–	Ardit et al. 2014
1031.24 ± 0.19[j]	980.08 ± 0.20[j]	2.180[j]	–	

[a]At 130 K; [b]at 180 K; [c]at 230 K; [d]at 298 K; [e]at 303 K; [f]at 413 K; [g]at 22.5 K; [h]at 301 K; [i]at ambient pressure; [j]at 19.5 GPa.

corresponding to a different thaumasite phase (thaumasite-I, thaumasite-II, and thaumasite-III) related by isosymmetric phase transitions.

The procedure for the synthesis of thaumasite consisted of mixing two solutions (10 mass%) that were cooled at 5°C (Aguilera et al. 2001). The first solution contained CaO or $Ca(OH)_2$, and the second one had Na_2SiO_3, Na_2SO_4, and Na_2CO_3. Different starting concentrations of reactants were used. The samples so obtained were kept at 0°C–5°C. After 4 months of reaction, thaumasite crystals were obtained in all the samples. The efficiency of the reaction is higher when CaO solution was used as a reagent rather than $Ca(OH)_2$. The thaumasite percentage in the samples increases with the time of reaction. Pure thaumasite was obtained when initial solutions contained less Na_2CO_3 and Na_2SiO_3 than the stoichiometric proportions after 15 months of reaction.

1.26 Silicon–Hydrogen–Calcium–Lead–Oxygen–Manganese–Sulfur

In the system containing these elements, the multinary compound $Pb_2Ca_6Mn^{2+}Si_6O_{18}(SO_4)_2(OH)_2{\cdot}4H_2O$ (mineral roeblingite), which crystallizes as a monoclinic structure with the lattice parameters a = 1320.8 ± 0.4, b = 828.7 ± 0.2, c = 1308.9 ± 0.9 pm, and β = 106.65 ± 0.06° (Moore and Shen 1984) [a = 1327 ± 3, b = 838 ± 2, c = 1309 ± 3 pm, and β = 103.86 ± 0.10° (Foit Jr 1966)], is formed.

1.27 Silicon–Hydrogen–Calcium–Phosphorus–Oxygen–Sulfur

In the system containing these elements, the multinary compound $Ca_{10}(SiO_4)(PO_4)_4(SO_4)(OH)_2$, which crystallizes as a hexagonal structure with the lattice parameters a = 944, c = 696 pm, and the calculated and experimental densities of 3.08 and 3.01 g·cm^{-3}, respectively, is formed (Dihn and Klement 1942).

1.28 Silicon–Hydrogen–Calcium–Oxygen–Sulfur

Three quinary compounds are formed in this system. $Ca_{10}(SiO_4)_3(SO_4)_3(OH)_2$ (mineral hydroxylellestadite) melts incongruently at 1260°C (Klement and Dihn 1941) and crystallizes as a hexagonal structure with the lattice parameters a = 949.6 ± 0.2 and c = 692.0 ± 0.2 pm (Onac et al. 2006) [a = 954, c = 699 pm, and the calculated and experimental densities of 3.00 and 3.07 g·cm^{-3}, respectively (Dihn and Klement 1942); a = 948.4, c = 629.7 pm, and the calculated and experimental densities of 3.080 and 3.018 g·cm^{-3}, respectively (Harada et al. 1971); in the monoclinic structure with the lattice parameters a = 947.6 ± 0.2, b = 950.8 ± 0.2, c = 691.9 ± 0.1 pm, and β = 119.53 ± 0.02° (Sudarsanan 1980)].

Hydroxylellestadite can be obtained by removing fluorine from $Ca_{10}(SiO_4)_3(SO_4)_3F_2$ and replacing it with hydroxyl by passing water vapor over this compound at 1000°C–1100°C (Klement and Dihn 1941; Dihn and Klement 1942).

$Ca_3Si(SO_4)_2(OH)_6{\cdot}12H_2O$ (mineral kottenheimite) also crystallizes as a hexagonal structure with the lattice parameters a = 1115.48 ± 0.03, c = 1057.02 ± 0.03 pm, and the calculated and experimental densities of 1.926 and 1.92 ± 0.02 g·cm^{-3}, respectively (Chukanov et al. 2011, 2012; Cámara et al. 2014b).

$Ca_3Si(SO_4)(SO_3)(OH)_6{\cdot}11H_2O$ (mineral hielscherite) also crystallizes as a hexagonal structure with the lattice parameters a = 1111.78 ± 0.02, c = 1053.81 ± 0.02 pm, and the calculated and experimental densities of 1.79 and 1.82 ± 0.03 g·cm^{-3}, respectively (Pekov et al. 2011, 2012; Cámara et al. 2015).

1.29 Silicon–Hydrogen–Calcium–Oxygen–Chlorine–Sulfur

According to Rouse and Dunn (1982), a hypothetical multicomponent compound $Ca_{10}(SiO_4)_3(SO_4)_3(OH,Cl)_2$, capable of crystallizing in a hexagonal structure with lattice parameters a = 954.3 and c = 691.7 pm, may exist in this system.

1.30 Silicon–Hydrogen–Barium–Oxygen–Manganese–Iron–Sulfur

In the system containing these elements, the multinary compound $Ba_4Mn_4Fe_2(Si_2O_7)_2(SO_4)_2O_2(OH)_2$ (mineral zinkgruvanite), which crystallizes as a triclinic structure with the lattice parameters $a = 539.82 \pm 0.01$, $b = 702.37 \pm 0.01$, $c = 1481.08 \pm 0.04$ pm, $\alpha = 98.256 \pm 0.002°$, $\beta = 93.379 \pm 0.002°$, $\gamma = 89.985 \pm 0.002°$, and a calculated density of 4.42 ± 0.01 g·cm^{-3}, is formed (Cámara et al. 2020a,b, 2021).

1.31 Silicon–Hydrogen–Zinc–Lead–Oxygen–Sulfur

In the system containing these elements, the multinary compound $ZnPb_{10}(SO_4)_6(SiO_4)_2(OH)_2$ (mineral raygrantite), which crystallizes as a triclinic structure with the lattice parameters $a = 931.75 \pm 0.04$, $b = 1119.73 \pm 0.05$, $c = 1083.18 \pm 0.05$ pm, $\alpha = 120.374 \pm 0.002°$, $\beta = 90.511 \pm 0.002°$, $\gamma = 56.471 \pm 0.002°$, and a calculated density of 6.374 g·cm^{-3}, is formed (Yang et al. 2013a, 2016; Belakovskiy et al. 2017c).

1.32 Silicon–Hydrogen–Aluminum–Carbon–Lead–Oxygen–Sulfur

In the system containing these elements, the multinary compound $Pb_8Al_4(Si_8O_{20})(SO_4)_2(CO_3)_4(OH)_8$ compound (mineral kegelite), which crystallizes as a monoclinic structure with the lattice parameters $a = 2104 \pm 1$, $b = 1555 \pm 1$, $c = 898.6 \pm 0.6$ pm, $\beta = 91.0 \pm 0.1°$, and an experimental density of 4.3–4.6 cm^{-3}, is formed (Dunn et al. 1990).

1.33 Silicon–Hydrogen–Aluminum–Lead–Oxygen–Manganese–Sulfur

In the system containing these elements, the multinary compound $Pb_{27}(Al_{1.6}Mn_{0.4})(Si_6O_{15})_2(SO_4)_2O_{10}(OH)_{24}$ (mineral substance), which crystallizes as a hexagonal structure with the lattice parameters $a = 920.3 \pm 0.4$, $c = 2572 \pm 2$ pm, and a calculated density of 6.37 g·cm^{-3}, is formed (Cipriani et al. 1995; Jambor et al. 1997a).

1.34 Silicon–Hydrogen–Lead–Oxygen–Chlorine–Sulfur

In the system containing these elements, the multinary compound $Pb_5(SiO_4)_{1.5}(SO_4)_{1.5}(Cl,OH)$ (mineral mattheddleite), which crystallizes as a hexagonal structure with the lattice parameters $a = 1000.56 \pm 0.06$, $c = 749.60 \pm 0.09$ pm, and a calculated density of 6.822 g·cm^{-3} (Steele et al. 2000) [$a = 996.3 \pm 0.5$, $c = 746.4 \pm 0.5$ pm, and a calculated density of 6.96 g·cm^{-3} (Livingstone et al. 1987; Jambor et al. 1988)], is formed.

1.35 Silicon–Hydrogen–Nitrogen–Oxygen–Sulfur

In the system containing these elements, the quinary compound $(NH_4)_2[Si(S_2O_7)_3$, which crystallizes as a trigonal structure with the lattice parameters $a = 956.2 \pm 0.6$ and $c = 1116.9 \pm 0.7$ pm, is formed (Logemann et al. 2012). The reaction for obtaining this compound was performed in thick-walled glass ampoule. The tube was loaded with $SiCl_4$ (1 mM), oleum (1 mL, 65% SO_3), and $(NH_4)_2SO_4$ (1 mM), torch-sealed under vacuum, and placed in a resistance furnace. The ampoule was maintained at a temperature of 250°C for 24 h and cooled down to room temperature at a rate of 1.8°C·h^{-1}. The colorless crystals are very moisture-sensitive and were separated from oleum in a glove box.

1.36 Silicon–Lithium–Silver–Indium–Sulfur

The $Li_xAg_{1-x}In_2SiS_6$ solid solutions are formed in this quinary system (Zhou et al. 2021a). They crystallize in the monoclinic structure with the lattice parameters $a = 1207.81 \pm 0.18$, $b = 716.33 \pm 0.11$, $c = 1209.86 \pm 0.18$ pm, $\beta = 109.6610 \pm 0.0004°$, a calculated density of 3.949 g·cm^{-3}, and an energy gap of 2.59 eV for $x = 0.79$, $a = 1206.42 \pm 0.05$, $b = 711.43 \pm 0.03$, $c = 1208.86 \pm 0.05$ pm, $\beta = 109.8320 \pm 0.0010°$, a calculated density of 3.759 g·cm^{-3}, and an energy gap of 2.69 eV for $x = 1.12$, and $a = 1206.96 \pm 0.06$, $b = 711.58 \pm 0.04$, $c = 1208.22 \pm 0.09$ pm, $\beta = 109.8050 \pm 0.0010°$, a calculated density of 3.538 g·cm^{-3}, and an energy gap of 2.89 eV for $x = 1.44$.

Li_2S, Ag, In, Si, S, and KI were the raw materials for the synthesis of these solid solutions. The total mass of the reactants was 900 mg, containing stoichiometric raw materials (500 mg) and flux KI (400 mg). The mixture was ground into a uniform powder and then pressed into pellets. Then they were placed in a graphite crucible. Next, they were sealed in quartz tubes with a vacuum pressure of 10^{-2} Pa. Finally, the tubes were transferred to a muffle furnace, using which the reactants were heated to 950°C at a rate of 30°C·h^{-1} and dwelled at several intermediate equilibrated temperatures with the aim of preventing breakage of the quartz tubes. The temperature of 950°C was maintained for 5 days, and the samples were subsequently cooled to 300°C at a rate of 5°C·h^{-1} before switching off the furnace. Yellow rod-like and block crystals were obtained. The crystals were then rinsed with deionized water and alcohol, followed by ultrasonication. The crystals are stable in air after several months and can be separated from the byproducts easily under an optical microscope because of their unique color and shape.

1.37 Silicon–Lithium–Boron–Oxygen–Sulfur

SiS_2–Li_2S–Li_3BO_3. Lithium ion conducting oxysulfide glasses were prepared along the (0.4SiS_2+ 0.6Li_2S)–Li_3BO_3 section of this quasiternary system by twin-roller rapid quenching (Tatsumisago et al. 1996). The glass-forming region is 0–25 mol% Li_3BO_3, and the addition of 5 mol% Li_3BO_3 improved the ionic conductivity and the glass stability against crystallization.

1.38 Silicon–Lithium–Germanium–Phosphorus–Sulfur

$Li_{10}(Ge_{1-x}Si_x)P_2S_{12}$ solid solutions are formed in this quinary system for the composition $0 \leq x \leq 1.0$ (Kato et al. 2014). These solid solutions crystallize in the trigonal structure with the lattice parameters $a = 867.99 \pm 0.02$ and $c = 1257.9 \pm 0.3$ pm for the $Li_{10}(Ge_{0.7}Si_{0.3})P_2S_{12}$ composition. To prepare the solid solutions, SiS_2, Li_2S, P_2S_5, and GeS_2 were mixed in an appropriate molar ratio in an Ar-filled glove box. The mixture was put into a ZrO_2 pot together with ZrO_2 balls, mechanically milled using a planetary ball milling apparatus at 370 rpm for 40 h, and then it was put into a quartz tube and heated at 550°C for 8 h.

1.39 Silicon–Lithium–Germanium–Oxygen–Sulfur

SiS_2–Li_2S–Li_4GeO_4. Lithium-ion-conducting oxysulfide glasses were prepared along the (0.4SiS_2+ 0.6Li_2S)–Li_4GeO_4 section of this quasiternary system by twin-roller rapid quenching (Tatsumisago et al. 1996). The glass-forming region is 0–15 mol% Li_4GeO_4, and the addition of 5 mol% Li_4GeO_4 improved the ionic conductivity and the glass stability against crystallization.

1.40 Silicon–Lithium–Tin–Phosphorus–Sulfur

Li_4SiS_4–Li_4SnS_4–Li_3PS_4. The monophasic region in this quasiternary system deviates from the tie line $Li_{10}SiP_2S_{12}$–$Li_{10}SnP_2S_{12}$ (Figure 1.1), and the composition of the solid solution was determined to be –0.1

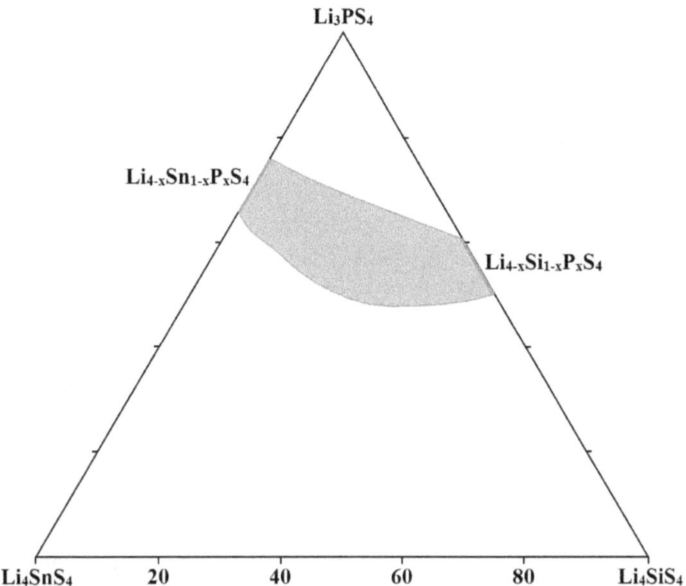

FIGURE 1.1 Region of the solid solution in the Li_4SiS_4–Li_4SnS_4–Li_3PS_4 quasiternary system. (From Sun, Y., et al., *Chem. Mater.*, **29**(14), 5858, 2017.)

$\leq \delta \leq 0.5$ and $0 \leq y \leq 1.0$ in formula $Li_{10+\delta}[Si_{1-y}Sn_y]_{1+\delta}P_{2-\delta}S_{12}$ or $0.5 \leq x \leq 0.7$ and $0 \leq y \leq 1.0$ in formula $Li_{4-x}[Si_{1-y}Sn_y]_{1-x}P_xS_4$ (Sun et al. 2017). This solid solution crystallizes as a tetragonal structure with the lattice parameters a = 867.822 ± 0.005 and c = 1257.44 ± 0.02 pm for $Li_{10.35}[Si_{1.08}Sn_{0.27}]P_{1.65}S_{12}$ or $Li_{3.45}[Si_{0.36}Sn_{0.09}]P_{0.55}S_4$, a = 869.204 ± 0.006 and c = 1261.66 ± 0.02 pm for $Li_{10.5}[Si_{1.2}Sn_{0.3}]P_{1.5}S_{12}$ or $Li_{4.5}[Si_{0.4}Sn_{0.1}]P_{0.5}S_4$, and a = 865.173 ± 0.006 and c = 1253.49 ± 0.02 pm for $Li_{10.2}[Si_{0.96}Sn_{0.24}]P_{1.8}S_{12}$ or $Li_{3.4}[Si_{0.32}Sn_{0.08}]P_{0.6}S_4$. The starting materials to synthesize this solid solution were Li_2S, SiS_2, SnS_2, and P_2S_5. They were weighed in stoichiometric ratios in an Ar-filled glove box and then mechanically mixed in a ZrO_2 pot with ZrO_2 balls using a planetary ball milling apparatus at 6.3 Hz for 20 h. The product mixture was pelletized at 120 MPa, sealed in a quartz tube at 10 Pa, followed by heating at 550°C for 24 h. Then, the samples were slowly cooled to room temperature.

1.41 Silicon–Lithium–Phosphorus–Oxygen–Sulfur

SiS_2–Li_2S–Li_3PO_4. The glasses $[(SiS_2)_{0.4}(Li_2S)_{0.6}]_{1-x}(Li_3PO_4)_x$ were prepared from the binary components up to x = 0.4 by twin-roller rapid quenching (Tatsumisago et al. 1993, 1996). The mixture in a vitreous carbon crucible was heated at 1000°C for 1 h. The molten mixture was poured into the rotating twin-roller to obtain flake-like samples with a thickness of about 20 μm. Brownish-orange transparent glasses were obtained in the range x < 0.4. Addition of small amounts (up to 5 mol%) of Li_3PO_4 improved the stability of glasses against crystallization. The values of glass-forming temperatures range from 290°C to 340°C.

1.42 Silicon–Sodium–Potassium–Calcium–Aluminum–Oxygen–Chlorine–Sulfur

Some multinary compounds are formed in this system. $Na_4K_2Ca_2(Al_6Si_6O_{24})(SO_4)Cl_2$ (mineral microsommite) crystallizes as a hexagonal structure with the lattice parameters a = 2216.0 ± 0.2 and c = 534.6 ± 0.1 pm (Bonaccorsi et al. 2001) [a = 2215.6 ± 0.5, c = 523.7 ± 0.2 pm, and the calculated and experimental densities of >2.30 and 2.37 ± 0.04 g·cm^{-3}, respectively (Leoni et al. 1979)]. The order–disorder transformation of this compound takes place at ca. 750°C with the activation energy of about 125 kJ·M^{-1} (Bonaccorsi et al. 2001).

TABLE 1.2

Crystallographic data for mineral afghanite, $(Na,K)_{22}Ca_{10}(Si_6Al_6O_{24})_4(SO_4)_6Cl_6$

		$d_{calc.}$	$d_{meas.}$	
a (pm)	*c* (pm)	g·cm^{-3}		References
1277 ± 3	2135 ± 4	2.65	2.55	Bariand et al. 1968; Fleischer 1968
1274 ± 4	2126 ± 8	–	2.517 ± 0.001	Ivanov and Sapozhnikov 1975
–	–	2.527	2.517	Hogarth 1979
1284.7 ± 0.3	2146.4 ± 0.5	2.56	2.44 ± 0.02	Leoni et al. 1979[a]
1279.6 ± 0.2	2136.1 ± 0.3	2.60	2.53 ± 0.02	
1276.1 ± 0.4	2141.6 ± 0.7	–	2.52	Pobedimskaya et al. 1991a
1279.97 ± 0.04	2140.62 ± 0.11	–	–	Ballirano et al. 1994
1280.13 ± 0.07	2141.19 ± 0.18	–	–	Ballirano et al. 1997; Jambor and Roberts 1997

[a] For two different samples.

$(Na,K,Ca)_{7\text{-}8}(Al_6Si_6O_{24})(SO_4,Cl)_{1\text{-}2}$ (mineral giuseppettite) also crystallizes as a hexagonal structure with the lattice parameters $a = 1285.6 \pm 0.2$, $c = 4225.6 \pm 0.8$ pm, and a calculated density of 2.32 g·cm^{-3} (Bonaccorsi 2004; Piilonen et al. 2005a) [$a = 1285.0 \pm 0.1$, $c = 4222 \pm 3$ pm, and the calculated and experimental densities of 2.365 and 2.35 g·cm^{-3}, respectively (Fleischer et al. 1982a)].

$(Na,K)_6Ca_2(Al_6Si_6O_{24})(Cl_2,SO_4)_2$ (mineral davyne) also crystallizes as a hexagonal structure with the lattice parameters $a = 1273.3 \pm 0.1$, $c = 533.4 \pm 0.1$ pm, and a calculated density of 2.46 g·cm^{-3} (Rozenberg et al. 2009) [$a = 1270.5 \pm 0.4$ and $c = 536.8 \pm 0.3$ pm (Bonaccorsi et al. 1990; Jambor and Burke 1990); $a = 1279.3 \pm 0.3$, 1285.4 ± 0.1 and $c = 536.7 \pm 0.3$, 535.7 ± 0.1 pm for two different samples (Hassan and Grundy 1990); $a = 1269.16 \pm 0.09$ and $c = 533.33 \pm 0.05$ pm (Ballirano et al. 1998)].

$(Na,K)_{16}Ca_8(Al_6Si_6O_{24})_3(SO_4)_5Cl_4$ (mineral liottite) also crystallizes as a hexagonal structure with the lattice parameters $a = 1285.75 \pm 0.03$ and $c = 1609.05 \pm 0.06$ pm (Ballirano et al. 1995) [$a = 1284.2 \pm 0.3$, $c = 1609.1 \pm 0.5$ pm, and the calculated and experimental densities of 2.61 and 2.56 ± 0.02 g·cm^{-3}, respectively (Merlino and Orlandi 1977); $a = 1287.0 \pm 0.1$ and $c = 1609.6 \pm 0.2$ pm (Ballirano et al. 1996)].

$(Na,K)_{22}Ca_{10}(Si_6Al_6O_{24})_4(SO_4)_6Cl_6$ (mineral afghanite) crystallizes as a trigonal structure with the lattice parameters given in Table 1.2.

1.43 Silicon–Sodium–Cesium–Uranium–Sulfur

Two quinary compounds, $Na_4Cs_2[U_2(SiS_4)_2(Si_2S_8)]$ and $Na_{3.88}Cs_{2.12}[U_2(SiS_4)_2(Si_2S_7)]$, are formed in this system (Klepov et al. 2019). Both compounds crystallize in the monoclinic structure with the lattice parameters $a = 1666.40 \pm 0.06$, $b = 971.15 \pm 0.03$, $c = 911.64 \pm 0.03$ pm, $\beta = 109.9330 \pm 0.0010°$ and a calculated density of 3.49 g·cm^{-3} for the first compound and $a = 1674.59 \pm 0.05$, $b = 967.76 \pm 0.03$, $c = 1700.08 \pm 0.05$ pm, $\beta = 97.5250 \pm 0.0010°$ and a calculated density of 3.50 g·cm^{-3} for the second one. To prepare these compounds, the starting reagents, US_2 (0.100 g), Na_2S (0.052 g), SiS_2 (0.122 g) in a 1:2:4 molar ratio and 1.00 g of eutectic CsI–NaI and CsI fluxes for the first and the second compound, respectively, were mixed inside a carbon-coated silica tube in a N_2-filled glove box. The reagent-filled silica tubes were flame sealed under a vacuum of $<10^{-2}$ Pa and placed vertically into a furnace. The furnace was ramped up to 600°C and 750°C, for CsI–NaI and CsI fluxes, respectively, in 1.5 h and maintained at this temperature for 20 h. Following this heating step, the furnace was cooled to below the melting point of the flux at a rate of 10°C·h^{-1} to effect the formation of crystallized products. Once at room temperature, the tube was cut open and the end of the tube containing the reaction products was placed into a beaker filled with DMF. After soaking the reaction products for several hours the crystals were isolated by filtration. A sample of $Na_4Cs_2[U_2(SiS_4)_2(Si_2S_8)]$, which was obtained from CsI–NaI flux, was slightly contaminated with US_3. CsI flux at 750°C gave rise to $Na_{3.88}Cs_{2.12}[U_2(SiS_4)_2(Si_2S_7)]$ single crystals along with a significant amount of UOS impurity.

1.44 Silicon–Sodium–Calcium–Barium–Titanium–Oxygen–Sulfur

In the system containing these elements, the multinary compound $Ba_4Ti_2Na(NaMn^{2+})Ti(Si_2O_7)_2[(SO_4)(PO_4)]O_2[O(OH)]$ (mineral innelite) (Bosi et al. 2023, 2024), which crystallizes as a triclinic structure with the lattice parameters a = 1476, b = 714, c = 538 pm, α = 90°, β = 95°, and γ = 99° (Chernov et al. 1971) [a = 538 ± 2, b = 714 ± 3, c = 1476 ± 10 pm, α ≈ 99°, β = 95°, γ ≈ 90°, and a calculated density of 3.96 g·cm^{-3} (Kravchenko et al. 1961; Fleischer 1962)], is formed.

1.45 Silicon–Sodium–Calcium–Aluminum–Carbon–Oxygen–Sulfur

In the system containing these elements, the multinary compound $(Ca,Na)_4(Al_6Si_6O_{24})(SO_4,CO_3)$ (mineral silvialite), which crystallizes as a tetragonal structure with the lattice parameters a = 1213.4 ± 0.2, c = 757.6 ± 0.2 pm, and the calculated and experimental densities of 2.77 and 2.75 g·cm^{-3}, respectively (Teertstra et al. 1999; Jambor and Roberts 2000), is formed.

1.46 Silicon–Sodium–Calcium–Aluminum–Oxygen–Sulfur

In the system containing these elements, the multinary compound $Na_6Ca_2(Al_6Si_6O_{24})(SO_4)_2$ (mineral haüyne), which crystallizes as a cubic structure with the lattice parameter a = 912.36 ± 0.01 pm at 33°C (Hassan et al. 2004) [a = 905.4 ± 0.9 pm, and a calculated density of 2.499 g·cm^{-3} (Tsuchiya and Takéuchi 1985); a = 911.6 ± 0.5 pm (Hassan and Buseck 1989); a = 910.97 ± 0.08 and 911.64 ± 0.05 pm and a calculated density of 2.42 and 2.41 g·cm^{-3} at 153 and 293 K, respectively (Hassan and Grundy 1991); a = 911.8 ± 0.2 pm (Evsyunin et al. 1996)], is formed. The a cell parameter for haüyne increases smoothly and non-linearly with temperature, but there is a discontinuity at 585°C (Hassan et al. 2004). Beyond 585°C, the rate of expansion is retarded. The a parameter expansion between 33°C and 576°C is given by the next least-squares fit (all temperatures are in °C): $a = 9.1213 + 9.5586{\cdot}10^{-5}T + 2.4947{\cdot}10^{-8}T^2$ A, and $a = 9.1229 + 1.5450{\cdot}10^{-4}T - 6.7698{\cdot}10^{-8}T^2$ A between 593°C and 1035°C.

1.47 Silicon–Sodium–Calcium–Aluminum–Oxygen–Chlorine–Sulfur

In the system containing these elements, the multinary compound $Na_7Ca(Al_6Si_6O_{24})(S_5)Cl$ (mineral bystrite), which crystallizes as a trigonal structure with the lattice parameters a = 1285.27 ± 0.06, c = 1069.07 ± 0.05 pm, and the calculated and experimental densities of 2.412 and 2.42 g·cm^{-3}, respectively (Belakovskiy et al. 2017b; Kaneva et al. 2017; Sapozhnikov et al. 2017; Chukanov et al. 2023) [a = 1285.9 ± 0.3, c = 1069.7 ± 0.8 pm, and the calculated and experimental densities of 2.45 and 2.43 ± 0.01 g·cm^{-3}, respectively (Pobedimskaya et al. 1991b; Sapozhnikov et al. 1991; Jambor and Puziewicz 1993a)], is formed.

1.48 Silicon–Sodium–Calcium–Phosphorus–Oxygen–Fluorine–Sulfur

The $Na_2Ca_9(SiO_4)(PO_4)_4(SO_4)F_2$ multinary compound, which crystallizes as a hexagonal structure with the lattice parameters a = 939, c = 689 pm, and the calculated and experimental densities of 3.24 and

3.10 g·cm^{-3}, respectively, is formed in this system (Dihn and Klement 1942). This compound can be synthesized by sintering the mixture of $Ca_3P_2O_8$, Na_2SO_4, $CaCO_3$, SiO_2, and CaF_2 (molar ratio 2:1:2:1:1) at 1250°C for 6 h.

1.49 Silicon–Sodium–Barium–Titanium–Phosphorus–Oxygen–Fluorine–Sulfur

In the system containing these elements, the multinary compound $Na_3Ba_4Ti_3(Si_4O_7)_2(PO_4,SO_4)_2O_2F$ (mineral phosphoinnelite), which crystallizes as a triclinic structure with the lattice parameters $a = 538 \pm 2$, $b = 710 \pm 2$, $c = 1476 \pm 5$ pm, $\alpha = 99.00 \pm 0.07°$, $\beta = 94.94° \pm 0.06$, $\gamma = 90.14° \pm 0.08$, and the calculated and experimental densities of 3.92 and 3.82 ± 0.05 g·cm^{-3}, respectively (Pekov et al. 2007; Ercit et al. 2009), is formed.

1.50 Silicon–Sodium–Aluminum–Manganese–Oxygen–Sulfur

In the system containing these elements, the multinary compound $Na_6Mn_2(Al_6Si_6O_{24})(SO_4)_2$, which crystallizes as a cubic structure with the lattice parameter $a = 897.80 \pm 0.02$, is formed (Warner et al. 2019). To prepare this compound, $MnSO_4{\cdot}H_2O$ and zeolite A, $Na_6(Al_6Si_6O_{24}){\cdot}2.2H_2O$, (molar ratio 2:1) were ground in a planetary mill with ZrO_2 balls using ZrO_2 pot at a rotation of 400 rpm for 30 min, with cyclohexane as the liquid medium. The resulting slurry was placed in a Petri-dish and left to evaporate in a fume cupboard overnight before dehydrating the contents in a furnace at 250°C for 20 h. This anhydrous powder mixture was compacted into monoliths using a stainless steel die and a uniaxial press under a load of 0.1 MPa. The monoliths were then placed on top of a sacrificial powder bed of anhydrous $MnSO_4$ in a small alumina crucible with a matching lid. This procedure was devised to create a stagnant atmosphere to limit the loss of SO_2 from the system. This crucible was in turn placed on top of a powder bed of anhydrous $CuSO_4$ inside a slightly larger alumina crucible with its matching lid secured. $CuSO_4$ is used to create an atmosphere sufficiently rich in SO_2/SO_3 to help prevent the thermal decomposition of the $MnSO_4$ in the inner crucible. This was then placed in a preheated chamber furnace at 650°C, and maintained there for 24 h before cooling to room temperature at a rate of 200°C·h^{-1}. The product material was removed from the alumina crucible for analysis.

1.51 Silicon–Sodium–Praseodymium–Oxygen–Sulfur

In the system containing these elements, the quinary compound $NaPr_9S_2(SiO_4)_6$, which crystallizes as a hexagonal structure with the lattice parameters $a = 981.05 \pm 0.04$ and $c = 689.68 \pm 0.02$ pm, is formed (Sieke and Schleid 1997). It was obtained as pale green single crystals from reaction of Pr, Pr_6O_{11}, S, and SiO_2 in suitable molar ratios and NaCl as a flux by heating in a fused evacuated silica ampoule at 850°C for 7 days.

1.52 Silicon–Sodium–Samarium–Oxygen–Sulfur

In the system containing these elements, the quinary compound $NaSm_9S_2(SiO_4)_6$, which crystallizes as a hexagonal structure with the lattice parameters $a = 975.32 \pm 0.09$ and $c = 676.46 \pm 0.07$ pm, is formed (Sieke and Schleid 1999). It was obtained as light yellow transparent single crystals from reaction of Sm, Sm_2O_3, S, and SiO_2 in suitable molar ratios and NaCl as a flux by heating in a fused evacuated silica ampoule at 850°C for 7 days.

1.53 Silicon–Potassium–Calcium–Aluminum–Carbon–Oxygen–Sulfur

In the system containing these elements, the multinary compound $(Ca,K)_4(Si,Al)_5O_{11}(SO_4,CO_3)$ (mineral latiumite), which crystallizes as a monoclinic structure with the lattice parameters a = 1206.4 ± 0.2, 1204.0 ± 0.9, 1203.6 ± 0.4, b = 510.7 ± 0.2, 510.5 ± 0.5, 510.6 ± 0.3, c = 1088.2 ± 0.3, 1087.4 ± 1.6, 1092.5 ± 0.9 pm, and β = 107.08 ± 0.03°, 106.98 ± 0.07°, 107.08 ± 0.04° for three different samples (Leoni et al. 1977) [a = 1212, b = 513, c = 1080 pm, β = 108°, and an experimental density of 2.93 cm^{-3} (Tilley and Henry 1953; Fleischer 1954); a = 1206 ± 1, b = 508 ± 2, c = 1081 ± 1 pm, β = 106.0° (Cannillo et al. 1973)], is formed.

1.54 Silicon–Potassium–Calcium–Germanium–Oxygen–Fluorine–Sulfur

$KF{\cdot}2[Ca_6(SO_4)(SiO_4)_2O$–$KF{\cdot}2[Ca_6(SO_4)(GeO_4)_2O$. The continuous series of the solid solutions is formed in this system (Navarro and Glasser 1985).

1.55 Silicon–Potassium–Calcium–Oxygen–Sulfur

The sulfate-rich portions of the K_2SO_4–$CaSO_4$–Ca_2SiO_4–CaO system were studied by the quenching technique, supplemented by DTA (Pliego-Cuervo and Glasser 1977). A quaternary eutectic point between $K_2Ca_2(SO_4)_3$, K_2SO_4, CaO, and Ca_2SiO_4 occurs at 824°C: the composition of the liquid in this point (in mass%) is CaO – 1.67, SiO_2–0.33, $CaSO_4$–39.20, and K_2SO_4–58.80.

1.56 Silicon–Potassium–Calcium–Oxygen–Fluorine–Sulfur

The $KCa_{12}(SiO_4)_4(SO_4)_2O_2F$ or $KF{\cdot}2[Ca_6(SO_4)(SiO_4)_2O$ multinary compound (mineral nabimusaite) is formed in this system (Triviño Vázquez 1982). It crystallizes as a trigonal structure with the lattice parameters a = 719.05 ± 0.04, c = 4125.1 ± 0.3 pm, and a calculated density of 3.119 cm^{-3} (Galuskin et al. 2013, 2015a; Belakovskiy et al. 2016b) [a = 719.7 ± 0.5, c = 4122.4 ± 2.8 pm, and the calculated and experimental densities of 2.975 and 2.92 $g{\cdot}cm^{-3}$, respectively (Fayos et al. 1985)]. Mineral nabimusaite is characterized by a variable chemical composition.

This compound was prepared by heating a stoichiometric mixture of $CaCO_3$, silica gel, K_2CO_3, and CaF_2 with 6 mass% excess of KF and 6 mass% excess of $CaSO_4$ at 1000°C for 1 h followed by crystallization of the melt at 950°C for 48 h to yield crystals in the form of hexagonal plates (Fayos et al. 1985).

1.57 Silicon–Copper–Silver–Oxygen–Iron–Sulfur

The experimental results for the gas/slag/matte/spinel system containing Ag as a minor element from 24 h equilibration at 1200°C, $p(SO_2)$ = 0.25 atm and $p(SO_2)$ = 0.25 atm and $p(O_2) = 10^{-8.1}$ atm are given in Hidayat et al. (2016). The concentration of Ag in matte of different particles was uniform around 0.15 mass%, while the concentration of Ag in slag of different locations had high variation between 2 and 8 ppm. The distribution coefficient of Ag was found to be approximately –2.51. This indicates that majority of Ag was accumulated in the matte phase.

1.58 Silicon–Copper–Silver–Iodine–Sulfur

Cu_7SiS_5I–Ag_7SiS_5I. In this system, the $(Cu_{0.5}Ag_{0.5})_7SiS_5I$ solid solution was synthesized from initial quaternary compounds in evacuated quartz ampoule (Studenyak et al. 2021). The synthesis included heating up to 750°C at a rate of 100°C·h^{-1} and aging for 48 h, and a further increase of temperature to 1200°C at a rate of 50°C·h^{-1} and aging at this temperature for 72 h. The annealing temperature was 600°C at an aging time of 120 h. Cooling to room temperature was performed in the furnace off mode.

1.59 Silicon–Copper–Magnesium–Calcium–Aluminum–Oxygen–Iron–Sulfur

Experimental measurements of gas/slag/matte/tridymite equilibria at 1200°C, 1300°C and $p(SO_2) = 0.25$ atm have provided the detailed data for the Si–Cu–Mg–Ca–Al–O–Fe–S system (Sineva et al. 2020). The effect of temperature on the multiphase equilibria has been measured after carrying out experiments and calculation. After providing a detailed analysis of experimental and calculated results it has been shown that: (1) increasing of temperature from 1200°C to 1300°C) at $p(SO_2) = 0.25$ atm results in lower matte grade for the fixed $p(O_2)$; (2) Increasing the Al_2O_3, CaO and MgO contents in slag does not affect the $p(O_2)$ and $p(S_2)$ in the system for the fixed matte grade; (3) The sulfur concentration in matte is inversely dependent on the matte grade. According to experimental points, increasing the Al_2O_3, CaO, and MgO concentrations in slag results in increasing S concentration in matte for a given matte grade; (4) Increasing the temperature from 1200°C to 1300°C decreases FeO concentration in slag by approximately 2 mass%. Increasing the Al_2O_3, CaO, and MgO concentrations in slag results in large decreases of FeO concentration in slag at fixed temperature and matte grade; (5) The corresponding Fe/SiO_2 ratio has a high sensitivity both to variation of temperature and Al_2O_3, CaO, and MgO concentrations in slag, resulting in decreasing of the ratio with increasing the temperature and increasing Al_2O_3, CaO, and MgO concentrations in slag; (6) The concentration of dissolved sulfur in slag is inversely related to the matte grade. It has been shown that increasing the temperature has no obvious effect on S concentration in the slag. Increasing the Al_2O_3, CaO, and MgO concentrations in slag results in the decreasing of S in slag at all matte grades; (7) Increasing the temperature has evident effect on dissolved Cu in slag, increasing it from 0.4 mass% at 1200°C to 0.6 mass% at 1300°C at fixed matte grade (60 mass% Cu). Increasing Al_2O_3, CaO and MgO concentrations in slag, results in decreasing Cu losses in slag. The calculated trend lines and experimental points are in good agreement; (8) The predicted Fe^{3+}/Fe ratio decreases with increasing both temperature and Al_2O_3, CaO, and MgO concentrations in slag at fixed matte grade.

1.60 Silicon–Copper–Magnesium–Oxygen–Iron–Sulfur

The effect of MgO on the phase equilibria of iron silicate slags in equilibrium with Fe_3O_4 spinel and matte of fixed 72 mass% Cu were identified at controlled $p(SO_2)$ 0.3 and 0.6 atm (Sun et al. 2020b). The experimental process includes equilibration, quenching, and electron probe microanalysis (EPMA). Spinel (Fe_3O_4) substrates were applied to confirm the equilibrium was achieved in the primary phase field of spinel. It was found that the presence of MgO increases the liquidus temperature of slag under a constant SiO_2 content, that is, more SiO_2 is required to be fluxed to keep a stable smelting temperature with MgO. The effects of CaO and MgO on the liquidus of the system were further compared and found that the influence of MgO is stronger than CaO at 1250°C under the same mass content while opposite situation occurs at 1200°C.

1.61 Silicon–Copper–Calcium–Oxygen–Iron–Sulfur

The phase equilibria in the Si–Cu–Ca–O–Fe–S system in equilibrium with matte at controlled $p(SO_2)$ of 0.3 and 0.6 atm and fixed matte grade of 72 mass% Cu were experimentally investigated in the spinel primary phase field (Sun et al. 2020a). The high-temperature equilibrations were realized in the spinel substrate and the sample after quenching were characterized using EPMA. The effect of CaO on the liquidus temperature of slags was quantified with varying CaO content from 0 to 6 mass%. It was found that the presence of CaO increased the liquidus temperature of slags and moreover, the increment effect was enhanced with increasing CaO content in the present range. The influence of $p(SO_2)$ was further clarified and it was found that the equilibrium SiO_2 content in the liquid phase, at the same temperature, remarkably increased with increasing $p(SO_2)$.

The solubility of oxygen in synthetic copper matte in the system Si–Cu–Ca–O–Fe–S saturated with silica and metallic iron under a N_2 atmosphere at 1200°C was found to decrease from 6.2 mass% in FeS to about zero in a matte containing 80 mass% Cu (Bor and Tarassoff 1971). Nearly one-third of the total oxygen is present in the form of Fe_3O_4 and the remainder as FeO. Within the range studied, the CaO and Al_2O_3 content of the slag was found to have no measurable effect on the oxygen solubility.

1.62 Silicon–Copper–Zinc–Germanium–Sulfur

Cu_2ZnSiS_4–Cu_2ZnGeS_4. This system is nonquasibinary section of the Si–Cu–Zn–Ge–S quinary system (Olekseyuk et al. 2019). At 400°C, Cu_2ZnSiS_4 dissolves 50 mol% Cu_2ZnGeS_4 and the solubility of Cu_2ZnSiS_4 in Cu_2ZnGeS_4 reaches 5 mol%. The alloys for the investigations were annealed at 400°C for 500 h followed by quenching into cold water.

1.63 Silicon–Copper–Zinc–Tin–Sulfur

Cu_2ZnSiS_4–Cu_2ZnSnS_4. This system is nonquasibinary section of the Si–Cu–Zn–Sn–S quinary system (Olekseyuk et al. 2019). At 400°C, the mutual solubility of both components each in other is 5 mol%. The alloys for the investigations were annealed at 400°C for 500 h followed by quenching into cold water.

According to Hamdi et al. (2014), the $Cu_2Zn(Si_xSn_{1-x})S_4$ ($0 \le x \le 0.5$ and $0.8 \le x \le 1$) solid solutions are formed in this system. To obtain them, Cu, Zn, Si, Sn, and S were weighted in stoichiometric amounts, sealed under vacuum in silica tubes, and heated at temperature ranging from 750°C to 900°C versus the target composition (the higher the Si content the higher the temperature) for one week. The as-obtained samples were then ground in an agate mortar and annealed at the synthesis temperature for 96 h and subsequently quenched in an iced bath. Materials crystallize with the kesterite structure type for $x \le 0.5$, but with the wurtz-stannite (enargite) structure type for $x \ge 0.8$. In between, a miscibility gap exists, and the two solid solutions $Cu_2ZnSi_{0.5}Sn_{0.5}S_4$ and $Cu_2ZnSi_{0.8}Sn_{0.2}S_4$ coexist. The optical band gap increases continuously with the Si content in the whole series.

1.64 Silicon–Copper–Aluminum–Oxygen–Iron–Sulfur

The solubility of oxygen in synthetic copper mattes in the system Si–Cu–Al–O–Fe–S saturated with silica and metallic iron under a N_2 atmosphere at 1200°C was found to decrease from 6.2 mass% in FeS to about zero in a matte containing 80 mass% Cu (Bor and Tarassoff 1971). Nearly one-third of the total oxygen is present in the form of Fe_3O_4 and the remainder as FeO. Within the range studied, the CaO and Al_2O_3 content of the slag was found to have no measurable effect on the oxygen solubility.

1.65 Silicon–Copper–Lanthanum–Palladium–Sulfur

In the system containing these elements, the quinary compound $La_6Pd_{0.5}CuSi_2S_{14}$, which crystallizes as a hexagonal structure with the lattice parameters a = 1029.73 ± 0.04, c = 578.78 ± 0.03 pm, and a calculated density of 4.537 g·cm^{-3} at 173 K, is formed (Akopov et al. 2022). To obtain this compound, first La–"Cu_2Pd"–Si precursor was prepared via arc-melting using pieces of La, "Cu_2Pd", and Si. A sample with a total mass of 0.6–1 g was weighed out in a ratio of La/"Cu_2Pd"/Si = 6:1:2.05. Single crystals were grown using arc-melted precursor, S powder, and KI flux. The ratio of (precursor + S)/KI was kept at 1:30 ratio by mass. The reactants and KI were added to a silica ampoule, which was sealed under vacuum. The reaction products were washed with deionized water to remove KI flux; the sample was then filtered under vacuum and allowed to dry. Bulk powder of the material was synthesized using the appropriate arc-melted precursor and sulfur powder. Sample was prepared by loading the silicide binary and sulfur (1:14 ratio) in a fused silica ampoule, which was then flame sealed under vacuum. For both crystals and bulk powder, the samples were heated over 12 h to 1000°C–1050°C, dwelled for 72–120 h and then cooled to room temperature over 8 h.

1.66 Silicon–Copper–Germanium–Phosphorus–Sulfur

Cu_2SiS_3–$CuGe_2P_3$. $CuGe_2P_3$ does not form any solid solution with Cu_2SiS_3 (Omar 1990).

1.67 Silicon–Copper–Germanium–Iodine–Sulfur

Cu_7SiS_5I–Cu_7GeS_5I. The compositional dependence of the lattice parameters is linear and corresponds to the Vegard's law, which is evidence for the formation of a continuous solid solution in this system (Studenyak et al. 2007). Due to the incongruent melting character of both the quaternary compounds, the alloys of this system were synthesized by a solid-state reaction method in evacuated (0.13 Pa) silica glass ampoules and thoroughly mixed. The synthesis was performed by the following method: heating at a rate 50°C·h^{-1} to 750°C, aging at this temperature for 240 h, and cooling to room temperature at the same rate. In order to achieve homogeneity after cooling, the sintered samples were ground to powder in an agate mortar, loaded once more in ampoules, and the solid-state synthesis cycle was repeated. The chemical interactions in this system was studied by XRD, metallography, and density determination.

1.68 Silicon–Copper–Tin–Oxygen–Iron–Sulfur

The distribution behavior of Sn between the matte and SiO_2-saturated slag was investigated under the controlled partial pressures of SO_2 at 1250°C (Koike and Yazawa 1994). From the results obtained, it was found that Sn was concentrated in the matte phase in general, but with increasing matte grade, tin became concentrated in the slag phase. The activity coefficient of SnO in the slag equilibrating with the matte was calculated by thermodynamic analysis.

1.69 Silicon–Copper–Tin–Iron–Sulfur

The $Cu_{5.5}Si_{1.5}Fe_4Sn_{12}S_{32}$ quinary compound, which crystallizes as a cubic structure with the lattice parameter a =1033.22 ± 0.06 pm, is formed in this system (Garg et al. 2001). The band gap was found to be 0.107 eV in the temperature range 170 K to 300 K. To prepare this compound, the stoichiometric amounts of Cu, Si, Sn, Fe, and S were loaded in the composition $Cu_{3.31}SiFe_4Sn_{12}S_{32}$ in a N_2-filled glove box. This was sealed in silica ampoules under high vacuum (10^{-3} Pa) and heated at 250°C for 2 days followed by annealing at 700°C for 6 days and then quenched at 680°C. After the reaction, the black cuboidal crystals of the title compound were formed.

1.70 Silicon–Copper–Oxygen–Iron–Sulfur

The gas/slag/matte/tridymite (SiO_2) equilibria in this system at 1200°C have been experimentally studied by Fallah-Mehrjardi et al. (2020a). An improved experimental technique was used, which includes high-temperature equilibration of sulfide–oxide mixtures on silica substrates at $p(SO_2) = 0.6$ atm, preservation of the equilibrium phases by rapid quenching, and direct compositional analysis of the phases using EPMA. The new information gives a more accurate description of the effect of $p(SO_2)$ on the phase equilibria of the system. Several key findings from the study are as follows: at a fixed mass% Cu in matte, (1) the corresponding $p(O_2)$ increases with increasing $p(SO_2)$; (2) the Fe/SiO_2 ratio at the slag in equilibrium with tridymite increases with increasing $p(SO_2)$; and (3) the concentrations of S in matte, S and Cu in slag are not sensitive to the changes in $p(SO_2)$ for the conditions considered in this study.

Experimental studies were also undertaken to determine the gas/slag/matte/tridymite equilibria in this system at 1200°C, $p(SO_2) = 0.25$ atm, and a range of $p(O_2)$ (Fallah-Mehrjardi et al. 2017). The experimental methodology involved high-temperature equilibration using a substrate support technique in controlled gas atmospheres ($CO/CO_2/SO_2/Ar$), rapid quenching of equilibrium phases, followed by direct measurement of the chemical compositions of the phases with EPMA. The experimental data for slag and matte were presented as a function of copper concentration in matte (matte grade). The data provided are essential for the evaluation of the effect of oxygen potential under controlled atmosphere on the matte grade, liquidus composition of slag and chemically dissolved copper in slag. The new data provide important accurate and reliable quantitative foundation for improvement of the thermodynamic databases for Cu-containing systems.

The gas/slag/matte/spinel (Fe_3O_4) equilibrium in the Si–Cu–O–Fe–S system at controlled gas atmosphere has been studied by Hidayat et al. (2016). New data on the system at 1200°C and $p(SO_2) = 0.25$ atm have been obtained. It was found that matte grade 63 and 75 mass% correspond to $p(O_2)$ from $10^{-8.5}$ to $10^{-8.2}$ atm, respectively. It was observed that an increase in matte grade led to a decrease in sulfur in matte and a decrease in "FeO" and S in slag. The chemically dissolved Cu in the range of matte grade under investigation was found to be around 0.8 and 1.1 mass%.

Experimental measurements of gas/slag/matte/tridymite equilibria in the Si–Cu–O–Fe–S system at 1250°C, $p(SO_2) = 0.25$ atm, and a range of $p(O_2)$'s were undertaken by Fallah-Mehrjardi et al. (2018). The comparison of data obtained at 1250°C and 1200°C shows that for the given matte grade, (1) S concentration in the matte and the slag phases is independent of temperature; (2) the Fe/SiO_2 ratio in the slag phase is inversely proportional to the temperature; (3) the Cu content in slag increases with the increase of temperature for the given matte grade; and (4) the Fe/SiO_2 ratio in slag is inversely proportional to the matte grade and is lower at higher temperature for the given matte grade. This ratio is used to control the flux addition in the pyrometallurgical copper plants.

The improved experimental methodology based on equilibration, rapid quenching of equilibrium phases, and EPMA analysis of the phases was used to investigate the gas/slag/matte/spinel (Fe_3O_4) equilibria in the Si–Cu–O–Fe–S system at 1200°C, $p(SO_2) = 0.25$ atm, and a range of $p(O_2)$ covering Cu concentrations in matte between 42 and 78 mass% (Hidayat et al. 2018). The four-point-test approach was employed to confirm the equilibrium achievement. The approach involved the evaluation of the effect of equilibration time, homogeneity of equilibrated phases, effect of direction of achievement of equilibrium, and analysis of reactions specific to the system. Confirmation by independent experiments and analysis of solid-state diffusion of solutes in spinel solid was also done. A set of key graphs was introduced to describe the equilibrium in the multinary, multiphase system Si–Cu–O–Fe–S. The graphs provide information on $p(O_2)$ in gas, sulfur, and oxygen in matte, and Fe/SiO_2 ratio, S, Cu, and Fe^{3+}/Fe_{total} ratio in slag, all presented as functions of Cu in matte.

Experimental measurements have been also made of the compositions of slag, matte, and tridymite phases in chemical equilibrium in this system at 1300°C, $p(SO_2) = 0.25$ atm and selected oxygen partial pressures (Fallah-Mehrjardi et al. 2020b). The high-temperature equilibration experiments were conducted using silica substrates under controlled $CO–CO_2–SO_2–Ar$ gas atmospheres. The resulting phases obtained from the equilibrations were retained at room temperature through rapid quenching of the samples. The condensed phase compositions from the equilibrium experiments were measured by EPMA. The data obtained have enabled the liquidus slag temperature to be accurately described as a function of Fe/SiO_2 ratio at $p(SO_2) = 0.25$ atm for temperatures between 1200°C and 1300°C and mattes containing from ~44 to ~78 mass% Cu.

The phase equilibria in the "FeO"–SiO_2 subsystem of the Si–Cu–O–Fe–S system in equilibrium with matte at controlled $p(SO_2)$ 0.3 and 0.6 atm and fixed matte grade 72 mass% Cu were experimentally investigated by Chen et al. (2019). The high-temperature equilibration using primary phase material as the substrate, quenching and EPMA were applied in the experiments where the $p(O_2)$ and $p(SO_2)$ were accurately controlled by $CO–CO_2–SO_2$ gas mixtures. The correlations between oxygen partial pressure and the matte grade were determined at $p(SO_2)$ 0.3 and 0.6 atm to obtain the target matte grade in the samples. The liquidus temperatures of the "FeO"–SiO_2 subsystem in equilibrium with matte at controlled $p(SO_2)$ 0.3 and 0.6 atm and fixed matte grade 72 mass% Cu were also reported.

1.71 Silicon–Silver–Indium–Selenium–Sulfur

The $Ag_2In_2SiSe_{2.94}S_{3.06}$ quinary compound, which crystallizes as a monoclinic structure with the lattice parameters a = 1240.90 ± 0.11, b = 731.47 ± 0.06, c = 1235.97 ± 0.11 pm, β = 109.274 ± 0.002°, a calculated density of 5.042 cm^{-3}, and an energy gap of 2.13 eV, is formed in the Si–Ag–In–Se–S system (Li et al. 2023b). It was obtained by the traditional solid-state reactions. For the typical reaction, a 500 mg mixture of Ag, In, Si, S, and Se with the molar ratios of 2:2:1:3:3 and 400 mg KI as a flux were weighed, ground into fine powder, and pressed into a pellet, followed by sealed into an evacuated quartz ampoule with the vacuum degree of 10^{-2} Pa. The tube was heated in a muffle furnace from room temperature to 890°C in 25 h and homogenized at several intermediate temperatures for several hours. The reaction was maintained at 890°C for about 120 h, and then cooled down to 300°C within 150 h. Finally, the tube was cooled down to room temperature naturally. The red rod-like single crystals of the title compound were obtained. The crystals were stable in air after several months and can be separated from the byproducts easily under an optical microscope because of their unique colors and shapes.

1.72 Silicon–Silver–Cerium–Chlorine–Sulfur

In the system containing these elements, the quinary compound $Ag_{0.63}Ce_3S_{2.63}Cl_{0.37}[SiS_4]$, which crystallizes as a hexagonal structure with the lattice parameters a = 1028.98 ± 0.05, c = 573.34 ± 0.03 pm, and a calculated density of 4.676 cm^{-3}, is formed (Hartenbach et al. 2007). It was prepared from a mixture Ag, S, SiS_2, and Ce_2S_3 with an excess of $CeCl_3$ as a flux. The annealing of the mixture at 850°C for 8 days in evacuated silica glass ampoule lead to dark red columnar single crystals of the title compound. They are air and water resistant.

1.73 Silicon–Silver–Germanium–Iodine–Sulfur

$Ag_7SiS_5I–Ag_7GeS_5I$. The phase diagram of this system (Figure 1.2), constructed through differential thermal analysis (DTA), XRD, and metallography using the alloy annealed at 460°C for 120 h, belongs to the type I according to the Rosebbom's classification (Pogodin et al. 2021).

1.74 Silicon–Beryllium–Calcium–Oxygen–Sulfur

In the system containing these elements, the quinary compound $Be_6Ca_8(SiO_4)_6(SO_4)_2$, which crystallizes as a cubic structure with the lattice parameter a = 877 pm, is formed (Kondo 1965). It was synthesized by means of substitution of the components of $Na_6Ca_2(Al_6Si_6O_{24})(SO_4)_2$, that is, Na by Ca and Al by Be.

1.75 Silicon–Beryllium–Zinc–Oxygen–Sulfur

In the system containing these elements, the quinary compound $Zn_4Be_3(SiO_4)_3S$ (mineral genthelvite), which crystallizes as a cubic structure with the lattice parameter a = 814.93 ± 0.05 and 810.91 ± 0.04 pm

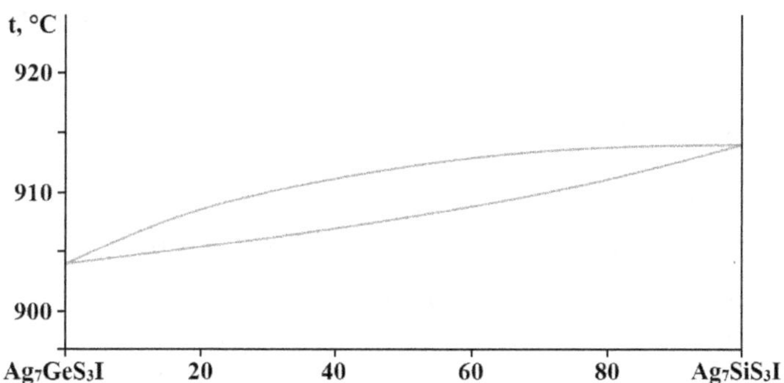

FIGURE 1.2 Phase diagram of the Ag_7SiS_5I–Ag_7GeS_5I system. (From Pogodin, A.I., et al., *Nauk. Visn. Uzhgorod. Univ., Ser. Khim.* [**1**(45)], 42, 2021.)

and a calculated density of 3.64 and 3.72 g·cm^{-3} for two different samples (Hassan and Grundy 1985) [a = 811.2 ± 0.6 and 816.6 ± 0.3 pm for two different samples (Lunts 1963); a = 813.3 ± 0.5 pm and an experimental density of 3.58 ± 0.01 g·cm^{-3} (Morgan 1967)], is formed.

This compound was synthesized trough hydrothermal synthesis at 450°C and 0.15 GPa for 7–10 days (1) using the mixture of ZnO, BeO, SiO_2, and ZnS with the solution of 10 mass% NaOH as a solvent, (2) using the same mixture with the solution of 10 mass% Na_2CO_3 or 10 mass% $NaHCO_3$ as a solvent, or (3) using the mixture of ZnO, BeO, and SiO_2, with the solution of 10 mass% Na_2S as a solvent (Mel'nikov et al. 1968). It can be also prepared at the same temperature and 0.2 GPa for 7–10 days (1) using the mixture of ZnO, BeO, SiO_2, and ZnS with the solution of 1 mass% NaOH or H_2O as a solvent or (2) using the mixture of ZnO, BeO, SiO_2, and S with the solution of 1 mass% NaOH as a solvent.

1.76 Silicon–Beryllium–Zinc–Oxygen–Iron–Sulfur

In the system containing these elements, the multinary compound $Zn_2Be_3Fe_2(SiO_4)_2S$, which crystallizes as a cubic structure with the lattice parameter a = 809.7 ± 0.1 pm, is formed (Armstrong et al. 2003). For obtaining this compound, BeO, SiO_2, powdered Fe and Zn, and S were homogeneously ground using an agate pestle and mortar in an Ar-filled glove box (1 ppm H_2O/1 ppm O_2). The mixture was placed in a quartz tube, which was evacuated and sealed. The tube was heated in the center of a tube furnace fitted with an inconel work tube in stages up to 1100°C and maintained at this temperature for 48 h. The sample tube was cooled slowly and transferred back to the glove box, where they were opened with a glass cutter.

1.77 Silicon–Beryllium–Cadmium–Oxygen–Sulfur

In the system containing these elements, the quinary compound $Cd_4Be_3(SiO_4)_3S$ is formed (Mel'nikov et al. 1968). This compound was synthesized trough hydrothermal synthesis at 450°C and 0.2 GPa for 7–10 days using the mixture of CdO, BeO, SiO_2, and CdS with the solution of 1 mass% NaOH as a solvent.

1.78 Silicon–Beryllium–Oxygen–Manganese–Sulfur

In the system containing these elements, the quinary compound $Mn_4Be_3(SiO_4)_3S$ (mineral helvine), which crystallizes as a cubic structure with the lattice parameter a = 829.13 ± 0.06 pm and a calculated density of 3.23 g·cm^{-3} and a = 823.65 ± 0.04 pm for two different samples (Hassan and Grundy 1985) [a = 852.5 pm and an experimental density of 3.2 g·cm^{-3} (Gottfried 1927); a = 825 pm (Pauling 1930); a = 829.4 ± 0.7 pm and a calculated density of 3.25 g·cm^{-3} (Holloway, Jr and Giordano 1972)], is formed.

This compound was synthesized through hydrothermal synthesis at 450°C and 0.15 GPa for 7–10 days (1) using the mixture of MnO, BeO, SiO_2, and S with the solution of 10 mass% $MnSO_4$ as a solvent, or at the same temperature and 0.2 GPa for 7–10 days or (2) using the same mixture with the solution of 8 mass% NH_4Cl or H_2O as a solvent (Mel'nikov et al. 1968).

1.79 Silicon–Beryllium–Oxygen–Manganese–Iron–Sulfur

In the system containing these elements, the multinary compound $Mn_2Be_3Fe_2(SiO_4)_2S$, which crystallizes as a cubic structure with the lattice parameter a = 822.3 ± 0.1 pm, is formed (Armstrong et al. 2003). It was prepared in the same way as $Zn_2Be_3Fe_2(SiO_4)_2S$ was synthesized using Mn instead of Zn.

1.80 Silicon–Beryllium–Iron–Oxygen–Sulfur

In the system containing these elements, the multinary compound $Fe_4Be_3(SiO_4)_3S$ (mineral danalite), which melts at 1274°C and crystallizes as a cubic structure with the lattice parameter a = 823.264 ± 0.005 pm at 33°C (Antao et al. 2003) [a = 819.6 pm (Thompson 1957); a = 823.17 ± 0.09 and 821.82 ± 0.02 pm a calculated density of 3.40 and 3.41 g·cm^{-3} for two different samples (Hassan and Grundy 1985); a = 821.27 ± 0.08 pm (Nimis et al. 1996); a = 820.3 ± 0.1 pm (Armstrong et al. 2003)], is formed.

The a cell parameter increases linearly with temperature without abrupt changes or discontinuities to 1035°C (Antao et al. 2003). The a parameter expansion is given by the least-squares fit: $a = 8.2307 + 4.4906 \cdot 10^{-5}$ T.

This compound was prepared in the same way as $Zn_2Be_3Fe_2(SiO_4)_2S$ was synthesized without using Zn (Armstrong et al. 2003).

1.81 Silicon–Calcium–Barium–Oxygen–Sulfur

Two quinary compounds are formed in this system. $Ca_6Ba(SiO_4)_2(SO_4)_2O$ (mineral gazeevite) crystallizes as a trigonal structure with the lattice parameters a = 715.40 ± 0.01, c = 2512.42 ± 0.05 pm, and a calculated density of 3.39 g·cm^{-3} (Galuskin et al. 2015b, 2017; Belakovskiy et al. 2017b).

$Ca_{12}Ba(SiO_4)_4(SO_4)_2O_3$ (mineral dargaite) also crystallizes as a trigonal structure with the lattice parameters a = 718.74 ± 0.04, c = 4129.2 ± 0.3 pm, and a calculated density of 3.235 g·cm^{-3} (Gfeller et al. 2015; Galuskina et al. 2019).

1.82 Silicon–Calcium–Aluminum–Oxygen–Sulfur

In the system containing these elements, the quinary compound $3CaAl_2Si_2O_8 \cdot CaSO_4$, which crystallizes as a tetragonal structure with the lattice parameters a = 1219.25 ± 0.10 and c = 760.45 ± 0.19 pm, is formed (Newton and Goldsmith 1976). This compound is a true high-pressure as well as a high-temperature phase, and its minimum temperature and pressure conditions for stability are 775°C at 1.7 GPa. Its melting temperature rises steeply from 1545°C at 1.0 GPa to 1555°C at 2.5 GPa. This compound can be obtained by the interaction of $CaAl_2Si_2O_8$ and $CaSO_4$ (molar ratio 3:1) at 1350°C and 1.5 GPa for 2 h.

1.83 Silicon–Calcium–Phosphorus–Oxygen–Fluorine–Sulfur

Some multinary compounds are formed in this system. $Ca_{9.5}(SiO_4)(PO_4)_3(SO_4)_2F_2$ crystallizes as a hexagonal structure with the lattice parameters a = 946, c = 691 pm, and the calculated and experimental densities of 3.04 and 3.10 g·cm^{-3}, respectively (Dihn and Klement 1942). It was synthesized by heating the mixture of $Ca_3(PO_4)_2$, $CaCO_3$, SiO_2, $CaSO_4$, and CaF_2 (molar ratio 3:4:2:4:2) at 1100°C for 7 h.

$Ca_{10}(SiO_4)(PO_4)_4(SO_4)F_2$ also crystallizes as a hexagonal structure with the lattice parameters $a = 945$, $c = 696$ pm, and the calculated and experimental densities of 3.09 and 3.13 g·cm^{-3}, respectively and can be prepared by heating the mixture of $Ca_{10}(PO_4)_6F_2$, $CaCO_3$, SiO_2, $CaSO_4$, and CaF_2 (molar ratio 2:6:3:3:1) (Dihn and Klement 1942).

$Ca_{10}(SiO_4)_2(PO_4)_2(SO_4)_2F_2$ also crystallizes as a hexagonal structure and can be obtained by heating the mixture of $Ca_3(PO_4)_2$, $CaCO_3$, SiO_2, $CaSO_4$, and CaF_2 (molar ratio 1:4:2:2:1) (Dihn and Klement 1942) or the mixture of $Ca_3(PO_4)_2$, Ca_2SiO_4, $CaSO_4$, and CaF_2 (molar ratio 1:2:2:1) (Klement 1939).

$Ca_{10.5}(SiO_4)_2(PO_4)_3(SO_4)F_2$ also crystallizes as a hexagonal structure with the lattice parameters $a = 940$, $c = 681$ pm, and the calculated and experimental densities of 3.23 and 3.20 g·cm^{-3}, respectively (Dihn and Klement 1942). It was synthesized by heating the mixture of $Ca_3(PO_4)_2$, $CaCO_3$, SiO_2, $CaSO_4$, and CaF_2 (molar ratio 3:8:4:2:2) at 1100°C for 7 h.

1.84 Silicon–Calcium–Phosphorus–Oxygen–Chlorine–Sulfur

The $Ca_{10}[(SiO_4)_x(PO_4)_{6-2x}(SO_4)_x]Cl_2$ solid solutions are formed in this system (Fang et al. 2014). They crystallize in the hexagonal structure with the lattice parameters $a = 961.82 \pm 0.01$ and $c = 679.20 \pm 0.01$ pm for $x = 0$, $a = 962.99 \pm 0.02$ and $c = 678.33 \pm 0.02$ pm for $x = 0.27$, $a = 962.57 \pm 0.04$ and $c = 678.92 \pm 0.03$ pm for $x = 0.66$, $a = 965.37 \pm 0.05$ and $c = 683.54 \pm 0.04$ pm for $x = 2.03$, $a = 966.22 \pm 0.03$ and $c = 684.38 \pm 0.02$ pm for $x = 2.4$, and $a = 966.91 \pm 0.03$ and $c = 685.31 \pm 0.03$ pm for $x = 3$. These solid solutions were prepared by a solid-state method where the nominal overall reaction was: $2x$CaO + $CaCl_2$ + $xSiO_2$ + $xCaSO_4$ + $(3–x)Ca_3(PO_4)_2$. The powders of the initial compounds were mixed in stoichiometric amounts. After grinding in acetone and drying, the mixtures were fired at 950°C in air for 9 h in alumina crucibles covered by an alumina slide to slow chlorine volatilization.

1.85 Silicon–Calcium–Oxygen–Fluorine–Sulfur

The $Ca_{10}(SiO_4)_3(SO_4)_3F_2$ quinary compound (mineral fluorellestadite), which melts congruently at 1250°C (Klement and Dihn 1941) and crystallizes as a hexagonal structure with the lattice parameters given in Table 1.3, is formed in this system.

The raw materials used for fluorellestadite synthesis were $CaCO_3$, $CaSO_4·2H_2O$, SiO_2, and CaF_2 in a stoichiometric ratio (Pajares et al. 2002). The ground mixture was heated at 1000°C for 2 h. Then, the sample was ground again in an agate mortar and heated at 1000°C for 32 h. This compound can be also synthesized at the sintering of the mixture of Ca_2SiO_4, $CaSO_4$, and CaF_2 (molar ratio 3:3:1) at 1150°–1200°C for a few hours in air (Klement 1939; Schwarz 1967a). Colorless crystals were obtained after quickly or slowly cooling.

1.86 Silicon–Calcium–Oxygen–Fluorine–Chlorine–Sulfur

The $Ca_{10}(SiO_4)_3(SO_4)_3F_xCl_{2-x}$ solid solutions, which crystallize in the hexagonal structure, are formed in this system (Fang et al. 2011). Their lattice parameters vary linearly with composition and show the expected shrinkage of unit cell volume as fluorine displaces chlorine: $a = 967.02 \pm 0.03$ and $c = 685.43 \pm 0.03$ pm for $x = 0$, $a = 962.39 \pm 0.03$ and $c = 687.749 \pm 0.003$ pm for $x = 0.4$, $a = 957.72 \pm 0.02$ and $c = 689.50 \pm 0.01$ pm for $x = 0.8$, $a = 953.12 \pm 0.02$ and $c = 691.81 \pm 0.02$ pm for $x = 1.2$, $a = 950.64 \pm 0.02$ and $c = 692.63 \pm 0.01$ pm for $x = 1.6$, and $a = 944.86 \pm 0.01$ and $c = 694.26 \pm 0.01$ pm for $x = 2.0$. These solid solutions were synthesized by solid-state reaction. The starting materials were prepared by preheating $CaCO_3$ at 1000°C for 12 h; SiO_2 and $CaSO_4·2H_2O$ at 600°C for 4 h; and $CaCl_2·2H_2O$ at 400°C for 4 h. Stoichiometric mixtures of CaO, $CaSO_4$, SiO_2, $CaCl_2$, and CaF_2 were manually ground with

TABLE 1.3

Crystallographic data for mineral fluorellestadite, $Ca_{10}(SiO_4)_3(SO_4)_3F_2$

		$d_{calc.}$	$d_{meas.}$	
a (pm)	*c* (pm)	g·cm^{-3}		References
953 ± 1	691 ± 1	3.046	3.068	McConnell 1937
954	699	3.00	3.06	Klement and Dihn 1941; Dihn and Klement 1942
944.7 ± 0.3	693.8 ± 0.3	3.106	3.08	Schwarz 1967a
948.5 ± 0.2	691.6 ± 0.2	3.09	3.03 ± 0.01	Chesnokov et al. 1987; Jambor and Puziewicz 1989
944.174 ± 0.009	693.964 ± 0.008	–	–	Pajares et al. 2002
944.86 ± 0.01	694.26 ± 0.01	–	–	Fang et al. 2011
944.71 ± 0.02[a]	693.11 ± 0.02[a]	3.131–3.179	–	Avdontceva et al. 2021
945.50 ± 0.02[b]	693.64 ± 0.02[b]	for 9 different samples	–	
946.80 ± 0.02[c]	694.58 ± 0.02[c]		–	
948.18 ± 0.02[d]	695.53 ± 0.02[d]		–	
949.48 ± 0.02[e]	696.52 ± 0.02[e]		–	
950.94 ± 0.02[f]	697.64 ± 0.02[f]		–	
952.42 ± 0.02[g]	698.79 ± 0.02[g]		–	
954.09 ± 0.02[h]	699.98 ± 0.02[g]		–	
955.94 ± 0.02[i]	701.26 ± 0.02[i]		–	

[a]At 20°C; [b]at 100°C; [c]at 200°C; [d]at 300°C; [e]at 400°C; [f]at 500°C; [g]at 600°C; [h]at 700°C; [i]at 800°C.

acetone and annealed at 900°C for 5 h in an alumina crucible and then reground and sintered at 950°C for a further 9 h.

1.87 Silicon–Calcium–Oxygen–Chlorine–Sulfur

Two quinary compounds are formed in this system. $Ca_4(SiO_4)(SO_4)Cl_2$ crystallizes as an orthorhombic structure with the lattice parameters a = 628, b = 700, c = 1090 pm and a calculated density of 2.90 g·cm^{-3} (Lampe von et al. 1989). It decomposes at ~950°C with formation of $Ca_{10}(SiO_4)_3(SO_4)_3Cl_2$ and $CaCl_2$. This compound forms, when a mixture of $CaCO_3$, SiO_2, and $CaSO_4$ blended in the ratio $CaO/SiO_2/CaSO_4$ = 46.16:16.18:37.75 was mixed with dried $CaCl_2$ in a mass proportion of 6:4, heated up to 800°C, annealed for 20 h and then cooled down slowly (20°C·h^{-1}). After dissolving excess $CaCl_2$ in methanol, one gets a white, finely crystallized powder containing a lot of very thin clear plate-like crystals of the title compound.

$Ca_{10}(SiO_4)_3(SO_4)_3Cl_2$ (mineral chlorellestadite) crystallizes in general in the hexagonal structure with the lattice parameters given in Table 1.4. Assuming the resulting composition $Ca_9(SiO_4)_3(SO_4)_3{\cdot}(1-x)CaCl_2$ for the nonstoichiometric $Ca_{10}(SiO_4)_3(SO_4)_3Cl_2$, $CaCl_2$ deficiency of about 30% could be achieved already after annealing for 12 h (Saint-Jean and Hansen 2005). The title compound begins to decompose at about 1200°C (Chen and Fang 1989) into three phases: $Ca_5(SiO_4)_2(SO_4)$, $CaSO_4$, and CaO (Saint-Jean and Hansen 2005).

$Ca_{10}(SiO_4)_3(SO_4)_3Cl_2$ with hexagonal structure was prepared by heating stoichiometric amounts of CaO, SiO_2, and $CaSO_4$ with $CaCl_2$ in 100% excess (Chen and Fang 1989; Saint-Jean et al. 2005; Saint-Jean and Hansen 2005). The starting materials used were prepared by preheating for 1 h: $CaCO_3$ at 1000°C; amorphous $SiO_2{\cdot}0.18H_2O$ and $CaSO_4{\cdot}2H_2O$ at 600°C; and $CaCl_2{\cdot}2H_2O$ at 400°C. After grinding in acetone and drying, the mixture was placed in boats of unglazed porcelain and heated in air at 1000°C for 4 h. Both quenching and slow cooling were attempated, but they gave the same results: a very fine

TABLE 1.4

Crystallographic data for mineral chlorellestadite, $Ca_{10}(SiO_4)_3(SO_4)_3Cl_2$

a (pm)	*c* (pm)	$d_{calc.}$	$d_{meas.}$	References
		g·cm^{-3}		
968.8	684.9	3.09	3.003	Chen and Fang 1989
967.73 ± 0.03	685.85 ± 0.01	–	–	Saint-Jean et al. 2005; Saint-Jean and Hansen 2005
956.08 ± 0.04	689.49 ± 0.02	–	–	Saint-Jean and Hansen 2005[a]
967.02 ± 0.03	685.43 ± 0.03	–	–	Fang et al. 2011
966.91 ± 0.03	685.31 ± 0.03	–	–	Fang et al. 2014
960.02 ± 0.02	689.92 ± 0.02	3.091	–	Środek et al. 2017a,b, 2018
(959.27 ± 0.06) – (964.32 ± 0.03)	(686.74 ± 0.03) – (690.08 ± 0.07)	–	–	Garbev et al. 2022[b]

[a]For $Ca_{9.4}(SiO_4)_3(SO_4)_3Cl_{0.8}$ composition; [b]for different samples prepared in various conditions.

white powder with a few needle-shaped crystals grown from the $CaCl_2$ flux. The prepared powders were washed with water to remove the excess $CaCl_2$ and dried. The product was then annealed in air for various periods of time between 2 and 48 h at 900°C, 950°C, 1000°C, and 1100°C, ground and analyzed by XRD. The obtained crystals were colorless and transparent.

According to Garbev et al. (2022), at lower temperatures $Ca_{10}(SiO_4)_3(SO_4)_3Cl_2$ crystallizes as a monoclinic structure with the lattice parameters changing from 955.48 ± 0.05 to 960.2 ± 0.1 for *a*, from 953.45 ± 0.04 to 956.4 ± 0.2 for *b*, from 689.8 ± 0.4 pm to 692.4 ± 0.3 for *c*, and from 119.95 ± 0.01° to 120.043 ± 0.007° for β for different samples prepared at various temperatures. To prepare monoclinic $Ca_{10}(SiO_4)_3(SO_4)_3Cl_2$, two samples from wastes of autoclaved aerated concrete with different sulfate contents were crushed and dried at 250°C. These samples were mixed with $CaCO_3$ and $CaCl_2$ with a targeted molar C/S-ratio of 2. $CaCl_2$ also was predried at 250°C. The mixtures were homogenized for 30 min in a planetary ball mill. About 20–30 mg of fine ground sample was weighed into a ceramic sample boat. Additionally, about 10–15 mg of tungsten oxide was added for better combustion of the sample. Amounts of 100 g of the samples were calcined in a furnace at temperatures between 700°C and 1200°C in air. The samples were heated at a rate of 10°C·min^{-1} from room temperature to the desired temperature followed by a dwelling time of 60 min. After that, the furnace was switched off and the samples were slowly cooled down to room temperature.

1.88 Silicon–Strontium–Phosphorus–Oxygen–Fluorine–Sulfur

The $Sr_{10}(SiO_4)_{x/2}(PO_4)_{6-x}(SO_4)_{x/2}F_2$ continuous series of solid solutions were formed in this system (Schwarz 1967a). They crystallize in the hexagonal structure with the lattice parameters a = 972.0 ± 0.3, c = 728.9 ± 0.3 pm, and the calculated and experimental densities of 4.132 and 4.09 g·cm^{-3}, respectively, for $x = 0$, a = 974.4 ± 0.3, c = 730.5 ± 0.3 pm, and a calculated density of 4.098 g·cm^{-3} for $x = 2$, and a = 978.7 ± 0.3, c = 733.0 ± 0.3 pm, and the calculated and experimental densities of 4.043 and 4.03 g·cm^{-3}, respectively, for $x = 4$.

1.89 Silicon–Strontium–Oxygen–Fluorine–Sulfur

In the system containing these elements, the quinary compound $Sr_{10}(SiO_4)_3(SO_4)_3F_2$, which crystallizes as a hexagonal structure with the lattice parameters a = 979.6 ± 0.3, c = 734.3 ± 0.3 pm, and the calculated and experimental densities of 4.023 and 4.00 g·cm^{-3}, respectively, is formed (Schwarz 1967a). This compound decomposes at temperatures above 1150°C into $SiSO_4$, Sr_2SiO_4, and SrF_2. It was prepared by sintering a mixture of Sr_2SiO_4, $SrSO_4$, and SrF_2 (molar ratio 3:3:1) twice at 1000°C–1100°C first for 16 and then for 8 h in air. Colorless crystals were obtained after quickly or slowly cooling.

1.90 Silicon–Barium–Gallium–Oxygen–Sulfur

In the system containing these elements, the quinary compound $Ba_5Ga_2SiO_4S_6$, which crystallizes as a monoclinic structure with the lattice parameters $a = 1178.5 \pm 0.3$, $b = 1382.9 \pm 0.4$, $c = 1162.3 \pm 0.3$ pm, $\beta = 115.847 \pm 0.004°$, and a calculated density of 4.327 $g{\cdot}cm^{-3}$, is formed (Shi et al. 2023). This compound was prepared from the mixture of BaS, Ga_2S_3, Ga_2O_3, and SiO_2 in a molar ratio of 15:1:2:3. All of the starting materials were weighed and dropped into quartz tube in an argon-filled glove box. The quartz tube was evacuated to 10^{-3} Pa, sealed, and then placed in a furnace. Subsequently, the quartz tube was heated from ambient temperature to 253°C within 20 h and maintained for 24 h, then heated to 1000°C for 35 h and maintained for 150 h, and finally cooled at a rate of 3.33°C·h^{-1} to 500°C before turning off the furnace. After washing the product with deionized water, light-yellow crystals of the title compound were separated. They are insoluble in water and ethanol and showed high durability toward air and moisture for more than one year.

1.91 Silicon–Barium–Gallium–Oxygen–Chlorine–Sulfur

In the system containing these elements, the multinary compound $[(Ba_{19}Cl_4)(Ga_6Si_{12}O_{42}S_8)]$, which crystallizes as an orthorhombic structure with the lattice parameters $a = 2502.8 \pm 0.2$, $b = 1076.0 \pm 0.7$, $c = 2507.6 \pm 0.2$ pm, a calculated density of 4.362 $g{\cdot}cm^{-3}$, and an energy gap of 4.65 eV, is formed (Shi et al. 2019a). Single crystals of this compound were found in a preparation based on the precursors $BaCl_2$, Ga_2O_3, SiO_2, BaO, BaS, and KCl in a 2:3:12:8:9:4 molar ratios by the high-temperature solid-state reaction. The starting materials were loaded into a carbon-coated silica crucible in a pressure of 10^{-3} Pa, and finally placed in a tube furnace. The reactants were gradually heated from room temperature to 400°C at a rate of 20°C·h^{-1} and maintained there for 12 h, and then heated to 800°C at a rate of 10°C·h^{-1} and kept at 800°C for 3 days. Finally, it was slowly cooled to 500°C at a rate of 4°C·h^{-1}, and then the furnace was switched off. The products were washed first with distilled water and then dried with ethanol. After such a treatment, colorless plates of the title compound were obtained.

1.92 Silicon–Barium–Indium–Oxygen–Sulfur

In the system containing these elements, the quinary compound $Ba_2In_2Si_3O_{10}S$, which crystallizes as an orthorhombic structure with the lattice parameters $a = 2306.0 \pm 0.6$, $b = 529.54 \pm 0.06$, $c = 877.9 \pm 0.2$ pm, a calculated density of 4.837 $g{\cdot}cm^{-3}$, and an energy gap of 4.64 eV, is formed (Guo et al. 2016). Single crystals of this compound were obtained by the high-temperature solid-state reaction employing K_2S (2.0 mM), S (2.0 mM), In (4.0 mM), SiO_2 (1.0 mM), and $BaCl_2$ (4.0 mM). The starting materials were ground into fine powders in an agate mortar and pressed into a pellet, which was loaded into a quartz tube, which was evacuated to 10^{-2} Pa and flame sealed. The tube was placed into a furnace, heated from a room temperature to 400°C at a rate of 35°C·h^{-1}, kept at 400°C for 10 h, and then heated to 800°C at a rate of 20°C·h^{-1} and kept at 800 °C for 4 days. Finally, it was slowly cooled to 400°C at a rate of 3°C·h^{-1}, and then the furnace was switched off. All the resultant materials were first carefully scraped from the tube and washed several times by ultrasonic cleaning in deionized water. Colorless flake-like crystals of the title compound were handpicked under a microscope.

1.93 Silicon–Barium–Lanthanum–Cerium–Sulfur

The $Ba(La_{1-x}Ce_x)_2Si_2S_8$ ($x \leq 0.10$) solid solution, which crystallizes as a trigonal structure with the lattice parameters $a = 917.344 \pm 0.016$ and $c = 2729.16 \pm 0.06$ pm for $x = 0.06$, is formed in this quinary system (Lee et al. 2014). The powder samples of this solid solution were prepared by solid-state reactions using La_2S_3, BaS, Ce_2S_3, and Si and S powders as raw materials. The stoichiometric amounts of the starting materials were thoroughly mixed and loaded into a vertically positioned fused silica ampoule, fully evacuated to 0.1 Pa, and sealed off under a dynamic vacuum. The fused silica ampoule was heated at 5°C·min^{-1} to 1000°C–1100°C for 6–8 h in a furnace and then cooled down slowly to room temperature.

1.94 Silicon–Barium–Uranium–Iron–Sulfur

In the system containing these elements, the quinary compound $Ba_8SiFeUS_{14}$, which crystallizes as a monoclinic structure with the lattice parameters $a = 2374.77 \pm 0.09$, $b = 706.68 \pm 0.03$, $c = 886.24 \pm 0.03$ pm, $\beta = 111.243 \pm 0.001°$, a calculated density of 4.479 g·cm^{-3} at 100 K, and an energy gap of 1.8 eV, is formed (Mesbah et al. 2015). This compound was obtained serendipitously in an attempt to synthesize new phases in the ternary Ba–U–S system. The reaction mixture consisted of U (0.084 mM), BaS (0.25 mM), and S (1.469 mM). It was heated to 950°C for 24 h, left to stand for 96 h, after which the tube was cooled to 200°C at a rate of 3°C·h^{-1}, and then the furnace was turned off. Black needles were isolated and analyzed by energy-dispersive x-ray (EDX) analysis to reveal Ba/Si/Fe/U/S ≈ 8:1:1:1:14. The presence of Fe was traced to contamination from another reaction loaded in the same set. Again the source of Si was the tube. A rational synthesis of $Ba_8SiFeUS_{14}$ was achieved by the reaction of Ba (0.840 mM), Fe (0.106 mM), U (0.105 mM), Si (0.224 mM), and S (1.469 mM). The reactants were first heated to 1000°C in 18 h and held for 8 h followed by quenching the tube in air. The reaction mixture was then annealed at 900°C for 99 h and then the furnace was switched off. The product contained black needles of the title compound. Primary phases included UOS, BaS, and Ba_2SiS_4.

1.95 Silicon–Barium–Tin–Oxygen–Sulfur

The $Ba_2SnSi_2SO_7$ quinary compound, which crystallizes as a tetragonal structure with the lattice parameters $a = 858.88 \pm 0.07$, $c = 546.75 \pm 0.05$ pm, a calculated density of 4.8881 g·cm^{-3}, and an energy gap of 2.4 eV (2.7 eV) assuming indirect (direct) transitions (Almoussawi et al. 2022) [$a = 859.15 \pm 0.03$, $c = 548.35 \pm 0.04$ pm, a calculated density of 4.871 g·cm^{-3}, and an energy gap of 4.05 eV (Shi et al. 2022)], is formed in the Si–Ba–Sn–O–S system. A single-phase material of this compound was prepared from a stoichiometric mixture of the precursors $BaO/SnO_2/SiO_2/Si/S$ (molar ratio of 2:1:1.5:0.5:1) (Almoussawi et al. 2022). Those precursors were mixed and thoroughly ground in an agate mortar before being pressed into pellets and heated in an evacuated sealed quartz tube. The heat treatment involved heating up to 500°C (at a rate of about 30°C·h^{-1}) for 5 h, then up to 750°C (at a rate of about 20°C·h^{-1}) at which the sample was kept for 24 h, and then the furnace was switched off.

This compound can be also prepared if a mixture of BaO (1 mM), BaS (1 mM), SnO_2 (1 mM), SiO_2 (2 mM), and KBr (2 mM) was sealed into an evacuated silica tube and subjected to the following heat treatment (Shi et al. 2022). After that, it was quickly heated to 650°C in 10 h, maintained at this temperature for 10 h, then, quickly heated to 850°C in 10 h, maintained at this temperature for 72 h, and then cooled to ambient conditions at 6°C·h^{-1}. Finally, the colorless single crystals of $Ba_2SnSi_2O_7S$ with a high yield of 90% (based on Sn) were obtained in air after washing with deionized water and drying with ethanol. All weighing processes were completed in an anhydrous and oxygen-free glove box.

1.96 Silicon–Zinc–Lead–Oxygen–Sulfur

In the system containing these elements, the quinary compound $Zn_2Pb_4(SiO_4)(Si_2O_7)(SO_4)$ (mineral queitite), which crystallizes as a monoclinic structure with the lattice parameters $a = 1136.2 \pm 0.3$, $b = 526.6 \pm 0.1$, $c = 1265.5 \pm 0.3$ pm, $\beta = 108.16 \pm 0.02°$, and a calculated density of 6.07 g·cm^{-3}, is formed (Fleischer et al. 1979a, 1980a; Hess and Keller 1980).

1.97 Silicon–Yttrium–Lanthanum–Lead–Sulfur

The $Y_xLa_{1-x}PbSi_2S_8$ solid solutions are formed in this system (Kaczorowski et al. 2020). They crystallize in the trigonal structure with the lattice parameters a = 900.30 ± 0.01 and c = 2675.10 ± 0.04 pm for x = 0.5, a = 894.96 ± 0.02, c = 2648.90 ± 0.08 pm, and a calculated density of 4.054 g·cm^{-3} for x = 1.0, and a = 889.49 ± 0.05 and c = 2620.30 ± 0.02 pm for x = 1.5. They were prepared by solid state syntheses. The nominal amounts of the elemental constituents were sealed in evacuated quartz tubes. All preparations were carried out in an inert atmosphere. The ampoules were first heated with a rate of 30°C·h^{-1} up to 1150°C and then kept at this temperature for 3 h. Afterwards, the samples were cooled slowly (10°C·h^{-1}) down to 500°C and annealed at this temperature for 720 h. Subsequently, the ampoules were quenched in cold water. These solid solutions were found to be stable against air and moisture.

1.98 Silicon–Yttrium–Cerium–Lead–Sulfur

The $Y_xCe_{1-x}PbSi_2S_8$ solid solutions are formed in this system (Kaczorowski et al. 2020). They crystallize in the trigonal structure with the lattice parameters a = 898.76 ± 0.09 and c = 2696.60 ± 0.03 pm for x = 0.5, a = 892.35 ± 0.02, c = 2637.40 ± 0.08 pm, and a calculated density of 4.102 g·cm^{-3} for x = 1.0, and a = 888.48 ± 0.06, c = 2616.90 ± 0.02 pm, and a calculated density of 4.0204 ± 0.0002 g·cm^{-3} for x = 1.5 (Kaczorowski et al. 2020; Melnychuk et al. 2020). They were prepared in the same way as $Y_xLa_{1-x}PbSi_2S_8$ was synthesized and also were found to be stable against air and moisture (Kaczorowski et al. 2020). In the case of a composition with x = 1.5, synthesis can also be carried out according to the following regime: heating to 600°C (36°C·h^{-1}), holding for 24 h, heating to 1170°C (12°C·h^{-1}), holding for 4 h, cooling to 500°C (6°C·h^{-1}), annealing for 500 h, and quenching in water together with the ampoule (Melnychuk et al. 2020).

1.99 Silicon–Yttrium–Praseodymium–Lead–Sulfur

The $Y_xPr_{1-x}PbSi_2S_8$ solid solutions are formed in this system (Kaczorowski et al. 2020). They crystallize in the trigonal structure with the lattice parameters a = 893.90 ± 0.01, c = 2650.40 ± 0.04 pm, and a calculated density of 4.2076 ± 0.0002 g·cm^{-3} for x = 0.5, a = 890.51 ± 0.05, c = 2631.20 ± 0.15 pm, and a calculated density of 4.133 g·cm^{-3} for x = 1.0, and a = 887.86 ± 0.06 and c = 2616.00 ± 0.02 pm for x = 1.5 (Kaczorowski et al. 2020; Melnychuk et al. 2020). They were prepared in the same way as $Y_xLa_{1-x}PbSi_2S_8$ was synthesized and also were found to be stable against air and moisture. In the case of a composition with x = 0.5, synthesis can also be carried out according to the following regime: heating to 600°C (36°C·h^{-1}), holding for 24 h, heating to 1170°C (12°C·h^{-1}), holding for 4 h, cooling to 500°C (6°C·h^{-1}), annealing for 500 h, and quenching in water together with the ampoule (Melnychuk et al. 2020).

1.100 Silicon–Yttrium–Europium–Lead–Sulfur

In the system containing these elements, the quinary compound $Y_{3.85}Eu_{0.15}(Si_2O_7)S_3$, which melts at 1545°C ± 15°C and crystallizes as a tetragonal structure with the lattice parameters a = 1165.06 ± 0.08, c = 1360.54 ± 0.16 pm, and a calculated density of 4.528 g·cm^{-3}, is formed (Tarasenko et al. 2018). It was synthesized when amorphous SiO_2 (7.5 mM), Si (2.5 mM), S (5 mM), Y_2O_2S (9.625 mM), Eu_2O_2S (0.375 mM), and CsCl (3 g) as a flux were ground in a mortar and placed in a glassy carbon crucible to prevent variation of stoichiometry due to possible reactivity of silica ampoule. The crucible was sealed under vacuum in silica ampoule and heated to 1100°C in 2 h, kept there for 12 h, and then cooled to 25°C in 5 h. The reaction product was washed by distilled water to remove CsCl. The crystals were orange well-shaped strangulated octahedrons.

1.101 Silicon–Yttrium–Terbium–Lead–Sulfur

In the system containing these elements, the quinary compound $Y_2Tb_2(Si_2O_7)S_3$, which crystallizes as a tetragonal structure with the lattice parameters a = 1167.94 ± 0.02, c = 1363.70 ± 0.04 pm, and a calculated density of 5.428 g·cm^{-3}, is formed (Tarasenko et al. 2018). The optical energy gaps of this compound remains unchanged with appearance of set of transitions in the region of 1.7–2.7 eV. It was synthesized in the same way as $Y_{3.85}Eu_{0.15}(Si_2O_7)S_3$ was prepared using 5 mM of Y_2O_2S and 5 mM of Tb_2O_2S instead of Eu_2O_2S. The crystals were colorless well-shaped strangulated octahedrons.

1.102 Silicon–Lanthanum–Cerium–Lead–Sulfur

The $La_xCe_{1-x}PbSi_2S_8$ solid solutions are formed in this system (Kaczorowski et al. 2020). They crystallize in the trigonal structure with the lattice parameters a = 900.46 ± 0.06 and c = 2677.90 ± 0.02 pm for x = 0.5, a = 901.20 ± 0.02, c = 2682.78 ± 0.09 pm, and a calculated density of 4.218 g·cm^{-3} for x = 1.0, and a = 903.09 ± 0.08 and c = 2688.80 ± 0.03 pm for x = 1.5. They were prepared in the same way as $Y_xLa_{1-x}PbSi_2S_8$ was synthesized and also were found to be stable against air and moisture.

1.103 Silicon–Lanthanum–Cerium–Iodine–Sulfur

The $La_{3-x}Ce_xI(SiS_4)_2$ quinary solid solution, which crystallizes as a monoclinic structure with the lattice parameters a = 1605.7 ± 0.1, b = 789.16 ± 0.06, c = 1091.16 ± 0.08 pm, and β = 97.939 ± 0.004° for x = 1, is formed in the Si–La–Ce–I–S system (Riccardi et al. 1999). This solid solution was prepared from SiS_2, iodine, La_2S_3, and Ce_2S_3 (molar ratio of La/Ce/Si/I/S = (3-x):x:3:1:10.5). The reactants were handled in a dry box under N_2 atmosphere and placed in a silica tube. The tube was evacuated, sealed, and heated for 10 days in a temperature gradient of 1000°C to 800°C. After cooling, the molten reaction mixture was finely ground and once more sealed in an evacuated quartz tube with some iodine (molar ratio approximately 0.15). The ampoule was placed in a furnace, following the temperature regimen described above. This procedure had to be repeated a second time to obtain the air-stable white crystals of $La_{3-x}Ce_xI(SiS_4)_2$. A higher crystalline quality for this solid solution was obtained by using the following slightly modified heating procedure: the samples were heated at 400°C for one day, then at 850°C for 7 days, and slowly cooled to room temperature afterward. The SiS_2 excess separated easily from the solid solution within the reaction tube.

1.104 Silicon–Lanthanum–Praseodymium–Lead–Sulfur

The $La_xPr_{1-x}PbSi_2S_8$ solid solutions are formed in this system (Kaczorowski et al. 2020). They crystallize in the trigonal structure with the lattice parameters a = 898.70 ± 0.06 and c = 2671.00 ± 0.02 pm for x = 0.5, a = 899.63 ± 0.01 and c = 2674.90 ± 0.04 pm for x = 1.0, and a = 901.74 ± 0.08 and c = 2683.90 ± 0.03 pm for x = 1.5. They were prepared in the same way as $Y_xLa_{1-x}PbSi_2S_8$ was synthesized and also were found to be stable against air and moisture.

1.105 Silicon–Lanthanum–Samarium–Lead–Sulfur

The $La_xSm_{1-x}PbSi_2S_8$ solid solutions are formed in this system (Kaczorowski et al. 2020). They crystallize in the trigonal structure with the lattice parameters a = 893.60 ± 0.01 and c = 2647.30 ± 0.04 pm for x = 0.5, a = 897.345 ± 0.015, c = 2665.25 ± 0.06 pm, and a calculated density of 4.337 g·cm^{-3} for x = 1.0, and a = 899.70 ± 0.01 and c = 2676.00 ± 0.04 pm for x = 1.5. They were prepared in the same way as $Y_xLa_{1-x}PbSi_2S_8$ was synthesized and also were found to be stable against air and moisture.

1.106 Silicon–Lanthanum–Terbium–Lead–Sulfur

The $La_xTb_{1-x}PbSi_2S_8$ solid solutions are formed in this system (Kaczorowski et al. 2020). They crystallize in the trigonal structure with the lattice parameters a = 890.11 ± 0.07 and c = 2629.60 ± 0.02 pm for x = 0.5, a = 894.85 ± 0.02, c = 2650.89 ± 0.09 pm, and a calculated density of 4.432 g·cm^{-3} for x = 1.0, and a = 899.45 ± 0.09 and c = 2672.40 ± 0.04 pm for x = 1.5. They were prepared in the same way as $Y_xLa_{1-x}PbSi_2S_8$ was synthesized and also were found to be stable against air and moisture.

1.107 Silicon–Lanthanum–Dysprosium–Lead–Sulfur

The $La_xDy_{1-x}PbSi_2S_8$ solid solutions are formed in this system (Kaczorowski et al. 2020). They crystallizes as a trigonal structure with the lattice parameters a = 889.38 ± 0.05 and c = 2623.90 ± 0.02 pm for x = 0.5, a = 894.44 ± 0.02, c = 2648.32 ± 0.09 pm, and a calculated density of 4.459 g·cm^{-3} for x = 1.0, and a = 900.05 ± 0.06 and c = 2673.60 ± 0.02 pm for x = 1.5. They were prepared in the same way as Y_xLa_{1-x} $PbSi_2S_8$ was synthesized and also were found to be stable against air and moisture.

1.108 Silicon–Lanthanum–Holmium–Lead–Sulfur

The $La_xHo_{1-x}PbSi_2S_8$ solid solutions are formed in this system (Kaczorowski et al. 2020). They crystallize in the trigonal structure with the lattice parameters a = 888.60 ± 0.04 and c = 2617.80 ± 0.01 pm for x = 0.5, a = 894.65 ± 0.03, c = 2645.37 ± 0.11 pm, and a calculated density of 4.476 g·cm^{-3} for x = 1.0, and a = 900.41 ± 0.06 and c = 2674.70 ± 0.02 pm for x = 1.5. They were prepared in the same way as $Y_xLa_{1-x}PbSi_2S_8$ was synthesized and also were found to be stable against air and moisture.

1.109 Silicon–Lanthanum–Erbium–Lead–Sulfur

The $La_xEr_{1-x}PbSi_2S_8$ solid solutions are formed in this system (Kaczorowski et al. 2020). They crystallize in the trigonal structure with the lattice parameters a = 888.19 ± 0.04 and c = 2610.80 ± 0.02 pm for x = 0.5, a = 893.60 ± 0.04, c = 2639.48 ± 0.19 pm, and a calculated density of 4.509 g·cm^{-3} for x = 1.0, and a = 899.70 ± 0.01 and c = 2671.70 ± 0.04 pm for x = 1.5. They were prepared in the same way as $Y_xLa_{1-x}PbSi_2S_8$ was synthesized and also were found to be stable against air and moisture.

1.110 Silicon–Cerium–Praseodymium–Lead–Sulfur

The $Ce_xPr_{1-x}PbSi_2S_8$ solid solutions are formed in this system (Kaczorowski et al. 2020). They crystallize in the trigonal structure with the lattice parameters a = 898.00 ± 0.01 and c = 2667.40 ± 0.01 pm for x = 0.5, a = 899.46 ± 0.03, c = 2675.31 ± 0.11 pm, and a calculated density of 4.257 g·cm^{-3} for x = 1.0, and a = 899.20 ± 0.01 and c = 2672.50 ± 0.04 pm for x = 1.5. They were prepared in the same way as $Y_xLa_{1-x}PbSi_2S_8$ was synthesized and also were found to be stable against air and moisture.

1.111 Silicon–Cerium–Samarium–Lead–Sulfur

The $Ce_xSm_{1-x}PbSi_2S_8$ solid solutions are formed in this system (Kaczorowski et al. 2020). They crystallize in the trigonal structure with the lattice parameters a = 892.18 ± 0.05 and c = 2643.20 ± 0.02 pm for x = 0.5, a = 893.74 ± 0.03, c = 2650.94 ± 0.12 pm and, a calculated density of 4.403 g·cm^{-3} for x = 1.0, and a = 896.17 ± 0.05 and c = 2660.00 ± 0.02 pm for x = 1.5. They were prepared in the same way as $Y_xLa_{1-x}PbSi_2S_8$ was synthesized and also were found to be stable against air and moisture.

1.112 Silicon–Cerium–Terbium–Lead–Sulfur

The $Ce_xTb_{1-x}PbSi_2S_8$ solid solutions are formed in this system (Kaczorowski et al. 2020). They crystallize in the trigonal structure with the lattice parameters $a = 889.13 \pm 0.04$, $c = 2627.50 \pm 0.02$ pm, and a calculated density of 4.582 ± 0.001 g·cm^{-3} for $x = 0.5$, $a = 891.70 \pm 0.02$, $c = 2638.47 \pm 0.10$ pm, and a calculated density of 4.491 g·cm^{-3} for $x = 1.0$, and $a = 896.54 \pm 0.06$ and $c = 2661.90 \pm 0.02$ pm for $x = 1.5$ (Kaczorowski et al. 2020; Melnychuk et al. 2020). They were prepared in the same way as $Y_xLa_{1-x}PbSi_2S_8$ was synthesized and also were found to be stable against air and moisture. In the case of a composition with $x = 0.5$, synthesis can also be carried out according to the following regime: heating to 600°C (36°C·h^{-1}), holding for 24 h, heating to 1170°C (12°C·h^{-1}), holding for 4 h, cooling to 500°C (6°C·h^{-1}), annealing for 500 h, and quenching in water together with the ampoule (Melnychuk et al. 2020).

1.113 Silicon–Cerium–Dysprosium–Lead–Sulfur

The $Ce_xDy_{1-x}PbSi_2S_8$ solid solutions are formed in this system (Kaczorowski et al. 2020). They crystallize in the trigonal structure with the lattice parameters $a = 888.58 \pm 0.05$ and $c = 2621.30 \pm 0.02$ pm for $x = 0.5$, $a = 892.19 \pm 0.03$, $c = 2637.22 \pm 0.10$ pm, and a calculated density of 4.507 g·cm^{-3} for $x = 1.0$, and $a = 896.34 \pm 0.05$ and $c = 2659.80 \pm 0.02$ pm for $x = 1.5$. They were prepared in the same way as $Y_xLa_{1-x}PbSi_2S_8$ was synthesized and also were found to be stable against air and moisture.

1.114 Silicon–Cerium–Holmium–Lead–Sulfur

The $Ce_xHo_{1-x}PbSi_2S_8$ solid solutions are formed in this system (Kaczorowski et al. 2020). They crystallize in the trigonal structure with the lattice parameters $a = 887.82 \pm 0.04$ and $c = 2615.40 \pm 0.01$ pm for $x = 0.5$, $a = 894.65 \pm 0.03$, $c = 2635.60 \pm 0.13$ pm, and a calculated density of 4.529 g·cm^{-3} for $x = 1.0$, and $a = 899.56 \pm 0.08$ and $c = 2673.10 \pm 0.03$ pm for $x = 1.5$. They were prepared in the same way as $Y_xLa_{1-x}PbSi_2S_8$ was synthesized and also were found to be stable against air and moisture.

1.115 Silicon–Cerium–Erbium–Lead–Sulfur

The $Ce_xEr_{1-x}PbSi_2S_8$ solid solutions are formed in this system (Kaczorowski et al. 2020). They crystallize in the trigonal structure with the lattice parameters $a = 886.04 \pm 0.08$, $c = 2604.50 \pm 0.03$ pm, and a calculated density of 4.7257 ± 0.0001 g·cm^{-3} for $x = 0.5$, $a = 891.68 \pm 0.03$, $c = 2631.21 \pm 0.09$ pm, and a calculated density of 4.549 g·cm^{-3} for $x = 1.0$, and $a = 896.54 \pm 0.08$ and $c = 2658.40 \pm 0.03$ pm for $x = 1.5$ (Kaczorowski et al. 2020; Melnychuk et al. 2020). They were prepared in the same way as $Y_xLa_{1-x}PbSi_2S_8$ was synthesized and also were found to be stable against air and moisture. In the case of a composition with $x = 0.5$, synthesis can also be carried out according to the following regime: heating to 600°C (36°C·h^{-1}), holding for 24 h, heating to 1170°C (12°C·h^{-1}), holding for 4 h, cooling to 500°C (6°C·h^{-1}), annealing for 500 h, and quenching in water together with the ampoule (Melnychuk et al. 2020).

1.116 Silicon–Praseodymium–Samarium–Lead–Sulfur

The $Pr_xSm_{1-x}PbSi_2S_8$ solid solutions are formed in this system (Kaczorowski et al. 2020). They crystallize in the trigonal structure with the lattice parameters $a = 891.40 \pm 0.02$ and $c = 2639.20 \pm 0.06$ pm for $x = 0.5$, $a = 893.06 \pm 0.04$, $c = 2649.28 \pm 0.12$ pm, and a calculated density of 4.416 g·cm^{-3} for $x = 1.0$, and $a = 895.20 \pm 0.01$ and $c = 2655.70 \pm 0.04$ pm for $x = 1.5$. They were prepared in the same way as $Y_xLa_{1-x}PbSi_2S_8$ was synthesized and also were found to be stable against air and moisture.

1.117 Silicon–Praseodymium–Terbium–Lead–Sulfur

The $Pr_xTb_{1-x}PbSi_2S_8$ solid solutions are formed in this system (Kaczorowski et al. 2020). They crystallize in the trigonal structure with the lattice parameters $a = 888.72 \pm 0.07$ and $c = 2625.30 \pm 0.03$ pm for $x = 0.5$, $a = 890.96 \pm 0.03$, $c = 2638.47 \pm 0.11$ pm, and a calculated density of 4.502 g·cm^{-3} for $x = 1.0$, and $a = 893.37 \pm 0.07$, $c = 2648.50 \pm 0.02$ pm, and a calculated density of 4.582 ± 0.001 g·cm^{-3} for $x = 1.5$ (Kaczorowski et al. 2020; Melnychuk et al. 2020). They were prepared in the same way as $Y_xLa_{1-x}PbSi_2S_8$ was synthesized and also were found to be stable against air and moisture. In the case of a composition with $x = 1.5$, synthesis can also be carried out according to the following regime: heating to 600°C (36°C·h^{-1}), holding for 24 h, heating to 1170°C (12°C·h^{-1}), holding for 4 h, cooling to 500°C (6°C·h^{-1}), annealing for 500 h, and quenching in water together with the ampoule (Melnychuk et al. 2020).

1.118 Silicon–Praseodymium–Dysprosium–Lead–Sulfur

The $Pr_xDy_{1-x}PbSi_2S_8$ solid solutions are formed in this system (Kaczorowski et al. 2020). They crystallize in the trigonal structure with the lattice parameters $a = 887.20 \pm 0.08$ and $c = 2615.30 \pm 0.03$ pm for $x = 0.5$, $a = 890.728 \pm 0.018$, $c = 2633.49 \pm 0.06$ pm, and a calculated density of 4.533 g·cm^{-3} for $x = 1.0$, and $a = 894.00 \pm 0.01$ and $c = 2639.70 \pm 0.04$ pm for $x = 1.5$. They were prepared in the same way as $Y_xLa_{1-x}PbSi_2S_8$ was synthesized and also were found to be stable against air and moisture.

1.119 Silicon–Praseodymium–Holmium–Lead–Sulfur

The $Pr_xHo_{1-x}PbSi_2S_8$ solid solutions are formed in this system (Kaczorowski et al. 2020). They crystallize in the trigonal structure with the lattice parameters $a = 886.61 \pm 0.06$ and $c = 2612.30 \pm 0.03$ pm for $x = 0.5$, $a = 890.26 \pm 0.03$, $c = 2630.94 \pm 0.12$ pm, and a calculated density of 4.556 g·cm^{-3} for $x = 1.0$, and $a = 893.73 \pm 0.06$ and $c = 2646.90 \pm 0.02$ pm for $x = 1.5$. They were prepared in the same way as $Y_xLa_{1-x}PbSi_2S_8$ was synthesized and also were found to be stable against air and moisture.

1.120 Silicon–Praseodymium–Erbium–Lead–Sulfur

The $Pr_xEr_{1-x}PbSi_2S_8$ solid solutions are formed in this system (Kaczorowski et al. 2020). They crystallize in the trigonal structure with the lattice parameters $a = 886.50 \pm 0.04$ and $c = 2605.60 \pm 0.01$ pm for $x = 0.5$, $a = 890.39 \pm 0.03$, $c = 2626.90 \pm 0.09$ pm, and a calculated density of 4.574 g·cm^{-3} for $x = 1.0$, and $a = 893.63 \pm 0.08$, $c = 2648.40 \pm 0.03$ pm, and a calculated density of 4.3563 ± 0.0001 g·cm^{-3} for $x = 1.5$ (Kaczorowski et al. 2020; Melnychuk et al. 2020). They were prepared in the same way as $Y_xLa_{1-x}PbSi_2S_8$ was synthesized and also were found to be stable against air and moisture. In the case of a composition with $x = 1.5$, synthesis can also be carried out according to the following regime: heating to 600°C (36°C·h^{-1}), holding for 24 h, heating to 1170°C (12°C·h^{-1}), holding for 4 h, cooling to 500°C (6°C·h^{-1}), annealing for 500 h, and quenching in water together with the ampoule (Melnychuk et al. 2020).

1.121 Silicon–Lead–Oxygen–Fluorine–Sulfur

In the system containing these elements, the quinary compound $Pb_{10}(SiO_4)_3(SO_4)_3F_2$, which crystallizes as a hexagonal structure with the lattice parameters $a = 989.0 \pm 0.3$, $c = 742.4 \pm 0.3$ pm, and the calculated and experimental densities of 7.061 and 7.03 g·cm^{-3}, respectively, is formed (Schwarz 1967a). It was synthesized by sintering a mixture of Pb_2SiO_4, $PbSO_4$, and PbF_2 (molar ratio 3:3:1) twice at 1000°C first for 14 and then for 9 h in air. Yellowish crystals were obtained after quickly cooling.

2

Systems Based on Silicon Selenides

2.1 Silicon–Hydrogen–Aluminum–Oxygen–Selenium

In the system containing these elements, the quinary compound, $Al_6(SeO_3)_3[SiO_3(OH)](OH)_9{\cdot}10H_2O$ (mineral petermegawite), which crystallizes in the orthorhombic structure with the lattice parameters a = 1623.92 ± 0.02, b = 1096.37 ± 0.01, c = 1533.67 ± 0.02 pm, and the calculated and experimental densities of 2.32 and 2.27 ± 0.5 $g{\cdot}cm^{-3}$, respectively, is formed (Yang et al. 2022g,h,2023h).

2.2 Silicon–Potassium–Gallium–Germanium–Selenium

In the system containing these elements, the quinary compound $K_3Ga_3(Ge_{4.95}Si_{2.05})Se_{20}$, which crystallizes in the monoclinic structure with the lattice parameters a = 701.01 ± 0.04, b = 3892.8 ± 0.2, c = 695.94 ± 0.04 pm, β = 90.531 ± 0.002°, a calculated density of 4.062 $g{\cdot}cm^{-3}$, and an energy gap of 2.12 eV, is formed (Li et al. 2019d). The title compound was first obtained using the stoichiometric mixture targeted to prepare $K_3Ga_3Ge_7Se_{20}$ and a minor amount of SnO_2/Si, together with abundant KI as a flux. After its chemical composition was determined, stoichiometric Si, Ga_2O_3, GeO_2, Se, B, and abundant KI were used to synthesize the compound with high yields. The optimized synthetic route is as follows. A 500 mg mixture of stoichiometric Si, Ga_2O_3, GeO_2, Se, and B and an additional 600 mg of KI were ground into a uniform powder using an agate mortar and then pressed into a pellet, which was then sealed into an evacuated quartz ampoule with a vacuum degree of ≈10^{-2} Pa. The ampoule was calcinated in a muffle furnace from room temperature to 950°C at a rate of 60°C$\cdot h^{-1}$ and homogenized at several intermediate temperatures for several hours. The reaction was maintained at 950°C for a week and then cooled down to 300°C at a rate of 5°C$\cdot h^{-1}$. Finally, the ampoule was cooled to room temperature naturally. The product was cleaned by distilled water, and yellow block single crystals were obtained.

2.3 Silicon–Potassium–Indium–Tin–Selenium

In the system containing these elements, the quinary compound $KInSi_{1.32}Sn_{0.68}Se_6$, which crystallizes in the monoclinic structure with the lattice parameters a = 1019.70 ± 0.06, b = 969.94 ± 0.06, c = 1249.92 ± 0.08 pm, β = 106.575 ± 0.002°, a calculated density of 4.179 $g{\cdot}cm^{-3}$, and an energy gap of 1.85 eV, is formed (Li et al. 2020c). The title compound was obtained by a solid state reaction, which was an unexpected product when attempted to obtain $KInSiSnSe_6$. A 500 mg mixture of Sn, In, Si, Se (molar ratio 1:1:1:6), and additional 400 mg KI as flux was ground into uniform powder using an agate mortar, and then pressed into a pellet, which was then sealed into an evacuated quartz ampoule with the vacuum degree of ≈10^{-2} Pa. The ampoule was heated in a furnace from room temperature to 950°C in 25 h and homogenized at several intermediate temperatures for several hours. The reaction was maintained at 950°C for about 120 h, and then cooled down to 300°C within 120 h. Finally, the tube was cooled down to room temperature naturally. The product was cleaned by distilled water and dimethylformamide (DMF),

DOI: 10.1201/9781003123460-2

and then the red block single crystals were obtained. The crystals are stable in air after several months and can be separated from the byproducts easily under an optical microscope because of their unique color and shape.

2.4 Silicon–Cesium–Gallium–Tin–Selenium

In the system containing these elements, the quinary compound $CsGaSiSnSe_6$, which crystallizes in the tetragonal structure with the lattice parameters a = 1041.93 ± 0.03, c = 943.97 ± 0.06 pm, a calculated density of 4.621 g·cm^{-3}, and an energy gap of 2.06 eV, is formed (Li et al. 2021a). This compound was obtained by solid state reaction. The mixture of CsCl as reactive flux, Ga, Si, Sn, and Se with the overall mass of 700 mg and the molar ratios of 2:1:1:1:6 was ground into uniform powder using an agate mortar and then pressed into a pellet, which was then sealed into an evacuated quartz ampoule with the vacuum degree of 10^{-2} Pa. The tube was heated to 800°C in 35 h, the reaction was kept at this temperature for 100 h, and then cooled down to 200°C within 200 h. Ultimately, the tube was cooled down to environmental temperature naturally. The product was washed DMF and distilled water, and the deep-red block crystals were obtained. These crystals are stable in air.

2.5 Silicon–Cesium–Indium–Tin–Selenium

In the system containing these elements, the quinary compound $CsInSiSnSe_6$, which crystallizes as a tetragonal structure with the lattice parameters a = 1049.79 ± 0.06, c = 952.52 ± 0.12 pm, a calculated density of 4.758 g·cm^{-3}, and an energy gap of 2.05 eV, is formed (Li et al. 2021a). It was prepared in the same way as $CsGaSiSnSe_6$ was synthesized using In instead of Ga. The crystals of this compound are also stable in air.

2.6 Silicon–Copper–Zinc–Germanium–Selenium

$Cu_2ZnSiSe_4$–$Cu_2ZnGeSe_4$. This system is nonquasibinary section of the Si–Cu–Zn–Ge–S quinary system (Olekseyuk et al. 2019). At 400°C, the mutual solubility of both components each in other is 8 mol%. The alloys for the investigations were annealed at 400°C for 500 h followed by quenching into cold water.

2.7 Silicon–Copper–Zinc–Tin–Selenium

$Cu_2ZnSiSe_4$–$Cu_2ZnSnSe_4$. This system is nonquasibinary section of the Si–Cu–Zn–Sn–S quinary system (Olekseyuk et al. 2019).At 400°C, $Cu_2ZnSiSe_4$ dissolves 8 mol% $Cu_2ZnSnSe_4$ and the solubility of $Cu_2ZnSiSe_4$ in $Cu_2ZnSnSe_4$ reaches 15 mol%. The alloys for the investigations were annealed at 400°C for 500 h followed by quenching into cold water.

2.8 Silicon–Copper–Germanium–Phosphorus–Selenium

Cu_2SiSe_3–$CuGe_2P_3$. $CuGe_2P_3$ does not form any solid solution with Cu_2SiSe_3 (Omar 1990).

2.9 Silicon–Beryllium–Zinc–Oxygen–Selenium

In the system containing these elements, the quinary compound $Zn_4Be_3(SiO_4)_3Se$ is formed (Mel'nikov et al. 1968). This compound was synthesized through a hydrothermal synthesis at 450°C and 0.2 GPa for 7–10 days using the mixture of ZnO, BeO, SiO_2, and Se with the solution of 1 mass% NaOH as a solvent.

2.10 Silicon–Beryllium–Cadmium–Oxygen–Selenium

The $Cd_4Be_3(SiO_4)_3Se$ quinary compound is formed in this system (Mel'nikov et al. 1968). This compound was synthesized through hydrothermal synthesis at 450°C and 0.2 GPa for 7–10 days using the mixture of CdO, BeO, SiO_2, and Se with the solution of 1 mass% NaOH as a solvent.

2.11 Silicon–Beryllium–Oxygen–Manganese–Selenium

The $Mn_4Be_3(SiO_4)_3Se$ quinary compound is formed in this system (Mel'nikov et al. 1968). This compound was synthesized tgrough hydrothermal synthesis at 450°C and 0.2 GPa for 7–10 days using the mixture of MnO, BeO, SiO_2, and Se with the solution of 8 mass% NH_4Cl as a solvent.

2.12 Silicon–Beryllium–Oxygen–Iron–Selenium

In the system containing these elements, the quinary compound $Fe_4Be_3(SiO_4)_3Se$, which crystallizes in the cubic structure with the lattice parameter a = 828.4 ± 0.1 pm, is formed (Armstrong et al. 2003). For obtaining it, BeO, SiO_2, powdered Fe, and Se were homogeneously ground using an agate pestle and mortar in an Ar-filled glove box (1 ppm H_2O/1 ppm O_2). The mixture was placed in a quartz tube, which was evacuated and sealed. The tube was heated in the center of a tube furnace fitted with an inconel work tube in stages up to 1100°C and maintained at this temperature for 48 h. The sample tube was cooled slowly and transferred back to the glove box, where they were opened with a glass cutter.

2.13 Silicon–Calcium–Erbium–Oxygen–Chlorine–Selenium

In the system containing these elements, the multinary compound $Ca_{0.25}Er_{3.75}(Si_2O_7)Se_{2.75}Cl_{0.25}$, which crystallizes as a tetragonal structure with the lattice parameters a = 1177.7 ± 0.2, c = 1376.5 ± 0.2 pm, and a calculated density of 7.177 g·cm^{-3}, is formed (Stöwe 1994). Well-shaped pink crystals of this compound were obtained as byproducts of the synthesis of erbium selenides from the elements in evacuated and sealed silica ampoules with graphite inlets at 980°C.

2.14 Silicon–Barium–Samarium–Oxygen–Selenium

In the system containing these elements, the quinary compound $BaSm_4(SiO_4)_3Se$, which crystallizes in the hexagonal structure with the lattice parameters a = 986.9 ± 0.1, c = 685.1 ± 0.1 pm, and a calculated density of 6.287 g·cm^{-3} at 153 K, is formed (Yang and Ibers 2000). The reaction was carried out at 850°C–900°C for 7 days in unprotected fused silica tubes with the use of a $BaBr_2$/KBr eutectic flux (molar ratio 1.1:1). Yellow needles were obtained in the reaction of BaSe, Ag, Sm, and Se (molar ratio 2:3:1:3).

2.15 Silicon–Zinc–Gallium–Phosphorus–Selenium

Si–ZnSe–GaP. Epitaxy layers of solid solutions $(ZnSe)_{1-x-y}(Si_2)_x(GaP)_y$ ($0 \leq x \leq 0.03$, $0 \leq y \leq 0.09$) were grown from the limited volume of tin solution, which is melted by a liquid-phase epitaxy procedure (Saidov et al. 2012). These solid solutions represent the stable phase.

3

Systems Based on Silicon Telluride

3.1 Silicon–Hydrogen–Zinc–Lead–Arsenic–Vanadium–Oxygen–Tellurium

In the system containing these elements, the multinary compound $Zn_3Pb_3(As,V,Si)_2Te(O,OH)_{14}$ (vanadian silician variety of mineral dugganite), which crystallizes as an orthorhombic structure with the lattice parameters $a = 857 \pm 3$, $b = 1484 \pm 5$, $c = 521 \pm 3$ pm, and a calculated density of 6.48 g·cm^{-3}, is formed (Kim et al. 1988; Jambor and Vanko 1991).

3.2 Silicon–Beryllium–Zinc–Oxygen–Tellurium

In the system containing these elements, the quinary compound $Zn_4Be_3(SiO_4)_3Te$ is formed (Mel'nikov et al. 1968). This compound was synthesized through hydrothermal synthesis at 450°C and 0.2 GPa for 7–10 days using the mixture of ZnO, BeO, SiO_2, and Te with the solution of 1 mass% NaOH as a solvent.

3.3 Silicon–Beryllium–Cadmium–Oxygen–Tellurium

In the system containing these elements, the quinary compound $Cd_4Be_3(SiO_4)_3Te$ is formed (Mel'nikov et al. 1968). This compound was synthesized through hydrothermal synthesis at 450°C and 0.2 GPa for 7–10 days using the mixture of CdO, BeO, SiO_2, and Te with the solution of 1 mass% NaOH as a solvent.

3.4 Silicon–Beryllium–Oxygen–Manganese–Tellurium

In the system containing these elements, the quinary compound $Mn_4Be_3(SiO_4)_3Te$ is formed (Mel'nikov et al. 1968). This compound was synthesized through hydrothermal synthesis at 450°C and 0.2 GPa for 7–10 days using the mixture of MnO, BeO, SiO_2, and Te with the solution of 8 mass% NH_4Cl as a solvent.

3.5 Silicon–Beryllium–Oxygen–Iron–Tellurium

In the system containing these elements, the quinary compound $Fe_4Be_3(SiO_4)_3Te$, which crystallizes as a cubic structure with the lattice parameter $a = 836.9 \pm 0.1$ pm, is formed (Armstrong et al. 2003). For obtaining it, BeO, SiO_2, powdered Fe, and Te were homogeneously ground using an agate pestle and mortar in an Ar-filled glove box (1 ppm H_2O/1 ppm O_2). The mixture was placed in a quartz tube, which was evacuated and sealed. The tube was heated in the center of a tube furnace fitted with an inconel work tube in stages up to 1100°C and maintained at this temperature for 48 h. The sample tube was cooled slowly and transferred back to the glove box, where they were opened with a glass cutter.

DOI: 10.1201/9781003123460-3

3.6 Silicon–Barium–Zinc–Oxygen–Tellurium

In the system containing these elements, the quinary compound $Ba_3Zn_6(TeO_6)(Si_2O_7)_2$, which is stable up to 950°C and crystallizes as a monoclinic structure with the lattice parameters a = 1597.5 ± 0.5, b = 1150.5 ± 0.4, c = 514.2 ± 0.2 pm, β = 107.437 ± 0.005°, a calculated density of 5.025 g·cm^{-3}, and an energy gap of 4.4 eV, is formed (Jiang and Mao 2006). It was prepared by a reaction mixture of $BaCO_3$, ZnO, TeO_2·H_2O, and SiO_2 (molar ratio 3:6:1:4) at 820°C for 7 days in a sealed silica tube.

3.7 Silicon–Zinc–Lead–Oxygen–Tellurium

In the system containing these elements, the quinary compound $Zn_4PbSiTeO_{10}$, which crystallizes as an orthorhombic structure with the lattice parameters a = 654.2 ± 0.5, b = 1562.4 ± 0.4, c = 828.0 ± 0.4 pm, is formed (Wedel et al. 1999). To prepare colorless crystals of this compound, PbO, SiO_2, and TeO_2 were intimately mixed, pressed into a tablet, and heated in a Pt crucible in air to 600°C for 96 h.

3.8 Silicon–Gallium–Lead–Oxygen–Tellurium

A variable-composition quinary phase $Pb_3Te_{1-x}Ga_{4-2x}Si_{1+3x}O_{14}$ exists in the Si–Ga–Pb–O–Te in the range of $0.15 \le x < 0.667$ (Mill 2010). This phase crystallizes as a trigonal structure with the lattice parameters a = 838.5 ± 0.4, c = 507.4 ± 0.2 pm, and a calculated density of 6.552 g·cm^{-3} for x = 0.333 and a = 837.3 ± 0.3, c = 507.3 ± 0.1 pm, and a calculated density of 6.408 g·cm^{-3} for x = 0.5. The starting compounds for synthesis of this phase were PbO, TeO_2, Ga_2O_3, and SiO_2. Thoroughly blended and compacted mixture was annealed in air in a platinum container at temperatures of 750°C–850°C to the melting or decomposition temperature with an interval of ≤50°C for 10–15 h. The heating rate was 250–300°C·h^{-1}.

3.9 Silicon–Lanthanum–Germanium–Oxygen–Tellurium

In the system containing these elements, the quinary compound $La_4(Si_{5.2}Ge_{2.8}O_{18})(TeO_3)_4$, which crystallizes as a triclinic structure with the lattice parameters a = 815.0 ± 0.5, b = 1289.2 ± 0.8, c = 1432.0 ± 0.8 pm, α = 106.132 ± 0.009°, β = 90.045 ± 0.005°, γ = 104.270 ± 0.005°, and an energy gap of 4.42 eV, is formed (Kong et al. 2008). Colorless plate-shaped single crystals of the title compound were initially obtained by solid state reaction of La_2O_3 (0.4 mM), GeO_2 (0.4 mM), TeO_2 (1.2 mM) with 0.4 mM of CsCl as a flux in an attempt to synthesize a La(III) germanate tellurite. The mixture was thoroughly ground and pressed into a pellet, which was then put into an evacuated quartz tube. The quartz tube was heated at 800°C for 6 days, and then cooled to 300°C at a rate of 10°C·h^{-1} before switching off the furnace. The incorporated Si was obviously extracted from the silica tube. After proper structural analyses, a single phase of this compound was prepared by reacting stoichiometric amounts of La_2O_3, GeO_2, SiO_2, and TeO_2 at 960°C for 6 days.

4

Systems Based on Germanium Sulfides

4.1 Germanium–Hydrogen–Lithium–Copper–Oxygen–Sulfur

In the system containing these elements, the multinary compound $Li_4Cu_8Ge_3S_{12}·H_2O$, which crystallizes as a cubic structure with the lattice parameters a = 1765.72 ± 0.03 pm and a calculated density of 2.786 g·cm^{-3} at 263 K, is formed(Wang et al. 2019a,b). To synthesize this compound, Cu powder (5 mM), Ge powder (2.5 mM), $LiOH·H_2O$ (95 mM), and thiourea powder (52 mM) were uniformly ground together and placed into a Teflon-lined stainless steel autoclave. The autoclave was tightly sealed and then heated at 200°C for 24 h. Then the autoclave was allowed to cool to ambient temperature. The product was first quickly washed with 80 mL of deionized water to wash out most of LiOH and thiourea, and then washed with anhydrous ethanol three times until light yellow crystals were separated. The product was dried in a vacuum at 60°C.

4.2 Germanium–Hydrogen–Sodium–Potassium–Copper–Oxygen–Sulfur

In the system containing these elements, the multinary compound $Na_{0.04}K_{3.96}Cu_8Ge_3S_{12}·2H_2O$, which crystallizes as a cubic structure with the lattice parameters a = 1772.80 pm, is formed (Zhang et al. 2018a). To prepare this compound, 40 mg of polycrystalline $Na_4Cu_8Ge_3S_{12}·2H_2O$ was added to 40 mL of KCl aqueous solution (1.0 M). The mixture then was kept at 45°C and stirred for 12 h. The exchanged products were isolated by filtration, washed several times with deionized water and ethanol, and then dried in air.

4.3 Germanium–Hydrogen–Sodium–Rubidium–Copper–Oxygen–Sulfur

In the system containing these elements, the multinary compound $Na_{0.32}Rb_{3.68}Cu_8Ge_3S_{12}·2H_2O$, which crystallizes as a cubic structure with the lattice parameters a = 1773.51 pm, is formed (Zhang et al. 2018a). It was prepared in the same way as $Na_{0.04}K_{3.96}Cu_8Ge_3S_{12}·2H_2O$ was obtained using RbCl instead of KCl.

4.4 Germanium–Hydrogen–Sodium–Copper–Oxygen–Sulfur

In the system containing these elements, the multinary compound $Na_4Cu_8Ge_3S_{12}·2H_2O$, which crystallizes as a cubic structure with the lattice parameters a = 1757.42 ± 0.15 pm, a calculated density of 2.944 g·cm^{-3}, and an energy gap of 2.4 eV, is formed (Zhang et al. 2018a). To obtain this compound, Cu powder (0.11 mM), GeO_2 powder (0.05 mM), Na_2CO_3 (0.1 mM), and S (1.0 mM) were placed in a Pyrex-glass tube (~10 mL in volume). Then, 300 mkL of diethyltriamine and 200 mkL of $CH_3OH:H_2O$ (volume ratio 1:1) were added as a mixed solvent and ultrasonically dispersed. The tube was sealed in an air atmosphere, placed in a stainless steel autoclave, and then heated at 160°C for 7 days. The products were washed

DOI: 10.1201/9781003123460-4

with ethylenediamine and ethanol several times. Light yellow cubic crystals of the title compound were obtained.

4.5 Germanium–Hydrogen–Sodium–Oxygen–Sulfur

Some quinary compounds are formed in this system. $Na_2GeS_2(OH)_2{\cdot}5H_2O$ crystallizes as an orthorhombic structure with the lattice parameters a = 1075.2 ± 0.2, b = 1378.7 ± 0.2, c = 1415.0 ± 0.2 pm, and the calculated and experimental densities of 1.943 and 1.95 ± 0.02 g·cm^{-3}, respectively (Krebs and Wallstab 1981a,b). It was obtained by reacting stoichiometric quantities of either GeO_2 and aqueous Na_2S, or GeS_2 and aqueous NaOH. Colorless highly soluble compound was isolated from the solution by adding a large excess of acetone.

$Na_3GeS_3(OH){\cdot}8H_2O$ crystallizes as a monoclinic structure with the lattice parameters a = 3179.4 ± 1.0, b = 667.8 ± 0.3, c = 1472.1 ± 0.5 pm, and β = 110.72 ± 0.03° (Krebs and Wallstab 1981b). It was prepared from a solution of GeS_2 in an aqueous mixture of NaOH + Na_2S, the three components taken in a 1:1:1 molar ratio according to the stoichiometric reaction $GeS_2 + Na_2S + NaOH + 8H_2O = Na_3GeS_3(OH){\cdot}8H_2O$. Colorless crystalline compound was isolated.

Four compounds, $Na_2GeS_3{\cdot}H_2O$, $Na_2Ge_2S_5{\cdot}11H_2O$, $Na_4GeS_4{\cdot}7H_2O$, and $Na_8GeS_6{\cdot}23H_2O$, were prepared by Sevryukov et al. (1969) from the stoichiometric quantities of GeO_2 and NaOH required to obtain a specific thiogermanate. The mixture was well stirred, placed in a quartz boat, and heated at ≈ 500°C in a tube furnace in a dry H_2S current for 30 min. After cooling, the cake obtained was dissolved in a small volume of water, and this solution was poured into acetone. The volume ratio of acetone and solution was ≈ 25:1. The thiogermanate crystals, which usually precipitated within 1 h, were separated and dried. According to Sevryukov and Salikova (1970), two more compounds, $Na_2GeS_3{\cdot}10H_2O$ and $Na_4GeS_4{\cdot}15H_2O$, were found in this system, but the formation the $Na_8GeS_6{\cdot}23H_2O$ compound could not be found.

$Na_4Ge_2S_6{\cdot}14H_2O$ crystallizes as a triclinic structure with the lattice parameters a = 997.8 ± 0.6, b = 702.0 ± 0.5, c = 960.1 ± 0.6 pm, α = 108.41 ± 0.04°, β = 92.39 ± 0.04°, γ = 91.69 ± 0.04°, and the calculated and experimental densities of 1.778 and 1.80 ± 0.02 g·cm^{-3}, respectively (Krebs et al. 1970a,b, 1972). To prepare this compound, precipitated GeS_2 (50 mM) was dissolved in portions in a solution of $Na_2S{\cdot}9H_2O$ (50 mM) in H_2O (20 mL). A pH value of 7.5–8 was maintained. The filtered, pale yellowish solution was poured into 500 mL of acetone, and a finely crystalline mass separated out. After decanting off the acetone, it was placed in a little water and crystallized over P_4O_{10} under a H_2S atmosphere. $Na_4Ge_2S_6{\cdot}14H_2O$, which crystallizes in colorless leaves, can be kept for a few hours in air. However, weathering and hydrolytic and oxidative decomposition gradually occur.

For obtaining $Na_6Ge_2S_7{\cdot}9H_2O$, freshly precipitated GeS_2 was dissolved in a saturated solution of Na_2S, with a Na_2S/GeS_2 molar ratio of 2:1 (Schwarz and Giese 1930; Ivanov-Emin and Kostrikin 1947). The resulting solution was poured into acetone, taken in a large excess, with vigorous stirring. First, a haze formed, which after 1 h gathered at the bottom of the vessel, forming a layer of a yellow oily substance. After separation of the latter from acetone and drying for 3 days in a desiccator over P_4O_{10} under a H_2S atmosphere, crystals of the title compound were isolated. According to Krebs et al. (1970a,b), this compound does not exist as a homogeneous phase: it consists of mixture of Na_4GeS_4 and $Na_4Ge_2S_6$.

4.6 Germanium–Hydrogen–Potassium–Cesium–Copper–Oxygen–Sulfur

In the system containing these elements, the multinary compound $(H_3O)K_{0.6}Cs_{0.4}Cu_6Ge_2S_8$, which crystallizes as a tetragonal structure with the lattice parameters a = 1464.5 ± 0.5, c = 734.8 ± 0.5 pm, a calculated density of 3.691 g·cm^{-3}, and an energy gap of 2.32 eV, is formed (Li et al. 2019e). It was synthesized using Cu powder (4.0 mg), GeO_2 powder (2.2 mg), S (20 mg), K_2CO_3 (2.5 mg), Cs_2CO_3 (3.5 mg), about 300 mg of ethylenediamine and 90 mg of deionized water, and then heated at 200°C for 12 days. Yellow needle crystals were obtained by washing with ethanol and distilled water.

4.7 Germanium–Hydrogen–Potassium–Copper–Oxygen–Sulfur

In the system containing these elements, the multinary compound $(H_3O)KCu_6Ge_2S_8$ [$(H_3O)KCu_6Ge_2S_8 \cdot nH_2O$ according to Zhang et al. (2021)], which crystallizes as a tetragonal structure with the lattice parameters a = 1029.30 ± 0.14, c = 366.97 ± 0.11 pm, a calculated density of 3.592 g·cm^{-3}, and an energy gap of 2.31 eV, is formed (Li et al. 2019e). To prepare this compound, a mixture of Cu powder (20.0 mg), GeO_2 powder (15.0 mg), K_2CO_3 (20.0 mg), excess S powder (92.52 mg), ethylenediamine (1.67 mL), and H_2O (0.45 mL) was sealed in a 23 mL Teflon-lined autoclave and heated at 200°C for 7 days (Zhang et al. 2021). The yellow, needle-shaped crystals of the title compound were harvested when the Teflon-lined autoclave was cooled to ambient temperature. Then, the crystals were separated by suction, washed with ethanol and distilled water (volume ratio 1:1) three times (10 mL per time), and dried in air.

4.8 Germanium–Hydrogen–Potassium–Silver–Oxygen–Sulfur

In the system containing these elements, the multinary compound $K_4Ag_2Ge_3S_9 \cdot H_2O$ which crystallizes as an orthorhombic structure with the lattice parameters a = 1445.24 ± 0.04, b = 963.39 ± 0.03, c = 1656.36 ± 0.09 pm, and a calculated density of 2.997 g·cm^{-3}, is formed (An et al. 2004). The synthesis of this compound was carried out as follows: GeS_2 (32.0 mg), $AgNO_3$ (9.0 mg), and K_2CO_3 (65.0 mg) were added into a glass tube, to which 0.4 mL of an ethanol/$HSCH_2CH(SH)CH_2OH$ (dimercaprol) mixed solvent with 2:1 volume ratio was also added. Then the glass tube was sealed (reagents filled about 10% of the tube), placed into a Teflon-lined stainless steel autoclave, and heated at 120°C for 5 days. The products were washed with ethanol and water, respectively, and cubic colorless crystals were obtained.

4.9 Germanium–Hydrogen–Potassium–Nitrogen–Sulfur

The quinary compound $K_2(GeS_2NH)$ is formed in this system (Behrens and Ostermeier 1962). This slightly soluble compound was prepared by reacting of KNH_2 with a solution of GeS_2 in liquid NH_3.

4.10 Germanium–Hydrogen–Potassium–Oxygen–Sulfur

Three quinary compounds are formed in this system. To obtain $K_2Ge_2S_5 \cdot 4H_2O$, germanium dioxide was dissolved in a concentrated solution of KOH (≈ 40 mass%), and then precipitated with glacial acetic acid, which was added until the acid reaction (Sevryukov et al. 1969). The suspension thus obtained was saturated with H_2S until a homogeneous viscous solution was obtained. Excess acetic acid was removed by extraction with acetone. The thiogermanate solution remaining after this reaction was poured into absolute alcohol, and crystals of $K_2Ge_2S_5 \cdot 4H_2O$ immediately precipitated.

$K_2Au_2GeS_4$ dissolves in water forming a light yellow clear solution (Davaasuren et al. 2017). The removal of water from such solution, by slow evaporation inside the glove box, resulted in the formation of the second quinary compound in this system, $K_3[GeS_3(OH)] \cdot H_2O$, as well as KHS, GeS_2, and Au_2S. This compound crystallizes as a triclinic structure with the lattice parameters a = 668.43 ± 0.13, b = 745.92 ± 0.15, c = 1000.3 ± 0.2 pm, α = 87.59 ± 0.03°, β = 81.81 ± 0.03°, γ = 69.52 ± 0.03°, a calculated density of 2.306 g·cm^{-3} at 203 K, and an energy gap of 3.70 eV (Melullis and Dehnen 2007). Colorless single crystals of this compound were prepared when K_4GeS_4 (0.100 mM) was dissolved in H_2O (10 mL) at 20°C. The solution was stirred for 30 min and filtered away from precipitating Ge powder. The filtrate was evaporated for several hours, whereupon $K_3[GeS_3(OH)] \cdot H_2O$ was obtained. All synthesis steps were performed with strong exclusion of air and external moisture (N_2 atmosphere at a high-vacuum, double-manifold Schlenk line or Ar atmosphere in a glove box). Water was degassed by applying dynamic vacuum (0.1 Pa) for several hours.

For obtaining the third quinary compound, $K_6Ge_2S_7{\cdot}9H_2O$, freshly precipitated GeS_2 was dissolved in a saturated solution of K_2S, with a K_2S/GeS_2 molar ratio of 2:1 (Schwarz and Giese 1930; Ivanov-Emin and Kostrikin 1947). The resulting solution was poured into acetone, taken in a large excess, with vigorous stirring. First, a haze formed, which after 1 h gathered at the bottom of the vessel, forming a layer of a yellow oily substance. After separation of the latter from acetone and drying for 3 days in a desiccator over P_4O_{10} under a H_2S atmosphere, crystals of the title compound were isolated. According to Krebs et al. (1970a,b), this compound does not exist as homogeneous phase: it consists of mixture of K_4GeS_4 and $K_4Ge_2S_6$.

4.11 Germanium–Hydrogen–Rubidium–Cesium–Copper–Oxygen–Sulfur

In the system containing these elements, the multinary compound $(H_3O)Rb_{0.75}Cs_{0.25}Cu_6Ge_2S_8$, which crystallizes as a tetragonal structure with the lattice parameters a = 1039.86 ± 0.12, c = 372.57 ± 0.08 pm, a calculated density of 3.694 g·cm^{-3}, and an energy gap of 2.33 eV, is formed (Li et al. 2019e). It was synthesized using Cu powder (4.0 mg), GeO_2 powder (2.2 mg), S (15 mg), Rb_2CO_3 (5.0 mg), Cs_2CO_3 (4.0 mg), about 300 mg of ethylenediamine and 90 mg of deionized water, and then heated at 200°C for 9 days. Yellow needle crystals were obtained by washing with ethanol and distilled water.

4.12 Germanium–Hydrogen–Rubidium–Copper–Oxygen–Sulfur

In the system containing these elements, the multinary compound $(H_3O)RbCu_6Ge_2S_8$, which crystallizes as a tetragonal structure with the lattice parameters a = 1042.9 ± 0.5, c = 369.2 ± 0.3 pm, a calculated density of 3.579 g·cm^{-3}, and an energy gap of 2.26 eV, is formed (Li et al. 2019e). It was synthesized using Cu powder (4.0 mg), GeO_2 powder (3.0 mg), S (13 mg), Rb_2CO_3 (7.5 mg), about 300 mg of ethylenediamine and 90 mg of deionized water, and then heated at 200°C for 12 days. Yellow needle crystals were obtained by washed with ethanol and distilled water.

4.13 Germanium–Hydrogen–Cesium–Oxygen–Sulfur

Two quinary compounds, $Cs_4Ge_4S_{10}{\cdot}3H_2O$ and $Cs_4Ge_4S_{10}{\cdot}4H_2O$, are formed in this system and both crystallize as a monoclinic structure with the lattice parameters a = 1255.8 ± 0.6, b = 1232.2 ± 0.6, c = 1669.8 ± 0.8 pm, and β = 92.20 ± 0.03°, and the calculated and experimental densities of 3.078 and 3.04 g·cm^{-3}, respectively, for the first compound (Pohl and Krebs 1976) and a = 1254.9 ± 0.5, b = 1230.5 ± 0.5, c = 1672.1 ± 0.6 pm, and β = 92.30 ± 0.04°, and the calculated and experimental densities of 3.128 and 3.11 g·cm^{-3}, respectively, for the second one (Krebs and Pohl 1971).

To obtain $Cs_4Ge_4S_{10}{\cdot}3H_2O$, in a solution of CsOH (20 mM) in carbonate-free H_2O (10 mL), H_2S was introduced with exclusion of air until saturation (Pohl and Krebs 1976). While stirring, a total of about 25 mM of freshly precipitated GeS_2 was added in portions at 70°C-80°C until no further dissolution occurs and after 1 h, a sediment of excess GeS_2 remains. After cooling, the solution was filtered and poured in acetone, whereupon the salt separates out as a white, crystalline precipitate. The acetone was decanted and the solid was taken up in little water and crystallized by evaporation in air. Colorless, compact to platelet shaped crystals were formed within a few hours. This compound is very easily water soluble and hardly sensitive to hydrolysis: it can be kept in air undecomposed for several days. $Cs_4Ge_4S_{10}{\cdot}3H_2O$ easily gives off part of the crystallization water when thoroughly dried.

4.14 Germanium–Hydrogen–Cesium–Oxygen–Iron–Sulfur

In the system containing these elements, the multinary compound $Cs_2FeGe_4S_{10}{\cdot}xH_2O$, which crystallizes as a tetragonal structure with the lattice parameters a = 841.3 ± 0.7, c = 1475 ± 2 pm, and a calculated density of 2.99 g·cm^{-3} at 173 K for x = 0.42, is formed (Bowes et al. 1996). It was prepared in a two-step reaction. Elemental Ge and S, in stoichiometric amounts, were hydrothermally digested in a 5% excess of aqueous CsOH (H_2O/Ge = 50; 150°C; tumbled; 16 h). From the resultant solution, $Cs_4Ge_4S_{10}$ was precipitated by adding acetone/ethanol. It was then recovered by filtration or centrifugation. Addition of aqueous Fe^{2+} to $Cs_4Ge_4S_{10}$ solution (molar ratio 1:1) resulted in the crystallization of the title compound. By applying diffusion crystal growth techniques and/or selective complexing, mineralizing and transporting agents, high-quality pseudo-tetrahedral single crystals as long as several hundred microns have been grown. $Cs_2FeGe_4S_{10}{\cdot}xH_2O$ dehydrated under N_2 atmosphere at around 100°C-200°C.

4.15 Germanium–Hydrogen–Calcium–Carbon–Oxygen–Sulfur

In the system containing these elements, the multinary compound $Ca_3(SO_4)[Ge(OH)_6](CO_3){\cdot}12H_2O$ (mineral carraraite), which crystallizes as a hexagonal structure with the lattice parameters a = 1105.6 ± 0.3, c = 1062.9 ± 0.6 pm, and a calculated density of 1.979 g·cm^{-3}, is formed (Mandarino 2001; Merlino and Orlandi 2001).

4.16 Germanium–Hydrogen–Calcium–Oxygen–Sulfur

In the system containing these elements, the quinary compound $Ca_3Ge(SO_4)_2(OH)_6{\cdot}3H_2O$ (mineral schaurteite), which crystallizes as a hexagonal structure with the lattice parameters a = 852.53 ± 0.04, c = 1080.39 ± 0.06 pm, and a calculated density of 2.642 g·cm^{-3} (Origlieri and Downs 2013) [a = 825.5 and c = 1080 pm (Fleischer 1967a); a = 852.5, c = 1080.3 pm, and the calculated and experimental densities of 2.64 and ≈ 2.65 g·cm^{-3}, respectively (Fleischer 1968a)], is formed.

4.17 Germanium–Hydrogen–Lead–Oxygen–Sulfur

Two quinary compounds, $GePb_3O_2(OH)_2(SO_4)_2$ and $GePb_3(OH)_6(SO_4)_2{\cdot}3H_2O$, are formed in this system (Fleischer 1960). First of them, mineral itoite, crystallizes as an orthorhombic structure with the lattice parameters a = 847, b = 538, c = 694 pm, and a calculated density of 6.67 g·cm^{-3} and the second one, mineral fleischerite, crystallizes as a hexagonal structure with the lattice parameters a = 889, c = 1086 pm, and the calculated and experimental densities of 4.59 and 4.2–4.4 g·cm^{-3}, respectively.

4.18 Germanium–Hydrogen–Nitrogen–Oxygen–Sulfur

In the system containing these elements, the quinary compound $(NH_4)_2[Ge(S_2O_7)_3]$, which crystallizes as a trigonal structure with the lattice parameters a = 966.17 ± 0.07 and c = 1110.3 ± 0.1 pm at 153 K, is formed (Logemann et al. 2012). The reaction for obtaining this compound was performed in thick-walled glass ampoule. The tube was loaded with $GeCl_4$ (1 mM), oleum (1 mL, 65% SO_3), and $(NH_4)_2SO_4$ (1 mM), torch-sealed under vacuum, and placed in a resistance furnace. The ampoule was held at a temperature of 250°C for 24 h and cooled down to room temperature at a rate of 1.8°C·h^{-1}. The colorless crystals are very moisture-sensitive and were separated from oleum in a glove box.

4.19 Germanium–Hydrogen–Nitrogen–Manganese–Sulfur

In the system containing these elements, the quinary compound $Mn_3Ge_2S_7(NH_3)_4$, which crystallizes as an orthorhombic structure with the lattice parameters $a = 910.68 \pm 0.02$, $b = 1392.3 \pm 0.3$, $c = 1275.0 \pm 0.3$ pm, a calculated density of 2.690 g·cm^{-3}, and an energy gap of ~2.65 eV, is formed (Zhang et al. 2014). To synthesize this compound, Mn (1.0 mM), GeS_2 (0.5 mM), S (2.5 mM), hydrazine monohydrate (0.4 mL, 98%), and polyethylene glycol-400 (1 g) were mixed together and sealed in a stainless steel autoclave with a 23 mL Teflon liner. The autoclave was heated at 160°C for 5 days and then taken out from the oven and cooled to room temperature naturally. The green block crystals of the title compound were collected by hand, washed with ethanol, and dried in air. This compound is stable under ambient conditions.

4.20 Germanium–Lithium–Cesium–Chlorine–Sulfur

In the system containing these elements, the quinary compound $Li_2Cs_4Ge_2S_5(S_2)Cl_2$, which crystallizes as an orthorhombic structure with the lattice parameters $a = 714.6 \pm 0.4$, $b = 1429.9 \pm 0.8$, $c = 1984.4 \pm 1.1$ pm, a calculated density of 3.230 g·cm^{-3}, and an energy gap of 2.92 eV, is formed (Li et al. 2019b). The crystals of this compound were synthesized by high-temperature flux method with Li metal, CsCl, Ge and S powder as the raw materials. The ratio of above raw materials was 1:1:1:3, and the highest temperature of the reaction was 850°C. All the reagents were weighted in the above ratio and then loaded into a graphite crucible, which was be sealed into a silica tube under the pressure of 10^{-3} Pa. The furnace was maintained at the highest temperature for about 33 h and cooled to room temperature at a rate of 5°C·h^{-1}. After the above reaction steps, the crystals of the title compound together with other impure phases were obtained. These crystals are air sensitive and transfer into liquid phase. Because of its moisture sensitivity, the pure phase has not been obtained.

4.21 Germanium–Lithium–Copper–Tin–Iron–Sulfur

In the system containing these elements, the multinary phases $Li_xCu_{3.31}GeSn_{12}Fe_4S_{32}$, which crystallize as a cubic structure with the lattice parameters $a = 1033.8 \pm 0.1$, 1033.5 ± 0.1, 1035.1 ± 0.2, and 1039.0 ± 0.3 for $x = 0$, 2, 4, and 6, respectively (Bousquet et al. 1998), are formed. The phase with $x = 0$ was synthesized from the pure elements, which were mixed and sealed in evacuated (< 10^{-3} Pa) quartz tubes. The mixtures were heated to 250°C at a rate of 50°C·h^{-1}. This temperature was kept constant for 48 h. Then the temperature was increased to 680°C for 7 days. The product was ground and stored in an Ar-filled glove box. The electrochemical lithium insertion was carried out in two-electrode cells.

4.22 Germanium–Lithium–Copper–Tin–Cobalt–Sulfur

The $Li_xCu_4GeCo_4Sn_{12}S_{32}$ multinary intercalation compounds are formed in this system (Vicente et al. 1999). To obtain them, electrochemical lithium insertion into $Cu_4GeCo_4Sn_{12}S_{32}$ was carried out in two-electrode cells. These compounds crystallize in the cubic structure with the lattice parameters $a = 1019.9 \pm 0.2$ for $x = 2$, $a = 1023.9 \pm 0.2$ for $x = 4$, $a = 1017.2 \pm 0.1$ for $Li_4Cu_{2.6}GeCo_4Sn_{12}S_{32}$, and $a = 1017.8 \pm 0.1$ for $Li_6Cu_{2.6}GeCo_4Sn_{12}S_{32}$.

4.23 Germanium–Lithium–Silver–Indium–Sulfur

In the system containing these elements, the quinary compound $LiAgIn_2GeS_6$, which crystallizes as a monoclinic structure with the lattice parameters $a = 1215.67 \pm 0.06$, $b = 752.25 \pm 0.04$, $c = 1217.57 \pm$

0.06 pm, $\beta = 109.9070 \pm 0.0010°$, a calculated density of 4.010 g·cm^{-3}, and an energy gap of 2.47 eV, is formed (Zhou et al. 2022a). Crystals of this compound were synthesized by traditional high-temperature solid state reactions using the following experimental process: a 500 mg stoichiometric mixture of Ag, Li_2S, In, Ge, and S and an additional 400 mg KI flux were ground in an agate mortar and loaded into a graphite crucible. Then, the graphite crucible was transferred into a silica tube, which was sealed by a hydrogen–oxygen flame in high vacuum (10^{-2} Pa). The samples were heated from an ambient temperature to 850°C within 9 h and held for 120 h, and finally cooled to 300°C at a rate of 5°C·h^{-1} before turning off the furnace. Moisture- and air-stable crystals of the title compound were obtained after washing with distilled water and alcohol.

4.24 Germanium–Lithium–Strontium–Cadmium–Sulfur

In the system containing these elements, the quinary compound $Li_2Sr_6CdGe_4S_{16}$, which crystallizes as a cubic structure with the lattice parameter a = 1391.59 ± 0.13, a calculated density of 3.587 g·cm^{-3}, and an energy gap of 3.80 eV, is formed (Lian et al. 2020). This compound was synthesized from a mixture of stoichiometric SrS, Li_2S, Cd, Ge, and S with a total mass of 0.4 g in a graphite crucible that was flame-sealed inside a silica jacket and preheated at 250°C for 6 h, and then heated to 900°C within 20 h, sintered at this temperature for 100 h, and subsequently cooled to 400°C at a rate of 5°C·h^{-1}. Block-shaped yellow crystals of the title compound were obtained. They are stable in the air for at least several months.

4.25 Germanium–Lithium–Gallium–Chlorine–Sulfur

GeS_2–LiCl–Ga_2S_3. The glass-forming region in this quasiternary system was determined by Tver'yanovich et al. (1996). It was shown that this region is situated on the GeS_2–Ga_2S_3 side near GeS_2. The ternary glassy alloys can include up to 20 mol% Ga_2S_3 and up to 18 mol% LiCl. To determine the glass-forming region, the samples were quenched in water from 800°C-900°C.

4.26 Germanium–Lithium–Tin–Phosphorus–Sulfur

The $Li_{10}(Ge_{1-x}Sn_x)P_2S_{12}$ solid solutions are formed in this quinary system for the composition $0 \le x \le 1.0$ (Kato et al. 2014). These solid solutions crystallize in the trigonal structure with the lattice parameters a = 873.00 ± 0.07 and c = 1265.9 ± 0.3 pm for the $Li_{10}(Ge_{0.7}Sn_{0.3})P_2S_{12}$ composition. To prepare the solid solutions, SnS_2, Li_2S, P_2S_5, and GeS_2 were mixed in an appropriate molar ratio in an Ar-filled glove box. The mixture was put into a ZrO_2 pot together with ZrO_2 balls, mechanically milled using a planetary ball milling apparatus at 370 rpm for 40 h, and then it was put into a quartz tube and heated at 550°C for 8 h.

4.27 Germanium–Sodium–Rubidium–Mercury–Sulfur

In the system containing these elements, the quinary compound $Na_3RbHg_2Ge_2S_8$, which melts congruently at 444°C and crystallizes as an orthorhombic structure with the lattice parameters a = 1315.72 ± 0.02, b = 656.280 ± 0.010, c = 1807.96 ± 0.03 pm, a calculated density of 4.073 g·cm^{-3}, and an energy gap of 2.76 eV, is formed (Tang et al. 2022). Single crystals of this compound were prepared by high-temperature solid state reaction of Na_2S (78 mg), HgS (233 mg), GeS_2 (137 mg), and RbCl (200 mg), which was added as cation exchanger and flux. The mixture was thoroughly ground in an agate mortar and loaded into silica tube. The silica tube was sealed under a high vacuum of 10^{-3} Pa and then placed in a furnace. The following temperature control program was executed: heating from room temperature to 800°C in 10 h, kept for 3 days, and then cooling slowly to room temperature with a rate of 3°C·h^{-1}. Greenish transparent block crystals were obtained.

A polycrystalline sample of $Na_3RbHg_2Ge_2S_8$ was obtained by reacting Na_2S (2 mM), HgS (2 mM), GeS_2 (2 mM), and RbCl (1 mM) in sealed evacuated fused-silica tube. The tube was placed in a furnace and heated to 400°C in 10 h, kept for a week, and cooled to room temperature. The crude product was washed using methanol to remove the by-product NaCl. A glove box or vacuum was used to complete all the preparation processes.

4.28 Germanium–Sodium–Silver–Indium–Sulfur

In the system containing these elements, the quinary compound $NaAgIn_2GeS_6$, which crystallizes as a monoclinic structure with the lattice parameters $a = 1207.62 \pm 0.09$, $b = 762.56 \pm 0.05$, $c = 1218.38 \pm 0.09$ pm, $\beta = 110.004 \pm 0.002°$, a calculated density of 3.940 g·cm^{-3}, and an energy gap of 2.40 eV, is formed (Zhou et al. 2022a). Crystals of this compound were synthesized in the same way as $LiAgIn_2GeS_6$ was prepared using Na instead of Li_2S. Moisture- and air-stable crystals of the title compound were obtained after washing with distilled water and alcohol.

4.29 Germanium–Sodium–Barium–Chlorine–Sulfur

In the system containing these elements, the quinary compound $NaBa_4Ge_3S_{10}Cl$, which crystallizes as a hexagonal structure with the lattice parameters $a = 976.53 \pm 0.02$, $c = 1205.81 \pm 0.03$ pm, a calculated density of 3.823 g·cm^{-3}, and an energy gap of 3.49 eV, is formed (Feng et al. 2014). To prepare this compound, a mixture of BaS, GeS_2, and NaCl (molar ratio of 1:1:1) was ground and loaded into a fused-silica tube under an Ar atmosphere in a glove box, which was sealed under vacuum of 10^{-3} Pa and then placed in a furnace. The sample was heated to 850°C over 20 h and kept at that temperature for 48 h, then cooled at a slow rate of 4°C·h^{-1} to 400°C, and finally cooled to room temperature.

4.30 Germanium–Sodium–Gallium–Selenium–Sulfur

In the system containing these elements, the quinary compound $Na_2Ga_2GeSSe_5$, which melts congruently at 606°C and crystallizes as an orthorhombic structure with the lattice parameters $a = 1298.7 \pm 1.0$, $b = 2365.3 \pm 1.7$, $c = 751.9 \pm 0.5$ pm, a calculated density of 3.939 g·cm^{-3}, and an energy gap of 1.56 eV, is formed (Li et al. 2018b). For the synthesis of the title compound, a stoichiometric mixture of the starting materials Na_2S, Ga, Ge, and Se (molar ratio 1:2:1:5) were loaded into a graphite crucible and placed in a quartz tube. The tube was flame-sealed under vacuum (~10^{-2} Pa), placed in a furnace, heated from room temperature to 800°C in 40 h, kept at that temperature for 96 h, and then cooled to room temperature at 4°C·h^{-1}. The product was washed with degassed DMF and dried with ethanol. Yellow crystals, which are stable in the air and water, were obtained.

4.31 Germanium–Sodium–Gallium–Chlorine–Sulfur

GeS_2–NaCl–Ga_2S_3. The liquidus surface of this quasiternary system was constructed by Nedoshovenko et al. (1986). In the system, there are fields of primary crystallization of the initial compounds, and the field of the Ga_2S_3 primary crystallization occupies the largest part of the concentration triangle. In the GeS_2–NaCl system, there is an immiscibility region, and the eutectic from the NaCl side is degenerate. The ternary eutectic E crystallizes at 580°C and contains 35 mol% GeS_2, 40 mol% NaCl, and 25 mol% Ga_2S_3. Stable ternary compounds were not found in the system; however, in the regime of sharp quenching, the formation of unstable ternary compounds is possible, which can be facilitated by the ability of the alloys to form glass.

The glass-forming region in this quasiternary system was also determined by Nedoshovenko et al. (1986). It was shown that the composition of glasses includes up to 45 mol% NaCl. Storage of the glasses in a humid atmosphere leads to clouding of their surface. The softening temperature of glasses varies from 416°C at 1 mol% NaCl to 280°C at a maximum content of NaCl. Glasses were obtained in the regime of melt quenching in cold water from 1000°C.

4.32 Germanium–Sodium–Europium–Chlorine–Sulfur

In the system containing these elements, the quinary compound $Na_{1.2}Eu_{3.4}Ge_3S_9Cl_2$, which crystallizes as a hexagonal structure with the lattice parameters a = 953.88 ± 0.03, c = 1171.78 ± 0.09 pm, and a calculated density of 4.034 g·cm^{-3}, is formed (Choudhury and Dorhout 2015). Single crystals of this compound were obtained from the reaction of the anhydrous $EuCl_3$ and Na_2GeS_3. Typically, $EuCl_3$ (0.36 mM) and Na_2GeS_3 (0.28 mM) were mixed and ground in an agate mortar inside a N_2-filled glove box. The mixture was transferred to a graphite crucible, which was then placed inside a fused-silica tube, evacuated, flame-sealed, placed vertically in a furnace, and heated at 825°C for 96 h. The reaction yielded red crystals of the title compound as the major product.

4.33 Germanium–Potassium–Rubidium–Mercury–Sulfur

In the system containing these elements, the quinary compound $K_{1.6}Rb_{0.4}Hg_3Ge_2S_8$, which crystallizes as a monoclinic structure with the lattice parameters a = 964.9 ± 0.5, b = 839.3 ± 0.2, c = 972.0 ± 0.2 pm, β = 95.08 ± 0.03°, and a calculated density of 4.658 g·cm^{-3}, is formed (Kanatzidis et al. 1997a,b; Liao et al. 2003). This compound was prepared by the heating of mixture HgS (0.51 mM), K_2S (0.37 mM), Rb_2S (0.12 mM), Ge (0.25 mM), and S (4.05 mM) at 400°C for 72 h with the next cooling to 385°C over 120 h and then to 165°C at a rate of 2°C·h^{-1}. Large plate-like yellow crystals of the title compound, which are stable in air, was obtained.

4.34 Germanium–Potassium–Calcium–Oxygen–Fluorine–Sulfur

In the system containing these elements, the multinary compound $KF{\cdot}2[Ca_6(SO_4)(GeO_4)_2O$, which crystallizes as a hexagonal structure with the lattice parameters a = 793.3 ± 0.1 and c = 4150.8 ± 1.1 pm, is formed (Navarro and Glasser 1985). This compound was synthesized from mechanically mixed $CaCO_3$, K_2CO_3, $CaCl_2$, CaF_2, and $CaSO_4{\cdot}2H_2O$. The mixture was heated in Pt foil container or crucible in a furnace in air atmosphere. It is formed at temperatures as high as 1200°C, where the compound finally decomposes upon prolonged heating owing to loss by evaporation of sulfur and fluorine species.

4.35 Germanium–Potassium–Barium–Chlorine–Sulfur

In the system containing these elements, the quinary compound $K_2Ba_3Ge_3S_9Cl_2$, which crystallizes as a hexagonal structure with the lattice parameters a = 979.68 ± 0.04, c = 1217.40 ± 0.12, a calculated density of 3.503 g·cm^{-3}, and an energy gap of 3.69 eV, is formed (Zhou et al. 2022c). This compound was synthesized utilizing the following experimental process: a 500 mg mixture of Ba (224.3 mg), Ge (118.6 mg), S (157.1 mg) in a molar ratio of 1:1:3 and an additional 400 mg KCl were ground uniformly and pressed into a pellet. The pellet was then loaded into a silica tube and sealed under vacuum (10^{-2} Pa). The tube was transferred into a muffle furnace, heated at 300°C for 5 h, raised to 950°C and stayed for 2 days, cooled to 400°C at a rate of 5°C·h^{-1}, and finally cooled to room temperature by switching off the furnace. The colorless crystals of $K_2Ba_3Ge_3S_9Cl_2$ were obtained and rinsed with deionized water and then dried with ethanol. The title compound can be steadily obtained when different ratios of raw materials and added different amounts of KCl were used.

4.36 Germanium–Potassium–Gallium–Tin–Sulfur

In the system containing these elements, the quinary compound $KGaGe_{1.37}Sn_{0.63}S_6$, which crystallizes as a tetragonal structure with the lattice parameters $a = 687.19 \pm 0.03$, $c = 2263.18 \pm 0.19$ pm, a calculated density of 2.955 g·cm^{-3}, and an energy gap of 3.50 eV, is formed (Li et al. 2023a). The title compound was obtained by a solid state reaction. A 500 mg mixture of KI as reactive flux to provide the K element, SnO, Ga_2O_3, GeO_2, S, and B with the molar ratios of 1:1:1:6:4 were weighed, then pressed into pellet, followed by its sealing into an evacuated quartz ampoule with the vacuum pressure of 10^{-2} Pa. The tube was heated in a muffle furnace from room temperature to 950°C in 25 h and homogenized at several intermediate temperatures for several hours. The reaction was maintained at 950°C for about 120 h, and then cooled down to 300°C within 120 h. Finally, the tube was cooled down to room temperature naturally. The faint yellow block single crystals of the title compound were obtained. The crystals are stable in air after several months, and can be separated from the byproducts easily under an optical microscope because of their unique colors and shapes.

4.37 Germanium–Potassium–Gallium–Chlorine–Sulfur

GeS_2–KCl–Ga_2S_3. The glass-forming region in this quasiternary system was determined by Tver'yanovich et al. (1996). It was shown that this region is situated on the GeS_2–Ga_2S_3 side near GeS_2. The ternary glassy alloys can include up to 58 mol% Ga_2S_3 and up to 62 mol% KCl. To determine the glass-forming region, the samples were quenched in water from 800°C to 900°C.

4.38 Germanium–Rubidium–Barium–Chlorine–Sulfur

In the system containing these elements, the quinary compound $RbBa_2GeS_4Cl$, which crystallizes as a monoclinic structure with the lattice parameters $a = 968.57 \pm 0.13$, $b = 859.55 \pm 0.13$, $c = 1226.9 \pm 1.8$ pm, $\beta = 90.118 \pm 0.008°$, a calculated density of 3.878 g·cm^{-3}, and an energy gap of 3.43 eV, is formed (Chu et al. 2018). In the preparation process of this compound, a graphite crucible was added into the vacuum-sealed silica tube evacuated to 10^{-3} Pa to avoid the reaction between RbCl and silica tube at the high temperature. $RbBa_2GeS_4Cl$ was prepared with a mixture under an Ar atmosphere in a glove box at the stoichiometric ratio of RbCl/BaS/Ge/S = 1:2:1:2. The temperature process was set as follows: firstly, the furnace was heated to 800°C in 40 h, and kept at this temperature for about 3 days, then slowly cooled down to 300°C at a rate of 5°C·h^{-1}, and eventually, swiftly cooled to room temperature. The obtained product was washed by DMF solvent. Near-colorless crystals were formed, which are stable in air.

4.39 Germanium–Cesium–Barium–Chlorine–Sulfur

In the system containing these elements, the quinary compound $CsBa_2GeS_4Cl$, which crystallizes as a monoclinic structure with the lattice parameters $a = 981.26 \pm 0.17$, $b = 861.07 \pm 0.13$, $c = 1241.6 \pm 0.2$ pm, $\beta = 90.738 \pm 0.012°$, a calculated density of 4.077 g·cm^{-3}, and an energy gap of 3.33 eV, is formed (Chu et al. 2018).

4.40 Germanium–Cesium–Gallium–Chlorine–Sulfur

GeS_2–CsCl–Ga_2S_3. The glass-forming region in this quasiternary system was determined by Tver'yanovich et al. (1996). It was shown that this region occupies the concentration triangle between GeS_2–Ga_2S_3 (50–100 mol% GeS_2) and Ga_2S_3–CsCl (55–72 mol% CsCl) system. To determine the glass-forming region, the samples were quenched in water from 800°C to 900°C.

4.41 Germanium–Copper–Silver–Strontium–Sulfur

In the system containing these elements, the quinary compound $Cu_2Ag_2Sr_6Ge_4S_{16}$, which crystallizes as a cubic structure with the lattice parameter a = 1403.0 ± 0.5, a calculated density of 4.021 g·cm^{-3}, and an energy gap of 2.76 eV, is formed (Yang et al. 2020a). The whole preparation process of this compound was completed in an Ar-filled glove box. It was obtained under a stoichiometric ratio of Cu_2S, Ag_2S, SrS, and GeS_2 in the vacuum-sealed silica tube. Firstly, the tube was heated to 600°C in 30 h and kept at this temperature for about 100 h, then quickly down to room temperature. The above product were ground and then loaded into the vacuum-sealed silica tube. It was further sintered again at 900°C for 10 days and cooled down to room temperature with 5°C·h^{-1}. Finally, many millimeter-level high-quality crystals with yellow color were formed in the tubes after washed with DMF. They are stable in the air within several weeks.

4.42 Germanium–Copper–Silver–Cadmium–Sulfur

Cu_2CdGeS_4–Ag_2CdGeS_4. A continuous series of solid solutions, which crystallize as an orthorhombic structure, is formed in this subsystem of the Ge–Cu–Ag–Cd–S system (Marushko 2012b). Small deviations from the linear concentration dependence exist in the change of a and b parameters, while the c parameter changes linearly. For obtaining these solid solutions, the ingots were annealed at 550°C for 300 h.

4.43 Germanium–Copper–Silver–Iodine–Sulfur

Cu_7GeS_5I–Ag_7GeS_5I. A continuous series of solid solutions, which crystallize as a cubic structure, is formed in this subsystem of the Ge–Cu–Ag–I–S system (Pogodin et al. 2016; Studenyak et al. 2019). The concentration dependence of the lattice parameter is characterized by a negative deviation from Vegard's law. Single crystals of these solid solutions in the entire range of concentrations were grown by the methods of vertical zone or directional crystallization.

4.44 Germanium–Copper–Barium–Magnesium–Sulfur

In the system containing these elements, the quinary compound $Ba_6(Cu_{1.9}Mg_{1.1})Ge_4S_{16}$, which crystallizes as a cubic structure with the lattice parameter a = 1430.8 ± 0.5, a calculated density of 4.030 g·cm^{-3}, and an energy gap of 2.92 ± 0.5 eV, is formed (Cicirello et al. 2021a). To prepare this compound, the reactant elements were loaded into carbonized silica ampoule at a molar ratio of Ba/Cu/Mg/Ge/S = 6:2:1:4:16. The ampoule was then flame-sealed under high vacuum and placed into a furnace, heated to 800°C in 10 h, held at that temperature for 168 h, and then cooled to 100°C in 30 h. The ampoule was then opened in the glove box and the sample was ground to a powder with a mortar. The sample was resealed under high vacuum and placed back into the furnace with the same temperature profile. After the second annealing, the sample appeared as polycrystalline powder.

For growing crystal samples of adequate size, a two-step procedure was followed. The solid state synthesis procedure was utilized, and the sample was ground in the glove box after the second annealing process. The pre-synthesized polycrystalline powder was loaded into carbonized silica ampoule and KI flux was added in a 1:1 mass ratio. The ampoule was sealed under high vacuum and heated to 800°C in 24 h, held at this temperature for 48 h, and slow cooled to 650°C at a rate of 1.5°C·h^{-1}. After annealing, the sample was washed by deionized water to remove the KI flux, yielding yellow-orange crystals of the title compound. All starting materials were stored and used in an Ar-filled glove box.

4.45 Germanium–Copper–Barium–Cadmium–Sulfur

In the system containing these elements, the quinary compound $Ba_6(Cu_{1.6}Cd_{1.4})Ge_4S_{16}$, which crystallizes as a cubic structure with the lattice parameter $a = 1433.6 \pm 0.2$, a calculated density of 4.249 $g \cdot cm^{-3}$, and an energy gap of 3.05 ± 0.5 eV, is formed (Cicirello et al. 2021a). This compound was synthesized in the same way as $Ba_6(Cu_{1.9}Mg_{1.1})Ge_4S_{16}$ was prepared but using Cd instead of Mg. Yellow-orange crystals of the title compound were obtained.

4.46 Germanium–Copper–Barium–Manganese–Sulfur

In the system containing these elements, the quinary compound $Ba_6(Cu_{2.2}Mn_{0.8})Ge_4S_{16}$, which crystallizes as a cubic structure with the lattice parameter $a = 1429.9 \pm 0.2$, a calculated density of 4.111 $g \cdot cm^{-3}$, and an energy gap of 2.70 ± 0.5 eV, is formed (Cicirello et al. 2021a). This compound was synthesized in the same way as $Ba_6(Cu_{1.9}Mg_{1.1})Ge_4S_{16}$ was prepared but using Mn instead of Mg. Red crystals of the title compound were obtained.

4.47 Germanium–Copper–Barium–Iron–Sulfur

In the system containing these elements, the quinary compound $Ba_6Cu_2FeGe_4S_{16}$, which crystallizes as a cubic structure with the lattice parameter $a = 1423.42 \pm 0.06$, a calculated density of 4.169 $g \cdot cm^{-3}$, and an energy gap of 1.72 eV, is formed (Cao et al. 2019).To prepare this compound, in an Ar-protected glove box, BaS, Cu, Fe, Ge, and S were mixed with the molar radio of 6:2:1:4:10. Next, the mixture was loaded into the graphite crucible. Then, it was put into the silica tube sealed under vacuum ($<10^{-3}$ Pa) and transferred to a muffle furnace. The sample was heated from room temperature to 900°C in 24 h and held at this temperature for 20 h. After the temperature was cooled to 300°C in 40 h, the furnace was shut down and the sample was cooled to ambient temperature.

For obtaining the single crystals of this compound, in an Ar-protected glove box, the prepared powder of $Ba_6Cu_2FeGe_4S_{16}$ was mixed with flux KI in a 1:1 mass ratio. The mixture was loaded into the graphite crucible, which was put into the silica tube, sealed the tube under vacuum ($<10^{-3}$ Pa), and transferred to a furnace. Afterward, it was heated to 900°C in 24 h and maintained at the current temperature for 48 h. Subsequently, the temperature was cooled to 300°C for 150 h. After that, the furnace was shut down and the sample was naturally cooled to ambient temperature. After the sample was taken out, it was cleaned with deionized water to obtain red crystals. These crystals are stable in air.

4.48 Germanium–Copper–Zinc–Gallium–Iron–Sulfur

In the system containing these elements, the multinary compound $Cu_2(Zn,Fe)(Ge,Ga)S_4$ (mineral zincobriartite), which crystallizes as a tetragonal structure with the lattice parameters $a = 534.33 \pm 0.04$ and $c = 1053.50 \pm 0.11$ pm, is formed (McDonald et al. 2016).

4.49 Germanium–Copper–Zinc–Tin–Sulfur

Cu_2ZnGeS_4–Cu_2ZnSnS_4. This system is nonquasibinary section of the Ge–Cu–Zn–Sn–S quinary system (Olekseyuk et al. 2019). At 400°C, Cu_2ZnGeS_4 dissolves 5 mol% Cu_2ZnSnS_4 and the solubility of Cu_2ZnGeS_4 in Cu_2ZnSnS_4 reaches 18 mol%. The alloys for the investigations were annealed at 400°C for 500 h followed by quenching into cold water.

4.50 Germanium–Copper–Zinc–Arsenic–Iron–Sulfur

In the system containing these elements, the multinary compound $(Cu,Zn)_{11}(Ge,As)_2Fe_4S_{16}$ (mineral renierite), which crystallizes as a tetragonal structure with the lattice parameters a = 1062.26 ± 0.05 and c = 1055.06 ± 0.08 pm (Bernstein et al. 1989) [in the cubic structure with the lattice parameter a = 1058.3 pm and an experimental density of 4.3 g·cm^{-3} (Murdoch 1953)], is formed. According to Bernstein (1986), the composition of renierite was found to be $Cu_{10}(Cu_xZn_{1-x})(Ge_{2-x}As_x)Fe_4S_{16}$, where $0 \le x \le 0.5$ for zincian renierite and $0.5 \le x \le 1$ for arsenian renierite: a = 1062.2 ± 0.1, c = 1055.1 ± 0.1 pm, and a calculated density of 4.414 g·cm^{-3} for $x = 0$ and a = 1060 ± 1, c = 1045 ± 1 pm, and a calculated density of 4.5 g·cm^{-3} for $x = 1$.

4.51 Germanium–Copper–Zinc–Selenium–Sulfur

The $Cu_2ZnGeS_{4-x}Se_x$ solid solutions are formed in this quinary system (Doverspike et al. 1990). To prepare them, the stoichiometric mass of the elements was introduced into a silica tube that was evacuated to 5×10^{-3} Pa. The tube was then sealed and enclosed in a tightly wound Kanthal coil to even out temperature gradients. It was then placed in a furnace and heated to 600°C at a rate of 30°C·h^{-1}. The temperature was then raised to 650°C for 24 h and then to 700°C for 72 h. The tube was then removed from the furnace and opened, and the product was ground under a N_2 atmosphere. The sample was then placed in another silica tube, evacuated, sealed, and placed back in the furnace for an additional 24 h at 700°C. Selenium can be substituted for sulfur up to 100% in the tetragonal structure of solid solution, prepared at 700°C and up to 75% in the orthorhombic structure prepared at 900°C. The temperature at which the phase transformation takes place was found to increase with higher selenium content.

4.52 Germanium–Copper–Zinc–Tungsten–Iron–Sulfur

In the system containing these elements, the multinary compound $Cu_{20}(Fe,Cu,Zn)_6W_2Ge_6S_{32}$ (mineral ovamboite), which crystallizes as a cubic structure with the lattice parameter a = 1068 ± 2 and a calculated density of 4.736 g·cm^{-3}, is formed (Spiridonov 2003; Jambor and Roberts 2004).

4.53 Germanium–Copper–Zinc–Manganese–Sulfur

The $Cu_2Zn_{1-x}Mn_xGeS_4$ solid solutions are formed in this quinary system (Honig et al. 1988). They were grown by chemical vapor transport using iodine as the transport agent. Optimum growth conditions were obtained when the maximum reaction temperature of the charge was 900°C for x = 0, 0.03, 0.05, 0.10, and 800°C for x = 1.0. A temperature gradient of 50°C between the charge and the growth zones was used for crystal growth in all the cases. The transport process was run for 10 days. All crystals were orange in color, although for x = 1.0 the crystals were more opaque. All samples crystallize with the wurtz-stannite structure, which has an orthorhombic unit cell.

4.54 Germanium–Copper–Zinc–Iron–Sulfur

Two quinary compounds, $Cu_2(Fe,Zn)GeS_4$ (mineral briartite) and $Cu_8ZnFe_2Ge_2S_{12}$ (mineral omariniite), are formed in this system. The first of them crystallizes as a tetragonal structure with the lattice parameters a = 532 ± 1 and c = 1051 ± 1 pm (Francotte et al. 1965; Fleischer 1966b).

The solid solution of $Cu_2Fe_{1-x}Zn_xGeS_4$ has been synthesized at 700°C with a composition varying between Cu_2FeGeS_4 and Cu_2ZnGeS_4 (Ottenburgs and Goethals 1972). The substitution of iron by zinc does not modify the structure of the solid solution. The Zn varieties show polymorphism: with increasing temperature, the tetragonal structure, derived from that of chalcopyrite, is transformed into an orthorhombic structure. With increasing Zn content, the temperature of polymorphic transformation is reduced, but in contrast the melting point rises.

Orthorhombic samples of $Cu_2Fe_{1-x}Zn_xGeS_4$ where $0 \le x \le 0.15$ can be also obtained by rapid quenching from 900°C (Doverspike et al. 1988). At $x = 0.20$, a small amount of the tetragonal phase was always present in the quenched sample. The phase transformation from tetragonal to orthorhombic could be reversed by either annealing the orthorhombic phase below the transition point or by applying elevated pressure.

$Cu_8ZnFe_2Ge_2S_{12}$ crystallizes as an orthorhombic structure with the lattice parameters $a = 1077.4 \pm 0.1$, $b = 539.21 \pm 0.05$, $c = 1608.5 \pm 0.2$ pm, and a calculated density of 4.319 g·cm^{-3} (Bindi et al. 2016b, 2017).

4.55 Germanium–Copper–Cadmium–Tin–Sulfur

Cu_2CdGeS_4–Cu_2CdSnS_4. This system is nonquasibinary section of the Ge–Cu–Cd–Sn–S quinary system (Olekseyuk et al. 2019). At 400°C, Cu_2CdGeS_4 dissolves 11 mol% Cu_2CdSnS_4 and the solubility of Cu_2CdGeS_4 in Cu_2CdSnS_4 reaches 16 mol%. The alloys for the investigations were annealed at 400°C for 500 h followed by quenching into cold water.

Cu_2GeS_3–Cu_2SnS_3–CdS. The isothermal section of this quasiternary system at 400°C is shown in the Figure 4.1 (Marushko et al. 2009a). The existence of four single-phase regions is established at this temperature. They are a continuous solution series of Cu_2GeS_3 and Cu_2SnS_3, solid solutions of the quaternary compounds stretched along the Cu_2CdGeS_4–Cu_2CdSnS_4 section, solid solution range of Cu_2CdGeS_4 with the orthorhombic structure that extends to 9 mol% Cu_2CdSnS_4, solid solution range of Cu_2CdSnS_4 with the stannite structure that extends to 14 mol% Cu_2CdGeS_4, and solid solution range of CdS with the wurtzite structure.

The liquidus surface of the Cu_2GeS_3–Cu_2SnS_3–CdS quasiternary system is shown in the Figure 4.2 (Marushko et al. 2009a). It consists of five fields of the primary crystallization of the solid solutions based on CdS (β), $Cu_2Cd_3GeS_6$ (ε), Cu_2CdGeS_4 (γ), and Cu_2CdSnS_4 (δ) quaternary compounds and $Cu_2Ge_xSn_{1-x}S_3$ (α) solid solutions. The field of the β solid solutions occupies the largest part of the concentration triangle. These fields of the primary crystallization are separated by seven monovariant lines and eight invariant

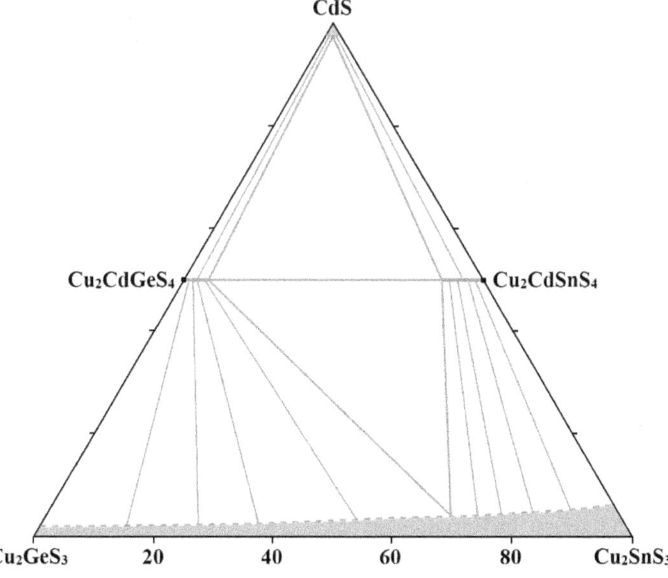

FIGURE 4.1 Isothermal section of the Cu_2GeS_3–Cu_2SnS_3–CdS quasiternary system at 400°C. (From Marushko, L.P., et al., *J. Alloys Compd.*, 484(1–2), 147, 2009.)

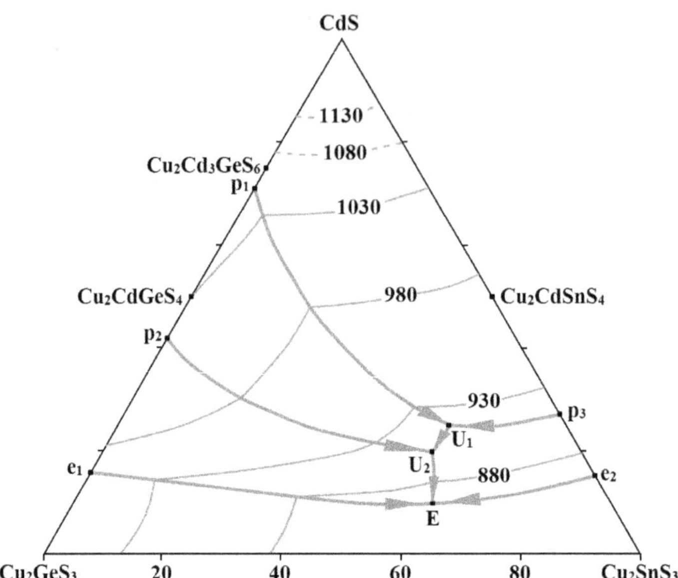

FIGURE 4.2 Liquidus surface of the Cu_2GeS_3–Cu_2SnS_3–CdS quasiternary system. (From Marushko, L.P., et al., *J. Alloys Compd.*, 484(1–2), 147, 2009.)

points: E (844°C) – L ⇔α + γ + δ; U_1 (914°C) – L + β ⇔δ + ε; and U_2 (903°C) – L + ε ⇔γ + δ. This system was investigated through DTA and XRD using the alloys annealed at 400°C for 500 h, followed by quenching into cold water.

4.56 Germanium–Copper–Cadmium–Selenium–Sulfur

Cu_2GeSe_3 + 3CdS + ⇔ Cu_2GeS_3 + 3CdSe. Five single-phase regions exist in this ternary mutual system at 400°C (Figure 4.3): the limited solid solution ranges of Cu_2GeS_3 and Cu_2GeSe_3, the limited solid solutions of the quaternary compounds (solid solution range of Cu_2CdGeS_4 with the orthorhombic structure and a small range of solid solution of low-temperature modification of $Cu_2CdGeSe_4$ with the stannite structure), and the continuous solid solution series of CdS and CdSe with the wurtzite structure (Marushko et al. 2009b). The samples were annealed at 400°C for 500 h and then quenched in cold water.

The liquidus surface of the Cu_2GeSe_3 + 3CdS + ⇔ Cu_2GeS_3 + 3CdSe ternary mutual system is shown in the Figure 4.4 (Marushko et al. 2009b). It consists of five fields of the primary crystallization that correspond to the $CdSe_xS_{1-x}$, $Cu_2Cd_3GeSe_{6(1-x)}S_{6x}$ and $Cu_2CdGeSe_{4(1-x)}S_{4x}$ solid solutions and solid solution based on the Cu_2GeS_3 and Cu_2GeSe_3 ternary compounds. The fields of the primary crystallization are separated by five monovariant lines and eight invariant points. Ternary transition point *U* takes place at 852°C. Neither of the sections that form the quadrangle diagonals is a quasibinary one.

4.57 Germanium–Copper–Cadmium–Iron–Sulfur

In the system containing these elements, the quinary compound $Cu_2(Cd,Fe)GeS_4$ (mineral barquillite), which crystallizes as a tetragonal structure with the lattice parameters $a = 545 \pm 4$, $c = 1060 \pm 10$ pm, and a calculated density of 4.53 g·cm^{-3}, is formed (Jambor and Roberts 1999, Mandarino 1999, Murciego et al. 1999). This mineral contains 2.1 at.% (2.2 mass%) Fe and its composition shows extensive Cd-for-Fe substitution and extends to at least Cd: Fe = 1: 1 ratio.

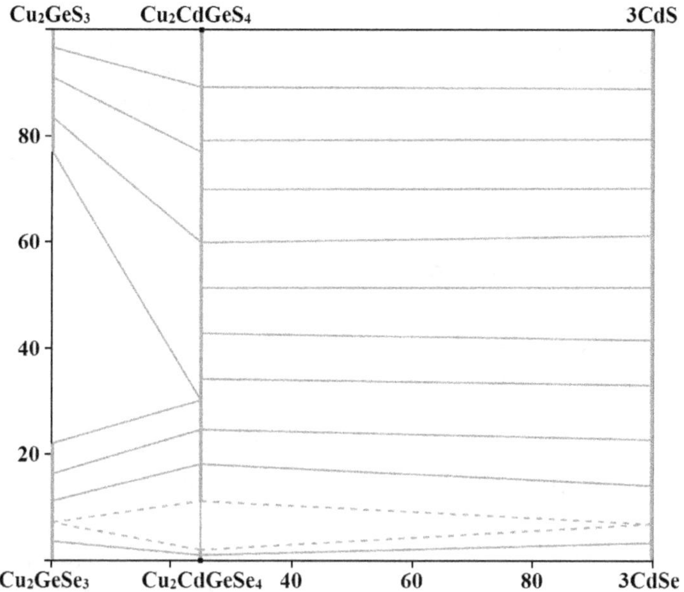

FIGURE 4.3 Isothermal section of the Cu_2GeSe_3 + 3CdS + ⇔ Cu_2GeS_3 + 3CdSe ternary mutual system at 400°C. (From Marushko, L.P., et al., *J. Alloys Compd.*, 473(1–2), 94, 2009.)

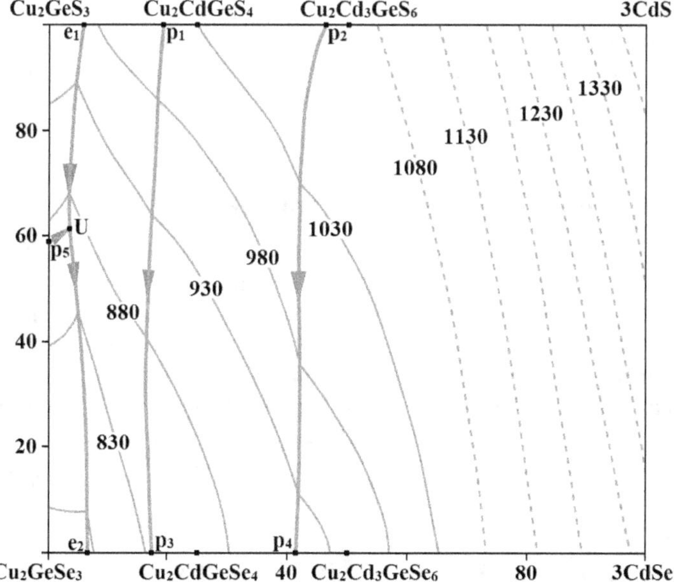

FIGURE 4.4 Liquidus surface of the Cu_2GeSe_3 + 3CdS + ⇔ Cu_2GeS_3 + 3CdSe ternary mutual system. (From Marushko, L.P., et al., *J. Alloys Compd.*, 473(1–2), 94, 2009.)

4.58 Germanium–Copper–Tin–Cobalt–Sulfur

In the system containing these elements, the quinary compound $Cu_4GeCo_4Sn_{12}S_{32}$, which crystallizes as a tetragonal structure with the lattice parameters $a = 715.7 \pm 0.1$ and $c = 1010.9 \pm 0.2$ pm or in the cubic structure with the lattice parameter $a = 1010.5 \pm 0.1$ ($a = 1016.5 \pm 0.1$ for $Cu_{2.6}GeCo_4Sn_{12}S_{32}$ composition), is formed (Vicente et al. 1999). To obtain this compound, a stoichiometric mixture of the pure elements was sealed under vacuum in a quartz tube. The mixture was heated to 250°C, at a constant rate of 50°C·h^{-1} and maintained at this temperature for 48 h, and then the temperature was increased to 680°C for a week and finally slowly cooled to room temperature. The copper extraction was carried out by treating the compound with a 0.1 M solution of I_2 in CH_3CN, as the oxidation reagent.

4.59 Germanium–Copper–Arsenic–Vanadium–Sulfur

In the system containing these elements, the quinary compound $Cu_{26}(Ge,As)_6V_2S_{32}$ (mineral germanocolusite), which crystallizes as a cubic structure with the lattice parameter $a = 1056.8 \pm 0.3$ pm and a calculated density of 4.55 g·cm^{-3}, is formed (Spiridonov et al. 1992b; Jambor and Grew 1994b)

4.60 Germanium–Copper–Selenium–Iodine–Sulfur

Cu_7GeS_5I–Cu_7GeSe_5I. The phase relations in this system were studied through DTA, XRD, metallography, and measurements of the density by Studenyak et al. (2007a). It was shown that the solid solutions are formed in the whole concentration range. The lattice parameter (cubic structure) and density change linearly with composition. All the alloys appeared to be stable in air.

4.61 Germanium–Copper–Molybdenum–Iron–Sulfur

In the system containing these elements, the multinary compound $Cu_{20}(Fe,Cu)_6Mo_2Ge_6S_{32}$ (mineral maikainite), which crystallizes as a cubic structure with the lattice parameter $a = 1064 \pm 1$ and a calculated density of 4.453 g·cm^{-3}, is formed (Spiridonov 2003; Jambor and Roberts 2004).

4.62 Germanium–Copper–Manganese–Cobalt–Sulfur

The $Cu_2Mn_xCo_{1-x}GeS_4$ solid solutions are formed in this quinary system (Bernert and Pfitzner 2006). To synthesize them, Cu_2MnGeS_4 and Cu_2CoGeS_4 were used as the starting materials. Mixed crystals were annealed at 800°C to 850°C for 5 days, homogenized in an agate ball mill, and then pressed to pellets. These were annealed again for 5 days at the same temperature. The melting points of the solid solutions decrease linearly with increasing Mn content. For $Cu_2Mn_xCo_{1-x}GeS_4$, from $x = 0$ to $x = 0.5$ the tetragonal structure dominates while from $x > 0.68$, solid solutions crystallize in the orthorhombic structure ($a = 757.7 \pm 0.2$, $b = 650.9 \pm 0.1$, $c = 623.3 \pm 0.1$ pm, and a calculated density of 4.162 g·cm^{-3} for $x = 0.68$).

4.63 Germanium–Silver–Cadmium–Gallium–Sulfur

Ag_2CdGeS_4–$AgCd_2GaS_4$. At 550°C, a continuous series of solid solutions, which crystallize as an orthorhombic structure, is formed in this subsystem of the Ge–Ag–Cd–Ga–S system (Marushko 2012a). The alloys for the investigations were annealed at 550°C for 300 h.

GeS_2–$AgGaS_2$–$CdGa_2S_4$. The glass-forming region in this quasiternary system is shown in the Figure 4.5 (Naydych et al. 2006). The glasses were obtained up to 15 mol% $AgGaS_2$ in the GeS_2–$AgGaS_2$ system and up to 48 mol% $CdGa_2S_4$ (65 mol% $0.5CdGa_2S_4$) in the GeS_2–$CdGa_2S_4$ system. To determine the glass-forming region, the samples were annealed at 1000°C for 6 h and quenched in 25 mass% NaCl aqueous solution.

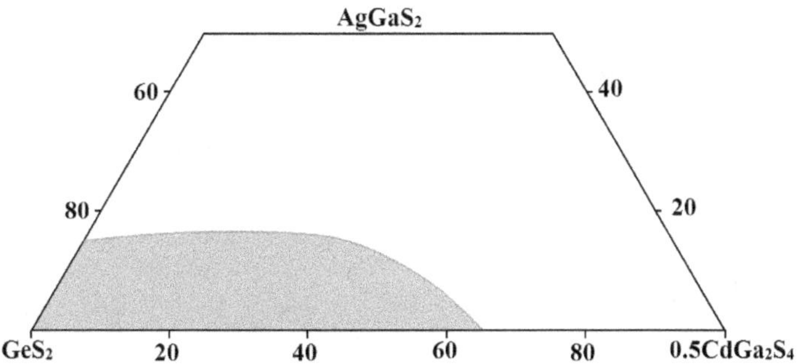

FIGURE 4.5 Glass-forming region in the GeS_2–$AgGaS_2$–$0.5CdGa_2S_4$ quasiternary system. (From Naydych, T.L., et al., *Nauk. Visnyk Volyns'k. Nats. Univ. im. Lesi Ukrainky. Ser. Khim. nauky*, (4), 80, 2006.)

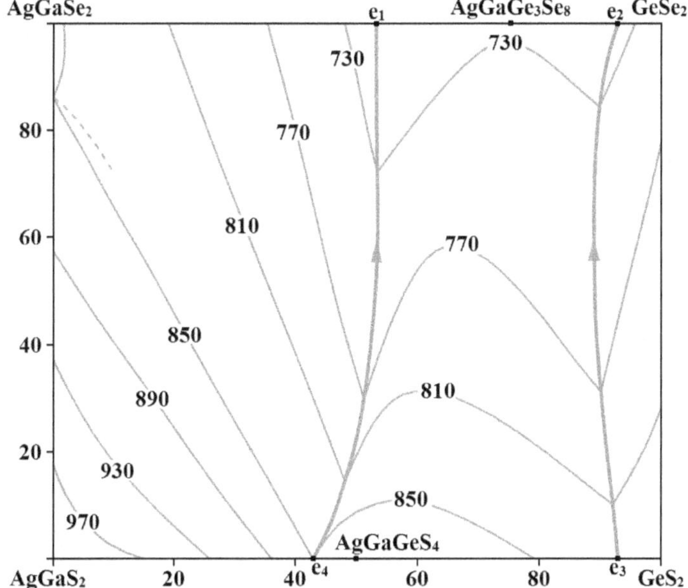

FIGURE 4.6 Liquidus surface of the $GeS_2 + AgGaSe_2 \Leftrightarrow GeSe_2 + AgGaS_2$ ternary mutual system. (From Shevchuk, M., *Nauk. Visnyk Volyns'k. Nats. Univ. im. Lesi Ukrainky. Ser. Khim. nauky*, [24(274)], 15, 2013.)

4.64 Germanium–Silver–Gallium–Selenium–Sulfur

$GeS_2 + AgGaSe_2 \Leftrightarrow GeSe_2 + AgGaS_2$. Three vertical sections, liquidus surface (Figure 4.6), and isothermal section at 450°C (Figure 4.7) of this ternary mutual system were constructed through DTA, XRD, and metallography using the ingots annealing at 450°C for 500 h (Shevchuk 2013c). The liquidus surface consists of three fields of primary crystallization of $AgGaS_{2x}Se_{2(1-x)}$, $AgGaGe_{1+2x}S_{4(1-x)}Se_{8x}$, and $GeS_{2x}Se_{2(1-x)}$ solid solutions, which are separated each other by two monovariant lines. The isothermal section of this system contains three single-phase regions. Both diagonal sections, GeS_2–$AgGaSe_2$ and $GeSe_2$–$AgGaS_2$, are nonquasibinary (Shevchuk 2013a,b,c). The solubility of GeS_2 in $AgGaSe_2$ is 39 mol% at 450°C, GeS_2 dissolves ≤ 2 mol% $AgGaSe_2$ and the solid solution region based on $AgGaS_2$ reaches 35 mol% $GeSe_2$ at the same temperature.

The glass-forming region was found during melt quenching from 1000°C in this system (Halyan et al. 2009). It is localized along the binary GeS_2–$GeSe_2$ system. The glass transition temperature of these alloys is in the interval from 390°C to 488°C.

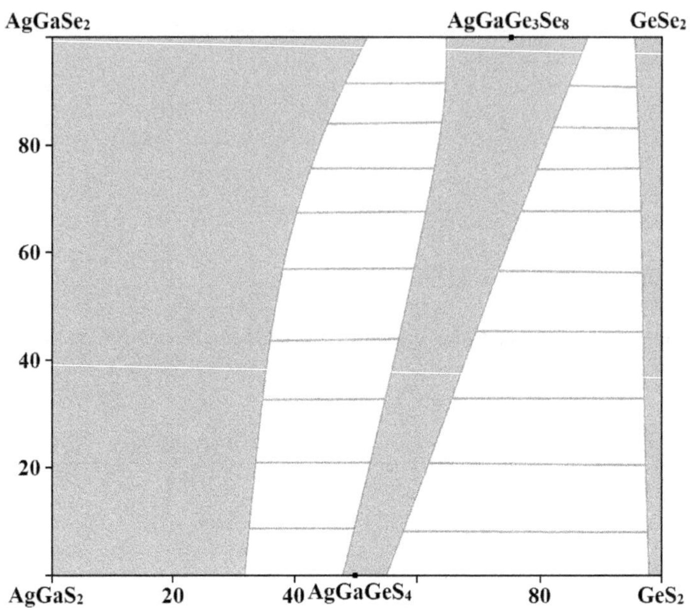

FIGURE 4.7 Isothermal section of the GeS_2 + $AgGaSe_2$ ⇔ $GeSe_2$ + $AgGaS_2$ ternary mutual system at 450°C. (From Shevchuk, M., *Nauk. Visnyk Volyns'k. Nats. Univ. im. Lesi Ukrainky. Ser. Khim. nauky*, [24(274)], 15, 2013.)

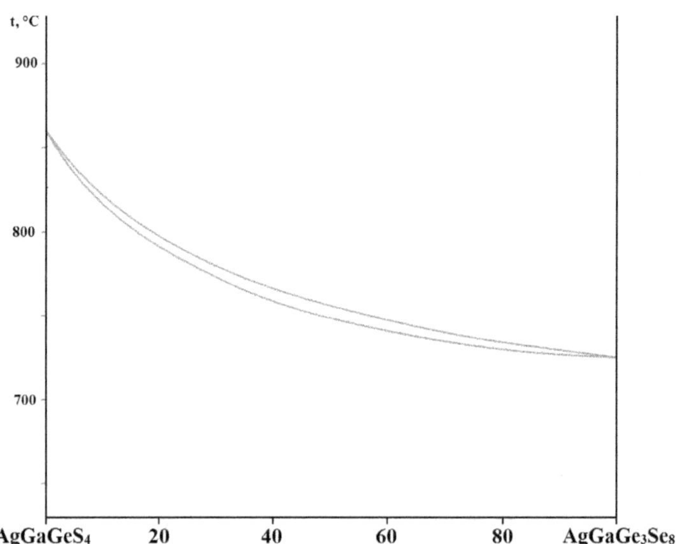

FIGURE 4.8 Phase diagram of the $AgGaGeS_4$–$AgGaGe_3Se_8$ system. (From Shevchuk, M.V., et al., *J, Cryst. Growth*, 318(1), 708, 2011.)

$AgGaGeS_4$–$AgGaGe_3Se_8$. The phase diagram of this system (Figure 4.8), constructed through DTA and XRD using the alloys annealed at 600°C for 500 h, belongs to the type I of Roozeboom's classification (Shevchuk and Olekseyuk 2006; Shevchuk et al. 2011). A continuous solid solution series exist in the system. The lattice parameters of the solid solutions change in accordance with Vegard's law and the band gap varies from 2.25 to 2.80 eV. Single crystals of the solid solutions were grown by the Bridgman–Stockbarger method.

4.65 Germanium–Silver–Gallium–Chlorine–Sulfur

GeS_2–AgCl–Ga_2S_3. The glass-forming region in this quasiternary system was determined by Aleksandrov (1990). It was shown that this region is situated on the GeS_2–Ga_2S_3 side near GeS_2. The ternary glassy alloys can include up to 45 mol% Ga_2S_3 and up to 42 mol% AgCl.

4.66 Germanium–Silver–Lead–Selenium–Sulfur

$Ag_{0.5}Pb_{1.75}GeS_4$–$Ag_{0.5}Pb_{1.75}GeSe_4$. This system is nonquasibinary section of the Ge–Ag–Pb–Se–S quinary system as $Ag_{0.5}Pb_{1.75}GeSe_4$ melts incongruently (Reshak et al. 2013). A continuous solid solution series, which crystallize as cubic structures, form in the system below 576°C with the linear variation of the lattice parameter. $Ag_{0.5}Pb_{1.75}GeSeS_3$ is *n*-type semiconductor with an energy gap of 1.9 eV at room temperature and 2.0 eV at 100 K, lattice parameter $a = 1415.68 \pm 0.01$ pm, and a calculated density of 6.2210 ± 0.0002 g·cm^{-3}.

4.67 Germanium–Silver–Arsenic–Selenium–Sulfur

$GeSe_2$-Ag_4SSe–As_2Se_3. The glass-forming region in this quasiternary system is presented in Figure 4.9 (Vassilev et al. 2005). It is situated entirely on the $GeSe_2$–As_2Se_3 side of the Gibbs diagram and partially on the As_2Se_3–Ag_4SSe (up to ~20 mol% Ag_4SSe) side. The glass transition temperature of these alloys varies in the range of 165°C-273°C. Glasses were prepared by alloying the initial binary components in evacuated (0.133 Pa) silica ampoules, using a rotary furnace.

4.68 Germanium–Silver–Phosphorus–Iodine–Sulfur

Ag_7GeS_5I–Ag_6PS_5I. This system is nonquasibinary section of the Ge–Ag–P–I–S quinary system as Ag_6PS_5I melts incongruently (Pogodin et al. 2021). A continuous series of solid solutions are formed in the solid state.

4.69 Germanium–Beryllium–Zinc–Oxygen–Sulfur

The $Zn_4Be_3(GeO_4)_3S$ quinary compound is formed in this system (Mel'nikov et al. 1968). This compound was synthesized through hydrothermal synthesis at 450°C and 0.2 GPa for 7–10 days using the mixture of ZnO, BeO, GeO_2, and ZnS with the solution of 1 mass% NaOH as a solvent.

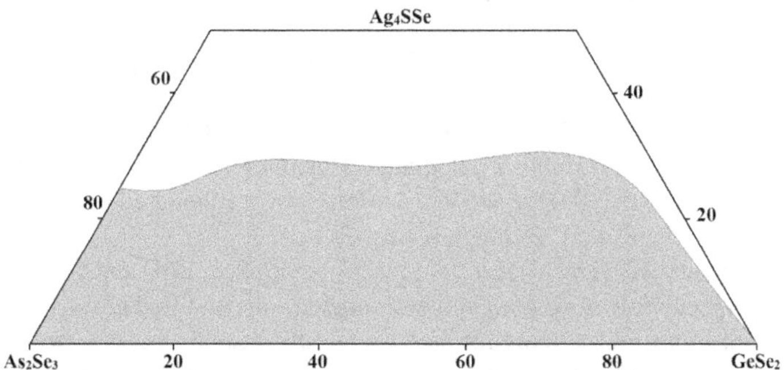

FIGURE 4.9 Glass-forming region in the Ag_4SSe–$GeSe_2$–As_2Se_3 quasiternary system. (From Vassilev, V., et al., *Mater. Lett.*, 59(1), 85, 2005.)

4.70 Germanium–Beryllium–Cadmium–Oxygen–Sulfur

The $Cd_4Be_3(GeO_4)_3S$ quinary compound is formed in this system (Mel'nikov et al. 1968). This compound was synthesized through hydrothermal synthesis at 450°C and 0.2 GPa for 7–10 days using the mixture of CdO, BeO, GeO_2, and CdS with the solution of 1 mass% NaOH as a solvent.

4.71 Germanium–Calcium–Gallium–Oxygen–Sulfur

In the system containing these elements, the quinary compound $Ca_2Ga_2GeOS_6$, which crystallizes as a tetragonal structure with the lattice parameters $a = 921.538 \pm 0.016$, $c = 600.44 \pm 0.03$ pm, a calculated density of 3.260 g·cm^{-3}, and an energy gap 3.15 eV, is formed (Wang et al. 2022c). Single crystals of this compound can be obtained as follows: CaO (1.0 mM), CaS (1.0 mM), Ga_2S_3 (1.0 mM), GeS_2 (1.0 mM), and the flux KI (10 mM) were weighed. These reagents were ground and loaded in a silica tube coated with carbon. The tube was flame-sealed, heated to 975°C for over 10 h, held for 48 h, and then cooled to 750°C over 48 h. The obtained product was sequentially washed with deionized water and acetone and then dried at 70°C.

4.72 Germanium–Calcium–Lanthanum–Chlorine–Sulfur

In the system containing these elements, the quinary compound $Ca_2LaGeS_4Cl_3$, which crystallizes as a hexagonal structure with the lattice parameters $a = 973.11 \pm 0.14$, $c = 633.66 \pm 0.13$ pm, and a calculated density of 3.363 g·cm^{-3}, is formed (Gitzendanner and DiSalvo 1996). It was first isolated from the reaction of equimolar amounts of La (200 mg), CaS (103.8 mg), Ge (209.0 mg), and S (253.9 mg). The reactants were mixed with an approximately equal mass of dry $CaCl_2$ (869.8 mg) to act as a flux. The mixture was loaded into an open graphite crucible and sealed in an evacuated fused-silica tube. It was then heated slowly (40°C·h^{-1}) to 600°C and held there for 24 h to allow for the complete reaction of the sulfur. The temperature was then raised to 875°C and maintained for 50 h. To encourage crystal growth, the mixture was slowly cooled back to 600°C (2.5°C·h^{-1}), the furnace was then shut off, and the reaction mixture was allowed to cool to room temperature with no external quenching. Excess $CaCl_2$ flux was removed by soaking the crucible in deionized water for several hours. The remaining product was filtered off, washed with more water, and allowed to dry overnight in a desiccator. This flux removal process ensures that the products isolated are water stable. The title compound has also been found to be air-stable on the time scale of at least months. All reagents were stored and handled in an Ar-filled glove box to prevent air and moisture exposure.

4.73 Germanium–Strontium–Zinc–Oxygen–Sulfur

In the system containing these elements, the quinary compound $Sr_2ZnGe_2OS_6$, which melts congruently at 1030°C and crystallizes as a tetragonal structure with the lattice parameters $a = 943.22 \pm 0.06$, $c = 618.13 \pm 0.05$ pm, a calculated density of 3.588 g·cm^{-3}, and an energy gap of 3.30 eV (Tian et al. 2022) [$a = 940.54 \pm 0.07$, $c = 617.18 \pm 0.07$ pm, and the calculated and experimental densities of 3.62 and 3.61 g·cm^{-3}, respectively (Teske 1985b); $a = 943.11 \pm 0.02$ and $c = 618.45 \pm 0.01$ pm (Endo et al. 2017); $a = 945.76 \pm 0.06$, $c = 619.81 \pm 0.06$ pm, a calculated density of 3.56 g·cm^{-3}, and an energy gap of 3.73 eV (Ran et al. 2022); $a = 941.05 \pm 0.03$, $c = 617.18 \pm 0.04$ pm, a calculated density of 3.610 g·cm^{-3}, and an energy gap of 3.00 eV (Wang et al. 2022c)], is formed.

This compound was synthesized by a solid state reaction with SrS, ZnO and GeS_2 in a 2:1:2 molar ratio (Tian et al. 2022). Raw materials were weighed and mixed in an Ar-filled glove box and vacuum-sealed quartz tube was placed in a furnace. This furnace was firstly heated to 900°C within 40 h, kept at this temperature for 80 h, and then slowly lowered to room temperature within 100 h. The colorless crystals, which are stable in the air, were obtained.

Single crystals of $Sr_2ZnGe_2OS_6$ were synthesized in the following way (Wang et al. 2022c). SrO (1.0 mM), SrS (1.0 mM), ZnS (1.0 mM), Ge (2.0 mM), S (4.0 mM), and KI as a flux (10 mM) were weighed. These reagents were ground and loaded in a silica tube coated with carbon. The tube was frame-sealed, heated to 850°C over 10 h, held for 48 h, and then cooled to 700°C over 48 h. The obtained product was sequentially washed with deionized water and acetone and then dried under 70°C.

4.74 Germanium–Strontium–Cadmium–Oxygen–Sulfur

The quinary compound $Sr_2CdGe_2OS_6$, which melts congruently at 941°C and crystallizes as a tetragonal structure with the lattice parameters a = 960.80 ± 0.02, c = 620.85 ± 0.02 pm, a calculated density of 3.715 g·cm^{-3}, and an energy gap of 3.13 eV (Tian et al. 2022) [a = 957.260 ± 0.010, c = 617.97 ± 0.02 pm, a calculated density of 3.76 g·cm^{-3}, and an energy gap of 3.62 eV (Ran et al. 2022); a = 959.186 ± 0.014, c = 619.79 ± 0.02 pm, a calculated density of 3.734 g·cm^{-3}, and an energy gap of 2.95 eV (Wang et al. 2022c)], is formed in the Ge–Sr–Cd–O–S system.

The pale yellow crystals of the title compound were obtained in high-temperature conditions (Ran et al. 2022). The mixture of SrS (1 mM), CdO (0.5 mM), GeS_2 (1 mM) was ground and heated from room temperature to 400°C in 20 h and then warmed to 850°C in 30 h after keeping for 20 h in 400°C. The furnace was then maintained for 120 h in the highest temperature, and cooled down to 350°C within 150 h, then to room temperature spontaneously with the heating stopped. After washing by hot water and ethanol, the crystals of $Sr_2CdGe_2OS_6$ were obtained. They are stable in air and moisture for more than one year. Single crystals of this compound were grown in the same way as single crystals of $Sr_2ZnGe_2OS_6$ were obtained but using ZnS instead of CdS (Wang et al. 2022c).

4.75 Germanium–Strontium–Gallium–Oxygen–Sulfur

In the system containing these elements, the quinary compound $Sr_2Ga_2GeOS_6$, which crystallizes as a tetragonal structure with the lattice parameters a = 939.35 ± 0.03, c = 617.45 ± 0.04 pm, a calculated density of 3.631 g·cm^{-3}, and an energy gap of 3.15 eV, is formed (Wang et al. 2022c). Single crystals of this compound were grown in the same way as in the case of the $Ca_2Ga_2GeOS_6$ compound using SrO and SrS instead of CaO and CaS, respectively.

4.76 Germanium–Strontium–Oxygen–Fluorine–Sulfur

In the system containing these elements, the quinary compound $Sr_{10}(GeO_4)_3(SO_4)_3F_2$, which crystallizes as a hexagonal structure with the lattice parameters a = 990.5 ± 0.3, c = 739.2 ± 0.3 pm, and the calculated and experimental densities of 4.262 and 4.23 g·cm^{-3}, respectively, is formed (Schwarz 1967a). It was synthesized by sintering a mixture of Sr_2GeO_4, $SrSO_4$, and SrF_2 (molar ratio 3:3:1) twice at 1000°C first for 14 and then for 9 h in air. Colorless crystals were obtained after quickly or slowly cooling.

4.77 Germanium–Strontium–Oxygen–Manganese–Sulfur

The quinary compound $Sr_2MnGe_2OS_6$, which crystallizes as a tetragonal structure with the lattice parameters a = 950.96 ± 0.02, c = 620.69 ± 0.03 pm, a calculated density of 3.45 g·cm^{-3}, and an energy gap of 3.51 eV (Ran et al. 2022) [a = 952.06 ± 0.02, c = 620.02 ± 0.01 pm, and an energy gap of 3.2 eV (Endo et al. 2017)], is formed in the Ge–Sr–O–Mn–S system.

4.78 Germanium–Strontium–Oxygen–Iron–Sulfur

The quinary compound $Sr_2FeGe_2OS_6$, which crystallizes as a tetragonal structure with the lattice parameters $a = 941.82 \pm 0.02$, $c = 614.70 \pm 0.02$ pm, a calculated density of 3.561 g·cm^{-3} at 100 K, and an energy gap of 2.24 eV, is formed in the Ge–Sr–O–Fe–S system (Yang et al. 2023a). This compound has good thermal stability up to 798°C under an N_2 atmosphere but starts to decompose at higher temperatures.

Single crystals of $Sr_2FeGe_2OS_6$ were grown by the flux method. A mixture of SrO (2 mM), $FeCl_2$ (1 mM), Ge (2 mM), S (6 mM), and Ba (1 mM) was weighed and an additional 300 mg of KI was added as the reactive flux. The mixture was placed in a silica crucible and then in a vacuum-sealed silica tube. The sample was gradually heated to 400°C, held for 15 h, and then heated to 800°C at a rate of 10°C·h^{-1}, held for 3 days, and then slowly cooled to 350°C at a rate of 3°C·h^{-1}. High-quality black-red block crystals were obtained after washing with 95% ethanol.

4.79 Germanium–Strontium–Oxygen–Cobalt–Sulfur

The quinary compound $Ba_2Ga_8GeSe_2S_{16}$, which melts congruently at 936°C and crystallizes as a tetragonal structure with the lattice parameters $a = 940.56 \pm 0.03$, $c = 617.41 \pm 0.04$ pm, a calculated density of 3.574 g·cm^{-3}, and an energy gap of 2.77 eV, is formed in the Ge–Sr–O–Co–S system (Zhang et al. 2022). This compound was synthesized via a solid state method. Sr, Co, GeO_2, and S were used. Raw materials were stoichiometrically weighed with a total mass of 0.5 g and ground thoroughly, followed by pressing it into a pellet. The mixture was put into a quartz tube, which was then evacuated to 10^{-2} Pa and sealed using oxygen/hydrogen flame. After heating at 850°C for 2.5 days and cooled to room temperature within 5 days, dark green crystals of the title compound were obtained, which are stable under atmosphere.

4.80 Germanium–Barium–Zinc–Oxygen–Sulfur

In the system containing these elements, the quinary compound $Ba_2ZnGe_2OS_6$, which crystallizes as a tetragonal structure with the lattice parameters $a = 963.59 \pm 0.22$, $c = 645.06 \pm 0.25$ pm, and the calculated and experimental densities of 3.85 and 3.74 g·cm^{-3}, respectively, is formed (Teske 1980). This compound could be obtained by the following reactions: $2BaS + ZnS + 2GeS_2 + 1/2O_2 = Ba_2ZnGe_2OS_6$ and $2BaS + ZnO + 2GeS_2 = Ba_2ZnGe_2OS_6$.

4.81 Germanium–Barium–Gallium–Selenium–Sulfur

In the system containing these elements, the quinary compound $Ba_2Ga_8GeSe_2S_{14}$, which melts congruently at 994°C and has an energy gap of 2.65 eV, is formed (Badikov et al. 2022b). Its crystals were synthesized in a graphitized quartz ampoule, filled with Ba, S, Ga_2S_3, GeS_2, and $GeSe_2$ in a stoichiometric ratio. The ampoule was evacuated to 10^{-4} Pa, sealed, placed in a horizontal furnace, and heated to 1050°C for 12 h. The melt obtained was kept at this temperature for 24 h and stirred for complete homogenization. Growth was performed in vertical furnace with a temperature gradient of 10°C–15°C·cm^{-1} in the crystallization zone, with melt heating to a temperature exceeding the melting point by 30°C-40°C. The crystal growth rate was 6 mm per day. The grown crystals were cooled to room temperature with the furnace switched off. High optical quality single crystals of this compound were grown by the vertical Bridgman–Stockbarger method,

4.82 Germanium–Barium–Tin–Bromine–Sulfur

In the system containing these elements, the quinary compound $Ba_4Ge_2SnS_8Br_2$, which crystallizes as an orthorhombic structure with the lattice parameters $a = 1198.45 \pm 0.06$, $b = 913.64 \pm 0.05$, $c = 884.37 \pm$

0.05 pm, a calculated density of 4.217 g·cm^{-3}, and an energy gap of 2.066 eV, is formed (Lin et al. 2014). To prepare this compound, reaction mixtures of BaS (2 mM), SnS (1 mM), GeS_2 (1 mM), and NaBr (1 mM) were ground and loaded into fused-silica tube under an Ar atmosphere in a glove box. The tube was flame-sealed under a high vacuum of 10^{-3} Pa and then placed in a furnace. The sample was heated to 850°C within 20 h, kept for 50 h, then slowly cooled to 400°C at a rate of 3°C·h^{-1}, and finally cooled to room temperature by switching off the furnace. Many dark red block-shaped crystals were obtained.

4.83 Germanium–Barium–Lead–Bromine–Sulfur

In the system containing these elements, the quinary compound $Ba_4Ge_2PbS_8Br_2$, which crystallizes as an orthorhombic structure with the lattice parameters a = 1200.81 ± 0.04, b = 917.44 ± 0.04, c = 880.42 ± 0.03 pm, a calculated density of 4.513 g·cm^{-3}, and an energy gap of 2.054 eV, is formed (Lin et al. 2014). This compound was synthesized in the same way as $Ba_4Ge_2SnS_8Br_2$ was prepared but using PbS instead of SnS.

4.84 Germanium–Barium–Oxygen–Fluorine–Sulfur

In the system containing these elements, the quinary compound $[Ba_2F_2][Ge_2O_3S_2]$, which crystallizes as an orthorhombic structure with the lattice parameters a = 918.13 ± 0.04, b = 918.06 ± 0.04, c = 2002.87 ± 0.09 pm, a calculated density of 4.485 g·cm^{-3}, and an energy gap of 4.45 eV, is formed (Liu et al. 2022a). This compound was synthesized by a solid state reaction. A mixture of BaF_2 (2 mM) GeO_2 (3 mM), Ba (2.15 mM), Ge (1.1 mM), and S (5 mM) was loaded into a graphite crucible, which was removed into a fused-silica tube. The tube was evacuated to about 10^{-3} Pa and then flame-sealed under an oxy-hydrogen flame. Further, the sealed tube was heated to 1030°C in 50 h, held for 30 h, and cooled to room temperature at a rate of 3°C·h^{-1}. After that, colorless bulk crystals of the title compound were obtained, which are stable under air conditions for at least 6 months. Ar-filled glove box was used to complete the preparation processes.

4.85 Germanium–Gallium–Thallium–Chlorine–Sulfur

GeS_2–TlCl–Ga_2S_3. The glass-forming region in this quasiternary system was determined by Aleksandrov (1990). It was shown that this region is situated on the GeS_2–Ga_2S_3 side near GeS_2. The ternary glassy alloys can include up to 35 mol% Ga_2S_3 and up to 80 mol% TlCl.

4.86 Germanium–Gallium–Lanthanum–Oxygen–Sulfur

The quinary compound $Ga_3La_3Ge_2S_3O_{10}$, which is thermally stable up to 800°C and crystallizes as a hexagonal structure with the lattice parameters a = 1017.01 ± 0.04, c = 751.98 ± 0.03 pm, a calculated density of 5.065 g·cm^{-3}, and an energy gap of 4.70 eV, is formed in the Ge–Ga–La–O–S system (Yan et al. 2021). Polycrystalline powder samples of this compound were synthesized from a stoichiometric mixture of La_2O_3 (3 mM), Ga_2O_3 (3 mM), GeO_2 (1 mM), Ge (3 mM), and S (6 mM). The mixture was ground thoroughly with an agate mortar and pestle, pressed into a pellet, sealed in a fused-silica tube under the vacuum value of 1 Pa, and then heated in a furnace at 850°C for 24 h.

Single crystals of the title compound were obtained by a flux crystal growth method using a eutectic $BaCl_2$–NaCl mixture. La_2S_3 (0.5 mM), Ga_2O_3 (0.5 mM), GeO_2 (0.5 mM), $BaCl_2$ (2.7 mM), and NaCl (4.0 mM) were loaded into an alumina crucible in an Ar-filled glove box, and the crucible was flame-sealed inside a fused-silica tube under the vacuum level of 1 Pa. The starting materials were heated in a muffle furnace to 850°C at 5°C·min^{-1}, held for 24 h, cooled to 550°C at 0.08°C·min^{-1}, and finally naturally cooled to room temperature by turning the heater off. The product was washed in sonicated water and extracted from the flux. Colorless transparent block crystals of $La_3Ga_3Ge_2S_3O_{10}$ were collected via vacuum filtration.

4.87 Germanium–Gallium–Neodymium–Oxygen–Sulfur

The quinary compound $Ga_3Nd_3Ge_2S_3O_{10}$, which crystallizes as a hexagonal structure with the lattice parameters a = 1001.590 ± 0.010, c = 733.730 ± 0.010 pm, and a calculated density of 5.435 g·cm^{-3}, is formed in the Ge–Ga–Nd–O–S system (Ran et al. 2023). The crystals of this compound were synthesized by high-temperature solid state reaction. Nd_2S_3 (0.066 mM), Nd_2O_3 (0.033 mM), Ga_2O_3 (1 mM) and GeO_2 (1 mM) were ground without further purification. The mixture was heated to 400°C from room temperature in 20 h, then warmed to 1000°C in 30 h after keeping 20 h at 400°C. The furnace was then maintained for 120 h in the highest temperature, and cooled down to 250°C within 200 h. Then it was brought down to room temperature spontaneously by stopping the heating. After washing with hot water and ethanol, the crystals of the title compound were obtained. These crystals are stable in air and moisture for more than one year. All weighing processes were completed in an anhydrous and oxygen-free glove box.

4.88 Germanium–Tin–Lead–Selenium–Sulfur

A solid solution of $GeSn_xPb_{2-x}Se_yS_{4-y}$ ($0 \le x \le 0.5$; $0 \le y \le 4$) is formed in the Ge–Sn–Pb–Se–S quinary system (Poduska et al. 2002). When $x > 0.5$, the resulting products were multiphase and did not contain the cubic high-temperature Pb_2GeS_4-type structure. This solid solution crystallizes as a cubic structure with the lattice parameter a = 1414.0 ± 0.1 pm for x = 0.5 and y = 1 and a = 1442.8 ± 0.1 pm for x = 0.5 and y = 3. The lattice constant increases monotonically with nominal Se content y.

4.89 Germanium–Lead–Oxygen–Fluorine–Sulfur

In the system containing these elements, the quinary compound $Pb_{10}(GeO_4)_3(SO_4)_3F_2$, which crystallizes as a hexagonal structure with the lattice parameters a = 1003.2 ± 0.3, c = 745.0 ± 0.3 pm, and the calculated and experimental densities of 7.180 and 7.17 g·cm^{-3}, respectively, is formed (Schwarz 1967a). It was synthesized by sintering a mixture of Pb_2GeO_4, $PbSO_4$, and PbF_2 (molar ratio 3:3:1) twice at 600°C first for 39 h and then for 48 h in a nitrogen atmosphere. Yellowish crystals were obtained after cooling quickly or slowly.

5

Systems Based on Germanium Selenides

5.1 Germanium–Hydrogen–Sodium–Oxygen–Selenium

Three quinary compounds are formed in this system. $Na_4GeSe_4 \cdot 14H_2O$ crystallizes as a monoclinic structure with the lattice parameters $a = 1286.2 \pm 0.4$, $b = 1120.5 \pm 0.3$, $c = 814.1 \pm 0.2$ pm, $\beta = 107.40 \pm 0.03°$ and the calculated and experimental densities of 2.173 and 2.15 ± 0.02 g·cm^{-3}, respectively (Krebs and Jacobsen 1976). To prepare this compound, H_2Se obtained by the reaction of 6 N HCl and Al_2Se_3 were introduced into a solution of NaOH (5 mM) in H_2O (5 mL) until a saturation was reached. Then solid NaOH (5 mM) was added to the pink aqueous solution of NaHSe. After removing H_2Se gas with suction, the solution containing Na_2Se (5 mM) was evaporated down to 10 mL and then added to a round bottom flask with $GeSe_2$ (2.5 mM). From the clear solution, which was under a N_2 atmosphere, the water was evaporated at reduced pressure at 25°C (approximately 48 h/10 mL). Colorless crystals were separated out and form long, flattened needles. The synthesis was carried out with the strictest exclusion of oxygen.

$Na_4Ge_2Se_6 \cdot 16H_2O$ also crystallizes as a monoclinic structure with the lattice parameters $a = 989.4 \pm 0.4$, $b = 1178.1 \pm 0.5$, $c = 1222.5 \pm 0.6$ pm, $\beta = 92.90 \pm 0.04°$ and the calculated and experimental densities of 2.33 and 2.31 ± 0.02 g·cm^{-3}, respectively (Krebs and Müller 1983). To synthesize this compound, a weak stream of H_2Se is introduced into a solution of NaOH (5 mM) in H_2O (50 mL) for about 1 h at 0°C, during which a crystal slurry of NaHSe separates out. Further, 200 mg of solid NaOH is then added. The resulting clear, colorless solution of Na_2Se (3 mM) was mixed with stirring at room temperature in portions with $GeSe_2$ (5 mM) (prepared by introducing H_2Se into an acidified germanate solution). At first, a reddish solution forms, which after a complete addition of $GeSe_2$ turns in orange-yellow, almost a clear solution. After filtration, the solvent was sucked off first with an oil pump and then with a water jet gun to allow slow crystallization. After the volume has reduced to a few milliliters, orange-colored crystals separate out in the form of leaves and needles.

For obtaining $Na_6Ge_2Se_7 \cdot 9H_2O$, GeO_2 (2 g) was dissolved in a solution of Na_2S (3.5 g in 20 mL of the solution with a GeO_2/NaOH molar ratio 1:4) (Schwarz and Giese 1930; Ivanov-Emin and Kostrikin 1947). The resulting solution mixed with Na_2Se was quickly poured in a beaker with acetone (500 mL). First, a yellowish oily substance is formed, which quickly crystallizes. The solution was filtered and the crystals were washed with acetone and then with dry ether. The crystals were stored in sealed test tubes in a hydrogen atmosphere. This compound rapidly decomposes with the formation of polyselenides, and then with the release of elemental selenium.

5.2 Germanium–Hydrogen–Potassium–Oxygen–Selenium

Three quinary compounds are formed in this system. $[K_4(H_2O)_4][GeSe_4]$ crystallizes as a triclinic structure with the lattice parameters $a = 786.95 \pm 0.16$, $b = 795.76 \pm 0.16$, $c = 1231.4 \pm 0.3$ pm, $\alpha = 80.24 \pm 0.03°$, $\beta = 87.66 \pm 0.03°$, $\gamma = 74.07 \pm 0.03°$, a calculated density of 2.803 g·cm^{-3} at 203 K, and an energy gap of 2.81 eV (Melullis and Dehnen 2007). Light yellow single crystals of this compound were prepared when K_4GeSe_4 (0.100 mM) was dissolved in H_2O (10 mL) at 20°C. The solution was stirred for 30 min

DOI: 10.1201/9781003123460-5

and filtered away from precipitating Ge powder. The filtrate was evaporated for several hours, whereupon the platelets of $[K_4(H_2O)_4][GeSe_4]$ were obtained. All synthesis steps were performed with a strong exclusion of air and external moisture (N_2 atmosphere at a high-vacuum, double-manifold Schlenk line or Ar atmosphere in a glove box). Water was degassed by applying a dynamic vacuum (10^{-1} Pa) for several hours.

$[K_4(H_2O)_3][Ge_2Se_6]$ crystallizes as a monoclinic structure with the lattice parameters $a = 1458.3 \pm 0.3$, $b = 838.58 \pm 0.17$, $c = 1417.0 \pm 0.3$ pm, $\beta = 96.27 \pm 0.03°$, a calculated density of 3.198 $g \cdot cm^{-3}$ at 203 K, and an energy gap of 2.65 eV (Melullis and Dehnen 2007). To synthesize this compound, stoichiometric amounts of K_2Se (25 mM), Ge (25 mM), and Se (50 mM) were mixed together and the mixture was heated in a sealed quartz ampoule under vacuum. 2 g of the resulting crude $K_4[Ge_2Se_6]$ was dissolved in H_2O (50 mL) at 20°C, stirred for 30 min and filtered away from precipitating Ge powder. The resulting yellow solution was slowly evaporated under vacuum to give yellow platelets of the title compound. All synthesis steps were performed with a strong exclusion of air and external moisture (N_2 atmosphere at a high-vacuum, double-manifold Schlenk line or Ar atmosphere in a glove box). Water was degassed by applying dynamic vacuum (10^{-1} Pa) for several hours.

$K_6Ge_2Se_7 \cdot 9H_2O$ was prepared in the same way as $Na_6Ge_2Se_7 \cdot 9H_2O$ was synthesized using K_2Se instead of Na_2Se (Schwarz and Giese 1930; Ivanov-Emin and Kostrikin 1947).

5.3 Germanium–Hydrogen–Rubidium–Silver–Oxygen–Selenium

In the system containing these elements, the multinary compound $Rb_3[AgGe_4Se_{10}] \cdot 2H_2O$, which crystallizes as a tetragonal structure with the lattice parameters $a = 876.7 \pm 0.3$, $c = 1595.6 \pm 0.7$ pm, and a calculated density of 4.009 $g \cdot cm^{-3}$, is formed (Loose and Sheldrick 1997). To synthesize this compound, the starting materials Rb_2CO_3 (0.7 mM), Ge (1.4 mM), Se (2.8 mM), and $Ag(CH_3COO)$ (0.7 mM) were slurried in H_2O (0.8 mL) and heated in a sealed glass tube to 160°C. After tempering for 4 days, the content was cooled to room temperature at $2°C \cdot h^{-1}$. Orange-yellow crystals of the title compound were obtained.

5.4 Germanium–Hydrogen–Rubidium–Oxygen–Manganese–Selenium

In the system containing these elements, the multinary compound $Rb_3[MnGe_4Se_{10}] \cdot 3H_2O$, which crystallizes as a tetragonal structure with the lattice parameters $a = 888.6 \pm 0.3$, $c = 1543.3 \pm 0.8$ pm, and a calculated density of 3.706 $g \cdot cm^{-3}$, is formed (Loose and Sheldrick 1997). To obtain this compound, the starting materials Rb_2CO_3 (0.67 mM), Ge (1.34 mM), Se (2.68 mM), and $Mn(CH_3COO)_2$ (0.67 mM) were slurried in H_2O (0.6 mL) and heated in a sealed glass tube to 175°C. After tempering for 4 days, the content was cooled to room temperature at $2°C \cdot h^{-1}$. Yellow-orange crystals of the title compound were prepared.

5.5 Germanium–Hydrogen–Cesium–Silver–Oxygen–Selenium

In the system containing these elements, the multinary compound $Cs_3[AgGe_4Se_{10}] \cdot 2H_2O$, which crystallizes as a tetragonal structure with the lattice parameters $a = 895.6 \pm 0.1$, $c = 1603.6 \pm 0.3$ pm, and a calculated density of 4.190 $g \cdot cm^{-3}$, is formed (Loose and Sheldrick 1997). To prepare this compound, the starting materials Cs_2CO_3 (0.7 mM), Ge (1.4 mM), Se (3.5 mM), and $Ag(CH_3COO)$ (0.7 mM) were slurried in H_2O (1.0 mL) and heated in a sealed glass tube to 175°C. After tempering for 4 days, the content was cooled to room temperature at $2°C \cdot h^{-1}$. Yellow-orange crystals of the title compound were obtained.

5.6 Germanium–Hydrogen–Cesium–Oxygen–Manganese–Selenium

In the system containing these elements, the multinary compound $Cs_3[MnGe_4Se_{10}]·3H_2O$, which crystallizes as a tetragonal structure with the lattice parameters *a* = 885.9 ± 0.1, *c* = 1544.8 ± 0.3 pm, and a calculated density of 3.985 g·cm^{-3}, is formed (Loose and Sheldrick 1997). To synthesize this compound, the starting materials Cs_2CO_3 (0.78 mM), Ge (1.56 mM), Se (3.12 mM), and $Mn(CH_3COO)_2$ (0.78 mM) were slurried in H_2O (0.8 mL) and heated in a sealed glass tube to 160°C. After tempering for 3 days, the content was cooled to room temperature at 2°C·h^{-1}. Yellow crystals of the title compound were obtained.

5.7 Germanium–Hydrogen–Barium–Oxygen–Selenium

In the system containing these elements, the quinary compound $[Ba_2(H_2O)_9][GeSe_4]$, which crystallizes as an orthorhombic structure with the lattice parameters *a* = 1391.6 ± 0.3, *b* = 1509.7 ± 0.3, *c* = 1611.6 ± 0.3 pm, and a calculated density of 3.238 g·cm^{-3} at 203 K, is formed (Melullis et al. 2005). For it obtaining, crude $Ba_2[GeSe_4]$ (1.5 mM) was suspended in water (50 mL) and heated to 100°C. A light orange solution was formed besides some gray powder which turned out to be excess Ge. The solution was filtered from the powder and layered by THF (50 mL), yielding pale yellow crystals of the title compound. All synthesis steps were performed with a strong exclusion of air and external moisture (N_2 atmosphere at a high-vacuum, double-manifold Schlenk line or Ar atmosphere in a glove box).

5.8 Germanium–Hydrogen–Barium–Oxygen–Manganese–Selenium

Two multinary compounds, $[Ba_3(H_2O)_{15}][Mn_6(H_2O)_6(\mu_6\text{-}Se)(GeSe_4)_4]·9H_2O$ and $[Ba_3(H_2O)_{16}][Mn_6(H_2O)_3(\mu_6\text{-}Se)(GeSe_4)_4]$, are formed in this system (Melullis et al. 2005). The first of them crystallizes as a rhombohedral structure with the lattice parameters *a* = 2046.3 ± 0.3, *c* = 2511.9 ± 0.5 pm, a calculated density of 3.188 g·cm^{-3} in hexagonal setting at 203 K, and an energy gap of 2.2 eV. To synthesize this compound, crude $[Ba_2(H_2O)_9][GeSe_4]$ (0.1·mM) was solved in H_2O (5 mL) producing a yellow solution. A colorless solution of $MnCl_2·4H_2O$ (0.1·mM) in H_2O (5 mL) was added, whereupon the reaction mixture turns intensely orange and traces of orange powder precipitate. After filtration of the precipitate, 10 mL of THF were allowed to slowly diffuse onto the reaction solution. After several days, orange-red rhombic dodecahedra of this compound crystallize.

The second compound crystallizes as a triclinic structure with the lattice parameters *a* = 1218.4 ± 0.2, *b* = 1508.9 ± 0.3, *c* = 1586.8 ± 0.3 pm, α = 88.73 ± 0.03°, β = 73.68 ± 0.03°, γ = 71.06 ± 0.03°, a calculated density of 3.412 g·cm^{-3} at 203 K, and an energy gap of 2.0 eV. To prepare it, crude $[Ba_2(H_2O)_9][GeSe_4]$ (0.1·mM) was solved in H_2O (5 mL) producing a yellow solution. A colorless solution of $MnCl_2·4H_2O$ (0.1·mM) in methanol (5 mL) was added, whereupon the reaction mixture turns intensely orange and some orange powder precipitates that can be re-dissolved in water and layered by THF to yield the first compound. After filtration of the precipitate, the solution was layered by 10 mL of THF. Crystallization of red parallelepipeds of the second compound were observed after 2 weeks.

All synthesis steps were performed with a strong exclusion of air and external moisture (N_2 atmosphere at a high-vacuum, double-manifold Schlenk line or Ar atmosphere in a glove box).

5.9 Germanium–Hydrogen–Gallium–Nitrogen–Selenium

In the system containing these elements, the quinary compound $(NH_4)_2Ga_2Ge_2Se_8$, which is stable up to 400°C under a N_2 atmosphere and crystallizes as a monoclinic structure with the lattice parameters *a* = 746.2 ± 0.3, *b* = 1230.2 ± 0.3, *c* = 1763.3 ± 0.5 pm, β = 92.40 ± 0.03°, and an energy gap of 1.83 eV, is formed (Wang et al. 2018). To prepare this compound, Ga (0.49 mM), GeO_2 (0.51 mM), Se (3.02 mM), 1-butylpyridinium bromide (1.48 mM), hydrazine monohydrate (1.0 mL, 80% aqueous solution), and ethanol (0.25 mL, 40% aqueous solution) were mixed and sealed in an autoclave equipped with a Teflon liner (28 mL), then heated at 190°C for 7 days and cooled to room temperature. The products were filtrated and washed several times with ethanol. Red rod crystals of the title compound were obtained.

5.10 Germanium–Hydrogen–Indium–Nitrogen–Selenium

In the system containing these elements, the quinary compound $(NH_4)_2In_2Ge_2Se_8$, which is stable up to 400°C under a N_2 atmosphere and crystallizes as a monoclinic structure with the lattice parameters $a = 764.16 \pm 0.12$, $b = 1246.98 \pm 0.19$, $c = 1822.2 \pm 0.3$ pm, $\beta = 95.214 \pm 0.006°$, and an energy gap of 1.95 eV, is formed (Wang et al. 2018). To synthesize this compound, In (0.49 mM), GeO_2 (0.51 mM), Se (3.04 mM), 1-butylpyridinium bromide (1.53 mM), hydrazine monohydrate (1.0 mL, 80% aqueous solution), and ethanol (0.25 mL, 40% aqueous solution) were mixed and sealed in an autoclave equipped with a Teflon liner (28 mL), then heated at 190°C for 10 days and cooled to room temperature. The products were filtrated and washed several times with ethanol. Orange block crystals of the title compound were obtained.

5.11 Germanium–Hydrogen–Nitrogen–Manganese–Selenium

In the system containing these elements, the quinary compound $Mn_2GeSe_4(N_2H_4)_8$ which crystallizes as a triclinic structure with the lattice parameters $a = 743.5 \pm 0.3$, $b = 839.0 \pm 0.4$, $c = 1731.3 \pm 0.8$ pm, $\alpha = 92.148 \pm 0.007°$, $\beta = 91.458 \pm 0.007°$, and $\gamma = 109.964 \pm 0.007°$, is formed (Yuan and Mitzi 2009). This compound was synthesized by the reaction of $GeSe_2$, MnS, and Se in a hydrazine solution.

5.12 Germanium–Potassium–Cesium–Gallium–Chlorine–Selenium

In the system containing these elements, the multinary compound $(Cs_5KCl)_2Cs_5[Ga_{15}Ge_9Se_{48}]$, which crystallizes as a tetragonal structure with the lattice parameters $a = 1433.09 \pm 0.06$, $c = 2779.7 \pm 0.2$ pm, and an energy gap of 2.76 eV, is formed (Huang-Fu et al. 2015). Crystals of this compound were grown by using single crystals of $(Cs_6Cl)_2Cs_5[Ga_{15}Ge_9Se_{48}]$ (ca. 25 mg) together with CsCl (100 mg) and KCl (44.3 mg) as mixed fluxes with a molar ratio of 5:5 (melt point 620°C). The reactants were sealed in silica tube under vacuum, heated at a rate of $15°C{\cdot}h^{-1}$ to 650°C, followed by a dwell time of 60 h, and then subsequently cooled at a rate of about $5.4°C{\cdot}h^{-1}$ to 200°C, before switching off the furnace. After washing with distilled water and ethanol to remove the flux, the clear, light yellow crystals were obtained.

5.13 Germanium–Potassium–Gallium–Tin–Selenium

Two quinary compounds, $KGaGe_{1.37}Sn_{0.63}Se_6$ and $K_3Ga_3(Ge_{6.17}Sn_{0.83})Se_{20}$, are formed in this system. The first of them crystallizes as a monoclinic structure with the lattice parameters $a = 733.89 \pm 0.06$, $b = 2406.40 \pm 0.19$, $c = 722.97 \pm 0.06$ pm, $\beta = 116.830 \pm 0.002°$, a calculated density of 4.413 $g{\cdot}cm^{-3}$, and an energy gap of 2.02 eV (Li et al. 2023a). The title compound was obtained by solid state reaction. A 500 mg mixture of KI as reactive flux and a source of the K element, SnO, Ga_2O_3, GeO_2, Se, and B with the molar ratios of 1:1:1:6:4 were weighed, then pressed into pellet, following by sealed into evacuated quartz ampoule with the vacuum degree of 10^{-2} Pa. The tube was heated in a muffle furnace from room temperature to 950°C in 25 h and homogenized at several intermediate temperatures for several hours. The reaction were maintained at 950°C for about 120 h, and then cooled down to 300°C within 120 h. Finally, the tube was cooled down to room temperature naturally. The red block single crystals of the title compound were obtained. The crystals are stable in air after several months, and can be separated from the byproducts easily under an optical microscope because of their unique colors and shapes.

$K_3Ga_3(Ge_{6.17}Sn_{0.83})Se_{20}$ also crystallizes as a monoclinic structure with the lattice parameters $a = 708.43 \pm 0.04$, $b = 3940.4 \pm 0.2$, $c = 702.12 \pm 0.04$ pm, $\beta = 90.567 \pm 0.002°$, a calculated density of 4.155 g·cm^{-3}, and an energy gap of 2.02 eV (Li et al. 2019d). The title compound was first obtained using a stoichiometric mixture targeting to prepare $K_3Ga_3Ge_7Se_{20}$ and a minor amount of SnO_2/Si, together with abundant KI as a flux. After its chemical composition was determined, stoichiometric SnO, Ga_2O_3, GeO_2, Se, B, and abundant KI were used to synthesize the compound with high yields. The optimized synthetic routine is as follows. A 500 mg mixture of stoichiometric SnO, Ga_2O_3, GeO_2, Se, and B and an additional 600 mg of KI were ground into a uniform powder using an agate mortar and then pressed into a pellet, which was then sealed into an evacuated quartz ampoule with a vacuum degree of ≈10^{-2} Pa. The ampoule was calcinated in a muffle furnace from room temperature to 950°C at a rate of 60°C·h^{-1} and homogenized at several intermediate temperatures for several hours. The reaction was maintained at 950°C for a week, and then cooled down to 300°C with the speed of 5°C·h^{-1}. Finally the ampoule was cooled to room temperature naturally. The product was cleaned by distilled water, and yellow block single crystals were obtained.

5.14 Germanium–Potassium–Gallium–Bromine–Selenium

$GeSe_2$–KBr–Ga_2Se_3. The glass-forming region in this quasiternary system is presented in the Figure 5.1 (Tang et al. 2008). At least 35 mol% KBr can be dissolved into the glasses. The glasses have relatively high glass transition temperatures (T_g = 290°C–350°C) and good thermal stability. To prepare the glasses, the mixtures of Ge, Ga, Se, and KBr were melted in evacuated (10^{-2} Pa) and flame-sealed silica ampoule at 900°C–950°C for 12 h in a rocking furnace. After that, the ampoules were quenched in cold water.

5.15 Germanium–Rubidium–Cesium–Gallium–Chlorine–Selenium

In the system containing these elements, the multinary compound $(Rb_6Cl)_2Cs_5[Ga_{15}Ge_9Se_{48}]$, which crystallizes as a tetragonal structure with the lattice parameters $a = 1421.93 \pm 0.06$, $c = 2741.0 \pm 0.2$ pm,

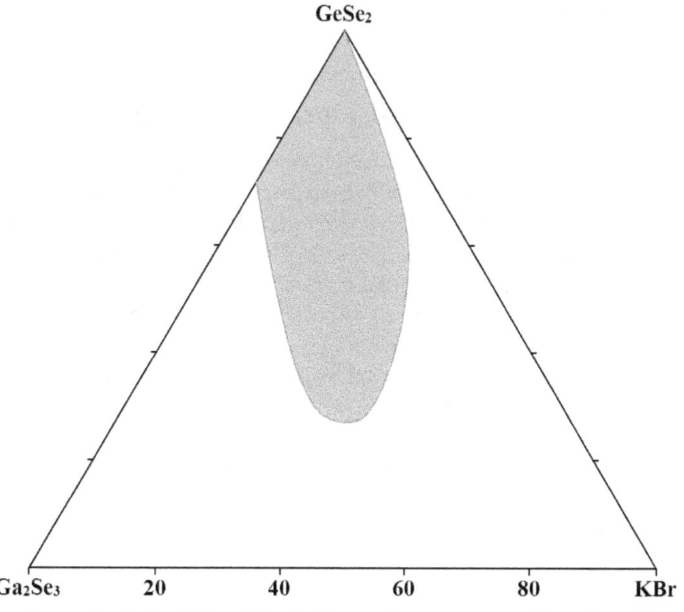

FIGURE 5.1 Glass-forming region in the $GeSe_2$–KBr–Ga_2Se_3 quasiternary system. (From Tang, G., et al., *J. Alloys Compd.*, 459(1–2), 472, 2008.)

and an energy gap of 2.73 eV, is formed (Huang-Fu et al. 2015). Crystals of this compound were prepared by using single crystals of $(Cs_6Cl)_2Cs_5[Ga_{15}Ge_9Se_{48}]$ (ca. 25 mg) together with RbCl (71.8 mg) and NaCl (10 mg) as mixed fluxes with a molar ratio of 2:8 (melt point 650°C). The reactants were sealed in a silica tube under vacuum, heated at a rate of 15°C·h^{-1} to 650°C, followed by a dwell time of 60 h, and then subsequently cooled at a rate of about 5.4°C·h^{-1} to 200°C, before switching off the furnace. After washing with distilled water and ethanol to remove the flux, the clear, light yellow crystals were obtained.

5.16 Germanium–Rubidium–Cesium–Gallium–Bromine–Selenium

In the system containing these elements, the multinary compound $(Rb_6Br)_2Cs_5[Ga_{15}Ge_9Se_{48}]$, which crystallizes as a tetragonal structure with the lattice parameters a = 1430.34 ± 0.05, c = 2759.8 ± 0.2 pm, and an energy gap of 2.61 eV, is formed (Huang-Fu et al. 2015). Crystals of this compound were grown by using single crystals of $(Cs_6Cl)_2Cs_5[Ga_{15}Ge_9Se_{48}]$ (ca. 25 mg) together with RbBr (63.7 mg) as flux. The reactants were sealed in silica tube under vacuum, heated at a rate of 15°C·h^{-1} to 675°C, followed by a dwell time of 60 h, and then subsequently cooled at a rate of about 5.4°C·h^{-1} to 200°C, before switching off the furnace. After washing with distilled water and ethanol to remove the flux, the clear, light orange-yellow crystals were obtained.

5.17 Germanium–Cesium–Gallium–Tin–Selenium

In the system containing these elements, the quinary compound $CsGaGeSnSe_6$, which crystallizes as a tetragonal structure with the lattice parameters a = 1042.45 ± 0.03, c = 944.34 ± 0.06 pm, a calculated density of 4.864 g·cm^{-3}, and an energy gap of 2.03 eV, is formed (Li et al. 2021a). This compound was obtained by solid state reaction. The mixture of CsCl as reactive flux, Ga, Si, Sn, and Se with the overall mass of 700 mg and the molar ratios of 2:1:1:1:6 was ground into uniform powder using an agate mortar and then pressed into a pellet, which was then sealed into an evacuated quartz ampoule at a vacuum of 10^{-2} Pa. The tube was heated to 800°C in 35 h and the reaction was maintained at this temperature for 100 h, before cooling down to 200°C within 200 h. Ultimately, the tube was cooled down to ambient temperature naturally. The product was washed DMF and distilled water, and the deep-red block crystals were obtained. These crystals are stable in air.

5.18 Germanium–Cesium–Gallium–Chlorine–Selenium

In the system containing these elements, the quinary compound $(Cs_6Cl)_2Cs_5[Ga_{15}Ge_9Se_{48}]$, which is stable up to 680°C, decomposes at 773°C, and crystallizes as a tetragonal structure with the lattice parameters a = 1438.70 ± 0.05, c = 2801.2 ± 0.2 pm, and an energy gap of 2.91 eV, is formed (Huang-Fu et al. 2015). A crystal obtained on attempting to synthesize "$Cs_2Ge_3Ga_6Se_{14}$" in an evacuated fused-silica tube with CsCl/Ga/Ge/Se in molar ratio 12:3:6:14 and total mass of ca. 600 mg was shown to have the formula of the title compound. In this reaction, CsCl acted as both a flux and a cesium source. Subsequently, in a typical reaction, a CsCl/Ga/Ge/Se mixture with a molar ratio of 11.3:8:1:15.6 and a total mass of 600 mg was heated at a rate of 18°C·h^{-1} to 950°C, followed by a dwell time of 60 h, and subsequently cooled at a rate of about 3.5°C·h^{-1} to 200°C, before switching off the furnace. After washing with distilled water and ethanol, light yellow, transparent crystals of $(Cs_6Cl)_2Cs_5[Ga_{15}Ge_9Se_{48}]$ with good quality were obtained; the only byproduct was nontransparent, red, block-shaped crystals of Ga_2Se_3. Because of their distinguishable shapes and color, it is possible to manually pick considerable numbers of crystals. The obtained crystals were relatively stable in air, to which they could be exposed for several weeks without significant change. However, the powder of this compound was relatively unstable, on the surface of which Se was observed after exposure to air for one day.

$GeSe_2$–CsCl–Ga_2Se_3. The glass-forming region in this quasiternary system is presented in Figure 5.2 (Calvez et al. 2007). The glass transition temperature of these alloys is within the interval from 292°C to

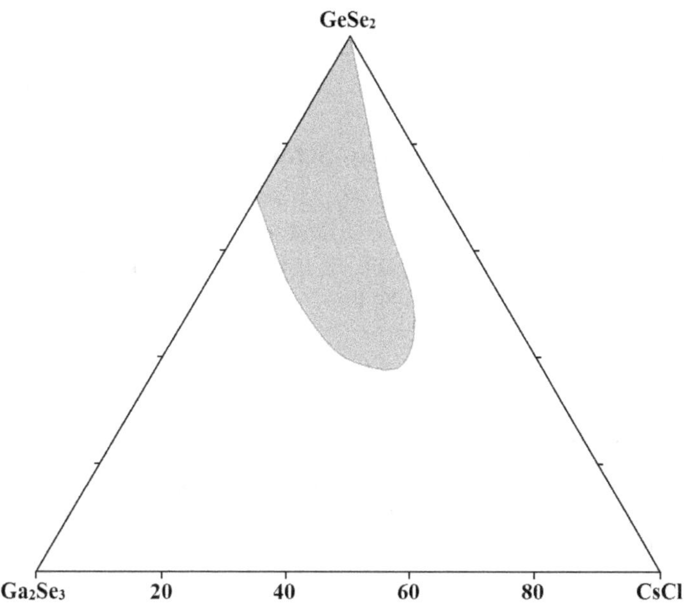

FIGURE 5.2 Glass-forming region in the $GeSe_2$–CsCl–Ga_2Se_3 quasiternary system. (From Calvez, L., et al., *Adv. Mater.*, 19(1), 129, 2007.)

370°C. Bulk glasses were prepared by melting the mixture of Ge, Ga, Se, CsCl and CsI under vacuum in a sealed silica tube. To establish the glassy region, the sealed ampoule containing the mixture was heated to 750°C in a rocking furnace for 12 h and quenched in water. All samples were annealed at 10°C below the glass transition temperature for 6 h to minimize inner tensions induced by the quenching step and were slowly cooled to room temperature.

5.19 Germanium–Cesium–Gallium–Chlorine–Iodine–Selenium

In the system containing these elements, the multinary compound $(Cs_6I_{0.6}Cl_{0.4})_2Cs_5[Ga_{15}Ge_9Se_{48}]$, which crystallizes as a tetragonal structure with the lattice parameters a = 1442.67 ± 0.05, c = 2808.2 ± 0.2 pm, and an energy gap of 2.80 eV, is formed (Huang-Fu et al. 2015). To grow the crystals of the title compound, mixed fluxes CsI (100 mg) and KI (42.6 mg) with molar ratio of 6:4 (melt point 571°C) together with single crystals of $(Cs_6Cl)_2Cs_5[Ga_{15}Ge_9Se_{48}]$ (ca. 25 mg) were sealed in silica tube under vacuum and then heated at a rate of 15°C·h^{-1} to 650°C, followed by a dwell time of 60 h, and subsequently cooled at a rate of 5.4°C·h^{-1} to 200°C, before switching off the furnace. After washing with distilled water and ethanol to remove the flux, the clear, light yellow crystals of this compound were obtained.

5.20 Germanium–Cesium–Gallium–Bromine–Selenium

In the system containing these elements, the quinary compound $(Cs_6Br)_2Cs_5[Ga_{15}Ge_9Se_{48}]$, which crystallizes as a tetragonal structure with the lattice parameters a = 1438.06 ± 0.04, c = 2793.4 ± 0.2 pm, and an energy gap of 2.75 eV, is formed (Huang-Fu et al. 2015). For obtaining the crystals of the title compound, mixed fluxes of CsBr (100 mg) and KBr (37.3 mg) with a molar ratio of 6:4 (melt point 571°C) together with single crystals of $(Cs_6Cl)_2Cs_5[Ga_{15}Ge_9Se_{48}]$ (ca. 25 mg) were sealed in silica tube under vacuum and then heated at a rate of 15°C·h^{-1} to 650°C, followed by a dwell time of 60 h, and subsequently cooled at a rate of 5.4°C·h^{-1} to 200°C, before switching off the furnace. After washing with distilled water and ethanol to remove the flux, the clear, light yellow crystals of this compound were obtained.

5.21 Germanium–Cesium–Indium–Tin–Selenium

In the system containing these elements, the quinary compound $CsInGeSnSe_6$, which crystallizes as a tetragonal structure with the lattice parameters $a = 1057.10 \pm 0.04$, $c = 960.65 \pm 0.08$ pm, a calculated density of 4.891 g·cm^{-3}, and an energy gap of 1.96 eV, is formed (Li et al. 2021a). It was prepared in the same way as $CsGaGeSnSe_6$ was synthesized but using In instead of Ga. The crystals of this compound are also stable in air.

5.22 Germanium–Copper–Silver–Gallium–Selenium

In the system containing these elements, the quinary phase $Cu_{0.02}Ag_{0.98}GaGe_3Se_8$, which crystallizes as an orthorhombic structure with the lattice parameters $a = 1244.97 \pm 0.08$, $b = 2383.6 \pm 0.2$, and $c = 714.37 \pm 0.04$ pm, is formed (Fedorchuk et al. 2011). Single crystals of this phase were grown by the Bridgman–Stockbarger technique.

5.23 Germanium–Copper–Silver–Tellurium–Selenium

In the system containing these elements, the quinary phase $Cu_{7.6}Ag_{0.4}GeSe_{5.1}Te_{0.9}$, which crystallizes as a trigonal structure with the lattice parameters $a = 737.156 \pm 0.007$, $c = 1805.82 \pm 0.03$ pm, and an energy gap of 0.758 eV, is formed (Jiang et al. 2017). To prepare this phase, Cu, Ag, Se, and Te shots and Ge pieces were weighed out in stoichiometric proportions and then sealed in silica tube under vacuum. The tube was heated to 1120°C and held at this temperature for 24 h before quenching into cold water. Then, the quenched ingot was annealed at 600°C for 5 days. Finally, the products were ground into a fine powder and sintered by spark plasma sintering at 410°C–440°C under a pressure of 75 MPa for 5 min.

5.24 Germanium–Copper–Silver–Iodine–Selenium

Cu_7GeSe_5I–Ag_7GeSe_5I. A continuous series of solid solutions, which crystallize in the face-centered cubic structure, is formed in this system (Pogodin et al. 2019). A linear increase in the lattice parameter with increasing Ag content was observed. Single crystals of these solid solutions were grown by zone crystallization from a melt according to a modified method.

5.25 Germanium–Copper–Zinc–Cadmium–Selenium

$Cu_2ZnGeSe_4$–$Cu_2CdGeSe_4$. This system is the nonquasibinary section of the Ge–Cu–Zn–Cd–Se quinary system (Olekseyuk et al. 2019). At 450°C, continuous solid solutions with tetragonal structure are formed in the system. The alloys for the investigations were annealed at 400°C for 500 h followed by quenching into cold water.

5.26 Germanium–Copper–Zinc–Indium–Arsenic–Selenium

Cu_2GeSe_3–$CuInSe_2$–$ZnGeAs_2$–InAs. A large volume of this segment of the Ge–Cu–Zn–In–As–Se multinary system extending from InAs was shown to be solid solutions with cubic structure of the sphalerite type (Pamplin and Hasoon 1985). No new compounds or superlattices were found, and only a nonequilibrium material was prepared in the center of the segment.

5.27 Germanium–Copper–Zinc–Tin–Selenium

$Cu_2ZnGeSe_4$–$Cu_2ZnSnSe_4$. This system is the nonquasibinary section of the Ge–Cu–Zn–Sn–Se quinary system (Olekseyuk et al. 2019). At 400°C, $Cu_2ZnGeSe_4$ dissolves 18 mol% $Cu_2ZnSnSe_4$ and the solubility of $Cu_2ZnGeSe_4$ in $Cu_2ZnSnSe_4$ reaches 16 mol%. The alloys for the investigations were annealed at 400°C for 500 h followed by quenching into cold water.

5.28 Germanium–Copper–Zinc–Manganese–Selenium

$Cu_2Zn_{1-x}Mn_xGeSe_4$. At room temperature, the XRD shows that samples in the range $0 < x < 0.375$ had the tetragonal stannite α structure, while for $0.725 < x \leq 1$ the wurtz-stannite δ structure (Caldera et al. 2014). The α and δ fields are separated by a relatively wide three-phase region (α + δ + $MnSe_2$). It was observed that undercooling effects occur for samples in the range $0.725 < x < 0.925$. All of the alloys were made by the usual melt and anneal technique. In each case, the components of 1 g sample were sealed under vacuum in a small quartz ampoule, which had previously been carbonized to prevent interaction of the components with the quartz. Then the components were heated up to 200°C and kept for about 1–2 h, and again the temperature was raised to 500°C using a rate of $40°C·h^{-1}$. Then, they were held at this temperature for 14 h. Then samples were heated from 500°C to 800°C at $30°C·h^{-1}$ and kept at this temperature for another 14 h. After that, the temperature was raised to 1150°C at $60°C·h^{-1}$, and the components were melted together at this temperature for about 1–2 h. During melting, the ampoules were periodically agitated to ensure good mixing of the constituent elements. In order to homogenize the material, annealing of samples was done in the range 700°C–500°C, and then samples were slowly and/or rapidly cooled to room temperature.

5.29 Germanium–Copper–Zinc–Iron–Selenium

It was found that at room temperature, a single-phase solid solution $Cu_2Zn_{1-x}Fe_xGeSe_4$ with the tetragonal stannite structure occurs across the whole composition range (Caldera et al. 2008). Undercooling effects occur for samples with $x > 0.9$. No variation of the lattice parameters with the composition was observed. All alloys were produced by the usual melt and anneal technique.

5.30 Germanium–Copper–Cadmium–Manganese–Selenium

$Cu_2CdGeSe_4$–$Cu_2MnGeSe_4$. This system is a nonquasibinary section of the Ge–Cu–Cd–Mn–Se quinary system as both compounds melt incongruently (Quintero et al. 2002, 2007). The $Cu_2Cd_{1-x}Mn_xGeSe_4$ solid solution is formed. In the composition range $0.05 < x < 0.2$, the alloys were found to be of two phases α + δ (the tetragonal stannite and orthorhombic wurtz-stannite structures), while that for the range $0.2 \leq x < 0.7$, the samples were found to be a single phase with the wurtz-stannite structure. For $0.7 \leq x \leq 1.0$, the samples were found to be of two phases δ + MnSe. The values of the crystal parameters a and c in the wurtz-stannite δ composition range decrease linearly with the composition x, while the values of b decrease nonlinearly. The experimental values of a and c were least squared fitted to a linear equation, but the b versus x curve was fitted to a quadratic form, and the obtained results are given by $a(x) = 0.8068 - 0.0086x$ (nm), $b(x) = 0.6869 - 0.0010x - 0.0060x^2$ (nm), and $c(x) = 0.6602 - 0.0057x$ (nm). All of the alloy samples were produced by the usual melt and anneal techniques.

The structure of the $Cu_2Cd_{0.5}Mn_{0.5}GeSe_4$ alloy was refined from the XRD pattern by Delgado et al. (2004). It crystallizes as an orthorhombic structure with the lattice parameters $a = 802.53 \pm 0.02$, $b = 685.91 \pm 0.02$, $c = 657.34 \pm 0.02$ pm, and a calculated density of $5.52\ g·cm^{-3}$.

5.31 Germanium–Copper–Cadmium–Iron–Selenium

$Cu_2CdGeSe_4$–$Cu_2CdFeSe_4$. This system is a nonquasibinary section of the Ge–Cu–Cd–Fe–Se quinary system as $Cu_2CdGeSe_4$ melts incongruently (Quintero et al. 2007). The $Cu_2Cd_{1-x}Fe_xGeSe_4$ solid solution is formed. There are only two single solid fields, namely, the tetragonal stannite α and the wurtz-stannite δ fields. The crystallographic parameter values appear, within the limits of experimental error, to follow the usual linear Vegard's behavior. The experimental values of a and c were least squared fitted to a linear equation, and the obtained results are given by $a(x) = 0.5746 - 0.0148x$ (nm) and $c(x) = 1.1057 - 0.5213 \cdot 10^{-4} x$ (nm). All of the alloy samples were produced by the usual melt and anneal techniques.

5.32 Germanium–Copper–Mercury–Tin–Selenium

$Cu_2HgGeSe_4$–$Cu_2HgSnSe_4$. The solid solution exists over the entire concentration range of this system (Hirai et al. 1967). Apparently, the phase diagram of the system belongs to the type I according to the Roozeboom's classification.

5.33 Germanium–Copper–Mercury–Tellurium–Selenium

$Cu_2HgGeSe_4$–$Cu_2HgGeTe_4$. The phase diagram of this system is a eutectic type (Figure 5.3; Hirai et al. 1967). The eutectic contains 10 mol% $Cu_2HgGeSe_4$ and crystallizes at 515°C. A wide region of solid solutions based on $Cu_2HgGeSe_4$ exists in the system.

5.34 Germanium–Copper–Gallium–Arsenic–Selenium

Cu_2GeSe_3–3GaAs. According to XRD and metallography, there is a continuous series of solid solutions in this system (Goryunova et al. 1965). The lattice parameter approximately corresponds to the Vegard's law. It was not possible to obtain completely homogeneous and sufficiently large crystalline samples. According to Averkieva et al. (1964, 1968), solid solutions based on GaAs contain up to 40 mol% Cu_2GeSe_3, and the lattice parameter changes linearly versus composition. A narrow region of the solid solution also exists on the base of GaAs.

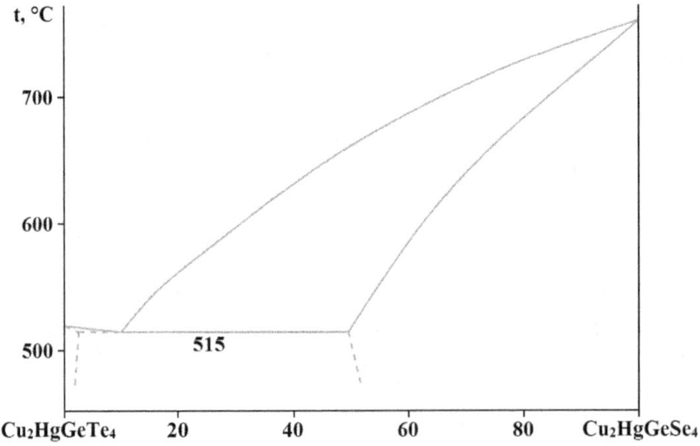

FIGURE 5.3 Phase diagram of the $Cu_2HgGeSe_4$–$Cu_2HgGeTe_4$ system. (From Hirai, T., et al., *Solid-State Electronics*, 10(10), 975, 1967.)

5.35 Germanium–Beryllium–Zinc–Oxygen–Selenium

In the system containing these elements, the quinary compound $Zn_4Be_3(GeO_4)_3Se$ is formed (Mel'nikov et al. 1968). This compound was synthesized through hydrothermal synthesis at 450°C and 0.2 GPa for 7–10 days using the mixture of ZnO, BeO, GeO_2, and Se with the solution of 1 mass% NaOH as a solvent.

5.36 Germanium–Beryllium–Cadmium–Oxygen–Selenium

The quinary compound $Cd_4Be_3(GeO_4)_3Se$ is formed in this system (Mel'nikov et al. 1968). This compound was synthesized through hydrothermal synthesis at 450°C and 0.2 GPa for 7–10 days using the mixture of CdO, BeO, GeO_2, and Se with the solution of 1 mass% NaOH as a solvent.

5.37 Germanium–Strontium–Yttrium–Antimony–Selenium

In the system containing these elements, the quinary compound $Sr_{2.63}Y_{0.37}Ge_{0.63}Sb_{2.37}Se_8$, which crystallizes as an orthorhombic structure with the lattice parameters a = 1259.4 ± 0.3, b = 433.23 ± 0.09, c = 2886.2 ± 0.6 pm, and a calculated density of 3.883 g·cm^{-3}, is formed (Chung and Lee 2013). In attempts to synthesize analogous compounds to $Sr_8YGe_2Bi_7Se_{24}$, Sr, Y, Ge, Sb, and Se were combined in the appropriate stoichiometric ratios (total mass approximately 0.5 g) under a dry nitrogen atmosphere. The reaction mixture was heated in a fused-silica tube in a furnace to 750°C over one day, maintained at that temperature for one day, slowly cooled to 400°C over a day, and finally cooled to room temperature by terminating the power. The product was polycrystalline ingot that exhibited a metallic luster.

5.38 Germanium–Strontium–Yttrium–Bismuth–Selenium

In the system containing these elements, the quinary compound $Sr_8YGe_2Bi_7Se_{24}$, which melts at 792°C and crystallizes as an orthorhombic structure with the lattice parameters a = 1283.5 ± 0.2, b = 2906.8 ± 0.3, c = 1284.8 ± 0.6 pm, and an energy gap of 0.22 eV, is formed (Chung and Lee 2012). In a typical reaction for obtaining this compound, stoichiometric proportions of the elements were mixed in an N_2-filled glove box (total mass ~0.5 g), placed in a carbon-coated silica tube, sealed under dynamic vacuum, and slowly heated to 750°C over 48 h. This temperature was maintained for one day, followed by slow cooling to 400°C at a rate of 15°C·h^{-1}, and finally to room temperature by simply terminating the power. This reaction yielded polycrystalline ingots with a metallic luster.

5.39 Germanium–Strontium–Lanthanum–Antimony–Selenium

In the system containing these elements, the quinary compound $Sr_{2.63}La_{0.37}Ge_{0.63}Sb_{2.37}Se_8$, which crystallizes as an orthorhombic structure with the lattice parameters a = 1253.7 ± 0.3, b = 434.8 ± 0.1, c = 2881.2 ± 0.6 pm, and a calculated density of 3.946 g·cm^{-3}, is formed (Chung and Lee 2013). This compound was synthesized in the same way as $Sr_{2.63}Y_{0.37}Ge_{0.63}Sb_{2.37}Se_8$ was prepared using La instead of Y.

5.40 Germanium–Barium–Lanthanum–Antimony–Selenium

In the system containing these elements, the quinary compound $Ba_4LaGe_3SbSe_{13}$, which is thermodynamically stable under the exclusion of air at least up to 720°C and crystallizes as a monoclinic structure with the lattice parameters a = 1633.30 ± 0.09, b = 1251.15 ± 0.07, c = 1303.21 ± 0.07 pm, β = 103.457 ±

0.002°, a calculated density of 5.27 g·cm^{-3}, and an energy gap of 1.5 eV, is formed (Assoud et al. 2004). This compound was prepared by heating the elements in an evacuated silica tube at 800°C over a period of 1 week, and then the furnace was slowly cooled to room temperature with a cooling rate of 100°C per day.

5.41 Germanium–Barium–Lead–Bromine–Selenium

In the system containing these elements, the quinary compound $Ba_4Ge_2PbSe_8Br_2$, which crystallizes as an orthorhombic structure with the lattice parameters a = 1250.37 ± 0.07, b = 949.63 ± 0.04, c = 906.71 ± 0.05 pm, a calculated density of 5.223 g·cm^{-3}, and an energy gap of 1.952 eV, is formed (Lin et al. 2014). To prepare this compound, reaction mixtures of BaSe (2 mM), PbSe (1 mM), $GeSe_2$ (1 mM), and NaBr (1 mM) were grounded and loaded into fused-silica tube under an Ar atmosphere in a glove box. The tube was flame-sealed under a high vacuum of 10^{-3} Pa and then placed in a furnace. The sample was heated to 850°C within 20 h, kept for 50 h, then slowly cooled to 400°C at a rate of 3°C·h^{-1}, and finally cooled to room temperature by switching off the furnace. Many dark red block-shaped crystals were obtained.

5.42 Germanium–Zinc–Antimony–Tellurium–Selenium

$GeSe_2$–ZnTe–Sb_2Se_3. The glass-forming region in this quasiternary system is presented in Figure 5.4 (Boycheva et al. 1999). The glassy phases of the system were found between the $GeSe_2$–Sb_2Se_3 and $GeSe_2$–ZnTe sides of the Gibbs diagram up to 70 mol% Sb_2Se_3 and 15 mol% ZnTe. The glass transition temperature varies in the range 200°C–280°C and depends on the concentration of ZnTe. To prepare the glasses, the required mass of the binary components were flame-sealed into a quartz ampoule under a vacuum of 10^{-3} Pa. The sealed ampoule was then placed in a rocking furnace and heated at 950°C. The final stage was water quenching with an average cooling rate of 2–5°C·s^{-1}.

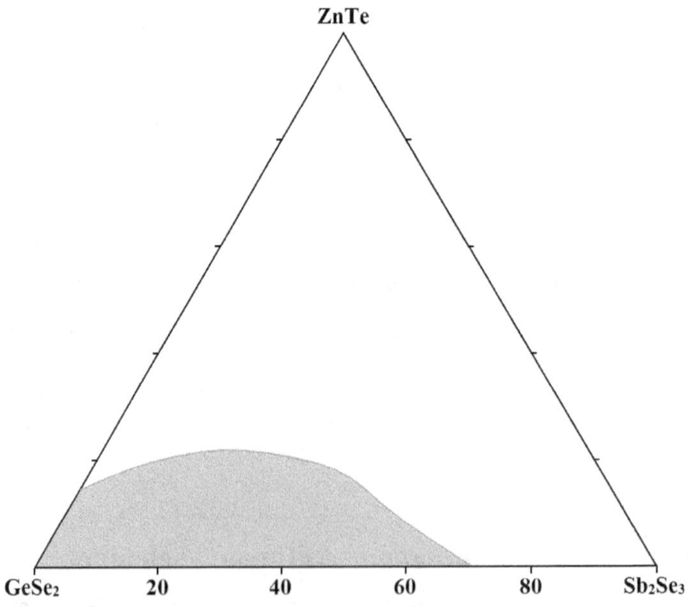

FIGURE 5.4 Glass-forming region in the $GeSe_2$–ZnTe–Sb_2Se_3 quasiternary system. (From Boycheva, S.V., et al., *J. Phys. D: Appl. Phys.*, 32(4), 529, 1999.)

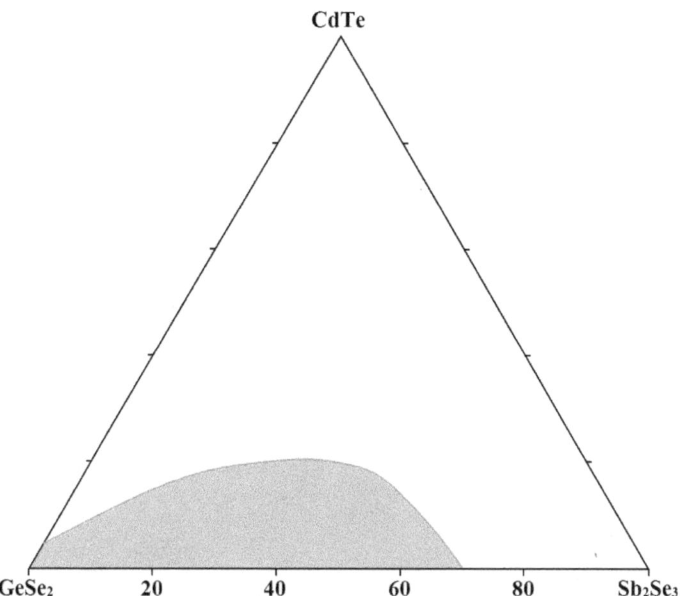

FIGURE 5.5 Glass-forming region in the $GeSe_2$–CdTe–Sb_2Se_3 quasiternary system. (From Vassilev, V.S., et al., *Mater. Lett.*, 52(1–2), 126, 2002.)

5.43 Germanium–Cadmium–Antimony–Tellurium–Selenium

$GeSe_2$–CdTe–Sb_2Se_3. The glass-forming region in this quasiternary system is shown in Figure 5.5 (Vassilev et al. 2002a). The glasses have been obtained in the $GeSe_2$-rich region. The maximum CdTe solubility in the glass ternary system is only 15 mol%. The glass transition temperature changes within the interval from 185°C to 250°C. The glass-forming region in this system was determined visually and through XRD, DTA, and EPMA.

5.44 Germanium–Cadmium–Bismuth–Oxygen–Iodine–Selenium

$GeSe_2$–CdI_2–Bi_2O_3. The glass-forming region in this quasiternary system is given in the Figure 5.6 (Vassilev et al. 2002b). The obtained glasses are located between $GeSe_2$–CdI_2 (55–100 mol% $GeSe_2$) and $GeSe_2$–Bi_2O_3 (92–100 mol% $GeSe_2$) sides of the Gibbs diagram. It should be mentioned that the CdI_2-rich glasses (above 20 mol% CdI_2) are not stable against moisture and hydrolize with time. The values of the glass transition temperature changes within the interval from 207°C to 228°C and increase at higher $GeSe_2$ contents. Bulk glasses were synthesized from the binary compounds. Mixtures of the starting materials with appropriate composition were melted up to 900°C–950°C in sealed silica ampoules by using a rocking furnace. Then the melts were quenched to room temperature in air or in ice water.

5.45 Germanium–Cadmium–Oxygen–Tellurium–Iodine–Selenium

$GeSe_2$–CdI_2–TeO_2. The glass-forming region in this quasiternary system is shown in the Figure 5.7 (Vassilev et al. 2002b). The obtained glasses are located between $GeSe_2$–CdI_2 (55–100 mol% $GeSe_2$) and $GeSe_2$–TeO_2 (86–100 mol% $GeSe_2$) sides of the Gibbs diagram. It should be mentioned that the CdI_2-rich glasses (above 20 mol% CdI_2) are not stable against moisture and hydrolyze with time. The values of the

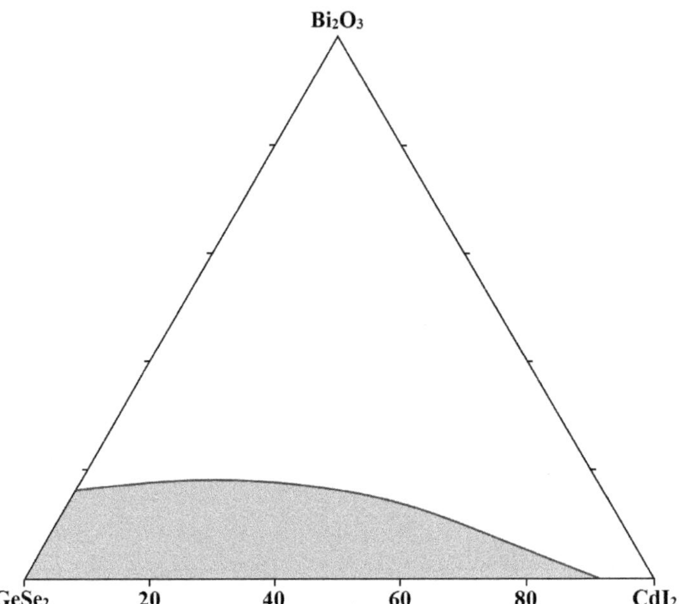

FIGURE 5.6 Glass-forming region in the $GeSe_2$–CdI_2–Bi_2O_3 quasiternary system. (From Vassilev, V.S., et al., *J. Phys. Chem. Solids*, 63(5), 815, 2002.)

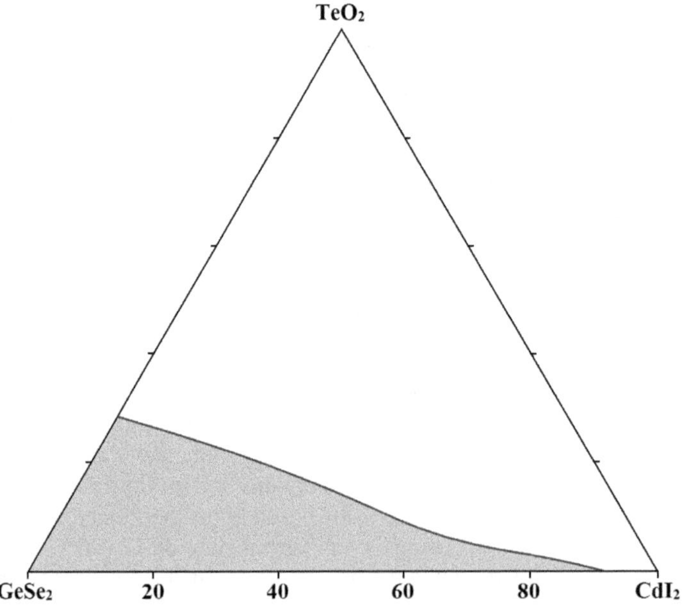

FIGURE 5.7 Glass-forming region in the $GeSe_2$–CdI_2–TeO_2 quasiternary system. (From Vassilev, V.S., et al., *J. Phys. Chem. Solids*, 63(5), 815, 2002.)

glass transition temperature change within the interval from 266°C to 293°C and increase at higher $GeSe_2$ contents. Bulk glasses were synthesized from the binary compounds. Mixtures of the starting materials with appropriate composition were melted up to 900°C–950°C in sealed silica ampoules by using a rocking furnace. Then the melts were quenched to room temperature in air or in ice water.

5.46 Germanium–Tin–Arsenic–Tellurium–Selenium

$GeSe_2$-SnTe–As_2Se_3. The glass-forming region in this quasiternary system is presented in Figure 5.8 (Vassilev et al. 2005). It is situated entirely on the $GeSe_2$–As_2Se_3 side of the Gibbs diagram and partially on the As_2Se_3–SnTe (up to ~30 mol% SnTe) side. The glass transition temperature of these alloys varies in the range of 200°C–280°C. Glasses were prepared by alloying the initial binary components in evacuated (0.133 Pa) silica ampoules, using a rotary furnace.

5.47 Germanium–Tin–Antimony–Tellurium–Selenium

The $Ge_7SnSb_2Te_7Se_4$ quinary compound, which has two polymorphic modifications and can be obtained in an amorphous state, is formed in the Ge–Sn–Sb–Te–Se system (Buller et al. 2012). The first modification of this compound crystallizes as a cubic structure with the lattice parameter a = 594.6 ± 0.1 pm and the second one crystallizes as a rhombohedral structure with the lattice parameters a = 418.1 ± 0.6 and a = 1047.6 ± 0.4 pm in the hexagonal setting. The experimental density and the energy gap of this compound are 5.63 and 5.94 g·cm^{-3} and 0.97 and 0.62 eV in an amorphous and in a crystalline state, respectively. The crystallization temperature of amorphous $Ge_7SnSb_2Te_7Se_4$ is ≈ 185°C. Thin films of the title compound were deposited by co-sputtering of $SnSb_2Se_4$ with diameters of 10 cm at a power of 5 W and GeTe.

5.48 Germanium–Bismuth–Oxygen–Tellurium–Selenium

$GeSe_2$–Bi_2O_3–TeO_2. The glass-forming region in this quasiternary system is given in Figure 5.9 (Vassilev et al. 2002b). The values of the glass transition temperature changes within the interval from 282°C to 334°C and increase at higher $GeSe_2$ contents. Bulk glasses were synthesized from the binary compounds.

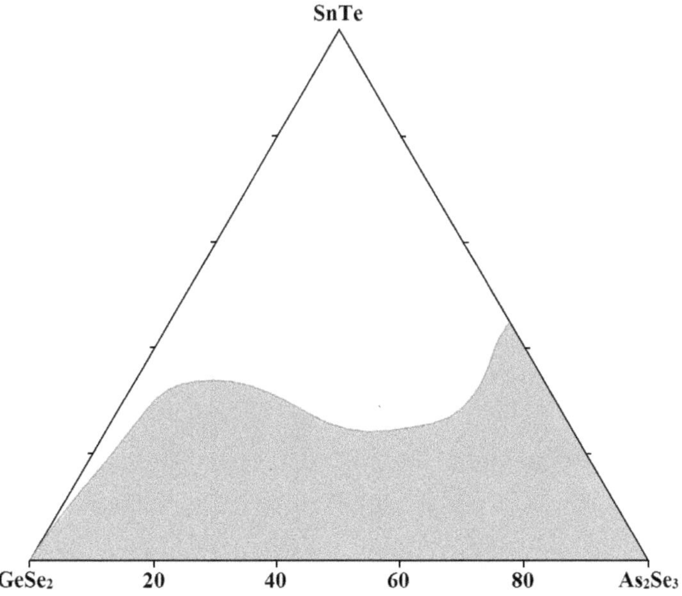

FIGURE 5.8 Glass-forming region in the $GeSe_2$-SnTe–As_2Se_3 quasiternary system. (From Vassilev, V., et al., *Mater. Lett.*, 59(1), 85, 2005.)

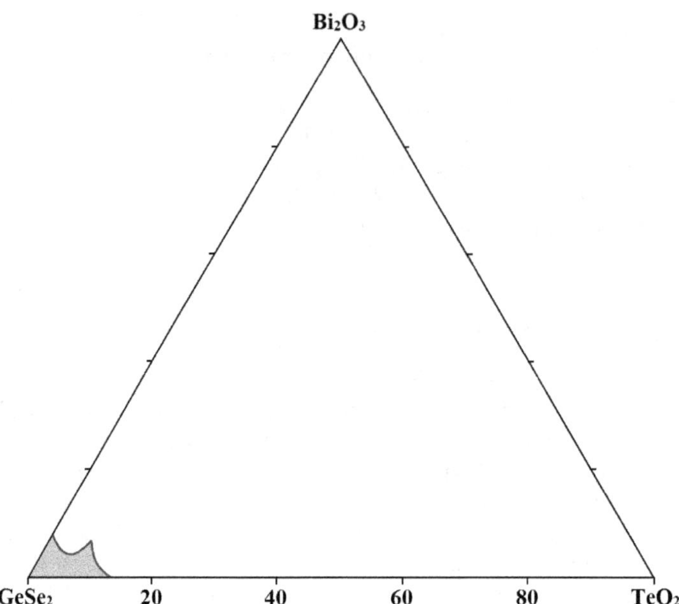

FIGURE 5.9 Glass-forming region in the $GeSe_2$–Bi_2O_3–TeO_2 quasiternary system. (From Vassilev, V.S., et al., *J. Phys. Chem. Solids*, 63(5), 815, 2002.)

Mixtures of the starting materials with appropriate composition were melted up to 900°C–950°C in sealed silica ampoules by using a rocking furnace. Then the melts were quenched to room temperature in air or in ice water.

6

Systems Based on Germanium Telluride

6.1 Germanium–Hydrogen–Potassium–Oxygen–Tellurium

In the system containing these elements, the quinary compound $[K_4(H_2O)_4][GeTe_4]$, which crystallizes as a monoclinic structure with the lattice parameters $a = 1943.0 \pm 0.4$, $b = 741.33 \pm 0.15$, $c = 1384.5 \pm 0.3$ pm, $\beta = 129.93 \pm 0.03°$, a calculated density of 3.368 g·cm^{-3} at 203 K, and an energy gap of 1.63 eV, is formed (Melullis and Dehnen 2007). Ruby red single crystals of this compound were prepared when K_6GeTe_6 (0.100 mM) was dissolved in H_2O (10 mL) at 20°C. The solution was stirred for 30 min and filtered away from precipitating Ge powder. The filtrate was evaporated for several hours, whereupon the blocks of $[K_4(H_2O)_4][GeTe_4]$ were obtained. All synthesis steps were performed with a strong exclusion of air and external moisture (N_2 atmosphere at a high-vacuum, double-manifold Schlenk line or Ar atmosphere in a glove box). Water was degassed by applying dynamic vacuum (10^{-1} Pa) for several hours.

6.2 Germanium–Beryllium–Zinc–Oxygen–Tellurium

The quinary compound $Zn_4Be_3(GeO_4)_3Te$ is formed in this system (Mel'nikov et al. 1968). This compound was synthesized through hydrothermal synthesis at 450°C and 0.2 GPa for 7–10 days using the mixture of ZnO, BeO, GeO_2, and Te with the solution of 1 mass% NaOH as a solvent.

6.3 Germanium–Beryllium–Cadmium–Oxygen–Tellurium

In the system containing these elements, the quinary compound $Cd_4Be_3(GeO_4)_3Te$ is formed (Mel'nikov et al. 1968). This compound was synthesized through hydrothermal synthesis at 450°C and 0.2 GPa for 7–10 days using the mixture of CdO, BeO, GeO_2, and Te with the solution of 1 mass% NaOH as a solvent.

6.4 Germanium–Strontium–Zinc–Oxygen–Tellurium

In the system containing these elements, the quinary compound $Sr_3Te(Zn_2Ge)Ge_2O_{14}$, which crystallizes as a trigonal structure with the lattice parameters $a = 832.23 \pm 0.05$, $c = 505.51 \pm 0.05$ pm, a calculated density of 5.274 g·cm^{-3}, and an energy gap of 4.03 eV, is formed (Chen et al. 2021a). This polycrystalline compound was synthesized by a solid state reaction. $Sr(NO_3)_2$, TeO_2, GeO_2, and ZnO were thoroughly ground in stoichiometric ratios and placed into a Pt crucible. An extra 5 mol% TeO_2 was added to compensate for volatilization losses. The mixture was heated from 30°C to 850°C at a rate of 50°C·h^{-1}. The crystals of $Sr_3Te(Zn_2Ge)Ge_2O_{14}$ were grown by the high-temperature solution method. The mixture of the polycrystalline sample, Na_2CO_3 and TeO_2 in the Pt crucible, was heated to 850°C for 24 h and then cooled to 700°C at a rate of 2°C·h^{-1} and to room temperature at a rate of 20°C·h^{-1}. Finally, colorless and transparent crystals of the title compound were obtained.

DOI: 10.1201/9781003123460-6

6.5 Germanium–Barium–Zinc–Oxygen–Tellurium

In the system containing these elements, the quinary compound $Ba_3Te(Zn_2Ge)Ge_2O_{14}$, which crystallizes as a trigonal structure with the lattice parameters $a = 861.6 \pm 0.3$, $c = 519.8 \pm 0.3$ pm, a calculated density of 5.527 $g \cdot cm^{-3}$, and an energy gap of 3.98 eV, is formed (Chen et al. 2021a). This polycrystalline compound was synthesized by a solid state reaction. $Ba(NO_3)_2$, TeO_2, GeO_2, and ZnO were thoroughly ground in stoichiometric ratios and placed into a Pt crucible. An extra 5 mol% TeO_2 was added to compensate for volatilization losses. The mixture was heated from 30°C to 850°C at a rate of 50°C·h^{-1}. The crystals of $Ba_3Te(Zn_2Ge)Ge_2O_{14}$ were grown by the high-temperature solution method. The mixture of the polycrystalline sample, Na_2CO_3 and TeO_2 in the Pt crucible, was heated to 850°C for 24 h and then cooled to 700°C at a rate of 2°C·h^{-1} and to room temperature at a rate of 20°C·h^{-1}. Finally, colorless and transparent crystals of the title compound were obtained.

6.6 Germanium–Zinc–Lead–Oxygen–Tellurium

In the system containing these elements, the quinary compound $Pb_3Te(Zn_2Ge)Ge_2O_{14}$, which crystallizes as a trigonal structure with the lattice parameters $a = 851.94 \pm 0.06$, $c = 512.35 \pm 0.06$ pm, a calculated density of 6.815 $g \cdot cm^{-3}$, and an energy gap of 3.43 eV, is formed (Chen et al. 2021a). The polycrystalline compound was synthesized by the solid state reaction. PbO, TeO_2, GeO_2, and ZnO were thoroughly ground in stoichiometric ratios and placed into a Pt crucible. An extra 5 mol% TeO_2 was added to compensate for volatilization losses. The mixture was heated from room temperature to 700°C at a rate of 50°C·h^{-1}. The crystals of $Pb_3Te(Zn_2Ge)Ge_2O_{14}$ were grown by the high-temperature solution method. The mixture of the polycrystalline sample, PbO and TeO_2 in the Pt crucible, was heated to 800°C for 24 h and then cooled to 650°C at a rate of 2°C·h^{-1} and to room temperature at a rate of 20°C·h^{-1}. Finally, colorless and transparent crystals of the title compound were obtained.

6.7 Germanium–Gallium–Lead–Oxygen–Tellurium

A variable-composition quinary phase $Pb_3Te_{1-x}Ga_{4-2x}Ge_{1+3x}O_{14}$ exists in the Ge–Ga–Pb–O–Te in the range of $0.15 \leq x < 1$ (Mill 2010). This phase crystallizes as a trigonal structure with the lattice parameters $a = 852.9 \pm 0.3$, $c = 507.3 \pm 0.1$ pm, and a calculated density of 6.797 $g \cdot cm^{-3}$ for $x = 0.333$ and $a = 849.3 \pm 0.3$, $c = 506.6 \pm 0.1$ pm, and a calculated density of 6.821 $g \cdot cm^{-3}$ for $x = 0.5$. The starting compounds for the synthesis of this phase were PbO, TeO_2, Ga_2O_3, and GeO_2. A thoroughly blended and compacted mixture was annealed in air in a platinum container at temperatures from 800°C to 840°C to the melting or decomposition temperature with an interval of ≤50°C for 10–15 h. The heating rate was 250–300°C·h^{-1}.

6.8 Germanium–Tin–Antimony–Bismuth–Tellurium

GeTe–SnTe–Sb_2Te_3–Bi_2Te_3. Samples of the system containing 77.4 mol% $(Sb_xBi_{1-x})_2Te_3$ and 22.6 mol% $Ge_ySn_{1-y}Te$ (the Sb/Bi and Ge/Sn ratios varied) have a fine lamellar structure of the eutectic type (Leonov and Popkov 1980).

7

Systems Based on Tin Sulfides

7.1 Tin–Hydrogen–Lithium–Oxygen–Sulfur

Two quinary compounds, $Li_4SnS_4{\cdot}13H_2O$ and $[Li_8(H_2O)_{29}[Sn_{10}O_4S_{20}]{\cdot}2H_2O$, are formed in this system. The first compound crystallizes as a cubic structure with the lattice parameter a = 1274.29 ± 0.04 pm and a calculated density of 1.63 g·cm^{-3} at 193 K (Kaib et al. 2012). To prepare the title compound, the crude product mixture from the synthesis of Li_4SnS_4 was pulverized. This powder (2 g) was solved in 20 mL of H_2O and stirred for 4 h. The starting materials and insoluble residues were removed by filtration, resulting in a clear, colorless solution. Colorless crystals of $Li_4SnS_4{\cdot}13H_2O$ were obtained upon very slow evaporation of the solvent by applying dynamic vacuum. All reaction steps were performed with a strong exclusion of air and external moisture (Ar atmosphere at a high-vacuum, double-manifold Schlenk line or Ar atmosphere in a glove box).

$[Li_8(H_2O)_{29}[Sn_{10}O_4S_{20}]{\cdot}2H_2O$ crystallizes as a triclinic structure with the lattice parameters a = 1123.2 ± 0.2, b = 1309.7 ± 0.3, c = 2373.5 ± 0.5 pm, α = 102.73 ± 0.03°, β = 90.43 ± 0.03°, and γ = 93.44 ± 0.03° at 173 K (Kaib at al. 2011). To synthesize this compound, $Li_4SnS_4{\cdot}13H_2O$ (3.93 mM) was dissolved in water (20 mL) and treated with HCl (1 mol%) to result in pH = 7. Then, the solvent was removed at 80°C in a constant Ar flow, thereby again lowering the pH value. The light-yellow residue was again dissolved in water (10 mL). Colorless needles of the title compound were obtained upon slow evaporation of the water applying dynamic vacuum. All reaction steps were performed with a strong exclusion of air and external moisture (Ar atmosphere at a high-vacuum, double-manifold Schlenk line or Ar atmosphere in a glove box).

7.2 Tin–Hydrogen–Sodium–Copper–Oxygen–Sulfur

In the system containing these elements, the multinary compound $Na_4Cu_{32}Sn_{12}S_{48}{\cdot}4H_2O$, which crystallizes as a cubic structure with the lattice parameter a = 1792.1 ± 0.2 pm, a calculated density of 2.978 g·cm^{-3}, and an energy gap of 2.0 eV, is formed (Zhang et al. 2015). For preparing this compound, starting materials of NaOH (10.0 g), Cu powder (5.0 mM), SnS_2 powder (5.0 mM), and thiourea powder (0.1 M) were weighed and ground uniformly, and placed into a Teflon-lined steel autoclave, which was tightly sealed and heated at 200°C for 3 days. Then the autoclave was pulled out of a hot oven and allowed to cool to ambient temperature. After cooling, the product was ultrasonically cleaned using distilled water until the black crystals were separated. Then the crystals were washed using distilled water several times in open air and dried by acetone. All the reagents were kept in a dry box.

7.3 Tin–Hydrogen–Sodium–Zinc–Oxygen–Sulfur

Two multinary compounds are formed in the Sn–H–Na–Zn–O–S system. $[Na_2Zn_{3.5}Sn_{3.5}S_{13}]{\cdot}6H_2O$ crystallizes as a cubic structure with the lattice parameter a = 1786.30 ± 0.03 pm, a calculated density of 2.993 g·cm^{-3} at 100 K, and energy gap of 2.9 eV (Wu et al. 2008). Its single crystals were obtained in a reaction containing Na_2S (0.50 mM), S (0.80 mM), Sn (0.25 mM), and $ZnCl_2$ 0.017 mM). The starting

DOI: 10.1201/9781003123460-7

materials were premixed and ground and then loaded in a thick-walled Pyrex tube. The sample was preheated at 90°C for 1 h and followed by the addition of 0.20 mL of CH_3OH and 0.20 mL of H_2O. The reaction mixture was heated at 150°C for 7 days. The products were washed with 80% alcohol followed by water. Pale yellow cubic crystals and crystalline powders of the title compound were obtained.

$[Na_{10}(H_2O)_{32}][Zn_5Sn(\mu_3\text{-}S)_4(SnS_4)_4]\cdot 2H_2O$ or $Na_{10}Zn_5Sn_5S_{20}\cdot 34H_2O$ crystallizes as a triclinic structure with the lattice parameters $a = 1484.57 \pm 0.10$, $b = 1487.72 \pm 0.09$, $c = 1751.01 \pm 0.11$ pm, $\alpha = 97.402 \pm 0.001°$, $\beta = 100.211 \pm 0.001°$, and $\gamma = 103.956 \pm 0.001°$, and a calculated density of 2.197 g·cm^{-3} at 203 K (Zimmermann et al. 2005). To synthesize this compound, $ZnCl_2$ (0.50 mM) was added to a solution of Na_4SnS_4 (0.50 mM) in H_2O (14 mL). After the mixture was stirred for 4 h, the colorless solution was layered with 10 mL of THF. The title compound crystallized as colorless needles, over the course of 2 days. Isolation by decanting the mother liquor and the loose precipitate and drying this compound leads to partial dehydration. Such dehydration always resulted in X-ray amorphous material with differing H_2O content. All synthesis steps were performed with a strong exclusion of air and external moisture (N_2 atmosphere on a high-vacuum, double-manifold Schlenk line or Ar atmosphere in a glove box).

7.4 Tin–Hydrogen–Sodium–Oxygen–Sulfur

Three quinary compounds are formed in the Sn–H–Na–O–S system. $Na_2SnS_3\cdot 8H_2O$ was obtained as follows. The aqueous solution of $Na_4SnS_4\cdot 18H_2O$ was boiled and treated with 1 N HCl until a spot test with bromocresol-green showed a slight excess (Jelley 1933). The liquid was gently boiled for 4 h and then set aside. The precipitated SnS_2 was washed ten times by decantation (each settling operation took 2 days) and then filtered off by suction (this took 5 days owing to the colloidal nature of the precipitate); it was dried at 120°C, and the resulting brownish substance was powdered and boiled with $Na_4SnS_4\cdot 18H_2O$ (50 g) in 100 mL of aqueous solution until it had dissolved. The addition of a granulated Sn prevented atmospheric oxidation of the solution. After concentration by boiling, the solution was set aside over calcium chloride. $Na_2SnS_3\cdot 8H_2O$ was separated as colorless prismatic crystals.

$Na_4SnS_4\cdot 18H_2O$ crystallizes as a monoclinic structure (Jelley 1933) with the lattice parameters $a = 862.1 \pm 0.2$, $b = 2354.3 \pm 0.5$, $c = 1135.8 \pm 0.2$ pm, $\beta = 110.58 \pm 0.03°$, and a calculated density of 1.875 g·cm^{-3} (Willett et al. 2000) [$a = 862.2 \pm 0.5$, $b = 2353.4 \pm 1.2$, $c = 1134.7 \pm 0.7$ pm, $\beta = 110.53 \pm 0.04°$, and the calculated and experimental densities of 1.821 and 1.81 ± 0.02 g·cm^{-3}, respectively (Krebs et al. 1970a,b; Schiwy et al. 1973)]. This compound was obtained by the reaction of $Na_2S\cdot 9H_2O$ with $SnCl_4\cdot 5H_2O$ in aqueous solution at 21°C (Willett et al. 2000). $Na_2S\cdot 9H_2O$ (75.0 g) was dissolved in deionized water (100 mL). $SnCl_4\cdot 5H_2O$ (18.4 g) was added to a separate volume of deionized water (20 mL). These two solutions were mixed with continuous stirring. An orange/yellow precipitate formed immediately, but it dissolved upon further stirring to leave a clear colorless solution. PCl_3 (6 mL) was added dropwise to the clear colorless solution. The initial drops caused no visible reaction, but the last drops caused a strong crackle when they entered the solution. PCl_3 turned the solution from colorless to gray. The gray solution was stirred in an ice bath for 20 min. It was then stirred for an additional hour at room temperature. The solution was then kept at 4°C for 18 h. Methanol was added to the solution after the 18-h cooling period, and a large amount of gray-white crystals precipitated out of solution. The crude product was dissolved in deionized distilled water and recrystallized using methanol.

$Na_4SnS_4\cdot 18H_2O$ was also prepared by adding $Na_2Sn(OH)_6$ (100 g) and $Na_2S\cdot 9H_2O$ (250 g) to boiling water (700 mL) (Jelley 1933). The mixture was kept at 90°C–100°C for 3 h. MgO (40 g) was added, and the heating was continued for 2–3 h. After filtration, the clear pale yellow liquid was concentrated to about 300 mL on the water bath, and then set aside. The crystals which separated were rinsed with a little ice-cold water and twice recrystallized.

According to Schiwy et al. (1973), this compound can be prepared as follows. Freshly precipitated SnS_2 (60 mM) was added in portions to a solution of $Na_2S\cdot 9H_2O$ (100 mM) while stirring, a pH of 11–12 being established. The undissolved matter was filtered off and the solution was poured into about 500 mL of acetone. $Na_4SnS_4\cdot 18H_2O$ occurs either in crystalline form or in the form of colorless flakes or as a highly viscous liquid, which is left to crystallize on a Petri dish in the absence of air and then filtered off.

To form single crystals, it is necessary to dissolve the compound in a little water and allow crystallizing out slowly. This compound is relatively stable and will last for some time in air without decomposition.

$Na_4Sn_2S_6{\cdot}14H_2O$ crystallizes as a triclinic structure with the lattice parameters a = 1011.4 ± 0.6, b = 702.7 ± 0.5, c = 980.1 ± 0.6 pm, α = 108.30 ± 0.04°, β = 92.18 ± 0.04°, γ = 91.11 ± 0.04°, and the calculated and experimental densities of 1.945 and 1.97 ± 0.02 g·cm^{-3}, respectively (Krebs et al. 1970a,b, 1972). This compound was prepared in the following two ways. (1) After adding SnS_2 (50 mM) to the solution of the sodium sulfide, heating was carried out until all SnS_2 has dissolved, a pH of 9.0–9.5 being established. When acetone was added, $Na_4Sn_2S_6{\cdot}14H_2O$ was obtained either in finely crystalline form or as a highly viscous liquid, which was allowed to crystallize on a Petri dish and filtered off. The residue was taken up in a little water and slowly crystallized over P_4O_{10}.

(2) $SnCl_2$ (20 mM) was slowly added dropwise to a solution of $Na_2S{\cdot}9H_2O$ (60 mM) in H_2O (100 mL) with stirring and ice-cooling. The mixture was stirred for some time with gentle warming until the reaction was ended. $SnCl_2$ can be added in an open vessel. When cooled, $Na_4Sn_2S_6{\cdot}14H_2O$ precipitates out in the form of colorless leaves after a while and was sucked off by mother liquor. This compound stays in air for some time without decomposition.

7.5 Tin–Hydrogen–Sodium–Cobalt–Oxygen–Sulfur

Two multinary compounds are formed in the Sn–H–Na–Co–O–S system (Zimmermann et al. 2005). $[Na_{10}(H_2O)_{32}][Co_5Sn(\mu_3\text{-}S)_4(SnS_4)_4]{\cdot}2H_2O$ crystallizes as a triclinic structure with the lattice parameters a = 1471.2 ± 0.2, b = 1437.7 ± 0.2, c = 1714.2 ± 0.2 pm, α = 97.298 ± 0.002°, β = 99.264 ± 0.002°, γ = 103.868 ± 0.003°, and a calculated density of 2.185 g·cm^{-3} at 203 K. For obtaining this compound, $[Co(en)_3]Cl_3$ or $[Co(en)_3]Br_3$ (0.46 mM), where *en* is ethylenediamine, was added to a solution of Na_4SnS_4 (0.50 mM) in 14 mL of H_2O; upon the addition, the reaction mixture immediately turned dark brown. After the mixture was stirred for 12 h, an insoluble black precipitate, containing CoS and S, was removed by filtration. The remaining solution was layered with 14 mL of THF. Over the course of 2 days, the solution became lighter in color upon crystallization of black blocks of the title compound together with further black precipitate, presumably further amounts of CoS and S.

$[Na_{10}(H_2O)_6][Co_5Sn(\mu_3\text{-}S)_4(SnS_4)_4]$ also crystallizes as a triclinic structure with the lattice parameters a = 1343.6 ± 0.2, b = 1357.0 ± 0.2, c = 1410.5 ± 0.2 pm, α = 95.770 ± 0.003°, β = 117.889 ± 0.003°, γ = 98.364 ± 0.003°, and a calculated density of 2.810 g·cm^{-3} at 203 K. To prepare this compound, the single crystals of $[Na_{10}(H_2O)_{32}][Co_5Sn(\mu_3\text{-}S)_4(SnS_4)_4]{\cdot}2H_2O$ were isolated by decanting both the mother liquor and the loose byproduct precipitate, and the washing procedure was repeated with a THF/H_2O (volume ratio 1:1) mixture. The solid material (0.157 g) was then pumped with dynamic vacuum for 5 h. The crystals changed appearance, compared to that of the starting material; a metallic luster appeared, and cracking of the crystal surfaces was observed. The yield of this compound is quantitative with respect to $[Na_{10}(H_2O)_{32}][Co_5Sn(\mu_3\text{-}S)_4(SnS_4)_4]{\cdot}2H_2O$.

7.6 Tin–Hydrogen–Potassium–Rubidium–Cesium–Zinc–Nitrogen–Sulfur

In the system containing these elements, the multinary compound $KRb_{1.29}Cs(NH_4)_{2.71}Zn_4Sn_5S_{17}$, which crystallizes as a tetragonal structure with the lattice parameters a = 1381.84 ± 0.04 and c = 1004.56 ± 0.07 pm, is formed (Manos et al. 2006). To prepare this compound, single crystals of $K_6Sn[Zn_4Sn_4S_{17}]$ were placed in H_2O (15 mL), and then, 0.24 mM each of CsCl, RbI, and NH_4I were added to the obtained suspension. The mixture was heated at 70°C for a week without stirring. The title compound was obtained.

7.7 Tin–Hydrogen–Potassium–Copper–Oxygen–Sulfur

In the system containing these elements, the multinary compound $K_{11}Cu_{32}Sn_{12}S_{48}{\cdot}4H_2O$, which crystallizes as a cubic structure with the lattice parameter a = 1805.59 ± 0.06 pm, a calculated density of 3.104

$g·cm^{-3}$, and an energy gap of 1.9 eV, is formed (Zhang et al. 2015). For obtaining this compound, starting materials of K_2CO_3 (17.3 g), Cu powder (5.0 mM), SnS_2 powder (5.0 mM), and S powder (25 mM) were weighed and ground uniformly, and placed into a Teflon-lined steel autoclave. Then 10 mL of distilled water was added. The autoclave was tightly sealed and heated at 200°C for 3 days. Then the autoclave was pulled out of a hot furnace and allowed to cool to ambient temperature. After cooling to room temperature, the product was ultrasonically cleaned using distilled water until the black crystals were separated. Then the crystals were washed using distilled water for several times in open air and dried by acetone. Black crystals were obtained. All the reagents were kept in a dry box.

7.8 Tin–Hydrogen–Potassium–Silver–Oxygen–Sulfur

In the system containing these elements, the multinary compound $K_4Ag_2Sn_3S_9·2H_2O$, which crystallizes as a monoclinic structure with the lattice parameters a = 780.71 ± 0.02, b = 2735.08 ± 0.11, c = 1050.08 ± 0.02 pm, β = 103.874 ± 0.011°, a calculated density of 3.445 $g·cm^{-3}$, and an energy gap of 2.4 eV, is formed (An et al. 2003, 2004; Baiyin et al. 2004). The synthesis of this compound is done as follows: Sn (26 mg), $AgNO_3$ (15 mg), K_2CO_3 (89 mg), and S (17 mg) were put into a glass tube, to which 0.4 mL of ethanol/$HSCH_2CH(SH)CH_2OH$ (volume ratio 2:1) mixed solvent was added. The glass tube was sealed (reagents filled about 10% of the tube), placed into a Teflon-lined stainless steel autoclave, and heated at 120°C for 5 days. The products were washed with ethanol and water, respectively, and yellowish chunky crystals were obtained.

7.9 Tin–Hydrogen–Potassium–Zinc–Nitrogen–Oxygen–Sulfur

The multinary compound $K_2(NH_4)_4Zn_4Sn_5S_{17}·3H_2O$ is formed in this system (Manos et al. 2006). For obtaining it, single crystals of $K_6Sn[Zn_4Sn_4S_{17}]$ were placed in a water solution of NH_4I (0.39 mM). Then, the mixture was heated at 70°C and kept undisturbed at this temperature for 2 weeks. The crystals of the title compound were isolated by filtration, washed several times with water, acetone, and ether.

7.10 Tin–Hydrogen–Potassium–Zinc–Nitrogen–Sulfur

Three quinary compounds are formed in the Sn–H–K–Zn–N–S system (Manos et al. 2006). $K(NH_4)_5Zn_4Sn_5S_{17}$ crystallizes as a tetragonal structure with the lattice parameters a = 1384.55 ± 0.06 and c = 1007.31 ± 0.09 pm. To synthesize this compound, single crystals of $K_6Sn[Zn_4Sn_4S_{17}]$ were placed in a water solution (10 mL) of NH_4I (0.24 mM). The mixture was then heated at 70°C and left undisturbed at this temperature for 1 week. The crystals were then isolated by filtration, washed several times with water, acetone, and ether.

$K_{4.6}(NH_4)_{1.4}Zn_4Sn_5S_{17}$ also crystallizes as a tetragonal structure with the lattice parameters a = 1382.9 ± 0.5 and c = 981.0 ± 0.7 pm (Manos et al. 2006). For obtaining it, single crystals of $K_6Sn[Zn_4Sn_4S_{17}]$ were placed in a water solution of NH_4I (0.39 mM). Then, the mixture was heated at 70°C for 48 h and kept undisturbed at this temperature for 2 weeks. The crystals of the title compound were then isolated by filtration, washed several times with water, acetone, and ether.

$K_5(NH_4)Zn_4Sn_5S_{17}$ also crystallizes as a tetragonal structure with the lattice parameters a = 1381.42 ± 0.09 and c = 968.24 ± 0.12 pm (Manos et al. 2006). For obtaining it, single crystals of $K_6Sn[Zn_4Sn_4S_{17}]$ were placed in a water solution of NH_4I (0.39 mM). Then, the mixture was heated at 70°C for 12 h and kept undisturbed at this temperature for 2 weeks. The crystals of the title compound were then isolated by filtration, washed several times with water, acetone, and ether.

7.11 Tin–Hydrogen–Potassium–Oxygen–Sulfur

Two quinary compounds, $K_2SnS_3{\cdot}2H_2O$ and $K_4SnS_4{\cdot}4H_2O$, are formed in the Sn–H–K–O–S system. The first of them crystallizes as an orthorhombic structure with the lattice parameters $a = 642.9 \pm 0.4$, $b = 1562.1 \pm 0.7$, $c = 1056.9 \pm 0.6$ pm, and the calculated and experimental densities of 2.059 and 2.06 $\pm$ 0.02 g·cm^{-3}, respectively (Schiwy et al. 1975). To prepare this compound, in a solution of 0.2 M KOH in H_2O (50 mL) was added H_2S up to saturation. The obtained solution was combined with the same amount of aqueous KOH solution and impurities were filtered off. The solution was heated to boiling while stirring constantly and then SnS_2 was added in small portions until sediment remains. The solution was boiled for another 5 min and then the mixture stirred for 1 h. The unsolved was filtered out and the solution was concentrated to about one-third of the original volume. The yellowish viscous liquid was allowed to crystallize in a desiccator in a flow of dry N_2 on a Petri dish. The mother liquor was then filtered off. The $K_2SnS_3{\cdot}2H_2O$, which is highly soluble in water, crystallizes in colorless to light-yellow rhombohedra and is in free air lasts for several days before it weathers.

$K_4SnS_4{\cdot}4H_2O$ crystallizes as a triclinic structure with the lattice parameters $a = 767.86 \pm 0.15$, $b = 778.53 \pm 0.16$, $c = 1220.2 \pm 0.2$ pm, $\alpha = 80.04 \pm 0.03°$, $\beta = 88.76 \pm 0.03°$, $\gamma = 73.78 \pm 0.03°$, and a calculated density of 2.290 g·cm^{-3} (Ruzin et al. 2008). It was prepared by the reaction of K (127.9 mM) with equimolar amounts of Sn in a quartz ampoule at ~700°C for 10 min under vacuum. After cooling to room temperature, equimolar amount of S was added and the mixture was re-melted for 20 min under vacuum. The crude "K_6SnS_6" was crushed to a fine powder, and the excess Sn was removed mechanically as a solid metallic plug. The ternary phase was dissolved in water (50 mL) at 20°C; the solution was stirred for 5 min and filtered off from newly precipitating Sn. The filtrate was evaporated for 24 h, whereupon $K_4SnS_4{\cdot}4H_2O$ was observed as colorless blocks. All synthesis steps were performed with a strong exclusion of air and external moisture (Ar atmosphere at a high-vacuum, double-manifold Schlenk line or N_2 atmosphere in a glove box).

7.12 Tin–Hydrogen–Potassium–Oxygen–Manganese–Sulfur

The $K_{2x}Mn_xSn_{3-x}S_6{\cdot}yH_2O$ ($x = 0.5$–0.95; $y = 2$–5) phases are formed in the Sn–H–K–O–Mn–S system (Manos et al. 2008). Their polycrystalline samples can be synthesized by hydrothermal reaction. Sn (60 mM), Mn (30 mM), K_2CO_3 (30 mM), S (180 mM), and H_2O (40 mL) were mixed in a 125-mL Teflon-lined stainless steel autoclave. The autoclave was sealed and placed in a box furnace with a temperature of 200°C for 4 days. Then the autoclave was allowed to cool to room temperature. A brown polycrystalline product was isolated by filtration, washed several times with H_2O, acetone, and ether, and dried under vacuum.

Single crystals of these phases, suitable for XRD, were obtained by a hydrothermal reaction of K_2S (0.40 mM), $MnCl_2$ (0.20 mM), Sn (0.40 mM), and S (0.40 mM), which were combined and loaded in a Pyrex tube along with 0.30 mL of H_2O under N_2 atmosphere in a glove box. The tube was then evacuated to <0.3 Pa, flame-sealed, and kept in an oven at ~220°C for 14 days. The product was isolated in air by filtration and washed with deionized water, ethanol, and ether. Under microscopic observation, the product consisted of dark-red/black hexagonal plate-like crystals plus unidentified material.

7.13 Tin–Hydrogen–Rubidium–Cadmium–Oxygen–Sulfur

The $Rb(H_3O)[CdSn_2S_6]$ multinary compound is formed in the Sn–H–Rb–Cd–O–S system (Pogu et al. 2019). This compound was prepared as a result of ion-exchange reactions, which were carried out at room temperature by stirring, for 10–12 h, 0.25 g of polycrystalline α-$K_2CdSn_2S_6$ with 20 mL aqueous solution of RbCl (molar ratio 1:5). The ion-exchanged solid product was filtered, washed with deionized water, and dried at room temperature in the open air.

7.14 Tin–Hydrogen–Rubidium–Oxygen–Sulfur

Two quinary compounds, $Rb_2Sn_3S_7·2H_2O$ and $Rb_4SnS_4·4H_2O$, are formed in this system. The first of them crystallizes as an orthorhombic structure with the lattice parameters $a = 369.9 \pm 0.1$, $b = 2588.2 \pm 0.5$, $c = 1717.2 \pm 0.3$ pm, and a calculated density of 3.19 g·cm^{-3} (Sheldrick and Schaaf 1994). To prepare this compound, H_2S was passed through a mixture of SnS_2 (1.25 mM) and Rb_2CO_3 (5.84 mM) in H_2O (1.5 mL) for 1.5 h. The mixture was then melted in a Duranglass ampoule (degree of filling 6%) and heated to 190°C; after 50 h, the ampoule was cooled (2°C·h^{-1}) to room temperature. Yellow crystals of $Rb_2Sn_3S_7·2H_2O$ were formed at the upper phase boundary.

$Rb_4SnS_4·4H_2O$ crystallizes as a triclinic structure with the lattice parameters $a = 789.14 \pm 0.09$, $b = 802.27 \pm 0.08$, $c = 1267.06 \pm 0.13$ pm, $\alpha = 80.091 \pm 0.012°$, $\beta = 89.716 \pm 0.013°$, $\gamma = 72.784 \pm 0.012°$, and a calculated density of 2.911 g·cm^{-3} (Ruzin et al. 2008). It was prepared by the reaction of Rb (58.5 mM) with equimolar amounts of Sn in a quartz ampoule at ~700°C for 10 min under vacuum. After cooling to room temperature, equimolar amount of S was added and the mixture was re-melted for 20 min under vacuum. The crude "Rb_6SnS_6" was crushed to a fine powder, and the excess Sn was removed mechanically as a solid metallic plug. The ternary phase was dissolved in water (50 mL) at 20°C; the solution was stirred for 5 min and filtered off from newly precipitating Sn. The filtrate was evaporated for 24 h, whereupon $Rb_4SnS_4·4H_2O$ was observed as colorless blocks. All synthesis steps were performed with a strong exclusion of air and external moisture (Ar atmosphere at a high-vacuum, double-manifold Schlenk line or N_2 atmosphere in a glove box).

7.15 Tin–Hydrogen–Cesium–Cadmium–Oxygen–Sulfur

The multinary compound $Cs(H_3O)[CdSn_2S_6]$ is formed in the Sn–H–Cs–Cd–O–S system (Pogu et al. 2019). This compound was prepared as a result of ion-exchange reactions, which were carried out at room temperature by stirring, for 10–12 h, 0.25 g of polycrystalline α-$K_2CdSn_2S_6$ with 20 mL aqueous solution of CsCl (molar ratio 1:5). The ion-exchanged solid product was filtered, washed with deionized water and dried at room temperature in the open air.

7.16 Tin–Hydrogen–Cesium–Oxygen–Sulfur

Three quinary compounds are formed in the Sn–H–Cs–O–S system. $Cs_4SnS_4·3H_2O$ crystallizes as an orthorhombic structure with the lattice parameters $a = 1167.4 \pm 0.2$, $b = 1567.1 \pm 0.3$, $c = 1700.2 \pm 0.3$ pm, a calculated density of 3.556 g·cm^{-3}, and an energy gap of 3.43 eV (Ruzin et al. 2008). It was prepared by the reaction of Cs (37.6 mM) with equimolar amounts of Sn in a quartz ampoule at ~700°C for 10 min under vacuum. After cooling to room temperature, equimolar amount of S was added and the mixture was re-melted for 20 min under vacuum. The crude "Cs_6SnS_6" was crushed to a fine powder, and the excess Sn was removed mechanically as a solid metallic plug. The ternary phase was dissolved in water (50 mL) at 20°C; the solution was stirred for 5 min and filtered off from newly precipitating Sn. The filtrate was evaporated for 24 h, whereupon $Cs_4SnS_4·4H_2O$ was observed as colorless blocks. All synthesis steps were performed with a strong exclusion of air and external moisture (Ar atmosphere at a high-vacuum, double-manifold Schlenk line or N_2 atmosphere in a glove box).

$Cs_4Sn_5S_{12}·2H_2O$ also crystallizes as an orthorhombic structure with the lattice parameters $a = 1412.1 \pm 0.5$, $b = 1311.7 \pm 0.2$, $c = 1627.2 \pm 0.7$ pm, and a calculated density of 2.832 g·cm^{-3} (Ko et al. 1994) [$a = 1311.0 \pm 0.3$, $b = 1623.9 \pm 0.3$, $c = 1423.2 \pm 0.2$ pm, and the calculated and experimental densities of 3.39 and 3.38 ± 0.02 g·cm^{-3}, respectively (Sheldrick 1988)]. For preparing this compound, SnS_2 (10.9 mM) and Cs_2CO_3 (10.9 mM) were suspended in H_2O (3 mL) and melted in a Duran glass ampoule (degree of filling 6%) (Sheldrick 1988). The suspension obtained in this way was heated to 130°C for 7 days, at 100°C for one day and then cooled to room temperature at a rate of 5°C·h^{-1}. In this way, yellow tabular air-stable crystals of the title compound were obtained at the upper phase boundary. This compound can be also synthesized from SnS_2, Cs_2CO_3 or CsHS and H_2O (molar ratio 1:1:20) at 200°C (Ko et al. 1994).

$Cs_8Sn_{10}O_4S_{20}{\cdot}13H_2O$ crystallizes as a monoclinic structure with the lattice parameters a = 1590.9, b = 2125.6, c = 1855.7 pm, β = 91.71°, and the calculated and experimental densities of 3.377 and 3.31 g·cm^{-3}, respectively (Schiwy and Krebs 1975a,b). For the preparation of this compound, CsOH (7.5 g) was dissolved in H_2O (30 mL) under nitrogen atmosphere. H_2S was passed into half of the solution until it was saturated. After filtration and combination with the rest of the solution SnS_2 (6 g) was added portion-wise at boiling temperature with the exclusion of air. After stirring for 1 h and subsequent filtration, colorless $Cs_8Sn_{10}O_4S_{20}{\cdot}13H_2O$ was induced to crystallize in a stream of nitrogen.

7.17 Tin–Hydrogen–Cesium–Oxygen–Manganese–Sulfur

In the system containing these elements, the multinary compound $[Cs_{10}(H_2O)_{18}][Mn_4(\mu_4\text{-}S)(SnS_4)_4]$, which crystallizes as a triclinic structure with the lattice parameters a = 1319.0 ± 0.3, b = 1493.6 ± 0.3, c = 1771.0 ± 0.4 pm, α = 77.07 ± 0.03°, β = 74.88 ± 0.03°, γ = 70.70 ± 0.03°, a calculated density of 3.057 g·cm^{-3}, and an energy gap of 2.73 eV, is formed (Ruzin et al. 2008). To prepare this compound, a solution of $Cs_4SnS_4{\cdot}3H_2O$ (0.1 mM) in H_2O (5 mL) was added to a solution of $MnCl_2{\cdot}4H_2O$ (0.1 mM) in MeOH (5 mL), whereupon the reaction solution immediately turned yellow. The reaction solution was stirred for 24 h, a precipitate of MnS was filtered off, and the filtrate was layered by 10 mL of THF. After 2 days, the title compound crystallized as yellow blocks. This compound remains stable under anaerobic, protic conditions. All synthesis steps were performed with a strong exclusion of air and external moisture (Ar atmosphere at a high-vacuum, double-manifold Schlenk line or N_2 atmosphere in a glove box).

7.18 Tin–Hydrogen–Copper–Oxygen–Sulfur

In the system containing these elements, the quinary compound $(H_3O)_4Cu_8Sn_3S_{13}$, which crystallizes as a cubic structure with the lattice parameter a = 1787.8 ± 0.2 pm and a calculated density of 1.452 g·cm^{-3}, is formed (Zhang et al. 2017b). A sample of this compound was prepared by a hydrothermal method in the reaction system of Cu–Sn–$(NH_2)_2CS$–NaOH–H_2O. Typically, a mixture of $(NH_2)_2CS$ (200 mM), Cu powder (10 mM), Sn powder (5 mM), NaOH (400 mM), and 20 mL of distilled H_2O was directly put into a Teflon liner (95 mL), which was sealed into a Teflon-lined stainless steel autoclave and heated at 200°C for 1 day after being stirred for 10 min by a glass bar. The crystal products were washed with distilled water and alcohol a couple of times and dried at 70°C for 6 h.

7.19 Tin–Hydrogen–Copper–Nitrogen–Oxygen–Sulfur

In the system containing these elements, the multinary compound $(H_3O)_{2.7}(NH_4)_{1.3}Cu_8Sn_3S_{13}$ is formed (Zhang et al. 2017b). It was obtained by soaking 0.2 g of $(H_3O)_4Cu_8Sn_3S_{13}$ crystals in 40 mL of 1 M NH_4Cl solution at 60°C for 2 h. The exchanged sample was washed sufficiently with distilled water and dried at 70°C for 6 h.

7.20 Tin–Hydrogen–Silver–Nitrogen–Sulfur

Two quinary compounds, $(NH_4)_2Ag_6Sn_3S_{10}$ and $(NH_4)_4Ag_{12}Sn_7S_{22}$, are formed in the Sn–H–Ag–N–S system. The first compound decomposes at 280°C and crystallizes as an orthorhombic structure with the lattice parameters a = 2385.79 ± 0.13, b = 648.75 ± 0.04, and c = 1339.92 ± 0.07 pm (Baiyin et al. 2005). It was synthesized as follows: Sn (23 mg), $AgNO_3$ (33 mg), $(NH_4)_2CO_3$ (126 mg), and S (25 mg) were put into a glass tube, to which 0.5 mL of mixed solvent with a volume ratio of pyridine/2,3-dimercapto-1-propanol = 2:1 was added. The glass tube was sealed, placed into a Teflon-lined stainless steel autoclave, and heated at 120°C for 7 days. The products were washed with ethanol and water, respectively, and dark-red needle-like crystals of the title compound were obtained.

The second compound crystallizes as a monoclinic structure with the lattice parameters a = 5463.4 ± 0.2, b = 681.75 ± 0.04, c = 1351.76 ± 0.05 pm, β = 99.955 ± 0.004°, a calculated density of 5.270 g·cm^{-3}, and an energy gap of 1.21 eV (Du et al. 2016). To prepare this compound, a mixture of Sn (40 mg), S (110 mg), 4-[4-(dimethylamino)styryl]-1-methylpyridinium iodide (45 mg), AgCl (99 mg), sodium *p*-toluenesulfonate (79 mg), and $N_2H_4 \cdot H_2O$ (0.5 mL) were sealed in a stainless steel reactor with a 20 mL Teflon liner and kept at 160°C for 5–7 days. After the reactor was cooled down to room temperature, black sheet-like crystals of the title compound were isolated.

7.21 Tin–Hydrogen–Calcium–Oxygen–Sulfur

In the system containing these elements, the quinary compound $Ca_3Sn(SO_4)_2(OH)_6 \cdot 3H_2O$ (mineral genplesite), which crystallizes as a hexagonal structure with the lattice parameters a = 851.39 ± 0.02, c = 1114.08 ± 0.03, and the calculated and experimental densities of 2.773 and 2.78 ± 0.01 g·cm^{-3}, respectively, is formed (Pekov et al. 2014, 2018; Belakovskiy and Cámara 2019).

7.22 Tin–Hydrogen–Strontium–Oxygen–Sulfur

Two quinary compounds, $[Sr_2Sn(OH)_6(H_2O)_6]S_4$ and $[Sr_2Sn(OH)_6(H_2O)_5]S_3 \cdot H_2O$, are formed in the Sn–H–Sr–O–S system (Wang et al. 2012). They crystallizes as a triclinic structure with the lattice parameters a = 593.5 ± 0.1, b = 844.0 ± 0.1, c = 946.4 ± 0.2 pm, α = 97.292 ± 0.002°, β = 104.957 ± 0.002°, γ = 110.328 ± 0.002°, and a calculated density of 3.265 g·cm^{-3} for the first compound and a = 1065.8 ± 0.2, b = 1108.2 ± 0.2, c = 1579.7 ± 0.2 pm, α = 73.303 ± 0.004°, β = 74.942 ± 0.004°, γ = 64.627 ± 0.003°, and a calculated density of 3.086 g·cm^{-3} for the second one. In a typical synthesis, granular Sn shot (2 mM, particle size ca. 2 mm), S powder (10 mM), $Sr(OH)_2 \cdot 8H_2O$ (1 mM), and $LiOH \cdot H_2O$ (7 mM) were mixed with deionized water (2 mL). The mixture was sealed in a Teflon-lined autoclave (23 mL) in air and heated at 220°C for 2 days. After cooled to room temperature in air, the product was vacuum-filtered, washed with water and ethanol, and dried in air. Dark brown ribbon-like crystals of the first compound and red rhombic plates of the second one were obtained with typical impurity crystals of $Sr_2Sn(OH)_8$, $SrCO_3$, and Se together with unidentified brown powder. In spite of extensive efforts to improve the synthesis by varying the reagent ratios, no single-phase product was obtained.

7.23 Tin–Hydrogen–Barium–Oxygen–Sulfur

The $Ba_2SnS_4 \cdot 11H_2O$ quinary compound, which crystallizes as a triclinic structure with the lattice parameters a = 763.15 ± 0.06, b = 791.87 ± 0.07, c = 1558.29 ± 0.12 pm, α = 102.851 ± 0.007°, β = 90.523 ± 0.007°, γ = 99.990 ± 0.007°, and a calculated density of 2.647 g·cm^{-3}, is formed in this system (Ruzin et al. 2008). This compound was prepared by the reaction of Ba (36.4 mM) with Sn (72.8 mM) in a quartz ampoule at ~700°C for 10 min under vacuum. To the melt, S (72.8 mM) was added and the mixture was heated for 20 more min under vacuum. The crude "$Ba_3[SnS_6]$" was crushed to a fine powder and the bulk of precipitated Sn was removed mechanically as a solid metallic plug. The ternary phase was dissolved in water (50 mL) at 20°C and the solution was stirred for 30 min and filtered off from precipitating tin. The filtrate was evaporated for 24 h, whereupon the title compound was observed as colorless blocks. All the synthesis steps were performed with a strong exclusion of air and external moisture (Ar atmosphere at a high-vacuum, double-manifold Schlenk line or N_2 atmosphere in a glove box).

7.24 Tin–Hydrogen–Cadmium–Thallium–Oxygen–Sulfur

The multinary compound $Tl(H_3O)[CdSn_2S_6]$ is formed in the Sn–H–Cd–Tl–O–S system (Pogu et al. 2019). This compound was prepared as a result of ion-exchange reactions, which were carried out at room temperature by stirring, for 10–12 h, 0.25 g of polycrystalline α-$K_2CdSn_2S_6$ with 20 mL aqueous solution of $TlNO_3$ (molar ratio 1:5). The ion-exchanged solid product was filtered, washed with deionized water, and dried at room temperature in the open air.

7.25 Tin–Hydrogen–Indium–Europium–Oxygen–Sulfur

At the interaction of $KInSn_2S_6$ with acidic radioactive waste containing Eu, the $[Eu(H_2O)_9]_{0.3}InSn_2S_6$ multinary compound is formed (Xiao et al. 2017).

7.26 Tin–Hydrogen–Nitrogen–Oxygen–Sulfur

Three quinary compounds, $(NH_4)_2[Sn(S_2O_7)_3$, $(NH_4)_4Sn_2S_6{\cdot}3H_2O$ and $(NH_4)_6Sn_3S_9{\cdot}1.3H_2O$, are formed. The first compound crystallizes as a trigonal structure with the lattice parameters a = 992.58 ± 0.10 and c = 1104.88 ± 0.10 pm at 153 K (Logemann et al. 2012). The reaction for obtaining this compound was performed in thick-walled glass ampoule. The tube was loaded with $SnCl_4$ (1 mM), oleum (1 mL, 65% SO_3), and $(NH_4)_2SO_4$ (1 mM), torch-sealed under vacuum, and placed in a resistance furnace. The ampoule was held at a temperature of 250°C for 24 h and cooled down to room temperature at a rate of 1.8°C·h^{-1}. The colorless crystals are very moisture-sensitive and were separated from oleum in a glove box.

$(NH_4)_4Sn_2S_6{\cdot}3H_2O$ crystallizes as a tetragonal structure with the lattice parameters a = 856.294 ± 0.008, c = 2277.03 ± 0.03 pm, and a calculated density of 2.212 g·cm^{-3} at 100 K and a = 859.2 ± 0.2 and c = 2299.6 ± 0.5 pm at room temperature (Nørby et al. 2014). The title compound is an excellent source for an aqueous ammonium thiostannate(IV) based solution. The thermal decomposition and corresponding condensation of crystalline $(NH_4)_4Sn_2S_6{\cdot}3H_2O$ to SnS_2 has been shown to follow a very complex route, including thermally stable and unstable intermediates. All thermal events involve the loss of ammonia, hydrogen sulfide, and water. To prepare this compound, a total of 10 mL of 96% ethanol was added to a total volume of 5 mL of 0.25 M ammonium thiostannate(IV) solution (based on the $[Sn_2S_6]^{4-}$ dimer) made by dissolving amorphous SnS_2 in aqueous $(NH_4)_2S$. Full crystallization occurred within 13 days. Aqueous $(NH_4)_2S$ can be replaced with $(NH_4)_2S_x$ (ammonium polysulfide). Crystals were dried by three times decantation of ethanol and subsequently two times with diethyl ether, which was evaporated under a flow of nitrogen gas.

$(NH_4)_6Sn_3S_9{\cdot}1.3H_2O$ crystallizes as a monoclinic structure with the lattice parameters a = 1698.72 ± 0.1, b = 1054.777 ± 0.007, c = 2108.72 ± 0.02 pm, β = 108.0389 ± 0.0009°, and a calculated density of 2.154 g·cm^{-3} at 100 K (Nørby et al. 2015). Single crystals of this compound were precipitated by the addition of DMF (10 mL) to 0.25 M ammonium thiostannate (IV) solution (5 mL). The crystals were formed after crystallization for 3–6 months at room temperature in a closed tube.

7. 27 Tin–Hydrogen–Nitrogen–Manganese–Sulfur

Three quinary compounds, $Mn_2SnS_4(N_2H_4)_2$, $Mn_2SnS_4(N_2H_4)_5$, and $Mn_2SnS_4(N_2H_4)_6$, are formed in the Sn–H–N–Mn–S system. The first of them is thermally stable up to ~215°C and crystallizes as a triclinic structure with the lattice parameters a = 816.03 ± 0.19, b = 1025.3 ± 0.3, c = 1269.3 ± 0.3 pm, α = 84.54 ± 0.02°, β = 72.840 ± 0.018°, γ = 72.328 ± 0.018°, a calculated density of 2.836 g·cm^{-3}, and an energy gap of ~2.2 eV (Manos et al. 2009). To prepare this compound, Sn (1 mM), Mn (2 mM), and S (4.5 mM), hydrazine monohydrate (2 mL, 98%), and distilled water (2 mL) were mixed in a 23 mL Teflon-lined stainless steel autoclave. The autoclave was sealed and placed in a temperature-controlled oven operated at 150°C. It remained undisturbed at this temperature for 4 days. Then, the autoclave was allowed to cool at room temperature. A yellow crystalline product was isolated by filtration, washed several times with water, acetone, and ether (in this order), and dried under vacuum.

$Mn_2SnS_4(N_2H_4)_5$ and $Mn_2SnS_4(N_2H_4)_6$ crystallize as a monoclinic structure with the lattice parameters $a = 853.5 \pm 0.8$ and 1291.9 ± 0.3, $b = 1330.0 \pm 1.2$ and 893.4 ± 0.2, $c = 1352.2 \pm 1.3$ and 1523.0 ± 0.3 pm, and $\beta = 90.588 \pm 0.016°$ and $106.992 \pm 0.004°$ for the first and the second compound, respectively (Yuan et al. 2007; Yuan and Mitzi 2009). Both compounds were synthesized in hydrazine by dissolving solid mixtures of SnS_2, S, and MnS under an inert atmosphere and stirring at room temperature. Single crystals were grown by diffusing 2-propanol into the filtered reaction mixtures over the time periods of 1–2 weeks. Hydrazine (6 or 4 mL for the first and the second compound, respectively, 98%, anhydrous) was added to a solid mixture of SnS_2 (1 mM), S (4 mM), and MnS (1.4 or 1 mM the first and the second compound, respectively). Gas bubbles were observed upon the addition of hydrazine, and the mixture was stirred at room temperature for 3 days, after which the reaction mixture was filtered through a 0.45 mkm syringe filter and the dark green filtrate was evenly distributed into three test tubes. In each test tube, 3 mL of 2-propanol/hydrazine mixture (volume ratio 1:1) and 12 mL of pure 2-propanol were layered consecutively on top of the filtrate solution for crystal growth. The single crystals were obtained at the bottom and the wall of each test tube after 2 weeks.

7.28 Tin–Lithium–Copper–Manganese–Sulfur

The intermediate lithiated phase $Li_xCu_2MnSn_3S_8$, which crystallizes as a cubic structure with the lattice parameter $a = 1040.5 \pm 0.9$ pm, is formed in the Sn–Li–Cu–Mn–S system (Lavela et al. 1996). It was obtained by discharging a lithium cell at a constant current of 100 $\mu A \cdot cm^{-2}$ for different periods of time. After cutting off the current, the cell voltage was allowed to subside for several days in order to achieve a constant potential and, consequently, a high homogeneity of the lithiated sample. The general stoichiometry of this compound can be expressed as $Li_xCu_{2-y}MnSn_3S_8$.

7.29 Tin–Lithium–Copper–Iron–Sulfur

The intermediate lithiated phase $Li_xCu_2FeSn_3S_8$, which crystallizes as a cubic structure with the lattice parameter $a = 1033.2 \pm 0.7$ pm, is formed in the Sn–Li–Cu–Fe–S system (Lavela et al. 1996). It was prepared in the same way as $Li_xCu_2MnSn_3S_8$ was synthesized. The general stoichiometry of this compound can be expressed as $Li_xCu_{2-y}FeSn_3S_8$. Electrochemical lithiated $Li_xCu_{1.8}FeSn_3S_8$ was also obtained.

7.30 Tin–Lithium–Copper–Cobalt–Sulfur

The intermediate lithiated phase $Li_xCu_2CoSn_3S_8$, which crystallizes as a cubic structure with the lattice parameter $a = 1029.5 \pm 0.6$ pm, is formed in the Sn–Li–Cu–Co–S system (Lavela et al. 1996). It was prepared in the same way as $Li_xCu_2MnSn_3S_8$ was synthesized. The general stoichiometry of this compound can be expressed as $Li_xCu_{2-y}CoSn_3S_8$. Electrochemical lithiated $Li_xCu_{1.1}CoSn_3S_8$ was also obtained.

7.31 Tin–Lithium–Copper–Nickel–Sulfur

The intermediate lithiated phase $Li_xCu_2NiSn_3S_8$, which crystallizes as a cubic structure with the lattice parameter $a = 1028.4 \pm 0.6$ pm, is formed in the Sn–Li–Cu–Ni–S system (Lavela et al. 1996). It was prepared in the same way as $Li_xCu_2MnSn_3S_8$ was synthesized. The general stoichiometry of this compound can be expressed as $Li_xCu_{2-y}NiSn_3S_8$.

7.32 Tin–Lithium–Silver–Zinc–Sulfur

The existence of two quinary phases $Li_xAg_{2-x}ZnSnS_4$ at x = 1.22 and 1.58 has been established in the Sn–Li–Ag–Zn–S system (Zhou et al. 2021b). They crystallize as an orthorhombic structure with the lattice parameters a = 807.25 ± 0.04, b = 676.41 ± 0.03, c = 647.83 ± 0.03 pm, a calculated density of 3.797 g·cm^{-3}, and an energy gap of 2.20 eV for the first phase and a = 802.77 ± 0.08, b = 673.80 ± 0.06, c = 642.76 ± 0.06 pm, a calculated density of 3.526 g·cm^{-3}, and an energy gap of 2.37 eV for the second one. It is possible that these phases are solid solution compositions.

The crystals of these phases were synthesized with the raw materials of Ag, Li_2S, Zn, Sn, S, and KI as a flux via a conventional high-temperature solid state routine. They were unexpected products when $LiAgZnSnS_4$ and $Li_{1.50}Ag_{0.50}ZnSnS_4$ were attempted as end products. For a typical reaction, the reagent mixture was ground thoroughly and transferred to a graphite crucible, which was then sealed into an evacuated quartz tube (10^{-2} Pa) by flame. Next, the reaction was heated to 850°C in 9 h in a muffle furnace, dwelling at that temperature for 72 h to make sure the mixture was completely melted. Then the sample was cooled slowly to 500°C at a rate of 5°C·h^{-1} and finally cooled naturally to room temperature. The products were cleaned by deionized water and ethanol. The red crystals of the first phase and yellow crystals of the second one were obtained, which are stable in air.

7.33 Tin–Lithium–Strontium–Cadmium–Sulfur

In the system containing these elements, the quinary compound $Li_2Sr_6CdSn_4S_{16}$, which crystallizes as a cubic structure with the lattice parameter a = 1423.68 ± 0.06 pm, a calculated density of 3.774 g·cm^{-3}, and an energy gap of 3.42 eV, is formed (Lian et al. 2020). This compound was synthesized from a mixture of stoichiometric SrS, Li_2S, Cd, Sn, and S with a total mass of 0.4 g in a graphite crucible that was flame-sealed inside a silica jacket and preheated at 250°C for 6 h, and then heat to 900°C within 20 h, sintered there for 100 h, and subsequently cooled to 400°C at a rate of 5°C·h^{-1}. Block-shaped saffron yellow crystals of the title compound were obtained. They are stable in the air for at least several months.

7.34 Tin–Lithium–Barium–Zinc–Sulfur

In the system containing these elements, the quinary compound $Li_2Ba_6ZnSn_4S_{16}$, which shows an excellent thermal stability before 900°C and crystallizes as a cubic structure with the lattice parameter a = 1459.24 ± 0.03 pm, a calculated density of 4.042 g·cm^{-3}, and an energy 3.30 eV, is formed (Duan et al. 2018). To prepare this compound, Ba, Li_2S, Zn, Sn, and S (molar ratio 6:1:1:4:15) were put in a graphite crucible. Under the high vacuum of 10^{-3} Pa, the crucible was sealed into a silicon tube, which was loaded in a heater. Then the heater was warmed up to 960°C in 50 h and kept at this temperature for 100 h. Next, the heater was refrigerated to 350°C through 100 h before closed. Finally, this compound was successfully synthesized. It can be stable for at least one year in the air.

7.35 Tin–Lithium–Barium–Cadmium–Sulfur

In the system containing these elements, the quinary compound $Li_2Ba_6CdSn_4S_{16}$, which is stable below 864°C and crystallizes as a cubic structure with the lattice parameter a = 1466.78 ± 0.05 pm, a calculated density of 4.079 g·cm^{-3}, and an energy gap of 3.02 eV, is formed (Duan et al. 2017). This compound was synthesized using a 6:1:1:4:15 molar ratio of Ba/Li_2S/Cd/Sn/S. These reagents were loaded in a graphite crucible that was flame-sealed under a high vacuum of 10^{-3} Pa. The mixture was heated to 950°C in 60 h and held at this temperature for 75 h. Subsequently, it was cooled down to 350°C in 75 h, and then the furnace was shut down. The light-yellow crystals of the title compound and no impure phase were observed. These crystals are stable in air for a long time. All manipulations were performed inside the glove box.

7.36 Tin–Lithium–Barium–Manganese–Sulfur

In the system containing these elements, the quinary compound $Li_2Ba_6MnSn_4S_{16}$, which crystallizes as a cubic structure with the lattice parameter a = 1460.80 ± 0.05 pm, a calculated density of 4.007 g·cm^{-3}, and an energy gap of 2.88 eV, is formed (Duan et al. 2020). To synthesize this compound, all reactants with the molar ratio of Li_2S/Ba/Mn/Sn/S = 1:6:1:4:15 were encased in a graphite crucible, and then sealed under the high vacuum of 10^{-3} Pa. The reactants were heated to 950°C in 70 h and held at this temperature for 120 h. Then the furnace was closed before slowly cooling to 250°C in 100 h. The light-yellow block crystals were obtained. The title compound can be kept in the air for a long time (more than two years). All of the elements were stored in the glove box, which was filled with Ar and controlled with oxygen and moisture levels below 0.1 ppm. All operations were performed inside the glove box.

7.37 Tin–Sodium–Copper–Oxygen–Sulfur

The quinary compound $Na_6CuSn(SO_4)_6$, which crystallizes as a rhombohedral structure with the lattice parameters a = 839.0 ± 0.5 and α = 107°49' ± 6', is formed in the Sn–Na–Cu–O–S system (Perret et al. 1974). To obtain this compound, the mixture of $CuSn(SO_4)_3$ and Na_2SO_4 (molar ratio 1:3) was heated in an evacuated sealed ampoule to 300°C in 40 h.

7.38 Tin–Sodium–Silver–Barium–Sulfur

The quinary compound $Na_{0.74}Ag_{1.26}BaSnS_4$, which crystallizes as a tetragonal structure with the lattice parameters a = 705.22 ± 0.03, c = 806.90 ± 0.06 pm, a calculated density of 4.446 g·cm^{-3}, and an energy gap of 3.70 eV, is formed in the Sn–Na–Ag–Ba–S system (Zhou et al. 2022b). This compound was an unexpected products when targeting to obtain $NaAgBaSnS_4$. Its single crystals were synthesized through a high-temperature solid state reaction. The element reactants were mixed with a stoichiometric ratios of 1:1:1:1:4 for Na:Ag:Ba:Sn:S (a total mass of 500 mg) with additional 400 mg KI flux. The Na and Ba metals were weighed in an Ar-filled glove box to prevent oxidation. The mixture was loaded into a graphite crucible and then transferred into a silica tube, followed by sealing under high vacuum with 10^{-2} Pa by hydrogen–oxygen flame. The tube was heated to 850°C in 9 h, kept for 5 days, and cooled to 300°C in another 5 days. Orange bulk single crystals were obtained.

7.39 Tin–Sodium–Magnesium–Oxygen–Sulfur

The quinary compound $Na_6MgSn(SO_4)_6$, which crystallizes as a rhombohedral structure with the lattice parameters a = 837.6 ± 0.5 and α = 107°36' ± 6', is formed in the Sn–Na–Mg–O–S system (Perret et al. 1974). To obtain this compound, the mixture of $MgSn(SO_4)_3$ and Na_2SO_4 (molar ratio 1:3) was heated in an evacuated sealed ampoule to 300°C in 40 h.

7.40 Tin–Sodium–Barium–Chlorine–Sulfur

In the system containing these elements, the quinary compound $NaBa_2SnS_4Cl$, which crystallizes as a tetragonal structure with the lattice parameters a = 821.620 ± 0.010, c = 1430.22 ± 0.05 pm, a calculated density of 3.991 g·cm^{-3}, and an energy gap of 2.28 eV, is formed (Li et al. 2015a). Polycrystalline sample of this compound was synthesized by a solid state reaction technique. The mixture of BaS, SnS_2, and NaCl according to the stoichiometric ratio was ground and loaded into a fused-silica tube under an Ar atmosphere in a glove box, which were sealed under 10^{-3} Pa and then placed in a furnace. The sample was heated to 900°C in 20 h, kept at that temperature for 48 h, and then the furnace was turned off.

To prepare single crystals of the title compound, the mixture of BaS, SnS_2, and NaCl (molar ratio 1:1:1) was ground and loaded into fused-silica tube under an Ar atmosphere in a glove box, which were sealed under 10^{-3} Pa and then placed in a furnace. The sample was heated to 900°C in 20 h and kept at that temperature for 48 h, then cooled at a slow rate of 4°C·h^{-1} to 400°C, and finally cooled to room temperature. The resultant dark-red crystals were manually selected.

7.41 Tin–Sodium–Zinc–Oxygen–Sulfur

The quinary compound $Na_6ZnSn(SO_4)_6$, which crystallizes as a rhombohedral structure with the lattice parameters a = 838.5 ± 0.5 and α = 107°46' ± 6', is formed in the Sn–Na–Zn–O–S system (Perret et al. 1974). To obtain this compound, the mixture of $ZnSn(SO_4)_3$ and Na_2SO_4 (molar ratio 1:3) was heated in an evacuated sealed ampoule to 300°C in 40 h.

7.42 Tin–Sodium–Cadmium–Oxygen–Sulfur

The quinary compound $Na_6CdSn(SO_4)_6$, which crystallizes as a rhombohedral structure with the lattice parameters a = 851.2 ± 0.5 and α = 108°30' ± 6', is formed in the Sn–Na–Cd–O–S system (Perret et al. 1974). To obtain this compound, the mixture of $CdSn(SO_4)_3$ and Na_2SO_4 (molar ratio 1:3) was heated in an evacuated sealed ampoule to 300°C in 40 h.

7.43 Tin–Sodium–Gallium–Selenium–Sulfur

In the system containing these elements, the quinary compound $Na_2Ga_2SnSSe_5$, which melts congruently at 659°C and crystallizes as an orthorhombic structure with the lattice parameters a = 1326.4 ± 0.4, b = 2393.6 ± 0.7, c = 751.4 ± 0.2 pm, a calculated density of 4.070 g·cm^{-3}, and an energy gap of 1.63 eV, is formed (Li et al. 2018b). For the synthesis of the title compound, a stoichiometric mixture of the starting materials Na_2S, Ga, Sn, and Se (molar ratio 1:2:1:5) were loaded into a graphite crucible and placed in a quartz tube. The tube was flame-sealed under vacuum (~10^{-2} Pa), placed in a furnace, heated from room temperature to 800°C in 40 h, kept at that temperature for 96 h, and then cooled to room temperature at 4°C·h^{-1}. The product was washed with degassed DMF and dried with ethanol. Orange crystals, which are stable in air and water, were obtained.

7.44 Tin–Sodium–Phosphorus–Chlorine–Sulfur

In the system containing these elements, the quinary compound $Na_{3.8}Sn_{0.9}P_{0.1}S_{3.9}Cl_{0.1}$, which crystallizes as a tetragonal structure with the lattice parameters a = 1380.4 ± 0.2 and c = 2744.0 ± 0.1 pm, is formed (Xiong et al. 2020). This phase was prepared via solid state reactions. The starting materials including Na_2S, SnS_2, P_2S_5, and NaCl were weighed and mixed with stoichiometric ratios. Na_2S was used with 5% excess to compensate the loss during heating. The mixture was ball-milled at 300 rpm for 16 h and then pressed into pellets with a diameter of 0.5 in. Each pellet weighted about 0.3 g. The pellets were loaded into a quartz tube in an Ar-filled glove box, and then the tube was sealed. The tube was then heated to 550°C (high-temperature synthesis) or 300°C (low-temperature synthesis) with a rate of 3°C·min^{-1} in a box furnace, left to stand for 24 h, and quenched on a copper plate to room temperature. The copper plate here serves as a good thermal conductor to facilitate the fast cooling process and the samples will typically be cooled to room temperature in 10–15 min. Then the tube was transferred into the glove box and the product powder was obtained through hand grinding with a mortar and a pestle after breaking the tube. The synthesized sample was stored in the Ar-filled glove box to avoid exposure to air and moisture.

7.45 Tin–Sodium–Phosphorus–Bromine–Sulfur

In the system containing these elements, the quinary compound $Na_{3.8}Sn_{0.9}P_{0.1}S_{3.9}Br_{0.1}$, which crystallizes as a tetragonal structure with the lattice parameters a = 1379.2 ± 0.7 and c = 2744.0 ± 0.1 pm, is formed (Xiong et al. 2020). This phase was synthesized and stored in the same way as $Na_{3.8}Sn_{0.9}P_{0.1}S_{3.9}Cl_{0.1}$ was prepared using NaBr instead of NaCl.

7.46 Tin–Sodium–Antimony–Chlorine–Sulfur

In the system containing these elements, the quinary phase $Na_{3.7}Sn_{0.8}Sb_{0.2}S_{3.9}Cl_{0.1}$, which crystallizes as a tetragonal structure with the lattice parameters a = 1383.9 ± 0.2 and c = 2759.0 ± 0.3 pm, is formed (Xiong et al. 2020). This phase was synthesized and stored in the same way as $Na_{3.8}Sn_{0.9}P_{0.1}S_{3.9}Cl_{0.1}$ was prepared by using Sb_2S_3 instead of P_2S_5.

7.47 Tin–Sodium–Antimony–Bromine–Sulfur

In the system containing these elements, the quinary phase $Na_{3.7}Sn_{0.8}Sb_{0.2}S_{3.9}Br_{0.1}$, which crystallizes as a tetragonal structure with the lattice parameters a = 1385.0 ± 0.4 and c = 2760.6 ± 0.4 pm, is formed (Xiong et al. 2020). This phase was synthesized and stored in the same way as $Na_{3.8}Sn_{0.9}P_{0.1}S_{3.9}Cl_{0.1}$ was prepared using NaBr instead of NaCl and Sb_2S_3 instead of P_2S_5.

7.48 Tin–Sodium–Manganese–Oxygen–Sulfur

The quinary compound $Na_6MnSn(SO_4)_6$, which crystallizes as a rhombohedral structure with the lattice parameters a = 842.8 ± 0.5 and α = 107°40′ ± 6′, is formed in the Sn–Na–Mn–O–S system (Perret et al. 1974). To obtain this compound, the mixture of $MnSn(SO_4)_3$ and Na_2SO_4 (molar ratio 1:3) was heated in an evacuated sealed ampoule to 300°C in 40 h.

7.49 Tin–Sodium–Cobalt–Oxygen–Sulfur

The quinary compound $Na_6CoSn(SO_4)_6$, which crystallizes as a rhombohedral structure with the lattice parameters a = 839.3 ± 0.5 and α = 107°44' ± 6', is formed in the Sn–Na–Co–O–S system (Perret et al. 1974). To obtain this compound, the mixture of $CoSn(SO_4)_3$ and Na_2SO_4 (molar ratio 1:3) was heated in an evacuated sealed ampoule to 300°C in 40 h.

7.50 Tin–Sodium–Nickel–Oxygen–Sulfur

The quinary compound $Na_6NiSn(SO_4)_6$, which crystallizes as a rhombohedral structure with the lattice parameters a = 837.5 ± 0.5 and α = 107°44' ± 6', is formed in the Sn–Na–Ni–O–S system (Perret et al. 1974). To obtain this compound, the mixture of $NiSn(SO_4)_3$ and Na_2SO_4 (molar ratio 1:3) was heated in an evacuated sealed ampoule to 300°C in 40 h.

7.51 Tin–Potassium–Rubidium–Cesium–Zinc–Sulfur

The multinary compound $K_{1.8}Rb_{3.1}Cs_{0.5}Zn_4Sn_{4.8}S_{17}$ is formed in the Sn–K–Rb–Cs–Zn–S system (Manos et al. 2006). To prepare this compound, in a suspension of $K_6Sn[Zn_4Sn_4S_{17}]$ (0.047 mM) in water (20 mL), CsCl (0.047 mM), and RbI (0.47 mM) were added as solids. The mixture was kept under magnetic stirring at room temperature for ~12 h. Then, the yellowish-white crystalline material was isolated by filtration and washed several times with water, acetone, and ether. The title compound was obtained.

7.52 Tin–Potassium–Rubidium–Zinc–Sulfur

Two quinary phases, $KRb_5Sn[Zn_4Sn_4S_{17}]$ and $K_5RbSn[Zn_4Sn_4S_{17}]$, are formed in the Sn–K–Rb–Zn–S system (Manos et al. 2005). Both phases crystallize as a tetragonal structure with the lattice parameters $a = 1383.54 \pm 0.10$ and $c = 988.93 \pm 0.09$ pm for the first phase and $a = 1376.87 \pm 0.10$, $c = 977.67 \pm 0.06$ pm, and an energy gap of ≈2.87 eV for the second one.

A typical ion-exchange experimental route for the preparation of first phase is as follows. An excess of solid RbI (0.2 mm) was added to a suspension of $K_6Sn[Zn_4Sn_4S_{17}]$ (0.01 mM) in water (20 mL). The mixture was stirred for ~12 h. Then, the yellowish-white crystalline material was isolated by filtration, washed several times with water, acetone, and ether (in this order), and dried in air.

For obtaining $K_5RbSn[Zn_4Sn_4S_{17}]$, a mixture of Sn (0.5 mM), Zn (0.4 mM), K_2S (0.9 mM), Rb_2S (0.1 mM), and S (8 mM) was sealed under vacuum (≈10^{-2} Pa) in a silica tube, heated (≈40°C·h^{-1}) to 400°C for 96 h, and then cooled to room temperature at a rate of 6°C·h^{-1}. The excess flux was removed with DMF to reveal a mixture of pale yellow polyhedral crystals of the title compound and orange crystals of $K_2Sn_2S_5$. The ternary phase was removed by washing with K_2CO_3 (pH = 9–10).

7.53 Tin–Potassium–Cesium–Zinc–Sulfur

Two quinary phases, $K_{4.4}Cs_{1.2}Zn_{4.8}Sn_5S_{17}$ and $K_5CsSn[Zn_4Sn_4S_{17}]$, are formed in the Sn–K–Cs–Zn–S system. The first phase was prepared as follows. In a suspension of $K_6Sn[Zn_4Sn_4S_{17}]$ (0.012 mM) in water (10 mL), excess CsCl (0.36 mM) was added as a solid (Manos et al. 2006). The mixture was kept under magnetic stirring at room temperature for ~12 h. Then, the yellowish-white crystalline material was isolated by filtration and washed several times with water, acetone, and ether. The title compound was obtained.

$K_5CsSn[Zn_4Sn_4S_{17}]$ crystallizes as a tetragonal structure with the lattice parameters $a = 1376.84 \pm 0.11$, $c = 956.30 \pm 0.09$ pm, and energy gap of ≈ 2.87 eV (Manos et al. 2005) [$a = 1384.3 \pm 0.2$ and $c = 969.0 \pm 0.3$ pm (Manos et al. 2006)]. For obtaining it, single crystals of $K_6Sn[Zn_4Sn_4S_{17}]$ were placed in a water solution of CsCl (0.39 mM) (Manos et al. 2006). Then, the mixture was heated at 70°C and kept undisturbed at this temperature for 2 weeks. The crystals of the title compound were then isolated by filtration, washed several times with water, acetone, and ether. This compound was also synthesized by the solid state reaction in the same way as $K_5RbSn[Zn_4Sn_4S_{17}]$ was prepared by using Cs_2S instead of Rb_2S (Manos et al. 2005).

7.54 Tin–Potassium–Copper–Oxygen–Sulfur

Two quinary compounds, $K_2CuSn(SO_4)_4$ and $K_6CuSn(SO_4)_6$, are formed in this system (Perret et al. 1974). Both compounds crystallize in the rhombohedral structure with the lattice parameters $a = 845.8 \pm 0.5$ and $\alpha = 33°05' \pm 6'$ for the first compound and $a = 899.6 \pm 0.5$ and $\alpha = 109°44' \pm 6'$ for the second one. $K_2CuSn(SO_4)_4$ and $K_6CuSn(SO_4)_6$ were prepared from the mixtures of $CuSn(SO_4)_3$ and K_2SO_4 (molar ratio 1:1 and 1:3, respectively), which were heated in an evacuated sealed ampoule to 300°C in 40 h.

7.55 Tin–Potassium–Magnesium–Oxygen–Sulfur

Two quinary compounds, $K_2MgSn(SO_4)_4$ and $K_6MgSn(SO_4)_6$, are formed in this system (Perret et al. 1974). Both compounds crystallize in the rhombohedral structure with the lattice parameters $a = 846.1$

± 0.5 and α = 33°09′ ± 6′ for the first compound and a = 898.9 ± 0.5 and α = 109°25′ ± 6′ for the second one. $K_2MgSn(SO_4)_4$ and $K_6MgSn(SO_4)_6$ were prepared from the mixtures of $MgSn(SO_4)_3$ and K_2SO_4 (molar ratio 1:1 and 1:3, respectively), which were heated in an evacuated sealed ampoule to 300°C in 40 h.

7.56 Tin–Potassium–Barium–Chlorine–Sulfur

In the system containing these elements, the quinary compound KBa_2SnS_4Cl, which crystallizes as a tetragonal structure with the lattice parameters a = 835.52 ± 0.07, c = 1445.4 ± 0.2 pm, a calculated density of 3.924 g·cm^{-3}, and an energy gap of 2.30 eV, is formed (Li et al. 2015a). Polycrystalline samples and single crystals of this compound were prepared in the same way as in the case of $NaBa_2SnS_4Cl$ using KCl instead of NaCl.

7.57 Tin–Potassium–Barium–Bromine–Sulfur

In the system containing these elements, the quinary compound KBa_2SnS_4Br, which crystallizes as an orthorhombic structure with the lattice parameters a = 1251.24 ± 0.04, b = 982.75 ± 0.02, c = 881.42 ± 0.03 pm, a calculated density of 3.926 g·cm^{-3}, and an energy gap 1.95 eV, is formed (Li et al. 2015a). Polycrystalline samples and single crystals of this compound were prepared in the same way as in the case of $NaBa_2SnS_4Cl$ using KBr instead of NaCl.

7.58 Tin–Potassium–Zinc–Oxygen–Sulfur

Two quinary compounds, $K_2ZnSn(SO_4)_4$ and $K_6ZnSn(SO_4)_6$, are formed in this system (Perret et al. 1974). Both compounds crystallize as a rhombohedral structure with the lattice parameters a = 846.5 ± 0.5 and α = 33°03' ± 6' for the first compound and a = 896.8 ± 0.5 and α = 109°33' ± 6' for the second one. $K_2ZnSn(SO_4)_4$ and $K_6ZnSn(SO_4)_6$ were prepared from the mixtures of $ZnSn(SO_4)_3$ and K_2SO_4 (molar ratio 1:1 and 1:3, respectively), which were heated in an evacuated sealed ampoule to 300°C in 40 h.

7.59 Tin–Potassium–Cadmium–Oxygen–Sulfur

Two quinary compounds, $K_2CdSn(SO_4)_4$ and $K_6CdSn(SO_4)_6$, are formed in this system (Perret et al. 1974). Both compounds crystallize in the rhombohedral structure with the lattice parameters a = 864.1 ± 0.5 and α = 32°39' ± 6' for the first compound and a = 905.4 ± 0.5 and α = 109°23' ± 6' for the second one. $K_2CdSn(SO_4)_4$ and $K_6CdSn(SO_4)_6$ were prepared from the mixtures of $CdSn(SO_4)_3$ and K_2SO_4 (molar ratio 1:1 and 1:3, respectively), which were heated in an evacuated sealed ampoule to 300°C in 40 h.

7.60 Tin–Potassium–Manganese–Oxygen–Sulfur

Two quinary compounds, $K_2MnSn(SO_4)_4$ and $K_6MnSn(SO_4)_6$, are formed in this system (Perret et al. 1974). Both compounds crystallize in the rhombohedral structure with the lattice parameters a = 851.8 ± 0.5 and α = 33°10' ± 6' for the first compound and a = 901.0 ± 0.5 and α = 109°21' ± 6' for the second one. $K_2MnSn(SO_4)_4$ and $K_6MnSn(SO_4)_6$ were prepared from the mixtures of $MnSn(SO_4)_3$ and K_2SO_4 (molar ratio 1:1 and 1:3, respectively), which were heated in an evacuated sealed ampoule to 300°C in 40 h.

7.61 Tin–Potassium–Cobalt–Oxygen–Sulfur

Two quinary compounds, $K_2CoSn(SO_4)_4$ and $K_6CoSn(SO_4)_6$, are formed in this system (Perret et al. 1974). Both compounds crystallize in the rhombohedral structure with the lattice parameters a = 845.3 ± 0.5 and α = 33°09' ± 6' for the first compound and a = 898.1 ± 0.5 and α = 109°40' ± 6' for the second one. $K_2CoSn(SO_4)_4$ and $K_6CoSn(SO_4)_6$ were prepared from the mixtures of $CoSn(SO_4)_3$ and K_2SO_4 (molar ratio 1:1 and 1:3, respectively), which were heated in an evacuated sealed ampoule to 300°C in 40 h.

7.62 Tin–Potassium–Nickel–Oxygen–Sulfur

Two quinary compounds, $K_2NiSn(SO_4)_4$ and $K_6NiSn(SO_4)_6$, are formed in this system (Perret et al. 1974). Both compounds crystallize as a rhombohedral structure with the lattice parameters a = 846.0 ± 0.5 and α = 32°55' ± 6' for the first compound and a = 898.0 ± 0.5 and α = 109°37' ± 6' for the second one. $K_2NiSn(SO_4)_4$ and $K_6NiSn(SO_4)_6$ were prepared from the mixtures of $NiSn(SO_4)_3$ and K_2SO_4 (molar ratio 1:1 and 1:3, respectively), which were heated in an evacuated sealed ampoule to 300°C in 40 h.

7.63 Tin–Rubidium–Copper–Oxygen–Sulfur

Two quinary compounds, $Rb_2CuSn(SO_4)_4$ and $Rb_6CuSn(SO_4)_6$, are formed in this system (Perret et al. 1974). Both compounds crystallize as a rhombohedral structure with the lattice parameters a = 884.5 ± 0.5 and α = 31°43' ± 6' for the first compound and a = 937.3 ± 0.5 and α = 110°21' ± 6' for the second one. $Rb_2CuSn(SO_4)_4$ and $Rb_6CuSn(SO_4)_6$ were prepared from the mixtures of $CuSn(SO_4)_3$ and Rb_2SO_4 (molar ratio 1:1 and 1:3, respectively), which were heated in an evacuated sealed ampoule to 300°C in 40 h.

7.64 Tin–Rubidium–Magnesium–Oxygen–Sulfur

Two quinary compounds, $Rb_2MgSn(SO_4)_4$ and $Rb_6MgSn(SO_4)_6$, are formed in this system (Perret et al. 1974). Both compounds crystallize as a rhombohedral structure with the lattice parameters a = 878.1 ± 0.5 and α = 32°06' ± 6' for the first compound and a = 930.5 ± 0.5 and α = 110°12' ± 6' for the second one. $Rb_2MgSn(SO_4)_4$ and $Rb_6MgSn(SO_4)_6$ were prepared from the mixtures of $MgSn(SO_4)_3$ and Rb_2SO_4 (molar ratio 1:1 and 1:3, respectively), which were heated in an evacuated sealed ampoule to 300°C in 40 h.

7.65 Tin–Rubidium–Barium–Chlorine–Sulfur

In the system containing these elements, the quinary compound $RbBa_2SnS_4Cl$, which crystallizes as a monoclinic structure with the lattice parameters a = 988.3 ± 0.6, b = 880.6 ± 0.5, c = 1243.7 ± 0.7 pm, β = 90.144 ± 0.007°, a calculated density of 3.943 g·cm^{-3}, and an energy gap of 2.82 eV, is formed (Chu et al. 2018). In the preparation process of this compound, a graphite crucible was added into the vacuum-sealed silica tube evacuated to 10^{-3} Pa to avoid the reaction between RbCl and silica tube at a high temperature. $RbBa_2SnS_4Cl$ was prepared with a mixture under an Ar atmosphere in a glove box at the stoichiometric ratio of RbCl/BaS/Sn/S = 1:2:1:2. The process was set as follows: firstly, the furnace was heated to 800°C in 40 h, and kept at this temperature for about 3 days, then slowly cooled down to 300°C with a rate of 5°C·h^{-1}, and eventually, swiftly cooled to room temperature. The obtained product was washed with DMF solvent. Yellow crystals of the title compound were formed, which are stable in air.

7.66 Tin–Rubidium–Zinc–Oxygen–Sulfur

Two quinary compounds, $Rb_2ZnSn(SO_4)_4$ and $Rb_6ZnSn(SO_4)_6$, are formed in this system (Perret et al. 1974). Both compounds crystallize in the rhombohedral structure with the lattice parameters $a = 879.6 \pm 0.5$ and $\alpha = 32°00' \pm 6'$ for the first compound and $a = 937.4 \pm 0.5$ and $\alpha = 110°15' \pm 6'$ for the second one. $Rb_2ZnSn(SO_4)_4$ and $Rb_6ZnSn(SO_4)_6$ were prepared from the mixtures of $ZnSn(SO_4)_3$ and Rb_2SO_4 (molar ratio 1:1 and 1:3, respectively), which were heated in an evacuated sealed ampoule to 300°C in 40 h.

7.67 Tin–Rubidium–Cadmium–Oxygen–Sulfur

Two quinary compounds, $Rb_2CdSn(SO_4)_4$ and $Rb_6CdSn(SO_4)_6$, are formed in this system (Perret et al. 1974). Both compounds crystallize in the rhombohedral structure with the lattice parameters $a = 890.3 \pm 0.5$ and $\alpha = 32°00' \pm 6'$ for the first compound and $a = 937.3 \pm 0.5$ and $\alpha = 110°15' \pm 6'$ for the second one. $Rb_2CdSn(SO_4)_4$ and $Rb_6CdSn(SO_4)_6$ were prepared from the mixtures of $CdSn(SO_4)_3$ and Rb_2SO_4 (molar ratio 1:1 and 1:3, respectively), which were heated in an evacuated sealed ampoule to 300°C in 40 h.

7.68 Tin–Rubidium–Manganese–Oxygen–Sulfur

Two quinary compounds, $Rb_2MnSn(SO_4)_4$ and $Rb_6MnSn(SO_4)_6$, are formed in this system (Perret et al. 1974). Both compounds crystallize as a rhombohedral structure with the lattice parameters $a = 881.5 \pm 0.5$ and $\alpha = 32°19' \pm 6'$ for the first compound and $a = 933.6 \pm 0.5$ and $\alpha = 110°09' \pm 6'$ for the second one. $Rb_2MnSn(SO_4)_4$ and $Rb_6MnSn(SO_4)_6$ were prepared from the mixtures of $MnSn(SO_4)_3$ and Rb_2SO_4 (molar ratio 1:1 and 1:3, respectively), which were heated in an evacuated sealed ampoule to 300°C in 40 h.

7.69 Tin–Rubidium–Cobalt–Oxygen–Sulfur

Two quinary compounds, $Rb_2CoSn(SO_4)_4$ and $Rb_6CoSn(SO_4)_6$, are formed in this system (Perret et al. 1974). Both compounds crystallize as a rhombohedral structure with the lattice parameters $a = 877.6 \pm 0.5$ and $\alpha = 32°04' \pm 6'$ for the first compound and $a = 931.5 \pm 0.5$ and $\alpha = 110°22' \pm 6'$ for the second one. $Rb_2CoSn(SO_4)_4$ and $Rb_6CoSn(SO_4)_6$ were prepared from the mixtures of $CoSn(SO_4)_3$ and Rb_2SO_4 (molar ratio 1:1 and 1:3, respectively), which were heated in an evacuated sealed ampoule to 300°C in 40 h.

7.70 Tin–Rubidium–Nickel–Oxygen–Sulfur

Two quinary compounds, $Rb_2NiSn(SO_4)_4$ and $Rb_6NiSn(SO_4)_6$, are formed in this system (Perret et al. 1974). Both compounds crystallize in the rhombohedral structure with the lattice parameters $a = 879.6 \pm 0.5$ and $\alpha = 31°52' \pm 6'$ for the first compound and $a = 931.7 \pm 0.5$ and $\alpha = 110°30' \pm 6'$ for the second one. $Rb_2NiSn(SO_4)_4$ and $Rb_6NiSn(SO_4)_6$ were prepared from the mixtures of $NiSn(SO_4)_3$ and Rb_2SO_4 (molar ratio 1:1 and 1:3, respectively), which were heated in an evacuated sealed ampoule to 300°C in 40 h.

7.71 Tin–Cesium–Copper–Oxygen–Sulfur

Two quinary compounds, $Cs_2CuSn(SO_4)_4$ and $Cs_6CuSn(SO_4)_6$, are formed in this system (Perret et al. 1974). Both compounds crystallize as a rhombohedral structure with the lattice parameters $a = 928.8 \pm$

0.5 and α = 30°35' ± 6' for the first compound and *a* = 972.2 ± 0.5 and α = 110°47' ± 6' for the second one. $Cs_2CuSn(SO_4)_4$ and $Cs_6CuSn(SO_4)_6$ were prepared from the mixtures of $CuSn(SO_4)_3$ and Cs_2SO_4 (molar ratio 1:1 and 1:3, respectively), which were heated in an evacuated sealed ampoule to 300°C in 40 h.

7.72 Tin–Cesium–Magnesium–Oxygen–Sulfur

Two quinary compounds, $Cs_2MgSn(SO_4)_4$ and $Cs_6MgSn(SO_4)_6$, are formed in this system (Perret et al. 1974). Both compounds crystallize as a rhombohedral structure with the lattice parameters *a* = 926.1 ± 0.5 and α = 30°45' ± 6' for the first compound and *a* = 971.4 ± 0.5 and α = 110°40' ± 6' for the second one. $Cs_2MgSn(SO_4)_4$ and $Cs_6MgSn(SO_4)_6$ were prepared from the mixtures of $MgSn(SO_4)_3$ and Cs_2SO_4 (molar ratio 1:1 and 1:3, respectively), which were heated in an evacuated sealed ampoule to 300°C in 40 h.

7.73 Tin–Cesium–Barium–Chlorine–Sulfur

The $CsBa_2SnS_4Cl$ quinary compound, which has two polymorphic modifications, is formed in this system. The first modification crystallizes as a monoclinic structure with the lattice parameters *a* = 992.4 ± 0.3, *b* = 879.0 ± 0.3, *c* = 1259.8 ± 0.4 pm, β =90.35 ± 0.02°, a calculated density of 4.170 g·cm^{-3}, and an energy gap of 2.60 eV (Chu et al. 2018) [*a* = 994.6 ± 0.2, *b* = 879.3 ± 0.2, *c* = 1253.2 ± 0.3 pm, β = 90.75 ± 0.03°, a calculated density of 4.182 g·cm^{-3}, and an energy gap of 2.06 eV (Li et al. 2015a)]. α-$CsBa_2SnS_4Cl$ was synthesized in the same way as $RbBa_2SnS_4Cl$ was prepared by using CsCl instead of RbCl (Chu et al. 2018). Polycrystalline samples and single crystals of this compound were also obtained in the same way as in the case of $NaBa_2SnS_4Cl$ using CsCl instead of NaCl (Li et al. 2015a).

β-$CsBa_2SnS_4Cl$ crystallizes as a tetragonal structure with the lattice parameters *a* = 848.7 ± 0.3, *c* = 1433.5 ± 0.9 pm, and a calculated density of 4.438 g·cm^{-3} (Chu et al. 2018). To obtain β-$CsBa_2SnS_4Cl$, the initial reagents were the mixture of CsCl, BaS, Sn, and S (molar ratio 1:4:3:6). The temperature process is similar to that of α-form. After washing with the DMF solvent, yellow crystals for this modification were found, which are also stable in the air for several months. Unfortunately, the α-form often coexists with the β-form and it is difficult to separate them. Adjusting the ratio of reactants and the reaction temperature is still difficult to increase the yield of the target compound.

7.74 Tin–Cesium–Zinc–Oxygen–Sulfur

Two quinary compounds, $Cs_2ZnSn(SO_4)_4$ and $Cs_6ZnSn(SO_4)_6$, are formed in this system (Perret et al. 1974). Both compounds crystallize in the rhombohedral structure with the lattice parameters *a* = 925.4 ± 0.5 and α = 30°33' ± 6' for the first compound and *a* = 970.6 ± 0.5 and α = 110°42' ± 6' for the second one. $Cs_2ZnSn(SO_4)_4$ and $Cs_6ZnSn(SO_4)_6$ were prepared from the mixtures of $ZnSn(SO_4)_3$ and Cs_2SO_4 (molar ratio 1:1 and 1:3, respectively), which were heated in an evacuated sealed ampoule to 300°C in 40 h.

7.75 Tin–Cesium–Cadmium–Oxygen–Sulfur

Two quinary compounds, $Cs_2CdSn(SO_4)_4$ and $Cs_6CdSn(SO_4)_6$, are formed in this system (Perret et al. 1974). Both compounds crystallize as a rhombohedral structure with the lattice parameters *a* = 934.7 ± 0.5 and α = 30°41' ± 6' for the first compound and *a* = 966.3 ± 0.5 and α = 110°26' ± 6' for the second one. $Cs_2CdSn(SO_4)_4$ and $Cs_6CdSn(SO_4)_6$ were prepared from the mixtures of $CdSn(SO_4)_3$ and Cs_2SO_4 (molar ratio 1:1 and 1:3, respectively), which were heated in an evacuated sealed ampoule to 300°C in 40 h.

7.76 Tin–Cesium–Manganese–Oxygen–Sulfur

Two quinary compounds, $Cs_2MnSn(SO_4)_4$ and $Cs_6MnSn(SO_4)_6$, are formed in this system (Perret et al. 1974). Both compounds crystallize as a rhombohedral structure with the lattice parameters a = 926.0 ± 0.5 and α = 30°53' ± 6' for the first compound and a = 964.9 ± 0.5 and α = 110°18' ± 6' for the second one. $Cs_2MnSn(SO_4)_4$ and $Cs_6MnSn(SO_4)_6$ were prepared from the mixtures of $MnSn(SO_4)_3$ and Cs_2SO_4 (molar ratio 1:1 and 1:3, respectively), which were heated in an evacuated sealed ampoule to 300°C in 40 h.

7.77 Tin–Cesium–Cobalt–Oxygen–Sulfur

Two quinary compounds, $Cs_2CoSn(SO_4)_4$ and $Cs_6CoSn(SO_4)_6$, are formed in this system (Perret et al. 1974). Both compounds crystallize as a rhombohedral structure with the lattice parameters a = 926.5 ± 0.5 and α = 30°38' ± 6' for the first compound and a = 972.3 ± 0.5 and α = 110°42' ± 6' for the second one. $Cs_2CoSn(SO_4)_4$ and $Cs_6CoSn(SO_4)_6$ were prepared from the mixtures of $CoSn(SO_4)_3$ and Cs_2SO_4 (molar ratio 1:1 and 1:3, respectively), which were heated in an evacuated sealed ampoule to 300°C in 40 h.

7.78 Tin–Cesium–Nickel–Oxygen–Sulfur

Two quinary compounds, $Cs_2NiSn(SO_4)_4$ and $Cs_6NiSn(SO_4)_6$, are formed in this system (Perret et al. 1974). Both compounds crystallize as a rhombohedral structure with the lattice parameters a = 927.8 ± 0.5 and α = 30°29' ± 6' for the first compound and a = 971.9 ± 0.5 and α = 110°40' ± 6' for the second one. $Cs_2NiSn(SO_4)_4$ and $Cs_6NiSn(SO_4)_6$ were prepared from the mixtures of $NiSn(SO_4)_3$ and Cs_2SO_4 (molar ratio 1:1 and 1:3, respectively), which were heated in an evacuated sealed ampoule to 300°C in 40 h.

7.79 Tin–Copper–Silver–Barium–Sulfur

Cu_2BaSnS_4–Ag_2BaSnS_4. The X-ray investigation of the $Cu_{2-x}Ag_xBaSnS_4$ alloys ($0 \leq x \leq 2$) yields a homogeneity range of $0 \leq x \leq 0.65$ for Cu_2BaSnS_4 (hexagonal structure) and $1.7 < x \leq 2$ for Ag_2BaSnS_4 (orthorhombic structure) (Teske and Vetter 1976). To prepare the samples required for the X-ray analysis of $Cu_{2-x}Ag_xBaSnS_4$, mixtures of Ag_2S and Cu_2S in the desired molar ratio were first annealed with the corresponding quantity of SnS_2 at the temperature < 400°C. The fine-grained and uniformly black powder obtained in this way was then triturated with BaS, and the mixtures were tempered at 700°C–740°C in alumina crucibles, which were inserted into sealed quartz ampoules. The annealing processes had to be interrupted several times for renewed grinding, and cooling was always very slow. Because the preparations are sensitive to moisture at higher temperatures, all operations must be carried out in a dry inert gas atmosphere.

7.80 Tin–Copper–Silver–Zinc–Indium–Iron–Sulfur

In the system containing these elements, the multinary compound $(Cu,Fe,Zn,Ag)_3(Sn,In)S_4$ (mineral petrukite), which crystallizes as an orthorhombic structure with the lattice parameters a = 767.7 ± 0.4, b = 644.2 ± 0.6, and c = 626.4 ± 0.3 pm (Radosavljevic et al. 2005) [a = 766.71 ± 0.81, 770.50 ± 0.27, 768.58 ± 1.01, b = 643.99 ± 0.32, 644.62 ± 0.14, 643.95 ± 0.60, and c = 626.05 ± 0.61, 627.57 ± 0.26, 629.47 ± 0.23 pm for three various samples (Kissin and Owens de 1989; Jambor and Burke 1990)], is formed.

7.81 Tin–Copper–Magnesium–Barium–Sulfur

In the system containing these elements, the quinary compound $Cu_{1.94}Mg_{1.06}Ba_6Sn_4S_{16}$, which crystallizes as a cubic structure with the lattice parameter a = 1456.49 ± 0.10 pm and a calculated density of 4.215 g·cm^{-3}, is formed (Ji et al. 2022b). For preparing this compound, the reactant elements were loaded in a molar ratio of Ba:Cu:Mg:Sn:S = 6:2:1:4:16 and mixed with KI salt flux in carbonized silica ampoule. Element mixtures of 0.4 g were mixed with 0.4 g of KI and loaded into carbonized silica tube. Then, the silica tube was flamed-sealed under a vacuum of 1 Pa and heated within a programmable box furnace. The sample was heated to 800°C in 10 h, held for 48 h at this temperature, and then slowly cooled to 650°C within 100 h. The sample was kept for 10 h at 650°C before the furnace was shut off. The KI flux was removed with deionized water and yellow compound was obtained. This compound is stable in air and water.

7.82 Tin–Copper–Strontium–Iron–Sulfur

In the system containing these elements, the quinary compound $Cu_2Sr_6FeSn_4S_{16}$, which crystallizes as a cubic structure with the lattice parameter a = 1413.49 ± 0.06 pm, a calculated density of 3.99 g·cm^{-3}, and an energy gap of 1.53 eV, is formed (Zhang et al. 2019a). The synthesis of the title compound involves two steps. In the first step, binary FeS was prepared. The reagents were loaded into silica tube, sealed under vacuum (<10^{-3} Pa), and heated to 600°C. In the second step, appropriate amounts of SrS, Cu, FeS, Sn, and S (molar ratio of 6:2:1:4:9) were loaded into the silica tube. The tube was sealed and placed into a furnace. The sample was heated to 800°C in 24 h, kept at that temperature for 24 h, and cooled at a rate of 5°C·h^{-1} to room temperature.

To prepare the single crystal of the title compound, a mixture of the as-prepared $Cu_2Sr_6FeSn_4S_{16}$ powder and KI (mass ratio 1:1) was mixed and loaded into the quartz tube, which was sealed (<10^{-3} Pa) and then placed into the furnace. Subsequently, the sample was heated to 800°C in 24 h and kept at that temperature for 48 h. After that, the sample was cooled at a rate of 1°C·h^{-1} to 650°C and then at a rate of 5°C·h^{-1} to room temperature. The products were washed with distilled water and the light-red plate crystals of the title compound, which are stable in air, were obtained. All operations were performed in an Ar-protected glove box.

7.83 Tin–Copper–Barium–Zinc–Sulfur

In the system containing these elements, the quinary compound $Cu_{1.87}Ba_6Zn_{1.06}Sn_4S_{16}$, which crystallizes as a cubic structure with the lattice parameter a = 1455.98 ± 0.10 pm, a calculated density of 4.450 g·cm^{-3}, and an energy gap of 2.5 eV, is formed (Ji et al. 2022b). This compound was prepared in the same way as $Cu_{1.94}Mg_{1.06}Ba_6Sn_4S_{16}$ was obtained using Zn instead of Mg. The red compound was obtained, which is also stable in air and water.

7.84 Tin–Copper–Barium–Cadmium–Sulfur

In the system containing these elements, the quinary compound $Cu_{2.47}Ba_6Cd_{0.52}Sn_4S_{16}$, which crystallizes as a cubic structure with the lattice parameter a = 1457.30 ± 0.06 pm and a calculated density of 4.352 g·cm^{-3}, is formed (Ji et al. 2022b). This compound was prepared in the same way as $Cu_{1.94}Ba_6Mg_{1.06}Sn_4S_{16}$ was obtained using Cd instead of Mg. An orange compound was obtained, which is also stable in air and water.

7.85 Tin–Copper–Barium–Indium–Sulfur

In the system containing these elements, the quinary compound $Cu_{2.46}Ba_6In_{0.54}Sn_4S_{16}$, which crystallizes as a cubic structure with the lattice parameter a = 1455.36 ± 0.05 pm and a calculated density of 4.375 g·cm^{-3}, is formed (Ji et al. 2022b). For preparing this compound, the reactant elements were loaded in a molar ratio of Ba:Cu:In:Sn:S = 6:2.5:0.5:4:16 and mixed with KI salt flux in carbonized silica ampoule. Element mixtures of 0.4 g were mixed with 0.4 g of KI and loaded into carbonized silica tube. Then, the silica tube was flamed-sealed under a vacuum of 1 Pa and heated within a programmable box furnace. The sample was heated to 800°C in 10 h, held for 48 h at this temperature, and then slowly cooled to 650°C within 100 h. The sample was kept for 10 h at 650°C before the furnace was shut off. The KI flux was removed with deionized water. A red compound was obtained, which is also stable in air and water.

7.86 Tin–Copper–Barium–Bismuth–Sulfur

In the system containing these elements, the quinary compound $Cu_{2.67}Ba_6Bi_{0.35}Sn_4S_{16}$, which crystallizes as a cubic structure with the lattice parameter a = 1456.55 ± 0.05 pm, a calculated density of 4.416 g·cm^{-3}, and an energy gap of 2.2 eV, is formed (Ji et al. 2022b). This compound was prepared in the same way as $Cu_{2.46}Ba_6In_{0.54}Sn_4S_{16}$ was obtained by using Bi instead of In. A dark-red compound was obtained, which is also stable in air and water.

7.87 Tin–Copper–Barium–Selenium–Sulfur

The solid solution $Cu_2BaSnSe_xS_{4-x}$ ($0 \le x \le 3$) is formed in the Sn–Cu–Ba–Se–S system (Shin et al. 2016). It crystallizes as a tetragonal structure with the lattice parameters a = 636.62 ± 0.01, c = 1582.87 ± 0.02 pm, and an energy gap of 1.95 eV for x = 0, a = 642.94 ± 0.03, c = 1600.21 ± 0.06 pm, and an energy gap of 1.80 eV for x = 1, a = 650.76 ± 0.01, c = 1620.18 ± 0.03 pm, and an energy gap of 1.63 eV for x = 2, and a = 656.99 ± 0.01, c = 1636.81 ± 0.02 pm, and an energy gap of 1.55 eV for x = 3.

Solid state reactions were employed to prepare bulk samples of this solid solution. Stoichiometric mixtures of BaS, SnS, CuS, BaSe, SnSe, and CuSe were carefully ground/homogenized and cold-pressed into pellets inside a nitrogen-filled glove box. The pellets were placed inside quartz tubes, and the quartz tubes were flame-sealed under dynamic vacuum (~10^{-5} Pa). The reaction mixtures were then heated to 650°C during 5–6 h inside a box furnace, and kept at this temperature for 10–15 h. After this step, the reactions were cooled to room temperature by switching off the furnace.

7.88 Tin–Copper–Barium–Manganese–Sulfur

In the system containing these elements, the quinary compound $Cu_{2.19}Ba_6Mn_{0.81}Sn_4S_{16}$, which crystallizes as a cubic structure with the lattice parameter a = 1459.14 ± 0.05 pm and a calculated density of 4.266 g·cm^{-3}, is formed (Ji et al. 2022b). This compound was prepared in the same way as $Cu_{1.94}Mg_{1.06}Ba_6Sn_4S_{16}$ was obtained by using Mn instead of Mg. An orange compound was obtained, which is also stable in air and water.

7.89 Tin–Copper–Barium–Iron–Sulfur

In the system containing these elements, the quinary compound $Cu_2Ba_6FeSn_4S_{16}$, which crystallizes as a cubic structure with the lattice parameter a = 1452.60 ± 0.07 pm, a calculated density of 4.323 g·cm^{-3}, and an energy gap of 1.20 eV, is formed (Zhang et al. 2019a). The synthesis of the title compound and preparation of its single crystals are the same as for $Cu_2Sr_6FeSn_4S_{16}$ compound using BaS instead of SrS.

7.90 Tin–Copper–Barium–Nickel–Sulfur

In the system containing these elements, the quinary compound $Cu_2Ba_6NiSn_4S_{16}$, which crystallizes as a cubic structure with the lattice parameter $a = 1451.14 \pm 0.18$ pm, a calculated density of 4.342 g·cm^{-3}, and an energy gap of 0.82 eV, is formed (Zhang et al. 2019a). The synthesis of the title compound and preparation of its single crystals are the same as for $Cu_2Sr_6FeSn_4S_{16}$ compound using BaS instead of SrS and NiS instead of FeS.

7.91 Tin–Copper–Zinc–Vanadium–Sulfur

The solid solution of $Cu_{26-x}Zn_xV_2Sn_6S_{32}$ ($0 \leq x \leq 2$) is formed in the Sn–Cu–Zn–V–S system (Bourgès et al. 2016). It crystallizes as a cubic structure with the lattice parameter from $a = 1075.7 \pm 0.1$ to 1079.1 ± 0.2 pm according to the data of XRD ($a = 1072.1 \pm 0.1$ pm for $x = 0$ according to the neutron diffraction data). XRD data refinements in the full series indicate a linear increase of the cell parameter a with zinc content from 1075.7 ± 0.6 pm for $x = 0$ to 1082.1 ± 0.4 pm for $x = 2$.

Mechanically alloyed powders of $Cu_{26-x}Zn_xV_2Sn_6S_{32}$ were prepared by milling 5 g of Cu, V, S, Sn, and Zn in stoichiometric proportions in a planetary ball mill. The milling was performed during 12 h at a speed of 600 rpm under Ar atmosphere in 45 mL tungsten carbide jar containing 10 balls of 10 mm in diameter. The obtained powders were then densified by spark plasma sintering for 45 min at 600°C under uniaxial pressure of 64 MPa with a cooling and heating rate of 50°C·min^{-1}. The final dimensions of the pellets were around 8 mm in thickness and 10 mm in diameter.

7.92 Tin–Copper–Zinc–Selenium–Sulfur

The calculated results revealed that the lattice constants variation of the $Cu_2ZnSn(Se_xS_{1-x})_4$ solid solutions obey Vegard's law (Zhao et al. 2015). The wurtzite-derived alloys have better solubility and uniform components compared with zinc-blende-derived alloys. In the whole range of x, the calculated lattice constants and band gaps are nearly linear varying with Se compositions.

7.93 Tin–Copper–Zinc–Iron–Sulfur

Two quinary phases are formed in the Sn–Cu–Zn–Fe–S system. $Cu_2(Zn,Fe)SnS_4$ (mineral kesterite) crystallizes as a tetragonal structure with the lattice parameters $a = 542.7 \pm 0.1$ and $c = 1087.1 \pm 0.5$ pm (Hall et al. 1978) [$a = 543$, $c = 1086$ pm, and an experimental density of 4.54–4.59 g·cm^{-3} (Fleischer 1959; Ivanov and Pyatenko 1959]. The structure of this mineral is characterized by a cell that is pseudo-cubic ($2a \approx c$).

$Cu_8(Fe,Zn)_3Sn_2S_{12}$ (mineral stannoidite) crystallizes as an orthorhombic structure with the lattice parameters $a = 1076.0 \pm 0.1$, $b = 541.6 \pm 0.1$, and $c = 1614.3 \pm 0.3$ pm (Ohtsuki et al. 1980) [$a = 1076 \pm 2$, $b = 540 \pm 1$, $c = 1609 \pm 4$ pm, and a calculated density of 4.29 g·cm^{-3} (Fleischer 1969b; Kato 1969; $a = 1076.7 \pm 0.1$, $b = 541.1 \pm 0.1$, $c = 1611.8 \pm 0.2$ pm, and a calculated density of 4.655 g·cm^{-3} (Kudoh and Takéuchi 1976)]. Synthetic stannoidite is stable below 410°C, at which it decomposes into a new unknown synthetic phase and Cu_5FeS_4 (Ohtsuki et al. 1980).

Cu_2FeSnS_4–ZnS. The phase diagram of this quasibinary system is shown in the Figure 7.1 (Moh 1975). A complete set of solid solutions are formed at elevated temperatures between β-Cu_2SnFeS_4 and α-ZnS, but only limited solid solutions exist below α-β-Cu_2SnFeS_4 transformation on both sides. α-Cu_2SnFeS_4 is converted at 706°C into β-Cu_2SnFeS_4, but when in stable coexistence with α-ZnS, this inversion

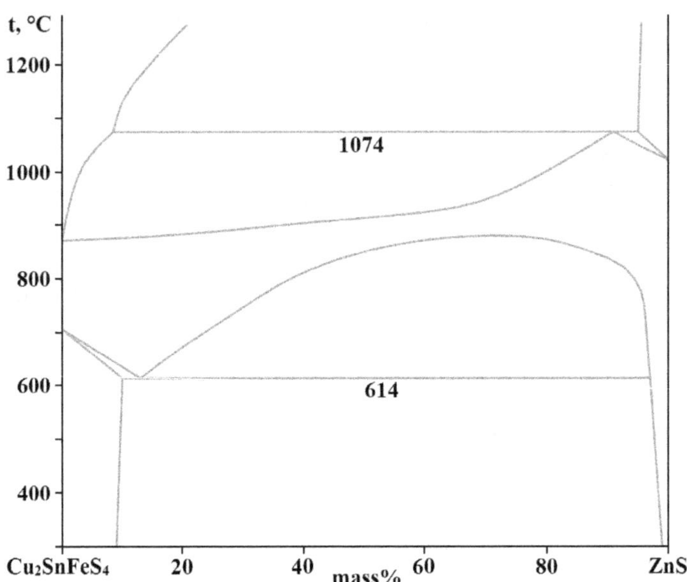

FIGURE 7.1 Phase diagram of the Cu_2SnFeS_4–ZnS quasibinary system. (From Moh, G.H., *Chem. Erde*, 34(1), 1, 1975.)

temperature is lowered to 614°C. At this temperature, α-Cu_2SnFeS_4 contains 10 mass% α-ZnS and β-Cu_2SnFeS_4 contains 13 mass% α-ZnS in the solid solution, whereas the solubility of Cu_2SnFeS_4 in α-ZnS reaches ~3 mass% and diminishes steadily with lower temperatures, for example 1.5 mass% at 400°C,. But β-Cu_2SnFeS_4 narrows only very little with decreasing temperature, and at 260°C, it still contains as much as 9 mass% α-ZnS. Complete α-Cu_2SnFeS_4–α-ZnS solid solutions become stable at ~870°C. The peritectic temperature of this system is 1074°C. At this temperature, the liquid dissolves 8.5 mass% α-ZnS, while the stable α-ZnS mixed crystal has a composition of 91 mass% α-ZnS and 9 mass% β-Cu_2SnFeS_4. Also at 1074°C, β-ZnS contains 5 mass% β-Cu_2SnFeS_4 in the solid solution, but the solubility decreases on further heating.

Cu_2ZnSnS_4–Cu_2FeSnS_4. The phase relations in this quasibinary system were determined between 300°C and the melting and decomposition temperatures that range from 878°C to 1002°C (Springer 1972). Quenching and DTA experiments were employed for investigations using silica tubes as sample containers. Unlimited solid solubility was found to occur above about 680°C, but below this temperature, there are two different phases separated by a miscibility gap.

This system was also investigated by Kissin (1989) in the temperature range from 500°C to 800°C. Phase relations in the system were established and it was shown that a solvus separating stannite and ferrokesterite–kesterite phases intersects the liquidus, indicating that no solid solution (β-$Cu_2Zn_xFe_{1-x}SnS_4$) extends entirely across the system. An inflection in the boundaries of the miscibility gap implies that an unquenchable high-temperature phase, presumably cubic $Cu_2Zn_xFe_{1-x}SnS_4$, exists. Below the inflection at ca. 700°C, the two-phase region slopes toward the iron end-member composition.

The 750°C and 550°C isotherms of the Cu_2ZnSnS_4–Cu_2FeSnS_4 quasibinary system were investigated by the evacuated silica tube technique using molten halide fluxes (Bernardini et al. 1990). The continuous variation of the *a* and *c* lattice parameters as a function of composition points to a complete solid solution between the two end-members, in spite of the different space groups. Although the possible mechanism of substitution between Fe and Zn and/or between (Fe,Zn) and Cu remains unclear, the space group transition is found to occur in the compositional range of 60–70 mol.% Cu_2ZnSnS_4.

Cu_2SnFeS_4–$CuFeS_2$–ZnS. The isothermal section of this quasiternary system at 600°C of the Sn–Cu–Zn–Fe–S quinary system is shown in the Figure 7.2 (Moh 1975).

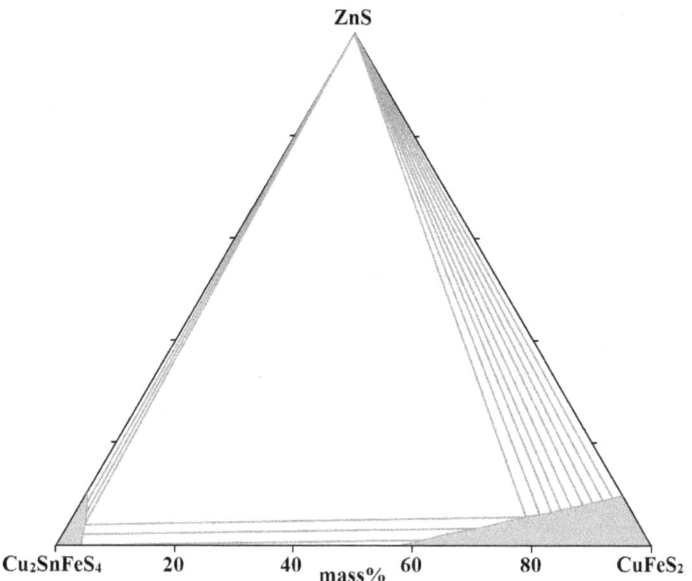

FIGURE 7.2 Isothermal section of the ZnS–$CuFeS_2$–Cu_2SnFeS_4 quasiternary system at 600°C. (From Moh, G.H., *Chem. Erde*, 34(1), 1, 1975.)

7.94 Tin–Copper–Zinc–Indium–Iron–Sulfur

In the system containing these elements, the multinary compound $(Cu,Zn,Fe)_3(In,Sn)S_4$ (mineral sakuraiite), which crystallizes as a cubic structure with the lattice parameter $a = 545.63 \pm 0.24$ pm (Kissin and Owens de 1986a; Jambor et al. 1988) [in the tetragonal (pseudo-cubic where $2a \approx c$) structure with the lattice parameters $a = 545.5$, $a = 10.9$ pm, and a calculated density of 4.45 g·cm^{-3} (Fleischer et al. 1968)], is formed.

7.95 Tin–Copper–Lead–Antimony–Bismuth–Sulfur

The multinary compound $Cu_3Sn_7Pb_8(Bi,Sb)_3S_{28}$ (mineral lévyclaudite), is formed in this system. The structure of synthetic lévyclaudite-(Sb) (the Sb-end member, a calculated density is 5.709 g·cm^{-3}) has been determined by single-crystal XRD on the basis of the (3+2)-dimensional superspace approach (Evain et al. 2006; Piilonen et al. 2007). This misfit-layer compound is of the cylindrite ($Sn_4Pb_3FeSb_2S_{14}$) type with a structure that results from the combination of two heavily modulated triclinic Q and H subcells with a common *q* wave-vector and only one shared reciprocal axis (stacking direction). The Q pseudo-tetragonal layer, ~($Pb_{0.7}Sb_{0.3}S$), is derived from the NaCl archetype: $a = 582.18 \pm 0.06$, $b = 586.45 \pm 0.05$, $c = 1214.43 \pm 0.13$ pm, $\alpha = 92.327 \pm 0.007°$, $\beta = 78.533 \pm 0.009°$, and $\gamma = 90.019 \pm 0.008°$; the H pseudo-hexagonal layer, ~($Sn_{0.85}Cu_{0.3}S_2$), is derived from the CdI_2 archetype: $a = 366.61 \pm 0.03$, $b = 631.38 \pm 0.05$, $c = 1190.28 \pm 0.10$ Å, $\alpha = 92.490 \pm 0.007°$, $\beta = 90.590 \pm 0.008°$, and $\gamma = 89.986 \pm 0.008°$.

According to earlier investigations (Moëlo et al. 1990; Jambor and Burke 1991), natural and synthetic samples show that lévyclaudite is isomorphous with cylindrite and has a noncommensurate lattice, with two monoclinic sublattice; the first subcell is pseudo-tetragonal (Q-type, $a = 1184 \pm 1$, $b = 582.5 \pm 1.0$, $c = 583.1 \pm 1.0$ pm, and $\beta = 92.6 \pm 0.2°$), and the second is pseudo-hexagonal (H-type, $a = 1184 \pm 1$,

$b = 367 \pm 1$, $c = 631 \pm 1$ pm, and $\beta = 92.6 \pm 0.2°$). The calculated density is 6.04 g·cm^{-3} for the global composition.

Dry syntheses of $Cu_3Sn_7Pb_8(Bi,Sb)_3S_{28}$ were carried out in evacuated glass tubes with either the elements or simple sulfides as reactants, in stoichiometric ratio (Evain et al. 2006; Piilonen et al. 2007). A small excess of sulfur was, however, introduced, to preclude the presence of divalent tin. Temperature was set at 600°C for 25–72 days.

7.96 Tin–Copper–Titanium–Niobium–Sulfur

The $Cu_{3.1}Nb_{0.125}Ti_{0.125}Sn_{0.9}S_4$ quinary phase, which crystallizes as a cubic structure with the lattice parameter $a = 543.2 \pm 0.2$ pm, is formed in the Sn–Cu–Ti–Nb–S system (Hirayama et al. 2022). To synthesize this phase, Cu wire, Nb grains, Ti powder, Sn grains, and S powder were sealed in an evacuated quartz tube and heated at 1100°C for 24 h. The obtained sample was then pulverized with an agate mortar and put into a carbon die with a 10-mm inner diameter. In a hot-press sintering furnace, the sample was sintered at 750°C for 40 min in a flowing N_2 environment at a uniaxial pressure of 50 MPa.

7.97 Tin–Copper–Selenium–Chromium–Sulfur

The solid solutions $CuCrSn(S_{1-x}Se_x)_4$ are formed in the Sn–Cu–Se–Cr–S quinary system (Riedel and Morlock 1978). Their polycrystalline samples have been prepared in the range $0 \leq x \leq 1$ from the elements by solid state reaction in evacuated quartz ampoules. The total reaction time was one or two weeks at temperatures between 600°C and 840°C. The lattice constants linearly increase with x.

7.98 Tin–Copper–Manganese–Iron–Sulfur

The solid solutions of $Cu_2Mn_xFe_{1-x}SnS_4$, which crystallize in the tetragonal structure with the lattice parameters $a = 545.80 \pm 0.10$, $c = 1082.70 \pm 0.10$ pm, and a calculated density of 4.422 g·cm^{-3} for $x = 0.4$, are formed in the Sn–Cu–Mn–Fe–S quinary system (López-Vergara et al. 2014). The polycrystalline solid solutions were prepared by a direct combination of powders of the corresponding chemical elements. The reaction mixture was sealed in evacuated quartz ampoules, manipulated under Ar atmosphere, and heated to 850°C for about 72 h in a furnace. Then, the reaction mixture was cooled by quenching in liquid nitrogen. The products appeared to be air and moisture stable over several weeks. These solid solutions are thermally stable up to approximately 400°C. After this temperature, they begin to decompose, a process, which occurs in two stages. The first stage of decomposition can be attributed to the partial sulfur loss. In a second stage, at 750°C, the phases decompose totally to metal sulfides.

7.99 Tin–Copper–Manganese–Cobalt–Sulfur

The solid solutions of $Cu_2Mn_xCo_{1-x}SnS_4$, which crystallize as a tetragonal structure with the lattice parameters $a = 542.88 \pm 0.08$ and $c = 1091.1 \pm 0.2$ pm for $x = 0.6$ and $a = 541.15 \pm 0.08$ and $c = 1084.7 \pm 0.2$ pm for $x = 0.8$, are formed in the Sn–Cu–Mn–Co–S quinary system (López-Vergara et al. 2013). The polycrystalline samples of these solid solutions were prepared by direct combination of powders of the corresponding elements in stoichiometric amounts. All manipulations were carried out under Ar atmosphere. The reaction mixtures were sealed in evacuated quartz ampoules, placed in a furnace, heated to 850°C for about 72 h, and then cooled by quenching in liquid nitrogen. The products appeared to be air and moisture stable over several weeks.

7.100 Tin–Copper–Tellurium–Iron–Palladium–Platinum–Sulfur

In the system containing these elements, the multinary compound $(Pd,Pt)_5(Cu,Fe)_4SnTe_2S_2$ (mineral oulankaite), which crystallizes as a tetragonal structure with the lattice parameters a = 904.4 ± 0.3, c = 493.7 ± 0.3 pm, and a calculated density of 10.27 ± 0.1 g·cm^{-3}, is formed (Barkov et al. 1996; Jambor et al. 1996a). The suggested generalized formula of this mineral is $(Pd,Pt)_{5+x}(Cu,Fe,Ag)_{4-x}SnTe_2S_2$ ($0 \le x < 1$) (Barkov et al. 2004; Jambor and Roberts 2005; Shvedov and Barkov 2015). The compositions extend from Ag-free to Ag-dominant samples (argentoan oulankaite).

7.101 Tin–Copper–Arsenic–Antimony–Vanadium–Sulfur

Three minerals are formed in the Sn–Cu–As–Sb–V–S multinary system. $Cu_{26}V_2(Sn,As,Sb)_6S_{32}$ (mineral nekrasovite) crystallizes as a cubic structure with the lattice parameter a = 1073 ± 5 pm and a calculated density of 4.62 g·cm^{-3} (Kovalenker et al. 1984a; Dunn et al. 1985d).

$Cu_{26}V_2(As,Sn,Sb)_6S_{32}$ (mineral colusite) also crystallizes as a cubic structure with the lattice parameter a = 1065.3 ± 0.3 pm (Frank-Kamenetskaya et al. 2002) [a = 530.4 ± 0.1 pm (Zachariasen 1933); an experimental density of 4.2 g·cm^{-3} (Landon and Mogilnor 1933); a = 1060 ± 1 pm and the calculated and experimental densities of 4.434 and 4.50 g·cm^{-3}, respectively (Berman and Gonyer 1939); a = 1060.8 pm and a calculated density of 4.43 g·cm^{-3} (Murdoch 1953); a = 1062.9 ± 0.3 pm (Dangel and Wuensch 1970); a = 1062.1 ± 0.6 pm (Orlandi et al. 1981); a = 1068 pm for ordered domains (Spry et al. 1994)]. Colusite is composed of ordered and disordered domains, which crystallize in the cubic structure (a = 534 pm for disordered domains) (Spry et al. 1994). The ordered domains consist of Sn-poor colusite.

$Cu_{26}V_2(Sb,Sn,As)_6S_{32}$ (mineral stibiocolusite) also crystallizes as a cubic structure with the lattice parameter a = 1070.5 ± 0.4 pm and a calculated density of 4.66 g·cm^{-3} (Spiridonov et al. 1992a; Jambor and Grew 1994a).

7.102 Tin–Copper–Arsenic–Antimony–Iron–Sulfur

In the system containing these elements, the multinary compound $Cu_{10}SnFe_4(As,Sb)S_{16}$ (mineral vinciennite), which crystallizes as a tetragonal (pseudo-cubic) structure with the lattice parameters $a = c$ = 1069.7 ± 0.3 pm and a calculated density of 4.29 g·cm^{-3}, is formed (Cesbron et al. 1985; Hawthorne et al. 1986).

7.103 Tin–Silver–Strontium–Barium–Sulfur

The $Sr_xBa_{6-x}[Ag_{4-y}Sn_{y/4}](SnS_4)_4$ quinary solid solutions are formed in the Sn–Ag–Sr–Ba–S system (Haynes et al. 2017). They crystallize as a cubic structure with the lattice parameters a = 1463.46 ± 0.04 pm, a calculated density of 4.326 g·cm^{-3}, and an energy gap of 1.44 eV for x = 1.9; a = 1456.46 ± 0.04 pm, a calculated density of 4.271 g·cm^{-3}, and an energy gap of 1.35 eV for x = 2.8; and a = 1451.42 ± 0.04 pm, a calculated density of 4.238 g·cm^{-3}, and an energy gap of 1.30 eV for x = 3.8. To prepare these solid solutions, the stoichiometric ratios of the chemical elements and KBr were loaded into a fused-silica tube in a dry N_2 atmosphere in a glove box and then sealed under a vacuum of < 0.01 Pa. The reactants were heated to 300°C for 4 h, which was raised to 800°C and maintained for 48 h, and then cooled to 300°C at a rate of 5°C·h^{-1}. The experiment was finished by turning off the power. Deionized water was used to remove the KBr flux in an ultrasonic cleaner. Dark-red and chunk-shaped crystals of the title solid solutions were formed. They are air and water stable.

7.104 Tin–Silver–Strontium–Iron–Sulfur

In the system containing these elements, the quinary compound $Ag_2Sr_6FeSn_4S_{16}$, which crystallizes as a cubic structure with the lattice parameter a = 1427.66 ± 0.07 pm, a calculated density of 4.075 g·cm^{-3}, and an energy gap of 1.87 eV, is formed (Zhang et al. 2019a). The synthesis of the title compound involves two steps. In the first step, binary FeS was prepared. The reagents were loaded into silica tube, sealed under vacuum (<10^{-3} Pa), and heated to 600°C. In the second step, appropriate amounts of SrS, Ag, FeS, Sn, and S (molar ratio of 6:2:1:4:9) were loaded in the silica tube. The tube was sealed and placed into a furnace. The sample was heated to 800°C in 24 h, kept at that temperature for 24 h, and cooled at a rate of 5°C·h^{-1} to room temperature.

To prepare the single crystal of the title compound, a mixture of the as-prepared $Ag_2Sr_6FeSn_4S_{16}$ powder and KI (mass ratio 1:1) was mixed and loaded into quartz tube, which was sealed (<10^{-3} Pa) and then placed into the furnace. Subsequently, the sample was heated to 800°C in 24 h and kept at that temperature for 48 h. After that the sample was cooled at a rate of 1°C·h^{-1} to 650°C and then at a rate of 5°C·h^{-1} to room temperature. The products were washed with distilled water and the deep-red block crystals of the title compound, which are stable in air, were obtained. All operations were performed in an Ar-protected glove box.

7.105 Tin–Silver–Barium–Zinc–Sulfur

In the system containing these elements, the quinary compound $Ag_2Ba_6ZnSn_4S_{16}$, which shows excellent thermal stability before 830°C and crystallizes as a cubic structure with the lattice parameter a = 1468.39 ± 0.04 pm, a calculated density of 4.391 g·cm^{-3}, and an energy 2.85 eV, is formed (Duan et al. 2018). To prepare this compound, Ba, Ag, Zn, Sn, and S (molar ratio 6:2:1:4:16) were put in a graphite crucible. Under the high vacuum of 10^{-3} Pa, the crucible was sealed into a silica tube, which was loaded in a heater. Then the reactants were heated up to 960°C in 50 h and kept at this temperature for 100 h. Next, the the reactants were cooled to 350°C for 100 h before turning off the heating. Finally, this compound was successfully synthesized. It stays stable for at least one year in the air.

7.106 Tin–Silver–Barium–Cadmium–Sulfur

In the system containing these elements, the quinary compound $Ag_2Ba_6CdSn_4S_{16}$, which is stable below 787°C and crystallizes as a cubic structure with the lattice parameter a = 1473.83 ± 0.06 pm, a calculated density of 4.440 g·cm^{-3}, and an energy gap of 2.07 eV (Duan et al. 2017) [a = 1472.53 ± 0.07 pm (Teske 1985a)], is formed. This compound was synthesized using a 6:2:1:4:16 molar ratio of Ba/Ag/Cd/Sn/S. These reagents were loaded in a graphite crucible that was flame-sealed under a high vacuum of 10^{-3} Pa. The mixture was heated to 950°C in 60 h and held at this temperature for 75 h. Subsequently, it was cooled down to 350°C in 75 h, and then the furnace was shut down. The yellow crystals of the title compound were obtained. These crystals are stable in air for a long time. All manipulations were performed inside the glove box.

7.107 Tin–Silver–Barium–Selenium–Sulfur

The $Ag_{3.57\pm0.06}Ba_6Sn_{4.11}Se_{6.47\pm0.02}S_{9.53}$ phase, which crystallizes as a cubic structure with the lattice parameters a = 1495.96 ± 0.02 pm and a calculated density of 4.986 g·cm^{-3}, is formed in this system (Lai et al. 2015). This phase was synthesized using a mixture of Ba (6.00 mM), Ag (3.57 mM), Sn (4.11 mM), S (9.53 mM), and KBr (4.20 mM), which was heated at 300°C for 4 h, raised to 800°C and kept at that temperature for 48 h, cooled to 300°C within 100 h, and finished by turning off the power. Dark-red crystals were obtained as a single product. DMF was used to remove the KBr flux since the obtained crystals dissolved completely in water giving black solutions. The surfaces of the crystals become ashy after a few days when exposed to external moisture.

7.108 Tin–Silver–Barium–Manganese–Sulfur

In the system containing these elements, the quinary compound $Ag_2Ba_6MnSn_4S_{16}$, which crystallizes as a cubic structure with the lattice parameter $a = 1470.64 \pm 0.06$ pm, a calculated density of 4.349 g·cm^{-3}, and an energy gap of 2.76 eV, is formed (Duan et al. 2020). To synthesize this compound, all reactants with the molar ratio of Ag/Ba/Mn/Sn/S = 2:6:1:4:16 were encased in a graphite crucible, and then sealed under the high vacuum of 10^{-3} Pa. The reactants were heated to 920°C in 70 h and held at this temperature for 120 h. Then the furnace was closed before slowly cooling to 250°C in 100 h. The yellow block crystals were obtained. The title compound can be kept in the air for a long time (more than two years). All of elements were stored in a glove box, which was filled with Ar and controlled with oxygen and moisture levels below 0.1 ppm. All operations were performed inside the glove box.

7.109 Tin–Silver–Zinc–Iron–Sulfur

In the system containing these elements, the quinary compound $Ag_2(Fe,Zn)SnS_4$ (mineral hocartite), which crystallizes as a tetragonal structure with the lattice parameters $a = 575.0 \pm 0.3$, $c = 1086.6 \pm 0.6$ pm, and a calculated density of 4.745 g·cm^{-3} (Johan and Picot 1982; Dunn et al. 1983b) [$a = 574 \pm 3$ and $c = 1096 \pm 5$ pm (Caye et al. 1968); Fleischer 1969a], is formed.

7.110 Tin–Silver–Gallium–Selenium–Sulfur

$SnS_2 + AgGaSe_2 \Leftrightarrow SnSe_2 + AgGaS_2$. The isothermal section of this ternary mutual system at 450°C (Figure 7.3) contains two single single-phase regions of the $AgGaSe_{2x}S_{2(1-x)}$ and $SnSe_{2x}S_{2(1-x)}$ solid

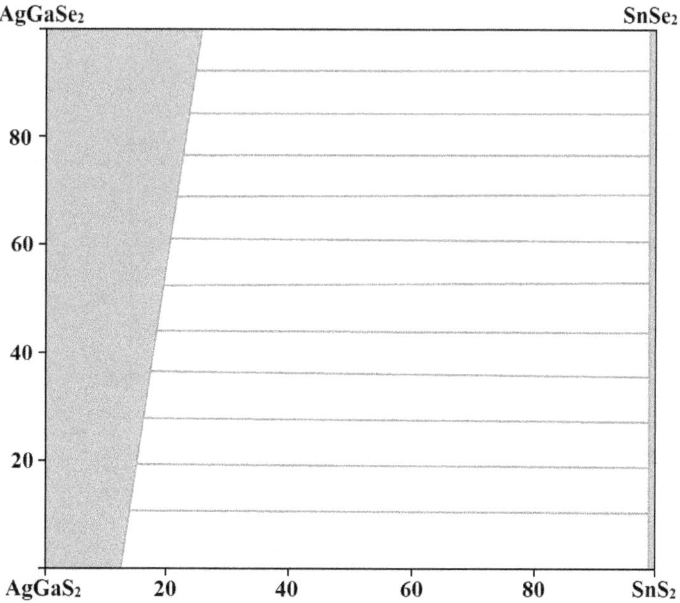

FIGURE 7.3 Isothermal section of the $SnS_2 + AgGaSe_2 \Leftrightarrow SnSe_2 + AgGaS_2$ ternary mutual system at 450°C. (From Shevchuk, M.V., and Olekseyuk I.D., *Fiz. khim. tv. tila*, 11(2), 386, 2010.)

solutions (Shevchuk and Olekseyuk 2010). The alloys for the investigations were annealed at 450°C for 480 h followed by quenching onto cold water. Liquidus surface of this system (Figure 7.4) was constructed using the literature data and the experimental results of the investigation of three vertical sections. It consists of two fields of the primary crystallization of the abovementioned solid solutions. The absence of a stable diagonal in the system indicates on its reversibility.

7.111 Tin–Silver–Lead–Antimony–Sulfur

$Ag_{1.2}Sn_{0.9}Sb_3S_6$–$AgPbSb_3S_6$. The phase diagram of this system, constructed through DTA, XRD, and metallography, is presented in the Figure 7.5 (Chang 1987). A complete solid solution series exists in this system with melting points ranging from 433°C to 506°C. The curvature of the concave liquidus is greater than that of the solidus, which in fact has a tendency to become concave, resulting in an inflected curve. This feature suggests that $AgPbSb_3S_6$ has a larger heat of fusion.

7.112 Tin–Silver–Selenium–Tellurium–Sulfur

SnTe–Ag_4SeS. The phase diagram of this quasibinary system, constructed through DTA, XRD, metallography, and measuring of the microhardness and density using the alloys annealed at 600°C for 3 h, is presented in Figure 7.6 (Vassilev et al. 2004). The eutectics contain 40 and 75 mol% SnTe and crystallize at 530°C and 550°C, respectively. The eutectoid point exists at 10 mol% SnTe and 70°C. Two quinary compounds, $3Ag_4SSe{\cdot}SnTe$ and $Ag_4SSe{\cdot}2SnTe$, exist in this system and both crystallize as a triclinic structure with the lattice parameters a = 785.1, b = 719.6, c = 629.6 pm, α = 101.32°, β = 85.90°, and γ = 111.36° for $3Ag_4SSe{\cdot}SnTe$ and a = 366.2, b = 330.3, c = 334.3 pm, α = 90.74°, β = 108.94°, and γ = 91.91° for α-$Ag_4SSe{\cdot}2SnTe$. $3Ag_4SSe{\cdot}SnTe$ melts incongruently at 540°C and $Ag_4SSe{\cdot}2SnTe$ melts congruently at 615°C and has a polymorphic transition at 110°C. The unit-cell parameters of the quinary compounds were determined for the samples previously kept at 90°C for 720 h and frozen in a mixture of H_2O+ice.

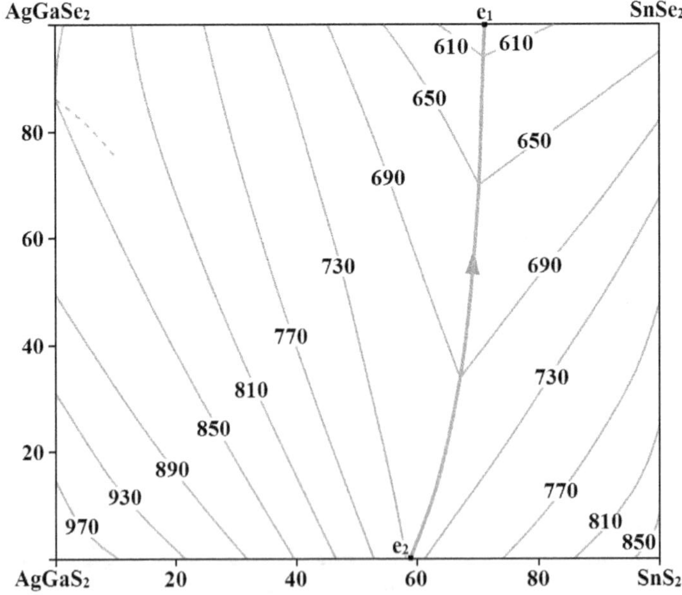

FIGURE 7.4 Liquidus surface of the $SnS_2 + AgGaSe_2 \Leftrightarrow SnSe_2 + AgGaS_2$ ternary mutual system. (From Shevchuk, M.V., and Olekseyuk I.D., *Fiz. khim. tv. tila*, 11(2), 386, 2010.)

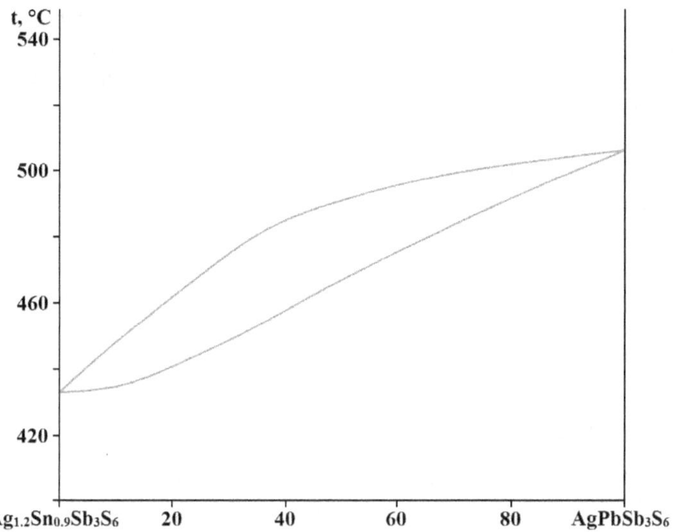

FIGURE 7.5 Phase diagram of the $Ag_{1.2}Sn_{0.9}Sb_3S_6$–$AgPbSb_3S_6$ quasibinary system. (From Chang, L.L.Y., *Mineralog. Mag.*, 51(363), 741, 1987.)

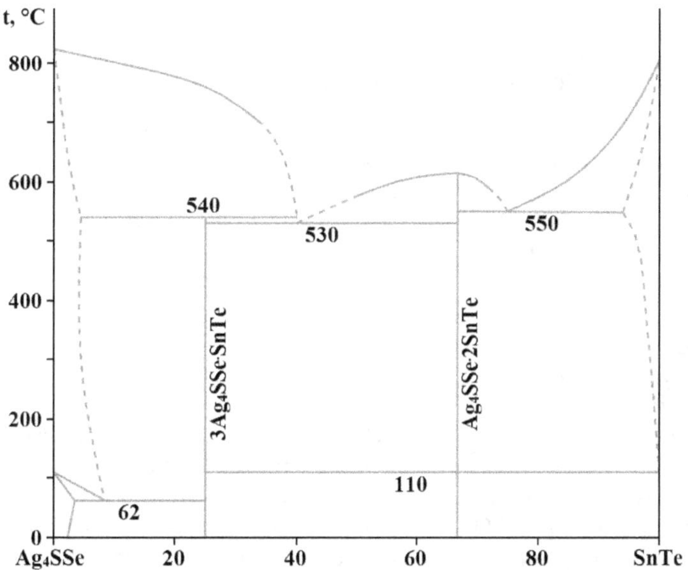

FIGURE 7.6 Phase diagram of the SnTe–Ag_4SeS quasibinary system. (From Vassilev, V., et al., *J. Therm. Anal. Calorim.*, 75(1), 63–71 (2004.)

7.113 Tin–Strontium–Zinc–Oxygen–Sulfur

In the system containing these elements, the quinary compound $Sr_2ZnSn_2OS_6$, which crystallizes as a tetragonal structure with the lattice parameters $a = 966.4$, $c = 633.0$ pm, a calculated density of 3.856 g·cm^{-3}, and an energy gap of 3.52 eV, is formed (Cheng et al. 2022). It is thermally stable until 750°C and after this temperature decomposes into SrS, SnS, and ZnS.

This compound was synthesized via the conventional solid state reaction method. First, SrS, ZnO, Sn, and S were mixed according to the molar ratio of 2:1:2:4 and then ground and transferred into fused-silica tube in a glove box. The mixture was heated to 700°C in 10 h, kept for 20 h, and then cooled down to 25°C in 10 h. The product was reground and heated again adopting the same procedure to improve its homogeneity and purity. The millimeter-size crystal of the title compound was grown by the solution method with SrS as the flux. A mixture of SrS (3 mM), ZnO (1 mM), Sn (2 mM), and S (4 mM) was loaded into a graphite crucible, which was put into a flame-sealed silica tube with a vacuum degree under 10^{-3} Pa. Then the tube was heated to 850°C, held at this temperature for one day, and then cooled it down to 450°C at a rate of 120°C per day. Finally, it was cooled to room temperature with the furnace shut down. The products were carefully washed with DMF to remove other byproducts and then dried in 50°C. Many colorless and flaky crystals with a small amount of unreacted SrS were discovered. $Sr_2ZnSn_2OS_6$ has the excellent chemical stability.

7.114 Tin–Indium–Lead–Bismuth–Sulfur

The $InSnPb_2BiS_7$ quinary compound (mineral abramovite) is formed in the Sn–In–Pb–Bi–S system (Yudovskaya et al. 2008; Ercit et al. 2009). This mineral is characterized by noncommensurate structure composed of regularly alternated pseudo-tetragonal and pseudo-hexagonal sheets. The pseudo-tetragonal subcell crystallizes as a triclinic structure with the lattice parameters $a = 2340 \pm 30$, $b = 577 \pm 2$, $c = 583 \pm 1$ pm, $\alpha = 89.1 \pm 0.5°$, $\beta = 89.7 \pm 0.7°$, $\gamma = 91.5 \pm 0.7°$. The pseudo-hexagonal subcell also crystallizes as a triclinic structure with the lattice parameters $a = 2360 \pm 30$, $b = 360 \pm 10$, $c = 620 \pm 10$ pm, $\alpha = 91 \pm 2°$, $\beta = 92 \pm 1°$, $\gamma = 90 \pm 2°$.

7.115 Tin–Thallium–Copper–Oxygen–Sulfur

Two quinary compounds, $Tl_2CuSn(SO_4)_4$ and $Tl_6CuSn(SO_4)_6$, are formed in this system (Perret et al. 1974). Both compounds crystallize in the rhombohedral structure with the lattice parameters $a = 886.3 \pm 0.5$ and $\alpha = 31°38' \pm 6'$ for the first compound and $a = 932.1 \pm 0.5$ and $\alpha = 110°42' \pm 6'$ for the second one. $Tl_2CuSn(SO_4)_4$ and $Tl_6CuSn(SO_4)_6$ were prepared from the mixtures of $CuSn(SO_4)_3$ and Tl_2SO_4 (molar ratio 1:1 and 1:3, respectively), which were heated in an evacuated sealed ampoule to 300°C in 40 h.

7.116 Tin–Thallium–Magnesium–Oxygen–Sulfur

Two quinary compounds, $Tl_2MgSn(SO_4)_4$ and $Tl_6MgSn(SO_4)_6$, are formed in this system (Perret et al. 1974). Both compounds crystallize as a rhombohedral structure with the lattice parameters $a = 877.5 \pm 0.5$ and $\alpha = 32°12' \pm 6'$ for the first compound and $a = 928.8 \pm 0.5$ and $\alpha = 110°38' \pm 6'$ for the second one. $Tl_2MgSn(SO_4)_4$ and $Tl_6MgSn(SO_4)_6$ were prepared from the mixtures of $MgSn(SO_4)_3$ and Tl_2SO_4 (molar ratio 1:1 and 1:3, respectively), which were heated in an evacuated sealed ampoule to 300°C in 40 h.

7.117 Tin–Thallium–Zinc–Oxygen–Sulfur

Two quinary compounds, $Tl_2ZnSn(SO_4)_4$ and $Tl_6ZnSn(SO_4)_6$, are formed in this system (Perret et al. 1974). Both compounds crystallize as a rhombohedral structure with the lattice parameters $a = 880.7 \pm 0.5$ and $\alpha = 31°58' \pm 6'$ for the first compound and $a = 929.6 \pm 0.5$ and $\alpha = 110°32' \pm 6'$ for the second one. $Tl_2ZnSn(SO_4)_4$ and $Tl_6ZnSn(SO_4)_6$ were prepared from the mixtures of $ZnSn(SO_4)_3$ and Tl_2SO_4 (molar ratio 1:1 and 1:3, respectively), which were heated in an evacuated sealed ampoule to 300°C in 40 h.

7.118 Tin–Thallium–Cadmium–Oxygen–Sulfur

Two quinary compounds, $Tl_2CdSn(SO_4)_4$ and $Tl_6CdSn(SO_4)_6$, are formed in this system (Perret et al. 1974). Both compounds crystallize as a rhombohedral structure with the lattice parameters a = 892.0 ± 0.5 and α = 31°54' ± 6' for the first compound and a = 936.3 ± 0.5 and α = 110°32' ± 6' for the second one. $Tl_2CdSn(SO_4)_4$ and $Tl_6CdSn(SO_4)_6$ were prepared from the mixtures of $CdSn(SO_4)_3$ and Tl_2SO_4 (molar ratio 1:1 and 1:3, respectively), which were heated in an evacuated sealed ampoule to 300°C in 40 h.

7.119 Tin–Thallium–Manganese–Oxygen–Sulfur

Two quinary compounds, $Tl_2MnSn(SO_4)_4$ and $Tl_6MnSn(SO_4)_6$, are formed in this system (Perret et al. 1974). Both compounds crystallize as a rhombohedral structure with the lattice parameters a = 882.4 ± 0.5 and α = 32°12' ± 6' for the first compound and a = 931.0 ± 0.5 and α = 110°20' ± 6' for the second one. $Tl_2MnSn(SO_4)_4$ and $Tl_6MnSn(SO_4)_6$ were prepared from the mixtures of $MnSn(SO_4)_3$ and Tl_2SO_4 (molar ratio 1:1 and 1:3, respectively), which were heated in an evacuated sealed ampoule to 300°C in 40 h.

7.120 Tin–Thallium–Cobalt–Oxygen–Sulfur

Two quinary compounds, $Tl_2CoSn(SO_4)_4$ and $Tl_6CoSn(SO_4)_6$, are formed in this system (Perret et al. 1974). Both compounds crystallize as a rhombohedral structure with the lattice parameters a = 878.5 ± 0.5 and α = 32°04' ± 6' for the first compound and a = 929.1 ± 0.5 and α = 110°10' ± 6' for the second one. $Tl_2CoSn(SO_4)_4$ and $Tl_6CoSn(SO_4)_6$ were prepared from the mixtures of $CoSn(SO_4)_3$ and Tl_2SO_4 (molar ratio 1:1 and 1:3, respectively), which were heated in an evacuated sealed ampoule to 300°C in 40 h.

7.121 Tin–Thallium–Nickel–Oxygen–Sulfur

Two quinary compounds, $Tl_2NiSn(SO_4)_4$ and $Tl_6NiSn(SO_4)_6$, are formed in this system (Perret et al. 1974). Both compounds crystallize in the rhombohedral structure with the lattice parameters a = 880.7 ± 0.5 and α = 31°52' ± 6' for the first compound and a = 928.8 ± 0.5 and α = 110°38' ± 6' for the second one. $Tl_2NiSn(SO_4)_4$ and $Tl_6NiSn(SO_4)_6$ were prepared from the mixtures of $NiSn(SO_4)_3$ and Tl_2SO_4 (molar ratio 1:1 and 1:3, respectively), which were heated in an evacuated sealed ampoule to 300°C in 40 h.

7.122 Tin–Lead–Arsenic–Bismuth–Chlorine–Sulfur

In the system containing these elements, the multinary compound $Sn_2Pb_{20}(Bi,As)_{22}S_{54}Cl_{16}$ (mineral vurroite), which crystallizes as a monoclinic structure with the lattice parameters a = 837.1 ± 0.2, b = 4550.2 ± 0.9, c = 2727.3 ± 0.6 pm, β =98.83 ± 0.03°, and a calculated density of 6.279 g·cm^{-3}, is formed (Pinto et al. 2008). According to Garavelli et al. (2005) and Piilonen et al. (2005), vurroite crystallizes as an orthorhombic structure with the lattice parameters a = 4582.4 ± 0.6, b = 836.8 ± 0.2, and c = 5399.0 ± 0.6 pm with a strong c-centered subcell: a = 2291.2 ± 0.3, b = 418.4 ± 0.1, and c = 2699.5 ± 0.3. However, the structural refinement points to the possible occurrence of twinning, and the real symmetry of vurroite could actually be monoclinic. The calculated density of $Sn_2Pb_{20}(Bi,As)_{22}S_{54}Cl_{16}$ is 6.15 g·cm^{-3}.

7.123 Tin–Lead–Arsenic–Iron–Sulfur

The quinary compound $(Pb,Sn)_{12.5}As_3FeSn_5S_{28}$ (mineral coiraite), which has two monoclinic subcells, is formed in the Sn–Pb–As–Fe–S system (Paar et al. 2008). The lattice parameters for the pseudo-tetragonal subcell are the next: $a = 584 \pm 1$, $b = 586 \pm 1$, $c = 1732 \pm 1$ pm, and $\beta = 94.14 \pm 0.01°$ and for the pseudo-hexagonal subcell – $a = 628 \pm 1$, $b = 366 \pm 1$, $c = 1733 \pm 1$ pm, and $\beta = 91.46 \pm 0.01°$. The calculated and experimental densities of this compound are 5.92 and 5.85–5.87 g·cm^{-3}, respectively.

7.124 Tin–Lead–Antimony–Manganese–Sulfur

The $Pb_{25.7}Sn_{8.3}Mn_{3.4}Sb_{6.4}S_{56.2}$ quinary compound (mineral ramosite), which has two monoclinic subcells, is formed in the Sn–Pb–Sb–Mn–S system (Keutsch et al. 2020). The lattice parameters for the pseudo-tetragonal subcell are the next: $a = 582$, $b = 592$, $c = 1765$ pm, and $\beta = 99.1°$ and for the pseudo-hexagonal subcell – $a = 623$, $b = 369$, $c = 1759$ pm, and $\beta = 93.3°$.

7.125 Tin–Lead–Antimony–Iron–Sulfur

Some multinary compounds are formed in the Sn–Pb–Sb–Fe–S system. $Sn_2(Pb,Sn)_6Sb_2FeS_{14}$ (mineral franckeite) is a layered misfit structure. The lattice parameters for the pseudo-tetragonal subcell are as follows: $a = 580.5 \pm 0.8$, $b = 585.6 \pm 1.6$, $c = 1733.8 \pm 0.5$ pm, $\alpha = 94.97 \pm 0.02°$, $\beta = 88.45 \pm 0.02°$, and $\gamma = 89.94 \pm 0.02°$ and for the pseudo-hexagonal subcell – $a = 366.5 \pm 0.8$, $b = 625.75 \pm 0.16$, $c = 1741.9 \pm 0.5$ pm, $\alpha = 95.25 \pm 0.02°$, $\beta = 95.45 \pm 0.02°$, and $\gamma = 89.97 \pm 0.02°$ (Makovicky et al. 2011). The calculated density of this mineral is 8.9802 g·cm^{-3}. According to Coulon et al. (1961), this mineral crystallizes as a triclinic structure with the lattice parameters $a = 4690 \pm 50$, $b = 582 \pm 4$, $c = 1730 \pm 50$ pm, $\alpha = \gamma = 90°$, $\beta = 94°40' \pm 15'$, and the calculated and experimental densities of 5.88 and 6.01 ± 0.15 g·cm^{-3}, respectively. Mottana et al. (1992) noted that franckeite crystallizes as a monoclinic structure with the lattice parameters $a = 585$, $b = 583$, $c = 1734$ pm, $\beta = 94°20'$.

$Sn_3Pb_6Sb_3FeS_{16}$ (mineral potosíite) is also a layered misfit structure and the cell parameters of two incommensurate triclinic subcells are the next: $a = 591.5 \pm 1.0$, $b = 593.8 \pm 1.3$, $c = 1723.9 \pm 1.7$ pm, $\alpha = 91.63 \pm 0.28°$, $\beta = 91.02 \pm 0.25°$, and $\gamma = 90.84 \pm 0.21°$ for the pseudo-tetragonal subcell and $a = 625.3 \pm 0.7$, $b = 373.4 \pm 0.8$, $c = 1722.9 \pm 1.9$ pm, $\alpha = 90.80 \pm 0.19°$, $\beta = 91.71 \pm 0.16°$, and $\gamma = 90.18 \pm 0.14°$ for the pseudo-hexagonal subcell (Kissin and Owens de 1986b) [$a = 588$, $b = 584$, $c = 1728$ pm, $\alpha = 90.0 \pm 0.2°$, $\beta = 92.2 \pm 0.2°$, and $\gamma = 90.0 \pm 0.2°$ for the pseudo-tetragonal subcell and $a = 626$, $b = 370$, $c = 1728$ pm, $\alpha = 90.0 \pm 0.2°$, $\beta = 92.2 \pm 0.2°$, and $\gamma = 90.0 \pm 0.2°$ for the pseudo-hexagonal subcell (Dunn et al. 1983)]. According to Shimizu et al. (1992), potosíite crystallizes as a triclinic structure with the lattice parameters $a = 619 \pm 2$, $b = 369 \pm 1$, $c = 1721 \pm 2$ pm, $\alpha = 89.8 \pm 0.3°$, $\beta = 92.2 \pm 0.2°$, and $\gamma = 89.2 \pm 0.2°$.

$Sn_4Pb_3FeSb_2S_{14}$ (mineral cylindrite) having a distinctive cylindrical morphology and a composite structure, is characterized by incommensurate structural modulations (Wang and Buseck 1992). It contains two types of sheets, and these have pseudo-hexagonal (*H*) and pseudo-tetragonal (*T*) structures with distinct lattice dimensions. The structure changes systematically from the core to periphery of the crystals. The *b* and *c* axes of the *H* and *T* sheets are almost parallel to each other near the crystal cores. Further from the cores, the angular divergence between the axes of the two types of sheets increases (i.e., the sheets are increasingly rotated relative to one another), and the wavelength of the structural modulation decreases. The crystal accommodates the dimensional misfit between the *H* and *T* sheets by a combination of this divergence, the variation of the structural modulation, and the curving of the layers.

$Sn_4Pb_4FeSb_2S_{15}$ (mineral incaite) crystallizes as a triclinic structure with the lattice parameters $a = 582 \pm 2$, $b = 578 \pm 1$, $c = 1726 \pm 2$ pm, $\alpha = 89.8 \pm 0.2°$, $\beta = 89.3 \pm 0.3°$, and $\gamma = 89.8 \pm 0.2°$ (Shimizu et al. 1992). According to Fleischer et al. (1975b), XRD shows that the structure of this mineral consists of two types of alternating layers, one pseudo-tetragonal with $a = 8623$, $b = 579$, $c = 3498$ pm, $\alpha = \gamma = 90°$, and $\beta = 90.28°$, and one pseudo-hexagonal with $a = 25870$, $b = 366$, $c = 6985$ pm, $\alpha = \gamma = 90°$, and $\beta = 90.28°$.

8

Systems Based on Tin Selenides

8.1 Tin–Hydrogen–Lithium–Oxygen–Selenium

Two quinary compounds, $Li_4SnSe_4{\cdot}13H_2O$ and $Li_4Sn_2Se_6{\cdot}14H_2O$, are formed in the Sn–H–Li–O–Se system (Kaib et al. 2013). The first compound crystallizes as an orthorhombic structure with the lattice parameters a = 1284.0 ± 0.3, b = 1293.0 ± 0.3, c = 1327.4 ± 0.3 pm, and a calculated density of 2.02 $g{\cdot}cm^{-3}$. $Li_4Sn_2Se_6{\cdot}14H_2O$ crystallizes as a triclinic structure with the lattice parameters a = 726.02 ± 0.15, b = 958.87 ± 0.19, c = 1049.6 ± 0.2 pm, α = 77.46 ± 0.03°, β = 71.10 ± 0.03°, γ = 74.00 ± 0.03°, and a calculated density of 2.501 $g{\cdot}cm^{-3}$. To prepare these compounds, ${}^{1}_{\infty}\{Li_2[SnSe_3]\}$ for the first compound and Li_4SnSe_4 for the second one were finely pulverized. Two grams of the powder was solved in 20 mL of H_2O and stirred for several hours. The solution was filtered to remove the starting materials and insoluble residues, and a clear yellow solution was obtained. Light yellow blocks of both compounds were obtained by very slow evaporation of the solvent under dynamic vacuum for several hours. All reaction steps were performed with a strong exclusion of air and external moisture (Ar atmosphere at a high-vacuum, double-manifold Schlenk line, or N_2 atmosphere in a glove box).

8.2 Tin–Hydrogen–Sodium–Oxygen–Selenium

Three quinary compounds are formed in the Sn–H–Na–O–Se system. $Na_4SnSe_4{\cdot}16H_2O$ crystallizes as a monoclinic structure with the lattice parameters a = 867.3 ± 0.3, b = 1656.3 ± 0.4, c = 864.7 ± 0.2 pm, β = 92.10 ± 0.02°, and the calculated and experimental densities of 2.180 and 2.20 ± 0.02 $g{\cdot}cm^{-3}$, respectively (Krebs and Hürter 1980). This compound was obtained from aqueous solution by the reaction of $SnSe_2$ with Na_2Se (molar ratio 1:2).

$Na_4Sn_2Se_6{\cdot}13H_2O$ crystallizes as a triclinic structure with the lattice parameters a = 710.6 ± 0.2, b = 1033.0 ± 0.2, c = 1900.9 ± 0.4 pm, α = 78.60 ± 0.02°, β = 85.66 ± 0.03°, γ = 72.85 ± 0.02°, and a calculated density of 2.637 $g{\cdot}cm^{-3}$ (Krebs and Uhlen 1987). To prepare this compound, a gentle stream of H_2S was introduced into a solution of NaOH (5 mM) in degasses H_2O (60 mL) under nitrogen at 0°C for about 1 h. A white crystal mass of NaHSe was separated out. After expelling the excess of H_2Se with N_2, another 5 mM of NaOH were added. $SnSe_2$ (5 mM) was added in portions to clear, colorless solution of Na_2Se (5 mM) while stirring at room temperature. The first portions of $SnSe_2$ were dissolved completely; later an undissolved residue remains, from which it was separated off after stirring overnight. The clear reddish solution was first concentrated under an oil pump vacuum and then, to enable slow crystallization, under a water jet vacuum. Compact, orange-red crystals were separated. They slowly weather at room temperature, releasing crystallization water.

$Na_6Sn_4Se_{11}{\cdot}22H_2O$ crystallizes as a monoclinic structure with the lattice parameters a =1532.5 ± 0.3, b = 1163.4 ± 0.2, c = 2673.9 ± 0.5 pm, β = 95.81 ± 0.03°, and a calculated density of 2.629 $g{\cdot}cm^{-3}$ (Loose and Sheldrick 2001). For obtaining this compound, solid NaOH (0.61 mM), Sn (0.64 mM), and Se (1.26 mM) were heated in a sealed glass tube to 130°C in a 0.3 mL H_2O + 0.6 mL CH_3OH mixture. After 3 days, the contents were cooled to room temperature. Leaving the mother liquor to stand at –8°C for a further 7 days led to crystallization of the title compound as orange-yellow needles.

DOI: 10.1201/9781003123460-8

8.3 Tin–Hydrogen–Potassium–Cadmium–Selenium

In the system containing these elements, the quinary compound $K_{14-x}H_xCd_{15}Sn_{12}Se_{46}$ ($x \approx 7$), which crystallizes as a cubic structure with the lattice parameter $a = 2308.97 \pm 0.05$ pm and a calculated density of 3.790 g·cm^{-3} for $x = 6.08$, is formed (Ding and Kanatzidis 2006a,b). After stirring $K_6Cd_4Sn_3Se_{13}$ in a concentrated aqueous solution of HI (pH = 1.1–1.8) for 1 h, the original cube-shaped crystals disappeared and new crystals of the title compound were formed.

8.4 Tin–Hydrogen–Potassium–Oxygen–Selenium

Two quinary compounds, $K_4Sn_4Se_{10}·4.5H_2O$ and $K_6Sn_4Se_{11}·8H_2O$, are formed in the Sn–H–K–O–Se system. The first compound crystallizes as a rhombohedral structure with the lattice parameters $a = 1532.0 \pm 0.3$, $c = 2101.9 \pm 0.7$, and a calculated density of 3.50 g·cm^{-3} in hexagonal setting (Loose and Sheldrick 1999a). For its preparation, a glass tube (~10-mL capacity) containing K_2CO_3 (1.08 mM), Sn (1.47 mM), and Se (2.85 mM) in 0.9 mL of a 1:2 H_2O/CH_3OH mixture was sealed under vacuum and heated to 130°C at a rate of 2°C·h^{-1}. After 72 h at this temperature, the glass tube was cooled to 20°C at 1°C·h^{-1} to afford bunches of orange-red prisms that were separated manually and washed with methanol.

$K_6Sn_4Se_{11}·8H_2O$ crystallizes as a monoclinic structure with the lattice parameters $a = 2800.4 \pm 0.6$, $b = 900.0 \pm 0.2$, $c = 1519.2 \pm 0.3$ pm, $\beta = 113.80 \pm 0.03°$, and a calculated density of 3.265 g·cm^{-3} (Loose and Sheldrick 1999b). To synthesize this compound, K_2CO_3 (1.06 mM), Sn (1.49 mM), and Se (2.91 mM) were heated to 130°C in a 0.5 mL H_2O/1.0 mL CH_3OH mixture. After tempering for 4 days, the contents were allowed to cool to room temperature at 1°C·h^{-1} and then left to stand for a further one day to optimize the yield of orange crystals of the title compound. Crystalline products were separated manually, washed, and characterized by X-ray structural analysis.

8.5 Tin–Hydrogen–Rubidium–Zinc–Oxygen–Selenium

In the system containing these elements, the multinary compound $[Rb_{10}(H_2O)_{14.5}][Zn_4(\mu_4\text{-}Se)(SnSe_4)_4]$ is formed (Ruzin and Dehnen 2006). It crystallizes as a triclinic structure with the lattice parameters $a = 1743.1 \pm 0.4$, $b = 1745.9 \pm 0.4$, and $c = 2273.0 \pm 0.5$ pm and $\alpha = 105.82 \pm 0.02°$, $\beta = 99.17 \pm 0.03°$, and $\gamma = 90.06 \pm 0.03°$, a calculated density of 1.646·cm^{-3}, and energy gap of 3.10 eV. To obtain this compound, $[Rb_4(H_2O)_4][SnSe_4]$ (0.2 mM) was drained in H_2O (5 mL) and added to a solution of $Zn(CH_3COO)_2$ (0.2 mM) in H_2O (5 mL), whereupon an yellow precipitate formed immediately. After stirring for 24 h, the precipitate was removed and the filtrate was layered by THF (10 mL). Yellow rhombuses of the targeted compound crystallized after 3 days. All reactions were performed under an inert atmosphere.

8.6 Tin–Hydrogen–Rubidium–Oxygen–Selenium

Three quinary compounds are formed in the Sn–H–Rb–O–Se system. $Rb_2Sn_4Se_9·H_2O$ crystallizes as an orthorhombic structure with the lattice parameters $a = 1329.8 \pm 0.3$, $b = 1241.8 \pm 0.3$, $c = 1256.3 \pm 0.3$ pm, and a calculated density of 4.400 g·cm^{-3} (Loose and Sheldrick 1998). For its synthesis, the starting compounds Rb_2CO_3 (1.03 mM), Sn (1.53 mM), and Se (2.90 mM) were slurried in 0.5 mL H_2O/1.0 mL CH_3OH solvent mixture and heated in a sealed glass tube to 115°C–130°C for 4 days. After tempering, the contents were allowed to cool to room temperature at 1°C·h^{-1}. The final pH value was 8. Black prisms of the title compound were obtained.

$Rb_4Sn_4Se_{10}·1.5H_2O$ crystallizes as a rhombohedral structure with the lattice parameters $a = 1553.9 \pm 0.2$, $c = 2137.3 \pm 0.4$, and a calculated density of 3.64 g·cm^{-3} in hexagonal setting (Loose and Sheldrick 1999a). For its preparation, a glass tube (~10-mL capacity) containing Rb_2CO_3 (1.06 mM), Sn (1.47 mM), and Se (2.85 mM) in 0.9 mL of a 1:2 H_2O/CH_3OH mixture was sealed under vacuum and heated to

130°C at a rate of 2°C·h^{-1}. After 96 h at this temperature, the glass tube was cooled to 20°C at 1°C·h^{-1} to afford a few black crystals of $Rb_2Sn_4Se_9{\cdot}H_2O$. Leaving the orange-yellow mother liquor to stand at -8°C for 72 h led to the crystallization of the title compound as orange prisms.

$Rb_6Sn_4Se_{11}{\cdot}xH_2O$ crystallizes as a monoclinic structure with the lattice parameters a =2810.1 ± 0.6, b = 901.2 ± 0.2, c = 1494.9 ± 0.3 pm, and β = 114.12 ± 0.03° (Loose and Sheldrick 1999b). To prepare this compound, Rb_2CO_3 (0.86 mM), Sn (1.41 mM), and Se (2.90 mM) were heated to 115°C in a 0.5 mL H_2O/1.0 mL CH_3OH mixture. After tempering for 8 days, the contents were allowed to cool to room temperature to afford black crystals of $Rb_2Sn_4Se_9{\cdot}H_2O$. Leaving the orange-yellow mother liquor to stand at –8°C for a further 24 h led to the crystallization of the title compound as yellow needles.

8.7 Tin–Hydrogen–Cesium–Oxygen–Selenium

Three quinary compounds are formed in the Sn–H–Cs–O–Se system. $Cs_2Sn_2Se_5{\cdot}H_2O$ crystallizes as an orthorhombic structure with the lattice parameters a = 1205.0 ± 0.2, b = 1240.5 ± 0.3, c = 942.6 ± 0.2 pm, and a calculated density of 4.318 g·cm^{-3} (Loose and Sheldrick 1999b). To synthesize this compound, Cs_2CO_3 (0.92 mM), Sn (1.32 mM), and Se (2.40 mM) were heated to 130°C in a 0.5 mL H_2O/1.0 mL CH_3OH mixture. After tempering for 4 days, the contents were allowed to cool to room temperature to afford light-red needles of the title compound (13% yield) and darker crystals of $Cs_2Sn_2Se_5$ (30% yield). The yield of $Cs_2Sn_2Se_5{\cdot}H_2O$ can be increased to 22% by reducing the reaction time to 2 days. Black prismatic crystals of $Cs_2Sn_4Se_9{\cdot}H_2O$ were formed as a product in a yield similar to $Cs_2Sn_2Se_5$.

$Cs_2Sn_4Se_9{\cdot}H_2O$ also crystallizes as an orthorhombic structure with the lattice parameters a = 1351.4 ± 0.3, b = 1254.0 ± 0.3, c = 1267.9 ± 0.3 pm, and a calculated density of 4.542 g·cm^{-3} (Loose and Sheldrick 1998). For its synthesis, the starting components Cs_2CO_3 (1.08 mM), Sn (1.47 mM), and Se (2.85 mM) were slurried in 0.5 mL H_2O/1.0 mL CH_3OH solvent mixture and heated in a sealed glass tube to 115°C–130°C for 4 days. After tempering, the contents were allowed to cool to room temperature at 1°C·h^{-1}. The final pH value was 8. Black prisms of the title compound were obtained.

$Cs_4Sn_4Se_{10}{\cdot}3.2H_2O$ crystallizes as a triclinic structure with the lattice parameters a = 723.5 ± 0.1, b = 946.8 ± 0.2, c = 1123.5 ± 0.2 pm, α = 68.77 ± 0.03°, β = 83.25 ± 0.03°, γ = 84.53 ± 0.03°, and a calculated density of 4.33 g·cm^{-3} (Loose and Sheldrick 1999a). For its preparation, a combination of Cs_2CO_3 (0.94 mM), Sn (1.25 mM), and Se (2.56 mM) was heated in a 1:2 H_2O/CH_3OH mixture in a sealed glass tube to 130°C at a rate of 2°C·h^{-1}. After tempering for 24 h at this temperature, the glass tube was cooled to 20°C at 1°C·h^{-1} to afford bunches of dark-red needles of the title compound that were separated manually from yellow plates of $Cs_4Sn_2Se_6$.

8.8 Tin–Hydrogen–Silver–Zinc–Nitrogen–Selenium

In the system containing these elements, the multinary compound $\{[Zn(NH_3)_6][Ag_4Zn_4Sn_3Se_{13}]\}_\infty$, which crystallizes as a cubic structure with the lattice parameter a = 1896.79 ± 0.04 pm and is an n-type semiconductor with an energy gap of 2.09 eV, is formed (Xiong et al. 2014). It was synthesized by mixing Ag (0.11 mM), $ZnCl_2$ (0.24 mM), SnSe (0.48 mM), Se (2.43 mM), octanedioic acid ($C_8H_{14}O_4$) (2.54 mM), and hydrazine monohydrate (98%, 3 mL). The mixture was sealed into an autoclave equipped with a Teflon liner (20 mL) and heated at 190°C for 3 days. After cooling to room temperature, the obtained product was filtrated and washed with EtOH. Red block crystals were selected by hand. This compound can also be prepared by the previously mentioned synthetic procedure without the addition of surfactant octanedioic acid, but the quality of single crystals were poor, and the yield of crystals were much lower.

8.9 Tin–Hydrogen–Silver–Nitrogen–Selenium

In the system containing these elements, the quinary compound $(NH_4)_3AgSn_3Se_8$, which crystallizes as a tetragonal structure with the lattice parameters a = 827.6 ± 0.3, c = 1346.1 ± 0.7 pm, a calculated density of 4.141·cm^{-3}, and energy gap of 1.82 eV, is formed (Zhang et al. 2017a). To synthesize this compound, freshly prepared AgCl (0.5 mM), Sn(1.3 mM), formamide (4.0 mL), and hydrazine hydrate (1.0 mL, 80%) was combined and sealed in a 23 mL Teflon-lined stainless steel autoclave at 160°C for 6 days, and then cooled to room temperature. After isolation, a mixture of unidentified black powders and dark-red sheet-like crystals of the title compound were obtained. Dark-red crystals were selected by hand and washed with distilled water and ethanol, respectively.

8.10 Tin–Hydrogen–Barium–Zinc–Oxygen–Selenium

In the system containing these elements, the multinary compound $[Ba_5(H_2O)_{32}][Zn_5Sn(\mu_3\text{-}Se)_4(SnSe_4)_4$, which crystallizes as a monoclinic structure with the lattice parameters a = 2523.1 ± 0.5, b = 2477.6 ± 0.5, c = 2539.6 ± 0.5 pm, β = 106.59 ± 0.03°, a calculated density of 3.285·cm^{-3}, and energy gap of 3.15 eV, is formed (Ruzin and Dehnen 2006). To obtain it, $ZnCl_2$ (0.1 mM) was added to $[Ba_2(H_2O)_5][SnSe_4]$ (0.1 mM) suspended in H_2O (10 mL), whereupon a yellow precipitate formed immediately. After stirring for 24 h, the precipitate was removed and the filtrate was layered by THF (10 mL). Yellow parallelepipeds of the title compound crystallize after 1 day. All reactions were performed under an inert atmosphere.

8.11 Tin–Hydrogen–Gallium–Nitrogen–Selenium

In the system containing these elements, the quinary compound $(NH_4)_2Ga_2Sn_2Se_8$, which is stable up to 300°C under a N_2 atmosphere and crystallizes as a monoclinic structure with the lattice parameters a = 758.18 ± 0.5, b = 1250.1 ± 0.3, c = 1837.7 ± 0.4 pm, β = 96.43 ± 0.03°, and an energy gap of 1.71 eV, is formed (Wang et al. 2018). To prepare this compound, Ga (0.53 mM), $SnCl_2$ (0.49 mM), Se (3.08 mM), 1-butylpyridinium bromide (4.63 mM), hydrazine monohydrate (1.0 mL, 80% aqueous solution), and ethanol (0.25 mL, 40% aqueous solution) were mixed and sealed in an autoclave equipped with a Teflon liner (28 mL), then heated at 190°C for 7 days and cooled to room temperature. The products were filtrated and washed several times with ethanol. Orange block crystals of the title compound were obtained.

8.12 Tin–Hydrogen–Indium–Nitrogen–Selenium

Two quinary compounds, $(NH_4)InSnSe_4$ and $(NH_4)_2In_2Sn_2Se_8$, are formed in the Sn–H–In–N–Se system (Wang et al. 2018). The first compound crystallizes as a tetragonal structure with the lattice parameters a = 837.31 ± 0.9 and c = 671.88 ± 0.10 pm. For its preparation, In (0.49 M), Sn (0.52 mM), Se (3.11 mM), hydrazine monohydrate (1.0 mL, 80% aqueous solution), and ethanol (0.25 mL, 40% aqueous solution) were mixed and sealed in an autoclave equipped with a Teflon liner (28 mL), then heated at 190°C for 7 days and cooled to room temperature. The products were filtrated and washed several times with ethanol. Yellow block crystals of $(NH_4)InSnSe_4$ were obtained.

$(NH_4)_2In_2Sn_2Se_8$ is stable up to 250°C under a N_2 atmosphere and crystallizes as a monoclinic structure with the lattice parameters a = 780.97 ± 0.35, b = 1246.49 ± 0.56, c = 1836.61 ± 0.84 pm, β = 96.59 ± 0.01°, and an energy gap of 1.83 eV. To synthesize this compound, In (0.53 mM), Sn (0.50 mM), Se (3.03 mM), 1-butylpyridinium bromide (1.49 mM), hydrazine monohydrate (1.0 mL, 80% aqueous solution), and ethanol (0.25 mL, 40% aqueous solution) were mixed and sealed in an autoclave equipped with a Teflon liner (28 mL), then heated at 190°C for 7 days and cooled to room temperature. The products were filtrated and washed several times with ethanol. Red rod crystals of the title compound were obtained.

8.13 Tin–Hydrogen–Nitrogen–Manganese–Selenium

Two quinary compounds, $Mn_2SnSe_4(N_2H_4)_7$, and $Mn_2SnSe_4(N_2H_4)_{10}$, are formed in the Sn–H–N–Mn–S system (Yuan et al. 2007; Yuan and Mitzi 2009). The first compound crystallizes as a triclinic structure with the lattice parameters $a = 810.3 \pm 0.2$, $b = 987.9 \pm 0.2$, $c = 1302.1 \pm 0.3$ pm, $\alpha = 95.008 \pm 0.004°$, $\beta = 94.423 \pm 0.004°$, and $\gamma = 113.425 \pm 0.004°$. The second compound crystallizes as a monoclinic structure with the lattice parameters $a = 1030.4 \pm 0.4$, $b = 1318.2 \pm 0.5$, $c = 1827.0 \pm 0.7$ pm, and $\beta = 97.657 \pm 0.007°$. Both compounds were synthesized in hydrazine by dissolving solid mixtures of SnSe, Se, and MnSe under an inert atmosphere and stirring at room temperature. Single crystals were grown by diffusing 2-propanol into the filtered reaction mixtures over a time period of 1–2 weeks. Hydrazine (8 or 16 mL for the first and the second compound, respectively, 98%, anhydrous) was added to a solid mixture of SnSe (5 or 1 mM for the first and the second compound, respectively), Se (10 or 5 mM for the first and the second compound, respectively), and MnSe (3.5 or 2 mM the first and the second compound, respectively). Gas bubbles were observed upon the addition of hydrazine, and the mixture was allowed to stir at room temperature for 2 or 10 days for the first and the second compound, respectively, after which the reaction mixture was filtered through a 0.45 mkm syringe filter and the dark green filtrate was evenly distributed into three test tubes. In each test tube, 3 mL of 2-propanol/hydrazine mixture (volume ratio 1:1) and 12 mL of pure 2-propanol were layered consecutively on top of the filtrate solution for crystal growth. Yellowish-green single crystals were obtained at the bottom and the wall of each test tube after 2 weeks.

8.14 Tin–Potassium–Rubidium–Zinc–Selenium

In the system containing these elements, the quinary compound $K_3Rb_3Zn_4Sn_3Se_{13}$, which crystallizes as a trigonal structure with the lattice parameters $a = 1465.0 \pm 0.2$, $c = 1574.4 \pm 0.3$, a calculated density of 3.435 g·cm^{-3}, and an energy gap of 2.6 eV, is formed (Wu et al. 2005). Single crystals of this compound were grown in a reaction containing K_2Se (0.5 mM), Se (0.6 mM), Sn (0.25 mM), RbCl (0.5 mM), and $ZnCl_2$ (0.017 mM) loaded in a thick wall Pyrex tube, followed by the addition of 0.2 mL of H_2O and 0.2 mL of CH_3OH. The reaction mixture sealed in a tube was heated at 130°C for 6 days. The product was washed with 80% EtOH followed by H_2O, and pure yellow crystals were obtained.

8.15 Tin–Potassium–Magnesium–Cadmium–Selenium

The quinary phases The $K_{14-2x}Mg_xCd_{15}Sn_{12}Se_{46}$ are formed in the Sn–K–Mg–Cd–Se system (Ding and Kanatzidis 2006a,b). To obtain them, K_2Se (0.6 mM), Cd (1.3 mM), Sn (1.0 mM), and Se (3.3 mM) were combined in an evacuated and flame-sealed fused-silica tube and melted at 900°C. The product was loaded into a 9 mm Pyrex tube along with $MgCl_2{\cdot}6H_2O$ (0.7 mM) and deionized water (0.2 mL). The tube was then evacuated to 0.4 Pa, flame-sealed and kept at 115°C for 18–24 h. The products, dark-orange polyhedral crystals, were collected by filtration and washed with water, ethanol, and diethyl ether. The obtained crystals are stable in air for months.

8.16 Tin–Potassium–Calcium–Cadmium–Selenium

The $K_{14-2x}Ca_xCd_{15}Sn_{12}Se_{46}$ quinary phases are formed in the Sn–K–Ca–Cd–Se system (Ding and Kanatzidis 2006a,b). For $x \approx 1.3$, it crystallizes as a cubic structure with the lattice parameter $a = 2328.07 \pm 0.04$ pm, a calculated density of 3.807 g·cm^{-3}, and an energy gap of ca. 2.00 eV. These phases were prepared by the similar experiment described for $K_{14-2x}Mg_xCd_{15}Sn_{12}Se_{46}$ using $CaCl_2$ (0.6 mM) instead of $MgCl_2{\cdot}6H_2O$.

8.17 Tin–Rubidium–Silver–Phosphorus–Selenium

In the system containing these elements, the multinary compound $Rb_4Ag_4Sn_2(P_2Se_6)_3$, which melts congruently at ca. 568°C and crystallizes as a monoclinic structure with the lattice parameters $a = 1118.9$

± 0.2, b = 768.8 ± 0.2, c = 2185.0 ± 0.3 pm, β = 94.31 ± 0.01°, and a calculated density of 4.639 g·cm^{-3} at 148 K and an energy gap of 2.15 eV, is formed (Chondroudis and Kanatzidis 1998). It was synthesized from a mixture of Sn (0.60 mM), Ag (0.30 mM), P_2Se_5 (0.60 mM), Rb_2Se (0.30 mM), and Se (3.00 mM), which was sealed under vacuum in a Pyrex tube and heated to 515°C for 4 days, followed by cooling to 150°C at 4°C·h^{-1}. The excess $Rb_xP_ySe_z$ flux was removed with degassed DMF, and the product was washed with tri-n-butylphosphine to remove elemental Se and to give black hexagonal crystals of $Ag_2SnP_3Se_9$ and red irregular-shaped crystals of $Rb_4Ag_4Sn_2(P_2Se_6)_3$. The latter are air and water stable.

8.18 Tin–Copper–Magnesium–Zinc–Selenium

The $Cu_{2-x}Mg_xZnSnSe_4$ (x = 0-0.4) solid solutions were fabricated by a liquid-phase reactive sintering technique at 600°C for 2 h (Kuo and Wubet 2014). Sintering the Mg-doped $Cu_2ZnSnSe_4$ ingots using Sb_2S_3 and Te sintering aids assisted the densification at 600°C and retained the atomic composition of the solid solutions very close to the favored composition design. Mg is a strong promoter of electrical mobility.

8.19 Tin–Copper–Magnesium–Indium–Selenium

The quinary phase $Cu_2MgIn_{0.1}Sn_{0.9}Se_4$, which crystallizes as a tetragonal structure with the lattice parameters a = 570.5 and c = 1139.9 pm, is formed in the Sn–Cu–Mg–In–Se system (Kumar et al. 2015). Powders of this phase were synthesized by solid state reaction of stoichiometric amounts of Mg metal, Cu, Sn, and Se powders in carbon-coated evacuated fused-silica quartz ampoules. The ampoules were then slowly heated up to 800°C with a heating rate of 2°C·min^{-1}, left to stand for 48 h, and then the furnace cooled to room temperature. The harvested powders were ground, sealed, and calcined for 96 h at 800°C. The final powders were densified in a spark plasma sintering furnace in vacuum at 550°C during 5 min in a graphite die (12 mm) under a pressure of 50 MPa.

8.20 Tin–Copper–Zinc–Aluminum–Selenium

The $Cu_{1.75}ZnAl_xSn_{1-x}Se_4$ (x = 0-0.6) solid solutions are formed in the Sn–Cu–Zn–Al–Se quinary system (Kuo and Tsega 2013). They were prepared by a liquid-phase reactive sintering method at 600°C with soluble sintering aids of Sb_2S_3 and Te. All ingots are semiconductors and exhibited p-type conductivity.

8.21 Tin–Copper–Zinc–Tellurium–Selenium

The $Cu_{2.2}Zn_{0.8}SnTe_xSe_{4-x}$ ($0.1 \leq x \leq 0.4$) solid solutions are formed in the Sn–Cu–Zn–Te–Se quinary system (Dong et al. 2015a). These solid solutions crystallize in the tetragonal structure with the lattice parameters a = 568.35 ± 0.01, c = 1133.11 ± 0.04 pm, and a calculated density of 5.720 g·cm^{-3} for x = 0.1, a = 569.70 ± 0.02, c = 1136.08 ± 0.07 pm, and a calculated density of 5.724 g·cm^{-3} for x = 0.2, a = 570.85 ± 0.03, c = 1138.23 ± 0.08 pm, and a calculated density of 5.746 g·cm^{-3} for x = 0.3, and a = 571.51 ± 0.08, c = 1139.7 ± 0.2 pm, and a calculated density of 5.750 g·cm^{-3} for x = 0.4. They were prepared by direct reaction of Cu, Sn, and Se powders, Zn shot, and Te ingot, which were loaded into silica ampoules in stoichiometric ratios. The reaction ampoules were sealed in quartz tubes, heated to 700°C and subsequently maintained at this temperature for 4 days. The furnace was turned off and the reaction tubes were quenched to room temperature in air. The products were then ground into fine powders, cold pressed into pellets, and annealed at 700°C for one week, with an additional grinding and annealing step performed in order to further promote homogeneity of the products. After appropriate annealing, the products were ground into fine powders and sieved (325 mesh) inside a glove box before loaded into graphite dies for hot pressing. Densification was accomplished by hot pressing at 400°C and 150 MPa for 3 h under a nitrogen flow.

8.22 Tin–Copper–Zinc–Iron–Selenium

The quinary phase $Cu_{2.2}Zn_{0.2}Fe_{0.6}SnSe_4$, which crystallizes as a tetragonal structure with the lattice parameters $a = 568.25 \pm 0.01$, $c = 1132.38 \pm 0.04$ pm, and a calculated density of 5.67 g·cm^{-3}, is formed in the Sn–Cu–Zn–Fe–Se quinary system (Dong et al. 2015b). Polycrystalline bulk specimens of this phase were prepared by the direct reaction of the constituent elements. Cu, Fe, Sn, and Se powders and Zn shot were loaded into silica ampoules in stoichiometric ratios. The reaction ampoules were sealed in quartz tubes under vacuum, heated to 750°C and subsequently maintained at this temperature for 5 days. The furnace was turned off and the reaction tubes were quenched to room temperature in air. The products were ground into fine powders and sieved (325 mesh) inside a glove box before being loaded into graphite dies for hot pressing. Densification was accomplished by hot pressing at 600°C and 150 MPa for 3 h under a nitrogen flow.

8.23 Tin–Copper–Cadmium–Manganese–Selenium

$Cu_2CdSnSe_4$–$Cu_2MnSnSe_4$. This system is a nonquasibinary section of the Sn–Cu–Cd–Mn–Se quinary system as both components melt incongruently (Moreno et al. 2009). The $Cu_2Cd_{1-x}Mn_xSnSe_4$ solid solutions are formed in this system (Sachanyuk est al. 2006, Moreno et al. 2009). Only two single solid phase fields, the tetragonal stannite α and the wurtz-stannite δ structures, were found to occur. In addition to the low-temperature α phase, extra XRD lines due to MnSe were observed for as-grown samples in the range $0.7 < x < 1.0$ (Moreno et al. 2009). However, it was found that the amount of the extra phase decreased for the compressed samples. The value of the crystal parameters decreases linearly with the composition x. The experimental values of a and c were least squared fitted to a linear equation, and the obtained results are given by $a(x) = 583.06–5.297x$ (pm), $c(x) = 1139.9–2.575x$ (pm), and $\rho_{calc.}(x) = 5.776–0.3757x$ (g·cm^{-3}). All of the alloy samples were produced by the usual melt and anneal techniques. In each case, the components of 1 g sample were sealed under vacuum in a small quartz ampoule, and then the components were melted together at 1150°C, initially, for about 1 h. The samples were annealed at 600°C for about 1 month, and finally slowly cooled to room temperature at a rate of 2°C·h^{-1}.

According to Sachanyuk et al. (2006), at 400°C, there is a continuous series of the substitution solid solutions between the quaternary chalcogenides. The solid solution of $Cu_2Cd_{0.5}Mn_{0.5}SnSe_4$ crystallizes as a tetragonal structure with the lattice parameters $a = 579.35 \pm 0.02$, $c = 1140.36 \pm 0.05$ pm, and a calculated density of 5.5951 ± 0.0007 g·cm^{-3}.

8.24 Tin–Copper–Cadmium–Iron–Selenium

$Cu_2CdSnSe_4$–$Cu_2FeSnSe_4$. This system is a nonquasibinary section of the Sn–Cu–Cd–Fe–Se quinary system as both components melt incongruently (Moreno et al. 2009). The $Cu_2Cd_{1-x}Fe_xSnSe_4$ solid solutions are formed in this system. Only two single solid phase fields, the tetragonal stannite α and the wurtz-stannite δ structures, were found to occur. In addition to the low-temperature α phase, extra XRD lines due to $FeSe_2$ were observed for as-grown samples in the range $0.7 < x < 1.0$. However, it was found that the amount of the extra phase decreased for the compressed samples. The value of the crystal parameter a decreases linearly with the composition x, while the values of c decreases nonlinearly. The experimental values of a were least squared fitted to a linear equation, but the c versus x curve was fitted to a quadratic form, and the obtained results are given by $a(x) = 583.06–10.6x$ (pm), $c(x) = 1139.98–4.264x – 8.45x^2$ (pm), and $\rho_{calc.}(x) = 5.776–0.254x + 0.065x^2$ (g·cm^{-3}). The alloy samples were produced by the usual melt and anneal technique. In each case, the components of 1 g sample were sealed under vacuum in a small quartz ampoule, and then the components were melted together at 1150°C, initially, for about 1 h. The samples were annealed at 600°C for about 1 month, and finally slowly cooled to room temperature at a rate of 2°C·h^{-1}.

8.25 Tin–Strontium–Lanthanum–Bismuth–Selenium

In the system containing these elements, the multinary compound $Sr_{2.71}La_{0.29}Sn_{0.77}Bi_{2.23}Se_8$, which crystallizes as an orthorhombic structure with the lattice parameters a = 1304.7 ± 0.4, b = 426.2 ± 0.1, c = 2922.0 ± 0.9 pm, and a calculated density of 5.177 g·cm^{-3}, is formed (Chung and Lee 2013). This compound was prepared in attempts to synthesize analogous compounds to $Sr_8YGe_2Bi_7Se_{24}$. Pure elements were combined in the appropriate stoichiometric ratios (total mass approximately 0.5 g) under a dry nitrogen atmosphere following the above formula. Reaction mixture was heated in a furnace to 750°C over one day, maintained at that temperature for another one day, slowly cooled to 400°C over a day, and finally cooled to room temperature by terminating the power. The product was polycrystalline that exhibited a metallic luster.

8.26 Tin–Strontium–Carbon–Oxygen–Selenium

In the system containing these elements, the quinary compound $Sr_3(SnOSe_3)(CO_3)$, which crystallizes as an orthorhombic structure with the lattice parameters a = 948.16 ± 0.09, b = 651.24 ± 0.06, c = 793.11 ± 0.06 pm, and an energy gap of 3.46 eV, is formed (Wang et al. 2022a). The pure polycrystalline samples of the title compound were synthesized by the traditional solid state reaction technology. The stoichiometric ratio (1:2:1:1) of $SrCO_3$, SrSe, Sn, and Se were weighed and thoroughly ground and then packed into a graphite crucible and removed into the silica tube and flame-sealed under a vacuum of 10^{-3} Pa. Subsequently, the silica tube was placed into a furnace and preheated from room temperature to 700°C at a rate of 3°C·min^{-1} and kept at this temperature for 60 h with several intermediate grindings to obtain the pure phase. All the initial materials were used and stored in a glove box filled with purified Ar (moisture and oxygen level is less than 0.1 ppm).

8.27 Tin–Barium–Lanthanum–Antimony–Selenium

In the system containing these elements, the quinary compound $Ba_{2.67}La_{0.33}Sn_{0.67}Sb_{2.33}Se_8$, which melts at 697°C and crystallizes as an orthorhombic structure with the lattice parameters a = 1263.0 ± 0.3, b = 462.6 ± 0.1, c = 2962.8 ± 0.8 pm, a calculated density of 4.050 g·cm^{-3}, and an energy gap of 0.81 eV, is formed (Chung and Lee 2013). This compound was prepared in the same way as $Sr_{2.71}La_{0.29}Sn_{0.77}Bi_{2.23}Se_8$.

8.28 Tin–Barium–Lanthanum–Bismuth–Selenium

In the system containing these elements, the quinary compound $Ba_{2.67}La_{0.33}Sn_{0.67}Bi_{2.33}Se_8$, which melts at 857°C and crystallizes as an orthorhombic structure with the lattice parameters a = 1306.2 ± 0.3, b = 445.96 ± 0.09, c = 2989.2 ± 0.6 pm, a calculated density of 4.609 g·cm^{-3}, and an energy gap of 0.52 eV, is formed (Chung and Lee 2013). This compound was prepared in the same way as $Sr_{2.71}La_{0.29}Sn_{0.77}Bi_{2.23}Se_8$.

8.29 Tin–Zinc–Gallium–Arsenic–Selenium

Sn–ZnSe–GaAs. Isothermal sections of this quasiternary system at 600°C, 650°C, 700°C, and 800°C as well as some verticals were constructed by Novikova et al. (1974) [the corresponding figures are given in Tomashyk (2015)]. The eutectic temperatures in all vertical sections are degenerated from the Sn-rich side and are equal to 218°C–222°C (the eutectic temperature increases with increasing ZnSe content).

8.30 Tin–Indium–Lead–Antimony–Bismuth–Selenium

In the system containing these elements, the multinary compound $In_{5.00}Sn_{6.13}Pb_{1.87}Sb_{10.12}Bi_{2.88}Se_{35}$ compound, which crystallizes as a monoclinic structure with the lattice parameters a = 3168.83 ± 0.05, b = 407.09 ± 0.01, c = 2654.06 ± 0.04 pm, β = 105.710 ± 0.001°, a calculated density of 6.335 g·cm^{-3}, and an energy gap of 0.62 eV, is formed (Chen et al. 2022a). A single-crystal sample of this compound was first found as one of the by-products of the reaction for synthesizing $In_5Sn_4Sb_9Se_{25}$. A pure phase was obtained with the following procedures. The mixture of Sn, In, Sb, Se, PbSe and Bi_2Se_3 were initially heated from room temperature to 800°C in 8 h; the temperature was maintained for 48 h, then cooled to 400°C over 24 h, and finally cooled down naturally by turning off the furnace. An additional annealing process was carried out at 580°C for 24 h to improve the crystallinity. The title compound is stable in air under ambient conditions.

9

Systems Based on Tin Telluride

9.1 Tin–Hydrogen–Potassium–Zinc–Oxygen–Tellurium

In the system containing these elements, the multinary compound $[K_{10}(H_2O)_{20}][Zn_4(\mu_4\text{-}Te)(SnTe_4)_4]$, which crystallizes as a cubic structure with the lattice parameter a = 1596.51 ± 0.18 pm, a calculated density of 2.984 g·cm^{-3} at 203 K, and an energy gap of 1.73 eV, is formed (Ruzin et al. 2006a). To synthesize this compound, $[K_4(H_2O)_{0.5}][SnTe_4]$ (0.1 mM) was dissolved in H_2O (5 mL) and added to a solution of $ZnCl_2$ (0.1 mM) in H_2O (5 mL), whereupon the reaction mixture immediately turned dark red. After stirring for 24 h, a reddish black precipitate of binary telluride was removed by filtration. The filtrate was layered by THF (10 mL). Black rhombic dodecahedra of the title compound, which are soluble in water, crystallized after two days. Its single crystals can be isolated under inert conditions from the mother liquor without decomposition. All synthesis steps were performed with a strong exclusion of air and moisture (nitrogen atmosphere at a high-vacuum, double-manifold Schlenk line or Ar atmosphere in a glove box).

9.2 Tin–Hydrogen–Potassium–Cadmium–Oxygen–Tellurium

In the system containing these elements, the multinary compound $[K_{10}(H_2O)_{20}][Cd_4(\mu_4\text{-}Te)(SnTe_4)_4]$, which crystallizes as a cubic structure with the lattice parameter a = 1614.23 ± 0.19 pm, a calculated density of 3.036 g·cm^{-3} at 203 K, and an energy gap of 1.68 eV, is formed (Ruzin et al. 2006a). This compound was prepared in the same way as $[K_{10}(H_2O)_{20}][Zn_4(\mu_4\text{-}Te)(SnTe_4)_4]$ was synthesized but using $CdCl_2$ instead of $ZnCl_2$. Black rhombic dodecahedra of the title compound, which are soluble in water, crystallized after two days. Its single crystals can be isolated under inert conditions from the mother liquor without decomposition. All synthesis steps were performed with a strong exclusion of air and moisture (nitrogen atmosphere at a high-vacuum, double-manifold Schlenk line or Ar atmosphere in a glove box).

9.3 Tin–Hydrogen–Potassium–Mercury–Oxygen–Tellurium

In the system containing these elements, the multinary compound $[K_{10}(H_2O)_{20}][Hg_4(\mu_4\text{-}Te)(SnTe_4)_4]$, which crystallizes as a cubic structure with the lattice parameter a = 1607.60 ± 0.19 pm, a calculated density of 3.355 g·cm^{-3} at 203 K, and an energy gap of 1.39 eV, is formed (Ruzin et al. 2006a). This compound was prepared in the same way as $[K_{10}(H_2O)_{20}][Zn_4(\mu_4\text{-}Te)(SnTe_4)_4]$ was synthesized but using $HgCl_2$ instead of $ZnCl_2$ and CH_3OH instead of H_2O as a solvent of $[K_4(H_2O)_{0.5}][SnTe_4]$. Black rhombic dodecahedra of the title compound, which are soluble in water, crystallized after two days. Its single crystals can be isolated under inert conditions from the mother liquor without decomposition. All synthesis steps were performed with a strong exclusion of air and moisture (nitrogen atmosphere at a high-vacuum, double-manifold Schlenk line or Ar atmosphere in a glove box).

DOI: 10.1201/9781003123460-9

9.4 Tin–Hydrogen–Potassium–Oxygen–Tellurium

In the system containing these elements, the multinary compound $[K_4(H_2O)_{0.5}][SnTe_4]$, which crystallizes as a monoclinic structure with the lattice parameters a = 2152.3 ± 0.4, b = 1151.8 ± 0.2, c = 1570.8 ± 0.3 pm, β = 130.53 ± 0.03°, a calculated density of 3.556 g·cm^{-3} at 203 K, and an energy gap of 2.1 eV, is formed (Ruzin et al. 2006b). This compound was prepared by reacting K (5.000 g) with equimolar amounts of Sn in a quartz ampoule at ca. 700°C for 10 min under vacuum. After cooling, equimolar amounts of elemental Te were added and the mixture was remelted for 20 min under vacuum. The ternary alloy was crushed to a fine powder and the bulk of precipitated tin was removed mechanically as a solid metallic plug. The crude "$K_6[SnTe_6]$" was dissolved in water (50 mL) at 20°C and the solution was stirred for 5 min and filtered away from precipitating Sn. The filtrate was evaporated for 24 h, whereupon the title compound was observed as dark red blocks. All synthesis steps were performed with a strong exclusion of air and moisture (nitrogen atmosphere at a high-vacuum, double-manifold Schlenk line or Ar atmosphere in a glove box).

9.5 Tin–Hydrogen–Potassium–Manganese–Oxygen–Tellurium

In the system containing these elements, the multinary compound $[K_{10}(H_2O)_{20}][Mn_4(\mu_4\text{-}Te)(SnTe_4)_4]$, which crystallizes as a cubic structure with the lattice parameter a = 1608.89 ± 0.19 pm, a calculated density of 2.883 g·cm^{-3} at 203 K, and an energy gap of 1.50 eV, is formed (Ruzin et al. 2006a). This compound was prepared in the same way as $[K_{10}(H_2O)_{20}][Zn_4(\mu_4\text{-}Te)(SnTe_4)_4]$ was synthesized but using $MnCl_2{\cdot}4H_2O$ instead of $ZnCl_2$. Black rhombic dodecahedra of the title compound, which are soluble in water, crystallized after two days. Its single crystals can be isolated under inert conditions from the mother liquor without decomposition. All synthesis steps were performed with a strong exclusion of air and moisture (nitrogen atmosphere at a high-vacuum, double-manifold Schlenk line or Ar atmosphere in a glove box).

9.6 Tin–Hydrogen–Rubidium–Barium–Oxygen–Tellurium

In the system containing these elements, the multinary compound $[Rb_2Ba(H_2O)_{11}][SnTe_4]$, which crystallizes as a monoclinic structure with the lattice parameters a = 852.32 ± 0.17, b = 1509.5 ± 0.3, c = 955.38 ± 0.19 pm, β = 97.03 ± 0.03°, a calculated density of 3.091 g·cm^{-3} at 203 K, and an energy gap of ≈ 1.6 eV, is formed (Ruzin et al. 2006b). For obtaining this compound, a solution of $BaCl_2$ (0.500 mM) in H_2O (5 mL) was added to a solution of $[Rb_4(H_2O)_2][SnTe_4]$ (0.500 mM) in CH_3OH (5 mL). The resulting deeply red mixture was stirred overnight. Upon filtration, the red solution was layered by 10 mL of THF. After one day, the targeted compound crystallized as black parallelepipeds.

9.7 Tin–Hydrogen–Rubidium–Cadmium–Oxygen–Tellurium

In the system containing these elements, the multinary compound $[Rb_{10}(H_2O)_{20}][Cd_4(\mu_4\text{-}Te)(SnTe_4)_4]$, which crystallizes as a cubic structure with the lattice parameter a = 1610.02 ± 0.19 pm and a calculated density of 3.429 g·cm^{-3} at 203 K, is formed (Ruzin et al. 2006a). This compound was prepared in the same way as $[K_{10}(H_2O)_{20}][Zn_4(\mu_4\text{-}Te)(SnTe_4)_4]$ was synthesized but using $Rb_4(H_2O)_{0.5}][SnTe_4]$ instead of $[K_4(H_2O)_{0.5}][SnTe_4]$ and $CdCl_2$ instead of $ZnCl_2$. Black rhombic dodecahedra of the title compound, which are soluble in water, crystallized after two days. Its single crystals can be isolated under inert conditions from the mother liquor without decomposition. All synthesis steps were performed with a strong exclusion of air and moisture (nitrogen atmosphere at a high-vacuum, double-manifold Schlenk line or Ar atmosphere in a glove box).

9.8 Tin–Hydrogen–Rubidium–Oxygen–Tellurium

In the system containing these elements, the quinary compound $[Rb_4(H_2O)_2][SnTe_4]$, which crystallizes as a monoclinic structure with the lattice parameters a = 1999.3 ± 0.4, b = 782.25 ± 0.16, c = 1422.0 ± 0.3 pm, β = 129.33 ± 0.03°, a calculated density of 3.888 g·cm^{-3} at 203 K, and an energy gap of ≈ 1.9 eV, is formed (Ruzin et al. 2006b). This compound was prepared in the same was as $[K_4(H_2O)_{0.5}][SnTe_4]$ was synthesized but using Rb instead of K. The title compound was observed as reddish black blocks. All synthesis steps were performed with a strong exclusion of air and moisture (nitrogen atmosphere at a high-vacuum, double-manifold Schlenk line or Ar atmosphere in a glove box).

9.9 Tin–Hydrogen–Cesium–Mercury–Oxygen–Tellurium

In the system containing these elements, the multinary compound $[Cs_{10}(H_2O)_{20}][Hg_4(\mu_4\text{-}Te)(SnTe_4)_4]$, which crystallizes as a trigonal structure with the lattice parameters a = 1588.0 ± 0.2, c =1736.7 ± 0.4 pm and a calculated density of 4.842 g·cm^{-3} is formed (Ruzin et al. 2006a). To prepare this compound, $[Cs_4(H_2O)_2][SnTe_4]$ (0.2 mM) was dissolved in H_2O (3 mL) and added to a solution of $HgCl_2$ (0.2 mM) in 3 mL of H_2O, whereupon the reaction mixture immediately turned dark red. After stirring for 24 h, traces of a black precipitate was removed by filtration. The filtrate was layered by 6 mL of THF. Small black cuboids of the title compound formed after two days. Its single crystals can be isolated under inert conditions from the mother liquor without decomposition. All synthesis steps were performed with a strong exclusion of air and moisture (nitrogen atmosphere at a high-vacuum, double-manifold Schlenk line or Ar atmosphere in a glove box).

9.10 Tin–Hydrogen–Cesium–Oxygen–Tellurium

In the system containing these elements, the quinary compound $[Cs_4(H_2O)_2][SnTe_4]$, which crystallizes as a monoclinic structure with the lattice parameters a = 2075.7 ± 0.4, b = 804.18 ± 0.16, c = 1462.4 ± 0.3 pm, β = 129.31 ± 0.03°, a calculated density of 4.209 g·cm^{-3} at 203 K, and an energy gap of ≈ 1.9 eV, is formed (Ruzin et al. 2006b). This compound was prepared in the same was as $[K_4(H_2O)_{0.5}][SnTe_4]$ was synthesized but using Cs instead of K. The title compound was observed as reddish black blocks. All synthesis steps were performed with a strong exclusion of air and moisture (nitrogen atmosphere at a high-vacuum, double-manifold Schlenk line or Ar atmosphere in a glove box).

9.11 Tin–Sodium–Lead–Antimony–Tellurium

$NaSn_{18}SbTe_{20}$–$NaPb_{18}SbTe_{20}$. The $NaSn_xPb_{18-x}SbTe_{20}$ (x = 3, 5, 9, 13, and 16) phases are formed in this system (Guéguen et al. 2009). They melts at 877°C, 888°C, 864°C, 838°C, and 820°C and crystallize as a cubic structure with the lattice parameter a = 644.1 ± 0.2, 642.2 ± 0.1, 639.5 ± 0.2, 637.1 ± 0.2, and 637.8 ± 0.2 pm and an energy gap of 0.16, 0.16, < 0.1, < 0.1, < 0.1 eV for x = 3, 5, 9, 13, and 16, respectively. All samples were prepared as polycrystalline ingots in silica tubes by mixing high-purity Na, Sn, Pb, Sb, and Te in the appropriate stoichiometric ratio. To prevent the reaction between the sodium metal and silica, the tubes were carbon-coated prior to use. For example, Na (0.050 g), Pb (5.8583 g), Sn (1.2909 g), Sb (0.2648 g), and Te (5.5503 g) were used to prepare $NaSn_5Pb_{13}SbTe_{20}$. All components (except Na) were loaded into silica tubes under ambient atmosphere and the corresponding amount of Na was later added under an inert atmosphere in a dry glove box. The silica tubes were then flame-sealed under a residual pressure of ~0.01 Pa, placed into a tube furnace (mounted on a rocking table), and heated at 980°C for 4 h to allow complete melting of all components. While molten, the furnace was allowed to rock for 2 h to facilitate complete mixing and homogeneity of the liquid phase. The furnace was finally immobilized at the vertical position and was cooled from 980°C to 550°C over 43 h followed by a faster cool (6–8 h) to room temperature. The resulting ingots generally were silvery-metallic in color with a smooth surface. These phases are not the composition of a solid solution.

9.12 Tin–Sodium–Lead–Bismuth–Tellurium

$NaSn_{18}BiTe_{20}$–$NaPb_{18}BiTe_{20}$. The $NaSn_xPb_{18-x}BiTe_{20}$ (x = 3, 5, 9, 13, and 16) phases are formed in this system (Guéguen et al. 2009; He et al. 2009). They melt at 902°C, 889°C, 864°C, 839°C, and 820°C and crystallize as a cubic structure with the lattice parameter $a = 644.6 \pm 0.2$, 643.6 ± 0.2, 641.2 ± 0.2, 636.5 ± 0.2, and 634.2 ± 0.2 pm for x = 3, 5, 9, 13, and 16, respectively. The energy gap for all phases is < 0.1 eV. These phases were prepared in the same way as $NaSn_xPb_{18-x}SbTe_{20}$ were synthesized but using Bi instead of Sb. They are not the composition of a solid solution.

9.13 Tin–Silver–Lead–Antimony–Tellurium

SnTe–$AgSbTe_2$–PbTe. According to DTA, XRD, and metallography, a continuous series of sold solutions, having mainly p-type conductivity, is formed in this system (Rodot 1961).

10

Systems Based on Lead Sulfide

10.1 Lead–Hydrogen–Sodium–Nitrogen–Oxygen–Chlorine–Sulfur

Using the methods of DTA, XRD, and chemical analyses, it has been established that when PbS interacts with NH_4NO_3 + NaCl at temperatures of 180°C–350°C, $PbSO_4$, $PbCl_2$, PbO, and Pb are formed in the system (Hîncu 1971; Hîncu and Golgoţiu 1973).

10.2 Lead–Hydrogen–Sodium–Oxygen–Sulfur

The quinary compound $Na_6[Pb(S_2O_3)_4]·6H_2O$ is formed in the Pb–H–Na–O–S system (Egorov 2010). It was synthesized using two methods. In method 1, exact mass portions of $Na_2S_2O_3·5H_2O$ and $Pb(NO_3)_2$ were dissolved in a minimum volume of water and combined by pouring a solution of $Pb(NO_3)_2$ to a solution of $Na_2S_2O_3$; the relevant reaction is $Pb(NO_3)_2 + xNa_2S_2O_3 \rightarrow Na_{2x-2}[Pb(S_2O_3)_x] + 2NaNO_3$, where $x = 2–4$. This reaction proceeded in two stages: poorly soluble PbS_2O_3 was first formed in the solution (the solubility product was 4×10^{-7}) and then dissolved in a $Na_2S_2O_3$ excess.

In method 2, PbS_2O_3 powder was dissolved in a saturated solution of $Na_2S_2O_3$; the relevant reaction is $PbS_2O_3 + yNa_2S_2O_3 \rightarrow Na_{2y}[Pb(S_2O_3)_{y+1}]$, where $y = 1–3$. PbS_2O_3 is formed and then the solutions were combined according to method 1 and the PbS_2O_3 powder used in method 2 is completely dissolved only when the molar ratio $Pb^{2+}/S_2O_3^{2-}$ = 1:4 or higher. At other ratios, insoluble PbS_2O_3 remains in the solution. On heating, the storage solutions of $Na_6[Pb(S_2O_3)_4]·6H_2O$ decompose to form PbS, which makes their isolation impossible by the evaporation of the solvent. The targeted compound can be isolated from solutions using alcohol. When ethanol is used, it is isolated as an oily liquid, which is separated and treated with absolute ethanol (without additional treatment, the liquid decomposes with time to form PbS). The oily liquid gradually transforms into a powder, which is filtered off, washed, and dried in air.

A drawback of method 1 is the necessity to re-precipitate the product because of its contamination with $NaNO_3$, which enhances the decomposition of $Na_6[Pb(S_2O_3)_4]·6H_2O$. In this case, the yield of the product decreases.

The product was isolated as a no hygroscopic white powder. The complete dehydration of $Na_6[Pb(S_2O_3)_4]·6H_2O$ occurs at 115°C and its complete decomposition occurs in the range 180°C-350°C.

10.3 Lead–Hydrogen–Copper–Calcium–Oxygen–Selenium–Tellurium–Sulfur

In the system containing these elements, the multinary compound $(Ca_{0.5}Pb_{0.5})Pb_3Cu_6Te^{6+}{}_2O_6(Te^{4+}O_3)_6(Se^{4+}O_3)_2(SO_4)_2·3H_2O$ (mineral tombstoneite), which crystallizes as a trigonal structure with the lattice parameters $a = 913.77 \pm 0.9$ and $c = 1227.97 \pm 0.09$ pm, is formed (Kampf et al. 2021c,d).

10.4 Lead–Hydrogen–Copper–Calcium–Oxygen–Tellurium–Sulfur

In the system containing these elements, the multinary compound $(Ca,Pb)_3CaCu_6[Te^{4+}{}_3Te^{6+}O_{12}]_2(Te^{4+}O_3)_2(SO_4)_2·3H_2O$ (mineral tlapallite), which crystallizes as a trigonal structure with the lattice parameters

DOI: 10.1201/9781003123460-10

$a = 912.19 \pm 0.17$, $c = 1193.20 \pm 0.09$ pm, and a calculated density of 5.035 g·cm^{-3}, is formed (Missen et al. 2019a). According to the earlier data (Williams and Duggan 1978; Fleischer et al. 1979c), this mineral crystallizes as a monoclinic structure with the lattice parameters a =1207 and 1197, b = 913 and 911, c = 1594 and 1566, β = 90°27' and 90°36', and the calculated and experimental densities of 5.465 and 5.05, and 5.38 and >4.65 g·cm^{-3}, respectively, for two different samples.

10.5 Lead–Hydrogen–Copper–Zinc–Oxygen–Sulfur

In the system containing these elements, the multinary compound $[Pb_8Zn_3Cu(OH)_{16}](SO_4)_4{\cdot}4H_2O$ (mineral cuprocherokeeite), which crystallizes as a monoclinic structure with the lattice parameters $a = 1268.28 \pm 0.05$, $b = 946.29 \pm 0.05$, $c = 1478.76 \pm 0.08$ pm, $\beta = 94.798 \pm 0.004°$, and a calculated density of 5.032 g·cm^{-3}, is formed (Kampf et al. 2022g, 2023a,c).

10.6 Lead–Hydrogen–Copper–Aluminum–Antimony–Chlorine–Oxygen–Sulfur

In the system containing these elements, the multinary compound $Cu_4AlPb_6SbO_2(SO_4)_2Cl_4(OH)_{16}$ (mineral mammothite), which crystallizes as a monoclinic structure with the lattice parameters $a = 1895.9 \pm 0.4$, $b = 733.98 \pm 0.19$, $c = 1136.3 \pm 0.3$ pm, and $\beta = 112.428 \pm 0.009°$ (Grice and Cooper 2014) [$a = 1893 \pm 3$, $b = 733 \pm 1$, $c = 1135 \pm 2$ pm, $\beta = 112.44 \pm 0.10°$, and a calculated density of 5.21 g·cm^{-3} (Effenberger 1985); $a = 1889 \pm 3$, $b = 722 \pm 1$, $c = 1131 \pm 2$ pm, $\beta = 112.43 \pm 0.18°$, and a calculated density of 5.25 g·cm^{-3} (Peacor et al. 1985; Dunn et al. 1986)], is formed.

10.7 Lead–Hydrogen–Copper–Aluminum–Oxygen–Sulfur

In the system containing these elements, the multinary compound $CuAl_2Pb(SO_4)_2(OH)_6$ (mineral osarizawaite), which crystallizes as a rhombohedral structure with the lattice parameters a = 704.5 pm and α = 60°03' or a = 705 and c = 1725 pm in hexagonal setting and the calculated and experimental densities of 4.114 and 4.037 g·cm^{-3}, respectively (Morris 1962, 1963) [a = 704 pm and α = 60°06' or a = 705 and c = 1723 pm in hexagonal setting and the calculated and experimental densities of 4.20 and 3.89–4.02 g·cm^{-3}, respectively (Taguchi 1961; Fleischer et al. 1962)], is formed.

10.8 Lead–Hydrogen–Copper–Carbon–Oxygen–Sulfur

In the system containing these elements, the multinary compound $Cu_2Pb_5(SO_4)_3(CO_3)(OH)_6$ (mineral caledonite), which crystallizes as an orthorhombic structure with the lattice parameters $a = 2008.5 \pm 0.3$, $b = 714.1 \pm 0.1$, and $c = 656.3 \pm 0.1$ pm (Schofield et al. 2009) [a = 714, b = 2006, and c = 655 pm (Palache and Richmond 1939); $a = 2008.9 \pm 0.7$, $b = 714.6 \pm 0.3$, $c = 656.0 \pm 0.5$ pm, and the calculated and experimental densities of 5.69 and 5.76 g·cm^{-3}, respectively (Giacovazzo et al. 1973); $a = 2008.8 \pm 0.2$, $b = 714.3 \pm 0.1$, and $c = 656.4 \pm 0.1$ pm (Giacovazzo et al. 1976); $a = 2004.3 \pm 0.8$, $b = 713.5 \pm 0.2$, and $c = 656.0 \pm 0.5$ pm (Zidarov et al. 2007)], is formed .

10.9 Lead–Hydrogen–Copper–Phosphorus–Oxygen–Sulfur

In the system containing these elements, the multinary compound $CuPb_2(PO_4)(SO_4)(OH)$ (mineral tsumebite), which crystallizes as a monoclinic structure with the lattice parameters a = 870, b = 580,

c = 785 pm, β = 111.5°, and an experimental density of 6.01 g·cm^{-3}, is formed (Bideaux et al. 1966; Nichols 1966).

10.10 Lead–Hydrogen–Copper–Arsenic–Oxygen–Sulfur

In the system containing these elements, the multinary compound $CuPb_2(AsO_4)(SO_4)(OH)$ (mineral arsentsumebite), which crystallizes as a monoclinic structure with the lattice parameters a = 780.4 ± 0.8, b = 589.0 ± 0.6, c = 896.4 ± 0.8 pm, β = 112.29 ± 0.06°, and a calculated density of 6.481 g·cm^{-3} (Zubkova et al. 2002) [a = 885, b = 592, c = 784 pm, β = 112.6°, and an experimental density of 6.46 g·cm^{-3} (Bideaux et al. 1966)], is formed .

10.11 Lead–Hydrogen–Copper–Oxygen–Sulfur

Some quinary compounds are formed in the Pb–H–Cu–O–S system. $CuPb(SO_4)(OH)_2$ (mineral linarite) crystallizes as a monoclinic structure with the lattice parameters a = 968.2 ± 0.2, b = 564.6 ± 0.1, c = 468.3 ± 0.6 pm, and β = 102.66 ± 0.01° (Schofield et al. 2009) [a = 980, b = 565, c = 468 pm, β = 105°04', and a calculated density of 5.319 g·cm^{-3} (Berry 1951); a = 981, b = 565, c = 470 pm, β = 104.7° (Bachmann and Zemann 1961); a = 968 ± 2, b = 565 ± 1, c = 468.5 ± 1.0 pm, and β = 102°36 ' ± 15' (Araki 1962); a = 970.1 ± 0.2, b = 565.0 ± 0.2, c = 469.0 ± 0.2 pm, β = 102.65 ± 0.02°, and a calculated density of 5.31 g·cm^{-3} (Effenberger 1987)].

$CuPb_4(SO_4)_2(OH)_6$ (mineral chenite) crystallizes as a triclinic structure with the lattice parameters a = 579.1 ± 0.1, b = 794.0 ± 0.1, c = 797.6 ± 0.1 pm, α = 112.02 ± 0.01°, β = 97.73 ± 0.01°, γ = 100.45 ± 0.01°, and the calculated and experimental densities of 6.044 and 5.98 ± 0.02 g·cm^{-3}, respectively (Paar et al. 1986; Hawthorne et al. 1987).

$CuPb_4(SO_4)O_2(OH)_4{\cdot}H_2O$ (mineral elyite) also crystallizes as a monoclinic structure with the lattice parameters a = 1423.3 ± 0.2, b = 1153.2 ± 0.1, c = 1461.1 ± 0.2 pm, β = 100.45 ± 0.01°, and a calculated density of 6.232 g·cm^{-3} (Kolitsch and Giester 2000) [a = 1428.8 ± 0.2, b = 576.8 ± 0.2, c = 730.9 ± 0.2 pm, β = 100°26' ± 1', and a calculated density of 6.321 g·cm^{-3} (Williams 1972); a = 1424.4 ± 0.1, b = 1153.6 ± 0.1, c = 1465.6 ± 0.1 pm, and β = 100.45 ± 0.01° (Miyawaki et al. 1997; Jambor and Roberts 1998)].

$Cu_4Pb(SO_4)_2(OH)_6{\cdot}3H_2O$ (mineral lautenthalite) also crystallizes as a monoclinic structure with the lattice parameters a = 2164.2 ± 0.8, b = 604.0 ± 0.2, c = 2254.4 ± 0.8 pm, β = 108.2 ± 0.1°, and a calculated density of 3.84 g·cm^{-3} (Jambor et al. 1991b).

$[Pb_4O_{1.5}(OH)_{2.5}]_2[Cu_5(S_2O_3)_4(S_2O_2OH)_2(H_2O)]{\cdot}4H_2O$ (mineral hayelasdiite) crystallizes as a triclinic structure with the lattice parameters a = 752.09 ± 0.01, b = 1493.45 ± 0.03, c = 1798.9 ± 0.1 pm, α = 106.727 ± 0.008, β = 90.966 ± 0.006°, and γ = 90.031 ± 0.006° (Kampf et al. 2022h,i).

$Pb_5(OH)_5[Cu(S^{6+}O_3S^{2-})_3](H_2O)_2$ (mineral steverustite) crystallizes as a monoclinic structure with the lattice parameters a = 1256.31 ± 0.07, b = 889.63 ± 0.05, c = 1801.32 ± 0.11 pm, β = 96.459 ± 0.001°, and a calculated density of 5.150 g·cm^{-3} (Cooper et al. 2009).

10.12 Lead–Hydrogen–Copper–Oxygen–Selenium–Sulfur

Two multinary compounds are formed in the Pb–H–Cu–O–Se–S system. $Pb_2Cu_2(SO_4)(SeO_3)(OH)_4$ (mineral munakataite) crystallizes as a monoclinic structure with the lattice parameters a = 980.23 ± 0.26, b = 567.51 ± 0.14, c = 928.11 ± 0.25 pm, β = 102.443 ± 0.006°, and a calculated density of 5.305 g·cm^{-3} (Kampf et al. 2010d) [a = 976.6 ± 0.8, b = 566.6 ± 0.5, c = 929.1 ± 1.0 pm, β = 102.40 ± 0.08°, and a calculated density of 5.526 g·cm^{-3} (Matsubara et al. 2008; Ercit et al. 2009)].

$Pb_5Cu_2(SO_4)_3(SeO_3)(OH)_6$ (mineral viskontite)) crystallizes as an orthorhombic structure with the lattice parameters a = 2062.65 ± 0.04, b = 720.40 ± 0.01, and c = 652.20 ± 0.01 pm (Pekov et al. 2023a,b).

10.13 Lead–Hydrogen–Copper–Oxygen–Tellurium–Sulfur

Two multinary compounds are formed in the Pb–H–Cu–O–Te–S system. $Pb_2Cu_4Te_2O_{10}(SO_4)(OH)_2{\cdot}H_2O$ (mineral bairdite) crystallizes as a monoclinic structure with the lattice parameters a = 1431.26 ± 0.10, b = 522.67 ± 0.03, c = 948.78 ± 0.05 pm, β = 106.815 ± 0.007°, and a calculated density of 6.021 g·cm^{-3} (Kampf et al. 2013c,e).

The multinary compound $Pb_4Cu_4Te_2O_{11}(SO_4)_2(OH)_2(H_2O)$ (mineral flaggite) crystallizes as a triclinic structure with the lattice parameters a = 956.10 ± 0.02, b = 997.55 ± 0.02, c = 1044.49 ± 0.03 pm, α = 74.884 ± 0.001, β = 89.994 ± 0.001°, γ = 78.219 ± 0.001°, and a calculated density of 6.137 g·cm^{-3} (Kampf et al. 2021a,b,2022a).

10.14 Lead–Hydrogen–Copper–Oxygen–Iron–Sulfur

In the system containing these elements, the multinary compound $Pb(Fe_2Cu)(SO_4)_2(OH)_6$ [mineral beaverite-(Cu)], which crystallizes as a trigonal structure with the lattice parameters a = 729.28 ± 0.13, c = 1724.2 ± 0.6 pm, and a calculated density of 4.208 g·cm^{-3} (Sato et al. 2009) [a = 722.88 ± 0.27 and c = 3440.7 ± 1.4 pm for natural mineral and a = 732.088 ± 0.026 and c = 1703.36 ± 0.07 pm for synthetic compound (Hudson-Edwards et al. 2008)], is formed .

Synthetic $Pb(Fe_2Cu)(SO_4)_2(OH)_6$ was prepared using a 1 L aqueous solution containing 0.054 M $Fe_2(SO_4)_3{\cdot}5H_2O$ and 0.02 M H_2SO_4, with the addition of 0.315 M $CuSO_4{\cdot}5H_2O$ (Hudson-Edwards et al. 2008). The solution was subsequently placed in a 2 L reaction vessel fitted with spiral condensers and then heated by means of a sand bath to 95°C (0.1 MPa) with constant stirring (400 rpm). When the solution temperature reached 95°C, 200 mL of 0.03 M $Pb(NO_3)_2$ was added, with stirring, to the solution at a rate of 6 mL·h^{-1}. Once all the $Pb(NO_3)_2$ had been added, the precipitates were stirred (400 rpm) for a further 5 h, after which they were allowed to settle and the residual supernatant solutions decanted. The precipitates were then washed several times with ultrapure water and dried at 110°C for 24 h.

$Pb(Fe_2Cu)(SO_4)_2(OH)_6$–$Pb_{0.5}Fe_3(SO_4)_2(OH)_6$. Solid-solution series was synthesized in this system using optimum procedures developed for $Pb_{0.5}Fe_3(SO_4)_2(OH)_6$ (Jambor and Dutrizac 1983, 1985). The experiments were done in a Ti autoclave for 24 h at a 600-rpm stirring speed. Various concentrations of Cu^{2+} (0 to 252 g·L^{-1}) were maintained by the addition of $CuSO_4$. To reduce the extent of hydronium ion incorporation, all solutions contained 2.0 M Li_2SO_4. At the completion of the experiments, soluble salts and excess $PbSO_4$ were selectively leached from the solution by washing with four 1-L portions of 10 mass% solution of NH_4CH_3COO at 25°C. The residue was then filtered, washed with water, and dried at 110°C.

10.15 Lead–Hydrogen–Magnesium–Carbon–Oxygen–Chlorine–Manganese–Sulfur

In the system containing these elements, the multinary compound $Pb_{12}O_6Mn_7(SO_4)(CO_3)_4Cl_4(OH)_{12}$ (mineral philolithite), which crystallizes as a tetragonal structure with the lattice parameters a = 1262.7 ± 0.9 and c = 1259.5 ± 0.9 pm, and a calculated density of 5.91·g cm^{-3}, is formed (Kampf et al. 1998; Jambor et al. 1999; Moore et al. 2000).

10.16 Lead–Hydrogen–Zinc–Arsenic–Oxygen–Iron–Sulfur

In the system containing these elements, the multinary compound $(Zn,Fe)Pb_2(AsO_4,SO_4)_2(OH,H_2O)$ (mineral feinglosite), which crystallizes as a monoclinic structure with the lattice parameters a = 897.3 ± 0.6, b = 595.5 ± 0.3, c = 776.6 ± 0.6 pm, β = 112.20 ± 0.06°, and a calculated density of 6.52·g cm^{-3}, is formed (Clark et al. 1997; Jambor et al. 1998).

10.17 Lead–Hydrogen–Zinc–Oxygen–Sulfur

Some quinary compounds are formed in the Pb–H–Zn–O–S system. $[Pb(H_2O)_{10}][Zn_{12}(OH)_{20}(H_2O)(SO_4)_3]$ (mineral haywoodite) crystallizes as a triclinic structure with the lattice parameters a = 835.298 ± 0.019, b = 1327.69 ± 0.07, c = 1827.44 ± 0.13 pm, α = 92.427 ± 0.008, β = 90.419 ± 0.006°, and γ = 108.214 ± 0.004°, and the calculated and experimental densities of 3.128 and 3.27 ± 0.02 g·cm^{-3}, respectively (Kampf et al. 2022j,k,2023e).

$[Pb_2Zn(OH)_4](SO_4)\cdot H_2O$ (mineral cherokeeite) crystallizes as a monoclinic structure with the lattice parameters a = 1716.97 ± 0.07, b = 647.173 ± 0.019, c = 1753.04 ± 0.12 pm, β = 115.440 ± 0.008°, and a calculated density of 5.011·g cm^{-3} (Kampf et al. 2022e,f,2023b).

$Pb_4Zn(OH)_6(SO_4)_2$ (mineral zincochenite) crystallizes as a triclinic structure with the lattice parameters a = 588.3 ± 0.4, b = 793.8 ± 0.5, c = 794.8 ± 0.5 pm, α = 110.511 ± 0.012, β = 98.497 ± 0.010°, and γ = 100.152 ± 0.008° (Kampf et al. 2022r,s).

$[Pb_8O_2Zn(OH)_6](S_2O_3)_4$ (mineral redmondite) crystallizes as a monoclinic structure with the lattice parameters a = 916.72 ± 0.04, b = 1065.76 ± 0.04, c = 1406.20 ± 0.10 pm, β = 101.173 ± 0.007°, and a calculated density of 5.757·g cm^{-3} (Kampf et al. 2022n,o,2023f).

$[Pb_8O_2Zn(OH)_6](S_2O_3)_4\cdot 2H_2O$ (mineral hydroredmondite) also crystallizes as a monoclinic structure with the lattice parameters a = 1259.91 ± 0.09, b = 928.19 ± 0.04, c = 1297.74 ± 0.09 pm, β = 90.443 ± 0.006°, and a calculated density of 5.124·g cm^{-3} (Kampf et al. 2022l,m,2023f).

$[Pb_8O_2Zn(OH)_6](SO_4)_4\cdot 6H_2O$ (mineral sulfatoredmondite) also crystallizes as a monoclinic structure with the lattice parameters a = 1729.4 ± 0.2, b = 736.68 ± 0.09, c = 1272.71 ± 0.18 pm, β = 110.622 ± 0.009°, and a calculated density of 5.173·g cm^{-3} (Kampf et al. 2022p,q,2023f).

10.18 Lead–Hydrogen–Zinc–Oxygen–Iron–Sulfur

In the system containing these elements, the multinary compound $Pb(Fe_2Zn)(SO_4)_2(OH)_6$ [mineral beaverite-(Zn)], which crystallizes as a trigonal structure with the lattice parameters a = 730.28 ± 0.02, c = 1705.17 ± 0.04 pm, and a calculated density of 4.25·g cm^{-3}, is formed (Sato et al. 2008, 2011; Belakovskiy et al. 2012).

10.19 Lead–Hydrogen–Boron–Oxygen–Sulfur

In the system containing these elements, the quinary compound $Pb_2[(BO_2)(OH)]SO_4$, which is thermally stable up to about 335°C and crystallizes as a monoclinic structure with the lattice parameters a = 698.17 ± 0.12, b = 539.17 ± 0.08, c = 816.66 ± 0.19 pm, β = 98.643 ± 0.019°, a calculated density of 6.231 g·cm^{-3}, and an energy gap of 4.08 eV, is formed (Ruan et al. 2018). Crystals of this compound were synthesized by hydrothermal reaction. Typically, $Pb(BO_2)_2\cdot H_2O$ (1 mM), H_2SO_4 solution (0.1 mL), $NH_3\cdot H_2O$ (0.25 mL), and H_2O (3 mL) were sealed in a 23 mL Teflon-lined stainless steel autoclave, and the pH value of the reaction mother liquid was about 5. The autoclave was heated at 220°C for three days, and then cooled to room temperature at a rate of 3°C·h^{-1}. The reaction products were washed with water and then dried in air. Colorless prism-shaped crystals of the targeted compound were obtained.

10.20 Lead–Hydrogen–Aluminum–Carbon–Oxygen–Manganese–Sulfur

The multinary compound $Al_4PbMn_3(CO_3)_4(SO_4)O_5\cdot 5H_2O$ (mineral nasledovite), which has an experimental density 3.069 g·cm^{-3}, is formed in the Pb–H–Al–C–O–Mn–S system (Enikeev 1958; Fleischer 1959).

10.21 Lead–Hydrogen–Aluminum–Phosphorus–Oxygen–Sulfur

In the system containing these elements, the multinary compound $PbAl_3(PO_4)(SO_4)(OH)_6$ (mineral hinsdalite) (Stanley 1987), which crystallizes as a rhombohedral structure with the lattice parameters a = 702.9 ± 0.4 and c = 1678.9 ± 0.4 pm in hexagonal setting (Kolitsch et al. 1999) [a = 690 pm, $\alpha = 61°08'$ and a calculated density of 4.06 g·cm^{-3} and a = 688 pm, $\alpha = 61°08'$ and a calculated density of 3.92·g cm^{-3} for two different samples, respectively (Baker 1963)], is formed .

The $H_6Pb_{10}Al_{20}(PO_4)_{12}(SO_4)_5(OH)_{40}\cdot 11H_2O$ multinary compound (mineral orpheite), which melts at about 1200°C and crystallizes as a trigonal structure with the lattice parameters a = 700 ± 2, c = 1672 ± 2 pm, and an experimental density of 3.75 ± 0.01 g·cm^{-3}, is also formed (Fleischer et al. 1976). This mineral needs further study. It is probably a variety of hinsdalite [$PbAl_3(PO_4)(SO_4)(OH)_6$].

10.22 Lead–Hydrogen–Aluminum–Arsenic–Oxygen–Sulfur

In the system containing these elements, the multinary compound $PbAl_3(AsO_4)(SO_4)(OH)_6$ (mineral hidalgoite), which crystallizes as a rhombohedral structure with the lattice parameters a = 711.42 ± 0.04 and c = 1709.73 ± 0.09 pm in hexagonal setting (Cooper and Hawthorne 2012) [a = 697 pm and α = 60°40' or a = 704 ± 2 and c = 1699 ± 2 pm in hexagonal setting and the calculated and experimental densities of 4.27 and 3.96 g·cm^{-3}, respectively (Smith et al. 1953)], is formed .

10.23 Lead–Hydrogen–Aluminum–Oxygen–Sulfur

Two quinary compounds, $[Pb_4O_2Al(OH)_5]_2(S_2O_3)_2\cdot H_2(S_2O_3)(H_2O)_5$ (mineral dinilawiite) and $Pb_3Al(OH)_6(SO_4)(OH)$ (mineral krivovichevite), are formed in this system. The first crystallizes as a monoclinic structure with the lattice parameters a = 1741.00 ± 0.05, b = 921.91 ± 0.02, c = 2169.8 ± 0.1 pm, and β = 107.276 ± 0.008° (Kampf et al. 2023b, 2024). The second compound crystallizes as a trigonal structure with the lattice parameters a = 774.2 ± 0.2, c = 3208.2 ± 0.9 pm, and a calculated density of 5.17 g·cm^{-3} (Krivovichev et al. 2009) [a = 769.3 ± 0.8, c = 3157 ± 9 pm, and a calculated density of 5.37 g·cm^{-3} (Poirier and Piilonen 2007; Yakovenchuk et al. 2007)].

10.24 Lead–Hydrogen–Gallium–Arsenic–Oxygen–Sulfur

In the system containing these elements, the multinary compound $PbGa_3(AsO_4,SO_4)_2(OH)_6$ (mineral gallobeudantite), which crystallizes as a rhombohedral structure with the lattice parameters a = 722.5 ± 0.4 and c = 1703 ± 2 pm in hexagonal setting and a calculated density of 4.61 g·cm^{-3}, is formed (Jambor et al. 1996b, 1997c).

10.25 Lead–Hydrogen–Carbon–Oxygen–Sulfur

In the system containing these elements, the quinary compound $Pb_4(CO_3)_2(SO_4)(OH)_2$, which has three polymorphic modifications, is formed . The first modification, mineral leadhillite, crystallizes as a monoclinic structure with the lattice parameters a = 910.4 ± 0.2, b = 2079.2 ± 0.6, c = 1157.7 ± 0.3 pm, and β = 90.50 ± 0.02° (Bindi and Menchetti 2005) [a = 908, b = 2076, c = 1156 pm, and β = 90°27.5' (Fleischer 1970); a = 909.9 ± 0.9, b = 2078.0 ± 8.6, c = 1156.4 ± 0.7 pm, and β = 90.255 ± 0.235° (Russell et al. 1983); a = 1158.2 ± 0.2, b = 2080.9 ± 0.3, c = 911.1 ± 0.3 pm, and β = 90.48° (Highcock et al. 1984)].

Reactions undergone by leadhillite on heating to 1000°C have been followed by DTA, thermogravimetry, differential scanning calorimetry, evolved gas analysis, continuous-heating XRD and IR, and hot stage microscopy by Milodowski and Morgan (1984). Intermediate decomposition products were identified by X-ray powder photography. At 80°C, biaxial leadhillite inverts to a uniaxial phase with properties similar to those of susannite, but this higher-temperature modification only partially reverts to the original structure on cooling (up to 24 h at room temperature is required for complete reversion). Between 250°C and 600°C the mineral undergoes two decomposition reactions: $PbO{\cdot}PbCO_3$ and $PbO{\cdot}PbSO_4$ form during the first reaction ($PbCO_3$ may form in the initial stages) and $4PbO{\cdot}PbSO_4$ during the second. α-$2PbO{\cdot}PbSO_4$ appears at 650°C due to solid state reaction between the other lead oxysulfate products. Melting occurs above 850°C.

The second modification, mineral macphersonite, crystallizes as an orthorhombic structure with the lattice parameters $a = 924.2 \pm 0.2$, $b = 2305.0 \pm 0.5$, $c = 1038.3 \pm 0.2$ pm, and a calculated density of 6.480 g·cm^{-3} (Steele et al. 1998) [$a = 909.4 \pm 0.3$, $b = 2078.9 \pm 0.9$, and $c = 1155.9 \pm 0.4$ pm (Russell et al. 1983); $a = 922.7 \pm 0.2$, $b = 2304.8 \pm 0.5$, and $c = 1036.8 \pm 0.4$ pm (Highcock et al. 1984); $a = 1037$, $b = 2310$, $c = 925$ pm, and the calculated and experimental densities of 6.60–6.65 and 6.50–6.55 g·cm^{-3}, respectively (Livingstone and Sarp 1984; Dunn et al. 1985c)].

The third modification, mineral susannite, crystallizes as a trigonal structure with the lattice parameters $a = 907.7 \pm 0.8$ and $c = 1161.1 \pm 0.9$ pm (Bindi and Menchetti 2005) [$a = 905$ and $c = 1154$ pm (Fleischer 1970); $a = 907.2$ and $c = 1153.9$ pm (Livingstone and Russel 1985); $a = 907.18 \pm 0.07$, $c = 1157.0 \pm 0.1$ pm, and a calculated density of 6.52 g·cm^{-3} (Steele et al. 1999)]. Susannite decomposes at about 300°C to $Pb_5(SO_4)O_4$.

The reversible phase transformation between leadhillite and susannite takes place at about 80°C (Steele et al. 1999). According to Fleischer (1970), single crystals of leadhillite are converted to susannite at 200°C and revert to leadhillite on cooling to room temperature.

10.26 Lead–Hydrogen–Nitrogen–Oxygen–Sulfur

PbS–NH_4NO_3. This system is a nonquasibinary section of the Pb–H–N–O–S quinary system (Hîncu 1971; Hîncu and Golgoţiu 1973). Using the methods of DTA, XRD, and chemical analyses, it has been established that when PbS interacts with NH_4NO_3 at temperatures of 180°C–380°C, $PbSO_4$, Pb, and $(NH_4)_2SO_4$ are formed.

The quinary compound $(NH_4)_2Pb(SO_4)_2$, which crystallizes as a rhombohedral structure with the lattice parameters $a = 796.3$ and $\alpha = 41°02'$ or $a = 558.1$ and $c = 2184.7$ pm in hexagonal setting and the calculated and experimental densities of 3.680 and 3.67 g·cm^{-3}, respectively (Schwarz 1966) [$a = 558 \pm 1$ and $c = 2184 \pm 5$ pm in hexagonal setting and an experimental density of 3.69 ± 0.03 g·cm^{-3} (Møller 1954)], is formed in this system.

To prepare this compound, 30 g of $(NH_4)_2SO_4$ in H_2O (100 mL) was slowly added to a solution of $Pb(CH_3COO)_2{\cdot}3H_2O$ (2.5 g) in H_2O (10 mL) (Schwarz 1966). After stirring the suspension for some days, the product was suctioned off sharply, pressed onto clay in a thin layer, and air dried. Large single crystals of the title compound were prepared in the following way (Møller 1954). Freshly precipitated $PbSO_4$ was boiled with a strong solution of $Pb(CH_3COO)_2$ and $(NH_4)_2SO_4$; upon filtering the hot solution, to get rid of excess of $PbSO_4$, and leaving the filtrate to cool slowly, nice crystals precipitated. They were filtered, washed with dilute alcohol, and dried at room temperature.

10.27 Lead–Hydrogen–Phosphorus–Oxygen–Iron–Sulfur

In the system containing these elements, the multinary compound $PbFe_3(PO_4)(SO_4)(OH)_6$ (mineral corkite), which crystallizes as a rhombohedral structure with the lattice parameters $a = 730.78 \pm 0.03$ and $c = 1685.5 \pm 0.2$ pm in hexagonal setting and a calculated density of 6.9243 ± 0.0007 g·cm^{-3} (Ferrenti et al. 2023) [$a = 728.0 \pm 0.1$ and $c = 1682.1 \pm 0.1$ pm in hexagonal setting and a calculated density of 4.309 g·cm^{-3} (Giuseppetti and Tadini 1987); $a = 730.65 \pm 0.05$ and $c = 1689.7 \pm 0.2$ pm in hexagonal setting and a calculated density of 4.281 g·cm^{-3} (Sato et al. 2009)], is formed .

Corkite samples were synthesized in Tefon-lined hydrothermal vessels at elevated temperatures (Ferrenti et al. 2023). Heating and cooling ramp rates for all syntheses were set at 100°C·h^{-1}. The precursor

$Pb_5(PO_4)_3Cl$ was synthesized in two steps. First, stoichiometric quantities of PbO, and $NH_4H_2PO_4$ were thoroughly ground and heated in an uncovered Pt crucible for 2 days at 1000°C. The product $Pb_3(PO_4)_2$ powder was then ground in a 3:2 molar ratio with $PbCl_2$ and heated for 2 days at 850°C in a covered alumina crucible. $FePO_4$ was also synthesized in two stages. First, equimolar amounts of $NH_4H_2PO_4$ and $Fe(NO_3)_3{\cdot}9H_2O$ were dissolved in deionized water, stirred until clear, then boiled until dry. The resulting powder was dried for 24 h at 400°C, and then for an additional 24 h at 700°C. Corkite samples were produced by the combination of a ground mixture of $Pb_5(PO_4)_3Cl$ (0.95 M), $FePO_4$ (2.00 M), and $Fe_2(SO_4)_3{\cdot}9H_2O$ (1.67 M) with $FeCl_3{\cdot}6H_2O$ (9.67 M). All powders were added to the Teflon liner, filled to ca. 70% with distilled H_2O, further acidified with 3 drops of ≈ 5M H_2SO_4 solution, sealed, and heated for 4 days at 98°C. Samples produced by the above method consistently resulted in large $Pb_5(PO_4)_3Cl$ and $PbSO_4$ impurities. A 5% deficiency of $Pb_5(PO_4)_3Cl$ was ultimately found to produce the purest product mixture (≈ 91% of corkite).

10.28 Lead–Hydrogen–Arsenic–Antimony–Oxygen–Sulfur

In the system containing these elements, the multinary compound $Pb_{0.59}Fe_3(AsO_4)_{0.18}(SO_4)_{1.82}(OH)_6$ (mineral arsenium plumbojarosite), which crystallizes as a trigonal structure with the lattice parameters $a = 731.43 \pm 0.01$ and $c = 3373.6 \pm 0.1$ pm, is formed (Mills et al. 2009b).

10.29 Lead–Hydrogen–Arsenic–Antimony–Oxygen–Iron–Sulfur

In the system containing these elements, the multinary compound $PbFe_3(AsO_4)(SO_4)(OH)_6$ (mineral beudantite), which crystallizes as a trigonal structure with the lattice parameters $a = 731.51 \pm 0.09$, $c = 1703.55 \pm 0.05$ pm, and the calculated and experimental densities of 4.49 and 4.48 ± 0.22 g·cm^{-3}, is formed (Szymański 1988).

10.30 Lead–Hydrogen–Arsenic–Oxygen–Iron–Sulfur

In the system containing these elements, the multinary compound $Pb_3Sb(SO_4)(AsO_4)(OH)_6{\cdot}3H_2O$ (mineral mallestigite), which crystallizes as a hexagonal structure with the lattice parameters $a = 893.8$, $c = 1109.8$ pm, and a calculated density of 4.91 g·cm^{-3}, is formed (Jambor and Roberts 2004).

10.31 Lead–Hydrogen–Oxygen–Tellurium–Sulfur

In the system containing these elements, the quinary compound $Pb_{10}Te_6O_{20}(OH)_{14}(SO_4){\cdot}5H_2O$ (mineral schieffelinite), which crystallizes as an orthorhombic structure with the lattice parameters $a = 965.81 \pm 0.03$, $b = 1958.33 \pm 0.07$, $c = 1050.27 \pm 0.07$ pm, and a calculated density of 6.055 g·cm^{-3} (Kampf et al. 2012c) [$a = 967$, $b = 1956$, $c = 1047$ pm, and the calculated and experimental densities of 5.15 and 4.98 ± 0.12 g·cm^{-3}, respectively (Williams 1980; Fleischer et al. 1981)], is formed .

10.32 Lead–Hydrogen–Oxygen–Chromium–Fluorine–Chlorine–Sulfur

In the system containing these elements, the multinary compound $Pb_6Cr^{3+}(Cr^{6+}O_4)_2(SO_4)(OH)_7FCl$ (mineral evanichite), which crystallizes as a trigonal structure with the lattice parameters $a = 776.51 \pm 0.14$, $c = 961.99 \pm 0.17$ pm, and a calculated density of 5.878 g·cm^{-3} (Yang et al. 2023b) [$a = 775.88 \pm 0.05$ and $c = 961.61 \pm 0.06$ pm (Yang et al. 2022c,d)], is formed .

10.33 Lead–Hydrogen–Oxygen–Chlorine–Sulfur

In the system containing these elements, the quinary compound $Pb_{10}(SO_4)O_7Cl_4{\cdot}H_2O$ (mineral symesite), which crystallizes as a triclinic structure with the lattice parameters a = 1972.7 ± 0.02, b = 879.6 ± 0.1, c = 1363.1 ± 0.2 pm, α = 82.21 ± 0.01, β = 78.08 ± 0.01°, γ = 100.04 ± 0.01°, and the calculated and experimental densities of 7.23 and 7.3 ± 0.2 g·cm^{-3}, respectively, is formed (Welch et al. 2000).

10.34 Lead–Hydrogen–Oxygen–Iron–Sulfur

In the system containing these elements, the quinary compound $Pb_{0.5}Fe_3(SO_4)_2(OH)_6$ (mineral plumbojarosite), which crystallizes as a trigonal structure with the lattice parameters a = 730.55 ± 0.07, c = 3367.5 ± 0.2 pm, and a calculated density of 3.618 g·cm^{-3} (Szymański 1985) [a = 720, c = 3360 pm, and the calculated and experimental densities of 3.71 and 3.67 g·cm^{-3}, respectively (Hendricks 1937); a = 731.5, c = 3375.8 pm, and the calculated and experimental densities of 3.60 and 3.64 g·cm^{-3}, respectively (Mumme and Scott 1966); a = 731.9 ± 0.4 and c = 3378 ± 2 pm (Jambor and Dutrizac 1983)], is formed.

$Pb_{0.5}Fe_3(SO_4)_2(OH)_6$ was prepared by reacting an excess of solid $PbSO_4$ with solutions containing $Fe_2(SO_4)_3$ in an autoclave at 130°C for 24 h (Mumme and Scott 1966; Jambor and Dutrizac 1983). Optimum operating conditions were attained by reacting a twofold excess of $PbSO_4$ in a well-stirred baffled reactor containing 1 L of 0.3 M Fe^{3+} as $Fe_2(SO_4)_3$ and 0.03 M H_2SO_4. Excess $PbSO_4$ was selectively leached from the precipitate by washing with four 1-L portions of 10 mass% solution of NH_4CH_3COO at 25°C. The residue was then filtered, washed with water, and dried at 110°C.

10.35 Lead–Sodium–Potassium–Copper–Magnesium–Aluminum–Oxygen–Iron–Sulfur

In the system containing these elements, the multinary compound $(K,Na,Pb)_4(Na,Ca)_2(Mg,Cu)_3(Fe_{0.5}Al_{0.5})(SO_4)_8$ (mineral philoxenite), which crystallizes as a triclinic structure with the lattice parameters a = 884.10 ± 0.03, b = 899.71 ± 0.03, c = 1618.61 ± 0.05 pm, α = 91.927 ± 0.003, β = 94.516 ± 0.003°, and γ = 90.118 ± 0.003°, is formed (Pekov et al. 2016a, 2020a).

10.36 Lead–Sodium–Potassium–Oxygen–Sulfur

In the system containing these elements, the quinary compound $(K,Na)_2Pb(SO_4)_2$ (mineral palmierite), which crystallizes as a rhombohedral structure with the lattice parameters a = 756 pm and α = 42°28' or a = 548 and c = 2061 pm in hexagonal setting (Strunz 1942) [an experimental density is 4.50 g·cm^{-3} (Zambonini 1921)], is formed. This compound was prepared by melting for 1 h at 1000°C a mixture of K_2SO_4 (5 g), $PbSO_4$ (7.5 g), Na_2SO_4 (9 g) and allowing the molten mass to cool slowly. The beautiful blades were obtained, which are easy to isolate, by treating the cooled mass with a solution of K_2SO_4 (2 mass%), and then washing the blades with a solution of K_2SO_4 (0.4 mass%). Finally, the blades were wiped between the papers.

10.37 Lead–Sodium–Copper–Arsenic–Sulfur

PbS–Na_2S–Cu_3As. The liquidus surface of this system was constructed using eight vertical sections and consists five fields of primary crystallization of PbS, Na_2S, $3PbS{\cdot}Na_2S$, Cu_3As, and Cu_xAs_y (Kopylov 1977). In the system, along with immiscibility region, the chemical interaction of the components occurs with the formation of new phases. Therefore, this system is a nonquasitrernary section of the Pb–Na–Cu–As–S quinary system. In addition to the thermal effects of equilibria related to the ternary system, the thermograms of vertical sections revealed effects corresponding to the phase transformations of new compounds formed in the system: for Cu_xAs_y within the interval 640°C–405°C; for Cu_2S and $mCu_2S{\cdot}Na_2S$ at 475°C, as well as formation of lead droplets at 380°C and their subsequent solidification at 285°C. The ternary eutectic *E* contains 55 mass% PbS, 32 mass% Na_2S, and 13 mass% Cu_3As, and the ternary transition point *U* exists at 450°C.

PbS–Na_2S–Cu_3As–$Cu_{2-x}S$. The liquidus surfaces of the isoconcentration sections of this system with 5, 10, and 20 mass% Na_2S were constructed by Kopylov (1977) and Kopylov et al. (1982). These sections are characterized by a complex phase interaction and the presence of a wide region of immiscibility in the liquid state into arsenide and sulfide phases. The minimum temperatures of primary crystallization (about 500ºC) of the isoconcentration section with 5 mass% Na_2S correspond to compositions with 40 wt.% PbS, 50 mass% $Cu_{2-x}S$, and no more than 5 mass% Cu_3As. The addition of Na_2S sharply reduces the solubility of sulfides in the arsenide phase of the melt.

An increase in the content of Na_2S to 10 mass% leads to a slight reduction in the separation area, a decrease in the field of $Cu_{2-x}S$ primary crystallization, the appearance of a field of $Na_2S{\cdot}4Cu_2S$ primary crystallization, and an increase in the fields of primary crystallization of Cu_3As (Cu_xAs_y) and PbS. The minimum temperature of primary crystallization (500°C–600°C) has a relatively narrow range of concentrations along the boundary lines of binary crystallization of PbS and Cu_3As (Cu_xAs_y).

The isoconcentration section with 20 mass% Na_2S is characterized by a significant decrease (up to 500°C–470°C) in the primary crystallization temperatures along the boundary line of binary crystallization of PbS and Cu_3As (Cu_xAs_y). The immiscibility region in the liquid state still occupies a large part of the liquidus surface, however, the addition of Na_2S to sulfide–arsenide melts somewhat reduces the possibility of the formation of an arsenide layer and increases the transition of arsenic to the sulfide melt. In this system, a partial interaction also occurs with the formation of new phases, the figurative points of the compositions of which do not lie in the volume of the tetrahedron. The interactions occurring in this system can be represented by the following schemes: $2PbS + 3Cu_3As \rightarrow 2Cu_{2-x}S + 2Pb + As$; $2PbS + 2Cu_3As \rightarrow 2Cu_{2-x}S + Cu_2As + 2Pb + As$; and $2Cu_3As \rightarrow Cu_5As_2 + Cu$. The resulting copper undergoes further sulfidation to form the $mCu_2S{\cdot}Na_2S$ complexes. This system was studied through DTA, metallography and EPMA.

10.38 Lead–Sodium–Copper–Arsenic–Iron–Sulfur

PbS–Na_2S–Cu_3As–Fe_2As. The immiscibity regions in the liquid state of this system at 950°C–1200°C was studied by Kopylov et al. (1988) and Kopylov and Toguzov (2002). The concentration volumes of the immiscibity region in this system were constructed at 1100°C and 1200°C. In the system, the interaction between sulfide and arsenide melts occurs with the formation of phases, the figurative points of which go beyond the limits of the concentration tetrahedron of this system. The immiscibility region occupies a significant part of the concentration tetrahedron. With an increase in temperature, it somewhat decreases, but remains dominant. A sufficiently large part of the concentration tetrahedron is occupied by the region of homogeneous composition based on PbS and Na_2S while the volume of homogeneous arsenide melt compositions based on Cu_3As and Fe_2As is insignificant and occupies a rather narrow section of the tetrahedron of compositions adjacent to the Cu_3As–Fe_2As system.

Pb–(PbS–Na_2S–Cu_3As–Fe_2As). It has been established that the presence of Na_2S in the Cu_2As melt, which is in contact with lead melt, reduces the solubility of copper in the latter and increases the solubility of arsenic (Kopylov et al. 1996). The presence of PbS in the sulfide–arsenide melt increases the solubility of copper and arsenic in the lead melt. Fe_2As decreases the solubility of Cu and As in Pb. At the same time, with the joint introduction into system Fe_2As and PbS to the mass ratio Fe_2As/PbS ≈ 1, the solubility of arsenic in the lead melt is determined mainly by the presence of Fe_2As in system, which leads to a sharp increase of arsenic solubility in lead with a simultaneous decrease in its copper content.

10.39 Lead–Sodium–Copper–Iron–Sulfur

PbS–Na_2S–Cu_2S–FeS. The liquidus surfaces of two isoconcentration sections of this system with 10 and 20 mass% Na_2S were constructed by Kopylov (1969).

10.40 Lead–Sodium–Arsenic–Iron–Sulfur

PbS–Na_2S–Fe_2As. At 1000°C and 1200°C, the immiscibility region in this system covers a significant part of the concentration triangle adjacent to the PbS–Fe_2As system (Kopylov and Toguzov 2002). The direction of the connodes tends to be fan-shaped from sulfide to arsenide melt. The curvature and kinks in their course indicate the complex nature of the components interaction in the melt with the formation of new phases, the figurative points of the compositions of which lie outside the concentration triangle of this system.

10.41 Lead–Sodium–Oxygen–Chlorine–Sulfur

In the system containing these elements, the quinary compound $Na_3Pb_2(SO_4)_3Cl$, which melts incongruently at 701°C and crystallizes as a hexagonal structure with the lattice parameters $a = 980.0 \pm 0.3$, $c = 710.0 \pm 0.3$ pm, and a calculated density of 4.544 g·cm^{-3} at room temperature and $a = 996.0 \pm 0.2$, $c = 726.0 \pm 0.3$ pm, and a calculated density of 4.299 g·cm^{-3} at 600°C (Knyazev et al. 2014) [$a = 981.5 \pm 0.4$, $c = 710.5 \pm 0.3$ pm, and the calculated and experimental densities of 4.519 and 4.50 g·cm^{-3}, respectively (Perret and Bouillet 1975)], is formed . The standard entropy of its formation is –1182.1 ± 1.9 J·(K·M)$^{-1}$ and the heat capacity c_p gradually increases with rising temperature and does not show any peculiarities until 63°C (Knyazev et al. 2014).

The title compound was prepared by the solid state reaction between $PbSO_4$, Na_2SO_4, and NaCl (Perret and Bouillet 1975; Knyazev et al. 2014). The synthesis was performed in a porcelain crucible, into which the reaction mixture with the atomic ratio Na/Pb/S/Cl = 3:2:3:1 was loaded. The mixture was calcined at 650°C for 10 h with dispersion in an agate mortar every 2 h.

10.42 Lead–Potassium–Copper–Oxygen–Sulfur

In the system containing these elements, the quinary compound $(K_2Pb)Cu_4O_2(SO_4)_4$ (mineral eleomelanite), which crystallizes as a monoclinic structure with the lattice parameters $a = 939.86 \pm 0.03$, $b = 489.11 \pm 0.01$, $c = 1822.93 \pm 0.05$ pm, $\beta = 104.409 \pm 0.003°$, and a calculated density of 3.790 g·cm^{-3}, is formed (Pekov et al. 2016b, 2020b).

10.43 Lead–Potassium–Gallium–Chlorine–Sulfur

In the system containing these elements, the quinary compound $[K_2PbCl][Ga_7S_{12}]$, which crystallizes as an orthorhombic structure with the lattice parameters $a = 724.09 \pm 0.03$, $b = 1051.88 \pm 0.05$, $c = 1463.51 \pm 0.05$ pm, a calculated density of 3.556 g·cm^{-3}, and an energy gap of 2.54 eV, is formed (Chen et al. 2023). For the synthesis of crystals of the title compound, the mixtures of K metal (0.81 mM), Ga metal (2.01 mM), S powder (3.62 mM), and $PbCl_2$ powder (0.40 mM) were weighed and carefully loaded into 11-mm (inner diameter) quartz tube. The tube was flame-sealed under a vacuum of 10^{-2} Pa and then placed in the furnace. Then, it was gradually heated from room temperature to 930°C over 25 h, maintained within 96 h, and cooled to 330°C at a rate of 3°C·h^{-1} before they reached ambient temperature naturally. The bright yellow crystals were mechanically separated from the products that were washed with distilled water and ethanol. This compound had unchanged weight and color at ambient temperature for half a year, which accounts for its physical and chemical stability.

10.44 Lead–Potassium–Gallium–Bromine–Sulfur

In the system containing these elements, the quinary compound $[K_2PbBr][Ga_7S_{12}]$, which crystallizes as an orthorhombic structure with the lattice parameters $a = 723.81 \pm 0.03$, $b = 1054.94 \pm 0.06$, $c = 1463.46 \pm 0.07$ pm, a calculated density of 3.679 g·cm^{-3}, and an energy gap of 2.49 eV, is formed (Chen et al. 2023). This compound was synthesized in the same way as $[K_2PbCl][Ga_7S_{12}]$ but using $PbBr_2$ instead of $PbCl_2$. The yellow crystals were mechanically separated from the products that were washed with distilled water and ethanol. This compound had also unchanged weight and color at ambient temperature for half a year which account for their physical and chemical stability.

10.45 Lead–Potassium–Gallium–Iodine–Sulfur

In the system containing these elements, the quinary compound $[K_2PbI][Ga_7S_{12}]$, which crystallizes as an orthorhombic structure with the lattice parameters $a = 724.34 \pm 0.03$, $b = 1061.22 \pm 0.03$, $c = 1466.25 \pm 0.05$ pm, a calculated density of 3.787 g·cm^{-3}, and an energy gap of 2.41 eV, is formed (Chen et al. 2023). This compound was synthesized in the same way as $[K_2PbCl][Ga_7S_{12}]$ but using PbI_2 instead of $PbCl_2$. The orange crystals were mechanically separated from the products that were washed with distilled water and ethanol. This compound had also unchanged weight and color at ambient temperature for half a year, which accounts for its physical and chemical stability.

10.46 Lead–Potassium–Phosphorus–Oxygen–Sulfur

In the system containing these elements, the quinary compound $K_4Pb_6(PO_4)_4(SO_4)_2$, which melts at 950°C and crystallizes as a hexagonal structure with the lattice parameters $a = 983.9 \pm 0.05$, $c = 733.5 \pm 0.5$ pm, and the calculated and experimental densities of 5.324 and 5.29 g·cm^{-3}, respectively, is formed (Schwarz 1967b). This compound was prepared by heating in air a mixture of $Pb_3(PO_4)_2$ and K_2SO_4 in the stoichiometric ratio first at 800°C for 5 h and then at 1000°C for 4 h with the next quickly cooling to room temperature.

10.47 Lead–Potassium–Arsenic–Oxygen–Sulfur

In the system containing these elements, the quinary compound $K_4Pb_6(AsO_4)_4(SO_4)_2$, which melts at 915°C and crystallizes as a hexagonal structure with the lattice parameters $a = 1013.0 \pm 0.05$, $c = 745.9 \pm 0.5$ pm, and the calculated and experimental densities of 5.379 and 5.36 g·cm^{-3}, respectively, is formed (Schwarz 1967b). This compound was prepared by heating in air a mixture of $Pb_3(AsO_4)_2$ and K_2SO_4 in the stoichiometric ratio first at 1050°C for 6 h and then at 1100°C for 3 h with the next quickly cooling to room temperature.

10.48 Lead–Potassium–Oxygen–Fluorine–Sulfur

The attempts to prepare the $K_4Pb_6(SO_4)_6F_2$ quinary compound were without success (Schwarz 1967b).

10.49 Lead–Copper–Silver–Mercury–Antimony–Sulfur

In the system containing these elements, the multinary compound $CuAg_7HgPb_7Sb_{24}S_{48}$ (mineral andreadiniite), which crystallizes as a monoclinic structure with the lattice parameters a = 1909.82 ± 0.14, b = 1700.93 ± 0.11, c = 1300.08 ± 0.10 pm, β = 90.083 ± 0.004° and a calculated density of 5.36 g·cm^{-3} (Biagioni et al. 2018a; Belakovskiy et al. 2019) [a = 1909.45 ± 0.13, b = 1702.55 ± 0.11, c = 1297.94 ± 0.09 pm, and β = 90.029 ± 0.001° (Biagioni et al. 2014b)], is formed .

10.50 Lead–Copper–Silver–Thallium–Arsenic–Sulfur

In the system containing these elements, the multinary phase $(Cu,Ag)TlPbAs_2S_5$ (mineral wallisite), which crystallizes as a triclinic structure with the lattice parameters a = 921.5 ± 1.0, b = 852.4 ± 1.0, c = 798.0 ± 1.0 pm, α = 55°59' ± 6', β = 62°30' ± 6', γ = 69°24' ± 6', and a calculated density of 5.71 g·cm^{-3} (Takéuchi et al. 1968; Fleischer 1969b), is formed . The reduced cell of this phase is also triclinic: a = 898.3 ± 1.0, b = 776.1 ± 1.0, c = 798.0 ± 1.0 pm, α = 65°33' ± 6', β = 65°30' ± 6', γ = 73°55' ± 6' (Takéuchi et al. 1968; Fleischer 1969b; Graeser and Guggenheim 1978) [a = 921.5, b = 852.4, c = 776 pm, α = 121°32', β = 100°54', γ = 110°36' (Fleischer 1966a)].

10.51 Lead–Copper–Silver–Arsenic–Antimony–Oxygen–Sulfur

In the system containing these elements, the multinary compound $(Ag,Cu)_{5.5}Pb_{42.4}(Sb,As)_{45.1}S_{112}O_{0.8}$ (mineral meerschautite), which crystallizes as a monoclinic structure with the lattice parameters a = 823.93 ± 0.01, b = 4360.15 ± 0.13, c = 2836.88 ± 0.08 pm, β = 94.128 ± 0.002° and a calculated density of 5.924 and 5.908 g·cm^{-3} for two samples, is formed (Biagioni et al. 2013a, 2016a; Gagne et al. 2017).

10.52 Lead–Copper–Silver–Arsenic–Antimony–Sulfur

Some multinary phases are formed in the Pb–Cu–Ag–As–Sb–S system. $Cu_{0.09}Ag_{1.04}Pb_{0.65}Sb_{2.82}As_{0.37}S_{6.08}$ (mineral roshchinite) crystallizes as an orthorhombic structure with the lattice parameters a = 1908.04 ± 0.01, b = 845.91 ± 0.02, and c = 1294.51 ± 0.03 pm (Makovicky et al. 2018a).

$Cu_{2.5}Ag_9Pb_{41}Sb_{36.5}As_7S_{112}$ (mineral sardashtite) crystallizes as a monoclinic structure with the lattice parameters a = 820.38 ± 0.03, b = 2710.02 ± 0.10, c = 2278.85 ± 0.09 pm, and β = 90.185 ± 0.001° (Topa et al. 2023m,n).

$Cu(Cu,Ag)_3Pb_{19}(Sb,As)_{22}(As_2)S_{56}$ (mineral ciriottiite) also crystallizes as a monoclinic structure with the lattice parameters a = 817.8 ± 0.2, b = 2822.3 ± 0.6, c = 4245.2 ± 0.5 pm, β = 93.55 ± 0.02° and a calculated density of 5.918 g·cm^{-3} (Bindi et al. 2015, 2016a; Belakovskiy et al. 2019).

$Cu(Ag,Cu)_3Pb_{19}(Sb,As)_{22}(As)_2S_{56}$ (mineral sterryite) also crystallizes as a monoclinic structure with the lattice parameters a = 818.91 ± 0.10, b = 2852.94 ± 0.13, c = 4298 ± 2 pm, β = 94.896 ± 0.008°, and a calculated density of 5.789 g·cm^{-3} (Moëlo et al. 2011, 2012; Belakovskiy et al. 2014) [in the orthorhombic structure with the lattice parameters a = 2840 ± 50, b = 4260 ± 60, c = 820 ± 5 pm, and a calculated density of 5.85 g·cm^{-3} (Jambor 1967; Fleischer et al. 1968; Pierrot 1968)].

$Cu_5Ag_{11}Pb_{76}Sb_{71}As_{17}(As^{2+})_8S_{224}$ (mineral hayyanite) also crystallizes as a monoclinic structure with the lattice parameters a = 819.3 ± 0.4, b = 4305 ± 2, c = 2863 ± 1 pm, and β = 90.08 ± 0.01 (Topa et al. 2023i,j).

10.53 Lead–Copper–Silver–Arsenic–Sulfur

In the system containing these elements, the quinary compound $Ag_4Cu_2Pb_{18}As_{12}S_{39}$ (mineral lengenbachite), which crystallizes as a monoclinic structure with the lattice parameters a = 3513, b = 1152, c = 3690 pm, and β = 92°36', is formed (Nowacki et al. 1960, 1961).

10.54 Lead–Copper–Silver–Antimony–Oxygen–Chlorine–Sulfur

In the system containing these elements, the multinary phase $(Cu,Ag)Pb_{10}Sb_{12}S_{27}O(Cl,S)_{0.6}$ (mineral pellouxite), which crystallizes as an orthorhombic structure with the lattice parameters $a = 5582.4 \pm 1.1$, $b = 408.92 \pm 0.08$, $c = 2412.8 \pm 0.5$ pm and a calculated density of 5.97 g·cm^{-3}, is formed (Orlandi et al. 2004; Palvadeau et al. 2004; Piilonen et al. 2005).

10.55 Lead–Copper–Silver–Antimony–Sulfur

In the system containing these elements, the quinary compound $CuAg_3Pb_4Sb_{12}S_{24}$ (mineral nakaséite), which crystallizes as a monoclinic structure with the lattice parameters $a = 1302 \pm 4$, $b = 1918 \pm 5$, $c = 10224 \pm 48$ pm, $\beta = 90°$, and the calculated and experimental densities of 5.379 and 5.30 g·cm^{-3}, respectively, is formed (Fleischer 1960; Ito et al. 1960).

10.56 Lead–Copper–Silver–Bismuth–Selenium–Sulfur

The multinary phase $Cu_2Ag_xPb_{6-2x}Bi_{8+x}(S,Se)_{19}$ (Ag-enriched felbertalite) with x ranging from 0.64 to 1.03 (probably up to 1.33), suggesting the presence of a solid-solution series in felbertalite, is formed in the Pb–Cu–Ag–Bi–Se–S system (Xiang-Ping et al. 2001). It crystallizes as a monoclinic structure with the lattice parameters $a = 2768 \pm 4$, $b = 404.6 \pm 0.1$, $c = 2067 \pm 3$ pm, and $\beta = 131.06 \pm 0.8°$ for the $Cu_{2.20}Ag_{0.70}Pb_{4.65}Bi_{8.45}S_{19.12}$ composition.

10.57 Lead–Copper–Silver–Bismuth–Sulfur

$PbS–Cu_2S–Ag_2S–Bi_2S_3$. Phase relations in this system were studied at 500°C on six planes, each representing a constant amount of Bi_2S_3: 75, 65, 50, 40, 25, and 15 mol% (Chang et al. 1988). Liquid phases present in the $Ag_2S–Cu_2S–PbS$ and $Cu_2S–PbS–Bi_2S_3$ systems become very extensive in the Bi_2S_3-poor region of the system, and they dominate phase relations. Nine solid-solution series have been established in the system. Among them, two complete series exist between PbS and $AgBiS_2$ and between Bi_2S_3 and $CuPbBiS_3$. Four extensive series can be represented by general formulas as follows: the pavonite series of $Ag_{1-(x+y)}Cu_{y+(n+1)z}Pb_{2x}Bi_{3-(x+z)}S_5$ ($0 < x < 0.36$, $0 < y < 0.36$, $0 < z < 0.53$, and $n \approx 2$; the benjaminite series of $Ag_{3.43-(x+y)}Cu_{y+(n+1)z}Pb_{2x}Bi_{6.85-(x+z)}S_{12}$ ($0 < x < 0.56$, $0 < y < 0.70$, $-0.21 < z < 0.16$, and $n \approx 2$; the $Cu_3Bi_5S_9$ series of $Ag_yCu_{3-(x+y)+(n+1)z}Pb_{2x}Bi_{5-(x+z)}S_9$ ($0 < x < 1.53$, $0 < y < 0.55$, $0 < z < 1.41$, and $n \approx 2$; and the $CuBi_3S_5$ series of $Ag_yCu_{1-(x+y)+(n+1)z}Pb_{2x}Bi_{3-(x+z)}S_5$ ($0 < x < 0.09$, $0 < y < 0.12$, $0 < z < 0.08$, and $n \approx 2$.

Some quinary minerals are formed in the Pb–Cu–Ag–Bi–S system. $(Cu,Ag)Pb_6Bi_7S_{17}$ (sulfosalts mineral) crystallizes as an orthorhombic structure with the lattice parameters $a = 881$, $b = 1306$, $c = 710.6$ pm, and a calculated density of 7.04 g·cm^{-3} (Jambor and Burke 1990).

$CuAg_2PbBi_6S_{13}$ (mineral mummeite) crystallizes as a monoclinic structure with the lattice parameters $a = 1347 \pm 1$, $b = 406 \pm 1$, $c = 2163 \pm 1$ pm, $\beta = 92.9 \pm 0.1°$, and a calculated density of 6.79 g·cm^{-3} (Jambor and Burke 1993).

$Cu_{1.12}Ag_{0.81}Pb_{0.27}Bi_{5.35}S_9$ (mineral makovickyite) also crystallizes as a monoclinic structure with the lattice parameters $a = 1323.9 \pm 0.3$, $b = 405.47 \pm 0.07$, $c = 1466.7 \pm 0.3$ pm, $\beta = 99.397 \pm 0.004°$, and a calculated density of 6.608 g·cm^{-3} (Topa et al. 2008) [a = 1337, 1337, 1383; b = 405, 405, 404; c = 1471, 1494, 1472 pm, β = 99.5°, 100.5°, 97.5° for three different samples and a calculated density of 6.66 g·cm^{-3} (Jambor et al. 1995d)].

$(Ag,Cu)_3(Bi,Pb)_7S_{12}$ (mineral benjaminite) [$(Ag,Cu)_2Pb_2Bi_4S_9$ according to Mintser (1967)] also crystallizes as a monoclinic structure with the lattice parameters $a = 1329.9 \pm 0.8$, $b = 407.0 \pm 0.8$, $c = 2020.9 \pm 1.0$ pm, and $\beta = 103.32 \pm 0.05°$ (Makovicky and Mumme 1979; Karup-Møller and Makovicky 1979) [$a = 1334$, $b = 406$, $c = 2025$ pm, and $\beta = 104°$ (Nuffield 1953); $a = 1324.6$, $b = 404.4$, $c = 2017.9$

pm, β = 103.21° and a calculated density of 6.7 g·cm^{-3} (Harris and Chen 1975); *a* = 1339, *b* = 406, *c* = 2021 pm, β = 103.27° and a calculated density of 6.68 g·cm^{-3} (Nuffield 1975); *a* = 1320, *b* = 400, *c* = 2020 pm, and β = 102° (Nenasheva et al. 1987)].

$Cu_2AgPbBiS_4$ (mineral ángelaite) crystallizes as an orthorhombic structure with the lattice parameters *a* = 1273.4 ± 0.5, *b* = 403.2 ± 0.1, *c* = 1463.3 ± 0.2 pm, and a calculated density of 6.89 g·cm^{-3} (Topa et al. 2010a) [*a* = 1273.3 ± 0.8, *b* = 402.2 ± 0.3, *c* = 1458.5 ± 1.0 pm, and a calculated density of 6.934 g·cm^{-3} (Topa et al. 2010d)].

$Cu_2AgPbBi_5S_{10}$ (mineral cupropavonite) crystallizes as a monoclinic structure with the lattice parameters *a* = 1344.5 ± 0.8, *b* = 402.3 ± 1.0, *c* = 3306.0 ± 1.0 pm, and β = 93.50 ± 0.05° (Karup-Møller and Makovicky 1979; Fleischer et al. 1980b) [*a* = 1340, *b* = 400, *c* = 3287 pm, and β = 93.87° (Nuffield 1980)].

$Cu_{3.44}Ag_{0.56}Pb_2Bi_6S_{13}$ (natural sulfide) crystallizes as an orthorhombic structure with the lattice parameters *a* = 397.3 ± 0.1, *b* = 1337.0 ± 0.2, *c* = 4218.2 ± 0.7 pm, and a calculated density of 7.008 g·cm^{-3} (Borisov et al. 2017).

$Cu_3Ag_2Pb_3Bi_7S_{16}$ (mineral berryite) crystallizes as a monoclinic structure with the lattice parameters *a* = 1270.3 ± 0.2, *b* = 403.05 ± 0.07, *c* = 2829.5 ± 0.5 pm, β = 102.484 ± 0.002°, and a calculated density of 6.899 g·cm^{-3} (Locock et. al. 2006a; Topa et al. 2006) [*a* = 1270.7, *b* = 402.1, *c* = 2892 pm, β = 102.06°, and a calculated density of 6.87 g·cm^{-3} (Karup-Møller 1966); *a* = 1272, *b* = 402, *c* = 5807 pm, β = 102.5°, and the calculated and experimental densities of 7.11 and 6.7 g·cm^{-3}, respectively (Nuffield and Harris 1966; Fleischer 1966a, 1967a)].

$Cu_7[(Cu,Ag)_{0.33}Pb_{1.33}Bi_{11.33}]S_{22}$ (mineral padĕraite) also crystallizes as a monoclinic structure with the lattice parameters *a* = 1758.5 ± 0.4, *b* = 393.86 ± 0.09, *c* = 2845.3 ± 0.7 pm, β = 105.41 ± 0.01°, and a calculated density of 6.694 g·cm^{-3} for the Ag-bearing material and *a* = 1757.3 ± 0.2, *b* = 394.26 ± 0.04, *c* = 2842.3 ± 0.3 pm, β = 105.525 ± 0.002°, and a calculated density of 6.691 g·cm^{-3} for the Ag-free material (Locock et al. 2006a; Topa and Makovicky 2006) [*a* = 2844 ± 8, *b* = 390 ± 3, *c* = 1755 ± 8 pm, β = 106.0 ± 0.2°, and a calculated density of 6.91 g·cm^{-3} (Mumme 1986; Hawthorne et al. 1987)].

$(Cu,Ag)_{21}(Pb,Bi)_2S_{13}$ (mineral larosite) crystallizes as an orthorhombic structure with the lattice parameters *a* = 2215, *b* = 2403, *c* = 1167 pm, and a calculated density of 6.18 g·cm^{-3} (Petruk 1972; Fleischer and Mandarino 1974).

$Cu_6Ag_2Pb_{25}Bi_{26}S_{68}$ (mineral neyite) crystallizes as a monoclinic structure with the lattice parameters *a* = 3752.7 ± 0.6, *b* = 407.05 ± 0.06, *c* = 4370.1 ± 0.7 pm, β = 108.801 ± 0.002°, and a calculated density of 7.037 g·cm^{-3} (Makovicky et al. 2001; Jambor and Roberts 2002) [*a* = 3750 ± 10, *b* = 407 ± 1, *c* = 4160 ± 10 pm, β = 96.8 ± 0.3°, and the calculated and experimental densities of 7.16 and 7.02 g·cm^{-3}, respectively (Drummond et al. 1969; Fleischer 1970)].

$Cu_8Ag_3Pb_4Bi_{19}S_{38}$ (mineral cupromakopavonite) also crystallizes as a monoclinic structure with the lattice parameters *a* = 1338.01 ± 0.19, *b* = 400.07 ± 0.06, *c* = 3108.3 ± 0.4 pm, β = 93.064 ± 0.002°, and a calculated density of 6.85 g·cm^{-3} (Topa et al. 2012) [*a* = 1340.5 ± 0.8, *b* = 401.6 ± 0.3, *c* = 2994.9 ± 1.9 pm, β = 99.989 ± 0.016°, and a calculated density of 6.78 g·cm^{-3} (Topa and Paar 2008; Piilonen et al. 2008; Topa et al. 2008)].

EPMA also identified the presence of another unknown mineral $Cu_2Ag_xPb_{10-2x}Bi_{12+x}S_{29}$ ($1.23 < x < 1.49$) (Xiang-Ping et al. 2001).

10.58 Lead–Copper–Barium–Iron–Nickel–Sulfur

In the system containing these elements, the multinary compound $(Ba,Pb)_6(Cu,Fe,Ni)_{25}S_{27}$ (mineral owensite), which crystallizes as a cubic structure with the lattice parameters *a* = 1037.3 ± 0.2 pm and a calculated density of 4.78 g·cm^{-3} (Laflamme et al. 1995) [*a* = 1034.9 ± 0.1 pm (Szymański 1995)], is formed .

10.59 Lead–Copper–Zinc–Iron–Sulfur

PbS–$Cu_{2-x}S$–ZnS–FeS. A part of this quinary system was investigated according to the isoconcentration sections with 5, 10, and 20 mass% ZnS, and each of these sections were studied according to six to eight vertical sections between ZnS–FeS quasibinary system and PbS–$Cu_{2-x}S$–ZnS quasiternary system (Toguzov et al. 1980). Liquidus surface of an isoconcentration section with 5 mass% ZnS includes the fields of ZnS, PbS, FeS, and Cu_5FeS_4 primary crystallization. The field of ZnS primary crystallization occupies the most part of this liquidus surface. The temperature of primary crystallization are minimum (760°C) near the composition of PbS–$Cu_{2-x}S$ quasibinary system at the mass ratio PbS/$Cu_{2-x}S$ = 1:1 and maximum (1180°C) near the FeS corner.

Liquidus surfaces of isoconcentration sections with 10 and 20 mass% ZnS include only the fields of ZnS and Cu_5FeS_4 primary crystallizations. The fields of PbS and FeS primary crystallizations disappear, and the field of Cu_5FeS_4 decreases noticeably as the ZnS concentration decreases. Increasing the ZnS and FeS concentrations in this system leads to an increase of the melting temperature. Just as in the case of isoconcentration section with 5 mass% ZnS, minimum melting temperatures (900°C and 1050°C, respectively) were found near the PbS–$Cu_{2-x}S$ quasibinary system and maximum melting temperatures (1300°C and 1370°C, respectively) are situated near FeS corner. This system was investigated through DTA and metallography.

10.60 Lead–Copper–Cadmium–Manganese–Iron–Sulfur

In the system containing these elements, the multinary compound (Mn,Pb,Cd)(Cu,Fe)$_8$S$_8$ (mineral manganoshadlunite), which crystallizes as a cubic structure with the lattice parameters a = 1073 pm, is formed (Evstigneeva et al. 1973; Fleischer 1973)].

10.61 Lead–Copper–Cadmium–Iron–Sulfur

In the system containing these elements, the quinary compound (Pb,Cd)(Cu,Fe)$_8$S$_8$ (mineral shadlunite), which crystallizes as a cubic structure with the lattice parameters a = 1091 pm, is formed (Evstigneeva et al. 1973; Fleischer 1973)].

10.62 Lead–Copper–Mercury–Antimony–Oxygen–Sulfur

In the system containing these elements, the multinary compound $Cu_2HgPb_{22}Sb_{28}S_{64}(O,S)_2$ (mineral rouxelite), which crystallizes as a monoclinic structure with the lattice parameters a = 4311.3 ± 0.9, b = 405.91 ± 0.08, c = 3787.4 ± 0.8 pm, β = 117.35 ± 0.03°, and a calculated density of 4.78 g·cm^{-3} (Orlandi et al. 2005; Locock et al. 2006) [a = 4310 ± 2, b = 406.0 ± 0.2, c = 3788 ± 2 pm, and β = 117.33 ± 0.02° (Biagioni et al. 2014a)], is formed .

10.63 Lead–Copper–Arsenic–Iron–Sulfur

PbS–Cu_3As–Fe_2As. The immiscibity regions in the liquid state of this system at 950°C–1200°C was studied by Kopylov et al. (1988) and Kopylov and Toguzov (2002). At 1000°C and 1200°C, the immiscibility region occupies a significant part of concentration triangle. The curvature and kinks in their course indicate the complex nature of the component interaction in the melt with the formation of new phases, the figurative points of the compositions of which lie outside the concentration triangle of this system. Therefore, this system is nonquasiternary section of the Pb–Cu–As–Fe–S quinary system.

10.64 Lead–Copper–Antimony–Bismuth–Sulfur

Some quinary minerals are formed in the Pb–Cu–Sb–Bi system. $Cu_xPb_{2+x}(Sb,Bi)_{2-x}S_5$ (x = 0.15–0.2) (mineral jaskólskiite) crystallizes as an orthorhombic structure with the lattice parameters a = 1131.2

± 0.2, b = 1982.9 ± 0.3, c = 408.8 ± 0.1 pm, and a calculated density of 6.551 ± 0.002 g·cm^{-3} for $x \approx 0.2$ (Makovicky and Nørrestam 1985) [a = 1131 ± 1, b = 1979 ± 2, c = 408.7 ± 0.4 pm, and a calculated density of 6.59 g·cm^{-3} for x = 0.15 (Harris et al. 1984); a = 1131, b = 1983, and c = 409 pm for x = 0.2 (Makovicky et al. 1984); a = 1133.1 ± 0.1, b = 1987.1 ± 0.2, c = 410.0 ± 0.1 pm, and a calculated density of 6.47 g·cm^{-3} for $Cu_{0.19}Pb_{2.22}Sb_{1.17}Bi_{0.62}S_5$ composition (Zakrzewski 1984; Dunn et al. 1985c)].

$Cu_{0.58}Pb_{12.72}(Sb_{7.04}Bi_{0.24})S_{24}$ (Cu-poor natural meneghinite) also crystallizes as an orthorhombic structure with the lattice parameters a = 2408.0 ± 0.5, b = 412.76 ± 0.08, c = 1136.9 ± 0.2 pm, and a calculated density of 6.393 g·cm^{-3} (Moëlo et al. 2002).

$CuPb_2(Bi,Sb,Pb)Bi_2S_7$ (mineral nuffieldite) also crystallizes as an orthorhombic structure with the lattice parameters a = 1449.49 ± 0.23, b = 2141.95 ± 0.49, c = 404.20 ± 0.15 pm, and a calculated density of 7.046 ± 0.005 g·cm^{-3} for $Cu_{1.4}Pb_2(Pb_{0.4}Bi_{0.4}Sb_{0.4})Bi_2S_7$ composition (Moëlo et al. 1997) [a = 1460.2 ± 0.6, b = 2134.4 ± 1.1, c = 402.6 ± 0.2 pm, and the calculated and experimental densities of 7.006 and 7.01 ± 0.07 g·cm^{-3}, respectively for $Cu_{1.4}Pb_2Bi_{2.4}Sb_{0.2}S_7$ composition (Kingston 1968; Fleischer 1969a); a = 1438.7 ± 0.7, b = 2101.1 ± 1.5, c = 404.6 ± 0.6 pm, and the calculated and experimental densities of 7.21 and 7.01 ± 0.07 g·cm^{-3}, respectively (Kohatsu and Wuensch 1972, 1973); a = 1460.2, b = 2137.5, and c = 402.6 pm (Mozgova et al. 1994)].

Syntheses of nuffieldite, $Cu_{1+x}Pb_2(Pb_xBi_{1-x-y}Sb_y)Bi_2S_7$ at $x \approx 0.37$ and $0.19 \le y \le 0.55$, were made from the mixtures of PbS, Cu_2S, Sb_2S_3, and Bi_2S_3 or $CuPbBiS_3$ and $Pb_3Bi_2S_6$ (Maurel and Moëlo 1990). The temperature was fixed at 435°C, for a duration of 90–110 days. The overall composition of the initial mixture corresponded to that of natural nuffieldite. At the end of the experiments, the samples were quenched.

$Cu_2Pb_{27}(Bi,Sb)_{19}S_{57}$ (mineral giessenite) crystallizes as a monoclinic structure with the lattice parameters a = 3434 ± 1, b = 3805 ± 1, c = 406 ± 1 pm, and β = 90.33 ± 0.05° (Makovicky and Karup-Møller 1986; Hawthorne et al. 1987) [a = 3451 ± 3, b = 3818 ± 5, c = 408.0 ± 0.8 pm, and β ≈ 90° (Graeser and Harris 1986); in the orthorhombic structure with the lattice parameters a = 3422.1 ± 0.8, b = 3793.3 ± 0.8, c = 406.3 ± 0.3 pm, and a calculated density of 6.61 g·cm^{-3} (Armbruster et al. 1984); a = 3450, b = 3830, and c = 408 pm (Graeser 1963; Fleischer 1965)].

$Cu_4Pb_{22}(Sb,Bi)_{30}S_{35}$ (mineral tintinaite) crystallizes as an orthorhombic structure with the lattice parameters a = 2230, b = 3400, c = 404 pm, and the calculated and experimental densities of 5.48 and 5.51 g·cm^{-3}, respectively (Harris et al. 1968; Fleischer 1969a).

$Cu_4Pb_{44}(Bi,Sb)_{30}S_{69}$ (mineral kobellite) also crystallizes as an orthorhombic structure with the lattice parameters a = 2257.5 ± 0.3, b = 3410.4 ± 0.4, and c = 403.8 ± 0.6 pm (Miehe 1971) [a = 2261.6, b = 3407.9, c = 401.8 pm, and an experimental density of 6.48 ± 0.05 g·cm^{-3} (Nuffield 1948)].

10.65 Lead–Copper–Antimony–Bismuth–Selenium–Sulfur

$CuPb_6(Bi_{3.8}Sb_{3.2})S_{1.9}Se_{15.6}$ (selenian kobellite), which crystallizes as an orthorhombic structure with the lattice parameters a = 2266.87 ± 0.06, b = 3425.77 ± 0.07, c = 404.091 ± 0.008 pm, and a calculated density of 6.44 g·cm^{-3}, is formed in this system (Mumme et al. 2013).

10.66 Lead–Copper–Antimony–Bismuth–Iron–Sulfur

The multinary compound $Pb_{27}(Cu,Fe)_2(Sb,Bi)_{19}S_{57}$ (mineral izoklakeite), which crystallizes as an orthorhombic structure with the lattice parameters a = 3406.7 ± 0.5, b = 3808.5 ± 0.4, and c = 405.6 ± 0.1 pm (Ozawa et al. 1998) [a = 3769, b = 3393, and c = 406 pm (Makovicky et al. 1984); a = 3388 ± 2, b = 3802 ± 2, c = 407.0 ± 0.2 pm, and the calculated and experimental densities of 6.505 and 6.47 g·cm^{-3}, respectively (Harris et al. 1986; Hawthorne et al. 1987); a = 3407 ± 1, b = 3798 ± 1, c = 407.2 ± 0.1 pm, and a calculated density of 6.68 g·cm^{-3} (Zakrzewski and Makovicky 1986; Hawthorne et al. 1987); a = 3422.1 ± 0.8, b = 3793.3 ± 0.8, and c = 406.3 ± 0.3 pm (Armbruster and Hummel 1987; a = 3427.2 ± 0.4, b = 3835.1 ± 0.5, and c = 409.8 ± 0.1 pm for Bi-rich izoklakeite (Orlandi et al. 2010)], is formed in the Pb–Cu–Sb–Bi–Fe–S system.

10.67 Lead–Copper–Antimony–Iodine–Sulfur

In the system containing these elements, the quinary compound $Cu_{0.507(5)}Pb_{8.73(9)}Sb_{8.15(8)}I_{1.6}S_{20.0(2)}$, which crystallizes as a monoclinic structure with the lattice parameters a = 498.01 ± 0.04, b = 411.32 ± 0.08, c = 2198.9 ± 0.1 pm, and β = 99.918 ± 0.006°, is formed in the Pb–Cu–Sb–I–S system (Kryukova et al. 2005). Nanowires of this iodine containing Pb–Sb-sulfosalt have been synthesized by chemical vapor transport with I_2 as a transport agent (0.3 mg I_2/mL) in a two-zone furnace with a temperature gradient ranging from 456°C to 380°C.

10.68 Lead–Copper–Bismuth–Selenium–Sulfur

Some quinary minerals are formed in the Pb–Cu–Bi–Se–S system. $CuPbBi(S,Se)_3$ (mineral součekite) crystallizes as an orthorhombic structure with the lattice parameters a = 817.6 ± 0.3, b = 852.9 ± 0.3, and c = 810.5 ± 0.3 pm (Johan et al. 1987) [a = 815.3, b = 849.8, c = 808.0 pm, and a calculated density of 7.60 g·cm^{-3} (Fleischer et al. 1980b)].

$Cu_{0\text{-}1}Pb_{7.5}Bi_{9.3\text{-}9.7}(S,Se)_{22}$ (mineral proudite) crystallizes as a monoclinic structure with the lattice parameters a = 3181.43 ± 0.09, b = 410.02 ± 0.02, c = 3656.0 ± 0.1 pm, β = 109.266 ± 0.001°, and a calculated density of 7.288 g·cm^{-3} (Mumme et al. 2009) [a = 3196 ± 1, b = 412 ± 1, c = 3669 ± 3 pm, β = 109.52 ± 0.03°, and a calculated density of 7.08 g·cm^{-3} (Mumme 1976)].

$CuPbBi_{11}(S,Se)_{18}$ (mineral pekoite) crystallizes as an orthorhombic structure with the lattice parameters a = 1147.2 ± 0.2, b = 3374.4 ± 0.2, c = 401.6 ± 0.1 pm, and a calculated density of 6.8 g·cm^{-3} (Mumme and Watts 1976a,b).

$CuPb_3Bi_7(S,Se)_{14}$ (mineral nordströmite) crystallizes as a monoclinic structure with the lattice parameters a = 1797 ± 8, b = 411 ± 2, c = 1762 ± 8 pm, and β = 94.3 ± 0.2° (Mumme 1980a,b).

$Cu_2PbBi_4(Se,S)_8$ (mineral watkinsonite) also crystallizes as a monoclinic structure with the lattice parameters a = 1295.2 ± 0.4, b = 415.23 ± 0.14, c = 1515.5 ± 0.5 pm, β = 108.931 ± 0.005°, and a calculated density of 7.76 g·cm^{-3} (Topa et al. 2010c) [a = 1292.1 ± 0.3, b = 399.7 ± 0.1, c = 1498.9 ± 0.3 pm, β = 109.2 ± 0.2°, and a calculated density of 7.82 g·cm^{-3} (Johan et al. 1987; Jambor and Vanko (1989)].

$Cu_2Pb_3Bi_8(S,Se)_{16}$ (mineral junoite) also crystallizes as a monoclinic structure with the lattice parameters a = 2671 ± 1, b = 406.0 ± 0.2, c = 1717.2 ± 0.7 pm, and β = 127.65 ± 0.03° (Pringle and Thorpe 1980) [a = 2666 ± 1, b = 406 ± 1, c = 1703 ± 1 pm, β = 127.20 ± 0.02°, and a calculated density of 6.77 g·cm^{-3} (Fleischer et al. 1975a; Large and Mumme 1975; Mumme 1975).

10.69 Lead–Copper–Bismuth–Iodine–Sulfur

The $Cu_{4-x}Pb_{1-x}Bi_{2+x}S_5I_2$ ($0 \leq x \leq 1$) quinary solid solution, which crystallizes as an orthorhombic structure with the lattice parameters a = 1323.6 ± 0.5, b = 2363.0 ± 1.6, c = 401.0 ± 0.5 pm, and a calculated density of 6.570 g·cm^{-3} for x = 0.88, is formed in this system (Ohmasa and Mariolacos 1974). It was synthesized by chemical transport reaction from a mixture of appropriate proportion of PbS, Bi_2S_3, and Cu_2S. The mixed sulfides and 0.1 N solution of HI were sealed in a silica glass tube and were kept at a temperature of 430°C with a gradient of about 15°C for 12 days. Needle crystals of this solid solution were obtained.

10.70 Lead–Copper–Bismuth–Iron–Sulfur

Three quinary compounds are formed in the Pb–Cu–Bi–Fe–S system. $(Cu,Fe)Pb_9Bi_{12}S_{28}$ (mineral eclarite) crystallizes as an orthorhombic structure with the lattice parameters a = 403.07 ± 0.01, b = 2270.11 ± 0.06, c = 5461.5 ± 0.1 pm, and a calculated density of 7.125 g·cm^{-3} (Topa and Makovicky 2012) [a = 5476

± 4, b = 403.0 ± 0.3, c = 2275 ± 3 pm, and the calculated and experimental densities of 6.88 and 6.85 g·cm^{-3}, respectively (Paar et al. 1983; Kupčik 1984; Dunn et al. 1985a); a = 5461.0 ± 0.5, b = 402.51 ± 0.04, and c = 2264.4 ± 0.2 pm (Pršek et al. 2008)].

$Cu_4PbBiFeS_6$ (mineral miharaite) also crystallizes as an orthorhombic structure with the lattice parameters a = 1088.0 ± 0.2, b = 1200.3 ± 0.4, c = 387.4 ± 0.1 pm, and a calculated density of 5.82 g·cm^{-3} (Petrova et al. 1988) [a = 1083.5, b = 1196.7, and c = 388 pm (Kovalenker et al. 1984b); a = 1085.4 ± 0.4 and 1086.9 ± 0.4, b = 1198.5 ± 0.4 and 1199.5 ± 0.6, c = 387.1 ± 0.1 and 387.3 ± 0.1 pm, and a calculated density of 6.06 and 6.04 g·cm^{-3}, respectively for two different samples (Sakharova et al. 1985); a = 1085.4 ± 0.4, b = 1198.5 ± 0.4, c = 387.1 ± 0.1 pm, and a calculated density of 6.06 g·cm^{-3} (Sugaki et al. 1980)]. Miharaite could not be synthesized in the dry experiments at 300°C (Sugaki et al. 1980). It is probably that this mineral is stable only to some temperature below 300°C.

$(Cu,Fe)Cu_{14}PbBi_{17}S_{35}$ (mineral pizgrischite) crystallizes as a monoclinic structure with the lattice parameters a = 3505.4 ± 0.2, b = 391.123 ± 0.001, c = 4319.2 ± 0.2 pm, and β = 96.713 ± 0.004°, and a calculated density of 6.58 g·cm^{-3} (Meisser et al. 2007).

10.71 Lead–Copper–Oxygen–Selenium–Sulfur

PbS–$CuSeO_3$. At a content of 50 mol% PbS, exothermic effects at 440°C and 550°C and an endothermic effect at 650°C were observed on heating thermograms (Angelova et al. 1975). It is assumed that the following chemical reactions take place in the system: PbS + $CuSeO_3$ = $PbSeO_3$ + CuS at <550°C and $PbSeO_3$ + CuS = PbSe + CuO + SO_2 at >650°C. Other reactions are also possible in the system, the final products of which are PbSe, CuO and SO_2.

10.72 Lead–Copper–Iron–Nickel–Platinum–Sulfur

In the system containing these elements, the multinary compound $(Cu,Pt,Pb,Fe,Ni)_9S_8$ (mineral kharaelakhite), which crystallizes as an orthorhombic structure with the lattice parameters a = 971.3 ± 0.5, b = 833.3 ± 0.5, c = 1450 ± 1 pm, and a calculated density of 7.78 g·cm^{-3}, is formed (Genkin et al. 1985; Jambor 1989).

10.73 Lead–Copper–Iron–Rhodium–Iridium–Platinum–Sulfur

In the system containing these elements, the multinary compound $(Pb,Cu,Fe)(Ir,Pt,Rh)_2S_4$ (mineral xingzhongite), which crystallizes as a cubic structure with the lattice parameter a = 997.0 pm (Dunn et al. 1984a) [a = 872 ± 1 pm (Fleischer et al. 1976); a = 1010 pm (Fleischer et al. 1980a)], is formed .

10.74 Lead–Copper–Rhodium–Iridium–Platinum–Sulfur

In the system containing these elements, the multinary compound $Cu_3Pb(Rh,Pt,Ir)_8S_{16}$ (mineral konderite), which crystallizes as a hexagonal structure with the lattice parameters a = 702.4 ± 0.2 and c = 1648 ± 2 pm, is formed (Rudashevskiy et al. 1984b; Dunn et al. 1986).

10.75 Lead–Copper–Iridium–Platinum–Sulfur

In the system containing these elements, the multinary compound $Cu_3Pb(Ir,Pt)_8S_{16}$ (mineral inaglyite), which crystallizes as a hexagonal structure with the lattice parameters a = 703 ± 1 and c = 1644 ± 1 pm, is formed (Rudashevskiy et al. 1984a; Dunn et al. 1986).

10.76 Lead–Silver–Cadmium–Arsenic–Sulfur

In the system containing these elements, the quinary phase $Ag(Cd,Pb)AsS_3$ (mineral quadratite), which crystallizes as a tetragonal structure with the lattice parameters a = 552.29 ± 0.04 and c = 3339.9 ± 0.5 pm (Bindi et al. 2013) [a = 549.9 ± 0.5, c = 3391 ± 4 pm, and a calculated density of 5.31 g·cm^{-3} (Graeser et al. 1998; Jambor and Roberts 2000)], is formed

10.77 Lead–Silver–Thallium–Arsenic–Antimony–Sulfur

Some multinary phases are formed in this system. $Ag_zTl_{8-x-z}Pb_{4+2x}As_ySb_{40-x-y}S_{68}$ ($0.00 \le x \le 0.40$; $16.15 \le y \le 19.11$; $0.04 \le z \le 0.11$) (mineral chabournéite) (Fleischer et al. 1979b; Miyawaki et al. 2021c,d) crystallizes as a triclinic structure with the lattice parameters a = 851.97 ± 0.04, b = 4246.1 ± 0.2, c = 1629.3 ± 0.8 pm, α = 83.351 ± 0.002, β = 90.958 ± 0.002°, and γ = 84.275 ± 0.002° for Pb-rich chabournéite (Biagioni et al. 2015) [a = 1632.0 ± 0.3, b = 4263.6 ± 0.6, c = 854.3 ± 0.2 pm, α = 83.98 ± 0.01, β = 89.06 ± 0.01°, γ = 83.20 ± 0.01°, and a calculated density of 4.88 g·cm^{-3} (Nagel 1979; Nagel and Nowacki 1979); a = 1634.6 ± 0.5, b = 4260.2 ± 1.0, c = 853.4 ± 0.3 pm, α = 95.86 ± 0.03, β = 86.91 ± 0.03°, γ = 96.88 ± 0.03°, and the calculated and experimental densities of 5.121 and 5.10 g·cm^{-3}, respectively (Johan et al. 1981; Fleischer et al. 1982b)].

The revised general formula of mineral rathite may be written as $Ag_xTl_yPb_{16-2(x+y)}As_{16+x+y-z}Sb_zS_{40}$ ($1.6 < x < 2$, $0 < y < 3$, $0 < z < 3.5$) [previously given as $(Pb,Tl)_3As_5S_{10}$] (Topa and Kolitsch 2018). The homogeneity range of rathite was determined to be unusually large, ranging from very Tl-poor compositions [0.16 mass%; refined single-crystal monoclinic unit-cell parameters: a = 847.1 ± 0.2, b = 792.6 ± 0.2, c = 2518.6 ± 0.5 pm, and β = 100.58 ± 0.03°] to very Tl-rich compositions [11.78 mass%; a = 852.1 ± 0.2, b = 800.5 ± 0.2, c = 2503.1 ± 0.5 pm, and β = 100.56 ± 0.03°] [a = 849.6 ± 0.1, b = 796.9 ± 0.1, c = 2512.2 ± 0.3 pm, and β = 100.704 ± 0.002° for $Ag_{2.01}Tl_{1.37}Pb_{9.25}As_{19.09}Sb_{0.83}S_{40}$ composition (Berlepsch et al. 2002)]. The Ag content is only slightly variable (3.1-4.1 mass%) with a mean value of 3.6 mass%. The Sb content is strongly variable (0.20-7.71 mass%) and not correlated with the Tl content. With increasing Tl content (0.16–11.78 mass%), a clear increase of the unit-cell parameters a and b, and a slight decrease of c was observed, although this is somewhat masked by the randomly variable Sb content. Depending on the composition, calculated density of rathite is within the interval from 4.986 to 5.446 g·cm^{-3}. According to Pring (2001), rathite is an ordered intermediate between $Pb_2As_2S_5$ and a theoretical end member $(Ag,Tl)_8Pb_8As_{24}S_{40}$.

$Ag_zTl_{10-x-z}Pb_{2x}Sb_{42-x-y}As_yS_{68}$ ($0.09 \le x \le 2.13$, $13.99 \le y \le 19.79$, $0.10 \le z \le 0.50$) (mineral dewitite) crystallizes as a triclinic structure with the lattice parameters a = 862.6 ± 0.2, b = 1635.1 ± 0.3, c = 2189.2 ± 0.4 pm, α = 74.96 ± 0.03, β = 83.59 ± 0.03°, and γ = 88.91 ± 0.03° (Topa et al. 2021a,b).

$AgTl_3Pb_4As_{11}Sb_9S_{36}$ (mineral écrinsite) also crystallizes as a triclinic structure with the lattice parameters a = 808.02 ± 0.01, b = 853.32 ± 0.02, c = 2261.22 ± 0.04 pm, α = 90.233 ± 0.001, β = 97.174 ± 0.001°, γ = 90.832 ± 0.001°, and a calculated density of 4.96 g·cm^{-3} (Topa et al. 2016a, 2017b).

$Ag_3Tl_{21.5}PbAs_{41.5}Sb_{63}S_{170}$ (mineral vallouiseite) also crystallizes as a triclinic structure with the lattice parameters a = 864.95 ± 0.02, b = 2807.28 ± 0.05, c = 3033.24 ± 0.07 pm, α = 88.687 ± 0.002, β = 86.759 ± 0.002°, and γ = 85.676 ± 0.002° (Topa et al. 2023k,l).

10.78 Lead–Silver–Thallium–Arsenic–Antimony–Iron–Sulfur

In the system containing these elements, the multinary compound $Ag_2(Ag_{0.5}Fe_{0.5})TlPb_{23.5}(Sb,As)_{33.5}S_{76}$ (mineral ginelfite), which crystallizes as a triclinic structure with the lattice parameters a = 836.92 ± 0.07, b = 2754.0 ± 0.2, c = 2917.7 ± 0.2 pm, α = 95.455 ± 0.003°, β = 93.852 ± 0.003°, and γ = 94.132 ± 0.003°, is formed (Biagioni et al. 2023a,b).

10.79 Lead–Silver–Thallium–Arsenic–Sulfur

In the system containing these elements, the quinary compound $AgTlPbAs_2S_5$ (mineral hatchite), which crystallizes as a triclinic structure with the lattice parameters $a = 922 \pm 1$, $b = 784 \pm 1$, $c = 806 \pm 1$ pm, $\alpha = 66°25' \pm 15'$, $\beta = 65°17' \pm 15'$, $\gamma = 74°54' \pm 15'$, and a calculated density of 5.8 g·cm^{-3} (Marumo and Nowacki 1967; Graeser and Guggenheim 1978) [$a = 927 \pm 9$, $b = 781 \pm 8$, $c = 801 \pm 7$ pm, $\alpha = 66°37' \pm 14'$, $\beta = 63°27' \pm 14'$, $\gamma = 85°06' \pm 14'$ (Nowacki et al. 1961)], is formed .

10.80 Lead–Silver–Thallium–Antimony–Sulfur

In the system containing these elements, the quinary compound $(Ag,Tl)_2Pb_8Sb_8S_{21}$ (mineral rayite), which crystallizes as a monoclinic structure with the lattice parameters $a = 1360 \pm 2$, $b = 1196 \pm 3$, $c = 2449 \pm 5$ pm, $\beta = 103.94 \pm 0.12°$, and a calculated density of 6.13 g·cm^{-3}, is formed (Basu et al. 1983; Dunn et al. 1984b).

10.81 Lead–Silver–Arsenic–Antimony–Sulfur

Some quinary phases are formed in the Pb–Ag–As–Sb–S system. $AgPb_9(Sb_{13-x}As_x)S_{29}$ (natural sulfosalt) crystallizes as a monoclinic structure with the lattice parameters $a = 1764.0 \pm 0.8$, $b = 1846.4 \pm 0.8$, $c = 396.7 \pm 0.3$ pm, $\beta = 90°48'$, and a calculated density of 5.51 g·cm^{-3} for $x = 4.2$ (Johan and Mantienne 2000).

$AgPb_{12}Sb_2(As,Sb)_8S_{20}$ (mineral carducciite) also crystallizes as a monoclinic structure with the lattice parameters $a = 849.09 \pm 0.03$, $b = 802.27 \pm 0.03$, $c = 2539.57 \pm 0.09$ pm, $\beta = 100.382 \pm 0.002°$, and a calculated density of 5.576 g·cm^{-3} (Biagioni et al. 2013b, 2014c; Belakovskiy et al. 2016a).

$AgPb_{16}As_{18}Sb_{27}S_{84}$ (mineral montpelvouxite) crystallizes as a the triclinic structure with the lattice parameters $a = 855.63 \pm 0.04$, $b = 2186.8 \pm 0.1$, $c = 2210.7 \pm 0.1$ pm, $\alpha = 119.106 \pm 0.002°$, $\beta = 100.079 \pm 0.002°$, and $\gamma = 91.000 \pm 0.002°$ (Topa et al. 2023o,p).

$AgPb_{46}As_{26}Sb_{23}S_{120}$ (mineral polloneite) also crystallizes as a monoclinic structure with the lattice parameters $a = 841.3 \pm 0.2$, $b = 2590.1 \pm 0.5$, $c = 2381.7 \pm 0.5$ pm, $\beta = 90.01 \pm 0.03°$, and a calculated density of 5.77 g·cm^{-3} (Topa et al. 2015, 2017a).

$Ag_3Pb_{10}Sb_8As_{11}S_{40}$ (mineral barikaite) also crystallizes as a monoclinic structure with the lattice parameters $a = 853.25 \pm 0.07$, $b = 807.49 \pm 0.07$, $c = 2482.8 \pm 0.2$ pm, $\beta = 99.077 \pm 0.006°$, and a calculated density of 5.34 g·cm^{-3} (Makovicky and Topa 2013; Topa et al. 2013f; Belakovskiy and Cámara 2016) [$a = 851.9 \pm 0.3$, $b = 805.7 \pm 0.3$, $c = 2490.5 \pm 0.8$ pm, and $\beta = 98.926 \pm 0.006°$ (Topa et al. 2013e)].

$Ag_4Pb_{20}Sb_{14}As_{10}S_{58}$ (mineral parasterryite) also crystallizes as a monoclinic structure with the lattice parameters $a = 839.65 \pm 0.05$, $b = 2759.40 \pm 0.07$, $c = 4388.40 \pm 0.13$ pm, $\beta = 90.061 \pm 0.012°$, and a calculated density of 5.753 g·cm^{-3} (Moëlo et al. 2010, 2011, 2012; Belakovskiy et al. 2014).

$Ag_{16}Pb_4Sb_{25}As_{15}S_{72}$ (mineral jasrouxite) crystallizes as a triclinic structure with the lattice parameters $a = 829.17 \pm 0.05$, $b = 1910.1 \pm 0.1$, $c = 1948.7 \pm 0.1$ pm, $\alpha = 89.731 \pm 0.001°$, $\beta = 83.446 \pm 0.001°$, $\gamma = 89.944 \pm 0.001°$, and a calculated density of 4.87 g·cm^{-3} (Topa et al. 2013a,b; Cámara et al. 2014a; Makovicky and Topa 2014).

$Ag_{17.6}Pb_{12.8}Sb_{38.1}As_{11.5}S_{96}$ (mineral arsenquatrandorite) crystallizes as a monoclinic structure with the lattice parameters $a = 1905.7 \pm 0.7$, $b = 1703.9 \pm 0.6$, $c = 1291.1 \pm 0.5$ pm, and $\beta = 89.993 \pm 0.006°$ (Topa et al. 2013d).

10.82 Lead–Silver–Arsenic–Antimony–Manganese–Sulfur

Two multinary phases are formed in the Pb–Ag–As–Sb–Mn–S system. $AgPb_{2.40}As_2Sb_3Mn_{1.60}S_{12}$ (mineral menchettiite) crystallizes as a monoclinic structure with the lattice parameters $a = 1923.3 \pm 0.2$,

$b = 1263.3 \pm 0.3$, $c = 847.6 \pm 0.2$ pm, $\beta = 90.08 \pm 0.02°$, and a calculated density of 5.146 g·cm^{-3} (Bindi et al. 2011, 2012; Tait et al. 2012).

$Ag_3Pb_4As_4Sb_7Mn_2S_{24}$ (mineral oyonite) also crystallizes as a monoclinic structure with the lattice parameters $a = 1918.06 \pm 0.18$, $b = 1277.55 \pm 0.14$, $c = 817.89 \pm 0.10$ pm, $\beta = 90.471 \pm 0.001°$, and a calculated density of 5.275 g·cm^{-3} (Bindi et al. 2018a,b,c; Belakovskiy et al. 2019).

10.83 Lead–Silver–Arsenic–Selenium–Tellurium–Sulfur

$PbTe–Ag_4SSe–As_2Se_3$. The region of glass formation in this system (Figure 10.1) was determined by Amova et al. (2013). It lays partially on the $As_2Se_3–Ag_4SSe$ (0–25 mol% Ag_4SSe) and $As_2Se_3–PbTe$ (50–100 mol% As_2Se_3) sides of the concentration triangle. Depending on the glass composition, the glass transition temperature changes between 113°C and 140°C, increases with the increase of Ag_4SSe and the decrease of PbTe content.

10.84 Lead–Silver–Antimony–Bismuth–Sulfur

$PbS–AgSbS_2–AgBiS_2$. The calculated isothermal sections of this system at 350°C and 400°C are shown in Figure 10.2 (Chutas et al. 2008). Above 441.7°C, the complete series of the solid solutions exists .

Some quinary phases are formed in the Pb–Ag–Sb–Bi–S system. $Ag_{0.70}Pb_{1.60}Bi_{1.35}Sb_{1.35}S_6$ (mineral staročeskéite) crystallizes as an orthorhombic structure with the lattice parameters $a = 425.39 \pm 0.08$, $b = 1330.94 \pm 0.08$, $c = 1962.53 \pm 0.12$ pm, and a calculated density of 6.191 g·cm^{-3} (Pažout and Dušek 2010; Pažout and Sejkora 2017, 2018; Belakovskiy et al. 2019).

$AgPb(Sb,Bi)_3S_6$ (mineral terrywallaceite) crystallizes as a monoclinic structure with the lattice parameters $a = 697.64 \pm 0.04$, $b = 1935.07 \pm 0.10$, $c = 838.70 \pm 0.04$ pm, $\beta = 107.519 \pm 0.002°$, and a calculated density of 6.005 g·cm^{-3} (Yang et al. 2011b, 2013b) [$a = 704.55 \pm 0.06$, $b = 1952.94 \pm 0.17$, $c = 834.12 \pm 0.11$ pm, and $\beta = 107.446 \pm 0.010°$ (Pažout and Dušek 2009)].

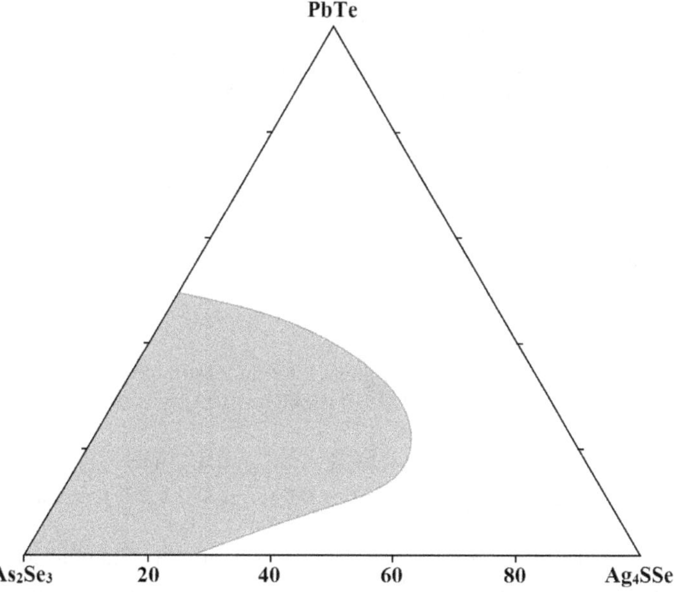

FIGURE 10.1 Glass forming region in the $PbTe–Ag_4SSe–As_2Se_3$ quasiternary system. (From Amova, A., et al., *J. Alloys Compd.*, 573, 32, 2013.)

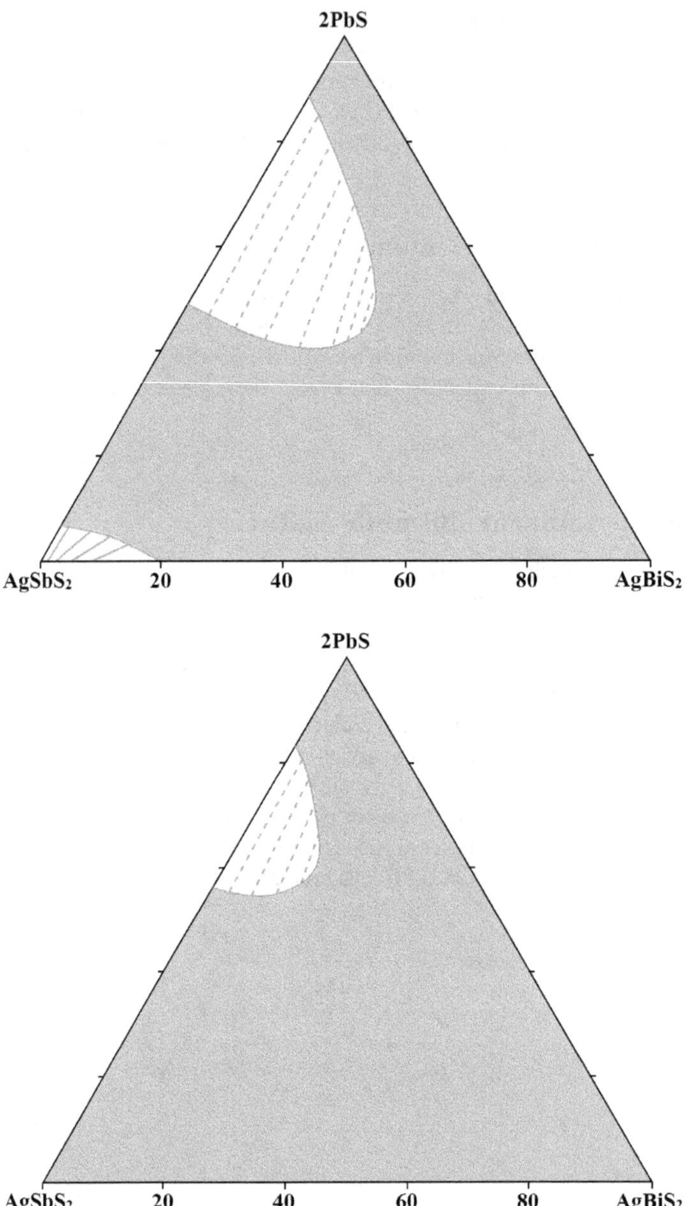

FIGURE 10.2 Isothermal section of 2PbS–$AgSbS_2$–$AgBiS_2$ quasiternary system at (a) 350°C and (b) 400°C. (Chutas, N.I., et al., *Amer. Mineralog.*, 93(10), 1630, 2008.)

$Ag_3Pb_6(Sb_8Bi_3)S_{24}$ (mineral holubite) crystallizes as a monoclinic structure with the lattice parameters $a = 1937.4 \pm 0.4$, $b = 1320.1 \pm 0.3$, $c = 865.1 \pm 0.2$ pm, $\beta = 90.112 \pm 0.018°$, and a calculated density of 5.899 g·cm^{-3} (Pažout et al. 2023a,cb).

$Ag_{3.75}Pb_{4.50}(Sb_{7.75}Bi_4)S_{24}$ (mineral lazerckerite) also crystallizes as a monoclinic structure with the lattice parameters $a = 1320.83 \pm 0.09$, $b = 1945.95 \pm 0.08$, $c = 840.5 \pm 0.1$ pm, and $\beta = 90.032 \pm 0.007°$ (Pažout et al. 2023d,e).

$Ag_{10}Pb_4(Sb_{17}Bi_9)S_{48}$ (mineral oscarkempffite) crystallizes as an orthorhombic structure with the lattice parameters $a = 1319.9 \pm 0.2$, $b = 1933.2 \pm 0.3$, $c = 829.4 \pm 0.1$ pm, and a calculated density of 5.8 g·cm^{-3} (Topa et al. 2011, 2016b; Belakovskiy et al. 2017a).

$Ag_{15}Pb_6Sb_{21}Bi_{18}S_{72}$ (mineral clino-oscarkempffite) crystallizes as a monoclinic structure with the lattice parameters a = 3981.1 ± 2.5, b = 1928.0 ± 1.2, c = 827.8 ± 0.5 pm, β = 96.195 ± 0.009°, and a calculated density of 6.04 g·cm^{-3} (Topa et al. 2013c; Makovicky et al. 2018b; Belakovskiy et al. 2019).

10.85 Lead–Silver–Antimony–Bismuth–Iron–Sulfur

In the system containing these elements, the multinary compound $Ag_5(Bi,Pb,Fe)_8(Sb,Bi)_2S_{17}$ (mineral borodaevite), which crystallizes as a monoclinic structure with the lattice parameters a = 1351.5 ± 0.7, b = 409.8 ± 0.3, c = 2600.0 ± 0.8 pm, β = 93.00 ± 0.04°, and a calculated density of 7.90 g·cm^{-3}, is formed (Nenasheva et al. 1992; Jambor et al. 1994).

10.86 Lead–Silver–Antimony–Chlorine–Sulfur

The existence of the quinary compound $AgPb_8Sb_8S_{20}Cl$ in this system was determined as mineral specie by Moëlo et al. (1989).

10.87 Lead–Silver–Antimony–Manganese–Sulfur

Two quinary compounds are formed in the Pb–Ag–Sb–Mn–S system. $AgPb_3MnSb_5S_{12}$ (mineral uchucchacuaite) has apparently two polymorphic modifications. The first of them crystallizes as a monoclinic structure with the lattice parameters a = 1936.45 ± 0.11 and 1934.62 ± 0.07, b = 1272.87 ± 0.08 and 1272.51 ± 0.05, c = 875.71 ± 0.06 and 874.72 ± 0.03 pm, β = 90.059 ± 0.003° and 90.017 ± 0.002°, and a calculated density of 5.487 and 5.498 g·cm^{-3} for two different samples (Yang et al. 2011a). The second modification crystallizes as an orthorhombic structure with the lattice parameters a = 1267, b = 1932, c = 438 pm, and a calculated density of 5.61 g·cm^{-3} (Moëlo et al. 1984; Dunn et al. 1985b); a = 1304 ± 1, b = 1946 ± 1, and c = 427 ± 1 pm for $AgPb_3MnSb_5S_{12}$; a = 1300, b = 1945, and c = 434 pm for $Ag_{1.7}Pb_{1.05}Mn_{1.45}Sb_{5.80}S_{12}$; a = 1291 ± 1, b = 1927 ± 1, and c = 421 ± 1 pm for $Ag_{2.14}Pb_{1.66}Mn_{0.50}Sb_{6.16}S_{12}$; and a = 1296 ± 1, b = 1932 ± 1, and c = 413 ± 1 pm for $Ag_{1.98}Pb_{1.24}Mn_{0.84}Sb_{6.04}S_{12}$ (Liu et al. 1994)]. This compound is readily synthesized from Ag_2S, MnS, PbS. And Sb_2S_3 in the temperature range 300°C–500°C (Liu et al. 1994).

$Ag_{12}Pb_{13}Mn_{11}Sb_{44}S_{96}$ (mineral lasmanisite) crystallizes as an orthorhombic structure with the lattice parameters a = 1305.07 ± 0.07, b = 1626.43 ± 0.09, and c = 1936.50 ± 0.10 pm (Topa et al. 2023e,f)

10.88 Lead–Silver–Antimony–Iron–Sulfur

In the system containing these elements, the quinary compound $Ag_4Pb_{12}Fe_4Sb_{20}S_{48}$ (mineral baiamareite), crystallizes as a monoclinic structure with the lattice parameters a = 871.50 ± 0.05, b = 1273.7 ± 0.1, c = 1935.0 ± 0.1 pm, and β = 90.062 ± 0.005°, is formed (Topa et al. 2023a,b).

10.89 Lead–Silver–Selenium–Tellurium–Sulfur

PbTe–Ag_4SSe. The phase diagram of this system, constructed through DTA, XRD, metallography, and measuring of the microhardness and density, is presented in Figure 10.3 (Amova et al. 2012). The eutectics contain 20 mol% and 78 mol% PbTe and crystallize at 590°C and 600°C, respectively. The immiscibility region exists at 630°C. The eutectoid contains 10 mol% PbTe and crystallizes at 70°C. The $Ag_4SSe{\cdot}(2{+}\delta)PbTe$ quinary compound with a wide homogeneity region is formed in the system. It

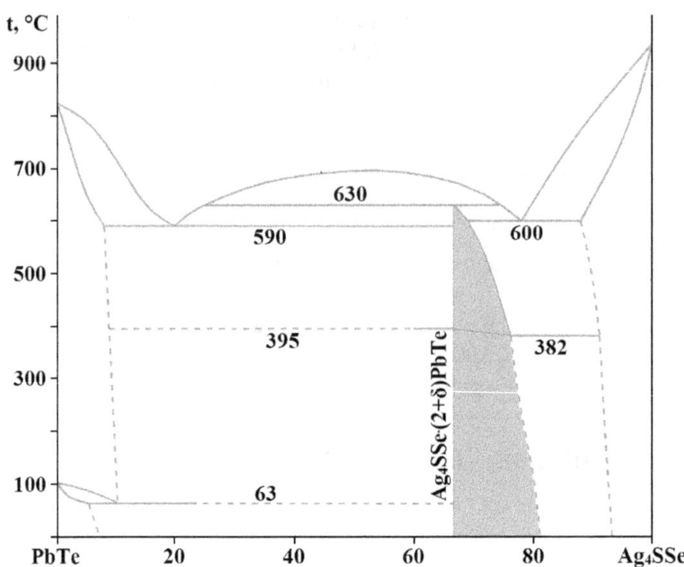

FIGURE 10.3 Phase diagram of the PbTe–Ag_4SSe quasibinary system. (From Amova, A., et al., *Thermochim. Acta*, 531, 42, 2012.)

melts incongruently at 630°C according to the syntactic reaction and crystallizes as a triclinic structure with the lattice parameters a = 351.0, b = 305.3, c = 295.1 pm, α = 99.07, β = 94.37°, and γ = 89.31°. This compound undergoes a phase transition at 395°C. No glassy phases were obtained in this system (Amova et al. 2013).

10.90 Lead–Gold–Antimony–Tellurium–Sulfur

Three quinary phases are formed in the Pb–Au–Sb–Te–S system. $Pb(Pb,Sb)S_2(Au,Te)$ crystallizes as a monoclinic structure with the lattice parameters a = 422.0 ± 0.1, b = 417.6 ± 0.1, c = 1511.9 ± 0.3 pm, β = 95.42 ± 0.03°, and a calculated density of 7.29 g·cm^{-3} (Effenberger et al. 1999). To prepare this compound, pieces of $AuTe_2$, Ag_2S, Sb, Pb, S, and Te were weighed according to the composition $AgAuSb_3Pb_{13}S_{16}Te_6$ and sealed in a Duran glass ampoule in nitrogen atmosphere at 20 kPa. Later synthesis runs were performed with variable ratios of elements. The total sample mass was chosen as ~1 g. The vertically positioned ampoule was heated to 480°C at a rate of 4°C·min^{-1} and kept there for 30 min. Reactions already started below 400°C and partial melting was visible; general surface melting was observed above 620°C. A temperature rise maintaining the rate of 4°C·min^{-1} produced a silvery melt with metallic luster containing numerous bubbles. Finally, the melt was kept at 780°C for 5 min and then cooled down at a rate of 2°C·min^{-1} in a vertical temperature gradient of approximately 0.5°C·cm^{-1}. Extremely thin-tabular crystals of chemically homogeneous synthetic $Pb(Pb,Sb)S_2(Au,Te)$, intergrown in fan-shaped aggregates were grown.

$Pb(Pb,Sb)S_2(Te,Au)$ (mineral nagyágite) crystallizes as an orthorhombic structure with the lattice parameters a = 836.3 ± 0.7, b = 3020 ± 1, c = 828.8 ± 0.7 pm, and a calculated density of 7.51 g·cm^{-3} (Stanley et al. 1994).

$AuPb_5SbTe_2S_{12}$ (mineral museumite) crystallizes as a monoclinic structure with the lattice parameters a = 436.1 ± 0.2, b = 661.8 ± 0.3, c = 2085.8 ± 0.9 pm, and β = 92.71 ± 0.05° (Bindi and Cipriani 2004b; Piilonen et al. 2005b).

10.91 Lead–Gold–Bismuth–Tellurium–Sulfur

In the system containing these elements, the quinary compound $AuPb_2BiTe_2S_3$ (mineral buckhornite), which crystallizes as an orthorhombic structure with the lattice parameters $a = 410.8 \pm 0.3$, $b = 1230.8 \pm 0.9$, $c = 933.1 \pm 0.6$ pm, and a calculated density of 8.25 g·cm^{-3} (Effenberger et al. 2000; Jambor and Roberts 2001) [$a = 409.2 \pm 0.2$, $b = 1224.5 \pm 0.4$, $c = 932.2 \pm 0.4$ pm, and a calculated density of 8.34 g·cm^{-3} (Francis et al. 1992; Jambor and Puziewicz 1993b); $a = 937.4 \pm 0.4$, $b = 1232.6 \pm 0.4$, $c = 407.3 \pm 0.2$ pm, and a calculated density of 8.347 g·cm^{-3} (Johan et al. 1994; Jambor and Roberts 1995a)], is formed .

To prepare the title compound, the mixture of $AuTe_2$, Ag_2S, Pb, Bi, and S according to the initial composition $Ag_{0.5}AuPb_2BiS_{3.25}Te_2$ served as input (Effenberger et al. 2000; Jambor and Roberts 2001). In total, ~1 g sample mass was sealed in a Duran glass ampoule. It was heated at a rate of 4°C·min^{-1} to 660° C and then at a rate of 2°C·min^{-1} to the maximum reaction temperature of 800°C. Finally, the melt was cooled down to room temperature at a rate of 2°C·min^{-1}.

10.92 Lead–Strontium–Antimony–Oxygen–Sulfur

In the system containing these elements, the quinary compound $Sr_{3.5}Pb_{2.5}Sb_6O_5S_{10}$, which crystallizes as a monoclinic structure with the lattice parameters $a = 755.56 \pm 0.04$, $b = 410.66 \pm 0.02$, $c = 1957.5 \pm 0.2$ pm, $\beta = 95.117 \pm 0.005°$, and an energy gap of 1.34 eV, is formed (Wang et al. 2020b). A polycrystalline sample of this compound was synthesized as follows: SrO (1.75 mM), PbO (0.75 mM), Sb_2S_3 (1.50 mM), and PbS (0.50 mM) were mixed and fully ground. The mixture was loaded into a carbon-coated silica tube, and then the tube was flame-sealed under vacuum (0.1 Pa). The tube was heated to 650°C in 10 h, held for 48 h, and slowly cooled to 400°C in 3 days. A black polycrystalline product was obtained. Black rod-shape single crystals of the title compound were grown by heating the reagents to 650°C, left to stand for 3 days, and cooled to 300°C in 4 days.

10.93 Lead–Zinc–Carbon–Nitrogen–Sulfur

In the system containing these elements, the quinary compound $ZnPb(SCN)_4$, which crystallizes as an orthorhombic structure with the lattice parameters $a = 916.9 \pm 0.4$, $b = 606.1 \pm 0.3$, $c = 2096.8 \pm 2.0$ pm, and the calculated and experimental densities of 2.95 and 2.97 g·cm^{-3}, respectively, is formed (Brodersen et al. 1987). The title compound was obtained from the binary $Zn(SCN)_2$ and $Pb(NCS)_2$ in stoichiometric ratio with traces of H_2O in an ultrasound bath for about 6 h at room temperature. It could also be obtained by dry milling of $Zn(SCN)_2$ and $Pb(SCN)_2$ or by shaking of $Pb(SCN)_2$ in a concentrated solution of $Zn(SCN)_2$ in H_2O (Dost et al. 1968).

10.94 Lead–Zinc–Antimony–Oxygen–Sulfur

$PbO–ZnS–Sb_2O_3$. Bulk glasses with starting compositions of $70Sb_2O_3–(30–x)PbO–xZnS$ ($x = 5–30$) were prepared in this system (Nouadji et al. 2013). The glass transition temperatures of these glasses are within the interval from 261°C to 272°C.

10.95 Lead–Cadmium–Arsenic–Bismuth–Chlorine–Sulfur

In the system containing these elements, the multinary compound $Cd_2Pb_{20}(As,Bi)_{22}S_{50}Cl_{10}$ (mineral tazieffite), which crystallizes as a monoclinic structure with the lattice parameters $a = 835.20 \pm 0.17$,

$b = 4552.90 \pm 0.92$, $c = 2726.10 \pm 0.55$ pm, $\beta = 98.84 \pm 0.03°$, and a calculated density of 6.077 g·cm^{-3}, is formed (Zelenski et al. 2009).

10.96 Lead–Mercury–Thallium–Antimony–Manganese–Sulfur

In the system containing these elements, the multinary compound $Mn_8Tl_8Hg_2(Sb_{21}Pb_2Tl)S_{48}$ (mineral tsygankoite), which crystallizes as a monoclinic structure with the lattice parameters $a = 2136.2 \pm 0.4$, $b = 385.79 \pm 0.10$, $c = 2713.5 \pm 0.4$ pm, and $\beta = 106.944 \pm 0.014°$, and a calculated density of 5.450 g·cm^{-3}, is formed (Kasatkin et al. 2018a,b,c; Belakovskiy et al. 2019).

10.97 Lead–Thallium–Arsenic–Antimony–Sulfur

Some quinary compounds are formed in the Pb–Tl–As–Sb–S system. $TlPbAs_2SbS_6$ (mineral jentschite) crystallizes as a monoclinic structure with the lattice parameters $a = 812.1 \pm 0.3$, $b = 2396.9 \pm 0.9$, $c = 584.7 \pm 0.3$ pm, $\beta = 107.68 \pm 0.03°$, and a calculated density of 5.24 g·cm^{-3} (Graeser and Edenharter 1997; Jambor and Roberts 1997) [$a = 809.58 \pm 0.05$, $b = 2391.7 \pm 0.2$, $c = 588.76 \pm 0.05$ pm, and $\beta = 108.063 \pm 0.008°$ (Berlepsch 1996; Jambor et al. 1997b)].

$TlPb_4(Sb_7As_2)S_{18}$ (mineral boscardinite) crystallizes as a triclinic structure with the lattice parameters $a = 810.17 \pm 0.04$, $b = 865.97 \pm 0.04$, $c = 2255.74 \pm 0.10$ pm, $\alpha = 90.666 \pm 0.002°$, $\beta = 97.242 \pm 0.002°$, and $\gamma = 90.850 \pm 0.002°$ (Biagioni and Moëlo 2017b; Belakovskiy et al. 2017d) [$a = 809.29 \pm 0.04$, $b = 876.10 \pm 0.05$, $c = 2249.71 \pm 0.11$ pm, $\alpha = 90.868 \pm 0.004°$, $\beta = 97.247 \pm 0.004°$, $\gamma = 90.793 \pm 0.004°$, and a calculated density of 5.355 g·cm^{-3} (Orlandi et al. 2012; Cámara et al. 2014c)].

$Tl_4Pb_2(As,Sb)_{20}S_{34}$ (mineral dalnegroite) also crystallizes as a triclinic structure with the lattice parameters $a = 1621.8 \pm 0.3$, $b = 4254.6 \pm 0.7$, $c = 855.8 \pm 0.1$ pm, $\alpha = 95.70 \pm 0.04°$, $\beta = 90.18 \pm 0.03°$, $\gamma = 96.38 \pm 0.04°$, and a calculated density of 9.515 g·cm^{-3} (Bindi et al. 2010) [$a = 1621.7 \pm 0.7$, $b = 4254.4 \pm 0.9$, $c = 855.7 \pm 0.4$ pm, $\alpha = 95.72 \pm 0.04°$, $\beta = 90.25 \pm 0.04°$, $\gamma = 96.78 \pm 0.04°$, and a calculated density of 4.819 g·cm^{-3} (Nestola et al. 2009; Piilonen and Poirier 2010).

$Tl_{4-x}Pb_{2+2x}Sb_{20-x-y}As_yS_{34}$ (mineral protochabournéite, $0.02 \le x \le 0.34$; $5.71 \le y \le 6.69$) (Miyawaki et al. 2021c,d) crystallizes as a triclinic structure with the lattice parameters $a = 815.0 \pm 0.2$, $b = 871.6 \pm 0.2$, $c = 2157.9 \pm 0.4$ pm, $\alpha = 85.18 \pm 0.01°$, $\beta = 96.94 \pm 0.01°$, $\gamma = 88.60 \pm 0.01°$, and a calculated density of 5.008 and 5.137 g·cm^{-3} for two different samples (Orlandi et al. 2011, 2013; Belakovskiy and Gagne 2015).

10.98 Lead–Thallium–Arsenic–Antimony–Oxygen–Sulfur

In the system containing these elements, the multinary compound $TlPb_{26}(Sb,As)_{31}S_{72}O$ (thallium-bearing chovanite) crystallizes as a monoclinic structure with the lattice parameters $a = 3428.0 \pm 0.3$, $b = 824.30 \pm 0.07$, $c = 4845.7 \pm 0.4$ pm, and $\beta = 106.290 \pm 0.004°$, is formed (Biagioni et al. 2018b).

10.99 Lead–Lanthanum–Bismuth–Oxygen–Sulfur

In the system containing these elements, the quinary compound $LaPbBiOS_3$, which crystallizes as a tetragonal structure with the lattice parameters $a = 409.676 \pm 0.001$ and $c = 1979.96 \pm 0.06$ pm (the neutron diffraction) or $a = 409.717 \pm 0.004$ and $c = 1979.33 \pm 0.02$ pm (the synchrotron XRD) (Mizuguchi et al. 2017) [$a = 409.82 \pm 0.01$, $c = 1977.54 \pm 0.06$ pm, and a calculated density of 6.6724 ± 0.0002 g cm^{-3} (Sun et al. 2014)], is formed . This compound is a narrow band gap semiconductor (Sun et al. 2014).

Polycrystalline samples of this compound were prepared by the solid state reaction method. The La_2O_3, La_2S_3, Bi_2S_3, and PbS powders were mixed, pelletized, sealed into an evacuated quartz tube, and heated at 670°C for 25 h (Mizuguchi et al. 2017). The obtained sample was ground, mixed to homogenize, pelletized, and heated at 750°C for another 25 h.

Polycrystalline samples can be also obtained using high-purity starting materials of La_2S_3, Bi_2S_3, and PbO (Sun et al. 2014). These starting materials in the stoichiometric ratio were accurately weighed first, and the mixture was placed into an alumina tube, which then was sealed in an evacuated quartz ampoule. After that, the ampoule was slowly heated to 670°C and annealed at this temperature for 25 h. The resultant product was then ground, pressed into pellets, and sintered under vacuum at 750°C for another 25 h, followed by cooling to room temperature. All the operations except for the ampoule sealing and heating were carried out in a glove box with the water and oxygen contents of <1 ppm. The pellet sample obtained was black in color.

10.100 Lead–Carbon–Nitrogen–Chlorine–Sulfur

The PbCl(SCN) quinary compound, which crystallizes as an orthorhombic structure with the lattice parameters $a = 1016.6 \pm 0.1$, $b = 425.5 \pm 0.1$, and $c = 958.9 \pm 0.1$ pm, is formed in this system (Gacemi et al. 2005). It was prepared from a mixture of $PbCl_2$ and KSCN: $PbCl_2$ (0.5 mM) was dissolved in water (100 mL) with the stoichiometric amount of KSCN. The mixture was refluxed for 1 h and the resulting clear solution was slowly evaporated at room temperature, yielding transparent needles of the title compound.

10.101 Lead–Arsenic–Antimony–Bismuth–Sulfur

In the system containing these elements, the quinary compound $Pb_{12}(Sb_3As_2Bi)S_{21}$ (mineral marcobaldiite), which crystallizes as a triclinic structure with the lattice parameters $a = 892.48 \pm 0.09$, $b = 2941.4 \pm 0.3$, $c = 853.01 \pm 0.08$ pm, $\alpha = 98.336 \pm 0.005°$, $\beta = 118.175 \pm 0.005°$, $\gamma = 90.856 \pm 0.005°$, and a calculated density of 6.56 g·cm^{-3}, is formed (Biagioni et al. 2016b, 2018c).

10.102 Lead–Arsenic–Antimony–Bismuth–Selenium–Sulfur

In the system containing these elements, the multinary compound $Pb_{10}(As,Bi)_6(S,Se)_{19}$ (Se-bearing variety of the mineral kirkiite) is formed (Borodaev et al. 1998). In this phase, the Se content ranges from 0.73 mass% to 1.16 mass% and As and Bi concentrations vary from 6.49 mass% to 7.98 mass% and from 14.32 to 16.84 mass%, respectively. The composition field of such kirkiite can be described by the formula $Pb_{10}(As_{6-x}Bi_x)_6(S_{19-y}Se_y)_{19}$, where $2.3 \le x \le 2.8$ and $0.3 \le y \le 0.5$, which is close to ideal formula $Pb_{10}(As,Bi)_6(S,Se)_{19}$.

10.103 Lead–Arsenic–Antimony–Oxygen–Sulfur

In the system containing these elements, the quinary phase $Pb_{28}(Sb,As)_{30}S_{72}O$ (mineral species), which crystallizes as a monoclinic structure with the lattice parameters $a = 3405.2 \pm 0.3$, $b = 820.27 \pm 0.07$, $c = 4807.8 \pm 0.4$ pm, and $\beta = 106.258 \pm 0.004°$, is formed (Biagioni and Moëlo 2017a).

10.104 Lead–Arsenic–Chlorine–Iodine–Sulfur

In the system containing these elements, the quinary compound $Pb_4As_2S_6ICl$ (mineral mutnovskite), which crystallizes as an orthorhombic structure with the lattice parameters $a = 1153.94 \pm 0.09$, $b = 667.32 \pm 0.05$, and $c = 934.54 \pm 0.07$ pm (Bindi et al. 2008) [$a = 1154.3 \pm 0.1$, $b = 667.54 \pm 0.07$, $c = 935.9 \pm$

0.1 pm, and a calculated density of 6.177 g·cm^{-3} (Zelenski et al. 2006)], is formed . At room temperature, mutnovskite possess a centrosymmetric structure type, space group *Pnma*, while at low temperature (110 K) it adopts a non-centrosymmetric orthorhombic structure type, space group $Pnm2_1$ (Bindi et al. 2008). Mutnovskite reconverts to the centrosymmetric-type upon returning to room temperature thus indicating that the phase transition is completely reversible in character.

10.105 Lead–Antimony–Bismuth–Iron–Sulfur

PbS–Sb_2S_3–Bi_2S_3–FeS. The phase equilibria in this system were studied through XRD, metallography, and EPMA (Chang et al. 1980). The evacuated glass capsule technique was used for investigations. Starting compositions were prepared from Pb, Sb, Bi, Fe, and S. The samples were heated in a horizontal muffle furnace. Generally, 50 days were used for homogenization at 500°C, 80 to 90 days at 450°C and below. At the end of heat treatment, the samples were quenched in air. Seven different phases were synthesized in the system. Among them, four have extensive solid solutions.

10.106 Lead–Antimony–Vanadium–Molybdenum–Sulfur

In the system containing these elements, the quinary compound $Pb_4Mo_4VSbS_{15}$ (mineral merelaniite) is formed (Jaszczak et al. 2016; Rumsey et al. 2016; Belakovskiy et al. 2019). This compound belongs to the cylindrite homologous series with alternating centered pseudo-tetragonal (*Q*) and pseudo-hexagonal (*H*) layers with respective PbS and MoS_2 structure types, and both layers are triclinic. The unit-cell parameters for the *Q* layer refined from single-crystal/powder XRD are a = (592.9 ±0.8)/(592.49 ± 0.08), b = (596.1 ± 0.5)/(598.7 ± 0.3), c = (1203 ± 1)/(1207.7 ± 0.6) pm, α = (91.33 ± 0.09°)/(90.61 ± 0.03°), β = (90.88 ± 0.05°)/(90.04 ± 0.02°), and γ = (91.79 ± 0.04°)/(89.95 ± 0.03°). For the *H* layer, a = (554.7 ± 0.9)/(555.03 ± 0.06), b = (315.6 ± 4)/(315.36 ± 0.08), c = (1191 ± 1)/(1187.7 ± 0.6) pm, α = (89.52 ± 0.09°)/(90.00 ± 0.01°), β = (92.13 ± 5°)/(90.05 ± 0.01°), and γ = (90.18 ± 0.04°)/(89.92 ± 0.02°). The calculated density of this compound is 4.895 g·cm^{-3}.

10.107 Lead–Antimony–Oxygen–Chlorine–Sulfur

In the system containing these elements, the quinary compound $Pb_9Sb_{10}S_{23}ClO_{0.5}$ (mineral pillaite), which crystallizes as a monoclinic structure with the lattice parameters a = 4949 ± 1, b = 412.59 ± 0.08, c = 2182.8 ± 0.4 pm, β = 99.62 ± 0.03°, and a calculated density of 5.87 g·cm^{-3} (Meerschaut et al. 2001) [a = 4965 ± 3, b = 415.0 ± 0.4, c = 2191 ± 1 pm, β = 99.76 ± 0.05°, and a calculated density of 5.77 g·cm^{-3} (Orlandi et al. 2001)], is formed .

10.108 Lead–Antimony–Oxygen–Chlorine–Manganese–Sulfur

In the system containing these elements, the multinary phases $(Mn_{1-x}Pb_x)Pb_{10+y}Sb_{12-y}S_{26-y}Cl_{4+y}O$, which crystallize in the monoclinic structure with the lattice parameters a = 3748.0 ± 0.8, b = 411.78 ± 0.08, c = 1816.7 ± 0.4 pm, β = 106.37 ± 0.03°, and a calculated density of 5.782 g·cm^{-3} for $Mn_{0.72}Pb_{10.97}Sb_{11.31}S_{25.31}Cl_{4.69}O$ composition, are formed (Doussier et al. 2007). The syntheses of these phases were performed using the mixtures of MnS, $PbCl_2$, PbS, and Sb_2S_3 (molar ratio 1:2:10:5 or 1:2:8:6), which were put into silica tubes, sealed under vacuum, and heated at 600°C for 10 and 15 days, respectively. An attempt performed to add oxygen as MnO (600°C; MnO, $PbCl_2$, PbS, and Sb_2S_3 with molar ratio 1:2:8:6 ratios; 10 days) failed to obtain these phases. All final products were complex and fine grained mixtures of chloro- and oxy-chloro-sulfosalts of Pb and Sb, with or without minor Mn.

11.109 Lead–Antimony–Selenium–Tellurium–Nickel–Sulfur

PbS–PbSe–PbTe–NiSb. When samples, containing 90 mol% PbS, 5 mol% PbSe, and 5 mol% PbTe, and NiSb interact up to 900°C, no changes were observed (Kuliev et al. 1974). During annealing of two-layer samples NiSb + (PbS+PbS+PbTe) at 600°C for 250 h, no diffusion layer was found in the contact area. The ingots were studied through DTA, XRD, and local X-ray spectral analysis.

10.110 Lead–Antimony–Manganese–Iron–Sulfur

$Pb_2MnSb_6S_{14}$–$Pb_2FeSb_6S_{14}$. A complete solid-solution series exists in this system with melting temperature varying continuously from 592°C ($Pb_2MnSb_6S_{14}$) to 583°C ($Pb_2FeSb_6S_{14}$) (Chang and Li 1987).

$Pb_4Sb_6S_{13}$–MnS–FeS. The $Pb_2Mn_xFe_{1-x}Sb_6S_{14}$ solid solutions are stable in this quasiternary system at 500°C (Chang and Li 1987). Solid solutions in equilibrium with FeS have compositions covering almost the entire series, whereas solid solutions in equilibrium with MnS are limited in their compositions to the Mn-rich end of the series. At 400°C, the ranges of the two-phase region vary in accordance with the decrease of mutual solubilities between FeS and MnS, but the composition of the $Pb_2Mn_xFe_{1-x}Sb_6S_{14}$ series that defines the three-phase region remains unchanged.

The starting compositions for the investigations of both systems were prepared from Pb, Sb, Fe, Mn, and S. The synthesis and heat treatment were made in muffle furnaces using the conventional technique of sealed, evacuated glass capsules. Generally, the duration of treatment ranged from 120 days at 400°C to 45 days at 500°C, and to 5 days at and above 600°C.

The $Pb_4(Mn,Fe)Sb_6S_{14}$ quinary phase (mineral benavidesite), which crystallizes as a monoclinic structure with the lattice parameters a = 1574, b = 1914, c = 406 pm, β = 91.50°, and a calculated density of 5.60 g·cm^{-3}, is formed in this system (Oudin et al. 1982; Fleischer and Pabst 1983).

10.111 Lead–Bismuth–Selenium–Tellurium–Sulfur

In the system containing these elements, the multinary compound $PbBi_2(Se,Te,S)_4$ (mineral poubaite), which crystallizes as a rhombohedral structure with the lattice parameters a = 427.1 ± 0.2 and c = 4027 ± 4 pm in hexagonal setting (Johan et al. 1987) [a = 425.2 and c = 4009.5 pm in hexagonal setting (Fleischer 1978)], is formed .

10.112 Lead–Bismuth–Selenium–Chlorine–Sulfur

In the system containing these elements, the quinary phases $Pb_{1+x}Bi_{2-x}(S_{4-x-y}Cl_xSe_y)$ ($0 < x < 0.3$, $0 < y < 0.2$) (chlorine-bearing galenobismutite), which crystallizes as an orthorhombic structure with the lattice parameters a = 1183.2 ± 0.1, b = 1464.0 ± 0.2, c = 408.17 ± 0.04 pm, and a calculated density of 7.10 g·cm^{-3} for x = 0.17 and y = 0.09; a = 1184.6 ± 0.4, b = 1466.7 ± 0.5, and c = 408.5 ± 0.1 pm for x = 0.20 and y = 0.20; and a = 1185.6 ± 0.6, b = 1468.7 ± 0.9, and c = 409.2 ± 0.3 pm for x = 0.22 and y = 0.09, are formed (Locock et al. 2006; Pinto et al. 2006).

10.113 Lead–Bismuth–Selenium–Platinum–Sulfur

The $Pt_{2-x}(Bi,Pb)_{11}(S,Se)_{11}$ or $(Pt,Pb)Bi_3(S,Se)_{4-x}$ (x = 0.4–0.8) quinary phase (mineral crerarite), which crystallizes as a cubic structure with the lattice parameter a = 586 ± 5 pm and a calculated density of 7.75 g·cm^{-3}, is formed in this system (Cook et al. 1994; Jambor et al. 1995b).

10.114 Lead–Oxygen–Tellurium–Chlorine–Iron–Sulfur

In the system containing these elements, the quinary compound $Pb_2Fe_6(Te^{4+}O_3)_3(SO_4)O_2Cl$ (mineral eztlite), which crystallizes as a monoclinic structure with the lattice parameters a = 1146.6 ± 0.2, b = 1977.5 ± 0.4, c = 1049.7 ± 0.2 pm, β = 102.62 ± 0.03°, and a calculated density of 5.458 g·cm^{-3} at 100 K (Missen et al. 2018) [a = 658, b = 968, c = 2052 pm, β = 90°15', and the calculated and experimental densities of 4.60 and 4.5 g·cm^{-3}, respectively (Williams 1982; Dunn et al. 1983a)], is formed . The chemical formula of eztlite has been revised (Hålenius et al. 2018c,d; Missen et al. 2018) from that stated previously as $Fe_6Pb_2(Te^{4+}O_3)_3(Te^{6+}O_6)(OH)_{10} \cdot nH_2O$ (Williams 1982).

10.115 Lead–Selenium–Tellurium–Iron–Sulfur

PbS–PbSe–PbTe–Fe. The samples, containing 90 mol% PbS, 5 mol% PbSe, and 5 mol% PbTe, begin to interact with Fe at 375°C (Kuliev et al. 1974). At the same time, iron sulfide was found in the system, and the main elements present in the diffusion layer after thermal annealing are iron and sulfur. The ingots were studied through DTA, XRD, and local X-ray spectral analysis.

10.116 Lead–Selenium–Tellurium–Cobalt–Sulfur

PbS–PbSe–PbTe–Co. The samples, containing 90 mol% PbS, 5 mol% PbSe, and 5 mol% PbTe, begin to interact with Co at 525°C (Kuliev et al. 1974). At the same time, cobalt sulfide was found in the system, and the main elements present in the diffusion layer after thermal annealing are lead, cobalt, and sulfur. The ingots were studied through DTA, XRD, and local X-ray spectral analysis.

10.117 Lead–Selenium–Tellurium–Nickel–Sulfur

PbS–PbSe–PbTe–Ni. The samples, containing 90 mol% PbS, 5 mol% PbSe, and 5 mol% PbTe, begin to interact with Ni at 620°C (Kuliev et al. 1974). At the same time, nickel sulfide and nickel telluride were found in the system. The ingots were studied through DTA, XRD, and local X-ray spectral analysis.

11

Systems Based on Lead Selenide

11.1 Lead–Hydrogen–Copper–Uranium–Oxygen–Selenium

In the system containing these elements, the multinary compound $Pb_2Cu_5(UO_2)_2(SeO_3)_6(OH)_6{\cdot}2H_2O$ (mineral demesmaekerite), which crystallizes as a triclinic structure with the lattice parameters a = 1195.5 ± 0.5, b = 1003.9 ± 0.4, c = 563.9 ± 0.2 pm, α = 89.78 ± 0.04°, β = 100.36 ± 0.04°, γ = 91.34 ± 0.04°, and the calculated and experimental densities of 5.42 ± 0.05 and 5.28 ± 0.04 $g{\cdot}cm^{-3}$, respectively (Ginderow and Cesbron 1983) [a = 1190 ± 3, b = 1002.3, c = 563 ± 2 pm, α = 90°11', β = 100°01', γ = 91°49', and the calculated and experimental densities of 5.45 and 5.28 ± 0.04 $g{\cdot}cm^{-3}$, respectively (Cesbron et al. 1965; Fleischer 1966b)], is formed.

11.2 Lead–Hydrogen–Copper–Nitrogen–Oxygen–Selenium

In the system containing these elements, the multinary compound $Cu_3Pb(OH)(NO_3)(SeO_3)_3{\cdot}0.5H_2O$, which crystallizes as a triclinic structure with the lattice parameters a = 776.1 ± 0.3, b = 947.8 ± 0.4, c = 951.4 ± 0.4 pm, α = 66.94 ± 0.02°, β = 69.83 ± 0.02°, γ = 81.83 ± 0.02°, and a calculated density of 4.76 $g{\cdot}cm^{-3}$, is formed (Effenberger 1986). For the synthesis of this compound, steel autoclave lined with Teflon and a reaction volume of ~6 cm^3 was used. An equimolar mixture consisting of Pb_3O_4 or $Pb(NO_3)_2$, CuO, $Cu(OH)_2$ or $Cu(NO_3)_2{\cdot}3H_2O$, and SeO_2 with or without HNO_3 was used, filled with H_2O to 80% of autoclave volume. The closed autoclave was heated between 150°C and 230°C for 48 h. After a cooling time of about 12 h, the crystals of the target compound were observed among other byproducts. The proportions of the individual phases formed vary relatively strongly in the individual synthesis approaches. The crystals of the title compound were colored light green, transparent, roughly isometric and very flat.

11.3 Lead–Hydrogen–Copper–Bismuth–Oxygen–Selenium

In the system containing these elements, the multinary compound $PbBiCu_6O_4(SeO_3)_4(OH){\cdot}H_2O$ (mineral favreauite), which crystallizes as a tetragonal structure with the lattice parameters a = 986.0 ± 0.4, c = 970.0 ± 0.5 pm, and a calculated density of 4.851 $g{\cdot}cm^{-3}$ at 100 K, is formed (Mills et al. 2014a,b; Belakovskiy et al. 2016c).

11.4 Lead–Hydrogen–Copper–Oxygen–Selenium

Two quinary compounds are formed in the Pb–H–Cu–O–Se system. $PbCu(SeO_4)(OH)_2$ (mineral franksousaite) crystallizes as a monoclinic structure with the lattice parameters a = 982.08 ± 0.03, b = 573.40 ± 0.02, c = 474.98 ± 0.01 pm, β = 102.683 ± 0.002°, and a calculated density of 5.59 g cm^{-3} (Yang et al. 2022i,j,k).

$Cu_2Pb_2(Se^{6+}O_4)(Se^{4+}O_3)(OH)_4$ (mineral schmiederite) also crystallizes as a monoclinic structure with the lattice parameters a = 992.2 ± 0.3, b = 571.2 ± 0.2, c = 939.6 ± 0.3 pm, β = 101.96 ± 0.03°, and a

DOI: 10.1201/9781003123460-11

calculated density of 5.63 g cm^{-3} (Effenberger 1987; Hawthorne et al. 1988) [a = 992 ± 1, b = 571 ± 2, c = 935 ± 1 pm, β = 102.0 ± 0.2°, and a calculated density of 5.62 g cm^{-3} (Sarp and Burri 1987)].

Synthetic crystals of schmiederite were prepared under hydrothermal conditions as follows (Effenberger 1987): 2 g of an equimolar mixture of PbO, CuO, and SeO_2 was put into a steel vessel lined with Teflon (~6 mL capacity). One mL each of H_2SeO_4 and H_2O_2 (perhydrol) were added and the vessel was filled up with H_2O to about 80 vol%. After a reaction time of 2 days at 240°C and after cooling to room temperature, the crystals of the title compound were obtained. The obtained crystals are light blue to greenish in color.

11.5 Lead–Hydrogen–Copper–Oxygen–Chlorine–Selenium

In the system containing these elements, the multinary compound $CuPb_4(SeO_3)_3(OH)Cl_3$ (mineral species), which crystallizes as a triclinic structure with the lattice parameters a = 829.0 ± 0.8, b = 1058.8 ± 1.3, c = 1358.7 ± 1.5 pm, α = 124.47 ± 0.08°, β = 110.60 ± 0.09°, γ = 63.26 ± 0.09°, and a calculated density of 5.25 g·cm^{-3}, is formed (Campostrini et al. 1999; Jambor et al. 2000).

11.6 Lead–Hydrogen–Cadmium–Oxygen–Iodine–Selenium

In the system containing these elements, the multinary compound $Cd_3Pb_2(SeO_3)_4I_2(H_2O)$, which is stable up to 380°C and crystallizes as a monoclinic structure with the lattice parameters a = 1120.0 ± 0.9, b = 543.9 ± 0.4, c = 1664.9 ± 1.0 pm, β = 119.24 ± 0.04°, a calculated density of 5.746 g cm^{-3}, and an energy gap of 3.42 eV, is formed (Zhang et al. 2012a). Brown rod-shaped crystals of the title compound were initially obtained along with crystals of $PbSeO_3$ and $CdSeO_3$ as impurities by hydrothermal reaction of a mixture containing PbO (0.1 mM), CdI_2 (0.1 mM), and SeO_2 (1.0 mM) in H_2O (4 mL), which was sealed in an autoclave equipped with a Teflon liner (23 mL) and heated at 230°C for 4 days, followed by slow cooling to room temperature at a rate of 6°C·h^{-1}. The final pH value of the reaction media was close to 1.0. After proper structural analysis, a single phase of this compound was obtained in a molar ratio of PbO, CdI_2, and SeO_2 of 1:2:12 under the same reaction conditions.

11.7 Lead–Hydrogen–Uranium–Oxygen–Selenium

In the system containing these elements, the quinary compound $Pb(UO_2)_3(SeO_3)_2O_2·3H_2O$ (mineral borzęckiite), which crystallizes as an orthorhombic structure with the lattice parameters a = 700 ± 2, b = 713 ± 1, and c = 1710 ± 4 pm, is formed (Siuda et al. 2022, 2023).

11.8 Lead–Hydrogen–Nitrogen–Oxygen–Selenium

The quinary compound $(NH_4)_2Pb(SeO_3)_3$ is formed in the Pb–H–N–O–Se system (Frydrych 1976). To prepare this compound, a 150-mL beaker equipped with a thermometer and glass electrode for pH measurements was placed on a heatable magnetic stirrer and filled with 50 mL of H_2O. Then, 13 g SeO_2 was dissolved (pH ≈ 0.8) and mixed with 10 g NH_4CH_3COO. The pH rises to around 3.6. The solution was then reacted with the appropriate alkali hydroxide or carbonate with heating to 60°C until the pH rises to 6.5. A 6 g of freshly prepared $Pb(SeO_3)_2$ was gradually added to this solution and the temperature was increased to 70°C. $Pb(SeO_3)_2$ dissolved within 2–3 min. If this is not completely successful, the remaining was sucked off in devices preheated to around 80°C. The title compound separated in a crystalline form from a clear yellow solution on cooling to the low temperature. The products separating out after 3 to 24 h were filtered off with suction, dried with a little of 12 mass% CH_3COOH, then washed with methanol and ether at 40°C.

11.9 Lead–Hydrogen–Vanadium–Oxygen–Selenium

In the system containing these elements, the quinary compound $Pb_4V_6O_{16}(SeO_3)_3(H_2O)$, which is stable up to 310°C and crystallizes as a monoclinic structure with the lattice parameters a = 713.620 ± 0.010, b = 2131.03 ± 0.04, c = 720.040 ± 0.010 pm, β = 94.887 ± 0.002°, a calculated density of 5.447 g cm^{-3}, and an energy gap of 2.54 eV, is formed (Cao et al. 2014). This compound was prepared by using hydrothermal techniques. A mixture of $PbCO_3$ (0.6 mM), V_2O_5 (0.45 mM), SeO_2 (1.5 mM), 0.10 mL of a 1 M KOH solution, and H_2O (6 mL) was sealed in an autoclave equipped with a Teflon liner (23 mL) and heated at 230°C for 4 days, followed by slow cooling to room temperature at a rate of 2°C·h^{-1}. The final pH value of the reaction media is close to 1.5. Yellow flake-shaped crystals of $Pb_4V_6O_{16}(SeO_3)_3(H_2O)$ were obtained as a single phase.

11.10 Lead–Hydrogen–Oxygen–Chlorine–Selenium

Three quinary compounds are formed in the Pb–H–O–Cl–Se system. $Pb_3(SeO_3)Cl_4{\cdot}H_2O$ (mineral orlandiite) crystallizes as a triclinic structure with the lattice parameters a = 813.6 ± 0.3, b = 843.0 ± 0.6, c = 923.3 ± 0.7 pm, α = 62.58 ± 0.07°, β = 71.84 ± 0.04°, γ = 75.13 ± 0.04°, and a calculated density of 5.699 g·cm^{-3} (Demartin et al. 2003) [a = 814.6 ± 1.2, b = 842.8 ± 2.2, c = 924.1 ± 2.2 pm, α = 62.32 ± 0.21°, β = 71.64 ± 0.17°, γ = 75.22 ± 0.19°, and a calculated density of 5.66 g·cm^{-3} (Campostrini et al. 1999; Jambor et al. 2000)].

$Pb_3(Se_4CO_3)(OH)Cl_3$ (mineral guangyuanite) crystallizes as an orthorhombic structure with the lattice parameters a = 1100.03 ± 0.05, b = 1064.60 ± 0.05, and c = 779.02 ± 0.03 pm (Yang et al. 2023f,g).

$Pb_3(SeO_3)(SeO_2OH)Cl_3$ crystallizes as a monoclinic structure with the lattice parameters a = 778.56 ± 0.09, b = 562.64 ± 0.06, c = 1241.30 ± 0.14 pm, and β = 99.057 ± 0.002° (Porter and Halasyamani 2001). This compound was synthesized by using a low-temperature aqueous method. H_2SeO_3 (3.74 mM) and $PbCl_2$ (1.87 mM) were dissolved in 5 mL of distilled water. The respective solution was refluxed overnight at 160°C and cooled to room temperature over a period of 2 h. The colorless parallelepiped-shaped crystals of the title compound were obtained. They were thoroughly washed with H_2O.

11.11 Lead–Hydrogen–Oxygen–Bromine–Selenium

Two quinary compounds are formed in the Pb–H–O–Br–Se system. $Pb_3(SeO_3)(OH)Br_2$ is stable up to 290°C and crystallizes as an orthorhombic structure with the lattice parameters a = 1030.59 ± 0.12, b = 1082.11 ± 0.12, c = 798.02 ± 0.09 pm, a calculated density of 5.247 g·cm^{-3}, and an energy gap of 3.51 eV (Shang and Halasyamani 2020). Crystals of the title compound were grown by hydrothermal methods, combining $PbBr_2$ (2 mM), SeO_2 (1 mM), PbO (1.5 mM), and K_2CO_3 (2 mM) with H_2O (8 mL). The solution was placed in 23-mL Teflon-lined autoclave that was closed. The autoclave was gradually heated to 220°C, maintained at the same temperature for 72 h, and then slowly cooled to room temperature at a rate of 6°C·h^{-1}. Colorless crystals, the only product from the reaction, were isolated from the mother liquor by vacuum filtration and washed with deionized water.

$Pb_3(SeO_3)(HSeO_3)Br_3$ is stable up to 200°C and crystallizes as a monoclinic structure with the lattice parameters a = 820.3 ± 0.2, b = 569.30 ± 0.10, c = 1257.1 ± 0.3 pm, β = 100.223 ± 0.003°, a calculated density of 6.417 g·cm^{-3}, and an energy gap of 3.11 eV (Shang and Halasyamani 2020). Crystals of this compound were grown by hydrothermal methods, combining $PbBr_2$ (2 mM), SeO_2 (4 mM), and PbO (1.5 mM) with H_2O (2 mL). The solution was placed in 23-mL Teflon-lined autoclave that was closed. The autoclave was gradually heated to 188°C, maintained at that temperature for 72 h, and then slowly cooled to room temperature at a rate of 6°C·h^{-1}. Colorless crystals, the only product from the reaction, were isolated from the mother liquor by vacuum filtration and washed with deionized water.

11.12 Lead–Hydrogen–Oxygen–Nickel–Selenium

In the system containing these elements, the quinary compound $PbNi_2(SeO_2OH)_2(SeO_3)_2$, which crystallizes as a monoclinic structure with the lattice parameters a = 1368.24 ± 0.10, b = 526.92 ± 0.05, c = 1934.76 ± 0.13 pm, β = 129.524 ± 0.004°, and a calculated density of 5.151 g·cm^{-3}, is formed (Kovrugin et al. 2015). This compound has been prepared by the hydrothermal techniques. After weighing and grinding PbO, SeO_2, and NiO, the reagents were mixed in 6 mL of distilled water. When necessary, a solution of NaOH was used to adjust the pH to its desired value. It has been experimentally established that only acidic conditions favor the crystallization. A large stoichiometric excess of SeO_2 was necessary to achieve reaction, where the amount of SeO_2 is multiplied by five. The pH value increases from ~1 to ~5.5–6.0 on decreasing the SeO_2 content. The chemical reactions were performed during 36 h in 23 mL Teflon-lined reaction vessels heated in an oven at 200°C. At the end of the experiment time, the vessels were cooled for 48 h. The precipitate was filtered through a filter paper. Green plates of the target compound were obtained.

11.13 Lead–Sodium–Carbon–Oxygen–Selenium

PbSe–Na_2CO_3–C. The interaction of PbSe with Na_2CO_3 in the carbon presence proceeds in two next stages: PbSe + Na_2CO_3 = PbO + Na_2Se + CO_2 and 2PbO + C = 2Pb + CO_2 (Shkodin and Malyshev 1969). The reduction of lead with carbon with simultaneous production of Na_2Se is most efficiently carried out at 1065°C–1200°C.

11.14 Lead–Potassium–Copper–Zinc–Oxygen–Chlorine–Selenium

In the system containing these elements, the multinary compound $KPb_{1.5}ZnCu_6O_2(SeO_3)_2Cl_{10}$ (mineral prewittite), which crystallizes as an orthorhombic structure with the lattice parameters a = 913.2 ± 0.2, b = 1941.5 ± 0.4, c = 1321.3 ± 0.3 pm, and the calculated and experimental densities of 3.889 and 3.90 ± 0.02 g·cm^{-3}, respectively, is formed (Shuvalov et al. 2013).

11.15 Lead–Potassium–Arsenic–Oxygen–Selenium

In the system containing these elements, the quinary compound $K_4Pb_6(PO_4)_4(SeO_4)_2$, which crystallizes as a hexagonal structure with the lattice parameters a = 984.3 ± 0.5, c = 733.6 ± 0.5 pm, and the calculated and experimental densities of 5.572 and 5.54 g·cm^{-3}, respectively, is formed (Schwarz 1967b). This compound was prepared by heating in oxygen current a mixture of $Pb_3(PO_4)_2$ and K_2SeO_4 in the stoichiometric ratio three times at 550°C for 90, 160, and 120 h with the next quickly cooling to room temperature after each heating.

11.16 Lead–Potassium–Oxygen–Fluorine–Selenium

The attempts to prepare the $K_4Pb_6(SeO_4)_6F_2$ quinary compound were without success (Schwarz 1967b).

11.17 Lead–Copper–Silver–Bismuth–Selenium

In the system containing these elements, the quinary compound $CuAgPbBi_4Se_8$ (mineral luxembourgite), which crystallizes as a monoclinic structure with the lattice parameters a = 1300.2 ± 0.1, b = 415.43 ± 0.03, c = 1531.2 ± 0.2 pm, β = 108.92 ± 0.01°, and a calculated density of 8.00 g·cm^{-3}, is formed (Philippo et al. 2019a,b, 2020).

11.18 Lead–Copper–Mercury–Bismuth–Selenium

Three quinary compounds are formed in the Pb–Cu–Hg–Bi–Se system. (Cu,Hg)(Bi,Pb)Se_2 (mineral hansblockite) crystallizes as a monoclinic structure with the lattice parameters $a = 685.3 \pm 0.1$, $b = 763.5 \pm 0.1$, $c = 726.4 \pm 0.1$ pm, $\beta = 97.68 \pm 0.01°$, and a calculated density of 8.26 g·cm^{-3} (Förster et al. 2016b, 2017).

$Cu_3HgPbBiSe_5$ (mineral petrovicite) crystallizes as an orthorhombic structure with the lattice parameters $a = 1617.6 \pm 0.5$, $b = 1468.4 \pm 0.5$, $c = 433.1 \pm 0.3$ pm, and a calculated density of 7.707 g·cm^{-3} (Johan et al. 1976; Fleischer et al. 1977a).

$Cu_6HgPb_2Bi_4Se_{12}$ (mineral quijarroite) also crystallizes as an orthorhombic structure with the lattice parameters $a = 924.13 \pm 0.08$, $b = 902.06 \pm 0.07$, $c = 962.19 \pm 0.08$ pm, and a calculated density of 5.771 g·cm^{-3} (Förster 2016a,c; Belakovskiy et al. 2019).

11.19 Lead–Copper–Nitrogen–Oxygen–Selenium

In the system containing these elements, the quinary compound $Cu_3Pb_2O_2(NO_3)_2(SeO_3)_2$, which crystallizes as an orthorhombic structure with the lattice parameters $a = 588.4 \pm 0.2$, $b = 1218.6 \pm 0.3$, $c = 1937.1 \pm 0.4$ pm, and a calculated density of 6.85 g·cm^{-3}, is formed (Effenberger 1986). For the synthesis of this compound, steel autoclave lined with Teflon and a reaction volume of ~6 cm^3 was used. An equimolar mixtures consisting of Pb_3O_4 or $Pb(NO_3)_2$, CuO, $Cu(OH)_2$ or $Cu(NO_3)_2{\cdot}3H_2O$, and SeO_2 with or without HNO_3 were used, filled with H_2O to 80% of autoclave volume. The closed autoclave was heated between 150°C and 230°C for 48 h. After a cooling time of about 12 h, the crystals of the target compound were observed among other byproducts. The proportions of the individual phases formed vary relatively strongly in the individual synthesis approaches. The crystals of the title compound were colored dark green, transparent, and platelet-shaped.

11.20 Lead–Copper–Oxygen–Chlorine–Selenium

Three quinary compounds are formed in the Pb–Cu–O–Cl–Se system. $CuPb_5(SeO_3)_4Cl_4$ (mineral sarrabusite) crystallizes as a monoclinic structure with the lattice parameters $a = 2491.7 \pm 0.3$, $b = 550.6 \pm 0.1$, $c = 1242.2 \pm 0.2$ pm, $\beta = 101.77 \pm 0.01°$, and a calculated density of 6.011 g·cm^{-3} (Gemmi et al. 2012; Tait et al. 2012) [in the triclinic structure with the lattice parameters $a = 829.0 \pm 0.8$, $b = 1058.8 \pm 1.3$, $c = 1358.7 \pm 1.5$ pm, $\alpha = 124.47 \pm 0.08°$, $\beta = 110.60 \pm 0.09°$, and $\gamma = 63.26 \pm 0.09°$ (Campostrini et al. 1999)].

$Cu_2Pb(SeO_3)_2Cl_2$ is stable up to 300°C and also crystallizes as a monoclinic structure with the lattice parameters $a = 1305.6 \pm 0.1$, $b = 955.67 \pm 0.09$, $c = 690.06 \pm 0.06$ pm, and $\beta = 90.529 \pm 0.007°$ (Berdonosov et al. 2013). Light green polycrystalline samples of this compound were prepared from stoichiometric mixtures of binary oxides and chlorides using one of the following reactions: $2CuO + 2SeO_2 + PbCl_2 = Cu_2Pb(SeO_3)_2Cl_2$ and $CuO + 2SeO_2 + PbO + CuCl_2 = Cu_2Pb(SeO_3)_2Cl_2$. However, only the last reaction – the one that does not involve $PbCl_2$ – resulted in a sufficiently pure product. A mixture of starting materials was ground, sealed in evacuated quartz tube, fired at 300°C for 24 h and at 500°C for another 60 h, and finally furnace was cooled. A single crystal suitable for the structure refinement was found in one of the samples. Reactants were handled in an Ar-filled glove box to avoid any contamination with air moisture.

$Cu^+Cu^{2+}{}_5PbO_2(SeO_3)_2Cl_5$ (mineral allochalcoselite) also crystallizes as a monoclinic structure with the lattice parameters $a = 1846.8 \pm 0.2$, $b = 614.75 \pm 0.08$, $c = 1531.4 \pm 0.2$ pm, $\beta = 119.284 \pm 0.002°$, and a calculated density of 4.606 g·cm^{-3} (Vergasova et al. 2005; Ercit et al. 2006; Krivovichev et al. 2006; Locock et al. 2006a).

11.21 Lead–Silver–Bismuth–Tellurium–Selenium

$AgPbBiSe_3$–$AgPbBiTe_3$. Continuous solid solutions series is formed in this system (Sportouch et al. 1998). The $AgPbBiSe_xTe_{3-x}$ solid solutions crystallize in the cubic structure with the lattice parameter $a = 627.4 \pm 0.1$, 624.7 ± 0.1, 619.1 ± 0.1, and 618.5 ± 0.1 for $x = 0.25$, 0.5, 0.75, and 1, respectively. The lattice parameter follows Vegard's law between the end members. The energy gap is 0.25 eV for $x = 0.5$ and 0.75 and 0.32 eV for $x = 1$. These solid solutions were synthesized by combining the chemical elements in stoichiometric ratios under static vacuum at 900°C for 3 days.

11.22 Lead–Strontium–Antimony–Oxygen–Selenium

The quinary compound $Sr_4Pb_{1.5}Sb_5O_5Se_8$, which is thermally stable up to approximately 730°C and crystallizes as a monoclinic structure with the lattice parameters $a = 1884.03 \pm 0.15$, $b = 423.35 \pm 0.03$, $c = 1533.64 \pm 0.12$ pm, $\beta = 95.068 \pm 0.003°$, and an energy gap of 0.920 eV, is formed in the Pb–Sr–Sb–O–Se system (Wang et al. 2020c). Single crystals of the title compound were synthesized by traditional high-temperature solid state reactions. A mixture of SrSe (1 mM), Pb (0.375 mM), Sb_2O_3 (0.42 mM), Sb_2Se_3 (0.21 mM), Se (0.375 mM), and CsI (5.77 mM) as a flux was ground and loaded into a carbon-coated fused-silica tube. The tube was flame-sealed under a high vacuum of 0.1 Pa and put into a furnace. The mixture was heated to 750°C in 10 h, kept at this temperature for 24 h, slowly cooled to 600°C in 80 h, and then followed by cooling to room temperature with the furnace. Many rod-like black single crystals were found after breaking the tube and washing the reaction product with deionized water.

11.23 Lead–Cadmium–Nitrogen–Oxygen–Fluorine–Selenium

In the system containing these elements, the multinary compound $CdPbF(SeO_3)(NO_3)$, which is stable up to 385°C and crystallizes as an orthorhombic structure with the lattice parameters $a = 1112.06 \pm 0.10$, $b = 1036.64 \pm 0.10$, $c = 539.50 \pm 0.05$ pm, a calculated density of 5.634 $g{\cdot}cm^{-3}$, and an energy gap of 4.42 eV, is formed (Ma et al. 2018). The single crystals of this compound were synthesized by a solid state reaction parts were as follows: the mixture of $Pb(NO_3)_2$ (1.0 mM), CdO (1.0 mM), NaF (1.0 mM), and SeO_2 (1.0 mM). The mixture was ground carefully. The evacuated quartz tube with a mixture was heated at 280°C for 48 h and then cooled to 30°C at a rate of 5°C$\cdot h^{-1}$. Colorless slab-shaped crystals of the title compound were achieved. Pure phases of the compound were synthesized by reacting stoichiometric amounts of CdO, PbF_2, $Pb(NO_3)_2$, and SeO_2 in the same way. The obtained colorless crystals are stable in air.

11.24 Lead–Cadmium–Oxygen–Chlorine–Bromine–Selenium

In the system containing these elements, the multinary compound $CdPb_8(SeO_3)_4Cl_4Br_6$, which is stable up to 470°C and crystallizes as an orthorhombic structure with the lattice parameters $a = 961.3 \pm 0.3$, $b = 1217.0 \pm 0.4$, $c = 1291.1 \pm 0.4$ pm, a calculated density of 6.374 $g{\cdot}cm^{-3}$, and an energy gap of 3.50 eV, is formed (Shang and Halasyamani 2020). Crystals of this compound were grown by hydrothermal methods, combining $PbBr_2$ (3 mM), SeO_2 (4 mM), PbO (1.5 mM), $CdCl_2$ (2 mM) with HBr (0.15 mL) and H_2O (2 mL). The solution was placed in 23-mL Teflon-lined autoclave that was closed. The autoclave was gradually heated to 220°C, maintained at that temperature for 72 h, and then slowly cooled to room temperature at a rate of 6°C$\cdot h^{-1}$. Colorless crystals, the only product from the reaction, were isolated from the mother liquor by vacuum filtration and washed with deionized water.

11.25 Lead–Gallium–Oxygen–Fluorine–Chlorine–Selenium

In the system containing these elements, the multinary compound $Pb_2GaF_2(SeO_3)_2Cl$, which starts to decompose above 400°C and crystallizes as a monoclinic structure with the lattice parameters $a = 832.67 \pm 0.08$, $b = 536.28 \pm 0.04$, $c = 1060.24 \pm 0.11$ pm, $\beta = 110.386 \pm 0.011°$, a calculated density of 6.073 g·cm^{-3}, and an energy gap of 4.32 eV, is formed (You et al. 2019). Single crystals of the title compound were prepared by hydrothermal reactions. $PbCl_2$ (0.5 mM), Ga_2O_3 (0.4 mM), SeO_2 (2 mM), 6 mL of H_2O, and 0.1 mL of HF were loaded into 23 mL Teflon-lined autoclave that was subsequently sealed. The autoclave was gradually heated to 230°C, maintained at this temperature for 4 days, and cooled slowly to room temperature at a rate of 3°C·h. After that, the autoclave was opened and the products were recovered by filtration, which were further washed with distilled water to remove soluble solids and dried in an oven at 60°C. Colorless blocks crystals of $Pb_2GaF_2(SeO_3)_2Cl$ were obtained.

11.26 Lead–Carbon–Oxygen–Tellurium–Selenium

In the system containing these elements, the quinary compound $Pb_5(SeO_4)_2(TeO_4)(CO_3)$, which crystallizes as an orthorhombic structure with the lattice parameters $a = 550.97 \pm 0.03$, $b = 1460.16 \pm 0.07$, $c = 1785.52 \pm 0.09$ pm, and a calculated density of 7.276 g·cm^{-3} at 100 K, is formed (Weil and Shirkhanlou 2017). This compound was prepared through hydrothermal synthesis, which was conducted in sealed Teflon containers (capacity ca. 10 mL) in steel autoclave at autogenous pressure. $PbCO_3{\cdot}Pb(OH)_2$ (0.205 g) was mixed in stoichiometric amounts with 80 mass% H_2SeO_4 (0.17 mL) and TeO_2 (0.084 g) (molar ratio 1.3:2:1. The mixture was placed in the Teflon container, filled up with water to about two-thirds of the containers volume and sealed with a Teflon lid. The sealed container was placed in the autoclave and heated at 210°C for one week. Afterward, the autoclave was cooled to room temperature overnight. The solid products were filtered off, washed subsequently with mother liquor, water, and ethanol. The reaction batch consisted of phase mixtures when inspected optically under a polarizing microscope.

11.27 Lead–Titanium–Oxygen–Fluorine–Chlorine–Selenium

In the system containing these elements, the multinary compound $Pb_2TiOF(SeO_3)_2Cl$, which is stable up to 389°C and crystallizes as a monoclinic structure with the lattice parameters $a = 828.63 \pm 0.05$, $b = 537.81 \pm 0.02$, $c = 1071.67 \pm 0.06$ pm, $\beta = 111.191 \pm 0.006°$, a calculated density of 5.867 g·cm^{-3}, and an energy gap of 3.34 eV, is formed (Cao et al. 2013). This compound was prepared by using hydrothermal techniques. The loaded composition was $PbCl_2$ (0.6 mM), TiO_2 (0.8 mM), SeO_2 (1.0 mM), HF (0.10 mL), and H_2O (6 mL). The reaction was carried out at 230°C for 6 days, and the final pH values of the reaction media was close to 0.5. Rod-shaped crystals of the title compound were obtained along with a small amount of single crystals of PbClF and $Pb_3(SeO_3)_2Cl_2$ as impurities, which were removed manually based on their different shapes.

11.28 Lead–Bismuth–Oxygen–Fluorine–Selenium

In the system containing these elements, the quinary compound $PbBi(SeO_3)_2F$, which crystallizes as an orthorhombic structure with the lattice parameters $a = 1296.72 \pm 0.09$, $b = 450.22 \pm 0.03$, $c = 1182.21 \pm 0.08$ pm, a calculated density of 6.632 g·cm^{-3}, and an energy gap of 3.75 eV, is formed (Jia et al. 2022). This compound was synthesized via facile hydrothermal reactions. PbF_2 (1 mM), Bi_2O_3 (0.5 mM), SeO_2 (2 mM), H_2O 4.5 (mL), and HNO_3 (1 M/L, 0.5 mL) were combined. The mixture solution was sealed in 23-mL Teflon-lined autoclave and gradually heated to 210°C, held for 4 days, and then cooled slowly to 30°C at a rate of 6°C·h^{-1}. The products were washed with absolute ethanol and then dried in air. The obtained crystals are highly stable in air for many months.

11.29 Lead–Bismuth–Oxygen–Chlorine–Selenium

In the system containing these elements, the quinary compound $Pb_2Bi(SeO_3)_2Cl_3$, which crystallizes as a monoclinic structure with the lattice parameters $a = 1322.70 \pm 0.09$, $b = 560.02 \pm 0.04$, $c = 770.21 \pm 0.05$ pm, $\beta = 115.0700 \pm 0.0010°$, a calculated density of 6.321 g·cm^{-3}, and an energy gap of 3.45 eV, is formed (Jia et al. 2022). This compound was synthesized in the same way as $PbBi(SeO_3)_2F$ using the mixture of PbO (1 mM), $BiCl_3$ (1 mM), SeO_2 (2 mM), and H_2O (5.0 mL). The crystals of the title compound are highly stable in air for many months.

11.30 Lead–Vanadium–Oxygen–Fluorine–Selenium

In the system containing these elements, the quinary compound $PbVO_2(SeO_3)F$, which is stable up to 265°C and crystallizes as an orthorhombic structure with the lattice parameters $a = 981.2 \pm 0.3$, $b = 814.7 \pm 0.3$, $c = 1302.6 \pm 0.4$ pm, a calculated density of 5.563 g·cm^{-3}, and an energy gap of 2.69 eV, is formed (Cao et al. 2014). To prepare this compound, a mixture of $PbCO_3$ (0.6 mM), V_2O_5 (0.25 mM), SeO_2 (1.5 mM), and H_2O (6 mL) with a pH value preadjusted to 1.0 with the addition of a few drops of hydrofluoric acid (~40%) was sealed in an autoclave equipped with a Teflon liner (23 mL) and heated at 230°C for 4 days. After washing with distilled water, large red, brick-like crystals of the title compound were obtained.

11.31 Lead–Vanadium–Oxygen–Chlorine–Selenium

In the system containing these elements, the quinary compound $Pb_2VO_2(SeO_3)_2Cl$, which is stable up to 322°C and crystallizes as a monoclinic structure with the lattice parameters $a = 833.3 \pm 0.3$, $b = 531.71 \pm 0.16$, $c = 1071.0 \pm 0.4$ pm, $\beta = 111.701 \pm 0.005°$, a calculated density of 5.926 g·cm^{-3}, and an energy gap of 2.41 eV, is formed (Cao et al. 2014). It was prepared by using hydrothermal techniques. The loaded compositions were $PbCl_2$ (0.4 mM), V_2O_5 (0.25 mM), SeO_2 (1.5 mM), 0.10 mL of HF acid (~40%), and H_2O (6 mL), and the resultant mixture was heated at 220°C for 4 days. The final pH value of the reaction media is close to 0.8. Garnet, brick-shaped single crystals of the title compound were obtained as a single phase.

$Pb_2VO_2(SeO_3)_2Cl$ can also be prepared by high-temperature solid state reactions. A stoichiometric mixture of $PbCl_2$ (1.5 mM), PbO (0.5 mM), V_2O_5 (0.5 mM), and SeO_2 (2 mM) was thoroughly ground and pressed into a pellet. The pellet was sealed into a silica tube under vacuum. The tube was gradually heated to 450°C, held for 20 h, and then cooled to room temperature. Defective crystals and excellent powders isolated were confirmed to be a single phase of the title compound.

11.32 Lead–Niobium–Oxygen–Chlorine–Selenium

In the system containing these elements, the quinary compound $Pb_2NbO_2(SeO_3)_2Cl$, which is stable up to 365°C and crystallizes as a monoclinic structure with the lattice parameters $a = 851.10 \pm 0.05$, $b = 541.48 \pm 0.02$, $c = 1073.00 \pm 0.06$ pm, $\beta = 111.753 \pm 0.007°$, a calculated density of 5.992 g·cm^{-3}, and an energy gap of 3.67 eV, is formed (Cao et al. 2013). This compound was prepared by using hydrothermal techniques. The loaded composition was PbCl (0.5 mM), Nb_2O_5 (0.4 mM), SeO_2 (2 mM), HF (0.10 mL), and H_2O (6 mL). The reaction was carried out at 210°C for 4 days and the final pH value of the reaction media was close to 0.5. Colorless rod-shaped crystals of this compound was recovered as single phase after the removal of the impurity $Pb_3(SeO_3)_2Cl_2$ crystals through manually picking.

11.33 Lead–Oxygen–Chlorine–Palladium–Selenium

In the system containing these elements, the quinary compound $Pb_2Pd(SeO_3)_2Cl_2$, which is stable up to 385°C and crystallizes as a monoclinic structure with the lattice parameters a = 837.8 ± 0.7, b = 525.5 ± 0.4, c = 1315.7 ± 0.8 pm, β = 124.70 ± 0.04°, a calculated density of 5.897 g·cm^{-3}, and an energy gap of 1.96 eV, is formed (Zhang et al. 2012b). For preparing this compound, a mixture containing $PbCO_3$ (0.2 mM), $PdCl_2$ (0.1 mM), SeO_2 (0.3 mM), and H_2O (4 mL), was sealed in an autoclave equipped with a Teflon liner (23 mL) and heated at 200°C for 4 days, followed by slow cooling to room temperature at a rate of 6°C·h^{-1}. The final pH values of the reaction media were close to 1.0. Red block-shaped crystals of the title were obtained along with some unknown brown powders. After proper structural analyses, $Pb_2Pd(SeO_3)_2Cl_2$ was obtained as a single phase by the reactions of $PbCl_2$ (0.1 mM), $PdCl_2$ (0.2 mM), SeO_2 (0.3 mM), and H_2O (4 mL) under the same reaction conditions as above.

12

Systems Based on Lead Telluride

12.1 Lead–Hydrogen–Copper–Calcium–Oxygen–Tellurium

In the system containing these elements, the multinary compound $(Ca,Pb)CuTeO_5(H_2O)$ (mineral eckhardite), which crystallizes as a monoclinic structure with the lattice parameters a = 816.06 ± 0.08, b = 530.76 ± 0.06, c = 1144.12 ± 0.15 pm, β = 101.549 ± 0.07°, and a calculated density of 4.671 g·cm^{-3}, is formed (Kampf et al. 2013d,f).

12.2 Lead–Hydrogen–Copper–Uranium–Oxygen–Tellurium

In the system containing these elements, the multinary phases $Cu_3Zn_6(TeO_3)_2O_6(OH)_6Ag_xPb_yCl_{x+2y}$ ($x + y \leq 2$) (mineral quetzalcoatlite), which crystallize as a trigonal structure with the lattice parameters a = 1014.5 ± 0.1, c = 499.25 ± 0.09 pm, and a calculated density of 4.82 g·cm^{-3} (Burns et al. 2000) [in the hexagonal structure with the lattice parameters a = 1009.7 ± 2.5, c = 494.4 ± 0.8 pm, and the calculated and experimental densities of 6.12 and 6.05 ± 0.03 g·cm^{-3}, respectively (Williams 1973; Fleischer 1974)], are formed.

12.3 Lead–Hydrogen–Copper–Carbon–Oxygen–Tellurium

In the system containing these elements, the multinary compound $CuPb_3TeO_5(OH)_2(CO_3)$ (mineral agaite), which crystallizes as an orthorhombic structure with the lattice parameters a = 1065.22 ± 0.07, b = 916.30 ± 0.05, c = 960.11 ± 0.07 pm, and a calculated density of 6.993 g·cm^{-3}, is formed (Kampf et al. 2012a, 2013a).

12.4 Lead–Hydrogen–Copper–Oxygen–Tellurium

Some quinary compounds are formed in the Pb–H–Cu–O–Te system. $CuPbTeO_5(H_2O)$ (mineral andychristyite) crystallizes as a triclinic structure with the lattice parameters a = 532.3 ± 0.2, b = 709.9 ± 0.2, c = 752.1 ± 0.2 pm, α = 83.611 ± 0.006°, β = 76.262 ± 0.007°, γ = 70.669 ± 0.008°, and a calculated density of 6.304 g·cm^{-3} from the powder XRD; a = 532.2 ± 0.3, b = 709.8 ± 0.4, c = 751.1 ± 0.4 pm, α = 83.486 ± 0.007°, β = 76.279 ± 0.005°, γ = 70.742 ± 0.005°, and a calculated density of 6.340 g·cm^{-3} from the single-crystal XRD (Kampf et al. 2015, 2016; Belakovskiy et al. 2017a).

$CuPb_6Te_4O_{18}(OH)_2$ (mineral housleyite) crystallizes as a monoclinic structure with the lattice parameters a = 785.52 ± 0.05, b = 1048.36 ± 0.07, c = 1104.26 ± 0.08 pm, β = 95.547 ± 0.002°, and a calculated density of 7.845 ± 0.001 g·cm^{-3} (Kampf et al. 2010b).

$Cu_3PbTeO_6(OH)_2$ has two polymorphic modifications. The first of them, mineral khinite, crystallizes as an orthorhombic structure with the lattice parameters a = 574.91 ± 0.10, b = 1001.76 ± 0.14, c = 2402.2 ± 0.3 pm, and a calculated density of 6.29 g·cm^{-3} (Cooper et al. 2008; Hawthorne and Cooper 2009) [a = 574.0 ± 0.5, b = 998.3 ± 0.9, c = 2396.0 ± 0.9 pm, and the calculated and experimental densities of

DOI: 10.1201/9781003123460-12

6.69 and 6.5–6.7 g·cm^{-3}, respectively (Williams 1978)]. The second modification, mineral parakhinite, crystallizes as a trigonal structure with the lattice parameters a = 576.5 ± 0.2, c = 1800.1 ± 0.9 pm, and a calculated density of 6.30 g·cm^{-3} (Cooper et al. 2008; Hawthorne and Cooper 2009) [a = 575.3 ± 0.9, c = 1795.8 ± 9.2 pm, and the calculated and experimental densities of 6.69 and 6.5–6.7 g·cm^{-3}, respectively (Williams 1978); a = 576.5 ± 0.2, c = 1800.1 ± 0.9 pm, and a calculated density of 6.302 g·cm^{-3} (Burns et al. 1995; Jambor and Roberts 1995b)].

$Cu_4Pb_2(TeO_6)_2(H_2O)_2$ (mineral paratimroseite) crystallizes as an orthorhombic structure with the lattice parameters a = 519.43 ± 0.04, b = 961.98 ± 0.10, c = 1167.45 ± 0.11 pm, and a calculated density of 6.557 g·cm^{-3} (Kampf et al. 2010c).

$Cu_5Pb_2(TeO_6)_2(OH)_2$ (mineral timroseite) also crystallizes as an orthorhombic structure with the lattice parameters a = 519.99 ± 0.02, b = 962.25 ± 0.04, c = 1153.40 ± 0.05 pm, and a calculated density of 6.982 g·cm^{-3} (Kampf et al. 2010c).

12.5 Lead–Hydrogen–Copper–Oxygen–Chlorine–Tellurium

In the system containing these elements, the multinary compound $Cu_6Pb_3TeO_6(OH)_7Cl_5$ [mineral fuettererite (Williams et al. 2012)], which crystallizes as a hexagonal structure with the lattice parameters a = 840.35 ± 0.12, c = 4468.1 ± 0.4 pm, and a calculated density of 5.552 g·cm^{-3}, is formed (Kampf et al. 2012b, 2013b).

12.6 Lead–Hydrogen–Calcium–Oxygen–Tellurium

In the system containing these elements, the quinary compound $H_6(Pb,Ca)_6(Te^{4+}O_3)_3(Te^{6+}O_6)_2 \cdot 2H_2O$ (mineral oboyerite), which crystallizes as a triclinic structure with the lattice parameters a = 1224.9 ± 0.8, b = 1511.3 ± 0.6, c = 686.8 ± 0.3 pm, α = 116.45 ± 0.04°, β = 98.58 ± 0.04°, γ = 85.82 ± 0.04°, and the calculated and experimental densities of 6.66 and 6.4 ± 0.6 g·cm^{-3}, respectively (Roberts 1980; Fleischer et al. 1981), was discredited as this substance is formed in at least two distinct phases, including Pb_2TeO_5 and $PbTeO_3$ (Missen et al. 2019b; Miyawaki et al. 2019a,b).

12.7 Lead–Hydrogen–Zinc–Arsenic–Antimony–Oxygen–Tellurium

In the system containing these elements, the multinary compound $Pb_3Zn_3(Sb^{5+},Te^{6+})As_2O_{13}(OH,O)$ (mineral joëlbruggerite), which crystallizes as a trigonal structure with the lattice parameters a = 848.03 ± 0.17, c = 523.34 ± 0.12 pm, and a calculated density of 6.73 g·cm^{-3}, is formed (Mills et al. 2009a).

12.8 Lead–Hydrogen–Carbon–Oxygen–Chlorine–Tellurium

In the system containing these elements, the multinary compound $Pb_6(Te_2O_{10})(CO_3)Cl_2(H_2O)$ (mineral thorneite), which crystallizes as a monoclinic structure with the lattice parameters a = 2130.5 ± 0.1, b = 1105.9 ± 0.1, c = 756.4 ± 0.1 pm, β = 101.112 ± 0.004°, and a calculated density of 6.829 g·cm^{-3}, is formed (Kampf et al. 2010a).

12.9 Lead–Hydrogen–Nitrogen–Oxygen–Tellurium

In the system containing these elements, the quinary compound $Pb_9Te_2O_{13}(OH)(NO_3)_3$, which is thermally stable up to about 400°C and crystallizes as a tetragonal structure with the lattice parameters a = 1272.68 ± 0.04, c = 1451.03 ± 0.05 pm, a calculated density of 7.153 g·cm^{-3}, and an energy gap of 3.62 eV, is formed (Chen et al. 2017b). This compound was prepared by hydrothermal synthesis. $Pb(NO_3)_2$

(4.0 mM), NaOH (4.0 mM), and TeO_2 (0.5 mM) were put in 4.0 mL of deionized water and sealed in a stainless steel bomb equipped with a Teflon liner (10 mL). The mixture was heated at 230°C for 4 days. After the mixture had cooled to room temperature at a rate of 4°C·h^{-1}, the transparent block crystals of the title compound were collected and washed with ethanol.

12.10 Lead–Hydrogen–Oxygen–Chromium–Tellurium

In the system containing these elements, the multinary compound $Pb_{10}Te_6O_{20}(OH)_{14}(CrO_4)(H_2O)_5$ (mineral chromschieffelinite), which crystallizes as an orthorhombic structure with the lattice parameters $a = 966.46 \pm 0.03$, $b = 1949.62 \pm 0.08$, $c = 1051.01 \pm 0.07$ pm, and a calculated density of 6.040 g·cm^{-3}, is formed (Kampf et al. 2012c; Tait et al. 2012).

12.11 Lead–Lithium–Lanthanum–Oxygen–Tellurium

In the system containing these elements, the quinary compound $LiLaPbTeO_6$, which crystallizes as a cubic structure with the lattice parameter $a = 798$ pm, is formed (Venevtsev et al. 1974).

12.12 Lead–Lithium–Bismuth–Oxygen–Tellurium

In the system containing these elements, the quinary compound $LiPbBiTeO_6$, which crystallizes as a cubic structure with the lattice parameter $a = 806$ pm, is formed (Venevtsev et al. 1974).

12.13 Lead–Sodium–Lanthanum–Oxygen–Tellurium

In the system containing these elements, the quinary compound $NaLaPbTeO_6$, which crystallizes as a cubic structure with the lattice parameter $a = 816$ pm, is formed (Venevtsev et al. 1974).

12.14 Lead–Sodium–Carbon–Oxygen–Tellurium

$PbTe–Na_2CO_3–C$. Interaction of PbTe with Na_2CO_3 in the presence of carbon proceeds in two next stages: $PbTe + Na_2CO_3 = PbO + Na_2Te + CO_2$ and $2PbO + C = 2Pb + CO_2$ (Shkodin and Malyshev 1969). The reduction of lead with carbon with simultaneously obtaining of Na_2Te is most efficiently carried out at 1065°C–1200°C.

12.15 Lead–Sodium–Bismuth–Oxygen–Tellurium

In the system containing these elements, the quinary compound $NaPbBiTeO_6$, which crystallizes as a cubic structure with the lattice parameter $a = 824$ pm, is formed (Venevtsev et al. 1974).

12.16 Lead–Copper–Gold–Iron–Tellurium

In the system containing these elements, the quinary phase $(Au,Te,Pb)_3(Cu,Fe)$ (mineral bogdanovite), which crystallizes as a cubic structure with the lattice parameter $a = 408.76 \pm 0.15$ pm (Bayliss 1990; Jambor and Burke 1991) [$a = 408.7$ pm and a calculated density of 14.12 g·cm^{-3} (Fleischer et al. 1979a; Spiridonov and Chvileva 1979)], is formed.

12.17 Lead–Copper–Calcium–Antimony–Oxygen–Chlorine–Tellurium

In the system containing these elements, the multinary compound $(Cu,Sb)_3(Pb,Ca)_3(TeO_3)_6Cl$ (mineral choloalite), which crystallizes as a cubic structure with the lattice parameter $a = 1260.10 \pm 0.08$ pm and a calculated density of 6.194 g·cm^{-3} (Weil et al. 2019) [a = 1251.9, 1257.6, and 1258.6 pm for three different samples and the calculated and experimental densities of 6.41 and 6.4 ± 0.1 g·cm^{-3}, respectively (Cabri et al. 1981; Williams 1981); a = 1251.4 pm and the calculated and experimental densities of 6.323 and 6.26 ± 0.08 g·cm^{-3}, respectively (Powell et al. 1994; Jambor et al. 1995a); $a = 1252.0 \pm 0.4$ pm for mineral choloalite of the same composition (Lam et al. 1999)], is formed. The ideal formula for this mineral is $CuPb(TeO_3)_2$ (Powell et al. 1994).

12.18 Lead–Copper–Carbon–Oxygen–Chlorine–Tellurium

In the system containing these elements, the multinary compound $Pb_8Cu(TeO_6)_2(CO_3)Cl_4$ (mineral hagstromite), which crystallizes as an orthorhombic structure with the lattice parameters $a = 2368.8 \pm 1.7$, $b = 902.6 \pm 0.8$, $c = 1046.1 \pm 0.8$ pm, and a calculated density of 7.074 g·cm^{-3}, is formed (Kampf et al. 2020a,b,c).

12.19 Lead–Copper–Nitrogen–Oxygen–Tellurium

In the system containing these elements, the quinary compound $[Cu_2Pb_2(Te_4O_{11})](NO_3)_2$, which crystallizes as a monoclinic structure with the lattice parameters $a = 2996.39 \pm 0.18$, $b = 542.22 \pm 0.04$, $c = 2422.11 \pm 0.18$ pm, $\beta = 129.015 \pm 0.004°$, and a calculated density of 5.874 g·cm^{-3}, is formed (Weil et al. 2019). The synthesis of this compound was carried out under hydrothermal conditions. The initial materials, $Cu(NO_3)_2{\cdot}2.5H_2O$ (0.187 g), $Pb(NO_3)_2$ (0.266 g), TeO_2 (0.257 g), and KOH (0.090 g) (molar ratio 1:1:2:2) were placed in a Teflon container that was filled up to three-fourth of its volume with water. The container was sealed and loaded into a stainless steel autoclave and heated in an oven at 210°C for one week. The plate-like blue crystals of the title compound were manually separated from the bulk.

12.20 Lead–Copper–Nitrogen–Oxygen–Chlorine–Tellurium

In the system containing these elements, the multinary compound $Cu_2Pb(TeO_3)_2Cl_2$, which is stable up to 500°C and crystallizes as a monoclinic structure with the lattice parameters $a = 724.01 \pm 0.02$, $b = 726.88 \pm 0.02$, $c = 828.46 \pm 0.02$ pm, and $\beta = 96.416 \pm 0.002°$, is formed (Berdonosov et al. 2013). Green polycrystalline samples of this compound were prepared using the following reaction: $CuO + 2TeO_2 + PbO + CuCl_2 = Cu_2Pb(SeO_3)_2Cl_2$. The reaction mixture was ground, sealed in an evacuated quartz tube, and fired at 430°C–460°C for 60 h and at 500°C for another 60 h with several intermediate grindings. Reactants were handled in an Ar-filled glove box to avoid any contamination with air moisture.

12.21 Lead–Copper–Phosphorus–Oxygen–Tellurium

In the system containing these elements, the quinary compound $Cu_3Pb_3TeP_2O_{14}$, which crystallizes as a trigonal structure with the lattice parameter $a = 813.0 \pm 0.5$, $c = 533.2 \pm 0.2$ pm, and a calculated density of 6.668 g·cm^{-3}, is formed (Mill' 2009). This polycrystalline compound was synthesized by conventional solid state method on air at 800°C–850°C with a heating rate of 250–300°C·h^{-1}.

12.22 Lead–Silver–Gold–Antimony–Bismuth–Tellurium

In the system containing these elements, the multinary compound (Au,Ag,Sb,Bi,Pb)$_{23}$(Te,Sb,Bi,Pb)$_{38}$ (mineral montbrayite), which crystallizes as a triclinic structure with the lattice parameters a = 1080.45 ± 0.06, b = 1214.70 ± 0.06, c = 1344.80 ± 0.07 pm, α = 108.091 ± 0.005°, β = 104.362 ± 0.005°, and γ = 97.471 ± 0.005° (Bindi et al. 2018) [a = 1210, b = 1346, c = 1080 pm, α = 104°30.5', β = 97°34.5', γ = 107°53.5' and an experimental density 9.94 g·cm^{-3} (Peacock and Thompson 1946); a = 1211, b = 1344, c = 1080 pm, α = 104°23', β = 97°30', γ = 107°56' (Bachechi 1971, 1972); a = 1212.3 ± 0.7, b = 1342.0 ± 0.8, c = 1078.3 ± 0.7 pm, α = 104.36 ± 0.05°, β = 97.45 ± 0.05°, and γ = 108.03 ± 0.04° (Edenharter et al. 1991)], is formed.

The previous formula of this mineral, (Au,Sb)$_2$Te$_3$, was changed due to redefinition of the nomenclature (Hålenius et al. 2018a,b).

12.23 Lead–Silver–Mercury–Antimony–Tellurium

In the system containing these elements, the quinary compound Ag$_3$HgPbSbTe$_5$ (mineral mazzettiite), which crystallizes as an orthorhombic structure with the lattice parameters a = 1649.5 ± 0.6, b = 1476.2 ± 0.7, c = 450.6 ± 0.2 pm, and a calculated density of 9.04 g·cm^{-3}, is formed (Bindi and Cipriani 2004a; Piilonen et al. 2005b).

12.24 Lead–Magnesium–Strontium–Oxygen–Tellurium

In the system containing these elements, the quinary compound MgSrPb$_2$TeO$_6$, which crystallizes as a cubic structure with the lattice parameter a = 795.5 ± 1.0 pm, is formed (Bayer 1963). A powder sample of this compound was prepared by solid state reaction of the oxides TeO$_2$, MgO and PbO and Sr(NO$_3$)$_2$. The stoichiometric mixture was compacted to disk, fired slowly to 700°C, and maintained at this temperature for about 6 h. During this heat treatment, most of the TeO$_2$ was oxidized to TeO$_3$. After regrinding, the specimens were subjected to the final heat treatment, namely 20 h at 1050°C.

12.25 Lead–Magnesium–Barium–Oxygen–Tellurium

In the system containing these elements, the quinary compound MgBaPb$_2$TeO$_6$, which crystallizes as a cubic structure with the lattice parameter a = 808 ± 1 pm, is formed (Bayer 1963). This compound was synthesized in the same way as MgSrPb$_2$TeO$_6$ but using Ba(NO$_3$)$_2$ instead of Sr(NO$_3$)$_2$.

12.26 Lead–Magnesium–Phosphorus–Oxygen–Tellurium

In the system containing these elements, the quinary compound Mg$_3$Pb$_3$TeP$_2$O$_{14}$, which melts incongruently, is thermally stable up to 1000°C, and crystallizes as a trigonal structure with the lattice parameter a = 840.72 ± 0.04, c = 521.58 ± 0.05 pm, a calculated density of 5.763 g·cm^{-3}, and an energy gap of 4.96 eV (Yu et al. 2016a,b) [a = 840.1 ± 0.2, c = 521.3 ± 0.1 pm, and a calculated density of 5.774 g·cm^{-3} (Mill' 2009)], is formed. This polycrystalline compound was synthesized by conventional solid state method on air (Yu et al. 2016a,b; Mill' 2009). Stoichiometric amounts of PbO, MgO, H$_2$TeO$_4$·2H$_2$O, and NH$_4$H$_2$PO$_4$ were ground thoroughly and pressed into a pellet. This pellet was placed on a platinum plate and heated to 400°C for 20 h to decompose H$_2$TeO$_4$·2H$_2$O and NH$_4$H$_2$PO$_4$, and then the temperature was raised to 850°C. The pellet was held at this temperature for 5 days with several intermittent grindings. Pure Mg$_3$Pb$_3$TeP$_2$O$_{14}$ was obtained. The single crystals of the title compound were grown from the high-temperature solution with TeO$_2$ as flux. Polycrystalline Mg$_3$Pb$_3$TeP$_2$O$_{14}$ was mixed thoroughly with TeO$_2$ in a molar ratio of 1:3. The mixture was heated to 950°C in a platinum crucible and kept at this temperature for 15 h to help the melt to become clear and homogeneous. A platinum wire was then dipped into the melt. The temperature was decreased to 900°C at a rate of 2°C·h^{-1}. Some small crystals were observed to nucleate on the platinum wire. Then the platinum wire was pulled out of the solution, and it was allowed to cool to room temperature at a rate of 10°C·h^{-1}. Some millimeter size and colorless crystals of $Mg_3Pb_3TeP_2O_{14}$ were obtained.

12.27 Lead–Magnesium–Arsenic–Oxygen–Tellurium

In the system containing these elements, the quinary compound $Mg_3Pb_3TeAs_2O_{14}$, which crystallizes as a trigonal structure with the lattice parameter $a = 853.7 \pm 0.2$, $c = 524.6 \pm 0.1$ pm, and a calculated density of 5.997 g·cm^{-3}, is formed (Mill' 2009). This polycrystalline compound was synthesized by conventional solid state method on air at 800°C–925°C with a heating rate of 250–300°C·h^{-1}.

12.28 Lead–Calcium–Zinc–Oxygen–Tellurium

In the system containing these elements, the quinary compound $(Ca,Pb)_3Zn_3(TeO_6)_2$ (mineral yafsoanite), which crystallizes as a cubic structure with lattice parameter $a = 1263.50 \pm 0.07$ pm and a calculated density of 5.441 g·cm^{-3} for $(Ca_{2.74}Pb_{0.26})Zn_{2.87}(TeO_6)_2$ composition (Mills et al. 2010) [$a = 631.5 \pm 0.2$ pm and a calculated density of 5.52 g·cm^{-3} for $(Ca_{1.36}Pb_{0.26})Zn_{1.38}(TeO_6)$ composition (for the ideal formula the calculated density is 5.55 g·cm^{-3}) (Kim et al. 1982a,b; Fleischer and Pabst 1983); $a = 1262.1 \pm 0.5$ pm and a calculated density of 4.83 g·cm^{-3} for $(Zn_{1.31}Ca_{1.32}Pb_{0.25})Te_{1.04}O_6$ composition (Rozhdestvenskaya et al. 1984); $a = 1263.2 \pm 0.2$ pm and the calculated nd experimental densities 5.54 and 5.40–5.55 g·cm^{-3}, respectively (Jarosh and Zemann 1989; Jambor and Grew 1990)], is formed.

12.29 Lead–Barium–Niobium–Oxygen–Tellurium

In the system containing these elements, the quinary compound $BaPb_{0.5}NbTe_2O_9$, which crystallizes as an orthorhombic structure with the lattice parameters $a = 1246.0 \pm 0.2$, $b = 993.6 \pm 0.2$, $c = 717.7 \pm 0.1$ pm, and a calculated density of 5.482 g·cm^{-3}, is formed (Launay et al. 1989a,b). This compound was prepared by solid state reaction from $BaTe_2O_6$ and $Pb(NbO_3)_2$. The stoichiometric mixture was heated between 600° and 750°C for 2 h under flowing argon. Yellow crystals of $BaPb_{0.5}NbTe_2O_9$ were obtained.

12.30 Lead–Zinc–Phosphorus–Oxygen–Tellurium

Two quinary compounds, $Zn_3Pb_3TeP_2O_{14}$ and $Zn_3Pb_3TeP_2O_{14}$, are formed in this system. $Zn_3Pb_2Te_2P_2O_{14}$ crystallizes as a monoclinic structure with the lattice parameters $a = 527.22 \pm 0.07$, $b = 794.32 \pm 0.10$, $c = 1521.3 \pm 0.2$ pm, $\beta = 90.058 \pm 0.006°$, and a calculated density of 6.003 g·cm^{-3} (Xia et al. 2016). Single crystals of this compound were grown by high-temperature solution method. The mixture of PbO, (30 mM), TeO_2 (53 mM), ZnO (30 mM), and $NH_4H_2PO_4$ (20 mM) were thoroughly ground and placed in a Pt crucible. It was slowly heated to 760°C and maintained at this temperature for several hours to homogenate thoroughly, then cooled down to 500°C at a rate of 5°C·h^{-1} before the furnace was switched off. After determining the crystal structure, the polycrystalline samples were synthesized by traditional solid state reaction method with stoichiometric chemical agents. Stoichiometric amounts of PbO, TeO_2, ZnO, and $NH_4H_2PO_4$ were mixed, and then the mixture was slowly preheated to 500°C and held on for 12 h. Despite several compositions and methods were attempted, pure samples of the title compound were not obtained.

$Zn_3Pb_3TeP_2O_{14}$ melts incongruently and is thermally stable up to 1000°C (Yu et al. 2016a). Apparently, this compound has two polymorphic modifications. The first compound crystallizes as a trigonal structure with the lattice parameter $a = 838.31 \pm 0.03$, $c = 519.30 \pm 0.04$ pm, a calculated density of 6.469 g·cm^{-3}, and an energy gap of 4.90 eV (Yu et al. 2016a) [a = 838.6 ± 0.1, c = 519.1 ± 0.1 pm, and a calculated density of 6.427 g·cm^{-3} (Mill' 2009); a = 839.2 ± 0.1 and c = 520.4 ± 0.1 pm for mineral kuksite with the same formula (Mills et al. 2010)]. According to the earlier investigations, mineral kuksite crystallizes as an orthorhombic structure with the lattice parameters a = 850 ± 3, b = 1472 ± 5, c = 519 ± 3 pm, and a calculated density of 6.21 g·cm^{-3} (Kim et al. 1990; Jambor and Vanko 1992).

The polycrystalline compound was synthesized by conventional solid state method on air (Yu et al. 2016a; Mill' 2009). Stoichiometric amounts of PbO, ZnO, $H_2TeO_4 \cdot 2H_2O$, and $NH_4H_2PO_4$ were ground thoroughly, packed tightly in a platinum plate, and heated to 400°C. The reaction was maintained at this temperature for 20 h to decompose $H_2TeO_4 \cdot 2H_2O$ and $NH_4H_2PO_4$. Then, the temperature was raised to 850°C and maintained for 5 days with several intermittent grindings. Pure $Zn_3Pb_3TeP_2O_{14}$ was obtained. The single crystals of the title compound were grown from the high-temperature solution with TeO_2 as flux. Polycrystalline $Zn_3Pb_3TeP_2O_{14}$ was mixed thoroughly with TeO_2 in a molar ratio of 1:2. The mixture was heated to 800°C in a platinum crucible and maintained at this temperature for 10 h. Then, the temperature was reduced to 650°C at a rate of 2°C·h^{-1} and cooled to room temperature at a rate of 10°C·h^{-1}. The sub-millimeter size crystals grew as regular hexagonal-shaped prims.

The second modification crystallizes as a monoclinic structure with the lattice parameters a = 1451.284 ± 0.003, b = 2513.85 ± 0.02, c = 518.74 ± 0.04 pm, and β = 90.0631 ± 0.0008° (Krizan et al. 2013). Polycrystalline samples of monoclinic $Zn_3Pb_3TeP_2O_{14}$ were prepared by typical solid state method with slight modification. The starting materials were PbO, TeO_2, ZnO, and $NH_4H_2PO_4$. The stoichiometric (by metal) mixture was ground and placed in high density, covered alumina crucible, and pre-reacted overnight at 450°C under flowing oxygen. The mixture was heated until shown to be pure by laboratory XRD.

12.31 Lead–Zinc–Arsenic–Oxygen–Tellurium

In the system containing these elements, the quinary compound $Zn_3Pb_3TeAs_2O_{14}$ (mineral dugganite), which crystallizes as a trigonal structure with lattice parameters a = 851.5 ± 0.2, c = 522.2 ± 0.1 pm, and a calculated density of 6.680 g·cm^{-3} (Mill' 2009) [a = 847.2 ± 0.5, c = 520.8 ± 0.5 pm, and the calculated and experimental densities of 6.33 and 6.33 ± 0.15 g·cm^{-3}, respectively (Williams 1978); a = 846.0 ± 0.2 and c = 520.6 ± 0.2 pm (Lam et al. 1998)], is formed. This polycrystalline compound was synthesized by conventional solid state method on air at 800°C-950°C with a heating rate of 250–300°C·h^{-1} (Mill' 2009).

12.32 Lead–Zinc–Vanadium–Oxygen–Tellurium

In the system containing these elements, the quinary compound $Zn_3Pb_3TeV_2O_{14}$, which melts incongruently, and crystallizes as a trigonal structure with the lattice parameter a = 860.8 ± 0.2, c = 518.6 ± 0.3 pm, a calculated density of 6.343 g·cm^{-3}, and an energy gap of 3.68 eV (Yu et al. 2016a) [a = 859.5 ± 0.2, c = 517.6 ± 0.1 pm, and a calculated density of 6.374 g·cm^{-3} (Mill' 2009)], is formed. According to the earlier investigations, mineral cheremnykhite with the same composition crystallizes as an orthorhombic structure with the lattice parameters a = 858 ± 3, b = 1486 ± 5, c = 518 ± 3 pm, and a calculated density of 6.44 g·cm^{-3} (Kim et al. 1990; Jambor and Vanko 1992).

This polycrystalline compound was synthesized by conventional solid state method on air (Mill' 2009; Krizan et al. 2013; Yu et al. 2016a). Stoichiometric amounts of PbO, ZnO, $H_2TeO_4 \cdot 2H_2O$ or TeO_2, and V_2O_5 were ground thoroughly, packed tightly in a platinum plate and heated to 400°C. The reaction was maintained at this temperature for 20 h to decompose $H_2TeO_4 \cdot 2H_2O$. Then the temperature was raised to 700°C and maintained at this temperature for 5 days with several intermittent grindings. Pure $Zn_3Pb_3TeV_2O_{14}$ was obtained. The single crystals of the title compound were grown from the high-temperature solution with TeO_2 as a flux. Polycrystalline $Zn_3Pb_3TeV_2O_{14}$ was mixed thoroughly with TeO_2 in a molar ratio of 1:2. The mixture was heated to 800°C in a platinum crucible and maintained at this temperature for 10 h. Then, the temperature was reduced to 650°C at a rate of 2°C·h^{-1} and cooled to room temperature at a rate of 10°C·h^{-1}. The sub-millimeter size crystals grew as regular hexagonal-shaped prims.

12.33 Lead–Cadmium–Phosphorus–Oxygen–Tellurium

In the system containing these elements, the quinary compound $Cd_3Pb_3TeP_2O_{14}$, which crystallizes as a trigonal structure with the lattice parameter a = 855.9 ± 1.0, c = 545.9 ± 0.6 pm, and a calculated density of 6.578 g·cm^{-3}, is formed (Mill' 2009). Polycrystalline compound was synthesized by conventional solid state method on air at 800°C with a heating rate of 250–300°C·h^{-1}.

12.34 Lead–Cadmium–Oxygen–Chlorine–Tellurium

Two quinary compounds, $CdPb_2Te_3O_8Cl_2$ and $Cd_{13}Pb_8Te_{14}P_{42}O_{14}$, are formed in the Pb–Cd–O–Cl–Te system (Bai et al. 2022). The first compound decomposes on heating under a N_2 atmosphere at 512°C and crystallizes as an orthorhombic structure with the lattice parameters a = 1825.26 ± 0.09, b = 813.57 ± 0.04, c = 794.08 ± 0.04 pm, a calculated density of 6.244 g·cm^{-3}, and an energy gap of ~3.89 eV. $Cd_{13}Pb_8Te_{14}P_{42}O_{14}$ decomposes on heating under a N_2 atmosphere at 618°C and crystallizes as a triclinic structure with the lattice parameters a = 874.87 ± 0.04, b = 901.58 ± 0.04, c = 2071.87 ± 0.09 pm, α = 89.824 ± 0.002°, β = 84.919 ± 0.002°, γ = 89.158 ± 0.002°, a calculated density of 6.196 g·cm^{-3}, and an energy gap of ~3.78 eV. The single crystals of these compounds were obtained by high-temperature solution method. The molar ratio of starting materials was $TeO_2/CdCl_2/PbCl_2/CdO$ = 1:1:1:1. The reactants were mixed thoroughly in an agate mortar and then sealed in a quartz tube. The sample was heated to 550°C, kept at this temperature for 24 h, and then the temperature was decreased to 400°C at a rate of 1°C·h^{-1}. Eventually, the furnace was turned off and the sample was cooled to room temperature naturally. It is worth noting that the two compounds were discovered simultaneously in the final products.

The pure phase polycrystalline powder samples of $CdPb_2Te_3O_8Cl_2$ and $Cd_{13}Pb_8Te_{14}P_{42}O_{14}$ were obtained by high-temperature solid state reaction with different starting materials, $CdCl_2/PbO/TeO_2$ = 1:2:3 (molar ratio) for the first compound and $CdCl_2/PbO/CdO/TeO_2$ = 7:8:6:14 (molar ratio) for the second one. The mixtures were placed into graphite crucibles and sealed in quartz tubes. After that, the samples were slowly heated to 600°C, kept at this temperature for 24 h, and then cooled to room temperature at a rate of 2.4°C·h^{-1}.

12.35 Lead–Aluminum–Oxygen–Fluorine–Tellurium

In the system containing these elements, the quinary compound $Al_3Pb_2F_3(Te_6F_2O_{16})$, which is stable up to about 665°C and crystallizes as a tetragonal structure with the lattice parameter a = 876.27 ± 0.04, c = 1123.52 ± 0.13 pm, a calculated density of 6.205 g·cm^{-3}, and an energy gap of 4.1 eV, is formed (Li et al. 2020a). This compound was obtained by a mild hydrothermal reaction. The component parts are as follows: TeO_2 (2 mM), Al_2O_3 (1 mM), PbF_2 (1 mM), HF (0.2 mL, 40%), and 3 mL of deionized water. The mixtures were sealed in an autoclave equipped with a Teflon liner (23 mL), heated at 230°C for 4 days, and then cooled down to 50°C at a rate of 3°C·h^{-1}. The product was washed with alcohol and dried in air at room temperature. Transparent crystals of the title compound were obtained as a single phase.

12.36 Lead–Aluminum–Oxygen–Chlorine–Tellurium

In the system containing these elements, the quinary compound $AlPb_2TeO_6Cl$ (mineral backite), which crystallizes as a trigonal structure with the lattice parameter a = 504.3 ± 0.1 and c = 934.6 ± 0.5 pm from the powder XRD, a = 504.41 ± 0.07 and c = 942.10 ± 0.05 pm from the single-crystal XRD, and a calculated density of 5.573 g·cm^{-3}, is formed (Tait et al. 2014a,b; Belakovskiy et al. 2016).

12.37 Lead–Gallium–Oxygen–Fluorine–Tellurium

In the system containing these elements, the quinary compound $Ga_3Pb_2F_3(Te_6F_2O_{16})$, which is stable up to about 590°C and crystallizes as a tetragonal structure with the lattice parameter a = 884.39 ± 0.03, c = 1158.02 ± 0.11 pm, a calculated density of 6.381 g·cm^{-3}, and an energy gap of 4.2 eV, is formed (Li et al. 2020a). This compound was prepared by a mild hydrothermal reaction. The component parts are as follows: TeO_2 (2 mM), Ga_2O_3 (1 mM), PbF_2 (1.5 mM), HF (0.3 mL, 40%), and 2 mL of deionized water. The mixture was sealed in an autoclave equipped with a Teflon liner (23 mL), heated at 230°C for 4 days, and then cooled down to 50°C at a rate of 3°C·h^{-1}. The product was washed with alcohol and dried in air at room temperature. Transparent crystals of the title compound were obtained as single phase.

12.38 Lead–Thallium–Neodymium–Bismuth–Tellurium

Tl_4PbTe_3–Tl_9NdTe_6–Tl_9BiTe_6. The Tl_4PbTe_3–Tl_9NdTe_6 and Tl_4BiTe_6–Tl_9NdTe_6 systems are the nonquasibinary sections of this ternary system as Tl_9NdTe_6 melts incongruently (Imamaliyeva et al. 2009). Continuous series of the solid solutions are formed in the systems below the decomposition temperature of Tl_9NdTe_6. The isothermal section of the ternary system at 580°C is given in Figure 12.1 and the liquidus and solidus surfaces are presented in Figure 12.2. The system was investigated through DTA, XRD, and measuring of the microhardness and EMF of the concentration chains using the alloys annealed at ca. 430°C for 800–1000 h.

12.39 Lead–Thallium–Samarium–Bismuth–Tellurium

Tl_4PbTe_3–Tl_9SmTe_6–Tl_9BiTe_6. The continuous series of solid solutions were found in this ternary system at the solidus temperatures and below (Imamaliyeva et al. 2018b). The isothermal sections of the system at 550°C and 570°C (Figure 12.3) include the region of liquid, two-phase region, and the region of $(Tl_4PbTe_3)_x(Tl_9BiTe_6)_{1-x}$ solid solutions. The liquidus surface (Figure 12.4) consists of the fields of the primary crystallization of $TlSmTe_2$ and $(Tl_4PbTe_3)_x(Tl_9BiTe_6)_{1-x}$ solid solutions. These fields are separated

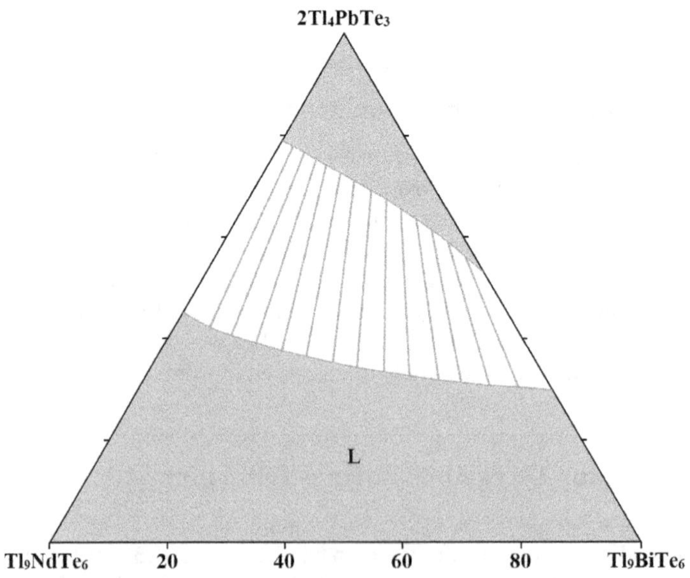

FIGURE 12.1 Isothermal section of the Tl_4PbTe_3–Tl_9NdTe_6–Tl_9BiTe_6 ternary system at 430°C. (From Imamaliyeva, S.Z., et al., *Azerb. khim. zhurn.*, (1), 49, 2009.) Open access.

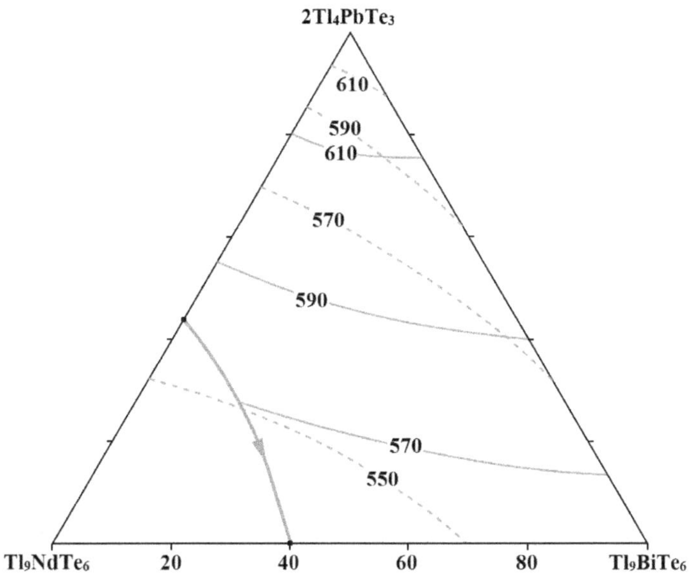

FIGURE 12.2 Liquidus (solid lines) and solidus (dashed lines) surface of the Tl_4PbTe_3–Tl_9NdTe_6–Tl_9BiTe_6 ternary system. (From Imamaliyeva, S.Z., et al., *Azerb. khim. zhurn.*, (1), 49, 2009.) Open access.

by a monovariant peritectic curve. The solidus isotherms are presented by dashed lines. The system was investigated through DTA, XRD, and measuring microhardness.

12.40 Lead–Thallium–Gadolinium–Bismuth–Tellurium

Tl_4PbTe_3–Tl_9GdTe_6–Tl_9BiTe_6. The continuous series of solid solutions were found in this ternary system at the solidus temperatures and below (Imamaliyeva et al. 2018c). The isothermal section of the system at 590°C is presented in Figure 12.5. It includes the region of liquid, two-phase region, and the region of the solid solutions based on Tl_4PbTe_3. The liquidus surface (Figure 12.6) consists of the fields of the primary crystallization of $TlGdTe_2$ and $(Tl_9GdTe_6)_x(Tl_9BiTe_6)_{1-x}$ solid solutions. These fields are separated by a monovariant peritectic curve. The solidus isotherms are presented by dashed lines. The system was investigated through DTA, XRD, and measuring the microhardness.

The analytical multi-3D model of the phase diagram of this ternary system was obtained and visualized for the quasiternary low-temperature and non-quasiternary high-temperature parts by Imamaliyeva et al. 2021b). For the Tl_4PbTe_3–Tl_9GdTe_6 and Tl_9GdTe_6–Tl_9BiTe_6 boundary systems, the analytical models of phase diagrams were obtained in forms of temperature dependence on the composition, and their graphs were constructed. The thermodynamic functions of solid solutions mixing were determined using the model of regular solutions of non-molecular compounds. It was found that in the entire concentration range in the temperature range from room temperature to 510°C, the second derivative of the integral mixing free energy is greater than zero. Therefore, the values of the stability function are also greater than zero in the entire range of concentrations, which indicates the thermodynamic stability of solid solutions.

12.41 Lead–Thallium–Terbium–Bismuth–Tellurium

Tl_4PbTe_3–Tl_9TbTe_6–Tl_9BiTe_6. Three vertical sections, the isothermal sections at 570°C and 590°C, and the liquidus surface of this system were constructed by Imamaliyeva et al. (2018a). The isothermal sections of the system at 570°C and 590°C (Figure 12.7) include the region of liquid, two-phase region, and the region of the solid solutions based on Tl_4PbTe_3. The liquidus surface (Figure 12.8) consists of the

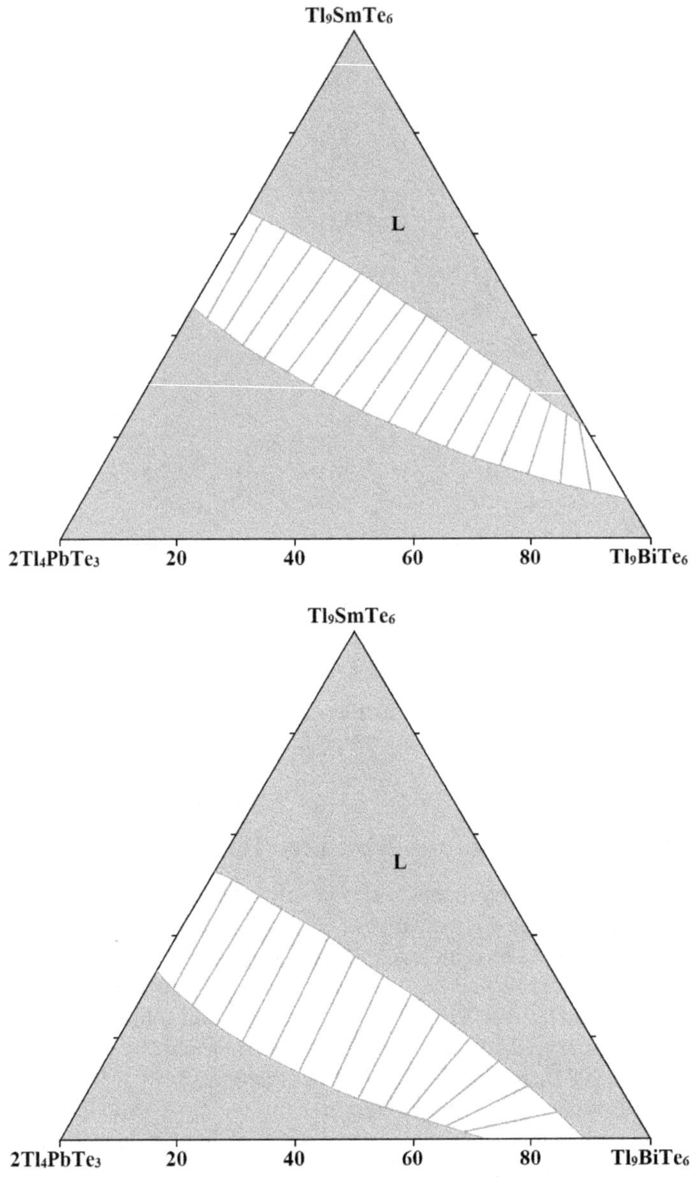

FIGURE 12.3 Isothermal section of the Tl_4PbTe_3–Tl_9SmTe_6–Tl_9BiTe_6 ternary system at (a) 550°C and (b) 570°C. (From Imamaliyeva, S.Z., et al., *Acta Chim. Slov.*, 65(2), 365, 2018.) Open access.

fields of the primary crystallization of $TlTbTe_2$ and $(Tl_4PbTe_3)_x(Tl_9BiTe_6)_{1-x}$ solid solutions. These fields are separated by a monovariant peritectic curve. The solidus isotherms are presented by dashed lines. The system was investigated through DTA, XRD, and measuring of the microhardness using the alloys annealed at 480°C for 1000 h.

The analytical multi-3D model of the phase diagram of this system was obtained as temperature dependence on the composition using the multipurpose genetic algorithm (Imamaliyeva et al. 2021a). It was determined that the system is characterized by the formation of continuous solid solutions (δ-phase) with tetragonal structure. The boundaries of the uncertainty band for the liquidus and solidus surfaces of the δ-phase and the $TlTbTe_2$ compound were determined. The thermodynamic functions of mixing of the solid solutions depending on the composition and temperature were determined based on the model

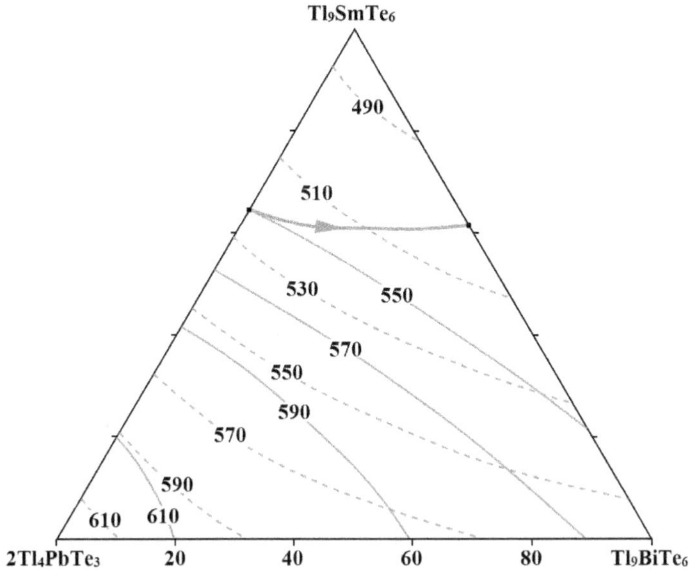

FIGURE 12.4 Liquidus (solid lines) and solidus (dashed lines) surface of the Tl_4PbTe_3–Tl_9SmTe_6–Tl_9BiTe_6 ternary system. (From Imamaliyeva, S.Z., et al., *Acta Chim. Slov.*, 65(2), 365, 2018.) Open access.

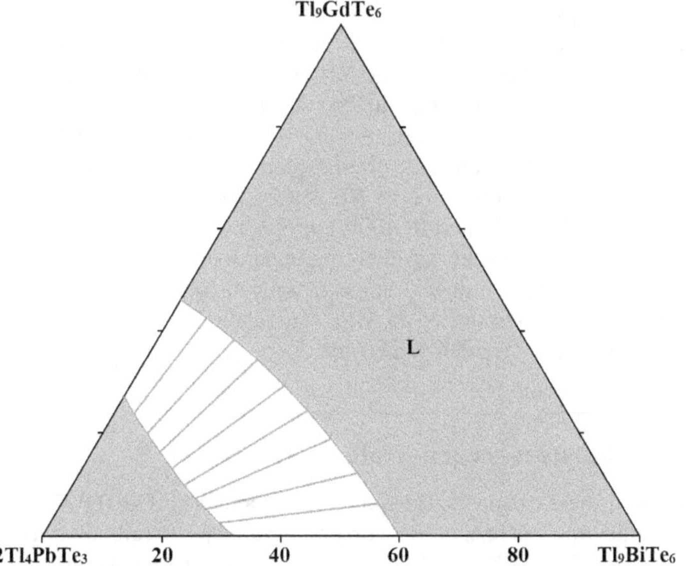

FIGURE 12.5 Isothermal section of the Tl_4PbTe_3–Tl_9GdTe_6–Tl_9BiTe_6 ternary system at 590°C. (From Imamaliyeva, S.Z., et al., *Chem. Probl.*, (4), 496, 2018.) Open access.

of regular solutions of non-molecular compounds. It was established that the δ-phase possesses thermodynamic stability in the entire concentration interval.

12.42 Lead–Phosphorus–Oxygen–Manganese–Tellurium

In the system containing these elements, the quinary compound $Pb_3Mn_3TeP_2O_{14}$, crystallizes as a trigonal structure with the lattice parameter $a = 847.956 \pm 0.001$ and $c = 533.153 \pm 0.001$ pm, a calculated density

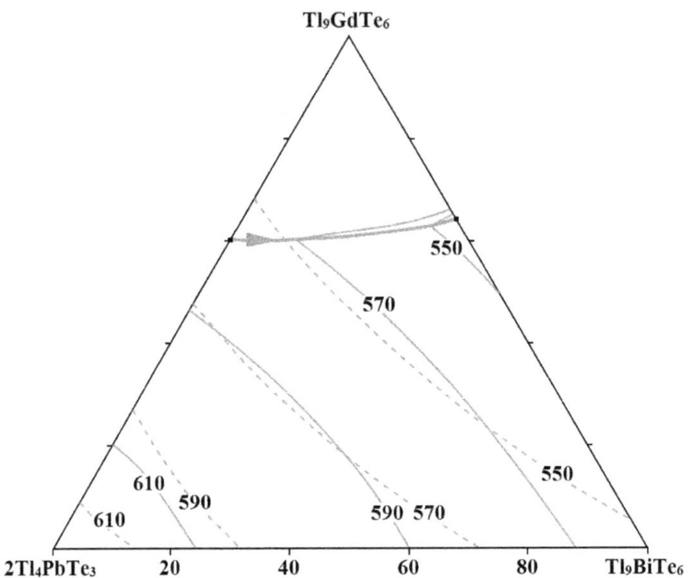

FIGURE 12.6 Liquidus (solid lines) and solidus (dashed lines) surface of the Tl_4PbTe_3–Tl_9GdTe_6–Tl_9BiTe_6 ternary system. (From Imamaliyeva, S.Z., et al., *Chem. Probl.*, (4), 496, 2018.) Open access.

of 5.763 g·cm^{-3}, and an energy gap of 4.96 eV (Silverstein et al. 2013a) [a = 846.5 ± 0.2, c = 532.6 ± 0.1 pm, and a calculated density of 6.029 g·cm^{-3} (Mill' 2009)], is formed. To synthesize this compound, the stoichiometric amounts of PbO, TeO_2, MnO, and $NH_4H_2PO_4$ were combined and ground by hand for 30 min for every 2 g of sample (Silverstein et al. 2013a). A pre-reaction stage is critical: simply heating the powders to temperatures greater than 500°C causes the reactants to melt and degrade. It was also found that pelleted samples at the pre-reaction stage often exploded during the reaction, most likely due to the release of NH_3 gas from the decomposition of $NH_4H_2PO_4$. For these reasons, the powder was loaded into an Al_2O_3 crucible and slowly heated to 400°C for 12 h. At this stage of the reaction, the powder appears light gray in color. The powder was then reground, loaded loose into a crucible, and annealed again at 550°C for another 12 h. The powder appears a much darker gray in color. Finally, the powder was reground, pelleted using a pressure of 25 MPa, and sintered at 800°C for 100 h with intermittent regrinding. The final product is a lavender-gray color.

12.43 Lead–Phosphorus–Oxygen–Cobalt–Tellurium

In the system containing these elements, the quinary compound $Pb_3Co_3TeP_2O_{14}$, which apparently has two polymorphic modifications, is formed. The first modification crystallizes as a trigonal structure with the lattice parameter a = 835.8 ± 0.2, c = 521.6 ± 0.1 pm, and a calculated density of 6.376 g·cm^{-3} (Mill' 2009). It was synthesized by conventional solid state method (Mill' 2009; Silverstein et al. 2013b). PbO, TeO_2, Co_3O_4, and $NH_4H_2PO_4$ were combined in stoichiometric amounts, ground by hand for 30 min for every 1.5 g of sample. Ammonia is already released during this process. The loose powder was loaded into an Al_2O_3 crucible and put into a horizontal tube furnace at 400°C for 6 h to convert $NH_4H_2PO_4$ into P_2O_5 and NH_3. The powder was then ground up, pelleted with a pressure of 20 MPa, and heated at 650°C in an open Pt crucible for 72 h with intermittent grindings every 16 h.

The second modification of $Pb_3Co_3TeP_2O_{14}$ crystallizes as a monoclinic structure with the lattice parameters a = 1447.371 ± 0.005, b = 2503.78 ± 0.004, c = 521.340 ± 0.009 pm, and β = 90.0304 ± 0.0004° at 4 K and a = 1447.600 ± 0.006, b = 2505.79 ± 0.001, c = 521.4222 ± 0.0016 pm, and β = 90.0304 ± 0.0004° at 100 K (Krizan et al. 2013). Polycrystalline samples of monoclinic $Pb_3Co_3TeP_2O_{14}$ were prepared by typical solid state methods, with slight modification. The starting materials were PbO, TeO_2,

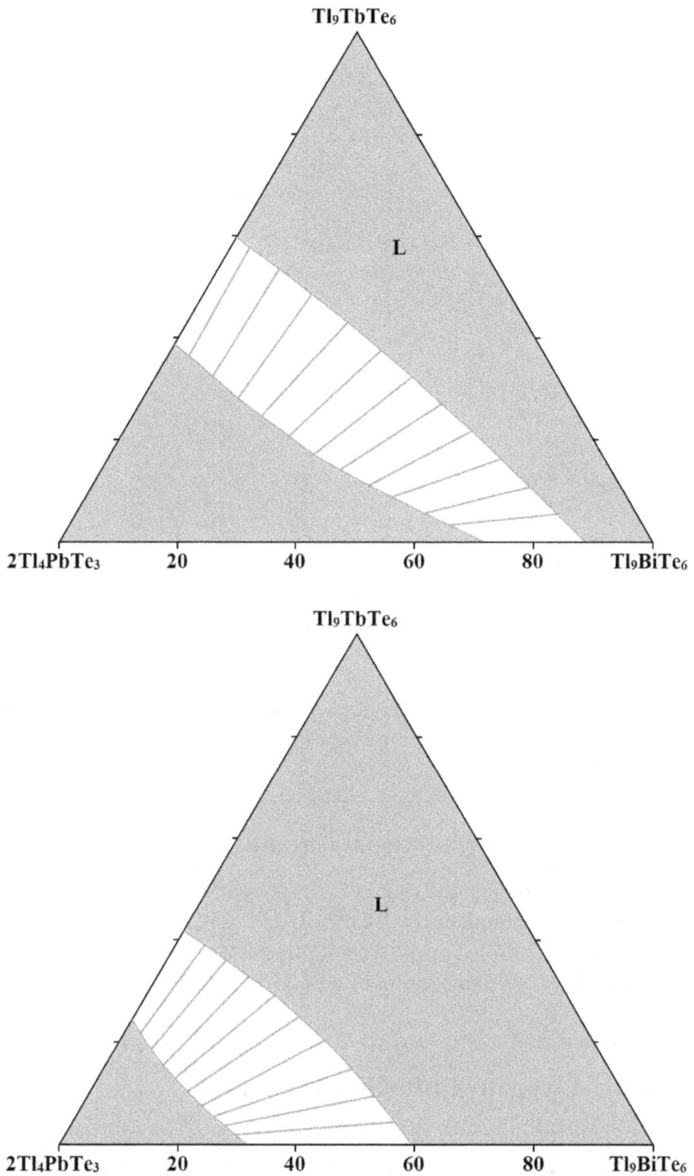

FIGURE 12.7 Isothermal section of the Tl_4PbTe_3–Tl_9SmTe_6–Tl_9BiTe_6 ternary system at (a) 550°C and (b) 570°C. (From Imamaliyeva, S.Z., et al., *Acta Chim. Slov.*, 65(2), 365, 2018.) Open access.

Co_3O_4, and $NH_4H_2PO_4$. The stoichiometric (by metal) mixture was ground and placed in high-density, covered alumina crucible and pre-reacted overnight at 450°C under flowing oxygen. To obtain this modification pure, additional heating at 700°C for 146 h in a closed crucible in flowing O_2 with periodic grinding is required. The powder of monoclinic $Pb_3Co_3TeP_2O_{14}$ has a very bright indigo color.

12.44 Lead–Arsenic–Oxygen–Manganese–Tellurium

In the system containing these elements, the quinary compound $Pb_3Mn_3TeAs_2O_{14}$, which crystallizes as a trigonal structure with the lattice parameter $a = 859.6 \pm 0.2$, $c = 536.2 \pm 0.1$ pm, and a calculated density

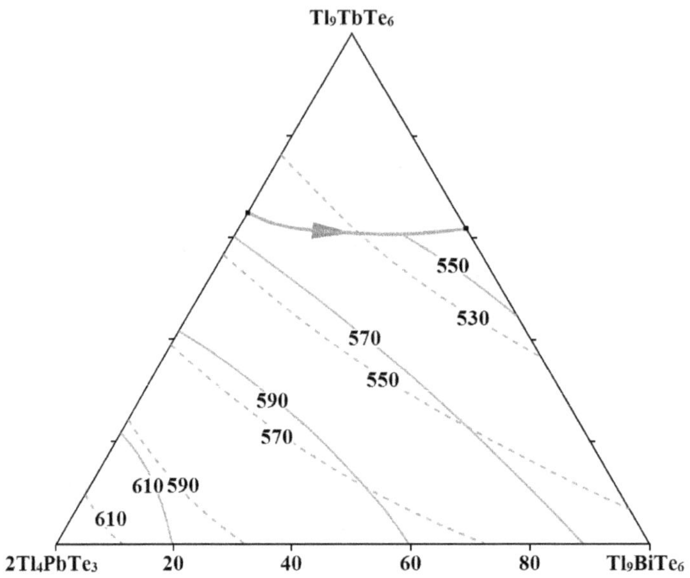

FIGURE 12.8 Liquidus (solid lines) and solidus (dashed lines) surface of the Tl_4PbTe_3–Tl_9SmTe_6–Tl_9BiTe_6 ternary system. (From Imamaliyeva, S.Z., et al., *Acta Chim. Slov.*, 65(2), 365, 2018.) Open access.

of 6.232 g·cm^{-3}, is formed (Mill' 2009). This polycrystalline compound was synthesized by conventional solid state method on air at 800°C with a heating rate of 250–300°C·h^{-1}.

12.45 Lead–Arsenic–Oxygen–Cobalt–Tellurium

In the system containing these elements, the quinary compound $Pb_3Co_3TeAs_2O_{14}$, which crystallizes as a trigonal structure with the lattice parameter a = 849.9 ± 0.2, c = 525.8 ± 0.1 pm, and a calculated density of 6.562 g·cm^{-3}, is formed (Mill' 2009). This polycrystalline compound was synthesized by conventional solid state method on air at 900°C with a heating rate of 250–300°C·h^{-1}.

12.46 Lead–Bismuth–Oxygen–Chlorine–Tellurium

$PbCl_2$–Bi_2O_3–TeO_2. The glass-forming region in this system was determined by Ersundu et al. (2021) and is given in Figure 12.9. This region is strongly adjacent to the region rich in TeO_2. All glasses possess pale yellow color. The values of the glass transition temperature vary between 256°C and 304°C and in general show a slight decrease with the equimolar substitution of TeO_2 by $PbCl_2$+Bi_2O_3. On the other hand, the glass transition temperatures clearly increase with increasing Bi_2O_3 content or decreasing $PbCl_2$ content in the samples containing 70 mol.% TeO_2.

12.47 Lead–Vanadium–Oxygen–Fluorine–Tellurium

In the system containing these elements, the quinary compound $PbVTeO_5F$, which melts incongruently at 378°C and crystallizes as an orthorhombic structure with the lattice parameters a = 989.76 ± 0.08, b = 806.85 ± 0.05, c = 1307.01 ± 0.10 pm, a calculated density of 6.169 g·cm^{-3}, and an energy gap of 2.22 eV, is formed (Yang et al. 2022l). The crystals of this compound were synthesized through a hydrothermal reaction. Red-brown columnar $PbVTeO_5F$ crystals were prepared by using a mixture of TeO_2 (1.5 mM), PbF_2 (2 mM), and V_2O_5 (0.5 mM) as well as 4 mL of deionized water. The mixture was placed in

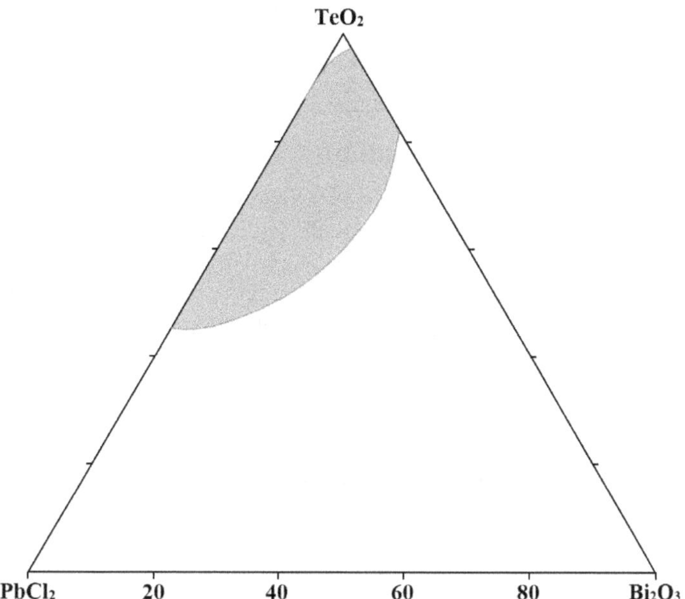

FIGURE 12.9 Glass forming region in the $PbCl_2$–Bi_2O_3–TeO_2 quasiternary system. (From Ersundu, A.E., et al., *J. Alloys Compd.*, 857, 158279, 2021.)

an autoclave equipped with a Teflon liner (25 mL), stirred at room temperature for 2 h, put into a stainless steel bomb, heated at 210°C for 4 days, and finally slowly cooled to room temperature at a rate of 6°C·h^{-1}. The obtained crystals were washed with deionized water and then dried in air.

12.48 Lead–Vanadium–Oxygen–Cobalt–Tellurium

In the system containing these elements, the quinary compound $Pb_3Co_3TeV_2O_{14}$, which apparently has two polymorphic modifications, is formed. The first modification crystallizes as a trigonal structure with the lattice parameter a = 855.95 ± 0.04 and c = 521.67 ± 0.03 pm (Silverstein et al. 2012) [a = 857.0 ± 0.2, c = 522.3 ± 0.1 pm, and a calculated density of 6.257 g·cm^{-3}] (Mill' 2009). It was synthesized using a standard solid state reaction (Mill' 2009; Silverstein et al. 2012, 2013b). The stoichiometric amounts of PbO, TeO_2, Co_3O_4, and V_2O_5 were combined, ground together, and pressed into pellets at a pressure of 20 MPa. The pellets were sintered at 650°C–800°C in an open Pt crucible for 15–24 h and then annealed at 600°C for up to 72 h with intermediate grindings every 16 h.

The second modification crystallizes as a monoclinic structure with the lattice parameters a = 1481.359 ± 0.009, b = 2564.683 ± 0.016, c = 521.137 ± 0.003 pm, and β = 90.0631 ± 0.0008° at 100 K (Krizan et al. 2013). Polycrystalline samples of monoclinic $Pb_3Co_3TeV_2O_{14}$ were prepared by typical solid state method, with slight modification. The starting materials were PbO, TeO_2, Co_3O_4, V_2O_5. The stoichiometric (by metal) mixture was ground and placed in high-density, covered alumina crucible and pre-reacted overnight at 550°C under flowing oxygen. The mixture was fully reacted after subsequent heating at 625°C in air.

12.49 Lead–Oxygen–Chlorine–Iron–Tellurium

In the system containing these elements, the quinary compound $Pb_2Fe(TeO_6)Cl$ (mineral müllerite), which crystallizes as a trigonal structure with the lattice parameter a = 520.40 ± 0.05, c = 896.54 ± 0.12

pm, and a calculated density of 5.812 g·cm^{-3} (Mills et al. 2020a,b) [a = 520.43 ± 0.03 and c = 896.3 ± 0.1 pm (Mills et al. 2019)], is formed.

12.50 Lead–Oxygen–Chlorine–Palladium–Tellurium

In the system containing these elements, the quinary compound $Pb_2Pd(TeO_3)_2Cl_2$, which is stable up to 590°C and crystallizes as a monoclinic structure with the lattice parameters a = 814.5 ± 0.2, b = 544.47 ± 0.15, c = 1314.0 ± 0.3 pm, β = 124.444 ± 0.011°, a calculated density of 6.516 g·cm^{-3}, and an energy gap of 1.96 eV, is formed (Zhang et al. 2012b). For preparing this compound, a mixture containing $PbCO_3$ (0.5 mM), $PdCl_2$ (0.5 mM), TeO_2 (1.0 mM), and H_2O (4 mL) was sealed in an autoclave equipped with a Teflon liner (23 mL) and heated at 200°C for 4 days, followed by slow cooling to room temperature at a rate of 6°C·h^{-1}. The final pH values of the reaction media were close to 3.0. Red block-shaped crystals of the title compound were obtained along with some unknown brown powders. After proper structural analyses, $Pb_2Pd(TeO_3)_2Cl_2$ was obtained as a single phase by the reactions of $PbCO_3$ (0.3 mM), $PdCl_2$ (0.2 mM), TeO_2 (0.3 mM), and H_2O (4 mL) under the reaction conditions, which is the same as above.

Appendix A

Ternary Alloys Based on IV-VI and IV-VI_2 Semiconductors

A.1 Systems Based on Silicon Sulfides

A.1.1 Silicon–Sodium–Sulfur

According to Harm et al. (2020), the ternary compound Na_4SiS_4 crystallizes as an orthorhombic structure with the lattice parameters $a = 1367.65 \pm 0.03$, $b = 878.39 \pm 0.02$, and $c = 688.940 \pm 0.015$ pm. This compound was obtained when stoichiometric amounts of Na_2S, Al_2S_3, Si (ball-milled), and S (sublimed in vacuum) were used as starting materials. An excess of 5 mass% S was added to the mixture to ensure an oxidizing atmosphere during the reaction. Samples were prepared by thoroughly mixing and grinding the starting materials in an agate mortar. The resulting fine powders were transferred into glassy carbon crucible, compacted and sealed under vacuum into quartz ampoule. The ampoule was subsequently transferred into a tube furnace and heated at a rate of 50°C·h^{-1} to 600°C and annealed for 3 days. Subsequently, the furnace was turned off. The ampoule was removed from the furnace when the temperature was below 100°C and transferred to a glove box. Na_4SiS_4 and Na_5AlS_4 compounds are off-white to yellow powders, probably from excess sulfur. Crystals of $Na_{8.5}(AlS_4)_{0.25}(SiS_4)_{0.75}$ were also obtained, which were colorless cuboids with a diameter of about 200 μm, immersed in an orange amorphous material, presumably a solidified melt of sodium polysulfide.

A.1.2 Silicon–Cadmium–Sulfur

The ternary compound Cd_4SiS_6, which is thermally stable up to ~710°C and crystallizes as a monoclinic structure with the lattice parameters $a = 1233.2 \pm 0.6$, $b = 705.0 \pm 0.4$, $c = 1235.6 \pm 0.7$ pm, $\beta = 110.470 \pm 0.009°$, a calculated density of 4.422 g·cm^{-3}, and an energy gap of 2.75 eV, is formed in the Si–Cd–S system (Chen et al. 2020a). To synthesize this compound, Cd, Si, and S were used as the starting materials. They were loaded into a silica ampoule with a molar ratio of Cd/Si/S = 4:1:6. The ampoule was sealed under a vacuum of 5×10^{-3} Pa, and then it was heated to 1050°C for 3 days and finally slowly cooled to room temperature in a muffle furnace. The polycrystalline sample of the title compound was obtained. High-quality single crystals of Cd_4SiS_6 were grown by a chemical transport reaction. A total of 1.674 g of Cd_4SiS_6 polycrystalline powder and 120 mg of iodine as a transport agent were loaded into a silica tube. The temperature of vaporization chamber and crystallization chamber were maintained at 950°C and 850°C for 4 days, respectively, and then cooled to 550°C and 450°C at 3°C·h^{-1}, and finally cooled to room temperature in 24 h. Millimeter-scale yellow crystals with good quality were obtained.

A.1.3 Silicon–Yttrium–Sulfur

Another ternary compound, $Y_4Si_3S_{12}$, which crystallizes as a trigonal structure with the lattice parameters $a = 1877.93 \pm 0.12$, $c = 768.54 \pm 0.07$ pm, a calculated density of 3.500 g·cm^{-3}, and an energy gap of 2.5 ± 0.1 eV, is formed in the Si–Y–S system (Bardelli et al. 2022). To synthesize this compound, Y pieces, Te broken ingot, Si and S powders, and I_2 resublimed crystals were loaded at a molar ratio of Y/Te/Si/S = 2:1:2:9 (total 0.5 g) mixed with two to three small pieces of I_2 into carbonized silica ampoule. The ampoule was slowly heated to 900°C over 10 h, maintained at that temperature for 96 h, and then slowly cooled to 500°C over 24 h; finally, the furnace was turned off and the ampoule was returned to room temperature. Powder XRD revealed that impurity TeI was mixed with the $Y_4Si_3S_{12}$ compound. Various synthesis attempts failed to make single phase of $Y_4Si_3S_{12}$. The quenching experiments at different temperatures also did not generate single phase of the title compound. All starting materials were stored and used in an Ar-filled glove box except I_2.

One more ternary compound, $Y_6Si_{2.5}S_{14}$, crystallizes as a hexagonal structure with the lattice parameters $a = 969.26 \pm 0.05$, $c = 572.95 \pm 0.03$ pm, a calculated density of 3.749 g·cm^{-3} at 173 K, and an energy gap of 2.4 ± 0.1 eV (Ye et al. 2021). To prepare this compound, the reactants were loaded at a molar ratio of Y/Si/S=6:2.5:14 (total 0.4 g) mixed with 0.4 g YCl_3/NaCl flux (molar ratio 1:1) into carbonized silica ampoule. The ampoule was slowly heated to 900°C over 10 h, held at that temperature for 96 h, then slowly cooled to 500°C over 24 h; finally, the furnace was turned off and the ampoule was returned to room temperature. After the reaction, the sample was washed with deionized water to remove the YCl_3/NaCl flux. After washing, the resulting yellow crystals were collected and separated. They are stable in the air for several months. All starting materials were stored and used in an Ar-filled glove box.

A.1.4 Silicon–Tin–Sulfur

The ternary compound $SiSn_2S_4$, which crystallizes as a monoclinic structure with the lattice parameters $a = 818.31 \pm 0.16$, $b = 828.51 \pm 0.17$, $c = 1117.4 \pm 0.4$ pm, $\beta = 116.84 \pm 0.02°$, a calculated density of 3.869 g·cm^{-3}, and an energy gap of 2.00 eV, is formed in the Si–Sn–S system (Li et al. 2015b). To synthesize this compound, reaction mixture of SnS (2 mM) and SiS_2 (1 mM) was ground and loaded into a fused-silica tube under an Ar atmosphere in a glove box. Then the tube was flame-sealed under a high vacuum of 10^{-3} Pa and placed in a computer-controlled furnace. The sample was heated to 900°C within 15 h, kept for 72 h, then slowly cooled to 400°C at a rate of 3°C·h^{-1}, and finally cooled to room temperature by switching off the furnace. Many block-shaped orange crystals of the title compound were found in the ampoule.

A.1.5 Silicon–Oxygen–Sulfur

The ternary compound $Si(SO_4)_2$ is formed in the Si–O–S system (Krüger et al. 1968). The interaction of Ag_2SO_4 with $SiCl_4$ (molar ratio 2:1) leads to the easily polymerizable $Si(SO_4)_2$.

A.2 Systems Based on Silicon Selenides

A.2.1 Silicon–Copper–Selenium

The enthalpy and entropy of the phase transition of the Cu_8SiSe_6 ternary compound at 52°C, determined through DSC, are 14.73 ± 0.59 kJ·M^{-1} and 45.32 ± 1.81 J·(M·K)$^{-1}$, respectively (Bayramova 2022a).

A.2.2 Silicon–Barium–Selenium

Another ternary compound, Ba_3SiSe_5, which crystallizes as an orthorhombic structure with the lattice parameters $a = 1243.6 \pm 0.3$, $b = 988.9 \pm 0.2$, $c = 895.64 \pm 0.19$ pm, a calculated density of 5.035 g·cm^{-3}, and an energy gap of 2.58 eV, is formed in the Si–Ba–Se system (Cheng et al. 2021). The polycrystalline samples of this compound were obtained by a solid state reaction in a closed system. Stoichiometric amounts of Ba, Si, and Se were mixed and placed into a fused-silica tube in a glove box filled with Ar, which was sealed under 10^{-3} Pa. Then, the mixed powders was heated up to 975°C in 30 h and kept at this temperature for 33 h. Next, the tube was slowly cooled down to 775°C in 13 h and to 500°C in another 13 h, respectively. Finally, the sample was cooled to 300°C in 10 h, and reached room temperature by closing the muffle furnace.

Single crystals of Ba_3SiSe_5 were synthesized by a solid state reaction technique. The raw materials, Ba, Si, and Se, with the molar ratio of 3:1:5 were placed into a fused-silica tube in a glove box filled with Ar, and the tube was sealed with under a high vacuum of 10^{-3} Pa and placed in a muffle furnace with program control; then, the mixed sample was heated up to 950°C in 55 h, kept at this temperature for 33 h, and then slowly cooled down with gradient to 550°C and to 300°C, respectively. Finally, yellow block-shaped single crystals were obtained at room temperature.

All the elementary substances were stored in a dry Ar-filled glove box with controlled oxygen and moisture levels below 0.1 ppm.

A.2.3 Silicon–Cadmium–Selenium

The ternary compound Cd_4SiSe_6, which is thermally stable up to ~362°C and crystallizes as a monoclinic structure with the lattice parameters $a = 1284.5 \pm 0.8$, $b = 736.0 \pm 0.4$, $c = 1283.4 \pm 0.8$ pm, $\beta = 110.009 \pm 0.009°$, a calculated density of 5.543 g·cm^{-3}, and an energy gap of 1.95 eV, is formed in the Si–Cd–Se system (Chen et al. 2020a). It was synthesized in the same way as Cd_4SiS_6 using Se instead of S. High-quality single crystals of Cd_4SiSe_6 were grown by a chemical transport reaction. A total of 2.379 g of Cd_4SiSe_6 polycrystalline powder and 120 mg of iodine as a transport agent were loaded into a silica tube. The temperature of vaporization chamber and crystallization chamber were maintained at 850°C and 750°C for 4 days, respectively, and then cooled to 550 and 450°C at 3°C·h^{-1}, and finally cooled to room temperature in 24 h. Millimeter-scale dark red crystals with good quality were obtained.

A.3 Systems Based on Germanium Sulfides

A.3.1 Germanium–Sodium–Sulfur

According to Yahia et al. (2023), the ternary compound Na_4GeS_4 melts incongruently at ≈ 660°C and crystallizes as a monoclinic structure with the lattice parameters $a = 1098.22 \pm 0.09$, $b = 2965.2 \pm 0.2$, $c = 2003.91 \pm 0.16$ pm, and $\beta = 105.77 \pm 0.01°$. Amorphous Na_4GeS_4 was prepared by mechanochemical synthesis from a stoichiometric mixture of Na_2S and GeS_2. The starting materials (0.5 g in total) were mixed in an agate mortar in a dry Ar glove box. The mixture was placed in a 45 mL zirconia milling pot with 90 g of zirconia balls of 4 mm diameter, sealed with O-ring and mechanically milled using a planetary ball mill apparatus. The rotational speed and milling time were set at 500 rpm and 20 h, respectively. The Na_4GeS_4 glass-ceramic was prepared by combining the mechanochemical process and a heat treatment. Pellets of the amorphous Na_4GeS_4 phase were sintered at 480°C for 12 h in alumina crucible under Ar flow. After an intermediate grinding, the pellets were heated again at 480°C for 48 h to ensure a total reaction. One sample also was heated at 620°C for 1 h and quenched in liquid nitrogen.

To avoid the decomposition of the Na_4GeS_4 after melting, the Na_2S-rich-phase Na_6GeS_5 was prepared. The Na_2S and GeS_2 reagents were ball-milled at 500 rpm for 20 h, and then the resulting powder was pelletized and sintered at 650°C for 36 h in alumina crucible under Ar flow. Although the pellet melted only partially, white opaque and colorless transparent crystals corresponding to the Na_4GeS_4 and Na_2S phases were formed.

A.3.2 Germanium–Potassium–Sulfur

Another ternary compound, K_4GeS_4, which crystallizes as a cubic structure with the lattice parameter $a = 1274.50 \pm 0.15$ pm, a calculated density of 2.292 g·cm^{-3} at 203 K, and an energy gap of 3.20 eV, is formed in the Ge–K–S system (Melullis and Dehnen 2007). The colorless hexagons of this compound were synthesized using pure K_2S that was prepared in THF solution from elemental K and S, applying naphthalene as a catalyst. K_2S was sealed in an evacuated quartz ampoule together with stoichiometric amounts of Ge and S under vacuum (0.1 Pa). The ampoule was heated at a rate of 1°C·min^{-1} up to 630°C, tempered for 2 h, and cooled at the same rate to give rise to single crystals. All synthesis steps were performed with a strong exclusion of air and external moisture.

A.3.3 Germanium–Copper–Sulfur

GeS_2–Cu_2S. The phase equilibria in this system were reinvestigated by Alverdiyev (2019) through DTA, XRD, and EMF measurements (Figure A.3.1). Two ternary compounds, Cu_2GeS_3 and Cu_8GeS_6, are formed in the system. The first compound melts congruently at 942°C, while the second one melts incongruently at 980°C and undergoes a polymorphic transformation at 55°C. The peritectic composition is 25 mol% GeS_2. The eutectics contain 43 and 90 mol% GeS_2 and crystallize at 932°C and 819°C, respectively. The lines at 102°C and 55°C correspond to polymorphic transformations of Cu_2S and Cu_8GeS_6, respectively. The existence of the Cu_4GeS_4 and $Cu_2Ge_2S_5$ ternary compounds were not confirmed.

According to Bagheri et al. (2015), Cu_2GeS_3 crystallizes as an orthorhombic structure with the lattice parameters $a = 1131.8 \pm 0.2$, $b = 378.02 \pm 0.05$, and $c = 521.66 \pm 0.08$ pm.

A.3.4 Germanium–Silver–Sulfur

According to Aliyeva et al. (2014), the high-temperature modification of Ag_8GeS_6 crystallizes as an orthorhombic structure with the lattice parameters $a = 1512.92 \pm 0.08$, $b = 745.63 \pm 0.03$, and $c = 1055.35 \pm 0.05$ pm and the low-temperature modification of this compound crystallizes as a cubic structure with the lattice parameter $a = 1070.56 \pm 0.07$ pm. The DSC data were used to determine the enthalpy and entropy of the phase transitions of Ag_8GeS_6 from the orthorhombic to the cubic modification at 222°C (Bayramova et al. 2022c). It was shown that $\Delta H_{tr} = 9.46 \pm 0.38$ kJ·M^{-1} and $\Delta S_{tr} = 19.11 \pm 0.76$ J·(M·K)$^{-1}$.

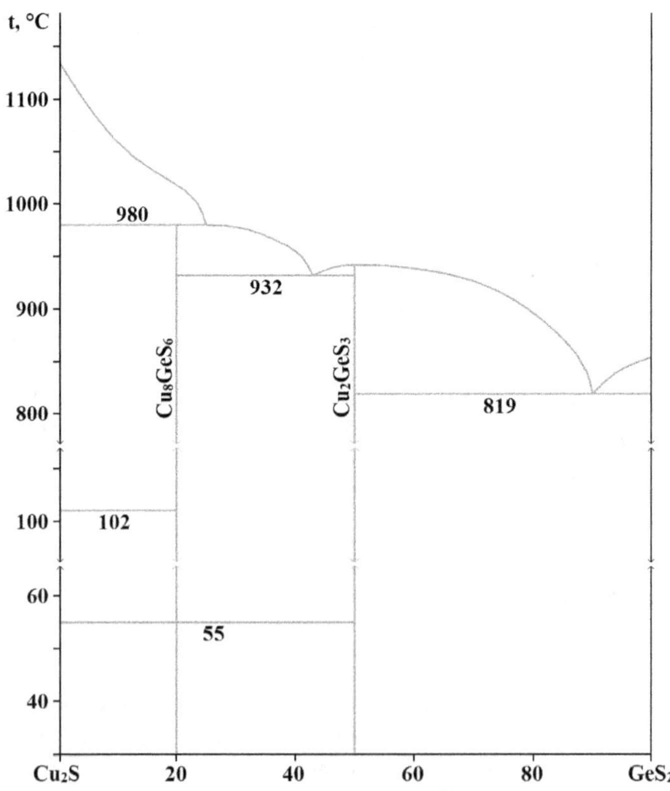

FIGURE A.3.1 Phase diagram of the GeS_2–Cu_2S quasibinary system. (From Alverdiyev, I.J., *Chem. Probl.*, (3), 423, 2019.) Open access.

A.3.5 Germanium–Cadmium–Sulfur

According to Li et al. (2018a), the ternary compound Cd_4GeS_6 is stable up to 700°C and is characterized by the energy gap of 2.52 eV. The crystals of this compound were synthesized using CdS and GeS_2 in the molar ratio 4:1. First, they were ground into fine powders in an agate mortar and pressed into pellet. Second, the pellet was loaded into quartz tube and flame-sealed under a high vacuum of 10^{-3} Pa. Third, the tube was placed into a computer-controlled furnace, heated from room temperature to 400°C in 10 h and kept at this temperature for 15 h, raised to 1000°C in 30 h and maintained for 100 h, and then slowly cooled to 300°C at a rate of $3°C \cdot h^{-1}$ before turning off the furnace. Finally, brown-yellow crystals of Cd_4GeS_6 were obtained which are stable in air for several months.

A.3.6 Germanium–Yttrium–Sulfur

According to Ye et al. (2021), the ternary compound $Y_6Ge_{2.5}S_{14}$, which crystallizes as a hexagonal structure, has an energy gap of 1.71 ± 0.1 eV. To prepare this compound, the reactants were loaded at a molar ratio of Y/Ge/S = 6:2.5:14 (total 0.4 g) mixed with 0.4 g YCl_3/NaCl flux (molar ratio 1:1) into carbonized silica ampoule. The ampoule was slowly heated to 900°C over 10 h, held at that temperature for 96 h, then slowly cooled to 500°C over 24 h; finally, the furnace was turned off and the ampoule was returned to room temperature. After the reaction, the sample was washed with deionized water to remove the YCl_3/NaCl flux. After washing, the resulting orange-red crystals were collected and separated. They are stable in the air for several months. All starting materials were stored and used in an Ar-filled glove box.

A.3.7 Germanium–Lanthanum–Sulfur

According to Cicirello et al. (2021b), the ternary compound $La_4Ge_3S_{12}$ crystallizes in the rhombohedral structure with the lattice parameter in the hexagonal setting a = 1942.7 ± 0.3, c = 807.53 ± 0.13 pm, and a calculated density of 4.372 $g \cdot cm^{-3}$ at 200 K. For preparing this compound, La powder, Ge pieces, S powder, I_2, $LaCl_3$, and NaCl were used. A two-step synthesis method was followed. The reactant elements were loaded into carbonized silica ampoule at a molar ratio of La/Ge/S = 4:3:12 with ~0.1 g of I_2 added as a transport agent. The ampoule was flame-sealed under high vacuum and placed in a programmable furnace. They were heated to 800°C in 10 h, held at that temperature for 96 h, and then cooled to 100°C in 24 h. The first step produced homogeneous powders. The as-synthesized powder from the first step was mixed with salt flux $LaCl_3$ and NaCl at a mass ratio of $LaCl_3$/NaCl = 0.68:0.32 in a glove box. The salt flux is of equal mass to the target compound. The admixture was sealed in a new ampoule heated by the same temperature profile as the first step. After the second annealing, the sample was washed with deionized water to remove the flux, yielding yellow-green crystals of $La_4Ge_3S_{12}$. This compound is stable in dry air for multiple months. All starting materials were stored and used in an Ar-filled glove box.

A.3.8 Germanium–Arsenic–Sulfur

The ternary compound Ge_2As_2S, which can exist in this system, were theoretically studied based on density function theory (DFT) calculation with generalized gradient approximation method (Yang et al. 2023i). It was shown that this compound is indirect band gap semiconductor with an energy gap of 0.822 eV, is dynamic and mechanically stable at 0 and 10 GPa, and crystallizes as an orthorhombic structure with the lattice parameters a = 658 and 619, b = 658 and 619, c = 640 and 596 pm, and a calculated density of 4.44 and 5.55 $g \cdot cm^{-3}$ at 0 and 10 GPa, respectively.

A.3.9 Germanium–Vanadium–Sulfur

The ternary compound GeV_4S_8, which exists in this system, undergoes successive structural phase transitions from cubic structure to rhombohedral at 33 K and to orthorhombic structure at 18 K with the decrease of temperature (Chudo et al. 2006). The antiferromagnetic ordering at 18 K [at 13 K (Müller et al. 2006)] is found to be accompanied with the structural transition. According to Bichler et al. (2008), the antiferromagnetic ordering occurs without symmetry change but with a certain increase in the distortion, and cubic structure at 30 K transforms into orthorhombic structure. The lattice parameters for the orthorhombic structure are as follows: $a = 684.28 \pm 0.01$, $b = 679.94 \pm 0.01$, $c = 964.76 \pm 0.01$ pm, and a calculated density of 3.942 g·cm^{-3} at 25 K (Bichler et al. 2008) and $a = 683.62 \pm 0.02$, $b = 1358.7 \pm 0.01$, $c = 963.84 \pm 0.04$ pm, and a calculated density of 3.953 g·cm^{-3} (Müller et al. 2006) [$a = 679.65$, $b = 684.05$, $c = 964.12$ pm (Chudo et al. 2006)] at 4 K. According to Chudo et al. (2006), this compound crystallizes at 25 K as a rhombohedral structure with the lattice parameters $a = 680.36$ pm and $\alpha = 60.299°$.

The powder sample of this compound was prepared by heating stoichiometric mixtures of the elements in an evacuated quartz tube (Chudo et al. 2006). The mixture was kept for 1 day at 750°C. Single crystals were grown by chemical transport method in a quartz tube.

A.3.10 Germanium–Oxygen–Sulfur

The ternary compound $Ge(SO_4)_2$ is formed in this system (Hayek and Hinterauer 1951). This compound can be prepared by reacting $GeCl_4$ with SO_3 in a sealed tube when heated without a solvent.

A.4 Systems Based on Germanium Selenides

A.4.1 Germanium–Potassium–Selenium

Another ternary compound, K_4GeSe_4, which crystallizes as a cubic structure with the lattice parameter $a = 1318.93 \pm 0.15$ pm, a calculated density of 3.155 g·cm^{-3} at 203 K, and an energy gap of 2.46 eV, is formed in the Ge–K–Se system (Melullis and Dehnen 2007). The light yellow platelets of this compound were synthesized using pure K_2Se, which was prepared in THF solution from elemental K and Se, applying naphthalene as a catalyst. K_2Se was sealed in an evacuated quartz ampoule together with stoichiometric amounts of Ge and Se under vacuum (0.1 Pa). The ampoule was heated at a rate of 1°C·min^{-1} up to 630°C, tempered for 2 h, and cooled at the same rate to give rise to single crystals. All synthesis steps were performed with a strong exclusion of air and external moisture.

A.4.2 Germanium–Rubidium–Selenium

Another ternary compound, $Rb_2Ge_4Se_{10}$, which crystallizes as a monoclinic structure with the lattice parameters $a = 683.4 \pm 0.3$, $b = 1127.6 \pm 0.5$, $c = 1245.0 \pm 0.5$ pm, $\beta = 92.517 \pm 0.008°$, a calculated density of 4.334 g·cm^{-3}, and an energy gap of 2.48 eV, is formed in the Ge–Rb–Se system (Liu et al. 2018). The polycrystalline sample of this compound was synthesized by the traditional solid state reactions. A stoichiometric mixture of Ba rod, Ge and Se powders, and RbCl acting as a reacting flux in a molar ratio of 1:4:10:10 was loaded into quartz vessel and then flame-sealed under a vacuum of 0.01 Pa. The vessel was placed into a resistance furnace and heated from room temperature to 850°C within 36 h, kept at that temperature for 96 h, and finally cooled to 350°C at 3°C·h^{-1} before turning off the power. Red crystals of the title compound were obtained after washing the products with deionized water and ethanol in an ultrasonic cleaner. Several attempts using the reactants not containing Ba failed to produce the compound, the Ba element may play the role of catalyst in the synthesis. All reagents were stored and operated in an Ar-filled glove box to prevent air and moisture exposure.

A.4.3 Germanium–Cesium–Selenium

Another ternary compound, $Cs_2Ge_4Se_{10}$, which crystallizes as a monoclinic structure with the lattice parameters $a = 699.1 \pm 0.1$, $b = 1136.0 \pm 0.3$, $c = 1259.4 \pm 0.2$ pm, $\beta = 91.697 \pm 0.006°$, a calculated density of 4.471 g·cm^{-3}, and an energy gap of 2.41 eV, is formed in the Ge–Cs–Se system (Liu et al. 2018). The polycrystalline red sample of this compound was synthesized in the same way as $Rb_2Ge_4Se_{12}$ but using CsCl instead of RbCl.

According to Yao et al. (2010), the $Cs_4Ge_2Se_6$ also crystallizes as a monoclinic structure with the lattice parameters $a = 1422.80 \pm 0.18$, $b = 740.61 \pm 0.10$, $c = 1031.1 \pm 0.2$ pm, $\beta = 124.121 \pm 0.004°$, and a calculated density of 4.248 g·cm^{-3}. This compound was prepared using $AgNO_3$ (0.15 mM), Ge (0.10 mM), Se (0.30 mM), Cs_2CO_3 (0.10 mM), about 0.3 mL of pyridine, and 0.2 mL of ethylenediamine. The mixture was sealed in a Pyrex glass tube with ca. 10% filling at air atmosphere, placed into a stainless steel autoclave, and then heated at 170°C for 7 days. After cooling naturally to ambient temperature, the products were washed with ethanol and water, respectively, and yellow block crystals of the title compound were obtained.

A.4.4 Germanium–Copper–Selenium

Based on the results the EMF measurements, the partial molar functions of copper in equilibrium alloys from the three-phase region of the Cu_2GeSe_3–$GeSe_2$–Se subsystem, as well as the standard thermodynamic functions of formation and the standard entropy of the Cu_2GeSe_3 were calculated (Yusibov et al. 2023). The following results were obtained: $\Delta G^0_{f,\,298} = -176.8 \pm 3.1$ kJ·M^{-1}, $\Delta H^0_{f,\,298} = -173.9 \pm 3.1$ kJ·M^{-1}, and $S^0_{298} = 233.3 \pm 5.1$ J·(M·K)$^{-1}$. Solid-phase equilibria in the Cu_2GeSe_3–$GeSe_2$–Se subsystem at 30°C–180°C are presented in Figure A.4.1.

According to Jiang et al. (2017), the Cu_8GeSe_6 ternary compound crystallizes as a hexagonal structure with the lattice parameters $a = 1266.01 \pm 0.04$, $c = 1176.98 \pm 0.03$ pm, and a calculated density of 6.545 g·cm^{-3} at room temperature and $a = 728.4 \pm 1.3$, $c = 1192.0 \pm 1.2$ pm, and a calculated density of 6.006 g·cm^{-3} at 200°C. To prepare this compound, Cu and Se shots and Ge pieces were weighed out in stoichiometric proportions and then sealed in silica tube under vacuum. The tube was heated to 1120°C and maintained at this temperature for 24 h before quenching into cold water. Then, the quenched ingot was annealed at 600°C for 5 days. Finally, the product was ground into fine powder and sintered by spark plasma sintering at 410°C–440°C under a pressure of 75 MPa for 5 min.

A.4.5 Germanium–Silver–Selenium

The partial and integral thermodynamic functions of α-Ag_8GeSe_6 ($\Delta G^0_{f,\,298} = -316.1 \pm 5.0$ kJ·M^{-1}, $\Delta H^0_{f,\,298} = -290.4 \pm 5.0$ kJ·M^{-1}, and $S^0_{298} = 711.6 \pm 10.0$ J·(M·K)$^{-1}$ were determined by Ibrahimova (2019b).

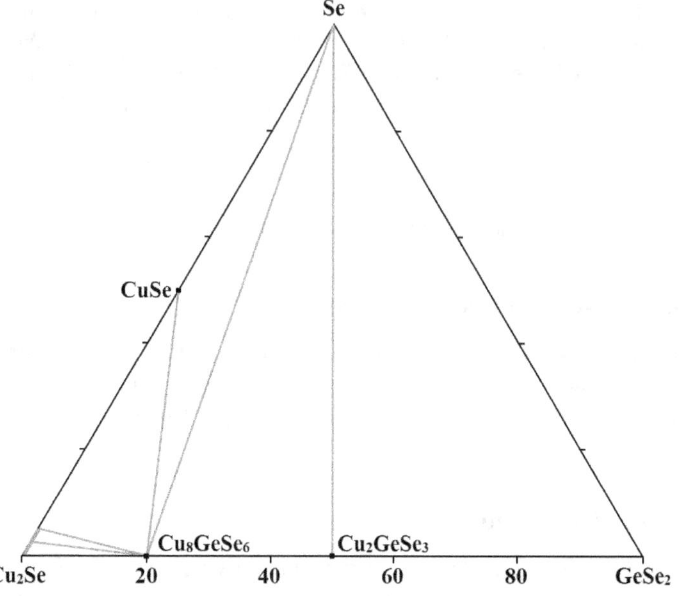

FIGURE A.4.1 Solid-phase equilibria in the $GeSe_2$–Cu_2Se–Se subsystem at 30°C-180°C. (From Yusibov, Yu.A., et al., *Azerb. Chem. J.*, (1), 108, 2023.) Open access.

These functions were also calculated by Alverdiev et al. (2017) based on EMF measurements of concentration circuits in the temperature range 17°C–157°C.

The DSC data were used to determine the enthalpy and entropy of the phase transitions of Ag_8GeSe_6 from the orthorhombic to the cubic modification at 48°C (Bayramova et al. 2022c). It was shown that $\Delta H_{tr} = 16.95 \pm 0.08$ kJ·M^{-1} and $\Delta S_{tr} = 52.80 \pm 2.11$ J·(M·K)$^{-1}$.

A.4.6 Germanium–Magnesium–Selenium

The structural, mechanical, electronic, dynamical, and optical properties of the Mg_2GeSe_4 compound were investigated using DFT calculations by Barde and Joubert (2020). It crystallizes as an orthorhombic structure with the lattice parameters calculated using PBE(*) and PBEsol(**) functionals a = 1364.1*(1343.0**), b = 793.7*(782.7**), c = 639.1*(631.6**) pm, and an energy gap of 2.58 eV. It is an indirect band gap semiconductor.

A.4.7 Germanium–Calcium–Selenium

The structural, mechanical, electronic, dynamical, and optical properties of the Ca_2GeSe_4 hypothetical compound were investigated using DFT calculations by Barde and Joubert (2020). It can crystallize as an orthorhombic structure with the lattice parameters calculated using PBE(*) and PBEsol(**) functionals a = 668.1*(654.6**), b = 865.8*(859.4**), c = 1445.8*(1413.6**) pm, and an energy gap of 2.30 eV. According to the calculations, this compound must be stable in this structure and must be an indirect band gap semiconductor.

A.4.8 Germanium–Strontium–Selenium

The structural, mechanical, electronic, dynamical, and optical properties of the γ-Sr_2GeSe_4 compound were investigated using DFT calculations by Barde and Joubert (2020). It crystallizes as an orthorhombic structure with the lattice parameters calculated using PBE(*) and PBEsol(**) functionals a = 1047.9*(1025.3**), b = 1075.6*(1048.1**), c = 748.5*(737.4**) pm, and an energy gap of 2.86 eV [2.00 eV (Menezes et al. 2020)]. It is a direct band gap semiconductor.

To prepare bulk γ-Sr_2GeSe_4 in high purity, the following procedure was developed (Menezes et al. 2020). Stoichiometric amounts of Sr granules, Ge chunks, and Se pellets were added to a glassy carbon crucible and placed at the bottom of a silica tube to be evacuated to 2.5 Pa and then flame-sealed. This ampoule was slowly heated at a rate of 100°C·h^{-1} to 800°C to enable complete reaction of the elements. After the mixture had reacted, the temperature was decreased to 650°C and held for 100 h to allow the elements to diffuse throughout the sample. The sample was finally slowly cooled to 200°C over 100 h.

A.4.9 Germanium–Yttrium–Selenium

$GeSe_2$–Y_2Se_3. No ternary compounds were found in this system (Lychmanyuk et al. 2006).

A.4.10 Germanium–Neodymium–Selenium

The ternary compound $Nd_6Ge_3Se_{14}$, which melts incongruently at 904°C and crystallizes as a hexagonal structure with the lattice parameters a = 1053 and c = 600 pm, is formed in this system (Guittard and Julien-Pouzol 1970). This compound was prepared by heating a stoichiometric mixture of Nd_2Se_3, Ge, and Se at 500°C.

A.4.11 Germanium–Dysprosium–Selenium

The ternary compound $Dy_6Ge_3Se_{14}$, which melts incongruently and crystallizes as a hexagonal structure with the lattice parameters a = 1018 and c = 609 pm, is formed in this system (Guittard and Julien-Pouzol 1970). This compound was prepared by heating a stoichiometric mixture of Dy_2Se_3, Ge, and Se at 700°C.

A.4.12 Germanium–Lead–Selenium

The ternary compound α-Pb_2GeSe_4, which crystallizes as a monoclinic structure with the lattice parameters a = 844.47 ± 0.03, b = 713.14 ± 0.03, c = 1380.79 ± 0.04 pm, β = 107.867 ± 0.002°, a calculated density of 6.74 g·cm^{-3}, and an energy gap of 1.42 eV, is formed in the Ge–Pb–Se system (Menezes et al. 2020). To synthesize this compound, stoichiometric amounts of the starting elements were added to a silica tube, evacuated, and flame-sealed. The sealed tube was placed into a furnace and heated to 900°C where a melt was achieved. The melt was quenched in an ice bath. Since powder XRD indicated the presence of the target phase in addition to the binaries $GeSe_2$ and PbSe, the sample was ground and then annealed at 500°C for 4 days. After 4 days at 500°C, only very minor amounts of $GeSe_2$ and PbSe as well as small unidentified peaks were present in the powder XRD pattern. This procedure gives the α-polymorph of Pb_2GeSe_4.

A.4.13 Germanium–Arsenic–Selenium

The Ge_2As_2Se ternary compound, which can exist in this system, was theoretically studied based on DFT calculation with generalized gradient approximation methods (Yang et al. 2023i). It was shown that this compound is indirect band gap semiconductor with an energy gap of 0.725 eV, is dynamic and mechanically stable at 0 and 10 GPa, and crystallizes as an orthorhombic structure with the lattice parameters a = 669 and 626, b = 669 and 626, c = 644 and 605 pm, and a calculated density of 4.91 and 6.08 g·cm^{-3} at 0 and 10 GPa, respectively.

A.5 Systems Based on Germanium Telluride

A.5.1 Germanium–Silver–Tellurium

The liquidus surface of the Ge–Ag–Te system, constructed through DTA, XTD, and metallography, is given in Figure A.5.1 (Ferhat et al. 1991). Only one ternary compound, Ag_8GeTe_6, which melts incongruently at 644°C, is formed in this system. There are also seven invariant points in the system; four ternary eutectics and three transition points: E_1 (649°C) – L ⇔ β-Ag_2Te + Ge + Ag; E_2 (595°C) – L ⇔ γ-Ag_8GeTe_6 + GeTe + Ge; E_3 (350°C) – L ⇔ γ-Ag_8GeTe_6 + Ag_5Te_3 + Te; E_4 (330°C) – L ⇔ γ-Ag_8GeTe_6 + GeTe + Te; U_1 (418°C) – L + $Ag_{1.9}Te$ ⇔ γ-Ag_8GeTe_6 + Ag_5Te_3; U_2 (453°C) – L + β-Ag_2Te ⇔ γ-Ag_8GeTe_6 + $Ag_{1.9}Te$; U_3 (610°C) – L + β-Ag_2Te ⇔ γ-Ag_8GeTe_6 + Ge. The system is also characterized by the existence of large immiscibility region near the Ag–Te system. A medium-sized ternary glass-forming region not presented in Figure A.5.1 was determined from the binary glass zone of the Ge–Te system.

The values of the standard thermodynamic functions of formation and the standard entropy of the γ-Ag_8GeTe_6 were calculated using the data of EMF measurements of concentration circuits in the temperature range 282°C–322°C (Moroz et al. 2014). The following results were obtained: $\Delta G^0_{f,\,298}$ = -266 ± 2 kJ·M^{-1}, $\Delta H^0_{f,\,298}$ = -221 ± 1 kJ·M^{-1}, and S^0_{298} = 150 ± 2 J·(M·K)$^{-1}$.

A.5.2 Germanium–Barium–Tellurium

Another ternary compound, $Ba_2GeTe_3(Te_2)$ or Ba_2GeTe_5, which is a metal (this was predicted from the DFT calculations) and crystallizes as an orthorhombic structure with the lattice parameters a = 679.00 ± 0.14, b = 2372.0 ± 0.5, c = 693.00 ± 0.14 pm, and a calculated density of 5.863 g·cm^{-3}, is formed in the Ge–Ba–Te system (Prakash et al. 2019). Black crystals of this compound were serendipitously obtained by combining the elements Ba, Ge, U, and Te. The aim of the reaction was to synthesize a new quaternary compound of Ba, Ge, U, and Te. The reactants were loaded into carbon-coated fused-silica tubes in an Ar-filled glove box, evacuated to 0.01 Pa, and flame-sealed. The reaction contained Ba (0.252 mM), U (0.042 mM), Ge (0.084 mM), and Te (0.756 mM). The sealed tube containing the reaction mixture was heated from room temperature to 780°C in 36 h, maintained for 15 h, further heated to 900°C in 24 h, and maintained for 192 h. The tube was then cooled to 720°C over 99 h and finally to room temperature in

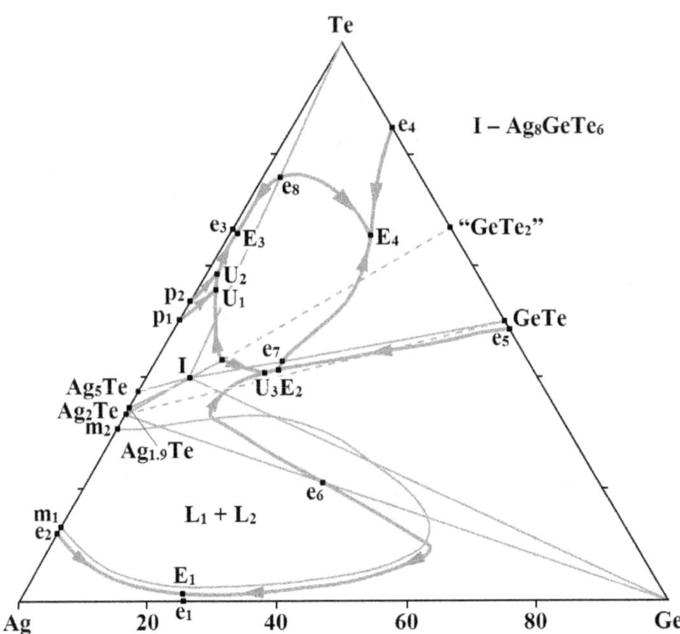

FIGURE A.5.1 Liquidus surface of the Ge–Ag–Te ternary system. (From Ferhat, A., et al., *J. Alloys Compd.*, 177(2), 337, 1991.)

a further 122 h. The reaction product also contained crystals of UOTe and BaTe. The crystals of the title compound are stable in air for at least one week as judged from unit-cell determinations.

According to Jana et al. (2022), the $Ba_2Ge_2Te_5$ ternary compound, which exists in the Ge–Ba–Te system, crystallizes as an orthorhombic structure with the lattice parameters a = 1338.79 ±0.04, b = 917.07 ±0.03, c = 994.41 ±0.03 pm, a calculated density of 5.755 g·cm^{-3}, and an energy gap of 0.6 eV [a = 1339.3 ±0.2, b = 917.69 ±0.15, c = 995.00 ±0.15 pm, a calculated density of 5.746 g·cm^{-3}, and an energy gap of 1.15 eV (Ran et al. 2021a)]. A polycrystalline sample of $Ba_2Ge_2Te_5$ was synthesized in a two-step method (Jana et al. 2022). First, the stoichiometric amounts of elements were heated at 850°C in a sealed evacuated carbon-coated fused silica tube. The product of this reaction was homogenized into a fine powder in the Ar-filled glove box and then compacted into a circular pellet that was further heated at 600°C inside a sealed evacuated fused silica tube. Black block-shaped crystals of this compound were synthesized by reaction of stoichiometric amounts of elements using the sealed tube solid state synthesis method. Since Ba is air and moisture sensitive, the reactants were handled inside an Ar-filled glove box.

The polycrystalline sample of this compound was also obtained by traditional solid state reactions in a stoichiometric mixture of Ba (0.1 mM), Ge (0.1 mM), and Te (0.25 mM) (Ran et al. 2021a). The mixture was loaded into a graphite crucible and transferred into a quartz tube. The tube was then evacuated to a vacuum of 10^{-3} Pa and sealed. Afterward, the sample was heated in a furnace to 950°C over 50 h and kept at that temperature for 100 h, and subsequently cooled to 350°C at a rate of 2°C·h^{-1} before switching off the furnace. Black crystals were obtained. They are stable in air for more than 5 months. All manipulations were performed in an Ar-filled glove box with H_2O and O_2 contents of less than 0.1 ppm.

A.5.3 Germanium–Arsenic–Tellurium

The Ge_2As_2Te ternary compound, which can exist in this system, was theoretically studied based on DFT calculation with generalized gradient approximation methods (Yang et al. 2023i). It was shown that this compound is an indirect band gap semiconductor with an energy gap of 0.461 eV, is dynamic and mechanically stable at 0 and 10 GPa, and crystallizes as an orthorhombic structure with the lattice parameters a = 690 and 636, b = 690 and 636, c = 651 and 618 pm, and a calculated density of 5.22 and 6.52 g·cm^{-3} at 0 and 10 GPa, respectively.

A.5.4 Germanium–Antimony–Tellurium

Quenching cubic high-temperature polymorphs of $(GeTe)_nSb_2Te_3$ ($n \geq 3$) yields metastable phases whose average structures can be approximated by the rock-salt type with $1/(n + 3)$ cation vacancies per anion (Urban et al. 2015). The phase with n = 5 crystallizes as a cubic structure with the lattice parameter a = 597.68 ± 0.12 pm and a calculated density of 6.34 g·cm^{-3} (Urban et al. 2015) and the phase with n = 8 crystallizes as a cubic structure with the lattice parameter a = 595.0 ± 0.3 pm or in the rhombohedral structure with the lattice parameters a = 420.99 ± 0.06 and c = 1024.7 ± 0.4 pm in hexagonal setting (Buller et al. 2012). The experimental density and the energy gap of the phase with n = 8 are 5.5 and 6.0 g·cm^{-3} and 0.82 and 0.58 eV in an amorphous and in a crystalline state, respectively. The glass transition temperature for $(GeTe)_8Sb_2Te_3$ is 165°C.

All samples were prepared by initially melting the elements Ge, Sb, and Te at 950°C in sealed silica glass ampoules under Ar atmosphere (Urban et al. 2015). Highly disordered crystals of $(GeTe)_2Sb_2Te_3 = Ge_{0.4}Sb_{0.4}Te$ were obtained by pouring a stoichiometric melt in liquid nitrogen. Multiply twinned crystal fragments were mechanically isolated from such quenched samples. Crystals with variable n in the range between 3 and 12 were obtained by physical vapor deposition. A melt with a molar ratio of Ge/Sb/Te = 1.44:2:5 was quenched by cooling the ampoule in water. About 100 mg of the powdered product were sealed in a silica glass ampoule (Ar atmosphere), kept at 628°C for 4 h, and then slowly cooled to 618°C within 6 h. After holding this temperature for 75 h, the ampoule containing octahedral crystals was quenched in air. Among others, crystals of $(GeTe)_4Sb_2Te_3 = Ge_{0.57}Sb_{0.29}Te$ were obtained. Besides the slow cooling of the sample also the natural temperature gradient of the tube furnace could have helped crystal growth.

Crystals of $(GeTe)_5Sb_2Te_3 = Ge_{0.63}Sb_{0.25}Te$ and $(GeTe)_{12}Sb_2Te_3 = Ge_{0.8}Sb_{0.13}Te$ were grown in the stability ranges of their high-temperature phases by chemical transport reactions and subsequently quenched. Quenched and powdered $(GeTe)_{12}Sb_2Te_3$ (typically 120 mg) and SbI_3 as transport agent [ca. 50 mg for $(GeTe)_5Sb_2Te_3$ and 10 mg for $(GeTe)_{12}Sb_2Te_3$] was sealed in silica glass ampoules. Chemical transport was carried out from 600°C to 520°C for n = 5 and 450°C for n = 12, respectively, within 1 day. Subsequently, the ampoule was removed from the furnace, that is, quenched in air. Crystals with n = 5 can also be obtained from $(GeTe)_3Sb_2Te_3$ (ca. 200 mg) and I_2 as transport agent (ca. 50 mg) under similar conditions.

Thin films of the phases from GeTe–Sb_2Te_3 quasibinary system crystallize under laser irradiation into an NaCl-type single phase as a metastable state, over a wide composition range of at least 50–100 mol% GeTe (Matsunaga and Yamada 2002).

A.5.5 Germanium–Bismuth–Tellurium

GeTe–Bi_2Te_3. The updated phase diagram of this system was constructed through DTA, XRD, and SEM and is given in Figure A.5.2 (Alakbarova et al. 2021, 2022a). After the fusion, the samples for investigations were quenched from a liquid state (680°C) in iced water and then subjected to thermal annealing at 530°C for 1000 h. The characteristic feature of the phase equilibria in this system is the incongruent nature of melting of the compounds and small difference (5–8°C) in the temperatures of peritectic reactions. The peritectic points have the compositions of 52, 60, 73, and 77 mol% Bi_2Te_3 and the eutectic contains 83 mol% Bi_2Te_3 and crystallizes at 565°C. The phase diagram includes six ternary compounds, which crystallize in the trigonal structure with the following lattice parameters: a = 426.38 ± 0.02 and c = 7327.1 ± 0.3 pm for $Ge_4Bi_2Te_7$; a = 427.30 ± 0.03 and c = 6263.4 ± 0.4 pm for $Ge_3Bi_2Te_6$; a = 429.86 ± 0.02 and c = 1733.5 ± 0.3 pm for $Ge_2Bi_2Te_5$ (incongruent melting at 590°C); a = 431.76 ± 0.03 and c = 4125.9 ± 0.5 pm for $GeBi_2Te_4$ (incongruent melting at 581°C); a = 435.56 ± 0.02 and c = 2392.8 ± 0.4 pm for $GeBi_4Te_7$ (incongruent melting at 575°C); and a = 435.72 ± 0.03 and c = 10191.1 ± 0.2 pm for $GeBi_6Te_{10}$ (incongruent melting at 570°C).

FIGURE A.5.2 Phase diagram of the GeTe–Bi_2Te_3 quasibinary system. (From Alakbarova, T.M., et al., *Condens. Matter Interphases*, 24(1), 11, 2022.) Open access.

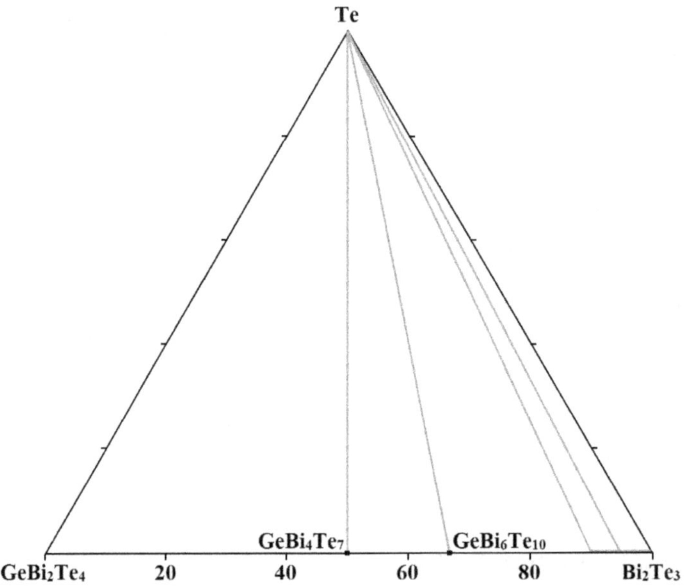

FIGURE A.5.3 Isothermal section of the $GeBi_2Te_4$–Bi_2Te_3–Te subsystem at 130°C. (From Alakbarova, T.M., et al., *Phys. Chem. Solid State*, 23(1), 25, 2022.) Open access.

$GeBi_2Te_4$–Bi_2Te_3–Te. The isothermal section of this quasiternary system, constructed through XRD and EMF measurements of concentration circuits, is presented in Figure A.5.3 (Alakbarova et al. 2022b). To construct this section, the samples after alloying were quenched from the liquid state by throwing ampoules into ice water to obtain highly dispersed cast alloys. Then, the samples were subjected to stepwise thermal annealing at 430°C for 1000 h and at 130°C for 100 h. The standard thermodynamic functions of the $GeBi_6Te_{10}$, $GeBi_4Te_7$, and $GeBi_2Te_4$ were calculated by the method of potential-forming reactions and the next values were obtained: $\Delta G^0_{f,\,298} = -313.4 \pm 0.8$ kJ·M^{-1}, $\Delta H^0_{f,\,298} = -312.8 \pm 2.5$ kJ·M^{-1}, and $S^0_{298} = 865.6 \pm 11.7$ J·(M·K)$^{-1}$ for $GeBi_6Te_{10}$; $\Delta G^0_{f,\,298} = -234.2 \pm 0.6$ kJ·M^{-1}, $\Delta H^0_{f,\,298} = -231.7 \pm 1.9$ kJ·M^{-1}, and $S^0_{298} = 611.0 \pm 8.7$ J·(M·K)$^{-1}$ for $GeBi_4Te_7$; and $\Delta G^0_{f,\,298} = -153.6 \pm 0.4$ kJ·M^{-1}, $\Delta H^0_{f,\,298} = -149.8 \pm 1.3$ kJ·M^{-1}, and $S^0_{298} = 354.2 \pm 5.6$ J·(M·K)$^{-1}$ for $GeBi_2Te_4$.

Two more ternary compounds are formed in the Ge–Bi–Te ternary system. $GeBi_3Te_4$ melts incongruently at 563°C and crystallizes as a trigonal structure with the lattice parameters $a = 436.25 \pm 0.05$ and $c = 3138.1 \pm 0.2$ pm (Aliyev et al. 2022). For synthesis this compound, Ge, Bi, and Te were melted in a quartz ampoule according to the stoichiometric amounts under the vacuum of 0.01 Pa at 800°C for 3 h and quenched in ice water. Then annealing was carried out at 450°C for 1000 h to reach an equilibrium state. A single-crystal sample with $GeBi_3Te_4$ composition was grown by the vertical Bridgman–Stockbarger method. For this, a quartz ampoule with a conical bottom was sealed (0.01 Pa) and passed through the hot zone (650°C) of the two-zone furnace into the cold zone (450°C) at a rate of 1.0 mm·h^{-1}.

The second compound, $GeBi_4Te_4$, melts incongruently at 538°C and crystallizes as a hexagonal structure with the lattice parameters $a = 440.71 \pm 0.16$ and $c = 1738.4 \pm 0.2$ pm (Aliyev et al. 2021). Ge, Bi, and Te were used to synthesize the polycrystalline sample of this compound and to prepare its single crystals. According to stoichiometric amounts, 1 g of sample was sealed under the vacuum (0.01 Pa) in quartz ampoule, heated up to 800°C, and kept at this temperature for 3 h. For better dissolving, the sample was mixed by shaking. The synthesized alloy was quenched in ice water and further annealed at 450°C for 1000 h. Single crystals with a starting composition of $GeBi_4Te_4$ were grown from the melt using the vertical Bridgman–Stockbarger method.

A.5.6 Germanium–Iron–Tellurium

The $Fe_{3-\delta}GeTe_2$ ternary compound, which is formed in this system, crystallizes as a hexagonal structure with the lattice parameters $a = 396.325 \pm 0.004$ and $c = 1638.57 \pm 0.02$ pm for $\delta = 0.1$ (Yuan et al. 2017) [$a = 402.99$ and $c = 1634.25$ pm for $\delta = 0$ (Chen et al. 2013); $a = 400.848 \pm 0.002$, $c = 1633.07 \pm 0.01$ pm, and a calculated density of 7.15 g·cm^{-3} at room temperature for $\delta = 0.112$; $a = 400.45 \pm 0.03$, $c = 1637.6 \pm 0.2$ pm, and a calculated density of 7.15 g·cm^{-3} at room temperature for $\delta = 0.1$; and $a = 399.12 \pm 0.03$, $c = 1629.6 \pm 0.2$ pm, and a calculated density of 7.23 g·cm^{-3} at 1.5 K for $\delta = 0.1$ (Verchenko et al. 2015)]. The homogeneity range of $Fe_{3-\delta}GeTe_2$ is $0 < \delta < 0.3$, where lattice parameters decrease monotonically with increasing δ (Verchenko et al. 2015). The ferromagnetic transition for this compound was observed at 220°C and the heat capacity at constant pressure at room temperature is about 200 J·(M·K)$^{-1}$ (Chen et al. 2013).

The synthesis of $Fe_{3-\delta}GeTe_2$ powder was performed by the standard ampoule technique (Verchenko et al. 2015). Fe powder, Ge chips, and Te pieces were used as starting materials. To examine the homogeneity range of this compound, the specimens with $0 \le \delta \le 0.3$ were synthesized. The mixtures of starting materials were placed in quartz ampoules, which were then sealed under a vacuum of 2 Pa. The ampoules were annealed at 625°C during 5 days, and furnace cooled to room temperature. After the first annealing, the specimens were ground in an argon glove box, sealed in quartz ampoules, annealed at 625°C for 5 more days, and furnace cooled again. Finally, the samples were ground in an argon glove box and stored under a pure argon atmosphere.

Single crystals of Fe_3GeTe_2 were grown by the chemical transport method with iodine as the transport agent (Chen et al. 2013). Fe, Ge, and Te were mixed in a stoichiometric molar ratio of 3:1:2. The mixture was placed in an evacuated quartz tube together with 2 mg·cm^{-3} iodine. Then, the quartz tube was set up in a temperature gradient of 750°C/700°C for one week. Single crystals with a tabular shape in the *ab*-plane were grown on the low-temperature side.

Two more ternary phases, $Fe_{5-x}GeTe_2$ and $Fe_{5-x}Ge_2Te_2$, are formed in the Ge–Fe–Te system. The first phase crystallizes as a hexagonal structure (Spirovski et al. 2006) with the lattice parameters $a = 403.20 \pm 0.01$ and $c = 2905.99 \pm 0.06$ pm for $x = 0$ (May et al. 2019b) [$a = 403.76 \pm 0.04$ and $c = 2919.4 \pm 0.6$ pm for $x = 0.377$ (Stahl et al. 2018)]. According to Ly et al. (2021), this phase belongs to the $Fe_{5-x}GeTe_2$ family ($x = 0, 1, 2$) where each layer consists of a Fe_xGe slab sandwiched by Te layers.

The thermal stability of quenched $Fe_{4.7\pm0.2}GeTe_2$ crystals was examined by in situ diffraction and magnetization measurements to 430°C as well as annealing studies, and a transition was observed at ≈280°C (May et al. 2019a). This is the critical temperature for a transition between a low- and high-temperature crystal structures, though the microscopic change in structure remains unclear. Samples quenched from above this temperature are thus metastable, and upon cooling to cryogenic temperatures, these crystals undergo a first-order transition below ≈100 K.

Polycrystalline samples of $Fe_{5-x}GeTe_2$ were prepared via solid state reaction from pure elements (Stahl et al. 2018; May et al. 2019b). Fe, Ge, and Te powders were mixed in molar ratio 4.5:1:2. The mixture was filled in alumina crucible and sealed in silica ampoule in an argon atmosphere. The sample was heated to 750°C for 100–120 h (heating and cooling rate: 100°C·h^{-1}). The product was metallic gray and stable at air.

Single crystals of this phase were synthesized by the chemical vapor transport method (Ly et al. 2021; May et al. 2019b). Fe, Ge, and Te powders were mixed in a molar ratio of 5:1:2. Iodine (5 mg·cm^{-3}) was added as a transport agent and the mixed precursor was sealed in an evacuated quartz glass ampoule, which was placed in a furnace. The temperature of the center region of the furnace was raised up to 700°C with a heating rate of 1°C·min^{-1} and maintained at 700°C for 96 h. To improve the crystallinity of crystals, the ampoule was slowly cooled down to 450°C for 10 days. Below 450°C, the furnace was rapidly cooled down to room temperature without any heating.

$Fe_{5-x}Ge_2Te_2$ crystallizes as a trigonal structure with the lattice parameters $a = 401.21 \pm 0.03$, $c = 1077.77 \pm 0.08$ pm, and a calculated density of 7.413 g·cm^{-3} for the composition of $Fe_{4.84}Ge_{1.96}Te_2$ (Jothi et al. 2020). Polycrystalline samples containing large single crystals of $Fe_{5-x}Ge_2Te_2$ were obtained by a solid state method. For this synthesis, stoichiometric amounts of Fe, Ge, and Te powders were mixed homogeneously and pressed into a pellet. The pellet was loaded in a quartz glass ampoule and sealed under vacuum. The quartz ampoule was transferred into a vertical tube furnace, slowly heated to 650°C, and kept at that temperature for 12 h. Then, the temperature was further raised to 760°C. After 2–3 days of reaction, the ampoule was taken out from the furnace and quenched in ice-cold water. Shiny silver-gray-colored plate-like micrometer-size crystals were obtained. The obtained polycrystalline material contains 76 mass% $Fe_{5-x}Ge_2Te_2$. In a second set of experiments, the temperature was raised to 800°C and slowly cooled to room temperature for 5 days. The obtained polycrystalline material contains 89 mass% $Fe_{5-x}Ge_2Te_2$.

A.5.7 Germanium–Nickel–Tellurium

The ternary compound Ni_5GeTe_2, which crystallizes as an orthorhombic structure, is formed in the Ge–Ni–Te system (Spirovski et al. 2006).

A.6 Systems Based on Tin Sulfides

A.6.1 Tin–Sodium–Sulfur

The ternary compound Na_4SnS_4, which is formed in this system, has two polymorphic modifications. They crystallize in the tetragonal structure with lattice parameters $a = 1384.13 \pm 0.02$, $c = 2742.32 \pm 0.04$ pm (Hartmann et al. 2022a,b) and $a = 784.19 \pm 0.06$, $c = 695.22 \pm 0.09$ pm, a calculated density of 2.633 g·cm^{-3}, and an energy gap of 3.25 eV (Gao et al. 2021b) for the first and second modifications, respectively. The first modification is formed by heating of $Na_4SnS_4 \cdot 14H_2O$ with a rate of 4°C·min^{-1} up to 170°C and is kinetically stable (Hartmann et al. 2022a,b). The whole preparation process of the second modification was completed in an Ar-filled glove box because of the instability for Na_2S in the air (Gao et al. 2021b). Their reaction process was achieved in the vacuum-sealed silica tube by the high-temperature muffle furnace. It was successfully synthesized under the sintering temperature at 500°C for several days and then cooled to room temperature at the rate of 5°C·h^{-1}. Pale yellow crystals, which are relatively stable in the air, were obtained.

A.6.2 Tin–Potassium–Sulfur

One more ternary phase, $K_{2x}Sn_{4-x}S_{8-x}$ ($x = 0.65$-1), which crystallizes as a monoclinic structure with the lattice parameters $a = 1309.2 \pm 0.3$, $b = 1688.2 \pm 0.2$, $c = 737.48 \pm 0.13$ pm, $\beta = 98.100 \pm 0.015°$, a calculated density of 2.722 g·cm^{-3}, and an energy gap of 2.38 eV for x = 1 with orthorhombic subcell (a = 368.31 ± 0.02, b = 2588.77 ± 0.19, c = 1681.55 ± 0.11 pm, and a calculated density of 2.729 g·cm^{-3}), is formed in the Sn–K–S system (Sarma et al. 2016). This phase remains stable up to 525°C, after which it starts to decompose into $K_2Sn_2S_5$ and SnS_2. For its preparation, K_2CO_3 (6 mM), Sn (9 mM), S (30 mM) were taken in a 23 mL polytetrafluoroethylene-lined stainless steel autoclave and deionized water (0.5 mL) was added dropwise until the mixture acquired dough-like consistency. The autoclave was sealed properly and maintained in a preheated oven at 220°C for 15 h under autogenous pressure. Then, the autoclave was allowed to cool to room temperature. The product was found to contain yellow rod-shaped crystals along with yellow polycrystalline powder. It was isolated by filtration, washed several times with water, and acetone and dried under vacuum. The product obtained is air and moisture stable.

According to Li et al. (2020d), the ternary compound $K_2Sn_2S_5$, which is formed in this system, also crystallizes as a monoclinic structure with the lattice parameters a = 1110.0 ± 0.5, b = 781.9 ± 0.4, c = 1154.3 ± 0.6 pm, β = 108.248 ± 0.005°, and a calculated density of 3.222 g·cm^{-3}. This compound can be synthesized at 150°C by mixing SnS (15.0 mg), K_2CO_3 (15.0 mg), about 500 mg mixed solvent of pyridine/thiophenol (volume ratio 3:1), and 150 mg of 1,3-diaminopropane in a Pyrex glass tube for 8 days.

A.6.3 Tin–Cesium–Sulfur

The ternary compound $Cs_4Sn_5S_{12}$, which crystallizes as an orthorhombic structure with the lattice parameters a = 1423.2 ± 0.4, b = 1316.1 ± 0.4, c = 1637.2 ± 0.5 pm, a calculated density of 3.270 g·cm^{-3}, and an energy gap of 2.3 eV, is formed in the Sn–Cs–S system (Baiyin et al. 2015). It was synthesized under solvothermal conditions.

A.6.4 Tin–Copper–Sulfur

Two more ternary compounds, $Cu_7Sn_3S_{10}$ and $Cu_{22}Sn_{10}S_{32}$, are formed in the Sn–Cu–S system. The first compound crystallizes as a tetragonal structure with lattice parameters a = 541.64 ± 0.03, c = 1083.2 ± 0.1 pm, and a calculated density of 4.65 g·cm^{-3} for $Cu_{5.64}Sn_{2.36}S_8$ composition (Deng et al. 2020). $Cu_7Sn_3S_{10}$ is stable up to 480°C and undergoes phase transformation at around 180°C. This phase transformation is expected to be caused by the Cu order–disorder transition rather than the change of structure symmetry. This compound was prepared using CuS and SnS powders, which were mixed in agate mortar by hands with different mole ratios. These mixtures were loaded into the graphite die. After loading each mixture into the graphite die, a pressure of 5 MPa was stressed on the powders to flatten the surface. After all the mixtures were loaded, they were sintered at 430°C for 15 min under the pressure of 60 MPa by spark plasma sintering. The sintered bulk samples were annealed at 400°C for 3 days to promote the reaction between CuS and SnS forming the target compound.

$Cu_{22}Sn_{10}S_{32}$ crystallizes as a cubic structure with the lattice parameter a = 1089.25 ± 0.18 pm and a calculated density of 4.639 g·cm^{-3} (Pavan Kumar et al. 2021b). It is a degenerate semiconductor. All sample preparation and handling of powder were performed in an Ar-filled glove box with oxygen contents of <1 ppm. Stoichiometric amounts of Cu, Sn, and S were loaded into a 25 mL tungsten carbide jar containing seven balls of 10 mm under an Ar atmosphere. High-energy ball-milling was performed in a planetary ball mill operating at room temperature at a disk rotation speed of 600 rpm for 6 h.

According to Pavan Kumar et al. (2021a), the ternary compounds Cu_2SnS_3 and $Cu_5Sn_2S_7$ crystallize in the monoclinic structure with the lattice parameters a = 665.6 ± 0.1, b = 1152.5 ± 0.2, c = 665.8 ± 0.1 pm, and β = 109.52 ± 0.01° for the first compound and a = 1205.80 ± 0.05, b = 540.63 ± 0.02, c = 805.26 ± 0.03 pm, and β = 98.163 ± 0.003° for the second one. Cu_2SnS_3 exhibits a degenerate semiconducting behavior. The polycrystalline sample of this compound was synthesized by a two-step solid state reaction in evacuated silica tube from Cu, Sn, and S. First, they were ground in an agate mortar in the appropriate stoichiometric ratio and compacted into a pellet, sealed in an evacuated silica tube, and heated to 750°C at a rate of 50°C·h^{-1}. The pellet was maintained at that temperature for 48 h followed by a cooling step down to room temperature at the same rate. Subsequently, the as-prepared Cu_2SnS_3 sample was crushed and the powder was compacted for a second firing carried out at 600°C for 24 h with an intermediate step at 400°C for 12 h. The same heating and cooling rates were used.

Powders of $Cu_5Sn_2S_7$ were synthesized by mechanical alloying (Pavan Kumar et al. 2021a). All sample preparations and handling of powders were performed in an argon-filled glove box with oxygen content <1 ppm. Stoichiometric amounts of Cu, Sn, and S were loaded in a 25 mL tungsten carbide jar containing seven balls of 10 mm under an argon atmosphere. High-energy ball-milling was performed in a planetary ball mill operating at room temperature at a disc rotation speed of 600 rpm for 6 h.

Cu_3SnS_4 crystallizes as a tetragonal structure with the lattice parameters a =539.28 ±0.29, c =1075.8 ±0.6 pm, and a calculated density of 4.645 g·cm^{-3} (Chen et al. 2020b).

According to Bourgès et al. (2015), $Cu_4Sn_7S_{16}$ crystallizes as a rhombohedral structure with the lattice parameters (in hexagonal setting) a = 735.0 ± 0.1 and c = 3606.2 ± 1.0 pm for as-ball-milled powder; a = 737.3 ± 0.1 and c = 3602.6 ± 0.1 pm for mechanical alloying and spark plasma sintering at 700°C; a = 737.4 ± 0.1 and c = 3602.8 ± 0.1 pm for mechanical alloying and spark plasma sintering at 600°C; and a = 736.7 ± 0.1 and c = 3599.0 ± 0.1 pm for sealed tube synthesis and spark plasma sintering at 700°C. The title compound was synthesized in two different ways. It was first prepared directly from the pure elements in a sealed fused silica tube using three dwell temperatures. Firstly, the container was heated at 450°C for 2 h to initiate sulfur reaction and prevent from tube blasting during the heating ramp. Secondly, it was heated up to 850°C for 2 h to reach the liquid state. It was then cooled down to 700°C and maintained for 48 h to crystallize the phase. The powder was then ground, pelletized, and heated again at 700°C for 72 h, in order to complete the crystallization and eliminate secondary phases. A rate of 100°C·h^{-1} was used for both heating and cooling ramps.

As a second synthesis approach, $Cu_4Sn_7S_{16}$ was obtained by mechanical alloying. A 7.5 g of pure elements in stoichiometric ratio was ball-milled in a WC jar of 45 ml containing 14 WC balls of 10 mm diameter at a speed of 600 rpm for 36 h. The XRD pattern of the ball-milled powder confirmed the formation of a single-phase sample. Both batches of powders were then compacted by spark plasma sintering using a dwell temperature of 600°C or 700°C during 30 min under a uniaxial pressure of 64 MPa. The atmosphere was under a slight argon over-pressure of 50 hPa.

A.6.5 Tin–Silver–Sulfur

According to Nagaoka et al. (2021), Ag_2SnS_3 melts at 645.5°C and one of its modification crystallizes as a monoclinic structure with the lattice parameters a = 659.1, b = 1152.6, c = 1306.4 pm, and β = 98.4 ± 0.7° [a = 663.23 ± 0.11, b = 1146.26 ± 0.13, c = 1323.81 ± 0.23 pm, β = 98.008 ± 0.014°, a calculated density of 5.7703 ± 0.0003 g·cm^{-3}, and an energy gap of 1.26 eV at room temperature and 1.37 eV at 77 K (Fedorchuk et al. 2012)]. The sample of the Ag_2SnS_3 composition was synthesized by co-melting calculated amounts of high-purity elements in a vacuum-sealed quartz ampoule (Fedorchuk et al. 2012). The preliminary stage was the synthesis in the oxygen-gas burner flame to complete bonding of the elementary sulfur under visual control. The ampoule was then heated in a furnace at a rate of 10°C·h^{-1} to 750°C, kept at this temperature for 6 h for the homogenization of the melt, and cooled at the same rate to 400°C. The sample was annealed at this temperature for 300 h, followed by quenching into cold water.

Slade et al. (2021) noted that the ternary compound Ag_8SnS_6 at low temperatures adopts two different orthorhombic structures with the lattice parameters a = 1529.93 ± 0.09, b = 754.79 ± 0.04, c = 1070.45 ± 0.06 pm, and a calculated density of 6.308 g·cm^{-3} at room temperature [a = 1533.38 ± 0.09, b = 756.20 ± 0.02, and c = 1072.44 ± 0.07 pm (Aliyeva et al. 2014)]; a = 1529.5 ± 0.2, b = 755.4 ± 0.1, c = 1068.5 ± 0.1 pm, and a calculated density of 6.317 g·cm^{-3} at 160 K; a = 1527.24 ± 0.09, b = 751.97 ± 0.05, c = 1065.69 ± 0.07 pm, and a calculated density of 6.372 g·cm^{-3} at 90 K; and a = 766.29 ± 0.05, b = 753.96 ± 0.05, c = 1063.00 ± 0.06 pm, and a calculated density of 6.349 g·cm^{-3} at 90 K. High-temperature modification of this compound crystallizes as a cubic structure with the lattice parameter a = 1085.83 ± 0.05 pm (Aliyeva et al. 2014).

On cooling from high temperature, Ag_8SnS_6 undergoes a cubic to orthorhombic phase transition at ≈190°C (Slade et al. 2021). The second structural transition exists at ≈120 K. This transition is first order and has a large thermal hysteresis.

The DSC data were used to determine the enthalpy and entropy of the phase transitions of Ag_8SnS_6 from the orthorhombic to the cubic modification at 173°C (Bayramova et al. 2022c). It was shown that $\Delta H_{tr} = 8.77 \pm 0.35$ kJ·M^{-1} and $\Delta S_{tr} = 19.66 \pm 0.79$ J·(M·K)$^{-1}$.

A.6.6 Tin–Strontium–Sulfur

According to Li et al. (2020b), the ternary compound Sr_2SnS_5, which exists in the Sn–Sr–S system, crystallizes as a monoclinic structure with the lattice parameters $a = 1591.71 \pm 0.03$, $b = 1154.73 \pm 0.02$, $c = 413.040 \pm 0.010$ pm, $\beta = 89.9998 \pm 0.0017°$, and an energy gap of 2.38 eV. To synthesize this compound, the mixture of SrS, Sn, S, and K_2S with a total mass of 0.5 g and a molar ratio of 2:1:4.5:0.6 was weighted and fully mixed before put into small graphite crucibles, and then were flame-sealed in high vacuum (10^{-3} Pa) in silica tube. The tube was put into a tube furnace, heated to 830°C in 60 h, maintained at 830° for 100 h, and then cooled down to 430°C in 100 h. In order to remove potassium polysulfides, the products were washed by DMF and then methanol thoroughly and dried at 80°C. Finally, orange yellow rod-like crystals of Sr_2SnS_5 were obtained after purification. The crystals can grow up to several millimeters. They are stable in the air for almost a year.

A.6.7 Tin–Barium–Sulfur

One more ternary compound, Ba_2SnS_5, which crystallizes as an orthorhombic structure with the lattice parameters $a = 1655 \pm 2$, $b = 1190 \pm 2$, $c = 438 \pm 10$ pm, and an energy gap of 2.38 eV, is formed in the Sn–Ba–S system (Li et al. 2020b). It was prepared in the same way as Sr_2SnS_5 was synthesized using BaS instead of SrS. Bright yellow rod-like crystals of this compound are also stable in the air for almost a year.

According to Gunatilleke et al. (2022), the ternary compound $BaSnS_2$, which exists in this system, crystallizes as a monoclinic structure with the lattice parameters $a = 609.21 \pm 0.13$, $b = 1216.17 \pm 0.19$, $c = 624.62 \pm 0.14$ pm, $\beta = 97.041 \pm 0.004°$, a calculated density of 4.6298 g·cm^{-3}, and an energy gap of 2.4 eV. This compound is stable up to 652°C but powder XRD of $BaSnS_2$ after heating to 530°C revealed $Ba_3Sn_2S_7$ and $BaSn_3S_2$ in addition to the target compound, indicating its partial decomposition above this temperature. To obtain this compound, Ba pieces, Sn powder, and ground S flakes were weighed in a 1:1:2 stoichiometric ratios inside a glove box under a nitrogen atmosphere. The elements were mixed by hand before being loaded onto a quartz crucible and sealed in an evacuated silica ampoule. The mixture was heated up to 800°C at a rate of 15°C·h^{-1}, maintained at this temperature for 30 h, and cooled to 500°C at a rate 2°C·h^{-1}, followed by cooling to room temperature at 10°C·h^{-1}.

A.6.8 Tin–Gallium–Sulfur

The ternary compound $Ga_2Sn_2S_5$, which exists in this system, melts congruently at 684°C (Li et al. 2019c) and crystallizes as an orthorhombic structure with the lattice parameters $a = 1241.87 \pm 0.07$, $b = 621.85 \pm 0.03$, $c = 1089.25 \pm 0.06$ pm, a calculated density of 4.241 g·cm^{-3}, and an energy gap of 2.02 eV (Shi et al. 2019b) [$a = 1242.8 \pm 0.5$, $b = 622.6 \pm 0.3$, $c = 1089.4 \pm 0.4$ pm, a calculated density of 4.23 g·cm^{-3}, and an energy gap of 2.14 eV (Li et al. 2019c)].

Single crystals of this compound were obtained by a solid state reaction (Shi et al. 2019b). The starting materials were Ga, S, and SnS. The sample had a total mass of 500 mg and the molar ratios of Sn/Ga/S = 2:2:5. The mixture was ground into fine powder in an agate mortar and pressed into a pellet, followed by its loading into a quartz tube. The tube was evacuated to a pressure of 0.01 Pa and flame-sealed. The sample was placed into a muffle furnace, heated from room temperature to 950°C in one day, homogenized for 5 days, and several intermediate equilibrated temperatures were set to homogenize the reactive system and prevent the breaking of the quartz tube. The red block crystals of the title compound were obtained, which are stable in moisture and air. The crystals of $Ga_2Sn_2S_5$ can be separated from the by-products easily under an optical microscope because of their unique color and shape.

The crystals of $Ga_2Sn_2S_5$ could be also obtained by a facile mid-temperature fluxing method using $SnCl_2$ as the flux (Li et al. 2019c). Typically, a graphite crucible was fitted with Sn (2 mM), Ga (2 mM),

S (5 mM), and $SnCl_2$ (2.8 mM) and then put inside a silica tube that was evacuated to 10^{-3} Pa and sealed. The tube was heated in a computer-controlled furnace to 450°C at a rate of 5°C·h^{-1}, the temperature was maintained for 72 h, and cooled to 200°C for 2 h, and the temperature of the furnace was shut off. The products were cleaned with distilled water to remove excess $SnCl_2$ and then dried by ethanol. Finally, red crystals of $Ga_2Sn_2S_5$ were obtained.

A.6.9 Tin–Lanthanum–Sulfur

One more ternary compound, $La_3Sn_{1.25}S_7$, which crystallizes as a hexagonal structure with the lattice parameters $a = 1027.70 \pm 0.15$, $c = 600.30 \pm 0.12$ pm, and a calculated density of 4.775 g·cm^{-3}, is formed in the Sn–La–S system (Zeng et al. 2008). This compound was obtained as follows. The mixture of La_2S_3, Na_2S, and SnS_2 (molar ratio 0.768:0.256:0.512) were mixed together and ground thoroughly within a N_2-filled glove box, pressed into pellets, and then sealed in evacuated quartz ampoules before being heated in resistant furnace to 750°C in 144 h, maintained at this temperature for 80 h, then heated to 1000°C in 250 h, annealed at this temperature for 144 h, and the furnace was turned off. The precursor obtained from the first-step solid state reaction was mixed with NaBr (0.80 g). Upon regrinding, repelleting, and resealing, the precursor/flux mixture was heated to 830°C in 64 h, maintained at this temperature for 212 h, then slowly cooled to 700°C in 65 h, and finally to room temperature by switching off the furnace powers. Transparent dark red crystals of the target compound were isolated manually from the residue after the flux was removed by washing with distilled water.

A.6.10 Tin–Ytterbium–Sulfur

SnS_2–YbS. The phase diagram of this system, constructed through DTA, XRD, metallography and measurement of microhardness, is presented in Figure A.6.1 (Aliyev and Gasymova 2009). The eutectic contains 20 mol% YbS and crystallizes at 585°C. $YbSnS_3$, $Yb_3Sn_2S_7$, $Yb_5Sn_3S_{12}$, and Yb_2SnS_4, which form in this system, melt incongruently at 752°C, 877°C, 977°C, and 1067°C, respectively. $YbSnS_3$ crystallizes as a triclinic structure with the lattice parameters $a = 830$, $b = 842$, $c = 1180$ pm, $\alpha = 111.30°$, $\beta = 99.70°$, $\gamma = 74.6°$, and the calculated and experimental density of 6.30 and 6.24 g·cm^{-3}, respectively. The rest

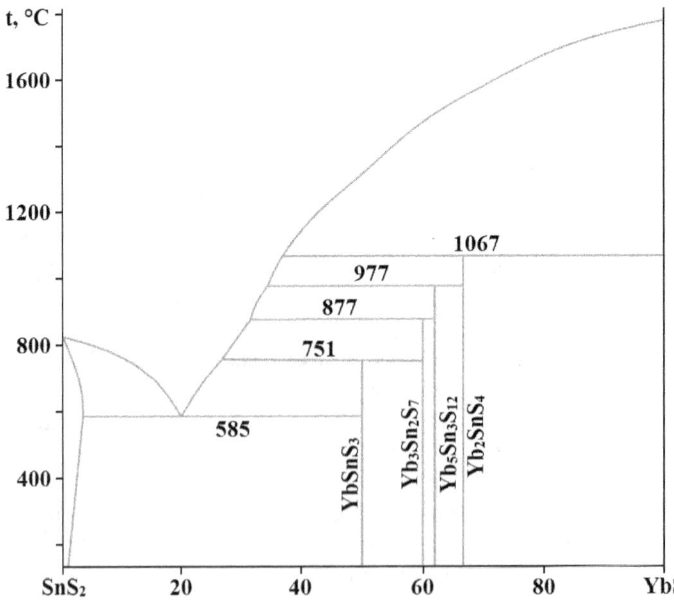

FIGURE A.6.1 Phase diagram of the SnS_2–YbS quasibinary system. (From Aliyev, O.M., and Gasymova, S.A., *Chem. Probl.*, (2), 387, 2009.) Open access.

of the four compounds crystallize in the orthorhombic structure with the lattice parameters a = 1140, b = 1264, c = 330 pm, and the calculated and experimental density of 5.30 and 5.30 g·cm^{-3}, respectively, for $Yb_3Sn_2S_7$, a = 1520, b = 886, c = 1128 pm, and the calculated and experimental density of 4.86 and 4.48 g·cm^{-3}, respectively for Yb_2SnS_4, a = 404, b = 1560, c = 1147 pm, and the calculated and experimental density of 6.10 and 5.48 g·cm^{-3}, respectively, for $Yb_4Sn_2S_9$, a = 340, b = 1145, c = 2020 pm, and the calculated and experimental density of 5.46 and 5.40 g·cm^{-3}, respectively for $Yb_5Sn_3S_{12}$. The alloys for the investigations were annealed for 185–200 h at 430°C (0–50 mol% YbS) and at 730°C (50–100 mol% YbS).

It should be noted that this system could not be quasibinary since the compositions of the $Yb_4Sn_2S_9$ and $Yb_5Sn_3S_{12}$ ternary compounds are shifted to sulfur corner of the Sn–Yb–S ternary system. Therefore, equilibria in the SnS_2–YbS system require reinvestigation.

A.6.11 Tin–Phosphorus–Sulfur

According to Shi et al. (2021), the ternary compound $SnPS_3$ or $Sn_2P_2S_6$, which exists in this system, crystallizes as a monoclinic structure with the lattice parameters a = 652.5 ± 0.3, b = 748.1 ± 0.3, c = 936.4 ± 0.4 pm, β = 91.19 ± 0.02°, a calculated density of 3.574 g·cm^{-3}, and an energy gap of 2.35 eV. Single crystals of this compound were synthesized by using Sn, SnS, and P_2S_5 (molar ratio 1:1:1). The mixture was loaded into a fused silica tube under an Ar atmosphere in a glove box, which was sealed under a vacuum of 0.01 Pa and then placed into a muffle furnace. The reactants were heated to 950°C within 25 h, kept at the temperature for 48 h, and then cooled to 400°C at a rate of 5°C·h^{-1} before the furnace was switched off. During this heating process, several intermediate equilibrated temperatures were set for safety. Black crystals stable in air and water were obtained.

A.6.12 Tin–Bromine–Sulfur

According to Li et al. (2022), the ternary compound $Sn_7S_2Br_{10}$, which exists in this system, crystallizes as a hexagonal structure with the lattice parameters a =1213.20 ± 0.09, c =437.65 ± 0.06 pm, a calculated density of 5.043 g·cm^{-3} at 100 K, and an energy gap of 2.56 eV. The title compound was obtained by solid state reaction. A 500 mg mixture of $SnBr_2$ and SnS at a molar ratio of 5:1 was ground into uniform powder using an agate mortar, and then pressed into a pellet, which was then sealed into an evacuated quartz ampoule to a vacuum pressure of 0.01 Pa. The tube was heated in a muffle furnace from room temperature to 950°C in 26 h and homogenized at several intermediate temperatures for several hours. The reaction was maintained at 950°C for about 48 h and then cooled down to 400°C within 110 h. Finally, the tube was cooled down to room temperature naturally and yellow single crystals were obtained.

A.6.13 Tin–Cobalt–Sulfur

The $Co_3Sn_2S_2$ ternary compound, which exists in the Sn–Co–S system, crystallizes as a rhombohedral structure with the lattice parameters a = 759.09 ± 0.03 pm and α = 89.919 ± 0.001° or a = 536.38 ± 0.04 and c = 1316.6 ± 0.1 pm in hexagonal settings (Rothballer et al. 2014) [a = 536.79 ± 0.03 and c = 1317.65 ± 0.06 pm in hexagonal settings (Weihrich and Anusca 2006)]. The preparation of this compound was done from the elements (Weihrich and Anusca 2006). Sealed quartz ampoule was heated to 920°C and cooled down within 7 days.

A.7 Systems Based on Tin Selenides

A.7.1 Tin–Sodium–Selenium

The ternary compound Na_4SnSe_4, which is formed in this system, crystallizes as a tetragonal structure with lattice parameters a = 816.74 ± 0.07, c = 726.65 ± 0.13 pm, a calculated density of 3.607 g·cm^{-3}, and an energy gap of 2.17 eV (Gao et al. 2021b). The preparation process of this compound was completed in an Ar-filled glove box because of the instability for Na_2Se in the air. Their reaction process was achieved in the vacuum-sealed silica tube by the high-temperature muffle furnace. It was successfully synthesized under the sintering temperature at 500°C for several days and then cooled to room temperature at a rate of 5°C·h^{-1}. Deep-yellow crystals of Na_4SnSe_4, which are hydroscopic and dissolve in water, were found in the silica tube.

A.7.2 Tin–Potassium–Selenium

The ternary compound $K_4Sn_3Se_8$, which is formed in this system, crystallizes as an orthorhombic structure with the lattice parameters a = 821.4, b = 2779.45 ±0.11, c =1875.11 ± 0.12 pm, and a calculated density of 4.053 g·cm^{-3} (Yao et al. 2010). This compound was prepared using $AgNO_3$ (0.15 mM), SnSe (0.05 mM), Se (0.20 mM), K_2CO_3 (0.10 mM), about 0.3 mL of pyridine, and 0.2 mL of triethylamine. The mixture was sealed in a Pyrex glass tube with ca. 10% filling at air atmosphere, placed in a stainless steel autoclave, and then heated at 170°C for 7 days. After cooling naturally to ambient temperature, the products were washed with ethanol and water, respectively, and yellow square crystals of the title compound were obtained.

A.7.3 Tin–Copper–Selenium

Using XRD and DTA, it was established that only one ternary compound, Cu_2SnSe_3, is formed in the Sn–Cu–Se system (Chorba et al. 2021). The formation of the Cu_2SnSe_4 compound has not been confirmed. According to the results of phase analysis in combination with the literature data, the triangulation of this system was carried out at 170°C (Figure A.7.1). The Cu_2Se–SnSe, Cu_2Se–$SnSe_2$, Cu_2SnSe_3–Se, Cu_2SnSe_3–$CuSe_2$, CuSe–SnSe, Cu_6Sn_5–SnSe, Cu_3Sn–SnSe, and Cu_3Sn–Cu_2Se sections are quasibinary.

Cu_2SnSe_3 has some polymorphic modifications. One of them crystallizes as a cubic structure with the lattice parameter a = 556.90 pm (Klymovych et al. 2013). The second modification (mineral okruginite) crystallizes as a monoclinic structure with the lattice parameters a = 699.06 ± 0.02, b = 1207.12 ± 0.03, c = 697.23 ± 0.02 pm, and β = 109.353 ± 0.010° (Vymazalová et al. 2023a,b). Nanocrystals of this compound crystallizes as a hexagonal structure with the lattice parameters a = 397.72 ± 0.06, c = 665.5 ± 0.2 pm, and an energy gap of 1.7 eV (Norako et al. 2012).

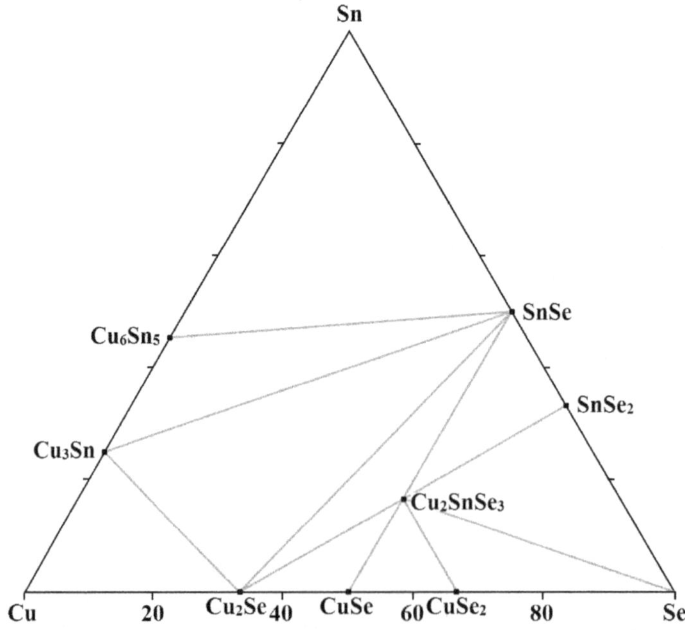

FIGURE A.7.1 Isothermal section of the Sn–Cu–Se ternary system at 170°C. (From Chorba, O.Y., et al., *Nauk. Visn. Uzhgorod. Univ., Ser. Khim.*, [2(46)], 22, 2021.) Open access.

Nanocrystal syntheses of Cu_2SnSe_3 were performed under N_2, in the absence of H_2O and O_2, using standard Schlenk techniques (Norako et al. 2012). In a typical synthesis, SnI_4 (0.200 mM) and CuCl (0.404 mM) were added to a two-neck round-bottom flask fitted with a reflux condenser, stir bar, and rubber septum and heated (10°C·min^{-1}) to 40°C. Dodecylamine (0.56 mL, 2.4 mM) and 1-dodecanethiol (0.10 mL, 0.42 mM) were added to the reaction flask via syringes and then cycled between vacuum and nitrogen three times. Then, the system was heated (10°C·min^{-1}) to 95°C, and was again cycled three times between vacuum and nitrogen to eliminate adventitious water and dissolved oxygen. The temperature was then increased (10°C·min^{-1}) to 180°C and di-*tert*-butyldiselenide (0.11 mL, 0.56 mM) was quickly injected into the system under flowing nitrogen, and allowed to react for 5 min with stirring. After being cooled to room temperature, the reaction mixture was dissolved in 5 mL of toluene and precipitated with 10 mL of ethanol, sonicated, and centrifuged (6000 rpm for 15 min) to yield a black solid. Dispersion/precipitation was repeated three times with toluene (1 mL) and ethanol (4 mL) to yield the as-prepared product.

A.7.4 Tin–Silver–Selenium

The isothermal sections of this ternary system at 250°C, 400°C, and 550°C were constructed by Zobac et al. (2022a) through DTA, XRD, and SEM-EDX (Figure A.7.2). Due to the low melting point of Sn and Se, two liquid phases were observed in this section at 250°C (Figure A.7.2a). Five binary phases, Ag_2Se, ζ-Ag_4Sn, ε-Ag_3Sn, SnSe and $SnSe_2$, are stable. All these binary phases have negligible solubility of the third element. Two ternary phases, $AgSnSe_2$ and Ag_8SnSe_6, are also stable at 250°C. $AgSnSe_2$ is a line compound, which contains a constant amount of Se and lies between $Ag_{24}Sn_{26}Se_{50}$ and $Ag_{22}Sn_{28}Se_{50}$. The ternary phase Ag_8SnSe_6 is almost a line compound with a homogeneity range between the approximate compositions $Ag_{52.5}Sn_{8.2}Se_{39.3}$ and $Ag_{35.5}Sn_{10.3}Se_{54.2}$. Ag_8SnSe_6 was exclusively found in the orthorhombic low-temperature form.

Two independent liquid phases and the same binary phases exist at 400°C (Figure A.7.2b). The binary phases have again very limited solubility of the third element. The two ternary phases $AgSnSe_2$ and Ag_8SnSe_6 are stable at this temperature. $AgSnSe_2$ is a line compound with a homogeneity range between

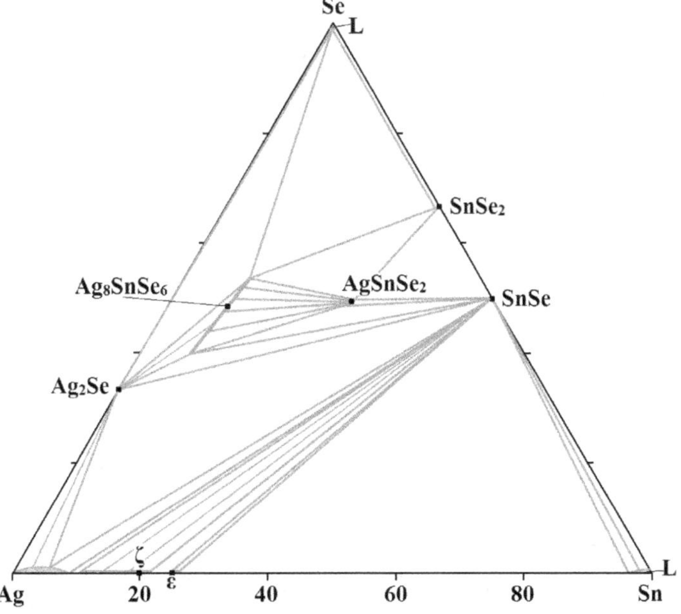

FIGURE A.7.2 Isothermal section of the Sn–Ag–Se ternary system at (a) 250, (b) 400°C, and (c) 550°C. (From Zobac, O., et al., *J. Phase Equilib. Diffus.*, 43(1), 32, 2022.)

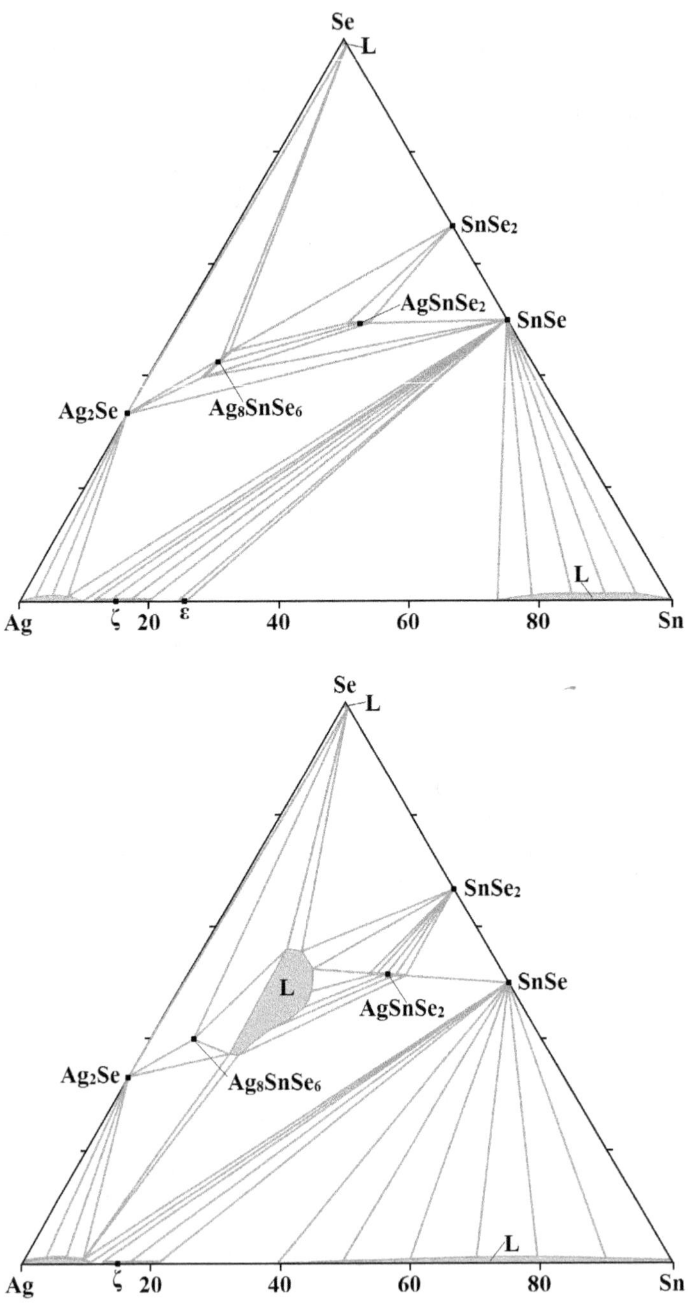

FIGURE A.7.2 Continued

$Ag_{25.2}Sn_{25.5}Se_{49.3}$ and $Ag_{21.9}Sn_{28.3}Se_{49.8}$, a slightly larger than the homogeneity range at 250°C. The ternary phase Ag_8SnSe_6 is almost a line compound with a homogeneity range between the approximate compositions $Ag_{52}Sn_8Se_{40}$ and $Ag_{45}Sn_{10}Se_{45}$, which is less pronounced than at 250°C.

Like the sections at 250°C and 400°C, two independent liquid phases exist at 550°C (Figure A.7.2c). An additional liquid region was found in the central part of the phase diagram. Four binary phases, Ag_2Se, ζ-Ag_4Sn, SnSe and $SnSe_2$, are stable. These binary phases have almost no solubility of third elements in their structures. Two ternary phases, $AgSnSe_2$ and Ag_8SnSe_6, are stable at 550°C. $AgSnSe_2$ exists between $Ag_{22.2}Sn_{28.2}Se_{49.6}$ and $Ag_{15.8}Sn_{33.4}Se_{50.7}$. This homogeneity range is wider in comparison to lower temperatures. The ternary phase Ag_8SnSe_6 is almost stoichiometric at 550°C.

Multi 3D visualization of the Ag_2Se, SnSe, $SnSe_2$, and Ag_8SnSe_6 crystallization surfaces and immiscibility surfaces in the Sn–Ag–Se system was carried out by Ibrahimova (2019a).

According to the data of Olekseyuk et al. (2009), the Ag_2SnSe_3 ternary compound is not formed in the $SnSe_2$–Ag_2Se system.

The SnSe–Ag_2Se–Se subsystem was studied by measuring EMF of concentration cells in the temperature range of 27°C–177°C (Alverdiev et al. 2019). The temperature of Ag_8SnSe_6 polymorphous transition (82°C) was determined and the partial molar functions of Ag in certain phase regions of this system were calculated. Standard thermodynamic functions of formation and standard entropies were estimated for ternary phases $AgSnSe_2$, $Ag_{0.84}Sn_{1.16}Se_2$, and two modifications of Ag_8SnSe_6 and also thermodynamic functions of Ag_8SnSe_6 polymorphous transition. The following values were obtained: $\Delta G^0_{f,\,298} = -144.1 \pm 2.6$ kJ·M^{-1}, $\Delta H^0_{f,\,298} = -145.0 \pm 4.8$ kJ·M^{-1}, and $S^0_{298} = 172.8 \pm 4.1$ J·(M·K)$^{-1}$ [$\Delta G^0_{f,\,298} = -133.9 \pm 1.6$ kJ·M^{-1}, $\Delta H^0_{f,298} = -124.9 \pm 1.3$ kJ·M^{-1}, and $\Delta S^0_{f,\,298} = 30.1 \pm 2.5$ J·(M·K)$^{-1}$ (Moroz and Prokhorenko 2015)] for $AgSnSe_2$; $\Delta G^0_{f,\,298} = -150.0 \pm 2.3$ kJ·M^{-1}, $\Delta H^0_{f,\,298} = -152.0 \pm 4.7$ kJ·M^{-1}, and $S^0_{298} = 182.7 \pm 6.0$ J·(M·K)$^{-1}$ for $Ag_{0.84}Sn_{1.16}Se_2$; $\Delta G^0_{f,\,298} = -335.3 \pm 2.9$ kJ·M^{-1}, $\Delta H^0_{f,\,298} = -320.4 \pm 6.4$ kJ·M^{-1}, and $S^0_{298} = 695.5 \pm 10.5$ J·(M·K)$^{-1}$ [$\Delta G^0_{f,\,298} = -352.5 \pm 1.9$ kJ·M^{-1}, $\Delta H^0_{f,\,298} = -323.1 \pm 1.6$ kJ·M^{-1}, and $\Delta S^0_{f,\,298} = 98.6 \pm 3.1$ J·(M·K)$^{-1}$ (Moroz and Prokhorenko 2015)] for α-Ag_8SnSe_6; and $\Delta G^0_{f,\,400} = -342.4 \pm 3.2$ kJ·M^{-1}, $\Delta H^0_{f,\,298} = -305.0 \pm 6.8$ kJ·M^{-1}, and $S^0_{298} = 738.8 \pm 10.6$ J·(M·K)$^{-1}$ for α-Ag_8SnSe_6. The DSC data were used to determine the enthalpy and entropy of the phase transitions of Ag_8SnS_6 from the orthorhombic to the cubic modification at 82°C (Bayramova et al. 2022c). It was shown that $\Delta H_{tr} = 19.67 \pm 0.60$ kJ·M^{-1} and $\Delta S_{tr} = 55.41 \pm 2.22$ J·(M·K)$^{-1}$ (Bayramova et al. 2022c) [$\Delta H_{tr} = 15.4 \pm 4.3$ kJ·M^{-1} and $\Delta S_{tr} = 43.4 \pm 12.1$ J·(M·K)$^{-1}$ (Alverdiev et al. 2019)].

A.7.5 Tin–Lanthanum–Selenium

According to the Kelly method, through summing up specified values of ion entropies, the values of standard entropy and formation entropy of La_2SnSe_4 were specified (Mamedov and Kulieva 2012). According to values of enthalpy and free energy of the formation of SnSe and La_2Se_3 when adjusted for the deviation from additiveness, enthalpy, and free energy of the formation of ternary compound were determined: $\Delta G^0_{f,\,298} = -1083.3$ kJ·M^{-1}, $\Delta H^0_{f,\,298} = -1094.6$ kJ·M^{-1}, $\Delta S^0_{f,\,298} = -37.95$ J·(M·K)$^{-1}$, and $S^0_{298} = 288.45$ J·(M·K)$^{-1}$.

A.7.6 Tin–Cerium–Selenium

According to the Kelly method, through summing up specified values of ion entropies, the values of standard entropy and formation entropy of Ce_2SnSe_4 were specified (Mamedov and Kulieva 2012). According to values of enthalpy and free energy of the formation of SnSe and Ce_2Se_3 when adjusted for the deviation from additiveness, enthalpy and free energy of the formation of ternary compound were determined: $\Delta G^0_{f,\,298} = -1081.0$ kJ·M^{-1}, $\Delta H^0_{f,\,298} = -1094.2$ kJ·M^{-1}, $\Delta S^0_{f,\,298} = -44.35$ J·(M·K)$^{-1}$, and $S^0_{298} = 307.25$ J·(M·K)$^{-1}$.

A.7.7 Tin–Praseodymium–Selenium

According to the Kelly method, through summing up specified values of ion entropies, the values of standard entropy and formation entropy of Pr_2SnSe_4 were specified (Mamedov and Kulieva 2012) According to values of enthalpy and free energy of the formation of SnSe and Pr_2Se_3 when adjusted for the deviation from additiveness, enthalpy and free energy of the formation of ternary compound were determined: $\Delta G^0_{f,\,298} = -1092.7$ kJ·M^{-1}, $\Delta H^0_{f,\,298} = -1103.4$ kJ·M^{-1}, $\Delta S^0_{f,\,298} = -35.95$ J·(M·K)$^{-1}$, and $S^0_{298} = 324.65$ J·(M·K)$^{-1}$.

A.7.8 Tin–Neodymium–Selenium

According to the Kelly method, through summing up specified values of ion entropies, the values of standard entropy and formation entropy of Nd_2SnSe_4 were specified (Mamedov and Kulieva 2012). According to values of enthalpy and free energy of the formation of SnSe and Nd_2Se_3 when adjusted for the deviation from additiveness, enthalpy and free energy of the formation of ternary compound were determined: $\Delta G^0_{f,\,298} = -1090.1$ kJ·M^{-1}, $\Delta H^0_{f,\,298} = -1103.2$ kJ·M^{-1}, $\Delta S^0_{f,\,298} = -44.35$ J·(M·K)$^{-1}$, and $S^0_{298} = 310.45$ J·(M·K)$^{-1}$.

A.7.9 Tin–Dysprosium–Selenium

SnSe–DySe. The phase diagram of this system, constructed through DTA, XRD, and measurement of microhardness, is presented in Figure A.7.3 (Huseynov et al. 2009). The eutectic contains 12 mol% DySe and crystallizes at 780°C. The $DySnSe_2$ ternary compound, which melts incongruently at 880°C, is formed in the system. At room temperature, SnSe dissolves 3.5 mol% DySe.

A.7.10 Tin–Lead–Selenium

The calculated isothermal sections of this ternary system at 350°C and 500°C (Figure A.7.4) (Zobac et al. 2022b) agrees reasonably with the experimental sections from Chen et al. (2020c), especially in the region with low selenium content. No ternary compound was found. The phase relationships are similar at 350°C and 500°C. At both temperatures, the binary compounds, PbSe and SnSe, have very significant ternary solubility, but the ternary solubility of Pb in $SnSe_2$ is negligible. There are still unresolved questions about the type and shape of the liquid phase at the Se corner.

The calculated liquidus surface of the Sn–Pb–Te system was also constructed and the calculated invariant reactions containing liquid phases in the system were defined (Zobac et al. 2022b). All key features of the experimental liquidus surface are in good qualitative agreement with calculated one. Theoretical calculations suggest the existence of a miscibility gap in the Se-rich corner up to high temperatures, but the equilibria at high temperatures are metastable at ambient pressure.

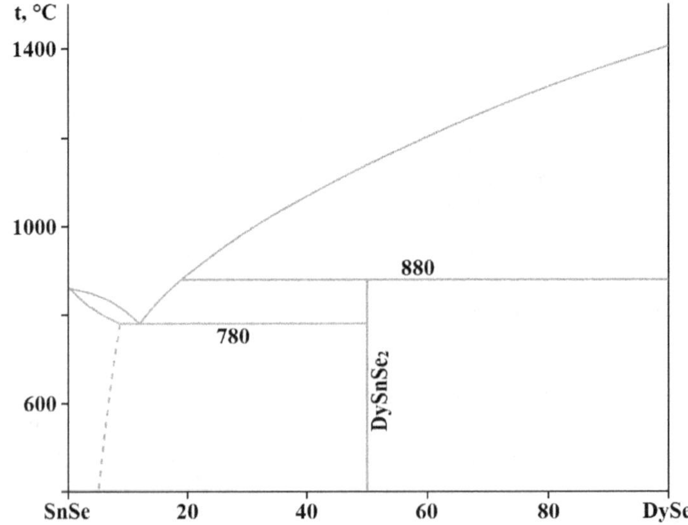

FIGURE A.7.3 Phase diagram of the SnSe–DySe quasibinary system. (From Huseynov, Dzh.I., et al., *Chem. Probl.*, (1), 110, 2009.) Open access.

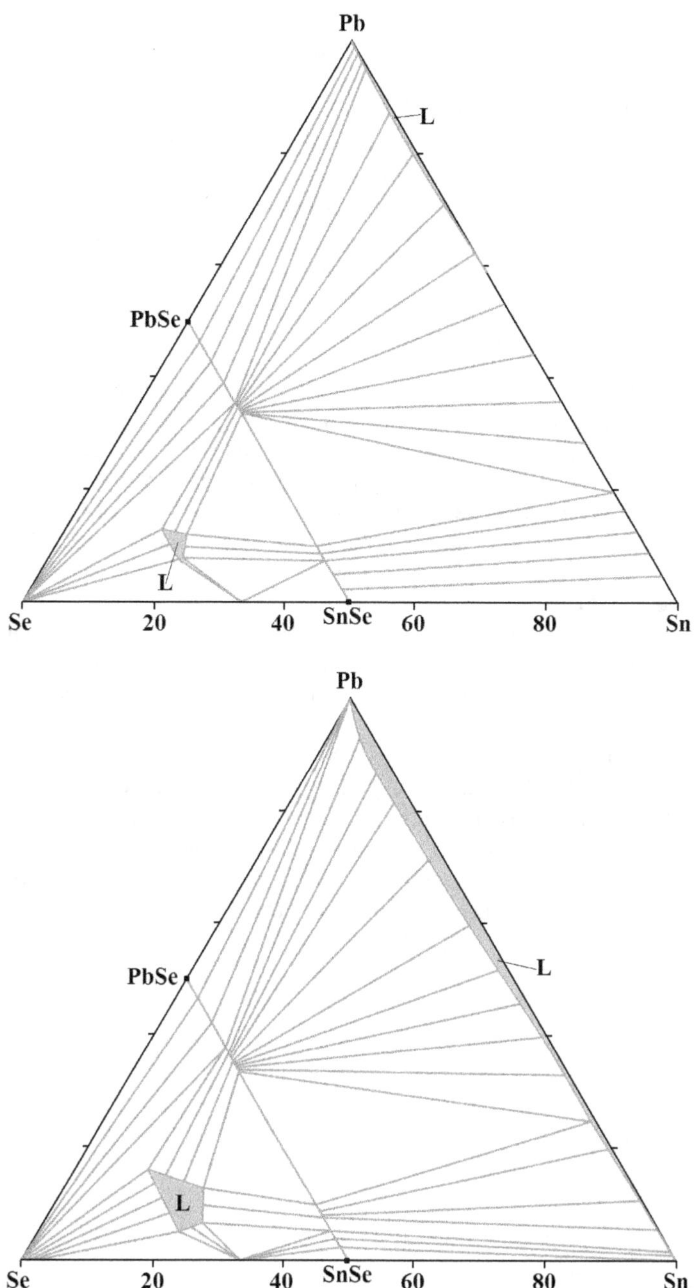

FIGURE A.7.4 Calculated isothermal sections of the Sn–Pb–Se ternary system at (a) 370°C and (b) 500°C. (From Zobac, O., et al., *J. Phase Equilib. Diffus.*, 43(2), 243, 2022.)

A.7.11 Tin–Phosphorus–Selenium

According to Shi et al. (2021), the $SnPSe_3$ or $Sn_2P_2Se_6$ ternary compound, which exists in this system, crystallizes as a monoclinic structure with the lattice parameters a = 681.6 ± 0.5, b = 772.6 ± 0.6, c = 964.4 ± 0.6 pm, β = 91.07 ± 0.02°, a calculated density of 5.056 g·cm^{-3}, and an energy gap of 1.73 eV [a = 681.45 ± 0.03, b = 771.70 ± 0.03, c = 1169.4 ± 0.1 pm, β = 124.549 ± 0.004°, and a calculated density of 5.069 g·cm^{-3} at 173 K and a = 680.8 ± 0.2, b = 768.2 ± 0.3, c = 1166.7 ± 0.7 pm,

$\beta = 124.75 \pm 0.06°$, and a calculated density of 5.121 g·cm^{-3} at room temperature (Israël et al. 1998)]. At 173°C, this compound is a ferroelectric phase and at room temperature, it is a paraelectric phase (Israël et al. 1998).

The single crystals of this compound were synthesized by using Sn, SnS, and P_2S_5 (molar ratio 1:1:1) (Shi et al. 2021). The mixture was loaded into a fused silica tube under an Ar atmosphere in a glove box, which was sealed under 0.01 Pa and then placed into a muffle furnace. The reactants were heated to 950°C within 25 h, kept at the temperature for 48 h, and then cooled to 400°C at a rate of 5°C·h^{-1} before the furnace was switched off. During this heating process, several intermediate equilibrated temperatures were set for safety. Brown block crystals stable in air and water were obtained.

A.7.12 Tin–Antimony–Selenium

$SnSe–Sb_2Se_3$. The refined version of the phase diagram was constructed through DTA and XRD by Ismailova et al. (2020) and is presented in Figure A.7.5. It was found that the $Sn_2Sb_2Se_5$ ternary compound and the intermediate γ-phase with a homogeneity region of 48–60 mol% Sb_2Se_3 are formed in the system. Both phases melt with decomposition by peritectic reactions at 598°C ($Sn_2Sb_2Se_5$) and 560°C (γ-phase). The area of γ-phase includes stoichiometric compositions of ternary compounds $SnSb_2Se_4$ and $Sn_2Sb_6Se_{11}$. The compositions of the peritectic point are 40 and 65 mol% Sb_2Se_3. The eutectic contains 72 mol% Sb_2Se_3 and crystallizes at 545°C. The ingots for the investigations were annealed at 450°C for ~ 500 h and quenched in cold water.

All phases crystallize as an orthorhombic structure with the lattice parameters $a = 2660.5 \pm 2.5$, $b = 2104.9 \pm 2.0$, and $c = 403.85 \pm 0.05$ pm for $SnSb_2Se_4$, $a = 3508 \pm 28$, $b = 2587 \pm 22$, and $c = 409 \pm 6$ pm for $Sn_2Sb_2Se_5$, and $a = 2660.4 \pm 2.5$, $b = 2106.8 \pm 2.5$, and $c = 382.65 \pm 0.05$ pm for $Sn_2Sb_6Se_{11}$.

The $SnSe–Sb_2Se_3–Se$ subsystem was studied by EMF and XRD, and the diagram of solid-phase equilibria at 130°C was constructed by Ismailova et al. (2021) and is shown in Figure A.7.6. The partial thermodynamic functions of SnSe in various phase regions of this subsystem were calculated from the measured EMF of concentration cells with respect to a SnSe electrode in the temperature range 30°C–180°C. These data together with the corresponding thermodynamic functions of SnSe and Sb_2Se_3 were used to calculate the partial molar functions of tin in alloys, and also the standard thermodynamic functions of formation

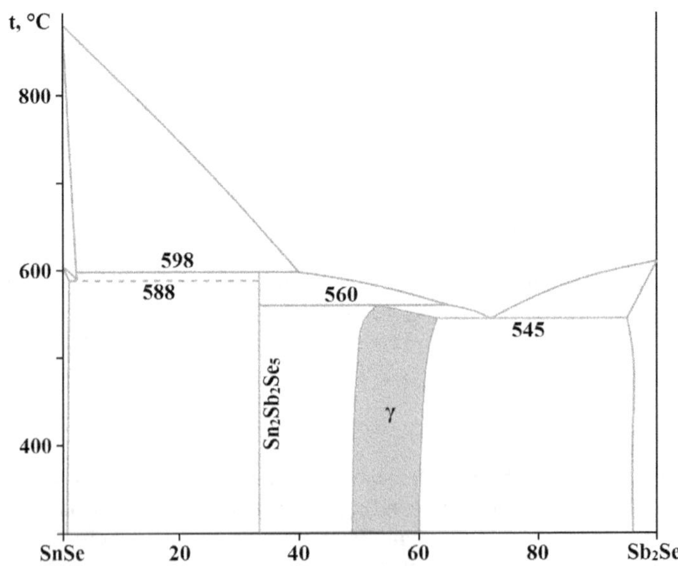

FIGURE A.7.5 Phase diagram of the $SnSe–Sb_2Se_3$ s quasibinary ystem. (From Ismailova, E.N., et al., *Chem. Probl.*, (2), 250, 2020.) Open access.

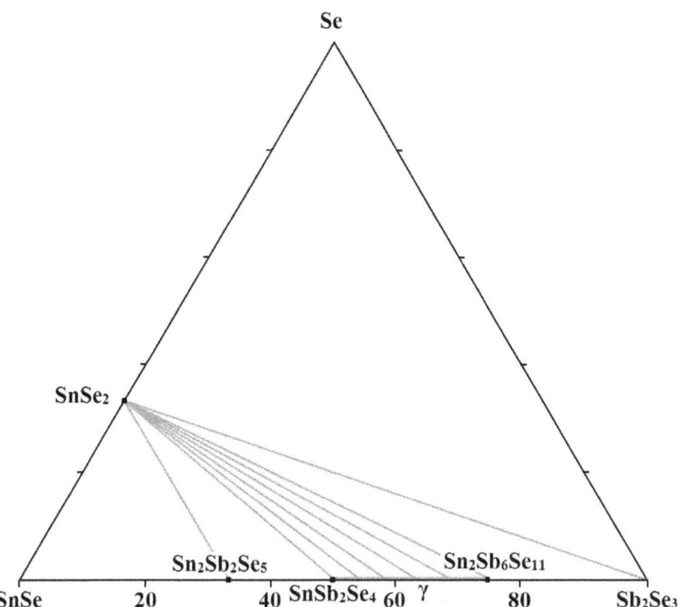

FIGURE A.7.6 Isothermal section of the SnSe–Sb_2Se_3–Se subsystem at 130°C. (From Ismailova, E.N., et al., *Russ. J. Inorg. Chem.*, 66(1), 96, 2021.)

and standard entropies of the ternary compounds: $\Delta G^0_{f,298} = -232.8 \pm 3.2$ kJ·M^{-1}, $\Delta H^0_{f,298} = -234.9 \pm 2.4$ kJ·M^{-1}, $\Delta S^0_{f,298} = -6.7 \pm 3.9$ J·(M·K)$^{-1}$, and $S^0_{298} = 306.5 \pm 6.0$ J·(M·K)$^{-1}$ for $SnSb_2Se_4$; $\Delta G^0_{f,298} = -335.4 \pm 3.4$ kJ·M^{-1}, $\Delta H^0_{f,298} = -338.6 \pm 3.3$ kJ·M^{-1}, $\Delta S^0_{f,298} = -3.5 \pm 6.1$ J·(M·K)$^{-1}$, and $S^0_{298} = 396.8 \pm 8.6$ J·(M·K)$^{-1}$ for $Sn_2Sb_2Se_5$; and $\Delta G^0_{f,298} = -592.9 \pm 9.4$ kJ·M^{-1}, $\Delta H^0_{f,298} = -596.9 \pm 5.2$ kJ·M^{-1}, $\Delta S^0_{f,298} = -13.1 \pm 5.5$ J·(M·K)$^{-1}$, and $S^0_{298} = 827.8 \pm 15.2$ J·(M·K)$^{-1}$ for $Sn_2Sb_6Se_{11}$.

The isothermal section of the Sn–Sb–Se ternary system at 400°C (Figure A.7.7) and the liquidus surface of this system (Figure A.7.8) were experimentally determined by Chang and Chen (2017). There are nine tie-triangles in the system: L + SnSe + SnSb, SnSe + SnSb + (Sb), SnSe + $Sn_2Sb_9Se_9$ + (Sb), $Sn_2Sb_9Se_9$ + (Sb) + Sb_2Se_3, SnSe + $Sn_2Sb_9Se_9$ + $SnSb_2Se_4$, $Sn_2Sb_9Se_9$ + $SnSb_2Se_4$ +Sb_2Se_3, SnSe + $SnSe_2$ + $SnSb_2Se_4$, $SnSe_2$ + $Sn_2Sb_9Se_9$ +$SnSb_2Se_4$, and $SnSe_2$ + Sb_2Se_3 + L. In addition, the possible new ternary compound, $Sn_2Sb_9Se_9$, was found. For the determination of the isothermal section, the quenched alloy was equilibrated at 400°C for six months.

Two immiscibility regions near the Sb–Se and Sn–Se side and the fields of ten primary crystallizations of (Sn), (Sb), (Se), Sn_3Sb_2, SnSb, Sb_2Se_3, SnSe, $SnSe_2$, $Sn_2Sb_9Se_9$, and $SnSb_2Se_4$ phases exist in the liquidus projection. There are ten invariant reactions in the Sn–Sb–Se ternary system, and seven of them were experimentally determined: U_1 (242.2°C) – L + Sn_3Sb_2 ⇔ SnSe + (Sn); U_2 (322.3°C) – L + SnSb ⇔ SnSe + Sn_3Sb_2; U_3 (423.1°C) – L + (Sb) ⇔ SnSb + SnSe; U_4 (613.4°C) – L + SnSe ⇔ (Sb) + $Sn_2Sb_9Se_9$; U_5 (523.5°C) – L + $Sn_2Sb_9Se_9$ ⇔ $SnSb_2Se_4$ + Sb_2Se_3; P_1 (569.4°C) – L + $Sn_2Sb_9Se_9$ + (Sb) ⇔Sb_2Se_3; and P_2 (615.1°C) – L + $Sn_2Sb_9Se_9$ + SnSe ⇔ $SnSb_2Se_4$.

A.7.13 Tin–Bismuth–Selenium

Some ternary compounds in this system along the SnSe–Bi_2Se_3 section were synthesized by Heinke et al. (2018): $SnBi_4Se_7$ crystallizes as a trigonal structure with the lattice parameters $a = 417.24 \pm 0.06$, $c = 3886.1 \pm 0.8$ pm, and a calculated density of 7.324 g·cm^{-3}; $Sn_2Bi_2Se_5$ crystallizes as an orthorhombic structure with the lattice parameters $a = 420.32 \pm 0.03$, $b = 1389.90 \pm 0.10$, $c = 3205.6 \pm 0.2$ pm, and a calculated density of 6.7046 g·cm^{-3}; $Sn_{2.2}Bi_{2.52}Se_6$ also crystallizes as an orthorhombic structure with the lattice parameters $a = 2118.6 \pm 0.4$, $b = 419.58 \pm 0.08$, $c = 1384.4 \pm 0.3$ pm, and a calculated density of 6.828 g·cm^{-3}; $Sn_3Bi_2Se_6$ crystallizes as a cubic structure with the lattice parameter $a = 595.007 \pm 0.008$

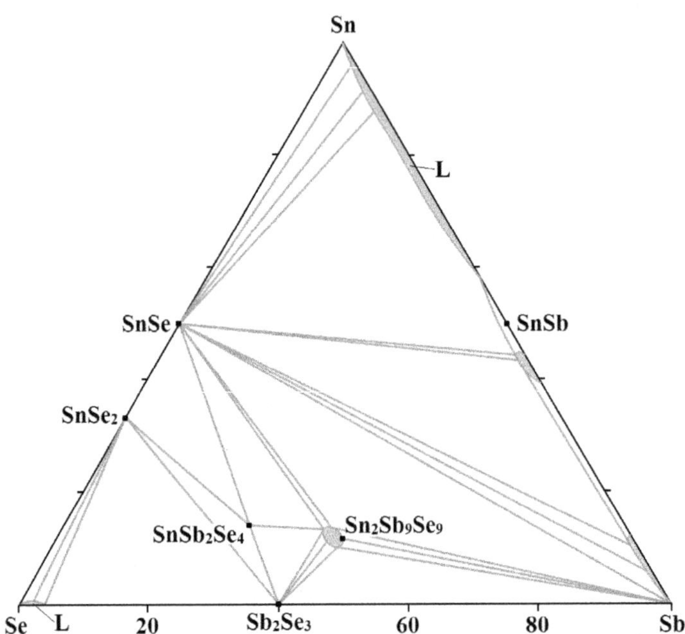

FIGURE A.7.7 Isothermal section of the Sn–Sb–Se ternary system at 400°C. (From Chang, J.-S., and Chen, S-W., *Metall. Mater. Trans. E*, 4(2–4), 89, 2017.)

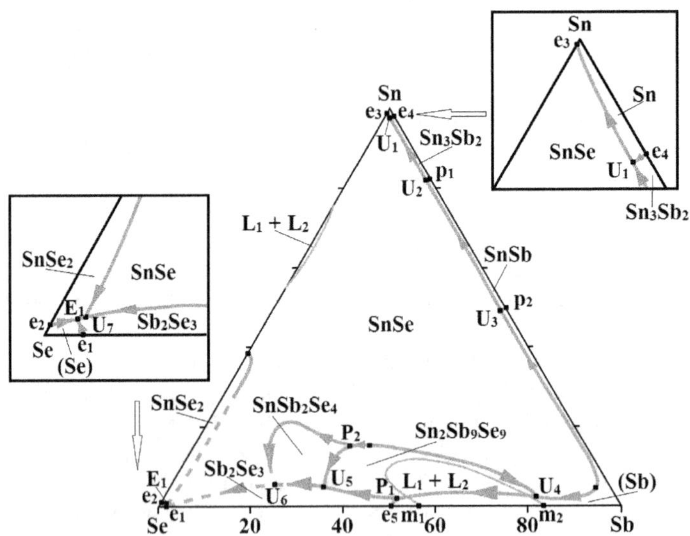

FIGURE A.7.8 Liquidus surface of the Sn–Sb–Se ternary system. (From Chang, J.-S., and Chen, S-W., *Metall. Mater. Trans. E*, 4(2–4), 89, 2017.)

and a calculated density of 6.5578 g·cm^{-3}; $Sn_4Bi_2Se_7$ also crystallizes as a cubic structure with the lattice parameter a = 593.59 ± 0.12 and a calculated density of 6.581 g·cm^{-3}; $Sn_{9.52}Bi_{10.96}Se_{26}$ crystallizes as a monoclinic structure with the lattice parameters a = 1384.7 ± 0.3, b = 419.61 ± 0.08, c = 2309.9 ± 0.5 pm, β = 98.78 ± 0.03°, and a calculated density of 6.791 g·cm^{-3}; and $Sn_{11.49}Bi_{12.39}Se_{30}$ also crystallizes as a monoclinic structure with the lattice parameters a = 1385.20 ± 0.10, b = 419.93 ± 0.03, c = 2669.8 ± 0.6 pm, β = 95.781 ± 0.006°, and a calculated density of 6.794 g·cm^{-3}. All these compounds were synthesized from stoichiometric mixtures of Sn, Bi, and Se. In typical reactions, mixtures with total weights of 250–300 mg were fused in sealed silica glass ampoules under dry Ar atmosphere. Some samples were annealed after previous melting at 900°C–950°C. Single crystals of $Sn_4Bi_2Se_7$ are accessible via the gas phase in the existence range of the cubic high-temperature phase (585°C).

Another ternary compound, $Sn_4Bi_{10}Se_{19}$, which melts congruently at 667°C and crystallizes as a monoclinic structure with the lattice parameters $a = 2815.2 \pm 0.6$, $b = 415.67 \pm 0.08$, $c = 2124.2 \pm 0.4$ pm, and $\beta = 131 \pm 3°$, was obtained by Lu et al. (2021). The structurally disordered but element-homogeneous sample of this compound was initially obtained through solid state reaction of high-purity elements sealed in an evacuated quartz tube followed by repeated melting and quenching of the product from solid state reaction using an induction melting furnace. This suggests the formation of a disordered, poorly crystalline $Sn_4Bi_{10}Se_{19}$ with congruent melting upon heating. The best crystalline samples were obtained by heating the induction-melted structurally disordered sample to 700°C (50°C above the melting temperature) to obtain a thick molten state, followed by a slow cooling across the crystallization window (700°C to 654°C in 48 h) in order to allow enough time for self-organization of the large building units. The sample was subsequently held at 654°C, the lower limit of the crystallization window, for another 48 h and finally cooled to room temperature in 12 h.

A.7.14 Tin–Vanadium–Selenium

A series of ferecrystalline compounds $([SnSe]_{1+\delta})_1(VSe_2)_1$ with varying Sn/V ratios were synthesized using the modulated elemental reactant technique (Falmbigl et al. 2015). The compounds form over a wide compositional range of Sn/V ratios within the interval of $0.89 \le \delta \le 1.37$. The lattice parameters of the compounds vary linearly with the Sn/V ratio. XRD reveals systematic changes in intensity and a slight decrease of the *c*-axis lattice parameter as the Sn/V ratio is increased. Temperature-dependent specific heat data reveal a phase transition at 102 K, where the heat capacity changes abruptly.

A.7.15 Tin–Nickel–Selenium

Three isothermal sections of the Sn–Ni–Se ternary system were experimentally investigated at 530°C, 730°C, and 830°C in the whole composition range by Zobac et al. (2023) and are shown in Figure A.7.9. The prepared samples underwent long-term annealing at selected temperatures afterward (1825 and 2784 h at 530°C, 1005, 1295, and 1775 h at 730°C, and 673, 1179, and 1222 h at 830°C). Eight binary phases were found to be stable in the isothermal section at 530°C (Figure A.7.9a). A metastable binary phase Ni_3Se_4, most likely stabilized by the third element, was observed. The amount of dissolved tin in this phase was only about 0.5 at% Sn, nevertheless it was enough for its stabilization. Only one stable ternary phase, $Ni_{5.62}SnSe_2$, was found with a small region of solubility. The other proposed ternary phases Ni_3SnSe and NiSnSe (Musa and Chen 2021) have not been found. Selenium and tin are already molten at 530°C.

The isothermal section at 730°C (Figure A.7.9b) looks like the isothermal section at 530°C. Binary phase NiSe contains about 15 at% of the tin in its structure. The existence of the Ni_3Se_4 phase cannot be confirmed at this temperature. The ternary phase $Ni_{5.62}SnSe_2$ is found to be stable at this temperature with a small but not negligible homogeneity region.

At 830°C (Figure A.7.9c), the Se-rich part of the phase diagram cannot be analyzed due to the high volatility of selenium. The Ni content is ≈17 at% in liquid Se and 8 at% in liquid Sn. The NiSe phase shows 8% tin solubility. The eutectic liquid at approximately 30 at% of Se according to the Ni–Se binary phase diagram dissolves up to 7 at% of tin. All other binary phases have negligible solubility for the third element and can be considered stoichiometric. No ternary phase was found at 830°C.

The isothermal section of the Sn–Ni–Se system at 300°C was also constructed by Musa and Chen (2021), where the existence of the Ni_3SnSe and NiSnSe metastable ternary compounds were determined. The interfacial reactions in the Ni/SnSe couple were systematically investigated at 250°C and 300°C. It was shown that the reaction rate decreases with lower reaction temperatures and the reaction zone grows thicker with longer reaction time.

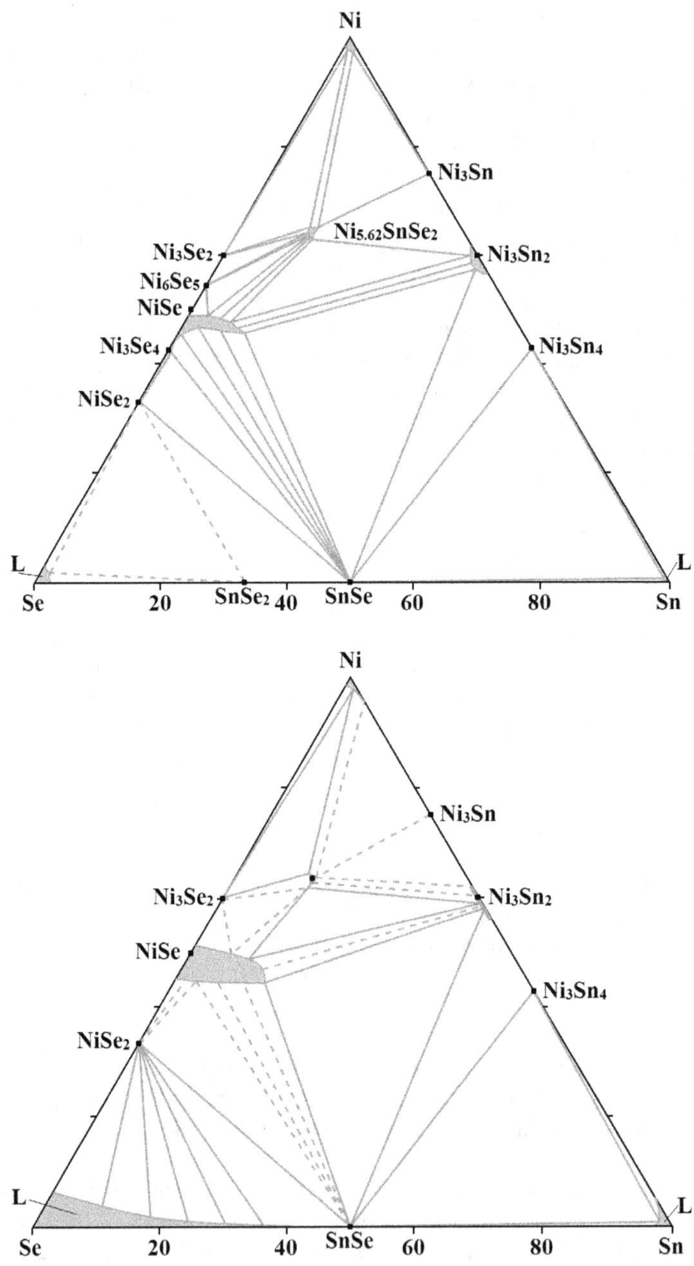

FIGURE A.7.9 Isothermal section of the Sn–Ni–Se ternary system at (a) 530, (b) 730°C, and (c) 830°C. (From Zobac, O., et al., *J. Phase Equilib. Diffus.*, 44(4), 594, 2023.)

A.8 Systems Based on Tin Telluride

A.8.1 Tin–Silver–Tellurium

The isothermal section of this system at 350°C is shown in Figure A.8.1a (Chen et al. 2022b). It was designed based on the experimental results and theoretical calculations. Besides the binary compounds, one ternary compound, $AgSnTe_2$, exists at this temperature. Its composition was found to be slightly shifted toward higher Sn content (≈30 at% Sn) in comparison with "ideal" composition ratio 1:1:2.

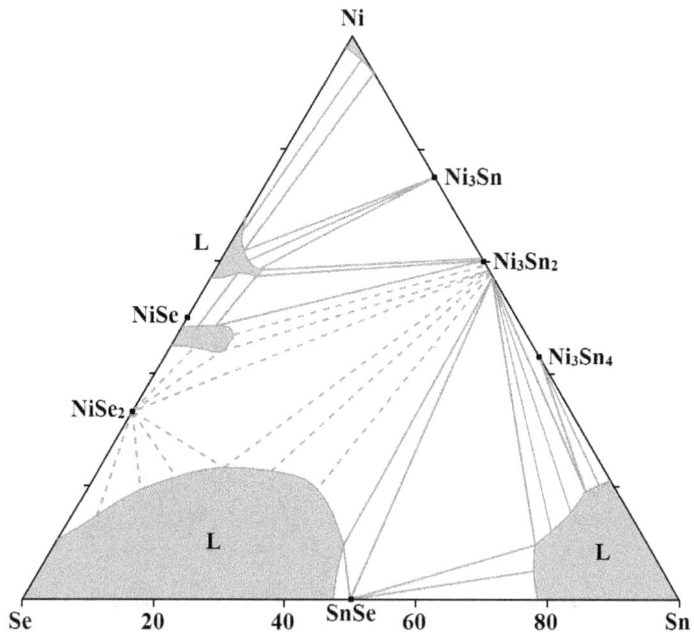

FIGURE A.7.9 Continued

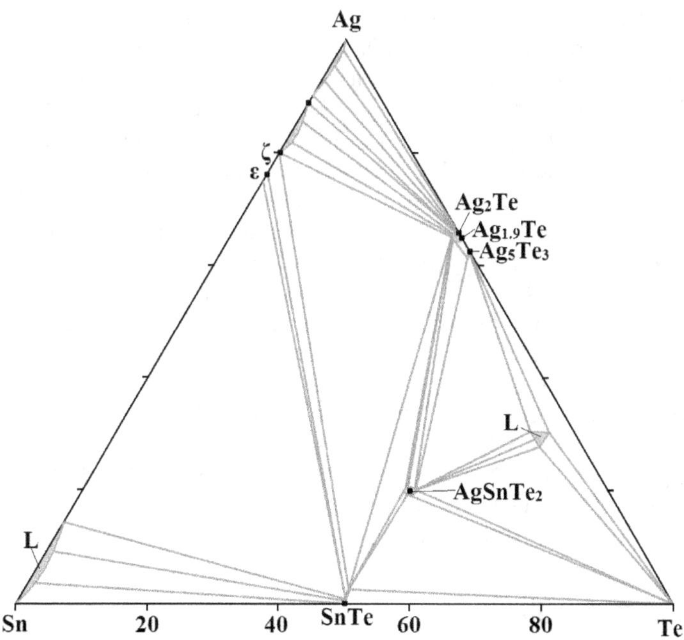

FIGURE A.8.1 Isothermal section of the Sn–Ag–Te ternary system at (a) 350, and (b) 500°C. (From Chen, S.-W., et al., *J. Phase Equilib. Diffus.*, 43(2), 139, 2022.)

Eleven tie-triangles were observed in the Ag–Sn–Te ternary system at 350°C. At 500°C (Figure A.8.1b), the ternary phase was not found in the isothermal section and it can be estimated that it melted incongruently around 430°C. There are four tie-triangles in the system. The alloys for the investigations were equilibrated at 350°C and 500°C for 180 days.

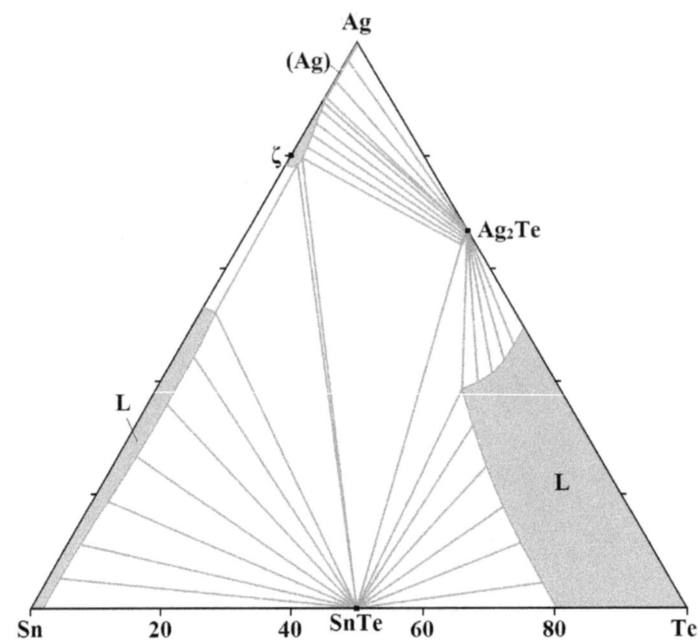

FIGURE A.8.1 Continued

The liquidus projection of the Ag–Sn–Te ternary system was determined experimentally by Chang et al. (2014). There are ten primary solidification regions: Ag, ζ-Ag_4Sn, ε-Ag_3Sn, Sn, SnTe, $AgSnTe_2$, Te, β-Ag_5Te_3, $Ag_{1.9}Te$, and γ-Ag_2Te, and eight ternary invariant reactions in the system. The calculated liquidus surface of this system is presented in Figure A.8.2 (Chen et al. 2022b).The results are compared with experimental data from Chang et al. (2014). The calculated liquidus surface correctly reproduced all important features of the experimentally established equilibria. There are three ternary eutectics, two ternary peritectics and seven transition points on the liquidus surface: U_1 (873.3°C) – L_1 + γ-Ag_2Te ⇔ L_2 + β-Ag_2Te; U_2 (813.5°C) – L + γ-Ag_2Te ⇔ (Ag) + β-Ag_2Te; U_3 (725.8°C) – L + (Ag) ⇔ β-Ag_2Te + ζ-Ag_4Sn; E_1 (583.9) – L_1 ⇔ L_2 + β-Ag_2Te + SnTe; U_4 (533.6°C) – L + β-Ag_2Te ⇔ ζ-Ag_4Sn + SnTe; P_1 (486.2°C) – L + ζ-Ag_4Sn + SnTe ⇔ ε-Ag_3Sn; P_2 (434.0°C) – L + β-Ag_2Te + SnTe ⇔ $AgSnTe_2$; U_5 (424.2°C) – L + β-Ag_2Te ⇔ β-$Ag_{1.9}Te$ + $AgSnTe_2$; U_6 (395.5°C) – L + β-$Ag_{1.9}Te$ ⇔ β-Ag_5Te_3 + $AgSnTe_2$; U_7 (386.8°C) – L + SnTe ⇔ (Te) + $AgSnTe_2$; E_2 (330.5) – L ⇔ (Te) + β-Ag_5Te_3 + $AgSnTe_2$; and E_3 (221.6) – L ⇔ SnTe + ε-Ag_3Sn + (Sn).

Standard thermodynamic functions $AgSnTe_2$ were calculated in the phase regions SnTe–Ag_2Te–$AgSnTe_2$ and SnTe–$AgSnTe_2$–Te of the Ag–Sn–Te system by Moroz et al. (2020). The following values were obtained: $\Delta G^0_{f,\,298} = -80.91 \pm 2.64$ kJ·M^{-1}, $\Delta H^0_{f,\,298} = -74.44 \pm 2.89$ kJ·M^{-1}, and $S^0_{298} = 214.58 \pm 6.78$ J·(M·K)$^{-1}$ in the first region and $\Delta G^0_{f,\,298} = -83.52 \pm 2.71$ kJ·M^{-1}, $\Delta H^0_{f,\,298} = -77.53 \pm 2.97$ kJ·M^{-1}, and $S^0_{298} = 212.97 \pm 6.93$ J·(M·K)$^{-1}$ in the second one.

A.8.2 Tin–Gallium–Tellurium

The liquidus surface of the Sn–Ga–Te ternary system was calculated by Pal et al. (2023) and is shown in Figure A.8.3. There is a liquid phase immiscibility region extrapolated from the Ga–Te system into the Sn–Ga–Te system.

A.8.3 Tin–Zirconium–Tellurium

The ZrSnTe ternary compound, which is formed in this system, crystallizes as a tetragonal structure with the lattice parameters $a = 405.98 \pm 0.03$ and $c = 870.49 \pm 0.09$ pm (Acharya et al. 2023). Single crystals

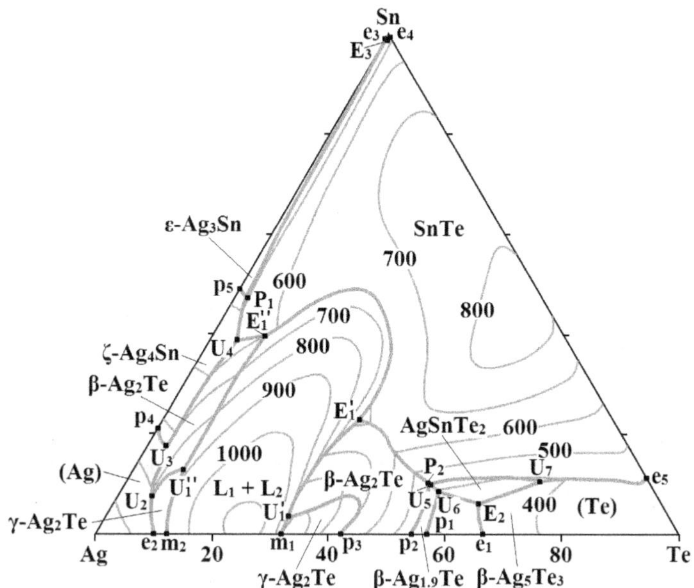

FIGURE A.8.2 Calculated liquidus surface of the Ag–Sn–Te ternary system. (From Chen, S.-W., et al., *J. Phase Equilib. Diffus.*, 43(2), 139, 2022.)

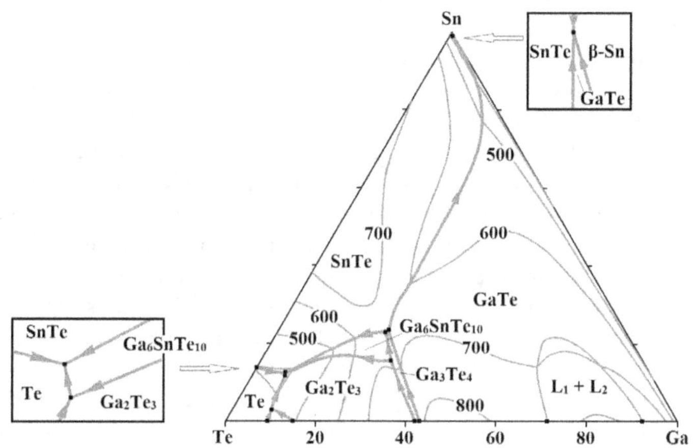

FIGURE A.8.3 Liquidus surface of the Sn–Ga–Te ternary system. (From Pal, V., et al., *J. Phase Equilib. Diffus.*, 44(5), 642, 2023.)

of this compound were successfully obtained by a flux method with Sn as the flux. Rapid cooling was found to be critical for minimizing secondary phases and obtaining ZrSnTe.

A.8.4 Tin–Antimony–Tellurium

$SnTe–Sb_2Te_3$. The phase diagram of this system was reconstructed by Seidzade et al. (2021) based on experimental data from DTA, XRD, and SEM-EDX using the alloys annealed at 450°C for 720°C and is presented in Figure A.8.4. Two ternary compounds, $SnSb_2Te_4$ and $SnSb_4Te_7$, are formed in this system. They melt incongruently at 595°C and 593°C, respectively. The coordinates of the peritectics were found to be 53 and 59 mol% Sb_2Te_3. The eutectic contains 58 mol% Sb_2Te_3 and crystallizes at 587°C. $SnSb_4Te_7$ has a very narrow primary crystallization area. This compound has a significant homogeneity area

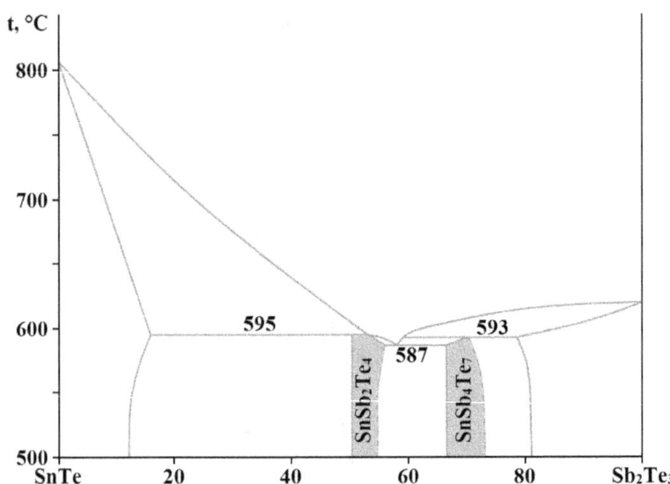

FIGURE A.8.4 Phase diagram of the SnTe–Sb_2Te_3 quasibinary system. (From Seidzade, A.E., et al., *J. Phase Equilib. Diffus.*, 42(3), 373, 2021.)

approximately from 66.5 to 72.7 mol% Sb_2Te_3. The homogeneity region of $SnSb_2Te_4$ has been detected approximately from 50 to 54 mol% Sb_2Te_3. The system has two more homogeneity areas on both initial compounds (for SnTe up to ≈11–13 mol% and for Sb_2Te_3 up to ≈18-19 mol% at room temperature). $SnSb_4Te_7$ crystallizes as a trigonal structure with the lattice parameters $a = 429.51 \pm 0.02$, $c = 2397.3 \pm 0.8$ pm, and a calculated density of 6.228 ± 0.003 g·cm^{-3}.

The SnTe–Sb_2Te_3 system was also investigated by using EMF measurements of reversible concentration cells relative to the SnTe electrode in the 30°C–130°C temperature range (Seyidzade et al. 2022). The formation of solid solutions (up to 20 mol%) based on Sb_2Te_3 was confirmed. Based on EMF measurements, the equations of the temperature dependences of the EMF for solid solutions based on Sb_2Te_3 were obtained and these results were used to calculate the partial thermodynamic functions of SnTe in alloys. The standard thermodynamic functions of the formation of solid solutions were also calculated.

SnTe–Sb_2Te_3–Te. The isothermal section of this subsystem at 130°C is given in Figure A.8.5 (Seidzade et al. 2022). It is seen that all telluride phases in this section are tie-lined with elementary tellurium. The subsystem was studied through XRD and EMF measurements and the alloys for the investigations were annealed at 380°C for 1000 h and at 130°C for 100 h. The standard integral thermodynamic properties of $SnSb_2Te_4$ and $SnSb_4Te_7$ were calculated and the following values were obtained: $\Delta G^0_{f,298} = -147.7 \pm 2.2$ kJ·M^{-1}, $\Delta H^0_{f,298} = -143.6 \pm 2.5$ kJ·M^{-1}, $\Delta S^0_{f,298} = 13.8 \pm 6.0$ J·(M·K)$^{-1}$ and $S^0_{298} = 361.6 \pm 11.0$ J·(M·K)$^{-1}$ for the first compound and $\Delta G^0_{f,298} = -208.2 \pm 3.3$ kJ·M^{-1}, $\Delta H^0_{f,298} = -202.4 \pm 3.4$ kJ·M^{-1}, $\Delta S^0_{f,298} = 19.5 \pm 9.0$ J·(M·K)$^{-1}$, and $S^0_{298} = 613 \pm 16$ J·(M·K)$^{-1}$ for the second one.

A.8.5 Tin–Bismuth–Tellurium

One more ternary compound, $SnBi_4Te_4$, which melts incongruently at 558°C and crystallizes as a trigonal structure with the lattice parameters $a = 443.306 \pm 0.057$, $c = 1773.96 \pm 0.57$ pm, and a calculated density of 8.0550 ± 0.0033 g·cm^{-3}, is formed in this system (Orujlu et al. 2020). To prepare this compound, the stoichiometric mixture of the elements was sealed in evacuated (0.01 Pa) quartz ampoule, heated up to 730°C, and kept at this temperature for 5 h with the next quenching in water. In order to achieve complete homogenization, the sample was annealed at 430°C for ~700 h.

The isothermal section of the SnTe–Sn–Bi at 160°C is presented in Figure A.8.6 (Chiu et al. 2012). The SnTe phase is very stable and has tie-lines with Sn, liquid, and Bi phases. No ternary compounds are observed in this region. Compared with the isothermal section at 160°C, all the Sn–Bi alloys are completely molten and only the liquid phase is stable along the Sn–Bi side at 500°C (Figure A.8.7). Four ternary compounds, $SnBiTe_2$, $SnBi_2Te_4$, $SnBi_4Te_7$, and $Sn_2Bi_2Te_5$, are found in the SnTe–Bi–Te

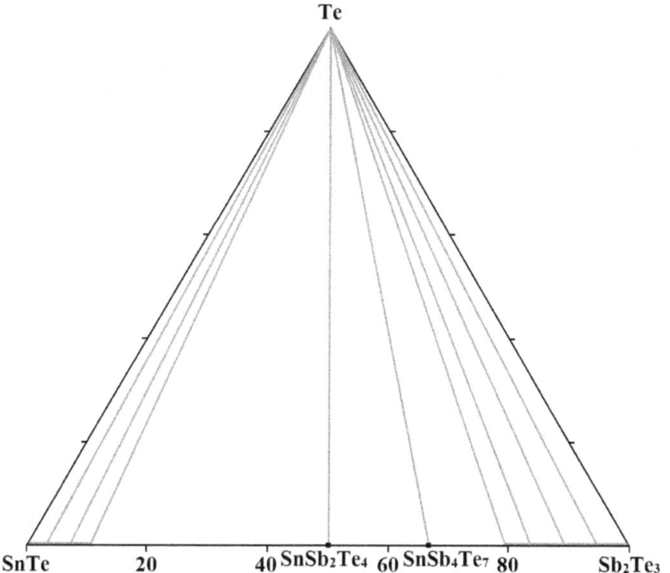

FIGURE A.8.5 Isothermal section of the SnTe–Sb_2Te_3–Te subsystem at 130°C. (From Seidzade, A.E., et al., *Russ. J. Inorg. Chem.*, 67(5), 683, 2022.)

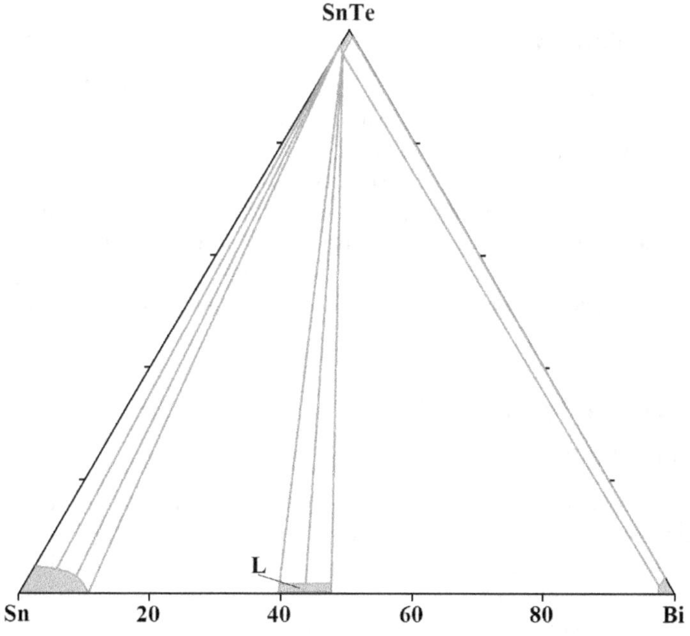

FIGURE A.8.6 Isothermal section of the SnTe–Sn–Bi subsystem at 160°C. (From Chiu, C.-N., et al., *J. Electron. Mater.*, 41(1), 22, 2012.)

region at 500°C. SnTe and Bi_2Te_3 have significant mutual solubility. The Bi solubility in SnTe is about 13.29 at%, and that of Sn in Bi_2Te_3 is 5.29 at%. There are about 3 at% of solubility range for both the $Sn_2Bi_2Te_5$ and $SnBi_2Te_4$ ternary phases. The samples for the investigations were placed in a furnace at 800°C for 24 h, quenched in ice water, equilibrated at 160°C or 500°C for 6 weeks, and then quenched in ice water again.

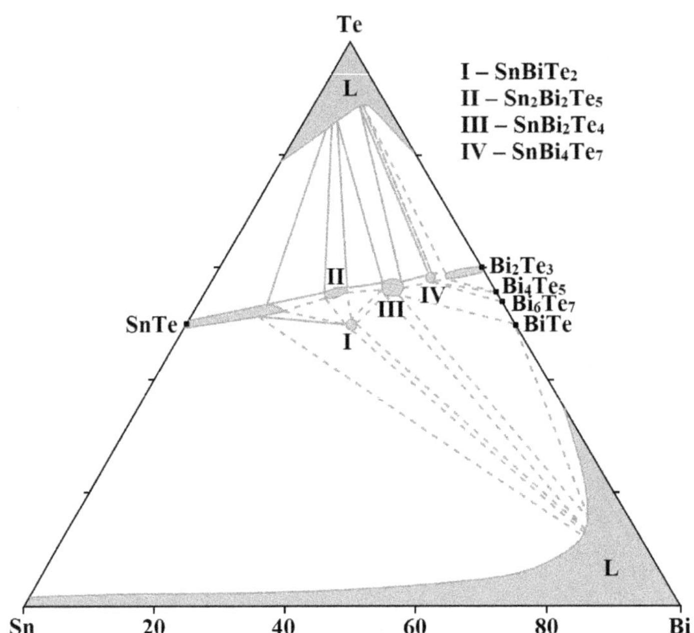

FIGURE A.8.7 Isothermal section of the Sn–Bi–Te ternary system at 160°C. (From Chiu, C.-N., et al., *J. Electron. Mater.*, 41(1), 22, 2012.)

A.8.6 Tin–Nickel–Tellurium

The isothermal section of the Sn–Ni–Te system at 400°C was constructed by Musa and Chen (2021) and is shown in Figure A.8.8. At 400°C, there are 12 thermodynamically stable phases; one liquid phase at the Sn corner, two terminal solid phases based on Ni and Te, seven binary phases, Ni_3Sn, Ni_3Sn_2, Ni_3Sn_4, Ni_3Te_2, $NiTe_{0.775}$, $NiTe_2$, and SnTe, and two ternary phases, Ni_3SnTe_2 and $Ni_{5.78}SnTe_2$. The as-prepared Ni/SnTe couple shows excellent adhesion without any reaction at the contact zone. Based on the interfacial reaction results, it can be concluded that Ni is the fastest diffusion species, Sn is second, and Te is relatively immobile. At 400°C, Ni atoms rapidly diffuse toward the SnTe substrate to form the Ni_3SnTe_2 ternary phase, and then some of Sn diffuses in the opposite direction to form Ni_3Sn and Ni_3Sn_2 leaving Te to form $Ni_{5.78}SnTe_2$.

A.9 Systems Based on Lead Sulfide

A.9.1 Lead–Copper–Sulfur

Using a membrane zero-manometer, the vapor pressure of S_2 over the surface of the PbS liquidus surface in the ternary system Pb–Cu–S was determined in the range 830°C–1130°C and 0–0.1 MPa (Babanly et al. 2021). Based on the thermodynamic calculation, the boundaries of the immiscibility of liquid alloys of the Cu–S, Pb–S, and Cu–Pb–S systems were determined and analytically described. Critical temperatures and pressures for immiscibility regions of sulfur-rich liquid alloys were characterized by high values: t_{cr} = 1250-1610°C and p_{cr} = 17.2-51.7 MPa. The crystallization surfaces of lead sulfide with electronic conductivity (*n*-type PbS) and with hole conductivity (*p*-type PbS) were calculated and analytically described, as well as the corresponding values of sulfur vapor pressure over the crystallization surface of PbS.

A.9.2 Lead–Gallium–Sulfur

The ternary compound $PbGa_2S_4$, which is formed in this system, has one more polymorphic modification which also crystallizes as an orthorhombic structure with the lattice parameters a = 1238.09 ± 0.02,

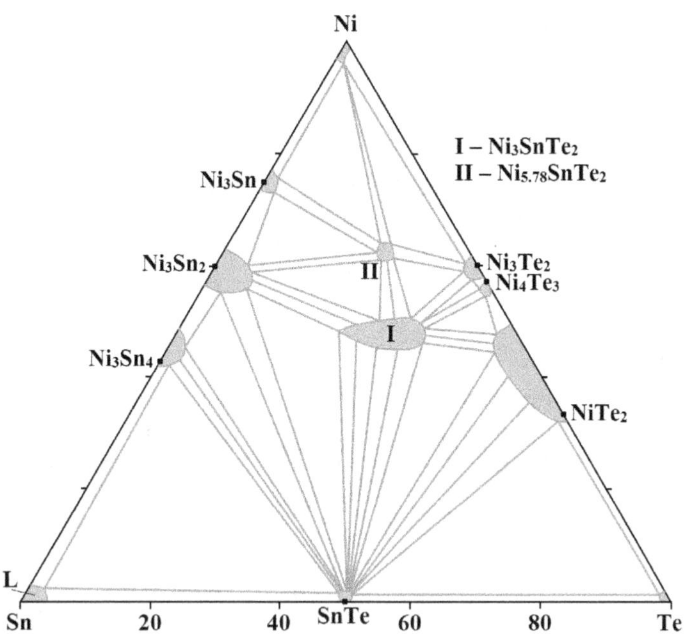

FIGURE A.8.8 Isothermal section of the Sn–Ni–Te ternary system at 400°C. (From Musa, A.F., et al., *J. Electron. Mater.*, 50(8), 4346, 2021.).

b = 1171.78 ± 0.02, c =1789.97 ± 0.04 pm, a calculated density of 4.858 g·cm^{-3}, and an energy gap of 2.46 eV (Chen et al. 2022c). This modification was obtained by the reaction with a mixture of K metal (0.74 mM), Ga metal (0.74 mM), S powder (1.48 mM), and $PbBr_2$ powder (0.74 mM). The weighing and mixing were performed in an argon-filled glove box with H_2O and O_2 levels below 0.1 ppm. The mixture was placed in a quartz tube and sealed with a vacuum of 0.01 Pa. Subsequently, the tube was heated in a computer-controlled furnace to 800°C in 25 h, kept at this temperature for 96 h, then cooled slowly to 300°C for 120 h, and finally the furnace was switched off. After washing the products with distilled water and ethanol, yellow crystals of the title compound were obtained.

A.9.3 Lead–Yttrium–Sulfur

The Y_2PbS_4 ternary compound, which is formed in this system, crystallizes as an orthorhombic structure with the lattice parameters a = 793.01 ± 0.03, b = 2869.66 ± 0.09, c =1205.11 ± 0.05 pm and a calculated density of 4.9719 g·cm^{-3} (Gulay et al. 2008; Marchuk and Smitiukh 2021). This compound was prepared by sintering the elements in evacuated silica ampoule. The synthesis was performed in a tube furnace. The ampoule was gradually heated (30°C·h^{-1}) to a maximal temperature of 1150°C. The sample was kept at this temperature for 4 h. Afterward, it was cooled slowly (10°C·h^{-1}) to 600°C and annealed at this temperature for 240 h. After annealing, the ampoule with the sample was quenched in cold water.

A.9.4 Lead–Dysprosium–Sulfur

The ternary compound Dy_2PbS_4, which is formed in this system, crystallizes as an orthorhombic structure with the lattice parameters a = 794.84 ± 0.08, b = 2872.1 ± 0.3, and c =1203.9 ± 0.1 pm (Gulay et al. 2008). This compound was prepared in the same way as Y_2PbS_4 was synthesized.

A.9.5 Lead–Holmium–Sulfur

The ternary compound Ho_2PbS_4, which is formed in this system, crystallizes as an orthorhombic structure with the lattice parameters a = 790.81 ± 0.02, b = 2862.22 ± 0.07, c =1202.20 ± 0.04 pm and a calculated density of 4.4951 g·cm^{-3} (Gulay et al. 2008). This compound was prepared in the same way as Y_2PbS_4 was synthesized.

A.9.6 Lead–Erbium–Sulfur

The ternary compound Er_2PbS_4, which is formed in this system, crystallizes as an orthorhombic structure with the lattice parameters a = 786.3 ± 0.2, b = 2852.5 ± 0.5, c =1199.5 ± 0.2 pm and a calculated density of 6.616 g·cm^{-3} (Gulay et al. 2008). This compound was prepared in the same way as Y_2PbS_4 was synthesized.

A.9.7 Lead–Thulium–Sulfur

The $Yb_2Pb_4S_7$ ternary compound, which is formed in this system, crystallizes as an orthorhombic structure with the lattice parameters a = 784.19 ± 0.03, b = 2841.84 ± 0.07, c =1196.55 ± 0.04 pm and a calculated density of 6.7078 g·cm^{-3} (Gulay et al. 2008). This compound was prepared in the same way as Y_2PbS_4 was synthesized.

A.9.8 Lead–Ytterbium–Sulfur

The ternary compound $Yb_2Pb_4S_7$, which crystallizes as an orthorhombic structure with the lattice parameters a = 1363.0, b = 2348.1, and c = 402.5 pm, is formed in this system (Gasymov and Hasanov 2020). The synthesis of this compound was carried out in a single-zone furnace in an evacuated quartz ampoule with simultaneous melting of binary sulfides. After keeping for 4 h at 950°C, the sample was annealed at 450°C for 172 h to obtain a homogenous sample.

A.9.9 Lead–Nitrogen–Sulfur

The $Pb(NS)_2$ ternary compound, which is formed in this system, crystallizes as an orthorhombic structure with the lattice parameters a = 437.5 ± 0.3, b = 765.4 ± 0.3, and c =1227.4 ± 0.7 pm (Martan and Weiss 1984). To prepare this compound, $Pb(NO_3)_2$ (250 mg) and S_4N_4 (100 mg) were mixed together, dry $CHCl_3$ was added and dry NH_3 was introduced in a slow stream. After 3–4 h the solution first turned red and after 3 days it was yellow. The residue left after decanting consisted of a white dispersed powder with a few, very small, red air stable crystals of the title compound.

A.9.10 Lead–Phosphorus–Sulfur

The ternary compound $Pb_2P_2S_6$, which is formed in this system, melts at 970°C (Becker et al. 1983) and crystallizes as a monoclinic structure with the lattice parameters a = 659.61 ± 0.06, b = 745.42 ± 0.07, c = 938.54 ± 0.10 pm, β = 91.515 ± 0.006°, and a calculated density of 4.814 g·cm^{-3} (Ji et al. 2022a) [a = 940.2 ± 0.4, b = 746.6 ± 0.3, c = 661.2 ± 0.3 pm, β = 91.53 ± 0.05°, and the calculated and experimental densities of 4.79 and 4.66 g·cm^{-3}, respectively (Becker et al. 1983); a = 660.3 ± 0.3, b = 745.6 ± 0.4, c = 940.1 ± 0.5 pm, β = 91.618 ± 0.006°, a calculated density of 4.800 g·cm^{-3}, and an energy gap of 2.460 eV (Wang et al. 2020a)].

The crystal growth of this compound was carried out in a two-temperature-zone gradient furnace (Becker et al. 1983; Ji et al. 2022a). A total of 2 g of elements mixed with a stoichiometric ratio of Pb/P/S = 2:2:6 were loaded into a long silica tube and to it was added ~0.04 g of I_2, which functioned as vapor transport agents. The long silica tube spanned two temperature zones with the reactants located at the hot zone. The temperature difference between the hot zone and cold zone was 100°C.

The program was set to control the temperature of the hot zone. The tube was heated up to 700°C in 40 h, dwelled at this temperature for 144 h, and cooled down to room temperature in 48 h. Yellow-orange crystal chunks were collected as the products when the tube was opened. $Pb_2P_2S_6$ exhibits extraordinary ambient stability. All reactants were stored in an argon-filled glove box with the moisture and oxygen levels below 0.3 ppm. All preparations were accomplished in the glove box or under vacuum.

Single crystals of this compound can be also obtained through the reaction of PbS, Zn, P, and S in the molar ratio of 1:2:6:14 (Wang et al. 2020a). The mixture with a total mass of ~0.25 g was weighed and loaded into a quartz silica tube under an Ar atmosphere in the glove box. This tube was evacuated to about 10^{-3} Pa and then flame-sealed. The sealed tube was heated to 950°C for 15 h and kept at this temperature for 48 h in a computer-controlled muffle furnace. After that, the furnace was cooled to 500°C in 48 h. Then, the temperature of the furnace was dropped slowly to room temperature. Finally, block-shaped crystals with sub-millimeter size and yellow color were obtained.

The ternary compound $Pb_3P_2S_8$, which is also formed in this system, crystallizes as a cubic structure with the lattice parameters a = 1092.53 ± 0.10 pm and a calculated density of 4.788 g·cm^{-3} (Ji et al. 2021). The direct and indirect band gaps of this compound are 2.6 ± 0.1 and 2.4 ± 0.1 eV, respectively. The millimeter-sized crystals of this compound were grown by a two-step vapor transport reaction. First, the reactants were loaded into carbonized silica ampoule at a molar ratio of Pb/P/S = 3:2:8 (total mass 2 g) to which ~0.1 g of I_2 was added as a vapor transport agent. The ampoule was flame-sealed under a high vacuum and placed in a programmable furnace. It was heated to 650°C in 48 h, maintained at that temperature for another 48 h, and then cooled to 100°C in 24 h. Second, the same tube from the first step was transferred to a gradient furnace with a 100°C difference between the hot and cold sides. The material side was placed in the hot temperature zone with the empty top side located in the cold zone. For the hot end, the tube was heated to 650°C in 48 h, held at that temperature for 120 h, and then cooled to 100°C in 120 h. Many yellow-orange crystals of $Pb_3P_2S_8$ were found after the second step. All starting materials were stored and used in an Ar-filled glove box.

A.9.11 Lead–Arsenic–Sulfur

The $Pb_2As_2S_5$ ternary compound, which is formed in this system, is thermally stable up to 493°C and has an energy gap of 1.64 eV (Chen et al. 2021b). This polycrystalline compound was prepared from a mixture of PbS and As_2S_3 (molar ratio 2:1) with a total mass of 0.5 g and the mixture was loaded in a dry and clean quartz tube that was evacuated to 10^{-3} Pa and flamed-sealed. The tube was placed in a high-temperature furnace and the temperature was controlled through the computer program that heated to 700°C and keeps them warm for 70 h and then slowly cooled to 250°C at a rate of 3°C·h^{-1}. Finally, the temperature controller was turned off, letting them cool to room temperature naturally. The products were taken out from the quartz tube, and it was found that the synthesized crystals are stable in air and moisture.

The ternary compound $Pb_3As_4S_9$ (mineral baumhauerite), which is also formed in this system, crystallizes as a triclinic structure with the lattice parameters a = 788.4 ± 0.4, b = 834.5 ± 0.4, c = 2281.1 ± 1.1 pm, α = 90.069 ± 0.008°, β = 97.255 ± 0.008°, γ = 90.082 ± 0.008°, and a calculated density of 5.4 g·cm^{-3} (Topa and Makovicky 2016).

A.9.12 Lead–Antimony–Sulfur

The ternary compound $Pb_6Sb_6S_{14}(S_3)$ (mineral moëloite), which crystallizes as an orthorhombic structure with the lattice parameters a = 1532.8 ± 0.3, b = 404.00 ± 0.08, c =2305.4 ± 0.5 pm, and a calculated density of 5.86 g·cm^{-3}, is formed in this system (Orlandi et al. 2002; Jambor and Roberts 2003; Mandarino 2003).

A.9.13 Lead–Bismuth–Sulfur

The ternary compound $PbBi_2S_4$, which is formed in this system, crystallizes as an orthorhombic structure with the lattice parameters a = 1178.9, b = 408.8, and c = 1457.2 pm (Ohta et al. 2014). For the preparation of this compound, a stoichiometric ratio of PbS and Bi_2S_3 with a total mass of 10 g was well mixed and then loaded into a fused silica tube. The tube was evacuated to a pressure of ~10^{-3} Pa and then flame-sealed. The mixture was heated to 700°C at a rate of ~340°C·h^{-1}, maintained at 700°C for 120 h, and then slowly cooled to room temperature at a rate of ~30°C·h^{-1}. Annealing was repeated at the same temperature after grinding/mixing the powder obtained to improve the homogeneity of the product. The total reaction time was 240 h.

The evaporation of mixture of PbS and Bi_2S_3 produced three types of $PbBi_4S_7$ crystals (Takéuchi et al. 1974). One type crystallizes as a monoclinic structure with the lattice parameters $a = 1324.7 \pm 0.3$, $b = 1204.2 \pm 0.2$, $c = 403.0 \pm 0.1$ pm, and $\beta = 105.02 \pm 0.02°$, and two another crystallize in the orthorhombic structure with the lattice parameters $a = 1330 \pm 10$, $b = 3680 \pm 20$, and $c = 403 \pm 1$ pm, and $a = 1334.3 \pm 0.2$, $b = 6002.2 \pm 0.9$, and $c = 403.3 \pm 0.1$ pm, respectively.

The ternary compound $Pb_3Bi_2S_6$, which is also formed in this system, also crystallizes as an orthorhombic structure with the lattice parameters $a = 411.6$, $b = 1357.1$, and $c = 2067.9$ pm (Ohta et al. 2014). This compound was obtained in the same way as $PbBi_2S_4$ was synthesized.

A.9.14 Lead–Selenium–Sulfur

There are no low-temperature phase transitions for PbS_xSe_{1-x} solid solutions (Lebedev and Sluchinskaya 1994).

A.10 Systems Based on Lead Selenide

A.10.1 Lead–Copper–Selenium

The liquidus surface of the Pb–Cu–Se ternary system was constructed through DTA and XRD and is presented in Figure A.10.1 (Mamedov et al. 2019). On the liquidus surface, there are the fields of primary crystallization of Cu, high-temperature modification of Cu_2Se and PbSe, as well as three areas of immiscibility. The crystallization surface of PbSe in this system at high pressure of selenium was also constructed.

A.10.2 Lead–Yttrium–Selenium

The ternary $Y_{4.2}Pb_{0.7}Se_7$ compound, which is formed in this system, crystallizes as a monoclinic structure with the lattice parameters $a = 1335.7 \pm 0.1$, $b = 404.69 \pm 0.03$, $c = 1223.56 \pm 0.08$ pm, $\beta = 104.529 \pm 0.003°$, and a calculated density of 5.556 g·cm^{-3} (Shemet et al. 2005).

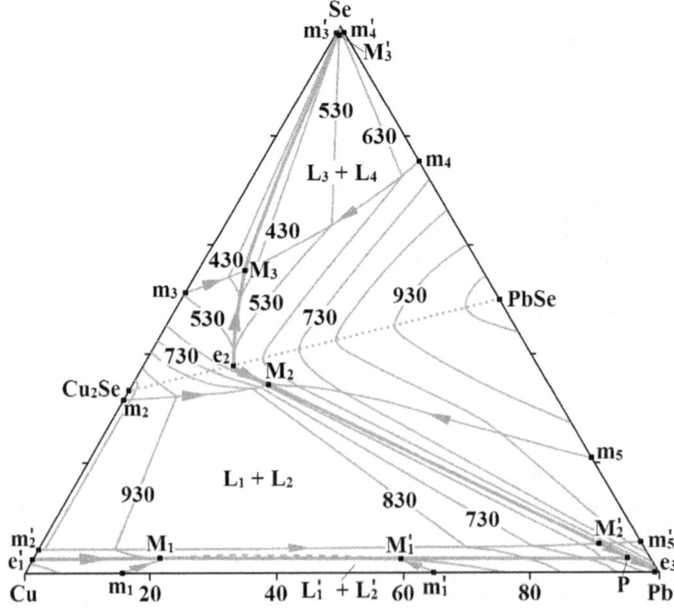

FIGURE A.10.1 Liquidus surface of the Pb–Cu–Se ternary system (From Mamedov, A.N., et al., *Chem. Probl.*, (1), 16, 2019.) Open access.

A.10.3 Lead–Praseodymium–Selenium

PbSe–PrSe. It was established that in this system the $PrPbSe_2$ ternary compound is formed, which melt incongruently at 1248°C (Shafagatova et al. 2012). According to the authors, continuous solid solutions are formed in the PbSe–$PrPbSe_2$ and $PrPbSe_2$–PrSe subsystems.

A.10.4 Lead–Arsenic–Selenium

PbSe–As_2Se_3. The phase diagram of this system constructed through DTA and XRD is given in Figure A.10.2 (Ostapyuk et al. 2014). The $PbAs_2Se_4$ ternary compound was not found in the system. The monotectic process $L_1 \Leftrightarrow L_2$ + PbSe takes place at 502°C and the degenerated eutectic crystallizes at 375°C.

A.10.5 Lead–Bismuth–Selenium

PbSe–Bi_2Se_3. The phase diagram of this system constructed through DTA and XRD is given in Figure A.10.3 (Shelimova et al. 2008, 2010; Zemskov et al. 2011). Three ternary compounds, $Pb_5Bi_6Se_{14}$, $Pb_5Bi_{12}Se_{23}$, and $Pb_5Bi_{18}Se_{32}$, which melt incongruently at 720°C, 700°C, and 675°C, are formed in this system. These compounds with monoclinic lattices belong to homologous series $[(PbSe)_5]_m[(Bi_2Se_3)_3]_n$. The eutectic contains 17.5 mol% PbSe and crystallizes at 657°C. The alloys for the investigations were annealed at 500°C for 200–400 h.

$Pb_5Bi_6Se_{14}$ crystallizes as a monoclinic structure with the lattice parameters a = 2153.7, b = 420.2, c = 1602.8 pm, and β = 97.34° (Ohta et al. 2014). For the preparation of this compound, a stoichiometric amount of PbSe and Bi_2Se_3 powders with a total mass of 10 g was well mixed and loaded into a fused silica tube. The tube was evacuated to a pressure of ~10^{-3} Pa and then flame-sealed. The mixture was heated to 600°C at a rate of ~290°C·h^{-1}, maintained at 600°C for 120 h, and then slowly cooled to room temperature at a rate of ~20°C·h^{-1}. To improve the homogeneity of the product, the powder obtained was thoroughly mixed and heated repeatedly under the same condition at 600°C for 120 h. The total reaction time was 240 h.

$Pb_5Bi_{12}Se_{23}$ and $Pb_5Bi_{18}Se_{32}$ also crystallize in the monoclinic structure with the lattice parameters a = 5316.8 ± 0.4, b = 417.86 ± 0.03, c = 2155.0 ± 0.2 pm, β = 107.491 ± 0.003°, a calculated density of

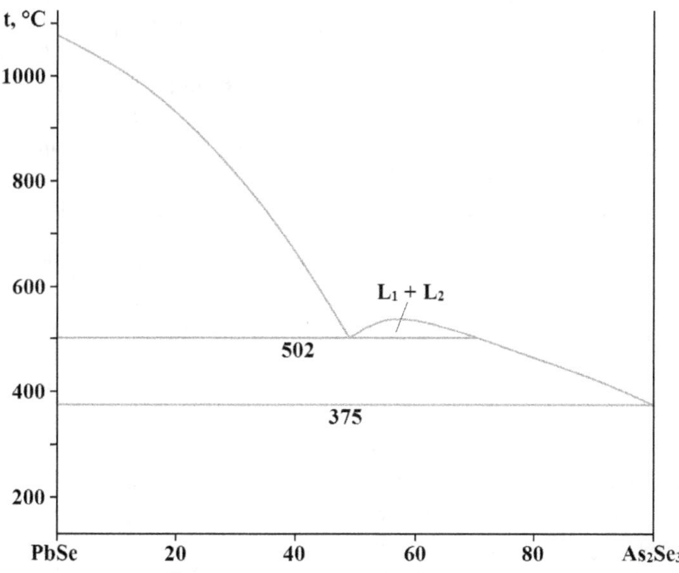

FIGURE A.10.2 Phase diagram of the PbSe–As_2Se_3 quasibinary system. (From Ostapyuk, T.A., et al., *Chem. Met. Alloys*, 7(1–2), 164, 2014.) Open access.

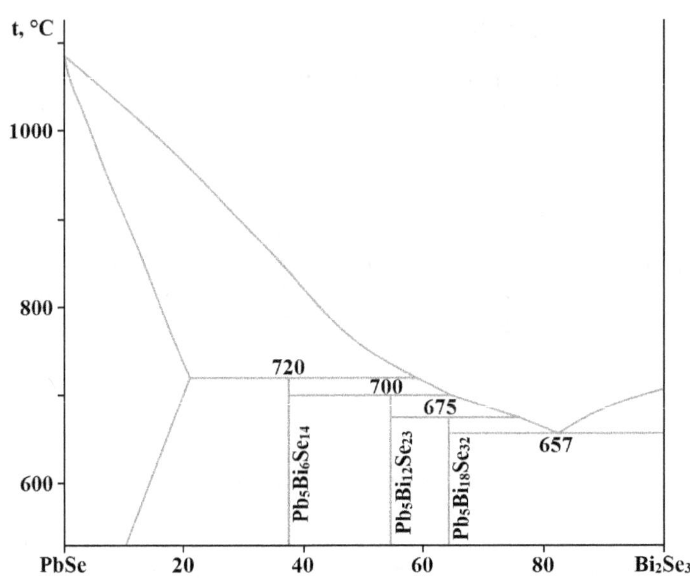

FIGURE A.10.3 Phase diagram of the PbSe–Bi_2Se_3 quasibinary system. (From Shelimova, L.E., et al., *Inorg. Mater.*, 46(2), 120, 2010.)

7.79 g·cm^{-3}, and an direct and indirect band gaps of 0.38 and 0.26 eV, respectively, for the first compound and a = 3566.7 ± 0.2, b = 416.70 ± 0.02, c = 2153.0 ± 0.1 pm, β = 102.275 ± 0.002°, a calculated density of 7.78 g·cm^{-3}, and an direct and indirect band gaps of 0.40 and 0.27 eV, respectively, for the second one (Sassi et al. 2018). To synthesize these compounds, appropriate stoichiometric ratios of PbSe and Bi_2Se_3 were ground into fine powders, pelletized, and loaded into evacuated quartz tubes sealed under secondary vacuum. The pellets were annealed for 10 days at 600°C. After the tubes were quenched in room-temperature water, the pellets were reground into fine powders and consolidated in a graphite die by spark plasma sintering at 400°C for 10 min under a pressure of 80 MPa. All manipulations were carried out in an argon-filled, dry glove box.

Two more ternary compounds, $Pb_6Bi_2Se_9$ and $Pb_7Bi_4Se_{13}$, were found in the PbSe–Bi_2Se_3 system. The first compound melts congruently at 990°C, is a narrow band gap (E_g = 0.58 eV) degenerate n-type semiconductor, and crystallizes as an orthorhombic structure with the lattice parameters a = 425.67 ± 0.09, b = 1410.5 ± 0.3, c = 3241.2 ± 0.7 pm, and a calculated density of 8.10 g·cm^{-3} (Casamento et al. 2017). A polycrystalline powder of this compound was synthesized from a solid state reaction of high-purity elemental powder in the desired ratio. A mixture of the starting reagents was loaded into fused quartz tube under an argon atmosphere, and the tube was flame-sealed under a residual pressure of 0.1 Pa. The sealed tube was loaded into a furnace, heated to 500°C over 12 h, and maintained at that temperature for 120 h to allow for complete reaction of constituent elements and to improve the crystallinity of the product. The furnace was finally cooled to room temperature over 24 h.

$Pb_7Bi_4Se_{13}$ melts congruently at 562°C, decomposes at temperatures above 700°C, is a narrow band gap semiconductor with E_g = 0.29 eV, and crystallizes as a monoclinic structure with the lattice parameters a = 1393.3 ± 0.2, b = 424.6 ± 0.1, c = 2329.6 ± 0.2 pm, β = 98.0 ± 0.1°, and a calculated density of 8.06 g·cm^{-3} at 100 K and a = 1399.1 ± 0.3, b = 426.2 ± 0.2, c = 2343.2 ± 0.5 pm, β = 98.3 ± 0.3°, and a calculated density of 7.96 g·cm^{-3} at room temperature (Olvera et al. 2015). Single crystals of this compound were obtained from a solid state reaction involving Pb powder, Bi pieces, and Se powder in the desired ratio (total mass was 5 g). A mixture of the starting reagents was loaded into a fused silica tube under an argon atmosphere, and the tube was flame-sealed under a residual pressure of 0.1 Pa. The sealed tube was loaded into a tube furnace, heated slowly to 300°C over 12 h, and maintained at that temperature for 48 h to allow complete melting of Bi and Se and reaction with the remaining lead powder. The temperature was then ramped up to 500°C and held constant for 72 h in order to improve the crystallinity of the product and also to promote crystal growth. Several needle-shaped black single crystals suitable for XRD were selected. The crystallinity of the final product was improved by annealing the ball-milled powder at 500°C for 14 days.

A.10.6 Lead–Oxygen–Selenium

The ternary compound $PbSeO_3$ (mineral paramolybdomenite), which is formed in this system, crystallizes as a monoclinic structure with the lattice parameters $a = 899.7 \pm 0.1$, $b = 815.9 \pm 0.1$, $c = 903.2 \pm 0.1$ pm, and $\beta = 103.33 \pm 0.01°$ (Pekov et al. 2023c,d).

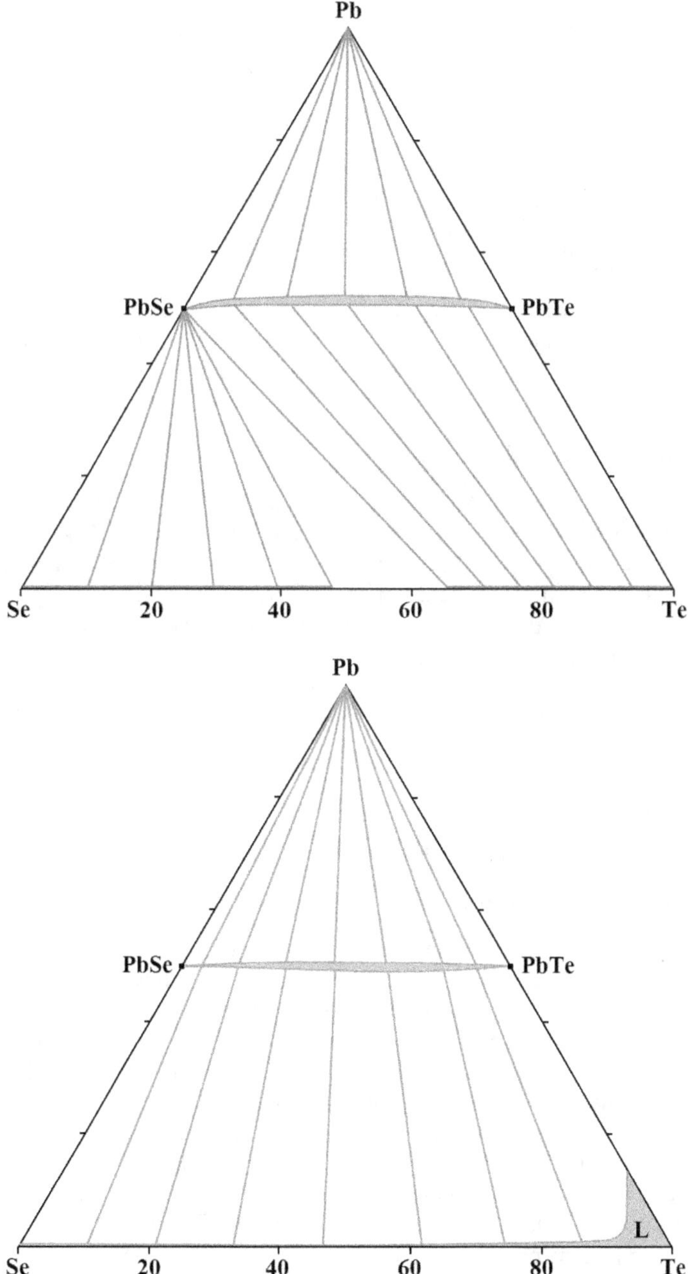

FIGURE A.10.4 Isothermal section of the Pb–Te–Se ternary system at (a) 350°C and (b) 500°C. (From Chen, S.-W., et al., *Calphad*, 74, 102310, 2021.)

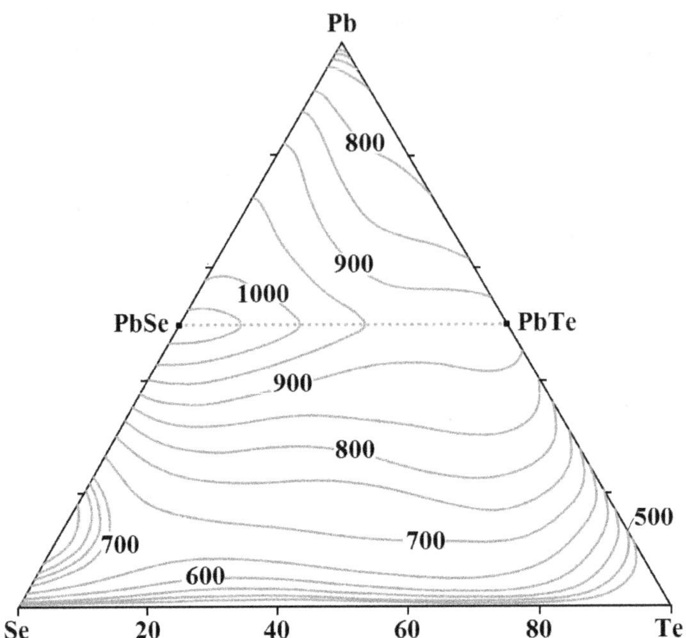

FIGURE A.10.5 Liquidus surface of the Pb–Te–Se ternary system (From Fikar, O., et al., *Calphad*, 74, 102309, 2021.)

A.10.7 Lead–Tellurium–Selenium

The isothermal sections of the Pb–Te–Se ternary system at 350°C and 500°C were proposed by Chen et al. (2021c) based on the experimental results and are presented in Figure A.10.4. No ternary compounds were found. There is one three-phase field, $PbSe_xTe_{1-x}$ + L + Se_xTe_{1-x}, at 350°C and no three-phase field at 500°C. At 350°C, there are also two liquid phases separated by the $PbSe_xTe_{1-x}$ solid solutions. PbSe and PbTe form a continuous solid solution at both temperatures. The calculated liquidus surface of this ternary system is given in Figure A.10.5 (Fikar et al. 2021). Practically, all liquidus surface is occupied by the field of the $PbTe_xSe_{1-x}$ solid solutions primary crystallization.

There are no low-temperature phase transitions for $PbTe_xSe_{1-x}$ solid solutions (Lebedev and Sluchinskaya 1994).

A.11 Systems Based on Lead Telluride

A.11.1 Lead–Bismuth–Tellurium

PbTe–Bi_2Te_3. A newly refined version of the phase diagram of this system was constructed through DTA and XRD by Gojayeva et al. (2022) and is presented in Figure A.11.1. It was established that three layered ternary compounds, $PbBi_2Te_4$, $PbBi_4Te_7$, and $PbBi_6Te_{10}$, which melt incongruently at 591°C, 583°C, and 578°C, are formed in this system. Earlier reported ternary compounds $Pb_3Bi_4Te_9$, $Pb_2Bi_6Te_{11}$, and $PbBi_8Te_{13}$ were not confirmed by the XRD studies. The eutectic contains 85 mol% Bi_2Te_3 and crystallizes at 575°C. The compositions of the peritectic points were found to be at 62, 70, and 78 mol% Bi_2Te_3. The existence of homogeneity areas based on all ternary compounds and the starting phases was detected. The homogeneity region based on PbTe reaches 18 mol% at 591°C. $PbBi_2Te_4$ and $PbBi_6Te_{10}$ crystallize as a rhombohedral structure with the lattice parameters $a = 443.82 \pm 0.5$ and $c = 4177.5 \pm 0.6$ and $a = 440.12 \pm 0.5$ and $c = 10253.8 \pm 0.2$ in hexagonal setting, respectively, and $PbBi_4Te_7$ crystallizes as a trigonal structure with the lattice parameters $a = 442.34 \pm 0.5$ and $c = 2387.6 \pm 0.8$.

The standard thermodynamic functions of the $PbBi_4Te_7$ and $PbBi_6Te_{10}$ were calculated using the potential-forming reactions method (Guseynov 2010).

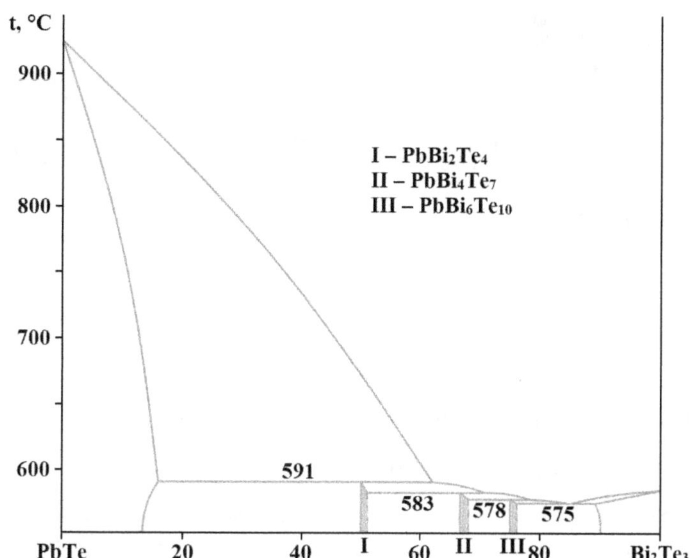

FIGURE A.11.1 Phase diagram of the PbTe–Bi_2Te_3 quasibinary system. (From Gojayeva, I.M., et al., *Azerb. Chem. J.*, (2), 47, 2022.) Open access.

A.11.2 Lead–Oxygen–Tellurium

One of the modifications of $PbTeO_3$ crystallizes as a monoclinic structure with the lattice parameters a = 2644.2 ± 0.5, b = 460.10 ± 0.09, c = 1789.0 ±0.4 pm, and β = 107.01 ±0.03° (Missen et al. 2019). The mineral matthiasweilite with the same chemical composition crystallizes as a triclinic structure with the lattice parameters a = 702.56 ± 0.08, b =1063.45 ± 0.06, c = 1199.65 ± 0.08 pm, α = 78.513 ±0.006°, β = 83.104 ±0.006°, γ = 84.083 ± 0.006°, and a calculated density of 7.314 g·cm^{-3} (Kampf et al. 2022b,c,d).

The $PbTeO_4$ ternary compound (mineral murphyite) also crystallizes as a monoclinic structure with the lattice parameters a = 1360.89 ± 0.03, b = 501.75 ± 0.01, c = 557.67 ±0.02 pm, β = 107.928 ±0.001°, and a calculated density of 7.579 g·cm^{-3} (Yang et al. 2022e,f,2023c).

Appendix B

Quaternary Alloys Based on IV-VI and IV-VI_2 Semiconductors

B.1 Systems Based on Silicon Sulfides

B.1.1 Silicon–Lithium–Aluminum–Sulfur

SiS_2–Li_2S–Al_2S_3. The glass-forming region in this quasiternary system is quite restricted and limited to compositions containing at least 40 mol% and at most 60 mol% SiS_2 (Martin and Sills 1991). The glass transition temperature shows a monotonic increase with substitution of Al_2S_3 for SiS_2 and Al_2S_3 for Li_2S. The highest values of the glass transition temperature were measured for glasses with maximum content of Al_2S_3. The glasses were prepared by weighing appropriate amounts of SiS_2, Li_2S, and Al_2S_3 into carbonized quartz tubes. The tubes were evacuated, sealed, and placed in a SiC tube furnace. They were heated to 1100°C and maintained there for 1 h after which they were quenched in a sand bath.

B.1.2 Silicon–Lithium–Phosphorus–Sulfur

The quaternary compound Li_7SiPS_8, which has two polymorphic modifications, is formed in this system (Harm et al. 2019). The first modification crystallizes as an orthorhombic structure with the lattice parameters $a = 1334.8 \pm 0.3$, $b = 797.03 \pm 0.16$, and $c = 613.43 \pm 0.12$ pm. It can be regarded as a solid solution of orthorhombic Li_4SiS_4 and Li_3PS_4. The second modification crystallizes as a tetragonal structure with the lattice parameters $a = 867.34 \pm 0.02$, $c = 1254.12 \pm 0.02$ pm, and a calculated density of 1.923 ± 0.009 g·cm^{-3}. Both modifications were prepared by heating Li_2S, Si, red P, and S (sublimed in vacuum) in stoichiometric amounts in a glassy carbon crucible sealed in a quartz glass ampoule under vacuum to 950°C for 2 h. Subsequently, the ampoule was quenched in ice water and annealed for 5 days at 575°C, 550°C, 525°C, 500°C, or 475°C. Cooling rates of the samples after annealing showed no significant influence on side-phase formation, and samples were therefore left in the oven after the end of the annealing program until the temperature was below 100°C. The resulting products were moisture-sensitive red or slightly yellow microcrystalline powders. All analytical procedures were therefore conducted in an Ar-filled glove box or in argon-filled sealed containers.

It was shown that annealing temperatures exceeding 500°C lead to the formation of tetragonal Li_7SiPS_8 and the tetragonal modification transforms to orthorhombic after annealing at 550°C for 3 h. Tetragonal Li_7SiPS_8 is a glass ceramic rather than a fully crystalline material since at least one amorphous side phase is present at all the explored temperatures.

B.1.3 Silicon–Lithium–Oxygen–Sulfur

SiS_2–Li_2S–Li_2SiO_4. Lithium-ion-conducting oxysulfide glasses were prepared along the ($0.4SiS_2$+ $0.6Li_2S$)–Li_2SiO_4 section of this quasiternary system by twin-roller rapid quenching (Tatsumisago et al. 1996). The glass-forming region is situated in the region of 0–20 mol% Li_2SiO_4, and the addition of 5 mol% Li_2SiO_4 improved the ionic conductivity and the glass stability against crystallization.

B.1.4 Silicon–Sodium–Aluminum–Sulfur

Na_4SiS_4–Na_5AlS_4. The $Na_{10-2x}(AlS_4)_{2-2x}(SiS_4)_{2x}$ solid solutions could be obtained in this system in the region of $0.5 \leq x \leq 0.75$ (Harm et al. 2020). These solid solutions crystallize as a monoclinic structure with the lattice parameters $a = 1756.73 \pm 0.06$, $b = 1354.08 \pm 0.05$, $c = 1425.43 \pm 0.05$ pm, and $\beta = 93.3683 \pm 0.0005°$ for $x = 0.75$. They were obtained when stoichiometric amounts of Na_2S, Al_2S_3, Si (ball-milled), and S (sublimed in vacuum) were used as starting materials. An excess of 5 mass% S was added to the mixture to ensure an oxidizing atmosphere during the reaction. Samples were prepared by thoroughly mixing and grinding the starting materials in an agate mortar. The resulting fine powders were transferred into glassy carbon crucibles, compacted, and sealed under vacuum into quartz ampoules. The ampoules were subsequently transferred into a tube furnace and heated at a rate of $50°C{\cdot}h^{-1}$ to 600°C and annealed for 3 days. Subsequently, the furnace was turned off. The ampoules were removed from the furnace when the temperature was below 100°C and transferred to a glove box. The colorless cuboids of $Na_{10-2x}(AlS_4)_{2-2x}(SiS_4)_{2x}$ with a diameter of about 200 μm, immersed in an orange amorphous material as well as the Na_4SiS_4 and Na_5AlS_4 compounds were obtained.

B.1.5 Silicon–Sodium–Gallium–Sulfur

The quaternary compound $Na_2Ga_2SiS_6$, which crystallizes as an orthorhombic structure with the lattice parameters $a = 1237.54 \pm 0.15$, $b = 2236.2 \pm 0.2$, $c = 728.33 \pm 0.09$ pm, a calculated density of 2.662 $g{\cdot}cm^{-3}$, and an energy gap of 3.93 eV, is formed in the Si–Na–Ga–S system (Xie et al. 2022b). The polycrystalline samples of this compound were manufactured by a high-temperature melt spontaneous crystallization method as follows. A total of 507 mg of starting materials Na (Na metal was weighed in an Ar-filled glove box to prevent oxidation), Ga, Si, and S was weighed in a molar ratio of 2:2:1:6, loaded into a graphite crucible, and then put into a clean silica tube under an Ar-filled glove box. The silicon tube was pumped to a high vacuum of 10^{-3} Pa and sealed with gas-oxygen flame. Then, the silica tube was put in a computer-controlled muffle furnace. The temperature of the furnace was gradually increased from room temperature to 830°C over two days, maintained for one day, and then cooled down from 830°C to 600°C over five days. After that, the furnace was cooled to room temperature in 10 h.

B.1.6 Silicon–Sodium–Phosphorus–Sulfur

SiS_2–Na_2S–P_2S_5. The calculated isothermal section of this quasiternary system at 0 K is characterized by the existence of seven quasibinary systems, namely Na_4SiS_4–Na_3PS_4, $Na_6Si_2S_7$–Na_3PS_4, Na_2SiS_3–Na_3PS_4, Na_2SiS_3–$Na_4P_2S_7$, $Na_2Si_2S_5$–$Na_4P_2S_7$, SiS_2–$Na_4P_2S_7$, and SiS_2–$NaPS_3$ (Richards et al. 2016). In the Na_4SiS_4–Na_3PS_4 section, a quaternary compound $Na_{10}SiP_2S_{12}$ can be formed, crystallizing in a tetragonal structure with lattice parameters $a = 960$ and $c = 1353$ pm (calculated values).

B.1.7 Silicon–Sodium–Oxygen–Sulfur

The quaternary compound $NaSi(S_2O_7)_3$, which crystallizes as an orthorhombic structure with the lattice parameters $a = 954.15 \pm 0.05$, $b = 1266.76 \pm 0.07$, $c = 1443.63 \pm 0.08$ pm, and a calculated density of 2.293 $g{\cdot}cm^{-3}$, is formed in the Si–Na–O–S system (Logemann et al. 2011). To synthesize this compound, $SiCl_4$ (1 mM), oleum (1 mL, 65% SO_3), and Na_2SO_4 (1 mM) were filled in a glass tube. The tube was sealed, placed in an oven, and heated to 250°C. After slow cooling ($1.8°C{\cdot}h^{-1}$), colorless and moisture-sensitive single crystals were obtained. During the reaction and even after cooling down to room temperature, the glass tube is under pressure. The tube has to be protected with an explosion shield during the reaction and should be cooled with liquid nitrogen before opening.

B.1.8 Silicon–Potassium–Gallium–Sulfur

The quaternary compound $K_{2.4}Ga_{2.4}Si_{1.6}S_8$, which crystallizes as a tetragonal structure with the lattice parameters $a = 789.10 \pm 0.07$, $c = 586.60 \pm 0.05$ pm, and a calculated density of 2.558 $g{\cdot}cm^{-3}$, is formed in the Si–K–Ga–S system (He et al. 2016). The crystals of the title compound were obtained through reactive flux method with K_2S_2 as the reactive flux. The starting materials K_2S_2 (2.25 mM), Ga_2S_3 (0.375 mM), Si (0.5 mM), and S (1.75 mM) were mixed and ground uniformly and loaded into a silica tube. The tube was then flame-sealed under vacuum (0.1 Pa) and slowly heated to 750°C in a programmable furnace. It was kept at this temperature for 3 days and then slowly cooled down to 300°C at a rate of 2°C·h^{-1}, and finally to room temperature by turning off the furnace. The melt was washed by DMF several times to remove the excessive flux and dried by acetone. Thin pink prisms up to 2–3 cm long were obtained. The powder sample was synthesized by stoichiometric combination reaction of the starting materials under the same reacting condition and by repeating the procedure three times. All operations were carried out in an Ar-protected glove box.

B.1.9 Silicon–Potassium–Praseodymium–Sulfur

The quaternary compound $KPrSiS_4$, which crystallizes as a monoclinic structure with the lattice parameters a = 649.41 ± 0.02, b = 659.91 ± 0.02, c = 865.83 ± 0.03 pm, β = 107.7620 ± 0.0010°, and a calculated density of 3.161 g·cm^{-3}, is formed in the Si–K–Pr–S system (Usman et al. 2019). Single crystals of this compound as green plates were grown by layering a mixture of Pr_2S_3 (0.5 mM), Si powder (1 mM), and S (2.5 mM) beneath a layer of 1 g of oven-dried KCl in an 8-in. long, finely carbon-coated fused silica tube. The loaded silica tube was evacuated to a pressure of $\lesssim 10^{-2}$ Pa, flame-sealed using an oxygen/methane torch, and placed in a furnace. A three-step reaction temperature profile was used that consisted of a ramp at 300°C·h^{-1} to 500°C, maintained at this temperature for 12 h, a ramp at 60°C·h^{-1} to 700°C, maintained at this temperature for 12 h, and finally a ramp at 60°C·h^{-1} to 900°C, maintained at this temperature for 24 or 48 h, and then fast cooling to room temperature by shutting off the furnace.

B.1.10 Silicon–Potassium–Neodymium–Sulfur

The quaternary compound $KNdSiS_4$, which crystallizes as a monoclinic structure with the lattice parameters a = 647.11 ± 0.06, b = 659.77 ± 0.06, c = 864.51 ± 0.08 pm, β = 107.857 ± 0.004°, and a calculated density of 3.211 g·cm^{-3}, is formed in the Si–K–Nd–S system (Usman et al. 2019). Single crystals of this compound as light blue needles were grown in the same way as in the case of $KPrSiS_4$ using Nd_2S_3 instead of Pr_2S_3.

B.1.11 Silicon–Potassium–Oxygen–Sulfur

The quaternary compound $K_2Si(S_2O_7)_3$ is formed in the Si–K–O–S system (Thilo and Winkler 1969). It was obtained by the reaction of K_2SiO_3 with excess of SO_3 in a tube at 95°C for about 24 h. The excess of SO_3 was distilled off at 65°C. When heated in a dry atmosphere, this compound remains unchanged up to 120°C and then gradually loses SO_3 at a heating rate of 1°C·min^{-1} up to 280°C.

B.1.12 Silicon–Rubidium–Zinc–Sulfur

The quaternary compound $Rb_2ZnSi_3S_8$, which melts at 631°C and crystallizes as a triclinic structure with the lattice parameters a = 707.98 ± 0.06, b = 748.67 ± 0.07, c = 1416.10 ± 0.12 pm, α = 88.319 ± 0.003°, β = 83.687 ± 0.003°, γ = 71.978 ± 0.003°, a calculated density of 2.701 g·cm^{-3}, and an energy gap of 3.66 eV, is formed in the Si–Rb–Zn–S system (Gao et al. 2021c). To prepare this compound, a mixture with ZnS (1 M), SiS_2 (3 M), S (1 M), and excess RbCl (3 M) was loaded into the graphite crucible and then put them into vacuum-sealed silica tube. Muffle furnace was used to complete the spontaneous crystallization and detailed setting temperature process was shown as following: firstly heated up to 750°C in 40 h to ensure the mixture melt completely and kept at this temperature within 5 days, and then slowly cooled down to room temperature in 6 days. Finally, many colorless crystals were found under the optical microscope after washing with DMF solution. Deionized water was also used to remove the excess RbCl. The title compound is stable in the air for several months.

B.1.13 Silicon–Rubidium–Cadmium–Sulfur

The quaternary compound $Rb_2CdSi_4S_{10}$, which is thermally stable up to 730°C and crystallizes as an orthorhombic structure with the lattice parameters a = 1863.78 ± 0.06, b = 913.09 ± 0.03, c = 1061.36 ± 0.03 pm, and a calculated density of 2.634 g·cm^{-3}, is formed in the Si–Rb–Cd–S system (Zhou et al. 2023). A single crystal of this compound was prepared by the flux method. The starting materials RbCl, CdS, Si and S were weighed and loaded into graphite crucible with the molar ratio of 2:1:3:8. The silica tube was sealed under a high vacuum of 10^{-3} Pa. After that, the sealed tube was put into a computer-controlled furnace and heating to 600°C in 20 h, kept at this temperature for 24 h, and after that the temperature was cooled to room temperature at a rate of 2.5°C·h^{-1}. Finally, the colorless single crystals of the title compound were harvested with a yield of about 90%. The pure phase of $Rb_2CdSi_4S_{10}$ powder sample was synthesized by solid state reaction at 630°C with stoichiometric ratio. All the raw reagents were stored in a dry Ar-filled glove box with limited oxygen and moisture levels below 0.1 ppm.

B.1.14 Silicon–Rubidium–Yttrium–Sulfur

The quaternary compound $RbYSiS_4$, which crystallizes as an orthorhombic structure with the lattice parameters a = 633.81 ± 0.04, b = 659.61 ± 0.04, c = 1688.68 ± 0.12 pm, and a calculated density of 3.111 g·cm^{-3}, is formed in the Si–Rb–Y–S system (Tarasenko et al. 2022). To prepare this compound, Y_2S_3 (1.4 mM), Si (2.1 mM), and S (4.2 mM) taken in a stoichiometric ratio were placed in a dry ampoule. RbCl (24.8 mM) was added to the mixture. The ampoule was evacuated to a residual pressure of ~ 1 Pa and sealed. It was placed into a programmable furnace and heated to 820°C in 10 h, kept at 820°C for 180 h, and cooled to room temperature in 12 h. The crystals are yellowish transparent rectangular plates. The title compound is completely hydrolyzed by water upon washing.

B.1.15 Silicon–Rubidium–Cerium–Sulfur

The quaternary compound $RbCeSiS_4$, which crystallizes as an orthorhombic structure with the lattice parameters a = 1716.8 ± 0.2, b = 664.89 ± 0.09, c = 647.14 ± 0.08 pm, and a calculated density of 3.434 g·cm^{-3}, is formed in the Si–Rb–Ce–S system (Usman et al. 2019). Single crystals of this compound as yellow plates were grown in the same way as in the case of $KPrSiS_4$ using Ce_2S_3 instead of Pr_2S_3 and RbCl instead of KCl.

B.1.16 Silicon–Rubidium–Praseodymium–Sulfur

The quaternary compound $RbPrSiS_4$, which crystallizes as an orthorhombic structure with the lattice parameters a = 646.11 ± 0.02, b = 664.85 ± 0.02, c = 1711.55 ± 0.06 pm, and a calculated density of 3.457 g·cm^{-3}, is formed in the Si–Rb–Pr–S system (Usman et al. 2019). Single crystals of this compound as green rods were grown in the same way as in the case of $KPrSiS_4$ using and RbCl instead of KCl.

B.1.17 Silicon–Rubidium–Neodymium–Sulfur

The quaternary compound $RbNdSiS_4$, which crystallizes as an orthorhombic structure with the lattice parameters a = 644.16 ± 0.02, b = 664.73 ± 0.02, c = 1706.76 ± 0.04 pm, and a calculated density of 3.509 g·cm^{-3}, is formed in the Si–Rb–Nd–S system (Usman et al. 2019). Single crystals of this compound as light blue plates were grown in the same way as $KPrSiS_4$ but using Nd_2S_3 instead of Pr_2S_3 and RbCl instead of KCl.

B.1.18 Silicon–Rubidium–Gadolinium–Sulfur

The quaternary compound $RbGdSiS_4$, which crystallizes as an orthorhombic structure with the lattice parameters a = 637.55 ± 0.02, b = 662.40 ± 0.02, c = 1698.20 ± 0.04 pm, and a calculated density of 3.696 g·cm^{-3}, is formed in the Si–Rb–Gd–S system (Usman et al. 2019). Single crystals of this compound as colorless rods were grown in the same way as $KPrSiS_4$ but using Gd_2S_3 instead of Pr_2S_3 and RbCl instead of KCl.

B.1.19 Silicon–Rubidium–Terbium–Sulfur

The quaternary compound $RbTbSiS_4$, which crystallizes as an orthorhombic structure with the lattice parameters a = 634.92 ± 0.02, b = 661.42 ± 0.02, c = 1696.43 ± 0.05 pm, and a calculated density of 3.736 g·cm^{-3}, is formed in the Si–Rb–Tb–S system (Usman et al. 2019). Single crystals of this compound as colorless blocks were grown in the same way as in the case of $KPrSiS_4$ using Tb_2S_3 instead of Pr_2S_3 and RbCl instead of KCl.

B.1.20 Silicon–Rubidium–Dysprosium–Sulfur

The quaternary compound $RbDySiS_4$, which crystallizes as an orthorhombic structure with the lattice parameters a = 633.55 ± 0.02, b = 660.58 ± 0.02, c = 1694.65 ± 0.05 pm, and a calculated density of 3.786 g·cm^{-3}, is formed in the Si–Rb–Dy–S system (Usman et al. 2019). Single crystals of this compound as colorless rods were grown in the same way as $KPrSiS_4$ but using Dy_2S_3 instead of Pr_2S_3 and RbCl instead of KCl.

B.1.21 Silicon–Rubidium–Holmium–Sulfur

The quaternary compound $RbHoSiS_4$, which crystallizes as an orthorhombic structure with the lattice parameters a = 632.18 ± 0.02, b = 659.96 ± 0.02, c = 1693.07 ± 0.05 pm, and a calculated density of 3.825 g·cm^{-3}, is formed in the Si–Rb–Ho–S system (Usman et al. 2019). Single crystals of this compound as yellow plates were grown in the same way as $KPrSiS_4$ but using Ho_2S_3 instead of Pr_2S_3 and RbCl instead of KCl.

B.1.22 Silicon–Cesium–Zinc–Sulfur

The quaternary compound $Cs_2ZnSi_3S_8$, which melts at 521°C and crystallizes as a triclinic structure with the lattice parameters a = 716.30 ± 0.03, b = 762.27 ± 0.03, c = 1424.37 ± 0.06 pm, α = 89.2550 ± 0.0010°, β = 85.4970 ± 0.0010°, γ = 73.2450 ± 0.0010°, a calculated density of 3.006 g·cm^{-3}, and an energy gap of 3.65 eV, is formed in the Si–Cs–Zn–S system (Gao et al. 2021c). This compound was obtained in the same way as $Rb_2ZnSi_3S_8$ but using CsCl instead of RbCl. Finally, many colorless crystals were found under the optical microscope after washing with DMF solution. Deionized water was also used to remove the excess CsCl. The title compound is stable in the air for several months.

B.1.23 Silicon–Cesium–Yttrium–Sulfur

The quaternary compound $CsYSiS_4$, which crystallizes as an orthorhombic structure with the lattice parameters a = 633.03 ± 0.04, b = 665.93 ± 0.05, c = 1757.95 ± 0.13 pm, a calculated density of 3.389 g·cm^{-3}, and an energy gap of 3.8 eV, is formed in the Si–Cs–Y–S system (Tarasenko et al. 2022). This compound was synthesized in the same way as $RbYSiS_4$ but using CsCl (17.8 mM) instead of RbCl (24.8 mM). The obtained product was rapidly washed from the CsCl melt with cold distilled water and dried with alcohol. In dry air, the crystals of the title compound are stable; in humid air, the crystals quickly become cloudy and disintegrate into a fine powder.

B.1.24 Silicon–Cesium–Lanthanum–Sulfur

The quaternary compound $CsLaSiS_4$, which crystallizes as an orthorhombic structure with the lattice parameters a = 1786.14 ± 0.11, b = 671.82 ± 0.04, c = 647.76 ± 0.04 pm, and a calculated density of 3.659 g·cm^{-3}, is formed in the Si–Cs–La–S system (Usman et al. 2019). Single crystals of this compound as yellow plates were grown in the same way as $KPrSiS_4$ but using La_2S_3 instead of Pr_2S_3 and CsCl instead of KCl.

B.1.25 Silicon–Cesium–Praseodymium–Sulfur

The quaternary compound $CsPrSiS_4$, which crystallizes as an orthorhombic structure with the lattice parameters $a = 643.70 \pm 0.02$, $b = 672.58 \pm 0.02$, $c = 1779.04 \pm 0.05$ pm, and a calculated density of 3.710 g·cm^{-3}, is formed in the Si–Cs–Pr–S system (Usman et al. 2019). Single crystals of this compound as green plates were grown in the same way as in the case of $KPrSiS_4$ using CsCl instead of KCl.

B.1.26 Silicon–Cesium–Neodymium–Sulfur

The quaternary compound $CsNdSiS_4$, which crystallizes as an orthorhombic structure with the lattice parameters $a = 641.79 \pm 0.02$, $b = 671.35 \pm 0.02$, $c = 1774.18 \pm 0.05$ pm, and a calculated density of 3.767 g·cm^{-3}, is formed in the Si–Cs–Nd–S system (Usman et al. 2019). Single crystals of this compound as green plates were grown in the same way as $KPrSiS_4$ but using Nd_2S_3 instead of Pr_2S_3 and CsCl instead of KCl.

B.1.27 Silicon–Cesium–Oxygen–Sulfur

The quaternary compound $CsSi(S_2O_7)_3$, which crystallizes as a trigonal structure with the lattice parameters $a = 1672.27 \pm 0.07$, $c = 1157.49 \pm 0.07$ pm, and a calculated density of 2.922 g·cm^{-3}, is formed in the Si–Cs–O–S system (Logemann et al. 2011). To synthesize this compound, $SiCl_4$ (1 mM), oleum (1 mL, 65% SO_3), and Cs_2SO_4 (1 mM) were filled in a glass tube. The tube was sealed, placed in an oven, and heated to 250°C. After slow cooling (1.8°C·h^{-1}), colorless and moisture-sensitive single crystals were obtained. During the reaction and even after cooling down to room temperature, the glass tube is under pressure. The tube has to be protected with an explosion shield during the reaction and should be cooled with liquid nitrogen before opening.

B.1.28 Silicon–Copper–Lanthanum–Sulfur

The quaternary compound $Cu_{1.97}La_6Si_2S_{14}$, which crystallizes as a hexagonal structure with the lattice parameters $a = 1032.31 \pm 0.14$, $c = 579.91 \pm 0.08$ pm, a calculated density of 4.541 g·cm^{-3} at 100 K, and an energy gap of ~2.6 eV, is formed in the Si–Cu–La–S system (Akopov et al. 2021). The powder of this compound was synthesized using pre-arc-melted precursor and sulfur without flux, while the crystals were grown using a salt flux synthesis from a pre-arc-melted precursor and sulfur. The precursor was prepared using pieces of La, Cu, and Si. A sample with a total mass of 0.6–1 g was weighed in a molar ratio of La/Cu/Si = 6:1:2.05. A slight excess of Si was used to account for its evaporation during arc melting. The sample was then placed in an arc-melter on a copper hearth along with oxygen getter material (zirconium metal). The arc-melter chamber was later sealed and evacuated for 20 min followed by purging with argon; this process was repeated three times to ensure no oxygen was present in the chamber. During arc melting, the getter was melted first to ensure the absorbance of any trace oxygen, and then the samples were heated to 1000°C in ~ 1 min at a current of ~ 80 A until molten, allowed to solidify, flipped, and re-arced two more times to ensure homogeneity under a current of ~ 100 A. The precursor ingots were then crushed into a powder using a steel Plattner-style diamond crusher.

The single crystals of $Cu_{1.97}La_3Si_2S_{14}$ were grown using pre-arc-melted precursor, S powder, and KI as a flux. The ratio of (precursor + sulfur) to KI flux was kept at 1 to 30 ratios by mass (for 0.1 g of reactants 3 g of KI was used). The reactants and KI were added to a silica ampoule, which was sealed under vacuum. The samples were heated to 950°C–1050°C over 12 h, allowed to dwell over 72–120 h, and then cooled down to room temperature over 8 h. The KI flux was washed with deionized water, and the sample was then filtered under vacuum and dried. The crystals have a transparent yellow-gold appearance.

Bulk powder of the compound was synthesized using pre-arc melted precursor and sulfur powder. The sample was prepared by loading the metal silicide precursor and sulfur in a fused silica ampoule in a 1:14 precursor to sulfur ratio by mass. The ampoule was then flame-sealed under vacuum. The sample was heated over 12 h to 950°C–1050°C, maintained for 72–120 h, and then allowed them to cool to room temperature over 8 h.

B.1.29 Silicon–Copper–Germanium–Sulfur

Cu_8SiS_6–Cu_8GeS_6. The continuous solid solutions based on low-temperature modifications of the starting compounds are formed in this system (Bayramova et al. 2023, 2022b). The solid solutions were prepared by alloying Cu_8SiS_6 and Cu_8GeS_6 in quartz ampoules under vacuum at ~1140°C, followed by homogenizing annealing at 550°C for 500 h with the cooling in the switched off furnace. The concentration dependence of the lattice parameters of the $Cu_8Si_xGe_{1-x}S_6$ obeys Vegard's law.

Cu_8SiS_6–Cu_8GeS_6–Cu_2S. The isothermal section of this system at room temperature consists only from one two-phase α-Cu_2S–$Cu_8Si_xGe_{1-x}S$ region (Bayramova et al. 2023). The liquidus surface of the Cu_8SiS_6–Cu_8GeS_6–Cu_2S subsystem consists of two fields of primary crystallization of α-Cu_2S and $Cu_8Si_xGe_{1-x}S$ solid solutions (Figure B.1.1). These fields are bounded by a monovariant eutectic curve, which further changes into peritectic equilibrium.

B.1.30 Silicon–Silver–Indium–Sulfur

The quaternary compound $Ag_2In_2SiS_6$, which crystallizes as a monoclinic structure with the lattice parameters a = 1214.00 ± 0.05, b = 716.83 ± 0.03, c = 1211.91 ± 0.05 pm, β = 109.3010 ± 0.0010°, a calculated density of 4.443 g·cm^{-3}, and an energy gap of 2.41 eV, is formed in the Si–Ag–In–S system (Zhou et al. 2021a). To prepare this compound, Ag, In, Si, S, and KI were used as the raw materials. The total mass of the reactants was 900 mg, containing stoichiometric raw materials (500 mg) and flux KI (400 mg). The mixture was ground into a uniform powder and then pressed into a pellet. Then it was placed in a graphite crucible. Next, it was sealed in quartz tube with a vacuum degree of 0.01 Pa. Finally, the tube was transferred to a muffle furnace, where it was heated to 950°C at a rate of 30°C·h^{-1} and maintained at several intermediate equilibrium temperatures in order to prevent destruction of the quartz tube. The temperature 950°C was maintained for 5 days, and the sample was subsequently cooled to 300°C at a rate of 5°C·h^{-1} before switching off the furnace. Saffron yellow crystals were obtained. They

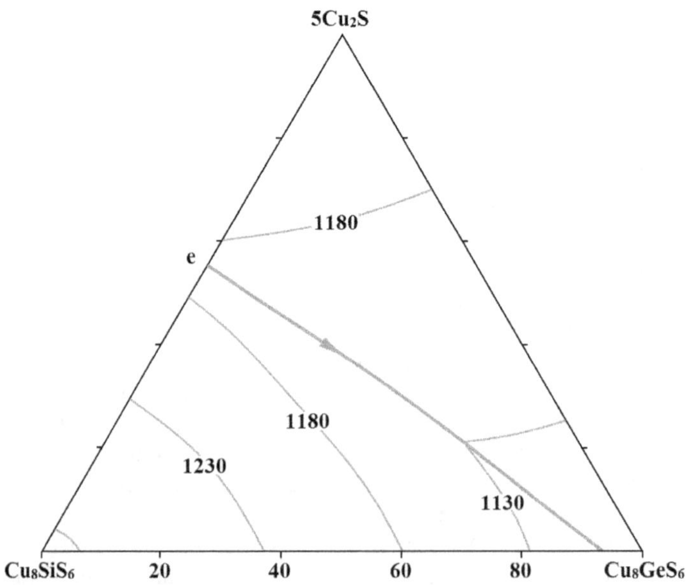

FIGURE B.1.1 Liquidus surface of the Cu_8SiS_6–Cu_8GeS_6–Cu_2S quasiternary system. (From Bayramova. U.R., et al., *J. Phase Equilib. Diffus.*,**44**(5), 509, 2023.)

were then rinsed with deionized water and alcohol, followed by ultrasonication. The crystals are stable in air after several months and can be separated from the byproducts easily under an optical microscope because of the unique color and shape.

B.1.31 Silicon–Silver–Germanium–Sulfur

Ag_8SiS_6–Ag_8GeS_6. The formation of continuous series of solid solutions between both crystalline modifications of the starting compounds was determined in this system (Figure B.1.2) (Ashirov et al. 2023a). The concentration dependence of lattice parameters of the solid solutions obeys Vegard's law. This system was studied through DTA and XRD. The ingots for the investigations were annealed at 630°C for 500 h.

Ag_8SiS_6–Ag_8GeS_6–Ag_2S. The liquidus surface of this quasiternary system consists of two fields corresponding to the primary crystallization of the high-temperature modifications of the $Ag_8Si_{1-x}Ge_xS_6$ solid solutions and high-temperature phase of Ag_2S (Figure B.1.3) (Ashirov et al. 2023a).

Two vertical sections and the isothermal section of the Ag_8SiS_6–Ag_8GeS_6–Ag_2S system were also constructed (Ashirov et al. 2023a). It was shown that the isothermal section at room temperature includes only one two-phase region: $Ag_8Si_{1-x}Ge_xS_6$ solid solutions and Ag_2S.

B.1.32 Silicon–Silver–Oxygen–Sulfur

Two more quaternary compounds, $Ag_2Si(S_2O_7)_3$ and $Si(AgOSO_3)_4$, are formed in the Si–Ag–O–S system. The first compound crystallizes as a trigonal structure with the lattice parameters a = 1594.90 ± 0.05 and c = 1031.34 ± 0.04 pm at 153 K (Logemann et al. 2012). The reaction for this compound obtaining was performed in a thick-walled glass ampoule. The tube was loaded with $SiCl_4$ (1 mM), oleum (1 mL, 65% SO_3), and Ag_2SO_4 (1 mM), torch-sealed under vacuum, and placed in a resistance furnace. The ampoule was maintained at the temperature of 250°C for 24 h and cooled down to room temperature at a rate of 1.8°C·h^{-1}. The colorless crystals are very moisture-sensitive and were separated from oleum in a glove box.

The interaction of $SiCl_4$ or $SiBr_4$ with excess Ag_2SO_4, in acetonitrile at –40°C yields, besides silver halide, $Si(AgOSO_3)_4$, which at room temperature decomposes into $Ag_2S_2O_7$ and polymeric $[(AgOSO_3)_2SiO^-]_n$ (Krüger et al. 1968).

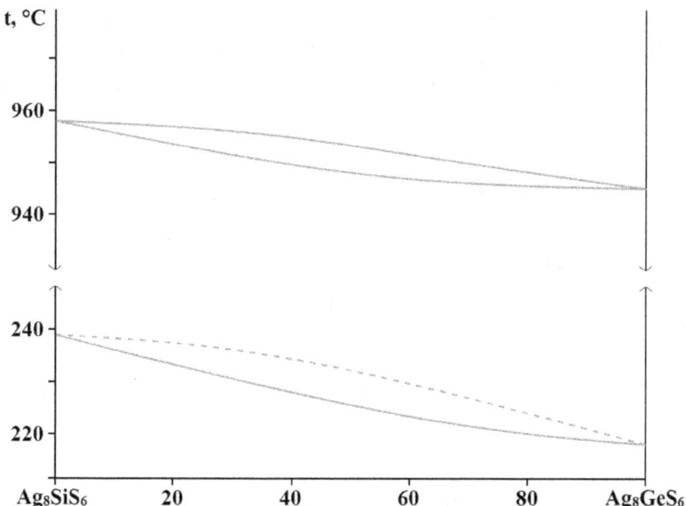

FIGURE B.1.2 Phase diagram of the Ag_8SiS_6–Ag_8GeS_6 system. (From Ashirov, G.M., et al., *Condens. Matter Interphases*, **25**(2), 292, 2023.) Open access.

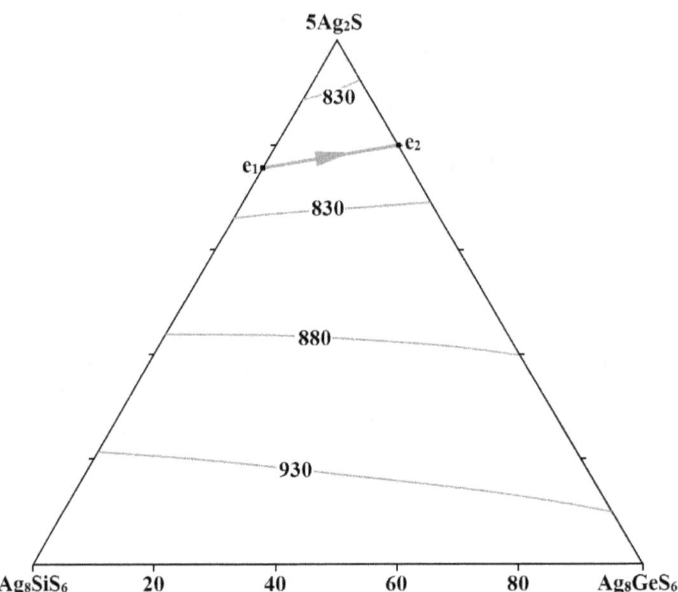

FIGURE B.1.3 Liquidus surface of the Ag_8SiS_6–Ag_8GeS_6–Ag_2S quasiternary system. (From Ashirov, G.M., et al., *Condens. Matter Interphases*, **25**(2), 292, 2023.) Open access.

B.1.33 Silicon–Silver–Tellurium–Sulfur

Ag_8SiS_6–Ag_8SiTe_6. The phase diagram of this system (Figure B.1.4) is characterized by the formation of a continuous series of solid solutions between Ag_8SiTe_6 and high-temperature Ag_8SiS_6 modification (Amiraslanova et al. 2023a). With the formation of solid solutions, the temperature of the polymorphic transition of the Ag_8SiS_6 decreases. This leads to the stabilization of the cubic phase in the range of compositions ≥30 mol% Ag_8SiTe_6 at room temperature and below. The homogeneity region based on room temperature Ag_8SiS_6 is 10 mol%. The system was studied through DTA, XRD, and SEM.

B.1.34 Silicon–Gold–Lanthanum–Sulfur

The quaternary compound $Au_{1.93}La_6Si_2S_{14}$, which crystallizes as a hexagonal structure with the lattice parameters a = 1038.04 ± 0.15, c = 582.50 ± 0.14 pm, a calculated density of 5.238 g·cm^{-3} at 100 K, and an energy gap of ~2.36 eV, is formed in the Si–Au–La–S system (Akopov et al. 2021). The single crystals and the bulk powder of this compound were obtained in the same way as in the case of $Cu_{1,97}La_3Si_2S_{14}$ but using Au instead of Cu.

B.1.35 Silicon–Calcium–Strontium–Sulfur

Ca_2SiS_4–Sr_2SiS_4. The solid solutions based on Sr_2SiS_4, which crystallize in the monoclinic structure, are formed in this system up to 40 mol% Ca_2SiS_4 and within the interval of 50–90 mol% Ca_2SiS_4 the phase separation was observed (Parmentier et al. 2010). The powders of these solid solutions were synthesized by sintering a homogeneous mixture of stoichiometric amounts of SrS, CaS, and Si in a continuous H_2S-flow at 875°C for 1 h. Hydrogen sulfide is used as a source of extra sulfur needed to form solid solutions. A slight surplus of Si (5%) was also used. After cooling, the samples are manually ground with a mortar and pestle in order to get a fine powder. Except for the sintering, the processing of the samples is done in ambient air.

B.1.36 Silicon–Strontium–Cadmium–Sulfur

The quaternary compound $SrCdSiS_4$, which exhibits thermal stability below 934°C under N_2 condition and decomposes to Sr_2SiS_4 and CdS at higher temperatures, is formed in the Si–Sr–Cd–S system (Yang

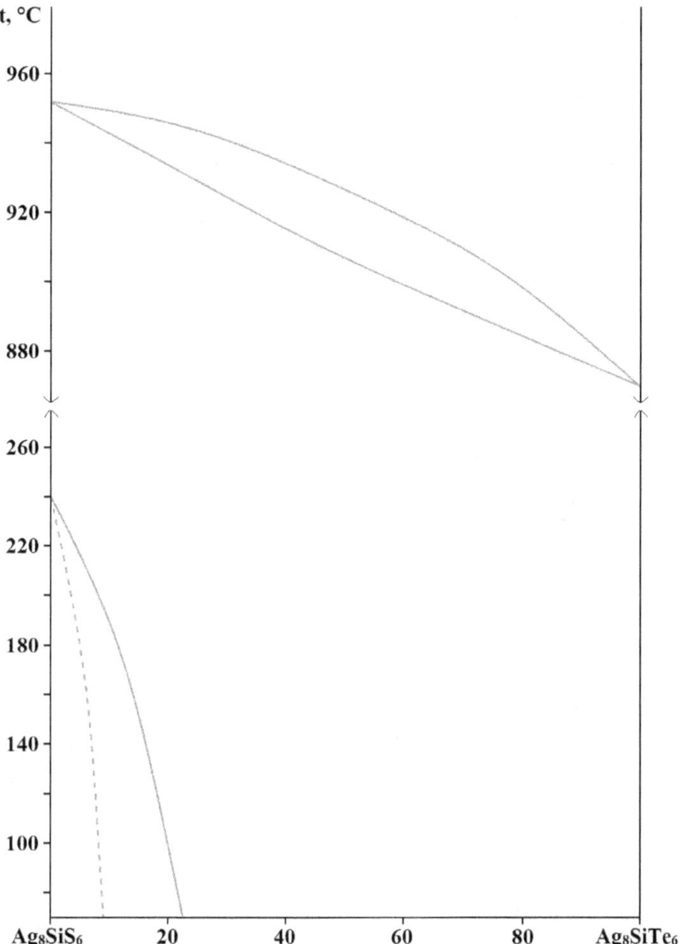

FIGURE B.1.4 Phase diagram of the Ag_8SiS_6–Ag_8SiTe_6 system. (From Amiraslanova, A.J., et al., *Azerb. Chem. J.*, (2), 169–177 (2023) Open access.

et al. 2022a). It crystallizes as an orthorhombic structure with the lattice parameters a = 1028.21 ± 0.07, b = 1015.51 ± 0.09, c = 636.99 ± 0.05 pm, a calculated density of 3.559 g·cm^{-3}, and an energy gap of 3.61 eV. The pale-yellow crystals of the title compound were obtained by flux method. SrS, CdS, Si, and S were weighed in the stoichiometric molar ratio and CsI as a flux. These raw materials (ca. 600 mg) were loaded into quartz silica tube and subsequently evacuated to 10^{-3} Pa and sealed. After that, it was heated to 400°C and kept 10 h, then cooled down to 350°C within 300 h after 120 h of insulation at 850°C, and finally the furnace was shut off. The single crystals were obtained after washing with hot distilled water and ethanol. They are insensitive to water and air.

B.1.37 Silicon–Barium–Lanthanum–Sulfur

The quaternary compound $BaLa_2Si_2S_8$, which crystallizes as a trigonal structure with the lattice parameters a = 918.109 ± 0.020 and c = 2729.41 ± 0.07 pm, is formed in this quinary system (Lee et al. 2014). The powder samples of this solid solution were synthesized by solid state reactions using La_2S_3, BaS, and Si and S powders as raw materials. The stoichiometric amounts of the starting materials were thoroughly mixed and loaded into a vertically positioned fused silica ampoule, fully evacuated to 0.1 Pa, and sealed off under a dynamic vacuum. The fused silica ampoule was heated at 5°C·min^{-1} to 1000°C-1100°C for 6–8 h in a furnace and then cooled down slowly to room temperature.

B.1.38 Silicon–Barium–Uranium–Sulfur

The quaternary compound $Ba_8Si_2US_{14}$, which crystallizes as a monoclinic structure with the lattice parameters a = 2392.08 ± 0.07, b = 708.18 ± 0.02, c = 885.49 ± 0.02 pm, β = 111.196 ± 0.001°, a calculated density of 4.374 g·cm^{-3} at 100 K, and an energy gap of 1.2 eV, is formed in the Si–Ba–U–S system (Mesbah et al. 2015). Black block-shaped single crystals of this compound were obtained in an attempt to synthesize the quaternary compound Ba_3RbUS_6. The reaction mixture comprised U (0.084 mM), Rb_2S_3 (0.074 mM), BaS (0.25 mM), and S (0.25 mM). It was heated to 950°C in 48 h, kept there for 192 h, then cooled to 200°C at a rate of 3°C·h^{-1}, and finally the furnace was turned off. A few black blocks were selected and analyzed by EDX to reveal Ba/Si/U/S = 8:2:1:14. A byproduct of the reaction turned out to be $Ba_{3.69}US_6$. The Si in the present compound certainly arises from the silica tube. Unfortunately, attempts to reproduce this synthesis steps or to effect a rational synthesis of $Ba_8Si_2US_{14}$ failed.

B.1.39 Silicon–Scandium–Lanthanum–Sulfur

The quaternary compound $Sc_{0.65}La_6Si_2S_{14}$, which crystallizes as a hexagonal structure with the lattice parameters a = 1033.83 ± 0.05, c = 574.59 ± 0.04 pm, and a calculated density of 4.269 g·cm^{-3} at 100 K, is formed in the Si–Sc–La–S system (Akopov et al. 2021). The single crystals and the bulk powder of this compound were obtained in the same way as $Cu_{1,97}La_3Si_2S_{14}$ but using Sc instead of Cu.

B.1.40 Silicon–Lanthanum–Titanium–Sulfur

The quaternary compound $Ti_{0.62}La_6Si_2S_{14}$, which crystallizes as a hexagonal structure with the lattice parameters a = 1029.70 ± 0.05, c = 577.20 ± 0.03 pm, and a calculated density of 4.287 g·cm^{-3} at 100 K, is formed in the Si–La–Ti–S system (Akopov et al. 2021). The single crystals and the bulk powder of this compound were obtained in the same way as $Cu_{1,97}La_3Si_2S_{14}$ but using Ti instead of Cu.

B.1.41 Silicon–Lanthanum–Zirconium–Sulfur

The quaternary compound $Zr_{0.52}La_6Si_2S_{14}$, which crystallizes as a hexagonal structure with the lattice parameters a = 1031.89 ± 0.06, c = 575.43 ± 0.04 pm, and a calculated density of 4.337 g·cm^{-3} at 100 K, is formed in the Si–La–Zr–S system (Akopov et al. 2021). The single crystals and the bulk powder of this compound were obtained in the same way as $Cu_{1,97}La_3Si_2S_{14}$ but using Zr instead of Cu.

B.1.42 Silicon–Lanthanum–Hafnium–Sulfur

The quaternary compound $Hf_{0.48}La_6Si_2S_{14}$, which crystallizes as a hexagonal structure with the lattice parameters a = 1031.88 ± 0.06, c = 576.12 ± 0.05 pm, and a calculated density of 4.451 g·cm^{-3} at 100 K, is formed in the Si–La–Hf–S system (Akopov et al. 2021). The single crystals and the bulk powder of this compound were obtained in the same way as $Cu_{1,97}La_3Si_2S_{14}$ but using Hf instead of Cu.

B.1.43 Silicon–Lanthanum–Vanadium–Sulfur

The quaternary compound $V_{0.77}La_6Si_2S_{14}$, which crystallizes as a hexagonal structure with the lattice parameters a = 1029.60 ± 0.04, c = 573.56 ± 0.03 pm, and a calculated density of 4.345 g·cm^{-3} at 100 K, is formed in the Si–La–V–S system (Akopov et al. 2021). The single crystals and the bulk powder of this compound were obtained in the same way as $Cu_{1,97}La_3Si_2S_{14}$ but using V instead of Cu.

B.1.44 Silicon–Lanthanum–Chromium–Sulfur

The quaternary compound $Cr_{0.66}La_6Si_2S_{14}$, which crystallizes as a hexagonal structure with the lattice parameters a = 1028.88 ± 0.12, c = 575.03 ± 0.09 pm, and a calculated density of 4.324 g·cm^{-3} at 100 K, is formed in the Si–La–Cr–S system (Akopov et al. 2021). The single crystals and the bulk powder of this compound were obtained in the same way as $Cu_{1,97}La_3Si_2S_{14}$ but using Cr instead of Cu.

B.1.45 Silicon–Lanthanum–Manganese–Sulfur

The quaternary compound $Mn_{0.98}La_6Si_2S_{14}$, which crystallizes as a hexagonal structure with the lattice parameters a = 1033.37 ± 0.04, c = 572.85 ± 0.03 pm, and a calculated density of 4.364 g·cm^{-3} at 100 K, is formed in the Si–La–Mn–S system (Akopov et al. 2021). The single crystals and the bulk powder of this compound were obtained in the same way as $Cu_{1,97}La_3Si_2S_{14}$ but using Mn instead of Cu.

B.1.46 Silicon–Lanthanum–Iron–Sulfur

The quaternary compound $Fe_{1.05}La_6Si_2S_{14}$, which crystallizes as a hexagonal structure with the lattice parameters a = 1029.91 ± 0.03, c = 573.89 ± 0.03 pm, a calculated density of 4.401 g·cm^{-3} at 100 K, and an energy gap of ~1.87 eV, is formed in the Si–La–Fe–S system (Akopov et al. 2021). The single crystals and the bulk powder of this compound were obtained in the same way as $Cu_{1,97}La_3Si_2S_{14}$ but using Fe instead of Cu.

B.1.47 Silicon–Lanthanum–Cobalt–Sulfur

The quaternary compound $Co_{1.01}La_6Si_2S_{14}$, which crystallizes as a hexagonal structure with the lattice parameters a = 1031.41 ± 0.11, c = 573.13 ± 0.10 pm, and a calculated density of 4.395 g·cm^{-3} at 100 K, and an energy gap of ~1.85 eV, is formed in the Si–La–Co–S system (Akopov et al. 2021). The single crystals and the bulk powder of this compound were obtained in the same way as $Cu_{1,97}La_3Si_2S_{14}$ but using Co instead of Cu.

B.1.48 Silicon–Lanthanum–Nickel–Sulfur

The quaternary compound $Ni_{0.99}La_6Si_2S_{14}$, which crystallizes as a hexagonal structure with the lattice parameters a = 1027.25 ± 0.06, c = 572.87 ± 0.04 pm, and a calculated density of 4.432 g·cm^{-3} at 100 K, is formed in the Si–La–Ni–S system (Akopov et al. 2021). The single crystals and the bulk powder of this compound were obtained in the same way as $Cu_{1,97}La_3Si_2S_{14}$ but using Ni instead of Cu.

B.1.49 Silicon–Lanthanum–Rhodium–Sulfur

The quaternary compound $Rh_{0.69}La_6Si_2S_{14}$, which crystallizes as a hexagonal structure with the lattice parameters a = 1025.07 ± 0.05, c = 574.90 ± 0.03 pm, a calculated density of 4.474 g·cm^{-3} at 100 K, and an energy gap of ~2.01 eV, is formed in the Si–La–Rh–S system (Akopov et al. 2021). The single crystals and the bulk powder of this compound were obtained in the same way as $Cu_{1,97}La_3Si_2S_{14}$ but using Rh instead of Cu.

B.1.50 Silicon–Lanthanum–Platinum–Sulfur

The quaternary compound $Pt_{0.53}La_6Si_2S_{14}$, which crystallizes as a hexagonal structure with the lattice parameters a = 1025.30 ± 0.03, c = 577.17 ± 0.03 pm, a calculated density of 4.557 g·cm^{-3} at 100 K, and an energy gap of ~1.8 eV, is formed in the Si–La–Pt–S system (Akopov et al. 2021). The single crystals and the bulk powder of this compound were obtained in the same way as $Cu_{1,97}La_3Si_2S_{14}$ but using Pt instead of Cu.

B.2 Systems Based on Silicon Selenides

B.2.1 Silicon–Lithium–Barium–Selenium

The quaternary compound $Li_2BaSiSe_4$, which crystallizes as a tetragonal structure with the lattice parameters $a = 692.1 \pm 0.3$, $c = 825.3 \pm 0.8$ pm, a calculated density of 4.160 g·cm^{-3}, and an energy gap of 2.47 eV, is formed in the Si–Li–Ba–Se system (Li et al. 2019a). This compound was obtained through high-temperature flux method. Li, Ba, Si, and Se were weighed with a molar ratio of 2:1:1:4, then loaded into the graphite crucible, and finally sealed in a silicon tube under ethane–oxygen flame. The tube was put in a program-computed furnace with the temperature program as follows: (1) heat from room temperature to 190°C in 3 h and keep at this temperature for 5 h; (2) heat from 190°C to 450°C in 10 h and keep at this temperature for 20 h; (3) heat from 450°C to 850°C in 10 h and keep at this temperature for 30 h; (4) cool from 850°C to 400°C in 100 h and turn down the furnace.

B.2.2 Silicon–Sodium–Mercury–Selenium

The quaternary compound $Na_2Hg_3Si_2Se_8$, which crystallizes as a tetragonal structure with the lattice parameters $a = 913.46 \pm 0.06$, $c = 946.39 \pm 0.09$ pm, a calculated density of 5.617 g·cm^{-3}, and an energy gap of 2.20 eV, is formed in the Si–Na–Hg–Se system (Gao et al. 2022b). To synthesize this compound, a mixture of Na (2 mM), HgSe (3 mM), $SiSe_2$ (2 mM), and Se (1 mM) was loaded into a graphite crucible that avoids the harmful reaction between Na metal and tube wall, and then put the graphite crucible into vacuum-sealed silica tube. Silica tube was firstly heated to 600°C in 50 h, kept at 600°C for about 100 h, and slowly cooled down to room temperature within 5 days. The final products were carefully washed with DMF and red microcrystals appeared in the tube. As for air-unstable Na block, an Ar-filled glove box was used to complete the whole preparation process.

B.2.3 Silicon–Potassium–Gallium–Selenium

The quaternary compound $K_{2.4}Ga_{2.4}Si_{1.6}Se_8$, which crystallizes as a tetragonal structure with the lattice parameters $a = 819.99 \pm 0.04$, $c = 614.60 \pm 0.07$ pm, and a calculated density of 3.746 g·cm^{-3}, is formed in the Si–K–Ga–Se system (He et al. 2016). The crystals of the title compound were obtained through reactive flux method with K_2Se_2 as the reactive flux. The starting materials K_2Se_2 (2.25 mM), Ga_2Se_3 (0.375 mM), Si (0.5 mM), and Se (1.75 mM) were mixed and ground uniformly and loaded into silica tube. The tube was then flame-sealed under vacuum (0.1 Pa) and slowly heated to 750°C in a programmable furnace. It was kept at this temperature for 3 days and then slowly cooled down to 300°C at a rate of 2°C·h^{-1}, and finally to room temperature by turning off the furnace. The melt was washed by DMF several times to remove the excessive flux and dried by acetone. Thin yellow prisms up to 2–3 cm long were obtained. The powder sample was synthesized by stoichiometric combination reaction of the starting materials under the same reacting condition and by repeating the procedure three times. All operations were carried out in an Ar-protected glove box.

B.2.4 Silicon–Silver–Gallium–Selenium

The quaternary compound $AgGaSiSe_4$, which crystallizes as an orthorhombic structure with the lattice parameters $a = 6323.89 \pm 0.15$, $b = 714.217 \pm 0.017$, $c = 1240.72 \pm 0.03$ pm, and an energy gap of 2.33 eV at 300 K and 2.42 eV at 100 K, is formed in the Si–Ag–Ga–Se system (Krymus et al. 2017). To prepare this compound, calculated amounts of substances were placed in the quartz ampoule that was previously graphitized. It was then evacuated to the residual pressure of 0.01 Pa and flame-sealed. The ampoule was placed in a shaft-type furnace and heated to 900°C with an intermediate exposure at 600°C for 12 h. The heating rate was 30°C·h^{-1} up to the intermediate stop temperature and it was 10°C·h^{-1} thereafter. Upon reaching the maximum temperature, the ampoule with the melt was maintained for 6 h and then it was cooled to ambient temperature over two days. The growth of the $AgGaSiSe_4$ crystals was done through Bridgman–Stockbarger method.

B.2.5 Silicon–Silver–Germanium–Selenium

Ag_8SiS_6–Ag_8GeSe_6. The continuous series of solid solution are formed based on high-temperature cubic modification and limited solid solution areas are formed based on low-temperature modification of initial compounds (Figure B.2.1) (Ashirov et al. 2023b). The formation of solid solutions leads to a sharp decrease of polymorphic transition temperatures of ternary compounds and stabilization of high-temperature phases at room temperature. The lattice parameters of solid solutions increase linearly with Ge substitution and follow Vegard's law. Phase equilibria in these systems were studied through DTA and XRD. To bring the samples to equilibrium, they were annealed at 630°C for 500 h.

Ag_8SiSe_6–Ag_8GeSe_6–Ag_2Se. The isothermal section at room temperature of this system (Figure B.2.2), the liquidus surface (Figure B.2.3), as well as some polythermal sections of the phase diagram were constructed by Ashirov et al. (2023b). The isothermal section consists of 10 heterogeneous regions: seven two-phase and three three-phase regions. The liquidus surface includes the regions of primary crystallization of $Ag_8Si_xGe_{1-x}Se_6$ solid solution and Ag_2Se.

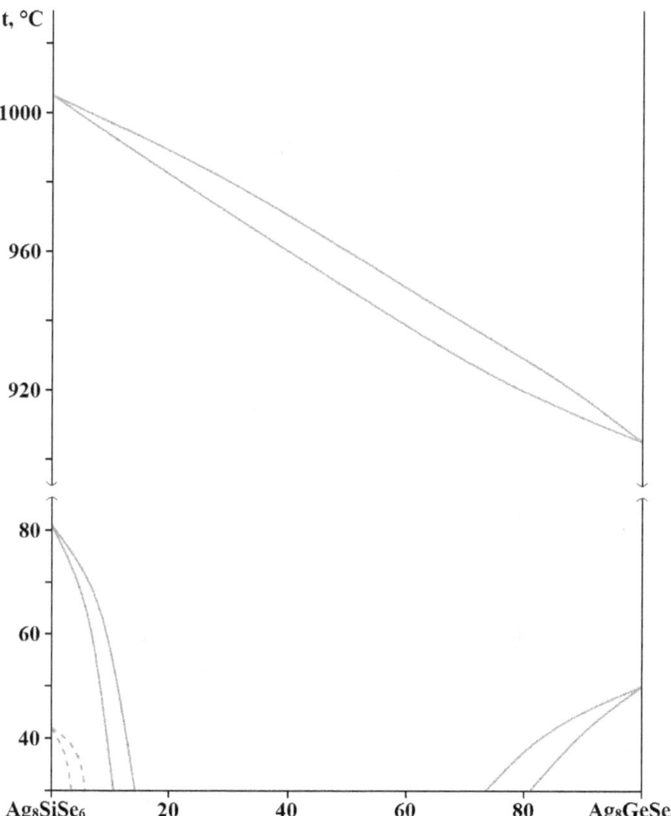

FIGURE B.2.1 Phase diagram of the Ag_8SiSe_6–Ag_8GeSe_6 system. (From Ashirov, G.M., et al., *Chem. Probl.*, (3), 229, 2023.) Open access.

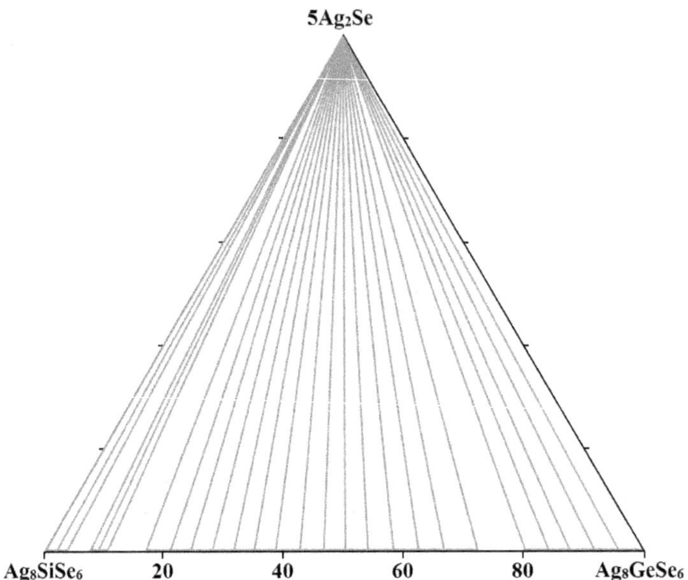

FIGURE B.2.2 Isothermal section of the Ag_8SiSe_6–Ag_8GeSe_6–Ag_2Se quasiternary system at room temperature. (From Ashirov, G.M., et al., *Chem. Probl.*, (3), 229, 2023.) Open access.

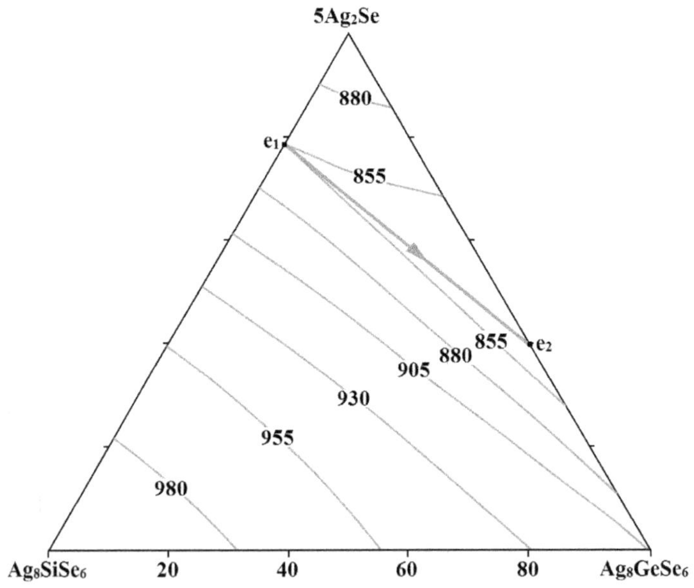

FIGURE B.2.3 Liquidus surface of the Ag_8SiSe_6–Ag_8GeSe_6–Ag_2Se quasiternary system. (From Ashirov, G.M., et al., *Chem. Probl.*, (3), 229, 2023.) Open access.

B.2.6 Silicon–Silver–Tellurium–Selenium

Ag_8SiSe_6–Ag_8SiTe_6. The phase diagram of this system is characterized by the formation of a continuous series of solid solutions between high-temperature Ag_8SiSe_6 and Ag_8SiTe_6 (Figure B.2.4) (Amiraslanova et al. 2023b). On the liquidus and solidus curves, the temperature changes monotonically between the melting points of initial compounds, and the melting temperature range does not exceed 15°C. The formation of solid solutions is accompanied by a strong decrease in the temperatures of polymorphic transitions

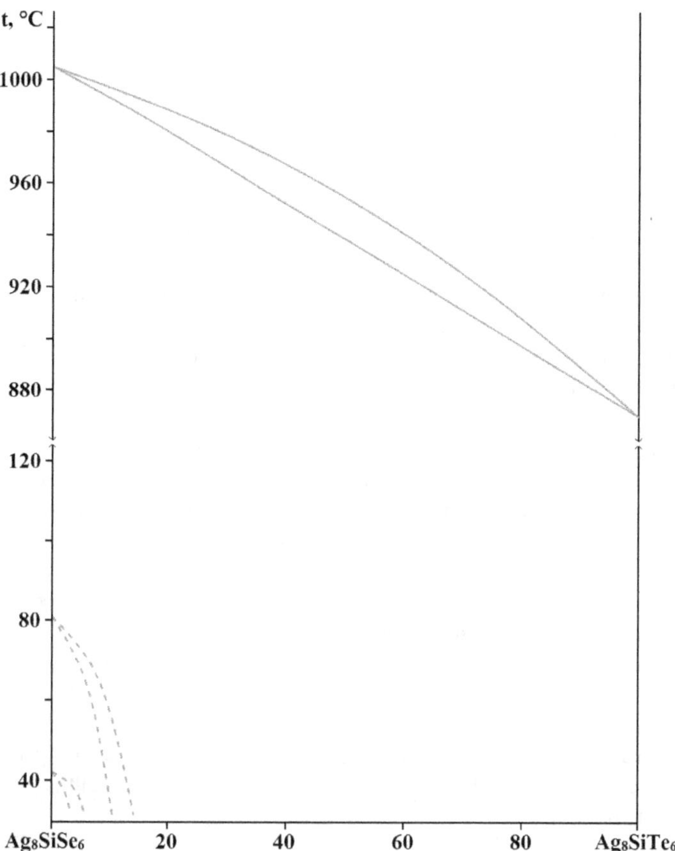

FIGURE B.2.4 Phase diagram of the Ag_8SiSe_6–Ag_8SiTe_6 system. (From Amiraslanova, A.J., et al., *Appl. Chem. Eng.*, **6**(2), 11–19, 2023.) Open access.

of Ag_8SiSe_6 (81°C and 42°C) and their transition to the temperature range below room temperature in the composition range of 0–10 mol% Ag_8SiTe_6.

The system was studied through DTA, XRD, and SEM. The alloys were annealed at 530°C for about 500 h. For some compositions, two series of alloys were prepared: the first series after annealing was quenched in ice water, and the second one was slowly cooled in a furnace to room temperature. For both series of alloys, the lattice parameter is an almost linear function of the composition, that is, Vegard's law is obeyed.

B.2.7 Silicon–Strontium–Zinc–Selenium

The quaternary compound $SrZnSiSe_4$, which crystallizes as an orthorhombic structure with the lattice parameters $a = 1086.73 \pm 0.13$, $b = 1071.81 \pm 0.13$, $c = 629.23 \pm 0.06$ pm, a calculated density of 4.503 g·cm^{-3}, and an energy gap of 1.93 eV, is formed in the Si–Sr–Zn–S system (Ma et al. 2023). To obtain this compound, SrSe, Zn, Si and Se powders were used as raw materials and put in a glove box filled with argon (the content of water and oxygen were both less than 0.1 ppm). These raw materials were mixed homogeneously and then sealed in a quartz tube under vacuum. The quartz tube was placed in a muffle furnace with program control, heated up to 900°C in 15 h, kept for 10 h at this temperature, then cooled to 500°C, and finally cooled naturally to room temperature. After the solid-phase synthesis, the sample was ground into powder and the powder was mixed in a 1:1 ratio with KI in an argon protective glove box; the mixture was sealed in a carbon-coated quartz tube and transferred to a muffle furnace with program control. Subsequently, the quartz tube was heated up to 950°C in 20 h and kept at the current temperature for 20 h, then cooled to 400°C at a rate of 4°C·h^{-1}, and finally the sample was cooled to room temperature. The product was cleaned with deionized water and stable $SrZnSiSe_4$ crystals were obtained.

B.2.8 Silicon–Yttrium–Iron–Selenium

The quaternary compound $FeY_6Si_2Se_{14}$, which crystallizes as a hexagonal structure with the lattice parameters a = 1019.71 ± 0.01, c = 597.99 ± 0.02 pm, a calculated density of 5.399 g·cm^{-3}, and an energy gap of 1.1 ± 0.1 eV, is formed in the Si–Y–Fe–S system (Panigrahi et al. 2023). A high-temperature reaction of corresponding elements produced the block-shaped, black-colored crystals of this compound. All the elements, Y (0.685 mM), Fe (0.114 mM), Si (0.224 mM), and Se (1.599 mM), were transferred into a carbon-coated fused silica tube, and then the tube was flame-sealed after evacuation to 0.01 Pa. The sealed ampoule was first heated to 650°C in 12 h inside a muffle furnace and annealed for 12 h before ramping the furnace's temperature to 950°C with a heating rate of 20°C·h^{-1}. The furnace's temperature was held constant for 72 h at 950°C and then cooled down to 300°C at a rate of 4°C·h^{-1}. Finally, the furnace was shut off, and the reaction mixture was cooled radiatively to room temperature. A black ingot was obtained after breaking the heat-treated ampoule under the ambient environment.

The high-temperature solid state sealed tube method was used to synthesize a polycrystalline sample of $FeY_6Si_2S_{14}$. In a carbon-coated fused silica tube, the reactants Y, Fe, Si, and Se were loaded with the molar ratio of 3:0.5:1:7. The total mass of the reactants was 1 g: Y (3.426 mM), Fe (0.571 mM), Si (1.142 mM), and Se (7.995 mM). The vacuum-sealed tube containing the reaction mixture was heat-treated inside the furnace. The temperature of the furnace was raised to 250°C in 12 h from room temperature and maintained for 12 h. Further, the temperature was ramped to 900°C in 24 h, and the reaction mixture was annealed for 96 h before switching off the furnace. The product formed in the first step of the reaction was homogenized into a fine powder inside the Ar-filled glove box and then compacted into a cylindrical pellet using a hydraulic press. The pellet was again sealed in a carbon-coated fused evacuated silica tube and reheated at 1000°C for four days.

B.3 Systems Based on Silicon Telluride

B.3.1 Silicon–Copper–Barium–Tellurium

The quaternary compound $CuBaSiTe_3$, which crystallizes as a monoclinic structure with the lattice parameters a = 758.24 ± 0.01, b = 884.40 ± 0.01, c = 1312.89 ± 0.02 pm, β = 122.022 ± 0.001°, a calculated density of 5.444 g·cm^{-3}, and an energy gap of 1.65 eV, is formed in the Si–Cu–Ba–S system (Jafarzadeh et al. 2019). This compound was prepared from the elements in the stoichiometric ratio (Ba pieces, Cu powder, Si powder, and Te broken ingots). The elements were added to carbon-coated silica tube in an Ar-filled glove box, and the tube was evacuated to ~0.1 Pa and sealed. In a programmable furnace, the sealed tube was heated to 700°C within 10 h, kept at this temperature for 2 h, and finally slowly cooled to room temperature. After grinding, the sample was reheated at 500°C to enhance the homogeneity and yield.

B.3.2 Silicon–Copper–Zinc–Tellurium

The quaternary compound $Cu_2ZnSiTe_4$, which crystallizes as a tetragonal structure with the lattice parameters a = 596.12 ±0.01, c = 1178.87 ± 0.04 pm, and a calculated density of 5.795 g·cm^{-3}, is formed in the Si–Cu–Zn–Te system (Levcenko et al. 2014). This compound was synthesized by a direct reaction of elements in evacuated and sealed quartz tube. Vacuum inside the tube was ~0.01 Pa. The synthesis was carried out in a vertical furnace with gradient 5°C·cm^{-1} along the furnace axes (higher temperature in upper part of furnace). The maximum temperature of synthesis was 950°C. After 24 h of maintaining this temperature, a slow (10°C·h^{-1}) cooling until 500°C was applied. At the end of the process a gray ingot with metallic luster was obtained.

B.3.3 Silicon–Barium–Antimony–Tellurium

The quaternary compound $Ba_{14}Si_4Sb_8Te_{32}(Te_3)$ compound, which crystallizes as a monoclinic structure with the lattice parameters a = 2422.0 ± 0.2, b = 1339.20 ± 0.16, c = 1361.29 ± 0.14 pm, β = 97.699 ± 0.004, a calculated density of 5.674 g·cm^{-3}, and an energy gap of 0.8 ± 0.2 eV, is formed in the Si–Ba–Sb–Te system (Jana et al. 2023). The following elemental reactants were used for the synthesis of the samples of this compound: Ba rod, Si granules, Sb lumps, and Te powder. This compound was obtained using the sealed tube solid state synthesis method. Several reactants with varying molar ratios were loaded, and heat-treated with diverse heating profiles inside evacuated fused silica tubes. One such reaction produced black block-shaped single crystals, which turned out to be the title compound. This reaction was loaded with the starting reactants of Ba (0.427 mM), Si (0.142 mM), Sb (0.569 mM), and Te (1.708 mM), which were placed into a carbon-coated fused silica tube inside the Ar-filled glove box. The loaded tube was flame-sealed after evacuating it to ~10^{-2} Pa and placed in a high-temperature furnace for heat treatment. At first, the furnace was heated to 400°C in 12 h, and the reaction mixture was soaked for 18 h before increasing the furnace's temperature to 800°C in 30 h. The furnace's temperature was held constant at 800°C for 72 h before cooling it to 700°C at a cooling rate of 4°C·h^{-1}. The reaction mixture was then annealed for 24 h at 700°C, followed by cooling it to 300°C with a rate of 5°C·h^{-1}. Finally, the furnace was shut off to cool it to room temperature radiatively. The heat-treated tube was cracked open to take out a homogeneous-looking black lump, which was further broken into small pieces. A few crystals were picked out.

A polycrystalline $Ba_{14}Si_4Sb_8Te_{32}(Te_3)$ was prepared by a two-step sealed tube solid state synthesis method. Initially, stoichiometric amounts of Ba (1.40 mM), Si (0.40 mM), Sb (0.80 mM), and Te (3.51 mM) were sealed inside a carbon-coated fused silica tube as discussed earlier. The sealed tube was first heated to 550°C in 15 h, where it was kept for 24 h before heating the furnace to 800°C with a heating rate of 20°C·h^{-1}. The tube was annealed at 800°C for 96 h before switching off the furnace to cool down the product to room temperature. The homogeneous-looking black lump obtained from the reaction was properly ground inside the Ar-filled glove box and compacted into an 8 mm diameter cylindrical disk under ~14 MPa pressure. The pellet was again sealed inside a fused carbon-coated silica tube and heated to 700°C in 10 h, followed by annealing for 72 h at 700°C before switching off the furnace.

B.3.4 Silicon–Barium–Manganese–Tellurium

The quaternary compound $Ba_4Mn_2Si_2Te_9$, which melts between 950°C and 1000°C and crystallizes as an orthorhombic structure with the lattice parameters a = 1346.90 ± 0.06, b = 872.23 ± 0.04, c = 1000.32 ± 0.04 pm, a calculated density of 5.267 g·cm^{-3}, and an energy gap of 0.6 ± 0.1 eV, is formed in the Si–Ba–Mn–Te system (Yadav et al. 2022). The methodology used for the synthesis of this compound was as follows: the reactants Ba (0.380 mM), Si (0.190 mM), Mn (0.380 mM), and Te (0.949 mM) were transferred into a carbon-coated fused silica tube inside the Ar-filled dry glove box. The tube was then evacuated to about 0.01 Pa, sealed with a flame torch, and kept in a programmable furnace for heat treatment. The furnace's temperature was ramped up to 1000°C from room temperature in 24 h and maintained for 48 h. After that, it was cooled to 800°C in 67 h and then maintained for 120 h. The reaction mixture was further cooled to 300°C in 48 h, and finally, the furnace was turned off. Under ambient conditions, the reaction ampoule was opened, and the product was examined under an optical microscope. The reaction product was a black ingot that was further broken into small pieces to reveal black-colored irregular-shaped crystals of the title compound and some red-colored crystals.

The synthesis of a polycrystalline sample of $Ba_4Mn_2Si_2Te_9$ was carried out using the sealed tube solid state method. Inside the Ar-filled glove box, stoichiometric amounts of Ba (2.14 mM), Si (1.07 mM), Mn (1.07 mM), and Te (4.82 mM) were loaded into a carbon-coated fused silica tube, which was then connected to a vacuum line. The tube was evacuated to ~0.01 Pa, followed by sealing with a flame torch. The sealed tube was then placed vertically inside a programmable muffle furnace, and the reaction mixture was heated to 1000°C in 12 h, and annealed for 48 h. It was then cooled to 800°C in 2 h and soaked for 96 h before cooling it to room temperature in 10 h. Under ambient conditions, the reaction tube was opened to reveal a homogeneous-looking black melted mass that was further crushed into a fine powder and homogenized by grinding inside the Ar-filled glove box using the agate mortar and pestle. The powder thus obtained was compressed into a circular disk by applying a pressure of about 10.3 MPa using a hydraulic press. This pellet was then placed carefully inside a carbon-coated fused silica tube, followed by evacuating it to ~0.01 Pa, and finally sealed with a flame torch. The fused silica ampoule containing the disk-shaped sample was heated to 1000°C in 10 h, annealed for 168 h at the same temperature, and then cooled to room temperature in 11 h. The resulting black pellet was ground into fine powder inside the Ar-filled glove box.

B.4 Systems Based on Germanium Sulfides

B.4.1 Germanium–Lithium–Cadmium–Sulfur

The quaternary compound $Li_4CdGe_2S_7$, which crystallizes as a monoclinic structure with the lattice parameters a = 1683.54 ± 0.10, b = 678.70 ± 0.04, c = 1014.99 ± 0.06 pm, β = 93.710 ± 0.003°, a calculated density of 2.926 g·cm^{-3}, and an energy gap of ~ 3.6 eV, is formed in the Ge–Li–Cd–S system (Craig et al. 2022). This compound was prepared by high-temperature, solid state synthesis. Stoichiometric quantities of the elemental reagents and a 10% excess of Li_2S were weighed in an argon-filled glove box before being briefly ground together in an agate mortar and pestle. This mixture was then placed in a loosely capped graphite tube that was next loaded into a fused-silica tube that was flame-sealed under ~0.1 Pa pressure before being placed in a box furnace. The sample was heated to 800°C in 12 h and allowed to dwell at this temperature for 120 h. After this, the reaction was cooled to 600°C over 100 h before being allowed to return to room temperature over 24 h. The reaction vessel was opened and the product was stored in the glove box because it was deemed to be somewhat air/moisture sensitive.

B.4.2 Germanium–Lithium–Yttrium–Sulfur

The quaternary compound LiY_3GeS_7, which crystallizes as a hexagonal structure with the lattice parameters a = 975.07 ± 0.03, c = 580.97 ± 0.04 pm, a calculated density of 3.962 g·cm^{-3}, and an energy gap of 1.80 eV, is formed in the Ge–Li–Y–S system (Gao et al. 2022a). The title compound was synthesized with the stoichiometric ratio of Li_2S, Ln_2S_3, and GeS_2 in vacuum-sealed silica tube.

The detailed temperature reaction process was caried out as follows: firstly, heat up to 900°C at a rate of 10°C·h^{-1} and then kept at this temperature within 10 days; secondly, cool down to room temperature at a rate of 5°C·h^{-1}. The synthesized crystals are stable in the air for several months.

B.4.3 Germanium–Lithium–Lanthanum–Sulfur

The quaternary compound $LiLa_3GeS_7$, which crystallizes as a hexagonal structure with the lattice parameters a = 1034.94 ± 0.04, c = 580.93 ± 0.05 pm, a calculated density of 4.442 g·cm^{-3}, and an energy gap of 3.02 eV, is formed in the Ge–Li–La–S system (Yang et al. 2021). The first-phase preparation process of this compound was completed in an Ar-filled glove box because of the instability for La_2S_3 and Li_2S in the air. In the synthesis process, first, the raw materials (La_2S_3, Li_2S, and GeS_2) were added into graphite crucible with cover and then put into graphite crucible within the silica tube to avoid the corrosion of silica tube by the raw materials. Their spontaneous crystallization processes were achieved in the vacuum-sealed silica tube with high-temperature muffle furnace. The title crystals were prepared under stoichiometric ratio of raw materials at 800°C. Their high yields (>95 %) were successfully achieved after the multiple grinding and calcination. Finally, many high-quality crystals were found in the silica tube after washing with DMF solvent. These crystals are relatively stable in the air for several weeks.

B.4.4 Germanium–Lithium–Praseodymium–Sulfur

The quaternary compound $LiPr_3GeS_7$, which crystallizes as a hexagonal structure with the lattice parameters a = 1015.64 ± 0.09, c = 576.58 ± 0.10 pm, and a calculated density of 4.685 g·cm^{-3}, is formed in the Ge–Li–Pr–S system (Yang et al. 2021). This compound was prepared in the same way as $LiLa_3GeS_7$ but using Pr_2S_3 instead of La_2S_3. These crystals are also relatively stable in the air for several weeks.

B.4.5 Germanium–Lithium–Neodymium–Sulfur

The quaternary compound $LiNd_3GeS_7$, which crystallizes as a hexagonal structure with the lattice parameters a = 1011.67 ± 0.18, c = 576.5 ± 0.2 pm, and a calculated density of 4.788 g·cm^{-3}, is formed in the Ge–Li–Nd–S system (Yang et al. 2021). This compound was prepared in the same way as $LiLa_3GeS_7$ but using Nd_2S_3 instead of La_2S_3. These crystals are also relatively stable in the air for several weeks.

B.4.6 Germanium–Lithium–Holmium–Sulfur

The quaternary compound $LiHo_3GeS_7$, which crystallizes as a hexagonal structure with the lattice parameters a = 972.74 ± 0.03, c = 582.65 ± 0.03 pm, and a calculated density of 5.556 g·cm^{-3}, is formed in the Ge–Li–Ho–S system (Yang et al. 2021). This compound was prepared in the same way as $LiLa_3GeS_7$ but using Ho_2S_3 instead of La_2S_3. These crystals are also relatively stable in the air for several weeks.

B.4.7 Germanium–Lithium–Erbium–Sulfur

The quaternary compound $LiEr_3GeS_7$, which crystallizes as a hexagonal structure with the lattice parameters a = 963.7 ± 0.3, c = 585.4 ± 0.3 pm, and a calculated density of 5.683 g·cm^{-3}, is formed in the Ge–Li–Er–S system (Yang et al. 2021). This compound was prepared in the same way as $LiLa_3GeS_7$ but using Er_2S_3 instead of La_2S_3. These crystals are also relatively stable in the air for several weeks.

B.4.8 Germanium–Sodium–Cadmium–Sulfur

The quaternary compound $Na_4CdGe_2S_7$, which crystallizes as a monoclinic structure with the lattice parameters a = 708.13 ± 0.02, b = 1190.07 ± 0.02, c = 1557.59 ± 0.03 pm, β = 90.791 ± 0.002°, a calculated density of 2.905 g·cm^{-3}, and an energy gap of 3.35 eV, is formed in the Ge–Na–Cd–S system (Zhang et al. 2020). The single crystals of this compound were obtained by high-temperature spontaneous crystallization using the mixture of NaS_2, CdS, Ge, and S in the molar ratio of 1:2:1:4. All the starting materials were ground and loaded into a fused silica tube under an Ar atmosphere in the glove box. The tube was sealed under a high vacuum of 10^{-3} Pa and placed in a computer-controlled furnace, subsequently heated to 900°C at a rate of 10°C·h^{-1} and kept at this temperature for 2 days before cooling to room temperature at a rate of 3°C·h^{-1}. The polycrystalline powder of $Na_4CdGe_2S_7$ has been successfully synthesized in high-temperature solid state method by heating a stoichiometric mixture of the elements. All elements were stored in a glove box filled with dry Ar without oxygen and moisture.

B.4.9 Germanium–Sodium–Phosphorus–Sulfur

Na_4GeS_4–Na_3PS_4. To prepare the glasses and glass-ceramics, Na_2S, P_2S_5, and GeS_2 powders were mixed together (Tanibata et al. 2018). The mixtures were mechanically milled using a planetary ball mill apparatus with a zirconia pot (45 mL) and 500 zirconia balls (4 mm diameter) at a rotation speed of 510 rpm. All the samples were mechanically milled for 5–15 h. The milled powders were compressed with a uniaxial press to prepare pellets of diameter 10 mm and thickness 1–1.5 mm. These pellets were crystallized by heating at an appropriate temperature between 210°C and 330°C (depending on composition) for 2 h to obtain glass-ceramics.

GeS_2–Na_2S–P_2S_5. The calculated isothermal section of this quasiternary system at 0 K is characterized by the existence of seven quasibinary systems, namely Na_4GeS_4–Na_3PS_4, $Na_6Ge_2S_7$–Na_3PS_4, Na_2GeS_3–Na_3PS_4, $Na_2Ge_2S_5$–Na_3PS_4, $Na_2Ge_2S_5$–$Na_4P_2S_7$, GeS_2–$Na_4P_2S_7$, and GeS_2–$NaPS_3$ (Richards et al. 2016). In the Na_4GeS_4–Na_3PS_4 section, a quaternary compound $Na_{10}GeP_2S_{12}$ can be formed, crystallizing as a tetragonal structure with lattice parameters $a = 962$ and $c = 1359$ pm (calculated values).

B.4.10 Germanium–Sodium–Chlorine–Sulfur

GeS_2–NaCl. The phase diagram of this system is a eutectic type with a degenerated eutectic from the NaCl side (Nedoshovenko et al. 1986). In the range of 2–38 mol% GeS_2, there is an immiscibility region at 800°C. The ingots for the investigations were annealed at 550°C and 800°C for 550 h.

B.4.11 Germanium–Potassium–Cadmium–Sulfur

The quaternary compound $K_2CdGe_3S_8$, which apparently has two polymorphic modifications, is formed in the Ge–K–Cd–S system. The first modification crystallizes as an orthorhombic structure with the lattice parameters $a = 737.58 \pm 0.03$, $b = 1210.51 \pm 0.05$, $c = 1686.58 \pm 0.06$ pm, a calculated density of 2.933 g·cm^{-3}, and an energy gap of 3.20 eV (Wu et al. 2023). For the preparation of this modification, $CdCl_2$ powder (0.10 mM), Ge metal (1.03 mM), S powder (2.59 mM), Li metal (1.05 mM), and KCl powder (1.55 mM) were weighed in an Ar-filled glove box. All reagents were loaded into a quartz tube with a carbon crucible inside and sealed under a vacuum with 0.1 Pa. The tube was then put into a furnace with the following temperature curves: first heating to 400°C for 5 h and holding there for 5 h, then heating to 700°C for 5 h and keeping at that temperature for 100 h; subsequently, cooling it down to 400°C for 96 h at a rate of 3°C·h^{-1} before turning down the furnace. Single crystals were finally obtained and were separated manually after washing the products with alcohol and deionized water.

The second modification crystallizes as a monoclinic structure with the lattice parameters $a = 731.139 \pm 0.010$, $b = 1205.817 \pm 0.018$, $c = 1692.74 \pm 0.02$ pm, $\beta = 95.3987 \pm 0.0012°$, a calculated density of 2.972 g·cm^{-3}, and an energy gap of 3.3 ± 0.1 eV (Ji et al. 2023). The crystals of monoclinic modification were synthesized using a solid state method in a stoichiometric molar ratio (K/Cd/Ge/S = 2:1:3:8). A total of 0.4 g of elements were loaded under vacuum in a flame-sealed silica ampoule and placed in a programmable furnace. The bottom of the silica ampoule was coated with amorphous carbon. The ampoule was heated from room temperature to 800°C in 20 h and kept at this temperature for 96 h. Then, the furnace was turned off and naturally cooled to room temperature. The crystals of monoclinic $K_2CdGe_3S_8$ were transparent light yellow. They are stable in dry air for many weeks with no detection of change. All starting materials were stored in an Ar-filed glove box with the oxygen level below 0.5 ppm.

B.4.12 Germanium–Potassium–Gallium–Sulfur

The quaternary compound $K_{2.4}Ga_{2.4}Ge_{1.6}S_8$, which crystallizes as a tetragonal structure with the lattice parameters $a = 788.24 \pm 0.09$, $c = 595.57 \pm 0.07$ pm, a calculated density of 2.844 g·cm^{-3}, and an energy gap of 3.5 eV for the crystalline compound and 3.0 eV for the glassy one, is formed in the Ge–K–Ga–S system (He et al. 2016). The crystals of the title compound were obtained through a reactive flux method with K_2S_2 as the reactive flux. The starting material K_2S_2 (2.25 mM), Ga_2S_3 (0.375 mM), Ge (0.5 mM), and S (1.75 mM) were mixed and ground uniformly and loaded into a silica tube. The tube was then flame-sealed under vacuum (0.1 Pa) and slowly heated to 750°C in a programmable furnace. It was kept at this temperature for 3 days and then slowly cooled down to 300°C at a rate of 2°C·h^{-1}, and finally to room temperature by turning off the furnace. The melt was washed with DMF several times to remove the excessive flux and dried by acetone. Thin yellow prisms up to 2–3 cm long were obtained. The powder sample was synthesized by a stoichiometric combination reaction of the starting materials under the same reacting condition and by repeating the procedure three times. All operations were carried out in an Ar-protected glove box. This compound can be also prepared in the glassy state with the glass transition temperature of 533°C. The glassy $K_{2.4}Ga_{2.4}Ge_{1.6}S_8$ is more stable in air than the crystalline form of this compound.

B.4.13 Germanium–Potassium–Yttrium–Sulfur

The quaternary compound $KYGeS_4$, which crystallizes as a monoclinic structure with the lattice parameters $a = 642.8 \pm 0.6$, $b = 664.1 \pm 0.6$, $c = 861.2 \pm 0.8$ pm, $\beta = 108.07°$, and an energy gap of 3.15 eV, is formed in the Ge–K–Y–S system (Mei et al. 2021). K_2S_3, Y, Ge, and S powders were used as raw materials for the synthesis of this compound. They were put in a glove box filled with argon, mixed homogeneously, and then sealed in a quartz tube under vacuum. The quartz tube was slowly heated to 800°C in 20 h and maintained for 10 h in a programmable furnace. The powder sample was obtained after this procedure. Single crystals were grown from the mixture of the raw materials, which was sealed in a carbon-coated quartz tube. The tube was heated to 850°C in 20 h, maintained at this temperature for 4 days, then slowly cooled to 300°C at a rate of $4°C{\cdot}h^{-1}$. After the cooling, the $KYGeS_4$ crystals were obtained.

B.4.14 Germanium–Rubidium–Zinc–Sulfur

The quaternary compound $Rb_2ZnGe_3S_8$, which crystallizes as a triclinic structure with the lattice parameters $a = 717.92 \pm 0.04$, $b = 747.90 \pm 0.04$, $c = 1443.05 \pm 0.07$ pm, $\alpha = 88.774 \pm 0.004°$, $\beta = 84.409 \pm 0.004°$, $\gamma = 72.211 \pm 0.005°$, a calculated density of 3.214 $g{\cdot}cm^{-3}$, and an energy gap of 3.24 eV, is formed in the Ge–Rb–Zn–S system (Chai et al. 2022). This compound was synthesized by high-temperature solid state reactions. The mixture of Mg, Zn, Ge, and S in a molar ratio 2:1:3:8 and RbCl as a flux was put in a glove box filled with argon, mixed homogeneously, and then sealed in a quartz tube under a vacuum of 0.01 Pa. The quartz tube was slowly heated in a furnace to 300°C in 5 h, maintained at this temperature for another 5 h, heated to 600°C in 9 h, maintained at 600°C for 9 h, heated to 900°C at a rate of $30°C{\cdot}h^{-1}$, maintained at this temperature for 96 h, and cooled at a rate of $8.4°C{\cdot}h^{-1}$ to 300°C. After that, the furnace was turned off. The powder sample was obtained after this procedure.

B.4.15 Germanium–Rubidium–Cadmium–Sulfur

The quaternary compound $Rb_2CdGe_3S_8$, which crystallizes as an orthorhombic structure with the lattice parameters $a = 734.59 \pm 0.02$, $b = 1222.34 \pm 0.04$, $c = 1691.63 \pm 0.07$ pm, a calculated density of 3.313 $g{\cdot}cm^{-3}$, and an energy gap of 3.16 eV, is formed in the Ge–Rb–Cd–S system (Chai et al. 2022). This compound was prepared in the same way as $Rb_2ZnGe_3S_8$ was synthesized using Cd instead of Zn.

B.4.16 Germanium–Rubidium–Mercury–Sulfur

The quaternary compound $Rb_4Hg_2Ge_2S_8$, which crystallizes as a monoclinic structure with the lattice parameters $a = 812.72 \pm 0.02$, $b = 1057.21 \pm 0.02$, $c = 1019.98 \pm 0.03$ pm, $\beta = 91.238 \pm 0.004°$, and a calculated density of 4.338 $g{\cdot}cm^{-3}$ is formed in the Ge–Rb–Hg–S system (Tang et al. 2022). The single crystals of this compound were prepared by high-temperature solid state reaction of Na_2S (78 mg,), HgS (233 mg), GeS_2 (137 mg), and RbCl (500 mg), which was added as cation exchanger and flux. The mixture was thoroughly ground in an agate mortar and loaded into silica tube. The silica tube was sealed under a high vacuum of 10^{-3} Pa, and then placed in a computer-controlled furnace. The following temperature control program was executed: heating from room temperature to 800°C in 10 h, kept for 3 days, then cooling slowly to room temperature at a rate of $3°C{\cdot}h^{-1}$. Greenish transparent block crystals were obtained. The synthesis of the polycrystalline sample by a similar strategy failed.

B.4.17 Germanium–Rubidium–Lanthanum–Sulfur

The quaternary compound $RbLaGeS_4$, which crystallizes as an orthorhombic structure with the lattice parameters a = 662.38 ± 0.04, b = 674.35 ± 0.04, c = 1725.95 ± 0.10 pm, and a calculated density of 3.663 g·cm^{-3}, is formed in the Ge–Rb–La–S system (Usman et al. 2019). Single crystals of this compound as yellow plates were grown by layering a mixture of La_2S_3 (0.5 mM), Ge powder (1 mM), and S (2.5 mM) beneath a layer of 1 g of oven-dried RbCl in an 8-in. long, finely carbon-coated fused silica tube. The loaded silica tube was evacuated to a pressure of not higher than ○0.01 Pa, flame-sealed using an oxygen/methane torch, and placed in a furnace. A three-step reaction temperature profile was used, which consisted of a ramp at 300°C·h^{-1} to 500°C, maintained at this temperature for 12 h, a ramp at 60°C·h^{-1} to 700°C, maintained at this temperature for 12 h, and finally a ramp at 60°C·h^{-1} to 900°C, maintained at this temperature for 24 or 48 h, and then fast cooling to room temperature by shutting off the furnace.

B.4.18 Germanium–Cesium–Lanthanum–Sulfur

The quaternary compound $CsLaGeS_4$, which crystallizes as an orthorhombic structure with the lattice parameters a = 660.95 ± 0.03, b = 679.71 ± 0.03, c = 1774.36 ± 0.07 pm, a calculated density of 3.938 g·cm^{-3} at 100 K, and an energy gap of 3.60 eV, is formed in the Ge–Cs–La–S system (Usman et al. 2019). Single crystals of this compound as yellow plates were grown in the same way as $RbLaGeS_4$ but using CsCl instead of RbCl.

B.4.19 Germanium–Cesium–Cerium–Sulfur

The quaternary compound $CsCeGeS_4$, which crystallizes as an orthorhombic structure with the lattice parameters a = 655.91 ± 0.02, b = 680.20 ± 0.02, c = 1784.21 ± 0.06 pm, and a calculated density of 3.954 g·cm^{-3}, is formed in the Ge–Cs–Ce–S system (Usman et al. 2019). Single crystals of this compound as yellow plates were grown in the same way as in the case of $RbLaGeS_4$ using Ce_2S_3 instead of La_2S_3 and CsCl instead of RbCl.

B.4.20 Germanium–Cesium–Praseodymium–Sulfur

The quaternary compound $CsPrGeS_4$, which crystallizes as an orthorhombic structure with the lattice parameters a = 654.07 ± 0.12, b = 679.29 ± 0.13, c = 1781.2 ± 0.3 pm, and a calculated density of 3.984 g·cm^{-3}, is formed in the Ge–Cs–Pr–S system (Usman et al. 2019). Single crystals of this compound as green plates were grown in the same way as $RbLaGeS_4$ but using Pr_2S_3 instead of La_2S_3 and CsCl instead of RbCl.

B.4.21 Germanium–Cesium–Neodymium–Sulfur

The quaternary compound $CsNdGeS_4$ quaternary compound, which crystallizes as an orthorhombic structure with the lattice parameters a = 651.95 ± 0.03, b = 678.00 ± 0.03, c = 1778.42 ± 0.07 pm, and a calculated density of 4.039 g·cm^{-3}, is formed in the Ge–Cs–Nd–S system (Usman et al. 2019). Single crystals of this compound as blue plates were grown in the same way as in the case of $RbLaGeS_4$ using Nd_2S_3 instead of La_2S_3 and CsCl instead of RbCl.

B.4.22 Germanium–Cesium–Europium–Sulfur

The quaternary compound $CsEuGeS_4$, which crystallizes as an orthorhombic structure with the lattice parameters a = 646.69 ± 0.02, b = 674.84 ± 0.02, c = 1771.48 ± 0.06 pm, and a calculated density of 4.173 g·cm^{-3}, is formed in the Ge–Cs–Eu–S system (Usman et al. 2019). Single crystals of this compound as red plates were grown in the same way as $RbLaGeS_4$ but using Eu_2S_3 instead of La_2S_3 and CsCl instead of RbCl.

B.4.23 Germanium–Cesium–Gadolinium–Sulfur

The quaternary compound $CsGdGeS_4$, which crystallizes as an orthorhombic structure with the lattice parameters $a = 646.57 \pm 0.02$, $b = 673.54 \pm 0.02$, $c = 1768.30 \pm 0.05$ pm, and a calculated density of 4.235 g·cm^{-3}, is formed in the Ge–Cs–Gd–S system (Usman et al. 2019). Single crystals of this compound as yellow plates were grown in the same way as $RbLaGeS_4$ but using Gd_2S_3 instead of La_2S_3 and CsCl instead of RbCl.

B.4.24 Germanium–Cesium–Terbium–Sulfur

The quaternary compound $CsTbGeS_4$, which crystallizes as an orthorhombic structure with the lattice parameters $a = 642.44 \pm 0.08$, $b = 672.47 \pm 0.09$, $c = 1768.9 \pm 0.2$ pm, and a calculated density of 4.282 g·cm^{-3}, is formed in the Ge–Cs–Tb–S system (Usman et al. 2019). Single crystals of this compound as yellow plates were grown in the same way as $RbLaGeS_4$ but using Tb_2S_3 instead of La_2S_3 and CsCl instead of RbCl.

B.4.25 Germanium–Copper–Barium–Sulfur

The quaternary compound Cu_2BaGeS_4, which crystallizes as a trigonal structure with the lattice parameters $a = 622.79 \pm 0.14$, $c = 1556.9 \pm 0.4$ pm, a calculated density of 4.432 g·cm^{-3}, and an energy gap of 2.49 eV, is formed in the Ge–Cu–Ba–S system (Liu et al. 2019). To prepare this compound, Cu powder (0.11 mM), Ge powder (0.08 mM), S powder (1.03 mM), $Ba(OH)_2 \cdot 8H_2O$ (0.04 mM), and about 300 mg ethylenediamine were mixed together and sealed in a Pyrex glass tube (about 10% filling volume of the tube) at ambient atmosphere. Finally, the glass tube was placed in a stainless steel autoclave, into which water was added to 80% filling to balance the pressure of the glass tube, and then heated in the furnace at 170°C for 6 days. After being cooled to room temperature naturally, the products were washed with water and ethanol, respectively, and yellow block crystals of the title compound were obtained.

B.4.26 Germanium–Copper–Phosphorus–Sulfur

Cu_8GeS_6–Cu_7PS_6. This system is a nonquasibinary section of the Ge–Cu–P–S system as Cu_8GeS_6 melts incongruently (Bereznyuk et al. 2022d). The system exhibits wide ranges of solid solutions at room temperature: up to 30 mol% based on α-Cu_7PS_6 and from ~9 to 32 mol% Cu_7PS_6 based on high-temperature modifications of both compounds. The solid solution based on the high-temperature modification Cu_2S primarily crystallizes as a range of 0–70 mol% Cu_7PS_6. The system was studied through DTA, XRD, and metallography. The ingots for the investigations were annealed at 230°C for 500 h.

B.4.27 Germanium–Copper–Arsenic–Sulfur

GeS_2–Cu_2S–As_2S_3. The region of the glass formation in this system was determined through XRD (Figure B.4.1) (Bereznyuk et al. 2021). Glassy samples were synthesized from elemental substances and As_2S_3. The mixtures were heated in evacuated quartz ampoules up to 830°C with exposure for 24 h at 400°C and 600°C. The samples were kept at 830°C for 10 h, followed by quenching in a 25% NaCl solution with crushed ice.

B.4.28 Germanium–Copper–Antimony–Sulfur

GeS_2–Sb_2S_3. The phase diagram of this system (Figure B.4.2), constructed through DTA, XRD, and metallography using the alloys annealed at 230°C for 500 h, is a eutectic type (Bereznyuk and Piskach 2023). The eutectic contains 35 mol% GeS_2 and crystallizes at 474°C. Samples with a content of 10–60 mol% GeS_2 are the glass or glass crystals.

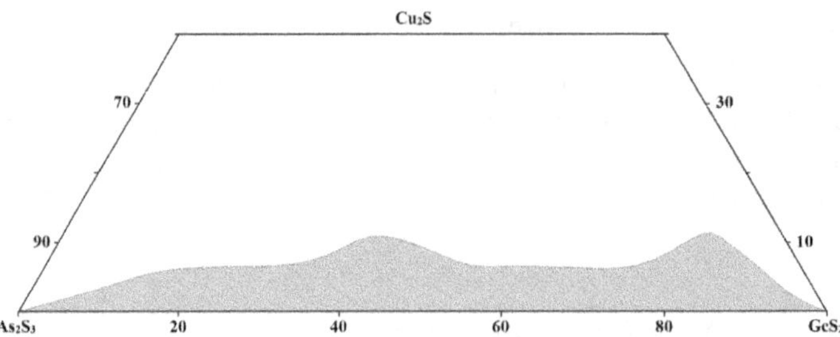

FIGURE B.4.1 Glass-forming region in the GeS_2–Cu_2S–As_2S_3 quasiternary system. (From Bereznyuk, O., et al., *Probl. khimii ta staloho rozvytku*, (4), 3, 2021.) Open access.

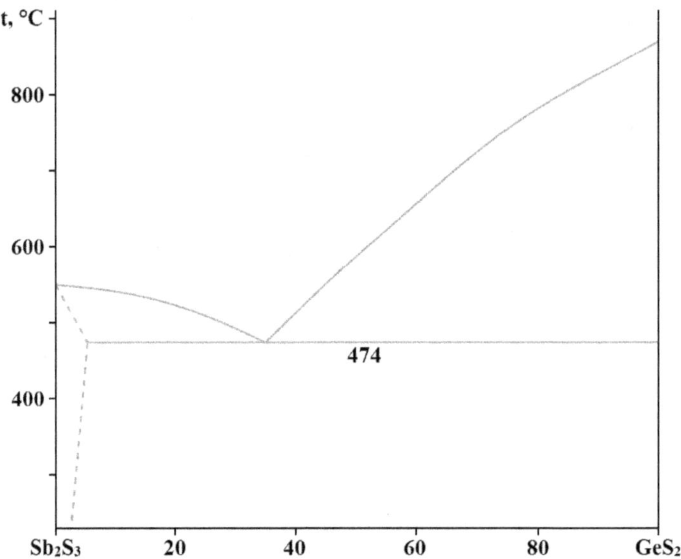

FIGURE B.4.2 Phase diagram of the GeS_2–Sb_2S_3 system. (From Bereznyuk, O., and Piskach, L., *Probl. khimii ta staloho rozvytku*, (3), 3, 2023.) Open access.

Cu_2GeS_3–Sb_2S_3. The phase diagram of this system (Figure B.4.3), constructed through DTA, XRD, and metallography using the alloys annealed at 230°C for 500 h, is a eutectic type (Bereznyuk and Piskach 2023). The eutectic contains 18 mol% Cu_2GeS_3 and crystallizes at 437°C. Solubility based on Sb_2S_3 and Cu_2GeS_3 is no more than 5 mol%.

Cu_2GeS_3–$Cu_3Sb_2S_3$. The phase diagram of this system (Figure B.4.4), constructed through DTA, XRD, and metallography using the alloys annealed at 230°C for 500 h, is a eutectic type (Bereznyuk and Piskach 2023). The eutectic contains 7 mol% Cu_2GeS_3 and crystallizes at 547°C. The horizontal line at 487°C corresponds to the peritectoid transformation, which is associated with polymorphism of Cu_3SbS_3.

GeS_2–Cu_2S–Sb_2S_3. The isothermal section of this quasiternary system (Figure B.4.5) contains eight single-phase, 13 two-phase, and six three-phase regions (Bereznyuk and Piskach 2023).

The region of the glass formation in this system was determined through XRD (Figure B.4.6) (Bereznyuk et al. 2021). Glassy samples were synthesized from elemental substances. The mixtures were heated in evacuated quartz ampoules up to 830°C with exposure for 24 h at 400°C and 600°C. The samples were kept at 830°C for 10 h, followed by quenching in a 25% NaCl solution with crushed ice.

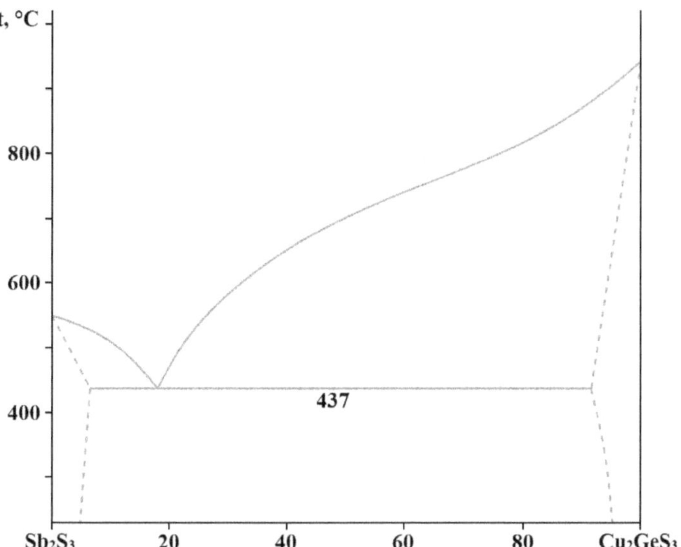

FIGURE B.4.3 Phase diagram of the Cu_2GeS_3–Sb_2S_3 system. (From Bereznyuk, O., and Piskach, L., *Probl. khimii ta staloho rozvytku*, (3), 3, 2023.) Open access.

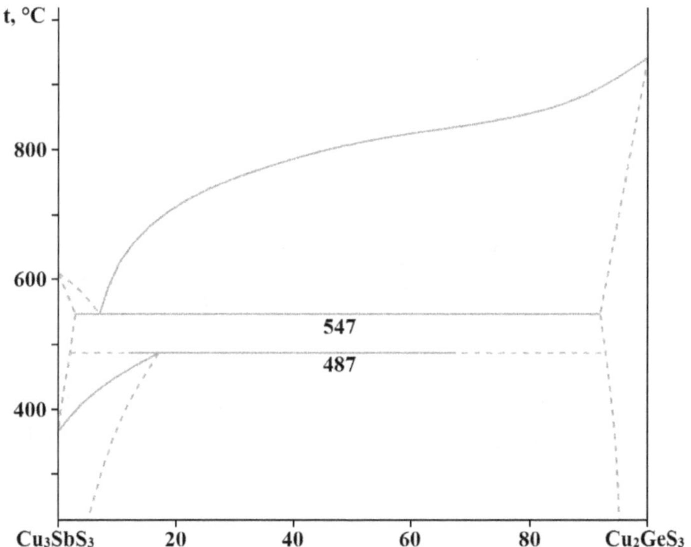

FIGURE B.4.4 Phase diagram of the Cu_2GeS_3–$Cu_3Sb_2S_3$ system. (From Bereznyuk, O., and Piskach, L., *Probl. khimii ta staloho rozvytku*, (3), 3, 2023.) Open access.

B.4.29 Germanium–Copper–Selenium–Sulfur

Cu_8GeS_6–Cu_8GeSe_6. This system is nonquasibinary due to the peritectic melting of the starting compounds (Bagheri et al. 2014). A high-temperature solid solution of $Cu_2S_xSe_{1-x}$ primarily crystallizes from liquid and forms a two-phase area in the system. The formation of these solid solutions decreases the temperature of polymorphic transitions of both components. The temperature of polymorphic transition of Cu_8GeS_6 within 0–20 mol% Cu_8GeSe_6 decreases from 55°C to 37°C and for 30 mol% Cu_8GeSe_6, this transition is not observed at higher than room temperature. The results show that the formation of the solid solutions extends the area of high-temperature cubic phase and exists at 30–90 mol% Cu_8GeSe_6 at

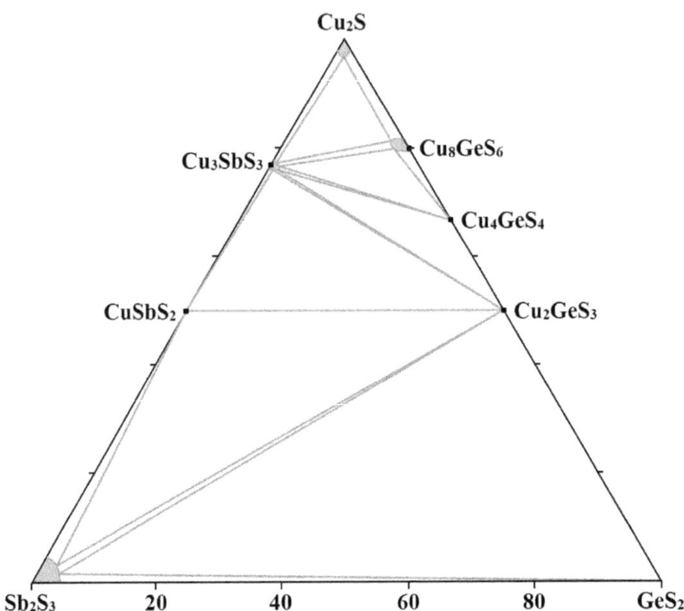

FIGURE B.4.5 Isothermal section of the GeS_2–Cu_2S–Sb_2S_3 quasiternary system at 230°C. (From (From Bereznyuk, O., and Piskach, L., *Probl. khimii ta staloho rozvytku*, (3), 3, 2023.) Open access.

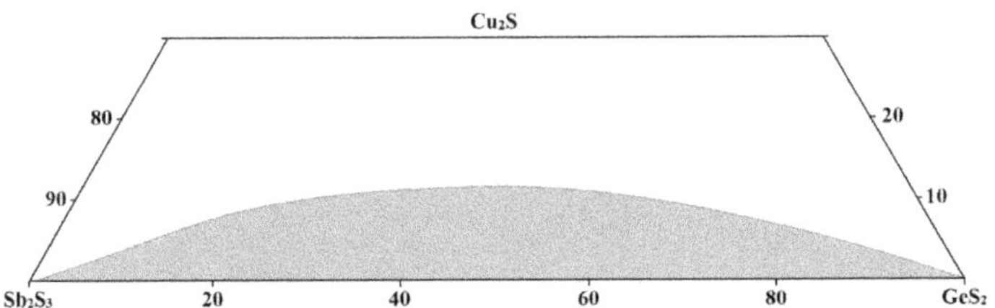

FIGURE B.4.6 Glass-forming region in the GeS_2–Cu_2S–Sb_2S_3 quasiternary system. (From Bereznyuk, O., et al., *Probl. khimii ta staloho rozvytku*, (4), 3, 2021.) Open access.

room temperature. The calculated partial molar thermodynamic function of Cu confirms that the solid solutions belong to the substitutional type of solid solutions. This system was studied through DTA and XRD, as well as EMF measurements.

B.4.30 Germanium–Copper–Iron–Sulfur

The quaternary compound $Cu_{22}Fe_8Ge_4S_{32}$, which crystallizes as a cubic structure with the lattice parameter a = 1059.5 pm, is formed in the Ge–Cu–Fe–S system (Paradis-Fortin et al. 2022). This compound was synthesized as follows. The powders were weighed in a stoichiometric ratio (22Cu:8Fe:4Ge:32S) and ground together in an agate mortar. Many 1.5 g batches of powder were then pressed into 5 mm diameter pellets. The pellets were placed in silica tubes and evacuated to 1 Pa. The sealed tubes were then placed in a vertical furnace and heated with a rate of 2°C·min^{-1} and annealed at 700°C for 24 h. The samples were cooled down at 500°C by switching off the heat power and using the thermal inertia of the furnace to limit the cooling rate and then air quenched. All starting element powders were stored and manipulated in a glove box under an Ar atmosphere.

B.4.31 Germanium–Silver–Gallium–Sulfur

The quaternary compound $AgGaGeS_4$, which crystallizes as an orthorhombic structure with the lattice parameters $a = 1202.15$, $b = 2291.78$, and $c = 687.38$ pm at room temperature and $a = 1211.53$, $b = 2300.73$, and $c = 693.23$ pm at 600°C, is formed in the Ge–Ag–Ga–S system (Wu et al. 2020).

B.4.32 Germanium–Silver–Indium–Sulfur

The quaternary compound $Ag_2In_2GeS_6$, which crystallizes as a monoclinic structure with the lattice parameters $a = 1222.14 \pm 0.11$, $b = 721.56 \pm 0.07$, $c = 1220.45 \pm 0.11$ pm, $\beta = 109.567 \pm 0.002°$, a calculated density of 4.653 g·cm^{-3}, and an energy gap of 2.30 eV, is formed in the Ge–Ag–In–S system (Zhou et al. 2022a). Crystals of this compound were synthesized by traditional high-temperature solid state reactions using the following experimental process: a 500 mg stoichiometric mixture of Ag, In, Ge, and S and an additional 400 mg KI flux were ground in an agate mortar and loaded into a graphite crucible. Then, the graphite crucible was transferred into a silica tube, which was flame-sealed in high vacuum (0.01 Pa). The samples were heated from an ambient temperature to 850°C within 9 h and held for 120 h, and finally cooled to 300°C at a rate of 5°C·h^{-1} before turning off the furnace. Moisture and air stable crystals of the title compound were obtained after washing with distilled water and alcohol.

B.4.33 Germanium–Silver–Tin–Sulfur

The quaternary compound $Ag_4SnGe_2S_7$, which crystallizes as a monoclinic structure with the lattice parameters $a = 1133.98 \pm 0.04$, $b = 697.06 \pm 0.02$, $c = 1548.85 \pm 0.04$ pm, $\beta = 91.2130 \pm 0.0010°$, a calculated density of 4.991 g·cm^{-3} at 153 K, and an energy gap of 2.40 eV, is formed in the Ge–Ag–Sn–S system (Wang et al. 2022b). The title compound was obtained by high-temperature solid state reaction. For crystal synthesis, the mixture of Ag_2S, CaS, and GeS_2, (molar ratio 3:2:2) with a total mass of ~ 0.487 g, as well as 0.3 g of $SnCl_2$ were weighed and loaded into a graphite crucible and then put it into a silica tube. The silica tube was flame-sealed under a high vacuum of 10^{-3} Pa, were then moved into computer-controlled furnace and heated to 450°C in 16.7 h, maintained at this temperature for another 16.7 h, heated to 650°C in 16.7 h, maintained at 650°C for 33.3 h, then cooled to 500°C in 100 h, cooled to 300°C for another 100 h, and cooled to room temperature after shutting down the furnace. After that, many yellow single crystals were crystallized in a carbon crucible.

The polycrystalline powders of $Ag_4SnGe_2S_7$ were also prepared by high-temperature solid state reaction. The mixtures of Ag, SnS, Ge and S (molar ratio 4:1:2:6) with a total mass of 3.75 g were sealed in silica tube and put in a horizontal tubular furnace, then heated to 450°C in 16.7 h, maintained at this temperature for another 16.7 h, heated to 750°C in 25 h, maintained 750°C for 33.3 h, then cooled to 650°C in 53.3 h, cooled to 500°C in 41.7 h, cooled to 300°C in 33.3 h, and cooled to room temperature in 10 h.

B.4.34 Germanium–Silver–Phosphorus–Sulfur

Ag_8GeS_6–Ag_7PS_6. This system is characterized by a peritectic interaction at 879°C (Figure B.4.7) (Berezniuk et al. 2022; Bereznyuk et al. 2022d). The solid solution based on Ag_7PS_6 extends to 43 mol% Ag_8GeS_6 and based on Ag_8GeS_6 reaches 35 mol% Ag_7PS_6. The $Ag_{7.9}Ge_{0.9}P_{0.1}S_6$ solid solution crystallizes as an orthorhombic structure with the lattice parameters $a = 1510.8 \pm 0.1$, $b = 746.09 \pm 0.07$, $c = 1057.1 \pm 0.1$ pm, and a calculated density of 6.204 ± 0.002 g·cm^{-3} and the $Ag_{7.1}Ge_{0.1}P_{0.9}S_6$ solid solution crystallizes as a cubic structure with the lattice parameters $a = 1040.38 \pm 0.07$ pm and a calculated density of 5.859 ± 0.001 g·cm^{-3}. The alloys achieved equilibrium by homogenizing annealing at 230°C for 450 h, followed by quenching of the ampoules into NaCl solution with crushed ice.

GeS_2–Ag_2S–P_2S_5. The isothermal section of this quasiternary system at 230°C is given in Figure B.4.8 (Berezniuk et al. 2022). The quasibinary sections separate this system into eight subsystems. Limited solid solutions were identified in the Ag_8GeS_6–Ag_7PS_6 section.

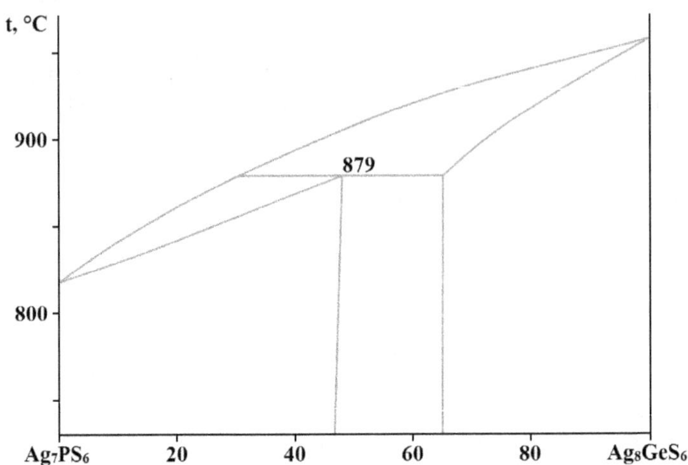

FIGURE B.4.7 Phase diagram of the Ag_8GeS_6–Ag_7PS_6 system. (From Berezniuk, O., et al., *J. Solid State Chem.*, **313**, 123340, 2022.)

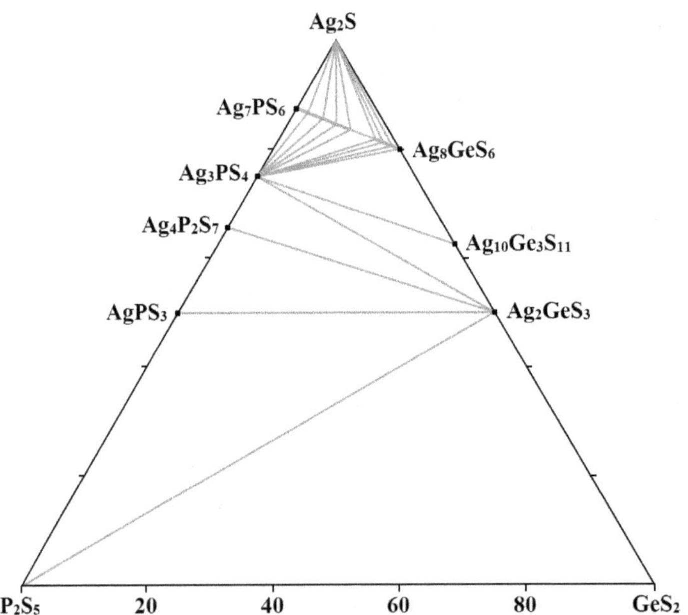

FIGURE B.4.8 Isothermal section of the GeS_2–Ag_2S–P_2S_5 quasiternary system at 230°C. (From Berezniuk, O., et al., *J. Solid State Chem.*, **313**, 123340, 2022.)

B.4.35 Germanium–Silver–Arsenic–Sulfur

GeS_2–Ag_2S–As_2S_3. The isothermal section of this quasiternary system at 230°C is given in the Figure B.4.9 (Bereznyuk et al. 2022b). No quaternary compounds were found in the system. The extent of solid solutions based on binary and ternary compounds at this temperature is negligible. The alloys for the investigations were annealed at 230°C for 500 h.

The glass-forming region in this system is presented in Figure B.4.10 (Bereznyuk et al. 2021, 2022c). The glass transition temperature of the glasses in this system is within the interval from 129°C to 148°C and increases with the increasing of the GeS_2 content (Bereznyuk et al. 2022b).

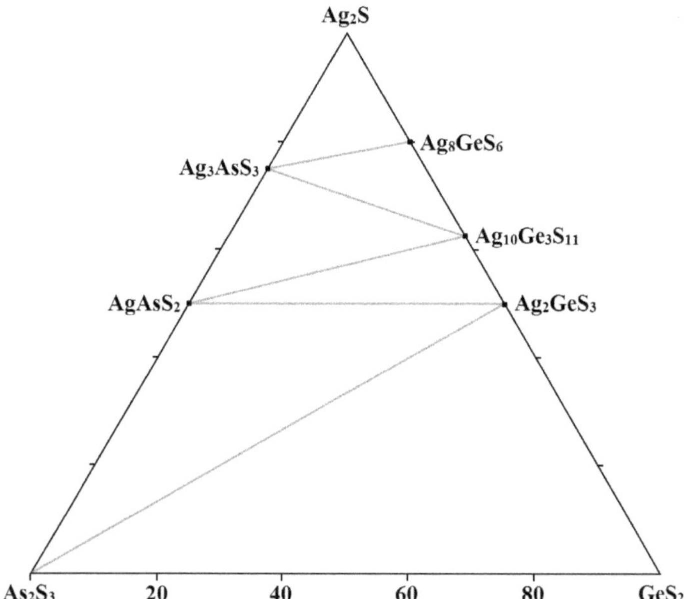

FIGURE B.4.9 Isothermal section of the GeS_2–Ag_2S–As_2S_3 quasiternary system at 230°C. (From Bereznyuk, O., et al., *Phys. Chem. Solid State*, **23**(1), 57, 2022.)

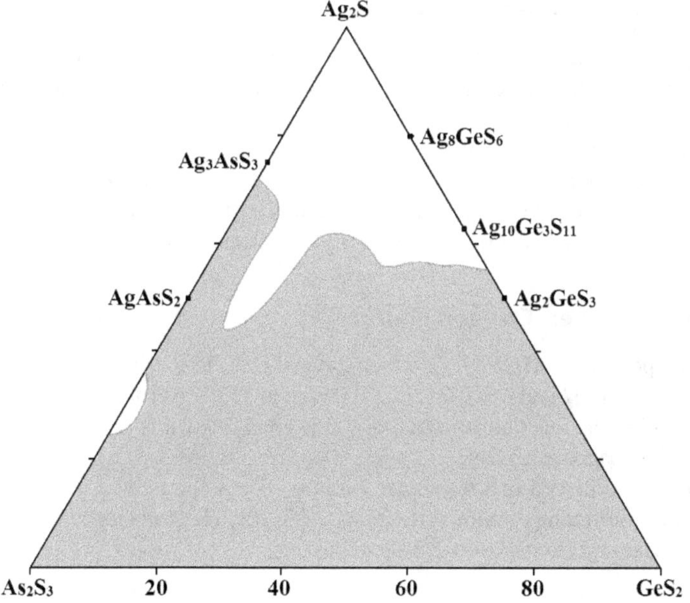

FIGURE B.4.10 Glass-forming region in the GeS_2–Ag_2S–As_2S_3 quasiternary system. (From Bereznyuk, O., et al., *Probl. khimii ta staloho rozvytku*, (4), 3, 2021.) Open access.

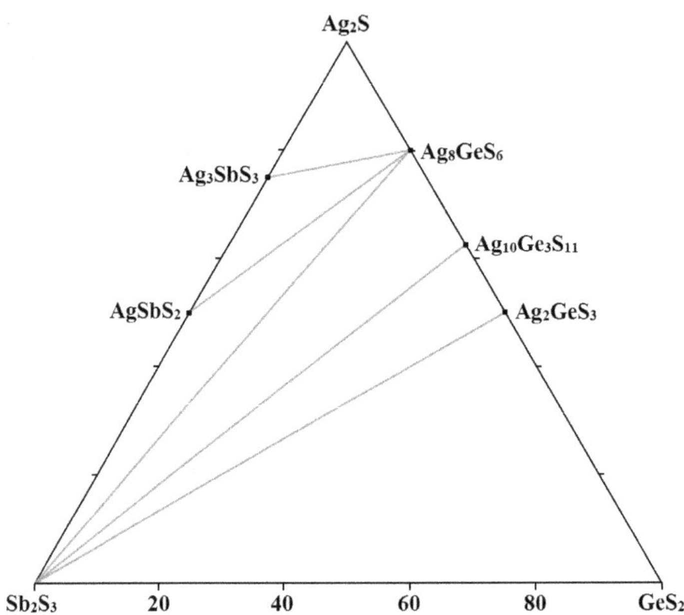

FIGURE B.4.11 Isothermal section of the GeS_2–Ag_2S–Sb_2S_3 quasiternary system at 230°C. (From Bereznyuk, O., et al., *Phys. Chem. Solid State*, **23**(1), 57, 2022.)

B.4.36 Germanium–Silver–Antimony–Sulfur

GeS_2–Ag_2S–Sb_2S_3. The isothermal section of this quasiternary system at 230°C is given in Figure B.4.11 (Bereznyuk et al. 2022b). No quaternary compounds were found in the system. The extent of solid solutions based on binary and ternary compounds at this temperature is negligible. The alloys for the investigations were annealed at 230°C for 500 h.

The glass-forming region in this system is shown in Figure B.4.12 (Bereznyuk et al. 2021, 2022b). The glass transition temperature of the glasses in this system is within the interval from 100°C to 165°C and increases with the increasing of the GeS_2 content (Bereznyuk et al. 2022b).

B.4.37 Germanium–Silver–Oxygen–Sulfur

The quaternary compound $Ag_2Ge(S_2O_7)_3$, which crystallizes as a trigonal structure with the lattice parameters $a = 1616.85 \pm 0.07$ and $c = 1031.83 \pm 0.04$ pm at 153 K, is formed in the Ge–Ag–O–S system (Logemann et al. 2012). The reaction for obtaining this compound was performed in thick-walled glass ampoule. The tube was loaded with $GeCl_4$ (1 mM), oleum (1 mL, 65% SO_3), and Ag_2SO_4 (1 mM), torch-sealed under vacuum, and placed in a resistance furnace. The ampoule was maintained at 250°C for 24 h and cooled down to room temperature at a rate of 1.8°C·h^{-1}. The colorless crystals are very moisture-sensitive and were separated from oleum in a glove box.

B.4.38 Germanium–Silver–Tellurium–Sulfur

Ag_8GeS_6–Ag_8GeTe_6. This system is nonquasibinary section of the Ge–Ag–Te–S system as Ag_8GeTe_6 melts incongruently (Amiraslanova et al. 2023c). It is characterized by the formation of continuous series of the solid solution between Ag_8GeTe_6 and high-temperature cubic modification of Ag_8GeS_6. When solid solutions are formed, the temperatures of the polymorphic phase transitions of Ag_8GeS_6 decreases. This leads to the stabilization of the cubic phase in the $\geq$ 30 mol% Ag_8GeTe_6 range of compositions at room temperature and below. The regions of homogeneity based on Ag_8GeS_6 is 15 mol%. The alloys for the investigations were annealed at 530°C for about 500 h.

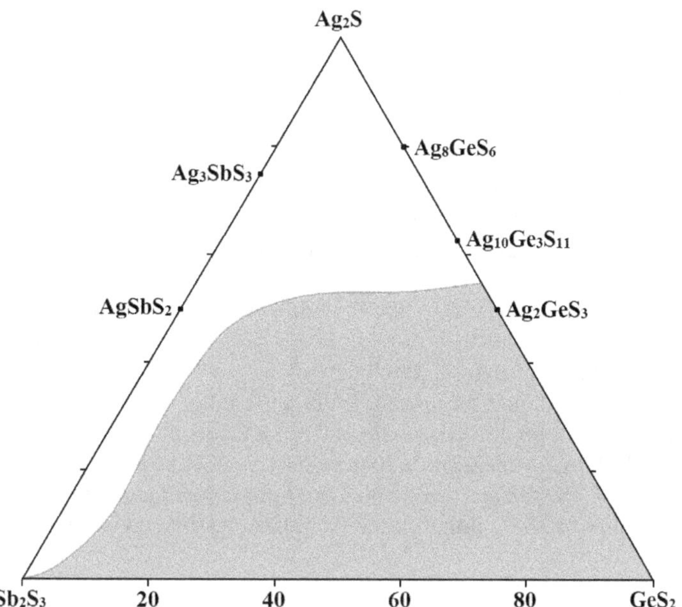

FIGURE B.4.12 Glass-forming region in the GeS_2–Ag_2S–Sb_2S_3 quasiternary system. (From Bereznyuk, O., et al., *Probl. khimii ta staloho rozvytku*, (4), 3, 2021.) Open access.

B.4.39 Germanium–Gold–Lanthanum–Sulfur

The quaternary compound $Au_{1.98}La_6Ge_2S_{14}$ compound, which crystallizes as a hexagonal structure with the lattice parameters a = 1039.05 ± 0.16, c = 588.48 ± 0.10 pm, and a calculated density of 5.479 g·cm^{-3} at 100 K, is formed in the Ge–Au–La–S system (Akopov et al. 2021). The powder compound was synthesized using pre-arc-melted precursor and sulfur without flux, while the crystals were grown using a salt flux synthesis from a pre-arc-melted precursor and sulfur. The precursor was prepared using pieces of La, Au, and Ge. A sample with a total mass of 0.6–1 g was weighed out in a molar ratio of La/Au/Ge = 6:1:2.05. A slight excess of Ge was used to account for its evaporation during arc melting. The sample was then placed in an arc-melter on a copper hearth along with oxygen getter material (zirconium metal). The arc-melter chamber was later sealed and evacuated for 20 min followed by purging with argon; this process was repeated three times to ensure no oxygen was present in the chamber. During arc melting, the getter was melted first to ensure the absorbance of any trace oxygen, and then the samples were heated for 1000°C in ~ 1 min at a current of ~ 80 A until molten, allowed to solidify, flipped and re-arced two more times to ensure homogeneity under a current of ~ 100 A. The precursor ingots were then crushed into a powder using a steel Plattner-style diamond crusher.

The single crystals of $Au_{1.98}La_6Ge_2S_{14}$ were grown using pre-arc-melted precursor, S powder, and KI as a flux. The ratio of (precursor + sulfur) to KI flux was kept at a 1: 30 ratio by mass (for 0.1 g of reactants, 3 g of KI was used). The reactants and KI were added to a silica ampoule, which was sealed under vacuum. The samples were heated to 950°C–1050°C over 12 h, allowed to maintain for 72–120 h, and then cooled down to room temperature over 8 h. The KI flux was washed with deionized water, and the sample was then filtered under vacuum and dried. The crystals have a transparent yellow-gold appearance.

The bulk powder of the compound was synthesized using pre-arc melted precursor and sulfur powder. The sample was prepared by loading the metal silicide precursor and sulfur in a fused silica ampoule in a 1:14 precursor to sulfur ratio by mass. The ampoule was then flame-sealed under vacuum. The sample was heated over 12 h to 950°C–1050°C, maintained for 72–120 h, and then allowing them to cool to room temperature over 8 h.

B.4.40 Germanium–Strontium–Zinc–Sulfur

The quaternary compound $SrZnGeS_4$, which is thermally stable below 710°C, decomposes at higher temperatures to SrS and ZnS, and crystallizes as an orthorhombic structure with the lattice parameters a = 2045.27 ± 0.15, b = 2080.75 ± 0.17, c = 1223.01 ± 0.09 pm, a calculated density of 3.612 g·cm^{-3}, and an energy gap of 3.63 eV, is formed in the Ge–Sr–Zn–S system (Liu et al. 2022b,c). The polycrystalline sample of this compound was synthesized through high-temperature solid state reactions. A mixture of SrS, Zn, Ge, and S was weighed in a stoichiometric ratio with a gross mass of 300 mg, and then loaded into a graphite crucible with an out cover of silica tubing. The whole assembly was sealed under a vacuum of 10^{-3} Pa and put into the tube furnace. In order to avoid latent explosion caused by vapor pressure of S, the tube was firstly heated to 400°C within 10 h, maintained for 12 h, and then heated to 890°C, maintained for 48 h, followed by a slow cooling to 400°C before turning off the furnace. Finally, the powder of $SrZnGeS_4$ was synthesized. The light yellow, nearly colorless, crystal was obtained using crystal seeds with KI as flux. Rather than the stoichiometric ratio, 400 mg mixture of SrS, Zn, Ge, and S with the molar ratio of 1:1:0.98:3, mixing with 600 mg KI, was put into graphite crucible. The sealed tube was heated to 800°C, maintained for 72 h, then slowly cooled to 500°C, and furnace closed. With the 0.98 ratio of Ge, $SrZnGeS_4$ crystals tend to be more light-colored and brighter. After deionized water washing to remove the flux and little impurities, light yellow crystal was obtained, being stable in air for at least 12 months.

B.4.41 Germanium–Strontium–Oxygen–Sulfur

Three quaternary compounds, $SrGeOS_2$, $SrGe_2O_3S_2$, and $Sr_3S(GeOS_3)$, are formed in the Ge–Sr–O–S system. The first compound possesses good thermal stability up to 830°C and crystallizes as an orthorhombic structure with the lattice parameters a = 466.24 ± 0.05, b = 805.45 ± 0.09, c = 1202.21 ± 0.12 pm, a calculated density of 3.536 g·cm^{-3}, and an energy gap of 4.36 eV (Yang et al. 2022b) [a = 466.27 ± 0.04, b = 806.22 ± 0.07, c = 1209.8 ± 0.1 pm, a calculated density of 3.510 g·cm^{-3}, and an energy gap of 3.9 eV (Zhang et al. 2019b)].

This compound was obtained via a flux method using stoichiometric mixture of SrO, Ge, and S with a total mass of 300 mg (Yang et al. 2022b). The mixture was placed in a quartz crucible and the same amount of CsI was added as the flux. Subsequently, they were vacuum-sealed under 10^{-3} Pa, gradually heated to 400°C and kept at this temperature for 20 h, then heated to 750°C and maintained for 72 h, and finally cooled at a rate of 3.5°C·h^{-1} to 350°C before turning off the furnace. Colorless and transparent rod-shaped crystals were obtained by washing with distilled water and ethanol, which are stable in air and water.

Single crystals of this compound were also prepared in the next way. SrO powder was annealed at 1000°C for 12 h before use (Zhang et al. 2019b). The mixture of SrO (3.0 mM), Ge (3.0 mM), S (6.0 mM), and KI (30 mM) was ground and loaded into the carbon-coated fused silica tube. The tube was then flame-sealed under vacuum (0.1 Pa) and slowly heated to 900°C in a furnace. The tube was maintained at this temperature for 3 days, followed by cooling to 500°C at a rate of 2°C·h^{-1}. Finally, the silica tube was quenched in air. The product was washed several times with distilled water to remove the KI flux and then dried with acetone. All operations were carried out in an Ar-filled glove box.

$SrGe_2O_3S_2$ is thermally stable up to at least 750°C under a nitrogen atmosphere and crystallizes as a monoclinic structure with the lattice parameters a = 737.8 ± 0.3, b = 915.6 ± 0.3, c = 876.5 ± 0.3 pm, β = 94.71 ± 0.04°, a calculated density of 3.882 g·cm^{-3}, and an energy gap of 3.73 eV (Xing et al. 2020). Single crystals of this compound were prepared from the equimolar mixture of SrO, GeO_2, and GeS_2, which was sealed into an evacuated quartz tube (<10^{-3} Pa) and subjected to the following heat treatment: heated to 1000°C in 1 day, maintained for 3 days, and cooled at a rate of 4°C·h^{-1} to room temperature. Finally, colorless transparent block crystals were obtained. The polycrystalline sample of $SrGe_2O_3S_2$ could be easily synthesized by heating the stoichiometric mixture sealed into an evacuated silica tube (1 mM of SrO, 1 mM of GeS_2, and 1 mM of GeO_2) at 850°C for 2 days. All the reactants were stored in an argon-filled glove box (H_2O and O_2 under 0.1 ppm).

$Sr_3S(GeOS_3)$ crystallizes as an orthorhombic structure with the lattice parameters $a = 1182.53 \pm 0.19$, $b = 598.24 \pm 0.09$, $c = 1166.32 \pm 0.19$ pm, a calculated density of 3.862 g·cm^{-3}, and an energy gap of 4.10 eV (Cui et al. 2023). Single crystals of this compound were synthesized using 3:0.5:0.5:1 molar ratios of SrS, GeO_2, Ge, and S with a total mass of 500 mg as the starting materials. After being ground to fine powder, the mixture was packed into quartz tube and further evacuated to 10^{-3} Pa, and then flame-sealed. The tube was put into a muffle furnace, heated from room temperature to 950°C within 6 h and kept at this temperature for 1 day, gradually cooled to 700°C at a rate of 5°C·h^{-1}, and finally the furnace was turned off. The single crystals of $Sr_3S(GeOS_3)$ were obtained. This compound could be synthesized in the range of 900°C–950°C, but it decomposes when the temperature was higher than 950°C.

B.4.42 Germanium–Barium–Tin–Selenium

The quaternary compound $Ba_4GeSn_3S_9$ compound, which crystallizes as an orthorhombic structure with the lattice parameters $a = 1246.3 \pm 0.3$, $b = 930.8 \pm 0.2$, $c = 1789.2 \pm 0.5$ pm, and energy gap of 1.21 eV, is formed in the Ge–Ba–Sn–Se system (Yuan et al. 2021b). The crystal of this compound was synthesized by a solid state reaction technique. A mixture of BaSe, Sn, GeSe, and Se (molar ratio 4:3:1:4) was ground and loaded into a graphite crucible sealed in the evacuated silica tube under 0.01 Pa pressure. The reactant was heated to 900°C within 25 h (held for 50 h) under the operation by a computer-controlled furnace, then the sample was slowly cooled to 800°C in 50 h and finally to room temperature in 25 h, obtaining black crystals, but there were always small amounts of yellow crystals of Ba_2GeSe_4 present. Since the crystals of Ba_2GeSe_4 are always associated with $Ba_4GeSn_3S_9$, the pure phase of the title compound is difficult to obtain. The pure phase of $Ba_4GeSn_3S_9$ was produced when the mixture was heated to 800°C in 15 h, kept for 50 h, and then the furnace was turned off.

B.4.43 Germanium–Barium–Oxygen–Sulfur

Two quaternary compounds, $BaGeOS_2$ and $Ba_3S(GeOS_3)$, are formed in the Ge–Ba–O–S system. The first compound possesses good thermal stability up to 830°C and crystallizes as an orthorhombic structure with the lattice parameters $a = 488.71 \pm 0.08$, $b = 847.28 \pm 0.17$, $c = 1209.1 \pm 0.2$ pm, a calculated density of 3.848 g·cm^{-3}, and an energy gap of 4.43 eV (Yang et al. 2022b) [$a = 484.41 \pm 0.01$, $b = 841.33 \pm 0.02$, $c = 1204.51 \pm 0.03$ pm, a calculated density of 3.925 g·cm^{-3}, and an energy gap of 4.1 eV (Zhang et al. 2019b)]. This compound was prepared in the same way as $SrGeOS_2$ but using BaO instead of SrO (Zhang et al. 2019b; Yang et al. 2022b).

$Ba_3S(GeOS_3)$ also crystallizes as an orthorhombic structure with the lattice parameters $a = 1235.22 \pm 0.07$, $b = 625.85 \pm 0.04$, $c = 1217.40 \pm 0.08$ pm, a calculated density of 4.438 g·cm^{-3}, and an energy gap of 3.63 eV (Cui et al. 2023). The title compound was obtained in the same way as $Sr_3S(GeOS_3)$ but using BaS instead of SrS.

B.4.44 Germanium–Barium–Tellurium–Sulfur

The quaternary compound Ba_3GeTeS_4, which crystallizes as an orthorhombic structure with the lattice parameters $a = 685.83 \pm 0.12$, $b = 1620.7 \pm 0.3$, $c = 962.39 \pm 0.19$ pm, and an energy gap of 2.5 ± 0.2 eV, is formed in the Ge–Ba–Te–S system (Yadav et al. 2023). Yellow-colored crystals of this compound were obtained by high-temperature reactions of Ba, Ge, Te, and S at 850°C. A small amount of black crystals of $Ba_2Ge_2Te_5$ were also found in the reaction product. All reactants were handled inside an Ar-filled glove box to avoid their oxidation.

B.4.45 Germanium–Zinc–Samarium–Sulfur

The quaternary compound $Zn_{0.5}Sm_3GeS_7$ compound, which crystallizes as a hexagonal structure with the lattice parameters $a = 1000.07 \pm 0.10$, $c = 573.92 \pm 0.11$ pm, a calculated density of 5.216 g·cm^{-3}, and an energy gap of 2.33 eV, is formed in the Ge–Zn–Sm–S system (Gao et al. 2022a). The title compound was synthesized with the stoichiometric ratio of Sm_2S_3, ZnS, and GeS_2 in vacuum-sealed silica tube. The detailed temperature reaction process was shown as follows: firstly, heat up to 900°C at a rate of 10°C·h^{-1} and then kept at this temperature for 10 days; secondly, cool down to room temperature at a rate of 5°C·h^{-1}. The synthesized crystals are stable in the air for several months.

B.4.46 Germanium–Cadmium–Yttrium–Sulfur

The quaternary compound $Cd_{0.5}Y_3GeS_7$, which crystallizes as a hexagonal structure with the lattice parameters $a = 981.83 \pm 0.13$, $c = 574.3 \pm 0.2$ pm, a calculated density of 4.294 g·cm^{-3}, and an energy gap of 1.62 eV, is formed in the Ge–Cd–Y–S system (Gao et al. 2022a). This compound was prepared in the same way as $Zn_{0.5}Sm_3GeS_7$ using Y_2S_3 instead of Sm_2S_3 and CdS instead of ZnS. It is also stable in the air for several months.

B.4.47 Germanium–Cadmium–Gadolinium–Sulfur

The quaternary compound $Cd_{0.5}Gd_3GeS_7$, which crystallizes as a hexagonal structure with the lattice parameters $a = 997.03 \pm 0.16$, $c = 573.27 \pm 0.14$ pm, a calculated density of 5.551 g·cm^{-3}, and an energy gap of 1.82 eV, is formed in the Ge–Cd–Gd–S system (Gao et al. 2022a). This compound was prepared in the same way as $Zn_{0.5}Sm_3GeS_7$ using Gd_2S_3 instead of Sm_2S_3 and CdS instead of ZnS. It is also stable in the air for several months.

B.4.48 Germanium–Lanthanum–Oxygen–Sulfur

The quaternary compound $La_4(GeS_2O_2)_3$, which is stable up to 800°C and crystallizes as a in the trigonal structure with the lattice parameters $a = 1682.83 \pm 0.03$, $c = 841.40 \pm 0.02$ pm, a calculated density of 5.126 g·cm^{-3}, and an energy gap of 3.67 eV, is formed in the Ge–La–O–S system (Yan et al. 2020). Single crystals of this compound were grown out of a eutectic $BaCl_2$–$CaCl_2$ flux. First, La_2S_3 (0.5 mM), GeO_2 (1.0 mM), $BaCl_2$ (3.13 mM), and $CaCl_2$ (3.13 mM) were loaded into a magnesia crucible in an Ar-filled glove box. The crucible was sealed inside a silica tube that was evacuated to 10^{-4} Pa. The tube containing the starting materials was heated in a muffle furnace to 800°C at 150°C·h^{-1}, maintained for 24 h, cooled to 500°C at 5°C·h^{-1}, and then cooled to room temperature by turning the heater off. The product was washed in sonicated water and extracted from the flux. Ivory-white block $La_4(GeS_2O_2)_3$ crystals, along with colorless transparent platelet LaOCl crystals, were collected via vacuum filtration.

B.4.49 Germanium–Lanthanum–Selenium–Sulfur

The solid solutions of $La_4Ge_3Se_xS_{12-x}$ are formed in the Ge–La–Se–S system (Cicirello et al. 2021b). This solid solutions crystallize in the rhombohedral structure with the lattice parameter in the hexagonal setting $a = 1955.75 \pm 0.06$, $c = 814.68 \pm 0.19$ pm, a calculated density of 4.605 g·cm^{-3}, and an energy gap of 2.4 ± 0.1 eV for $x = 2$, $a = 1966.79 \pm 0.12$, $c = 819.50 \pm 0.06$ pm, a calculated density of 4.922 g·cm^{-3}, and an energy gap of 2.2 ± 0.1 eV for $x = 4$, $a = 1977.66 \pm 0.11$, $c = 824.48 \pm 0.04$ pm, a calculated density of 5.078 g·cm^{-3}, and an energy gap of 2.1 ± 0.1 eV for $x = 6$, $a = 1989.76 \pm 0.12$, $c = 829.84 \pm 0.04$ pm, a calculated density of 5.328 g·cm^{-3}, and an energy gap of 2.0 ± 0.1 eV for $x = 8$, and $a = 2007.23 \pm 0.12$, $c = 834.60 \pm 0.05$ pm, a calculated density of 5.530 g·cm^{-3}, and an energy gap of 1.9 ± 0.1 eV for $x = 10$. The solution limits for replacing S atoms with Se is $x = 10$.

For preparing these solid solutions, La powder, Ge pieces, S powder, Se shot, I_2, $LaCl_3$, and NaCl were used. A two-step synthesis method was followed. The reactant elements were loaded into carbonized silica ampoules at a molar ratio of La/Ge/Se/S = 4:3:x:(12 – x) (x = 2, 4, 6, 8, and 10) with ~0.1 g of I_2 added as a transport agent. The ampoules were flame-sealed under high vacuum and placed in a programmable furnace. They were heated to 800°C in 10 h, maintained at that temperature for 96 h, and then cooled to 100°C in 24 h. The first step produced homogeneous powders. The as-synthesized powders from the first step were mixed with salt flux $LaCl_3$ and NaCl at a mass ratio of $LaCl_3$/NaCl = 0.68:0.32 in a glove box. The salt flux is of equal mass to the target compounds. The admixtures were sealed in a new ampoules heated by the same temperature profile as the first step. After the second annealing, the samples were washed with deionized water to remove the flux, yielding red crystals ($x = 8$ and $x = 10$), red-orange crystals ($x = 6$), yellow-orange crystals ($x = 4$), and yellow crystals ($x = 2$). The solid solutions are stable in dry air for multiple months. All starting materials were stored and used in an Ar-filled glove box.

B.5 Systems Based on Germanium Selenides

B.5.1 Germanium–Hydrogen–Nitrogen–Selenium

Two quaternary compounds, $Ge_2Se_6(N_2H_5)_4$ and $Ge_2Se_6(N_2H_4)_3(N_2H_5)_4$, are formed in the Ge–H–N–Se system (Mitzi 2005; Yuan and Mitzi 2009). The first compound crystallizes as a tetragonal structure with the lattice parameters $a = 1270.8 \pm 0.1$, $c = 2195.5 \pm 0.2$ pm, and a calculated density of 2.814 g·cm^{-3} and the second one crystallizes as a triclinic structure with the lattice parameters $a = 659.7 \pm 0.2$, $b = 938.1 \pm 0.3$, $c = 974.5 \pm 0.3$ pm, $\alpha = 93.67 \pm 0.01°$, $\beta = 100.34 \pm 0.01°$, and $\gamma = 104.97 \pm 0.01°$. A multiphase yellow product was obtained by reacting (under a nitrogen atmosphere) of $GeSe_2$ (2 mM) and Se (2 mM) in 1 mL of freshly distilled hydrazine. The highly toxic hydrazine was handled using appropriate protective equipment to prevent physical contact with either vapor or liquid. The reaction produces vigorous bubbling and rapidly yielded a clear light-yellow solution. The solution was evaporated overnight to dryness under a N_2 flow and transferred into a dry box where it was allowed to further dry in the dry box antechamber for several hours, yielding a bright-yellow crystalline product containing both compounds.

B.5.2 Germanium–Lithium–Magnesium–Selenium

The quaternary compound $Li_2MgGeSe_4$ which crystallizes as an orthorhombic structure with the lattice parameters $a = 829.61 \pm 0.07$, $b = 700.69 \pm 0.05$, $c = 661.16 \pm 0.06$ pm, and a calculated density of 3.686 g·cm^{-3} at 153 K, is formed in the Ge–Li–Mg–Se system (Gao et al. 2021a). Single crystals of this compound were prepared using a melting method in a sealed quartz tube. The starting mixture (molar ratio of Li/Mg/Ge/Se = 2:1:1:4) was packaged in graphite crucible in a glove box. After that, the graphite crucible was moved into quartz tube which was flame-sealed under a vacuum about 10^{-3} Pa. Then, the sample was heated to 880°C in 46 h, kept at 880°C for 50 h, and then cooled to room temperature in 48 h. Breaking the tube, the yellow $Li_2MgGeSe_4$ single crystals were harvested in the graphite crucible. It is worth mentioning that these crystals show strong moisture absorptions in air.

The syntheses of $Li_2MgGeSe_4$ powder sample was tried at a higher temperature. The mixtures of Li, Mg, Ge, and Se with a molar stoichiometric ratio were first weighed, ground, and sealed in quartz tube. The sealed sample was slowly heated to 900°C in 60 h in a muffle furnace, and kept at this temperature for 100 h, then cooled to room temperature in 100 h.

B.5.3 Germanium–Lithium–Mercury–Selenium

The quaternary compound $Li_2HgGeSe_4$, which crystallizes as an orthorhombic structure with the lattice parameters $a = 1415.0 \pm 0.3$, $b = 826.05 \pm 0.19$, $c = 669.24 \pm 0.14$ pm, a calculated density of 5.119 g·cm^{-3}, and an energy gap of 2.27 eV, is formed in the Ge–Li–Hg–Se system (Gao et al. 2022b). To prepare this compound, a mixture of Li (2 mM), HgSe (1 mM), $GeSe_2$ (1 mM), and Se (1 mM) was loaded into a graphite crucible that avoids the harmful reaction between Li metal and tube wall, and then the graphite crucible was put into a vacuum-sealed silica tube. The silica tube was firstly heated to 700°C in 50 h, kept at 700°C for about 100 h, and slowly cooled down to room temperature within 5 days. DMF solvent was used to wash the final products. Many deep-yellow crystals of the title compound were found in the tube. As for air-unstable Li, an Ar-filled glove box was used to complete the whole preparation process.

B.5.4 Germanium–Sodium–Mercury–Selenium

The quaternary compound $Na_2Hg_3Ge_2Se_8$, which crystallizes as a tetragonal structure with the lattice parameters $a = 919.38 \pm 0.03$, $c = 954.37 \pm 0.05$ pm, a calculated density of 5.865 g·cm^{-3}, and an energy gap of 1.98 eV, is formed in the Ge–Na–Hg–Se system (Gao et al. 2022b). To synthesize this compound, a mixture of Na (2 mM), HgSe (3 mM), $GeSe_2$ (2 mM), and Se (1 mM) was loaded into a graphite crucible that avoids the harmful reaction between Na metal and tube wall, then the graphite crucible was put into vacuum-sealed silica tube. Silica tube was firstly heated to 600°C in 50 h, kept at 600°C for about 100 h, and slowly cooled down to room temperature within 5 days. The final products were carefully washed with DMF and red microcrystals appeared in the tube. As for air-unstable Na block, an Ar-filled glove box was used to complete the whole preparation process.

B.5.5 Germanium–Potassium–Silver–Selenium

The quaternary compound $K_2Ag_2GeSe_4$, which crystallizes as an orthorhombic structure with the lattice parameters $a = 661.29 \pm 0.13$, $b = 1377.6 \pm 0.3$, $c = 2182.2 \pm 0.4$ pm, a calculated density of 4.560 g·cm^{-3}, and an energy gap of 2.0 eV, is formed in the Ge–K–Ag–Se system (Yao et al. 2010). This compound was prepared using $AgNO_3$ (0.15 mM), Ge (0.10 mM), Se (0.30 mM), K_2CO_3 (0.10 mM), 0.3 mL of mixed solvent of thiophenol/pyridine (volume ratio 1:3), and 0.2 mL of ethylenediamine. The mixture was sealed in a Pyrex glass tube with ca. 10% filling at air atmosphere, placed into a stainless steel autoclave, and then heated at 170°C for 7 days. After cooling naturally to ambient temperature, the products were washed with ethanol and water, respectively. Yellow trigonal bipyramidal crystals were obtained.

B.5.6 Germanium–Potassium–Cadmium–Selenium

The quaternary compound $K_2CdGe_3Se_8$, which crystallizes as a monoclinic structure with the lattice parameters $a = 760.686 \pm 0.011$, $b = 1240.155 \pm 0.019$, $c = 1765.85 \pm 0.03$ pm, $\beta = 95.3063 \pm 0.0014°$, a calculated density of 4.165 g·cm^{-3}, and an energy gap of 2.3 ± 0.1 eV, is formed in the Ge–K–Cd–Se system (Ji et al. 2023). The crystals of this compound were synthesized using a solid state method in a stoichiometric molar ratio (K/Cd/Ge/Se = 2:1:3:8). A total of 0.4 g of elements were loaded under vacuum in a flame-sealed silica ampoule and placed in a programmable furnace. The bottom of silica ampoule was coated with amorphous carbon. The ampoule was heated from room temperature to 800°C in 20 h and kept at this temperature for 96 h. Then, the furnace was turned off and naturally cooled to room temperature. The crystals of $K_2CdGe_3Se_8$ were yellow-orange. They are stable in dry air for many weeks with no detection of change. All starting materials were stored in an Ar-filed glove box with the oxygen level below 0.5 ppm.

B.5.7 Germanium–Potassium–Gallium–Selenium

The quaternary compound $K_{2.4}Ga_{2.4}Ge_{1.6}Se_8$, which crystallizes as a tetragonal structure with the lattice parameters $a = 812.08 \pm 0.08$, $c = 620.80 \pm 0.08$ pm, a calculated density of 4.093 g·cm^{-3}, and an energy gap of 2.6 eV for the crystalline compound and 2.4 eV for the glassy one, is formed in the Ge–K–Ga–Se system (He et al. 2016). The crystals of the title compound were obtained through the reactive flux method with K_2Se_2 as the reactive flux. The starting materials K_2Se_2 (2.25 mM), Ga_2Se_3 (0.375 mM), Ge (0.5 mM), and Se (1.75 mM) were mixed and ground uniformly and loaded into the silica tube. The tube was then flame-sealed under vacuum (0.1 Pa) and slowly heated to 750°C in a programmable furnace. It was kept at this temperature for 3 days and then slowly cooled down to 300°C at a rate of 2°C·h^{-1}, and finally to room temperature by turning off the furnace. The melt was washed with DMF several times to remove the excessive flux and dried with acetone. Thin brown prisms up to 2–3 cm long were obtained. The powder sample was synthesized by stoichiometric combination reaction of the starting materials under the same reacting condition and by repeating the procedure three times. All operations were carried out in an Ar-protected glove box. This compound can be also prepared in the glassy state. The glassy $K_{2.4}Ga_{2.4}Ge_{1.6}S_8$ is also stable in air.

B.5.8 Germanium–Rubidium–Silver–Selenium

The quaternary compound $Rb_2Ag_2GeSe_4$, which crystallizes as an orthorhombic structure with the lattice parameters $a = 659.43 \pm 0.03$, $b = 1397.40 \pm 0.07$, $c = 2262.74 \pm 0.12$ pm, and an energy gap of 1.83 eV, is formed in the Ge–Rb–Ag–Se system (Jiang et al. 2020). This compound was prepared by the starting materials of Rb_2CO_3, $AgNO_3$, Ge, Se, and mixed DMF/ethylenediamime solvent.

B.5.9 Germanium–Cesium–Copper–Selenium

The quaternary compound $Cs_8Cu_2GeSe_{14}$, which crystallizes as a monoclinic structure with the lattice parameters $a = 782.31 \pm 0.09$, $b = 2395.28 \pm 0.17$, $c = 601.97 \pm 0.06$ pm, $\beta = 113.747 \pm 0.011°$, and an energy gap of 1.46 eV, is formed in the Ge–Cs–Cu–S system (Jiang et al. 2020). This compound was prepared using the starting materials Cs_2CO_3, $CuCl_2{\cdot}2H_2O$, Ge, Se, and mixed DMF/ethylenediamine as the solvent.

B.5.10 Germanium–Cesium–Cadmium–Selenium

The quaternary compound $Cs_2CdGe_3Se_8$, which crystallizes as an orthorhombic structure with the lattice parameters $a = 768.9 \pm 0.3$, $b = 1258.2 \pm 0.5$, and $c = 1778.8 \pm 0.6$ pm at 150 K, is formed in the Ge–Cs–Cd–Se system (Hahn et al. 2013). The title compound was prepared by the reaction of elements with the use of the reactive halide-flux technique. A combination of Cd, Ge, and Se powders was mixed in a fused silica tube in a molar ratio of Cd/Ge/Se = 1:1:4 and then CsCl was added. The mass ratio of the reactants and CsCl was 1:2. The tube was evacuated to 0.133 Pa, sealed, and heated gradually (14°C·h^{-1}) to 650°C, where it was kept for 5 days. The tube was cooled to 100°C at a rate 5°C·h^{-1}. The excess halide was removed with distilled water and dark green block-shaped crystals were obtained. They are stable in air and water.

B.5.11 Germanium–Copper–Tellurium–Selenium

The solid solutions of $Cu_8GeSe_{6-x}Te_x$, which crystallize as a cubic structure, are formed in the Ge–Cu–Te–Se system (Schwarzmüller et al. 2018). The lattice parameter a increases with temperature in two linear regimes from 1046.82 ± 0.18 pm at 50°C to 1055.6 ± 0.3 pm at 400°C and to 1060.9 ± 0.3 pm at 500°C with a superionic transition at 400°C. At the same time, the calculated density decreases from 6.597 g·cm^{-3} at 50°C to 6.435 g·cm^{-3} at 400°C and to 6.339 g·cm^{-3} at 500°C. While in $Cu_8GeSe_{6-x}Te_x$ with $0 \leq x \leq 2$, thermodynamically stable solid solutions were obtained, and phase segregation occurs for $x > 2$.

Samples of these solid solutions were synthesized by fusing the elements Cu, Ge, Se, and Te in silica glass ampoules. Mixtures of the starting materials were heated stepwise for several hours at 300°C, 550°C, and 950°C, followed by quenching in air. Crystals were grown by annealing at 750°C for 2 weeks in a tube furnace.

According to Jiang et al. (2017), the quaternary phase $Cu_8GeSe_{5.1}Te_{0.9}$ crystallizes as a trigonal structure with the lattice parameters $a = 731.96 \pm 0.05$, $c = 1792.2 \pm 0.2$ pm, and a calculated density of 6.143 g·cm^{-3} at 100 K and $a = 734.356 \pm 0.001$ and $c = 1799.08 \pm 0.05$ pm at room temperature. To prepare this phase, Cu, Se, and Te shots and Ge pieces were weighed in stoichiometric proportions and then sealed in a silica tube under vacuum. The tube was heated to 1120°C and maintained at this temperature for 24 h before quenching in cold water. Then, the quenched ingot was annealed at 600°C for 5 days. Finally, the products were ground into fine powder and sintered by spark plasma sintering at 410°C–440°C under a pressure of 75 MPa for 5 min.

B.5.12 Germanium–Copper–Tungsten–Selenium

The quaternary compound Cu_6GeWSe_8, which has two polymorphic modifications, is formed in the Ge–Cu–W–Se system (Zhou et al. 2023). The hexagonal phase is a semiconductor and the cubic phase ($a = 1113.72 \pm 0.02$ pm and a calculated density of 6.104 g·cm^{-3}) is a semimetal. The phase transformation from cubic to hexagonal structures happens at around 440°C upon rising temperature. The transition starts at ~440°C and is completed at ~470°C. The wide transition temperature range and long transition time indicate that the phase transition is very slow.

The polycrystalline sample of the cubic Cu_6GeWSe_8 was synthesized by solid state reaction using stoichiometric amounts of Cu powder, Ge pieces, W powder, and Se shot. The well-ground ingredients with a total mass of ~5 g were thoroughly mixed, and then pelletized and sealed in an evacuated quartz tube. The initial heating cycle was done at 420°C for 3 days. Subsequently, the as-prepared sample was fully ground in a mortar, pressed into cylindrical pellets with diameter 10 mm under an uniaxial pressure of 165 MPa, and resealed in an evacuated quartz tube, which was annealed at 420°C for 7 days and naturally cooled down to room temperature.

B.5.13 Germanium–Silver–Gallium–Selenium

The solid solutions of $AgGaGe_nSe_{2(n+1)}$ are formed in the Ge–Ag–Ga–Se system. At $n = 1$, the solid solution crystallizes as a tetragonal structure with the lattice parameters $a = 580.56 \pm 0.05$, $c = 1034.88 \pm 0.10$ pm, a calculated density of 5.389 g·cm^{-3}, and an energy gap of 2.27 eV (Dang et al. 2022). To prepare this phase, appropriate amounts of $AgGaSe_2$ and $GeSe_2$ (molar ratio 1:1) in a glove box filled with argon, were put into a silica glass tube, and sealed with hydrogen under a vacuum of less than 10^{-3} Pa. Then, the tube was put into a computer-controlled muffle furnace, heated to 950°C for 20 h, maintain at this temperature for 10 h, and then cool it to room temperature. The reaction was repeated three times and ground intermittently to synthesize the quaternary phase $AgGaGeSe_4$. For preparing single crystals, $AgGaSe_2$ and $GeSe_2$ (molar ratio 1:1) were heated under a vacuum less than 10^{-3} Pa to 900°C for 20 h, maintained at this temperature for 24 h, cooled to 600°C at a rate of 3°C·h^{-1}, and then cooled to room temperature to obtain the single crystal of $AgGaGeSe_4$.

The solid solutions of $AgGaGe_nSe_{2(n+1)}$ with $n = 1.5$–9 crystallizes as an orthorhombic structure with the lattice parameters $a = 1247.266$, $b = 2384.884$, $c = 712.598$ pm, and a calculated density of 5.122 g·cm^{-3} for $n = 1.5$, $a = 1247.403$, $b = 2386.119$, $c = 712.723$ pm, and a calculated density of 5.048 g·cm^{-3} for $n = 1.75$, $a = 1249.103$, $b = 2390.375$, $c = 714.621$ pm, and a calculated density of 4.959 g·cm^{-3} for $n = 2$, $a = 1244.891$, $b = 2384.304$, $c = 714.377$ pm, and a calculated density of 4.826 g·cm^{-3} for $n = 3$, $a = 1242.978$, $b = 2380.369$, $c = 714.570$ pm, and a calculated density of 4.741 g·cm^{-3} for $n = 4$, $a = 1237.885$, $b = 2369.461$, $c = 712.276$ pm, and a calculated density of 4.731 g·cm^{-3} for $n = 5$, and $a = 1236.671$, $b = 2365.151$, $c = 712.432$ pm, and a calculated density of 4.609 g·cm^{-3} for $n = 9$ (Liu et al. 2022d). The band gap of $AgGaGe_nSe_{2(n+1)}$ changes along with the n value, increasing from 2.049 eV for $n = 1.5$ to 2.144 eV for $n = 9$, and the band gap of $AgGaGe_5Se_{12}$ is 2.127 eV [2.14 eV (Ni et al. 2015)].

To prepare these solid solutions, Ag, Ga, Ge, and Se were weighed and then loaded into quartz ampoules (Ni et al. 2015; Liu et al. 2022d). The ampoules were evacuated to 10^{-4} Pa and sealed. The polycrystalline synthesis was conducted in a two-zone furnace. The lower zone was first heated to 600°C and then equilibrated for 20 h to consume most of the selenium. Then, it was heated to 960°C and maintained for 24 h, while the upper zone was kept at 300°C all the time. After this period, the temperature of the two zones would increase to 1080°C and 1040°C, respectively, and right after this, the lower zone would undergo a temperature oscillation process between 960°C and 1040°C to ensure that the raw materials react completely. At last, considering the different contents of $GeSe_2$ in different solid solutions, the cooling rate of the furnace was changed from 30 to 16°C·h^{-1}. The synthetic product was ground into powder and reloaded into a quartz ampoule. Then, the ampoule was evacuated, sealed, and put into a six-zone Bridgman growth furnace to prepare the single crystals.

B.5.14 Germanium–Silver–Tellurium–Selenium

Ag_8GeSe_6–Ag_8GeTe_6. This system is a nonquasibinary section of the Ge–Ag–Te–Se system as Ag_8GeTe_6 melts incongruently (Amiraslanova et al. 2023c). It is characterized by the formation of continuous series of the solid solution between Ag_8GeTe_6 and high-temperature cubic modification of Ag_8GeSe_6. When solid solutions are formed, the temperatures of the polymorphic phase transitions of Ag_8GeSe_6 decrease.

This leads to the stabilization of the cubic phase in the ≥ 40 mol% Ag_8GeTe_6 range of compositions at room temperature and below. The regions of homogeneity based on Ag_8GeSe_6 is 10 mol%. The alloys for the investigations were annealed at 530°C for about 500 h.

B.5.15 Germanium–Strontium–Lead–Selenium

The $Sr_{2-x}Pb_xGeSe_4$ and $Sr_{19-x}Pb_xGe_{11}Se_{44}$ phases are formed in the Ge–Sr–Pb–Se system. $Sr_{2-x}Pb_xGeSe_4$ crystallizes as an orthorhombic structure with the lattice parameters a = 1031.220 ± 0.001, b = 1039.320 ± 0.001, c = 742.140 ± 0.001 pm, and a calculated density of 5.39 g·cm^{-3} for x = 0.688 and in the cubic structure with the lattice parameter a = 1461.77 ± 0.03 pm and a calculated density of 6.63 g·cm^{-3} for x = 1.81 (Menezes et al. 2020). The optical band gaps of the phases with x = 0.7 and 1.75 were determined to be 1.65 and 1.48 eV, respectively. The sample with x = 0.7 was prepared in the following way. Stoichiometric amounts of Sr, Pb, Ge, and Se were added to a glassy carbon, placed at the bottom of a silica tube to be evacuated to 0.25 Pa, and then flame-sealed. This ampoule was slowly heated at a rate of 100°C·h^{-1} to 800°C to enable complete reaction of the elements. After the mixture had reacted, the temperature was decreased to 650°C and maintained for 100 h to allow the elements to diffuse throughout the sample. The sample was finally slowly cooled to 200°C over 100 h. Then, the ampoule was opened and the content was ground to make a homogeneous powder before being resealed in the ampoule and annealed at 500°C for 1 week. The result was a homogeneous dark brown powder of nominal composition $Sr_{1.3}Pb_{0.7}GeSe_4$.

To prepare the phase with x = 1.75, stoichiometric amounts of the starting elements were added to a silica tube, evacuated, and flame-sealed. The sealed tube was placed into a furnace and heated to 900°C where a melt was achieved. The melt was quenched in an ice bath. Since XRD indicated the presence of the target phase in addition to the binaries $GeSe_2$ and PbSe, the sample was ground and then annealed at 500°C for 4 days. After 4 days at 500°C, only very minor amounts of $GeSe_2$ and PbSe as well as small unidentified peaks were present in the XRD pattern.

$Sr_{19-x}Pb_xGe_{11}Se_{44}$ crystallizes as a hexagonal structure with the lattice parameters a = 2001.2 ± 0.2, c = 1205.5 ± 0.1 pm, and a calculated density of 5.17 g·cm^{-3} for x = 5.0 and a = 1998.67 ± 0.02, c = 1205.46 ± 0.01 pm, and a calculated density of 5.34 g·cm^{-3} for x = 6.4 (Assoud and Kleinke 2017). The synthesis of these phases started from the elements stored in an Ar-filled glove box. Sr block, Pb powder, Ge powder, and Se powder were placed into a silica tube in the stoichiometric ratio, which was sealed under dynamic vacuum and then placed in a furnace. The furnace was heated to 800°C within 24 h, then cooled to 650°C within 10 h, and then kept at that temperature for 96 h. Finally, the furnace was slowly cooled down to room temperature with a rate of 2.5°C·h^{-1}.

B.5.16 Germanium–Strontium–Oxygen–Selenium

Three quaternary compounds, $SrGeOSe_2$, $Sr_3Se(GeOSe_3)$, and $Sr_3Ge_2O_4Se_3$, are formed in the Ge–Sr–O–Se system. The first compound possesses good thermal stability up to 780°C and crystallizes as an orthorhombic structure with the lattice parameters a = 465.19 ± 0.03, b = 830.92 ± 0.04, c = 1242.15 ± 0.06 pm, a calculated density of 4.6 g·cm^{-3}, and an energy gap of 3.16 eV (Ran et al. 2020). This compound was prepared from SrO, Ge, and Se (molar ratio 1:1:2) and CsI as a flux. These raw materials were loaded into a quartz silica tube that was subsequently evacuated to 10^{-3} Pa, sealed in a glove box, maintained at 400°C for 20 h, and then heated to 750°C at a rate of 3°C h^{-1}, maintained for 3 days, finally cooled at 3.5°C h^{-1} to 350°C, and later the furnace was shut off. The colorless single crystals of $SrGeOSe_2$ were obtained by washing with distilled water and dried by ethanol.

$Sr_3Se(GeOSe_3)$ also crystallizes as an orthorhombic structure with the lattice parameters a = 1223.66 ± 0.13, b = 621.52 ± 0.06, c = 1202.63 ± 0.12 pm, a calculated density of 4.846 g·cm^{-3}, and an energy gap of 3.52 eV (Cui et al. 2023). Single crystals of this compound were synthesized using 3:0.5:0.5:1 ratios of SrSe, GeO_2, Ge, and Se with a total mass of 500 mg as the starting materials. After being ground to a fine powder, the mixture was packed into a quartz tube and further evacuated to 10^{-3} Pa, and then flame-sealed. The tube was put into a muffle furnace, heated from room temperature to 900°C within 6 h and kept at this temperature for 1 day, gradually cooled to 700°C at a rate of 5°C·h^{-1}, and finally the furnace was turned off. The single crystals of $Sr_3Se(GeOSe_3)$ were obtained. This compound could be synthesized in the range of 850°C–900°C, but it decomposes when the temperature was higher than 900°C.

$Sr_3Ge_2O_4Se_3$ is thermally stable up to at least 750°C under a nitrogen atmosphere and crystallizes as a trigonal structure with the lattice parameters a = 1024.34 ± 0.06, c = 844.88 ± 0.05 pm, a calculated density of 4.600 g·cm^{-3}, and an energy gap of 2.96 eV (Xing et al. 2020). Single crystals of the title compound were initially found in the reaction products based on the precursors SrSe, $GeSe_2$, Sb_2Se_3, and Sb_2O_3 in a molar ratio of 2:1:1:1. Sb_2Se_3 may act as a fluxing agent and Sb_2O_3 as an oxygen agent. After being mixed and ground in a glove box, the reactants were sealed into an evacuated silica tube (<10^{-3} Pa) and subjected to the following heat treatment: heated to 1000°C in 1 day, maintained for 3 days, and cooled at a rate of 4°C·h^{-1} to room temperature. Finally, colorless transparent block crystals of $Sr_3Ge_2O_4Se_3$ were obtained. The polycrystalline sample of this compound could be easily synthesized by heating the stoichiometric mixture sealed into an evacuated silica tube (3 mM of SrSe and 2 mM of GeO_2) at 850°C for 2 days. All of the reactants were stored in an argon-filled glove box (content of H_2O and O_2 < 0.1 ppm).

B.5.17 Germanium–Barium–Antimony–Selenium

The quaternary compound $Ba_4GeSb_2Se_{11}$, which crystallizes as an orthorhombic structure with the lattice parameters a = 937.0 ± 1.1, b = 2585.0 ± 0.0, c = 879.8 ± 1.0 pm, a calculated density of 5.404 g·cm^{-3}, and an energy gap of 1.35 eV, is formed in the Ge–Ba–Sb–Se system (Yuan et al. 2021a). The crystals of this compound were synthesized by high-temperature solid state reactions. BaSe, Ge, Sb, and Se were mixed according to the molar ratio of 4:1:2:7 and loaded into a graphite crucible, which is sealed in a silica tube under vacuum (0.01 Pa). The mixture was heated to 750°C within 25 h and kept at this temperature for 15 h. Then, the sample was slowly cooled to 650°C in 100 h (holding for 10 h) and then to room temperature in 15 h. All operations were done through a computer-controlled furnace and finally black crystals were obtained. The pure phase of $Ba_4GeSb_2Se_{11}$ was synthesized by a stoichiometric mixture of the BaSe, Ge, Sb, and Se, which was heated to 670°C in 20 h, kept for 30 h, and then cooled to room temperature naturally.

B.5.18 Germanium–Barium–Oxygen–Selenium

The quaternary compound $Ba_3Se(GeOSe_3)$, which crystallizes as an orthorhombic structure with the lattice parameters a = 1276.1 ± 0.2, b = 649.37 ± 0.09, c = 1220.58 ± 0.19 pm, a calculated density of 5.233 g·cm^{-3}, and an energy gap of 3.52 eV, is formed in the Ge–Ba–O–Se system (Cui et al. 2023). The title compound was obtained in the same way as $Sr_3Se(GeOSe_3)$ was synthesized using BaS instead of SrS.

B.5.19 Germanium–Gallium–Lead–Selenium

PbSe–$Ga_2PbGeSe_6$. The phase diagram of this system, constructed through DTA, XRD, and metallography, is presented in Figure B.5.1 (Bellagra et al. 2023). The eutectic contains 54 mol% PbSe and crystallizes at 636°C. The phase transition of the solid solution based on $Ga_2PbGeSe_6$ takes place at 522°C. The quaternary compound $Ga_2Pb_3GeSe_8$, which melts incongruently at 656°C and crystallizes as a tetragonal structure with the lattice parameters a = 1300.5 ± 0.7 and c = 625.7 ± 0.9 pm, is formed in this system. This compound has a homogeneity region within the interval of 64–77 mol% PbSe at 400°C. The composition of the peritectic point is 62 mol% PbSe. The solid solubility at 400°C based on $Ga_2PbGeSe_6$ is ~ 3 mol% and is negligible from the PbSe side.

$Ga_2PbGeSe_6$–Ga_2Se_3. The phase diagram of this system is a eutectic type (Bellagra et al. 2023). The eutectic contains 13 mol% Ga_2Se_3 and crystallizes at 664°C. The phase transition of the solid solution based on $Ga_2PbGeSe_6$ takes place at 522°C. The solid solubility of the components in each other at 400°C is below 5 mol%.

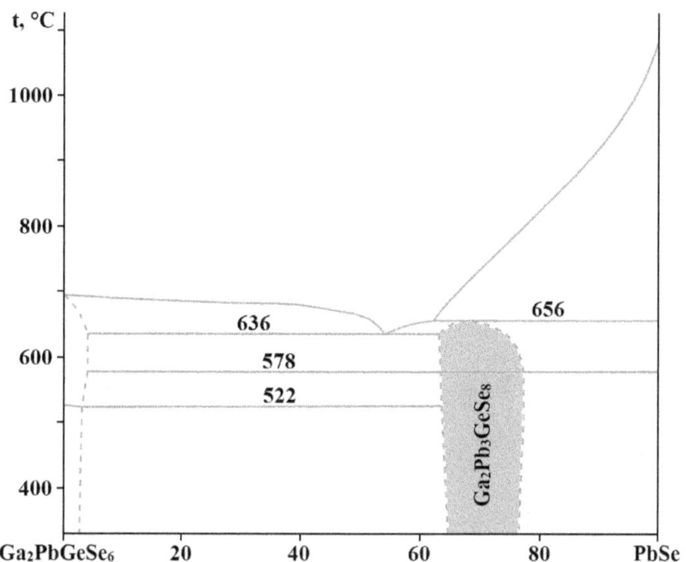

FIGURE B.5.1 Phase diagram of the PbSe–$Ga_2PbGeSe_6$ system. (From Bellagra, H.K., et al., *J. Phase Equilib. Diffus.*, **44**(1), 3, 2023.)

$GeSe_2$–Ga_2PbSe_4. This section is nonquasibinary due to the incongruent melting of Ga_2PbSe_4 (Bellagra et al. 2023). The quaternary compound $Ga_2PbGeSe_6$ (orthorhombic structure, $a = 471.42 \pm 0.09$, $b = 758.1 \pm 0.3$, and $c = 1215.8 \pm 0.5$ pm) is formed in this section. It melts congruently at 687°C and has a homogeneity region of ~4–5 mol% to either side of the stoichiometric composition at 400°C. Two eutectics in this section crystallize at 589°C and 617°C and contain 33 and 80 mol% $GeSe_2$, respectively.

$GePb_2Se_4$–Ga_2PbSe_4. This section is also nonquasibinary as both ternary compounds melt incongruently (Bellagra et al. 2023). The $Ga_2Pb_3GeSe_8$ quaternary compound, which melts incongruently at 656°C and has a phase transition at 578°C, is formed in this section.

PbSe–$Ga_4Pb_4GeSe_{12}$. This section is also nonquasibinary (Bellagra et al. 2023). The solid solution containing up to 25 mol% PbSe is formed based on $Ga_4Pb_4GeSe_{12}$.

$GeSe_2$–Ga_2Se_3–PbSe. The isothermal section of this quasiternary system is shown in Figure B.5.2 (Bellagra et al. 2023). The solid solubility based on PbSe and $GeSe_2$ is negligible (~1–2 mol%). The largest solid solution is based on Ga_2Se_3 (α) and reaches ~6 mol% $Ga_2PbGeSe_6$ in the $Ga_2PbGeSe_6$–Ga_2Se_3 system. Solid solutions based on Ga_2PbSe_4 and $GePb_2Se_4$ are within 3–5 mol%. Three quaternary compounds, $Ga_2PbGeSe_6$, $Ga_4Pb_4GeSe_{12}$, and $Ga_2Pb_3GeSe_8$, are formed. A large γ-solid solution exists in the Ga_2PbSe_4–$GePb_2Se_4$ section (~22 mol%) and both sides of it. Expressed by the formula $Ga_{2+2x}Pb_{3+x}GeSe_{8+4x}$, this includes the compositions $Ga_2Pb_3GeSe_8$ and $Ga_4Pb_4GeSe_{12}$. The β-solid solution based on $Ga_2PbGeSe_6$ (~7 mol%) is significantly smaller.

The liquidus surface of this system (Figure B.5.3) includes the fields of primary crystallization of solid solutions based on Ga_2Se_3, α- and β-$Ga_2PbGeSe_6$, α- and β-$Ga_{2+2x}Pb_{3+x}GeSe_{8+4x}$, Ga_2PbSe_4, and $GePb_2Se_4$ separated by 24 monovariant curves and 24 invariant points, of which five correspond to binary and 19 to ternary invariant reactions. There are five ternary eutectics and eight transition points on the liquidus surface: U_1 (595°C) – L + PbSe ⇔ (Ga_2PbSe_4) + (β-$Ga_{2+2x}Pb_{3+x}GeSe_{8+4x}$); U_2 (537°C) – L + PbSe ⇔ (α-$Ga_{2+2x}Pb_{3+x}GeSe_{8+4x}$) + ($GePb_2Se_4$); U_3 (a, 522°C) – L + (β-$Ga_2PbGeSe_6$) ⇔ (α-$Ga_2PbGeSe_6$) + (α-$Ga_{2+2x}Pb_{3+x}GeSe_{8+4x}$); U_4 (b, 522°C) – L + (β-$Ga_2PbGeSe_6$) ⇔ (α-$Ga_2PbGeSe_6$) + ($GePb_2Se_4$); U_5 (c, 578°C) – L + (β-$Ga_{2+2x}Pb_{3+x}GeSe_{8+4x}$) ⇔ (α-$Ga_{2+2x}Pb_{3+x}GeSe_{8+4x}$) + PbSe; U_6 (d, 578°C) – L + (β-$Ga_{2+2x}Pb_{3+x}GeSe_{8+4x}$) ⇔ (α-$Ga_{2+2x}Pb_{3+x}GeSe_{8+4x}$) + (β-$Ga_2PbGeSe_6$); U_7 (f, 578°C) – L + (β-$Ga_{2+2x}Pb_{3+x}GeSe_{8+4x}$) ⇔ (α-$Ga_{2+2x}Pb_{3+x}GeSe_{8+4x}$) + ($Ga_2PbSe_4$); U_8 (g, 578°C) – L + (β-$Ga_{2+2x}Pb_{3+x}GeSe_{8+4x}$) ⇔ (α-$Ga_{2+2x}Pb_{3+x}GeSe_{8+4x}$) + (α-$Ga_2PbGeSe_6$); E_1 (570°C) – L ⇔ (Ga_2PbSe_4) + (β-$Ga_2PbGeSe_6$) + (α-$Ga_{2+2x}Pb_{3+x}GeSe_{8+4x}$); E_2

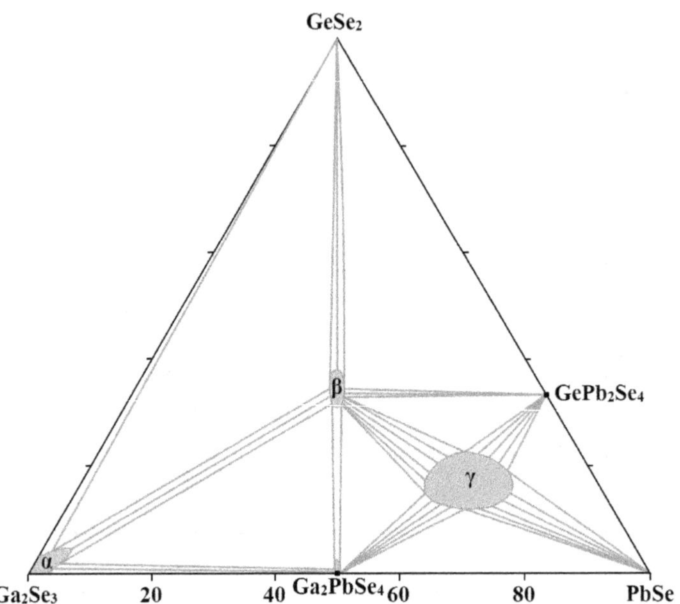

FIGURE B.5.2 Isothermal section of the $GeSe_2$–Ga_2Se_3–PbSe quasiternary system at 400°C. (From Bellagra, H.K., et al., *J. Phase Equilib. Diffus.*, **44**(1), 3, 2023.)

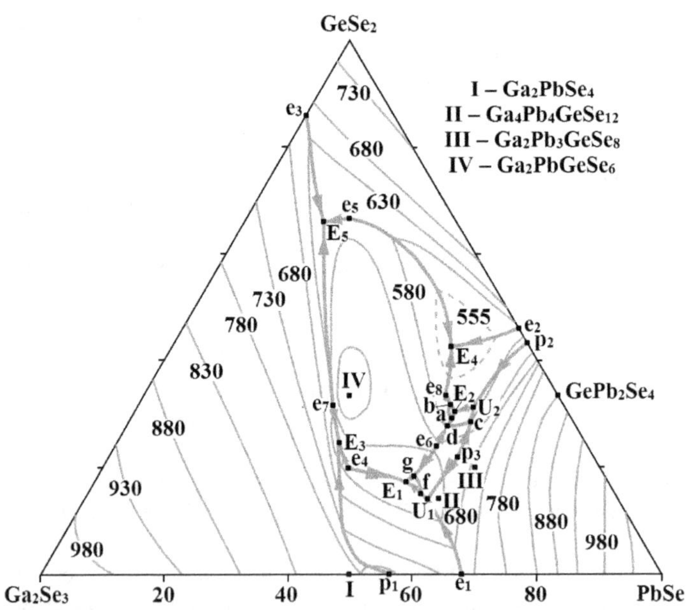

FIGURE B.5.3 Liquidus surface of the $GeSe_2$–Ga_2Se_3–PbSe quasiternary system. (From Bellagra, H.K., et al., *J. Phase Equilib. Diffus.*, **44**(1), 3, 2023.)

(514°C) – L ⇔ (α-$Ga_2PbGeSe_6$) + (α-$Ga_{2+2x}Pb_{3+x}GeSe_{8+4x}$) + ($GePb_2Se_4$); E_3 (552°C) – L ⇔ (Ga_2Se_3) + (β-$Ga_2PbGeSe_6$) + (Ga_2PbSe_4); E_4 (540°C) – L ⇔ (β-$Ga_2PbGeSe_6$) + ($GePb_2Se_4$) + $GeSe_2$; and E_5 (581°C) – L ⇔ (Ga_2Se_3) + (β-$Ga_2PbGeSe_6$) + $GeSe_2$.

The alloys for the investigation of these systems were annealed at 400°C for 250 h followed by quenching in a 20% NaCl solution.

B.5.20 Germanium–Antimony–Tellurium–Selenium

The quaternary compound $Ge_8Sb_2Te_6Se_5$, which has two polymorphic modifications and can be obtained in an amorphous state, is formed in the Ge–Sb–Te–Se system (Buller et al. 2012). The first modification of this compound crystallizes as a cubic structure with the lattice parameter a = 598.68 ± 0.08 pm and the second one crystallizes as a rhombohedral structure with the lattice parameters a = 415.74 ± 0.05 and c = 1037.0 ± 0.4 pm in the hexagonal setting. The experimental density and the energy gap of this compound are 5.56 and 5.89 g·cm^{-3} and 0.89 and 0.66 eV in an amorphous and in a crystalline state, respectively. The crystallization temperature of amorphous $Ge_8Sb_2Te_6Se_5$ is ≈ 210°C. Thin films of the title compound were deposited by co-sputtering of $Ge_3Sb_2Te_6$ with a diameter of 10 cm at a power of 10 W and GeSe.

B.6 Systems Based on Germanium Telluride

B.6.1 Germanium–Sodium–Antimony–Tellurium

The quaternary compound $Na_9Sb(Ge_2Te_6)_2$, which crystallizes as a monoclinic structure with the lattice parameters a = 954.1 ± 0.2, b = 2625.3 ± 0.7, c = 758.20 ± 0.18 pm, β = 122.233 ± 0.015°, and a calculated density of 4.445 g·cm^{-3}, is formed in the Ge–Na–Sb–Te system (Schwarzmüller et al. 2019). Single crystals of this compound were initially found in heterogeneous samples with the nominal composition $Na_2Ge_6Sb_2Te_{10}$. These were obtained by fusing the elements (Na, Ge, Sb, and Te) in graphitized silica glass ampoules. They were kept at 800°C for 1 h, followed by quenching in air. Subsequently, the sample was melted again at 700°C, cooled to 500°C with a cooling rate of 5°C·h^{-1}, and then cooled to room temperature by switching off the furnace. Dark gray crystals were obtained. As they are sensitive to moisture, they were selected under paraffin oil, transferred into silica glass capillaries and sealed in dry argon atmosphere. Bulk samples of $Na_9Sb(Ge_2Te_6)_2$ were obtained by pre-reacting Sb, Ge and Te in stoichiometric ratio at 950°C for 30 min in sealed silica glass ampoules under dry argon atmosphere. After quenching at air, Na was added and the mixture was fused in a graphitized silica glass ampoule by heating from room temperature to 700°C within 30 min. Subsequently, it was quenched at air, re-melted at 800°C for 30 min, quenched at air again, then annealed at 500°C for 5 days, and cooled by switching off the furnace. A gray ingot with metallic luster was obtained.

B.6.2 Germanium–Barium–Indium–Tellurium

The quaternary compound $Ba_6In_2Ge_2Te_{15}$, which melts at 760°C and crystallizes as a monoclinic structure with the lattice parameters a = 962.01 ± 0.06, b = 1342.25 ± 0.07, c = 2758.45 ± 0.16 pm, β = 92.469 ± 0.002°, a calculated density of 5.810 g·cm^{-3}, and an energy gap of 0.92 eV, is formed in the Ge–Ba–In–Te system (Sun et al. 2022b). Single crystals of this compound were grown spontaneously by crystallization. BaTe (0.944 mM), In_2Te_3 (0.135 mM), GeTe (0.674 mM), and Te (0.270 mM) were thoroughly ground and loaded into a quartz tube. The tube containing the raw materials was sealed under a vacuum of 10^{-3} Pa and placed in a tube furnace. Then, set the heating program as follows: the tube was heated from room temperature to 900°C in 20 h, kept for 50 h, then slowly cooled to 500°C at a rate of 3°C·h^{-1}, and finally cooled to room temperature within 5 h. The metallic plate-like crystals of $Ba_6In_2Ge_2Te_{15}$ were obtained. These crystals are not sensitive to oxygen and water in the air. All manipulations are carried out in an Ar-filled glove box.

B.6.3 Germanium–Barium–Gallium–Tellurium

The quaternary compound $Ba_5Ga_2Ge_3Te_{12}$, which melts incongruently at 817°C and crystallizes as a monoclinic structure with the lattice parameters a = 1365.40 ± 0.03, b = 967.05 ± 0.02, c = 2311.34 ± 0.07 pm, β = 91.829 ± 0.002°, a calculated density of 5.607 g·cm^{-3}, and an energy gap of <0.56 eV, is formed in the Ge–Ba–Ga–Te system (Sun et al. 2021). Single crystals of the title compound were obtained from BaTe (5% mass was added to compensate for the reaction with the quartz tube), Ga_2Te_3, GeTe, and Te (molar ratio 5:1:3:1), which were mixed and ground and placed in a quartz tube. Then, the tube was sealed under high vacuum and placed in the furnace, heated to 900°C in 36 h and maintained for 100 h, then cooled to 350°C in 133 h, and finally the furnace was turned off. A large number of black crystals were observed in the ampoule. The compound can be kept in air without degradation for 3 months. All operations were carried out in a glove box filled with Ar and containing <0.1 ppm of H_2O and O_2.

B.6.4 Germanium–Barium–Oxygen–Tellurium

The quaternary compound $Ba_3Ge_2O_4Te_3$, which melts incongruently at 1018°C and crystallizes as a trigonal structure with the lattice parameters $a = 1095.18 \pm 0.04$, $c = 897.40 \pm 0.03$ pm, a calculated density of 5.366 g·cm^{-3}, and an energy gap of 2.08 eV, is formed in the Ge–Ba–O–Te system (Sun et al. 2022a). Single crystals of this compound were grown via spontaneous crystallization. BaTe (1 mM), BaO (1 mM), Ge (1 mM), and Te (2 mM) were thoroughly ground and loaded into a quartz tube. This tube was slowly heated to 850°C within 20 h and maintained for 25 h, then heated to 950°C within 17 h and maintained for 34 h, gradually cooled to 650°C at a rate of 4°C·h^{-1}, and finally the furnace was turned off. Dark red block-like crystals of $Ba_3Ge_2O_4Te_3$ were obtained. These crystals are not sensitive to oxygen and water in the air.

The polycrystalline powder of $Ba_3Ge_2O_4Te_3$ was synthesized using a solid state reaction. A mixture of BaTe (0.396 g) and GeO_2 (0.104 g) was evenly ground and placed into quartz tubes, which were sealed under a high vacuum of 10^{-3} Pa. These samples were gradually heated to 730°C in 22 h, maintained for 70 h, and finally cooled to room temperature at a rate 4°C·h^{-1}. All manipulations were carried out in an Ar-filled glove box.

B.6.5 Germanium–Arsenic–Iron–Tellurium

The $Fe_{3-y}Ge_{1-x}As_xTe_2$ solid solution, which contains up to 85 at% As, is formed in the Ge–As–Fe–Te system (Yuan et al. 2017). Polycrystalline samples with $y = 0.1$ and $x = 0$, 0.25, 0.5, 0.75 and 0.85, were synthesized by conventional solid state reactions using powder of Fe, Ge, As, and Te as starting materials. They were thoroughly ground, pressed into pellets, loaded in alumina crucibles, sealed in evacuated quartz tubes and then slowly heated to 800°C for 48 h, and finally furnace-cooled to room temperature.

B.6.6 Germanium–Iron–Cobalt–Tellurium

The solid solution of $Fe_{5-x}Co_xGeTe_2$, which crystallizes as a trigonal structure with the lattice parameters $a = 401.96 \pm 0.02$ and $c = 980.00 \pm 0.06$ pm at 220 K, is formed in the Ge–Fe–Co–Te system (May et al. 2020). The crystals were grown using initial compositions of $Fe_{5-x}Co_xGeTe_2$. The raw elements were sealed in evacuated silica ampoules and heated to 750°C at 120°C h^{-1}, followed by an isothermal step for 1–2 weeks at this temperature. The ampoules were quenched in ice water. The single crystals were grown using an iodine-assisted reaction. Iodine was washed from the surfaces of the crystals with ethanol and/or isopropanol with an acetone rinse.

B.6.7 Germanium–Iron–Nickel–Tellurium

The solid solution of $Fe_{5-\delta-x}Ni_xGeTe_2$, which crystallizes as a trigonal structure, is formed in the Ge–Fe–Ni–Te system (Stahl et al. 2018). The lattice parameter a decreases from 403.85 ± 0.01 pm at $x = 0.8$ and $\delta = 0.38$ to 402.94 ± 0.01 pm at $x = 1.29$ and $\delta = 0.75$ while c increases from 2916.3 ± 0.2 to 2923.2 ± 0.2 pm at the same values of x and δ. This solid solution was successfully synthesized up to $x = 1.3$. Polycrystalline samples were prepared via solid state reaction from pure elements. Fe, Ni, Ge, and Te powder were mixed in a molar ratio (4.5–x): x:1:2 with x = 0.1, 0.25, 0.5, 0.75, 1.0, 1.1, 1.25, 1.5. The mixtures were filled in alumina crucibles and sealed in silica ampoules in an Ar atmosphere. The samples were heated to 750°C for 100–120 h (heating and cooling rate – 100°C·h^{-1}). The products were metallic gray and stable at air. Single crystals show a hexagonal plate-like shape and a pronounced layer character. All samples contain solid solution as a side phase with 9–18 mol%.

B.7 Systems Based on Tin Sulfides

B.7.1 Tin–Hydrogen–Nitrogen–Sulfur

The quaternary compound $Sn_2S_6(N_2H_5)_4$, which crystallizes as an orthorhombic structure with the lattice parameters a = 852.20 ± 0.05, b = 1369.91 ± 0.08, and c = 1441.02 ± 0.09 pm, is formed in the Sn–H–N–S system (Mitzi et al. 2004; Yuan and Mitzi 2009). Pale-yellow crystals of this compound were formed by dissolving (in a nitrogen atmosphere, over several hours, with stirring) of SnS_2 (1 mM) in 2 mL of hydrazine and S (2 mM), ultimately yielding a yellow solution. Vigorous bubbling during dissolution supports the generation of gas. The highly toxic hydrazine was handled using appropriate protective equipment to prevent physical contact with either vapor or liquid. Evaporation of the solution to dryness under flowing nitrogen gas over a period of several hours, as well as under vacuum in the dry-box antechamber, leads to the formation of the crystalline yellow product.

B.7.2 Tin–Lithium–Lanthanum–Sulfur

The quaternary compound $LiLa_3SnS_7$, which crystallizes as a hexagonal structure with the lattice parameters a = 1031.22 ± 0.09, c = 597.05 ± 0.10 pm, a calculated density of 4.631 g·cm^{-3}, and an energy gap of 2.40 eV, is formed in the Sn–Li–La–S system (Yang et al. 2021). The first-phase preparation process of this compound was completed in an Ar-filled glove box because of the instability for La_2S_3 and Li_2S in the air. In the synthesis process, firstly, the raw materials (La_2S_3, Li_2S, and SnS_2) were added into a graphite crucible with a cover and then graphite crucible was put within the silica tube to avoid the corrosion of silica tube by raw materials. Their spontaneous crystallization processes were achieved in the vacuum-sealed silica tube with high-temperature muffle furnace. The title crystals were prepared under a stoichiometric ratio of raw materials at 800°C. Their high yields (>95 %) were successfully achieved after multiple grinding and calcination. Finally, many high-quality crystals were found in the silica tube after washing with DMF solvent. These crystals are relatively stable in the air within several weeks.

B.7.3 Tin–Lithium–Praseodymium–Sulfur

The quaternary compound $LiPr_3SnS_7$, which crystallizes as a hexagonal structure with the lattice parameters a = 1010.86 ± 0.14, c = 599.63 ± 0.16 pm, and a calculated density of 4.837 g·cm^{-3}, is formed in the Sn–Li–Pr–S system (Yang et al. 2021). This compound was prepared in the same way as $LiLa_3GeS_7$ but using Pr_2S_3 instead of La_2S_3. These crystals are also relatively stable in air within several weeks.

B.7.4 Tin–Lithium–Neodymium–Sulfur

The quaternary compound $LiNd_3SnS_7$, which crystallizes as a hexagonal structure with the lattice parameters a = 1003.97 ± 0.04, c = 604.10 ± 0.05 pm, and a calculated density of 4.930 g·cm^{-3}, is formed in the Sn–Li–Nd–S system (Yang et al. 2021). This compound was prepared in the same way as $LiLa_3GeS_7$ was synthesized using Nd_2S_3 instead of La_2S_3. These crystals are also relatively stable in air within several weeks.

B.7.5 Tin–Sodium–Silver–Sulfur

The quaternary compound Na_3AgSnS_4, which crystallizes as a monoclinic structure with the lattice parameters $a = 810.9 \pm 0.4$, $b = 648.3 \pm 0.3$, $c = 1594.1 \pm 0.8$ pm, $\beta = 103.713 \pm 0.009°$, a calculated density of 3.458 g·cm^{-3}, and an energy gap of 2.70 eV, is formed in the Sn–Na–Ag–S system (Yang et al. 2020b). To synthesize this compound, Na and Ag blocks and Sn and S powders were used. In order to ensure the reliability of raw material ratio, vacuum glove box was chosen to complete the weighing process and eliminate the effect of air oxidation. The raw mixture with a stoichiometric proportion of elements was firstly loaded into the graphite crucible and the the crucible was put into the silica tube. Using the flame gun and air extractor, silica tube was carefully vacuum-sealed with the internal vacuum of ~ 10^{-3} Pa in an ampoule. Muffle furnace was used to complete the crystallization reaction and the setting temperature process was shown as follows: firstly heated up to 850°C in 30 h to ensure the mixture melts completely and kept at this temperature within 4 days, then slowly cooled down to 200°C in 100 h, finally quickly down to the room temperature in one day. Finally, many sub-millimeter-level red crystals were found under the optical microscope after washing with DMF solution. They are relatively stable in the air over several days.

B.7.6 Tin–Sodium–Phosphorus–Sulfur

SnS_2–Na_2S–P_2S_5. The calculated isothermal section of this quasiternary system at 0 K is characterized by the existence of five quasibinary systems, namely Na_4SnS_4–Na_3PS_4, Na_2SnS_3–Na_3PS_4, SnS_2–Na_3PS_4, SnS_2–$Na_4P_2S_7$, and SnS_2–$NaPS_3$ (Richards et al. 2016). In the Na_4SnS_4–Na_3PS_4 section, a quaternary compound $Na_{10}SnP_2S_{12}$ can be formed, crystallizing in a tetragonal structure with lattice parameters $a = 968$ and $c = 1363$ pm (calculated values). This compound was synthesized by mixing stoichiometric amounts of Na_2S, P_2S_5, and SnS_2 with a planetary ball mill (380 rpm for 17 h). The pelletized mixture was wrapped in a gold foil and heated at 700°C for 12 h in an evacuated quartz tube and slowly cooled down to room temperature for 99 h (approximately 0.1°C·min^{-1}).

Two more quaternary compounds, $Na_6Sn_3P_4S_{16}$ and $Na_{11}Sn_2PS_{12}$, are formed in the Sn–Na–P–S system. $Na_6Sn_3P_4S_{16}$ crystallizes as a trigonal structure with the lattice parameters $a = 1930.4 \pm 0.4$, $c = 618.1 \pm 0.2$ pm, a calculated density of 2.824 g·cm^{-3}, and an energy gap of 2.52 eV (Zhao et al. 2023). Microcrystals of this compound were synthesized in a stoichiometric proportion by the high-temperature solid state method. The detailed steps are as follows: First, a mixture of Na (6 mM), SnS (3 mM), P (4 mM), and S (13 mM) was weighed and loaded into a graphite crucible and then put into a silica tube. Then the silica tube was pumped to a 10^{-3} Pa vacuum, sealed with a flame, and placed in a muffle furnace. A rising temperature program was set to heat the tube to 750°C at 15°C·h^{-1} and it was kept at this temperature for about 100 h, then slowly cooled down to room temperature over 3 days. Finally, many millimeter-level yellow crystals of $Na_6Sn_3P_4S_{16}$ were obtained. They are stable in air for several months. The storage and weighing processes were carried out in an Ar-filled glove box (the oxygen and water vapor contents were lower than 0.1 ppm) to prevent oxidation of the Na metal.

$Na_{11}Sn_2PS_{12}$ crystallizes as a tetragonal structure with the lattice parameters $a = 1360.12 \pm 0.04$ and $c = 2718.72 \pm 0.14$ pm at 100 K, $a = 1364.368 \pm 0.013$ and $c = 2727.15 \pm 0.04$ pm at 295 K, and $a = 1367.656 \pm 0.013$ and $c = 2733.47 \pm 0.44$ pm at 373 K (Duchardt et al. 2018a,b) [$a = 1361.48 \pm 0.03$, $c = 2722.44 \pm 0.07$ pm, and a calculated density of 2.373 g·cm^{-3} at 280 K (Ramos et al. 2018; Zhang et al. 2018b); $a = 1358.76 \pm 0.12$ and $c = 2715.5 \pm 0.5$ pm (Yu et al. 2018)].

Single crystals of this compound were synthesized from Na_2S, P_2S_5, and SnS_2 which were mixed together in a mortar at a molar ratio of 5.5:0.5:2 (Zhang et al. 2018b). The mixture was pelletized and sealed in a carbon-coated quartz tube under vacuum. After a heat treatment at 700°C for 2 h followed by a slow cooling step of 99 h from 700°C to 250°C, the sample was cooled to room temperature. The resultant material consisted of an agglomerate of crystals that were gently separated from each other to give single crystals.

This compound can be also prepared by heating stoichiometric amounts of Na_2S, phosphorus lump, Sn, and S in an evacuated quartz tube to 700°C at a rate of 0.5°C·min^{-1} (Yu et al. 2018). It was kept at

700°C for 48 h before cooling down to room temperature at a rate of 0.1°C·min^{-1}. Brown-colored single crystals were obtained.

$Na_{11}Sn_2PS_{12}$ was also obtained from corresponding mixtures of Na_3PS_4 and Na_4SnS_4 at 600°C with heating and cooling rates of 40°C·h^{-1} and 10°C·h^{-1}, respectively (Duchardt et al. 2018a,b). All synthetic steps were performed under argon atmosphere.

B.7.7 Tin–Sodium–Antimony–Sulfur

The solid solution of $Na_{4-x}Sn_xSb_xS_4$, which crystallizes as a tetragonal structure with the lattice parameters a = 1383.99 ± 0.03 and c = 2742.73 ± 0.08 pm for x = 0.01 and a = 1382.84 ± 0.03 and c = 2746.79 ± 0.08 pm for x = 0.1 (Hartmann et al. 2022a,b) [a = 1383.62 ± 0.01 and c = 2744.99 ± 0.03 pm for x = 0.05, a = 1382.89 ± 0.01 and c = 2751.79 ± 0.04 pm for x = 0.25, and a = 1382.78 ± 0.01 and c = 2754.94 ± 0.04 pm for x = 0.33 (Heo et al. 2018); a = 1376.81 ± 0.10, c = 2744.2 ± 0.2 pm, and a calculated density of 2.557 g·cm^{-3} for $Na_{11.20}Sn_2SbS_{12}$ or $Na_{3.73}Sn_{0.67}Sb_{0.33}S_4$ at 200 K and a = 1379.73 ± 0.08, c = 2750.26 ± 0.19 pm, and a calculated density of 2.534 g·cm^{-3} for $Na_{11.08}Sn_2SbS_{12}$ or $Na_{3.69}Sn_{0.67}Sb_{0.33}S_4$ at 280 K (Ramos et al. 2018)], is formed in the Sn–Na–Sb–S system.

The powders of this solid solution were prepared by heat treatment of a stoichiometric mixture of Na_2S, SnS_2, Sb_2S_3, and S at 450°C or 550°C for 12 h in a fused silica ampoule sealed under vacuum (Heo et al. 2018). For the solution process, the as-prepared $Na_{4-x}Sn_xSb_xS_4$ powders were fully dissolved in deionized water under dry Ar. After the powders were dried under vacuum at room temperature, further heat treatment was carried out in a sealed glass ampoule at 450°C.

The solid solution with x = 0.33 was also synthesized by solid state reaction (Ramos et al. 2018). Na_2S, SnS_2, Sb_2S_3, and S powder were mixed together in a molar ratio of 5.5:2:0.5:1. The mixture was loaded into a glassy carbon crucible, which was then vacuum-sealed in a quartz tube. The mixture was heated at 700°C with a heating rate of 1°C·min^{-1}, kept for 5 h, followed by a slow cooling step of 99 h from 700°C to room temperature. The sample resulted in agglomerates of crystals that were gently separated from each other.

$Na_{4-x}Sn_xSb_xS_4$ can be also prepared by thermal co-decomposition of stoichiometric amounts of $N_4SnS_4{\cdot}14H_2O$ and $Na_3SbS_4{\cdot}9H_2O$ at 170°C (Hartmann et al. 2022a,b).

B.7.8 Tin–Potassium–Copper–Sulfur

Two quaternary phases, $K_{2.1}Cu_{7.9}Sn_{3.1}S_{12}$ and $K_{3.9}Cu_{8.1}Sn_{2.8}S_{12}$, are formed in the Sn–K–Cu–S system (Zhang et al. 2017b). They were obtained by soaking for 2 h 0.2 g of $(H_3O)_4Cu_8Sn_3S_{13}$ crystals in 40 mL of 1 M KCl solution at 20°C for the first compound and in 40 mL of 1 M KCl solution at 60°C for the second one. The exchanged samples were washed sufficiently with distilled water and dried at 70°C for 6 h.

B.7.9 Tin–Potassium–Silver–Sulfur

Two quaternary compounds, $K_2Ag_2SnS_4$ and $K_2Ag_2Sn_2S_6$, are formed in the Sn–K–Ag–S system (Li et al. 2020d). Both compounds crystallize as an orthorhombic structure with the lattice parameters a = 1152.95 ± 0.15, b = 2695.3 ± 0.4, c = 584.83 ± 0.08 pm, a calculated density of 3.954 g·cm^{-3}, and an energy gap of 2.66 eV for the first compound and a = 992.33 ± 0.07, b = 3027.4 ± 0.2, c = 855.37 ± 0.06 pm, a calculated density of 3.741 g·cm^{-3}, and an energy gap of 2.43 eV for the second one.

$K_2Ag_2SnS_4$ was synthesized at 210°C by mixing $AgNO_3$ (10.0 mg), SnS (15.0 mg), K_2CO_3 (15.0 mg), about 500 mg of mixed solvent of pyridine/thiophenol (volume ratio 3:1), and 150 mg of 1,3-diaminopropane in a Pyrex glass tube for 8 days. After washing with ethanol and deionized water, yellow plate-like crystals were obtained. $K_2Ag_2Sn_2S_6$ was prepared in the same way as $K_2Ag_2SnS_4$ was synthesized using the temperature of 150°C instead of 210°C.

B.7.10 Tin–Potassium–Indium–Sulfur

The quaternary compound $KInSn_2S_6$, which melts congruently at ~ 743°C and crystallizes as a trigonal structure with the lattice parameters a = 369.3 ± 0.1, c = 2346.6 ± 0.3 pm, a calculated density of 3.498 g·cm^{-3}, and an energy gap of 2.17 ± 0.05 eV (Friedrich et al. 2022) [a = 368.030 ± 0.010, c = 2612.76 ± 0.10 pm, and a calculated density of 3.162 g·cm^{-3} at 100 K (Xiao et al. 2017)], is formed in the Sn–K–In–S system.

The title compound was synthesized using two experimental methods (Friedrich et al. 2022). According to the first method, a batch of 1.0 g containing stoichiometric amounts of K_2S (0.85 mM), In (1.71 mM), Sn (3.43 mM), and S (9.42 mM) were weighed in a fused silica tube in a nitrogen-filled glove box. The tube containing the starting mixture was evacuated to ~0.01 Pa, flame-sealed, and placed in a programmable furnace. The sample was heated to 800°C with a heating rate of 1°C·min^{-1}, annealed at 800°C for 24 h, and cooled to room temperature at a cooling rate of 1°C·min^{-1}. The tube contained a soft, yellow, sintered polycrystalline material that could easily be crushed in an agate mortar while falling apart into plate-like crystallites. They are stable in moisture and air.

$KInSn_2S_6$ can be also prepared below its melting points. In a batch of 1.0 g, K_2S (0.85 mM), In_2S_3 (0.85 mM), Sn (3.43 mM), and S (6.85 mM) were used. These amounts were ground together in an agate mortar and pestle until it becomes homogeneous. The resultant mixture was loaded into a fused silica tube that was internally lined with an Al foil. The Al foil is necessary to protect the region to be sealed from the powder; without its use, the reactant residue on the glass will react with silica in the high heat of the torch, producing high-melting-point silicate in the sealing region, making effective sealing of the ampoule difficult. After loading, the Al foil was removed, and the tube was evacuated to ~0.01 Pa, flame-sealed, and placed in a programmable tube furnace. The sample was heated to 700°C at a heating rate of 50°C·h and annealed at 700°C for 24 h, after which the furnace was turned off, and the sample was allowed to cool to room temperature.

$KInSn_2S_6$ can be also prepared using another two different methods (Xiao et al. 2017). (1) K_2CO_3 (22 mM), Sn (60 mM), In (30 mM), and S (195 mM) were combined and loaded in a 50 mL grinding jar under N_2 atmosphere in a glove box. The mixture was ball-milled at 100 rpm for 1 min and at 250 rpm for 30 min. Then 3 g of the ball-milled mixture was placed in a carbon-coated fused silica tube under N_2 atmosphere. A rubber balloon was attached at the end of the reaction tube in order to accommodate the created pressure of the CO_2 evolution. The mixture was heated gradually to 200°C where it was kept for 5 h before being successfully brought to 800°C. It was kept at 800°C for 8 h. Shiny yellow plate shape crystals were obtained at cooling at a rate of 40°C·h^{-1} to room temperature. (2) A mixture of K_2S_6 (1 mM), S (6 mM), Sn (4 mM), and In (2 mM) was sealed under vacuum (0.01 Pa) in a carbon-coated fused silica tube and heated (80°C·h^{-1}) to 800°C. It was kept there for 24 h, followed by cooling to room temperature at 40°C·h^{-1}.

B.7.11 Tin–Rubidium–Silver–Sulfur

The quaternary compound $Rb_2Ag_2SnS_4$, which crystallizes as an orthorhombic structure with the lattice parameters a = 1167.87 ± 0.18, b = 2752.6 ± 0.4, c = 594.72 ± 0.09 pm, a calculated density of 4.403 g·cm^{-3}, and an energy gap of 2.70 eV, is formed in the Sn–Rb–Ag–S system (Li et al. 2020d). This compound was synthesized at 210°C by mixing $AgNO_3$ (10.0 mg), SnS (15.0 mg), Rb_2CO_3 (33.0 mg), about 500 mg of mixed solvent of pyridine/thiophenol (volume ratio 3:1), and 150 mg of 1,3-diaminopropane in a Pyrex glass tube for 8 days. After washing with ethanol and deionized water, yellow plate-like crystals were obtained.

B.7.12 Tin–Rubidium–Indium–Sulfur

The quaternary compound $RbInSn_2S_6$, which melts congruently at ~ 735°C and crystallizes as a trigonal structure with the lattice parameters a = 369.3 ± 0.1, c = 2466.7 ± 0.7 pm, a calculated density of 3.591 g·cm^{-3}, and an energy gap of 2.31 ± 0.05 eV, is formed in the Sn–Rb–In–S system (Friedrich et al. 2022).

The title compound was prepared using two experimental methods. According to the first method, a batch of 1.0 g of stoichiometric amounts of Rb_2S (0.79 mM), In (1.58 mM), Sn (3.17 mM), and S (8.73 mM) were weighed in a fused silica tube in a nitrogen-filled glove box. The further process of preparing this compound is similar to that used in the synthesis of $KInSn_2S_6$. $RbInSn_2S_6$ can be also obtained by combining Rb_2S (0.79 mM), In_2S_3 (0.79 mM), Sn (3.17 mM), and S (6.35 mM). The following preparation scheme is similar to that used in the synthesis of $KInSn_2S_6$.

B.7.13 Tin–Cesium–Silver–Sulfur

The quaternary compound $Cs_2Ag_2SnS_4$, which crystallizes as an orthorhombic structure with the lattice parameters $a = 601.1 \pm 0.2$, $b = 1187.6 \pm 0.4$, $c = 1413.3 \pm 0.6$ pm, a calculated density of 4.796 $g \cdot cm^{-3}$, and an energy gap of 2.72 eV, is formed in the Sn–Cs–Ag–S system (Li et al. 2020d). This compound was synthesized at 210°C by mixing $AgNO_3$ (10.0 mg), SnS (15.0 mg), Cs_2CO_3 (580 mg), about 500 mg of mixed solvent of pyridine/thiophenol (volume ratio 3:1), and 150 mg of 1,3-diaminopropane in a Pyrex glass tube for 8 days. After washing with ethanol and deionized water, yellow plate-like crystals were obtained.

B.7.14 Tin–Cesium–Indium–Sulfur

The quaternary compound $CsInSn_2S_6$, which melts congruently at ~ 726°C and crystallizes as a trigonal structure with the lattice parameters $a = 369.0 \pm 0.1$, $c = 2591.0 \pm 0.9$ pm, a calculated density of 3.681 $g \cdot cm^{-3}$, and an energy gap of 2.47 ± 0.05 eV, is formed in the Sn–Cs–In–S system (Friedrich et al. 2022).

The title compound was prepared using two experimental methods. According to the first method, a batch of 1.0 g of stoichiometric amounts of Cs_2S (0.79 mM), In (1.58 mM), Sn (3.17 mM), and S (8.73 mM) were weighed in a fused silica tube in a nitrogen-filled glove box. The further process of preparing this compound is similar to that used in the synthesis of $KInSn_2S_6$. $CsInSn_2S_6$ can be also obtained by combining Cs_2S (0.79 mM), In_2S_3 (0.79 mM), Sn (3.17 mM), and S (6.35 mM). The following preparation scheme is similar to that used in the synthesis of $KInSn_2S_6$.

B.7.15 Tin–Copper–Barium–Sulfur

Two quaternary compounds, Cu_2BaSnS_4 and $Cu_{2.93}Ba_6Sn_{4.4}S_{16}$, are formed in the Sn–Cu–Ba–S system. The first compound crystallizes as a trigonal structure with the lattice parameters $a = 637.11 \pm 0.03$, $c = 1584.25 \pm 0.12$ pm, a calculated density of 4.574 $g \cdot cm^{-3}$, and an energy gap of 1.88 eV (Liu et al. 2019) [$a = 636.62 \pm 0.01$, $c = 1582.87 \pm 0.02$ pm, and an energy gap of 1.95 eV (Shin et al. 2016)]. To prepare this compound, Cu powder (0.11 mM), Sn powder (0.08 mM), S powder (1.03 mM), $Ba(OH)_2 \cdot 8H_2O$ (0.04 mM), and about 300 mg of ethylenediamine were mixed together and sealed in a Pyrex glass tube (about 10% filling volume of the tube) at air atmosphere (Liu et al. 2019). Finally, the glass tube was placed in a stainless steel autoclave, into which water was added to 80% filling to balance the pressure of the glass tube, and then heated in the furnace at 170°C for 6 days. After being cooled to room temperature naturally, the products were washed with water and ethanol, respectively, and red block crystals of the title compound were obtained.

Cu_2BaSnS_4 can also be prepared as follows (Shin et al. 2016). A stoichiometric mixture of BaS, SnS, and CuS was carefully ground/homogenized and cold-pressed into pellet inside a nitrogen-filled glove box. The pellet was placed inside a quartz tube, which was flame-sealed under dynamic vacuum (~7×10^{-5} Pa). The reaction mixture was then heated to 650°C for 5–6 h inside a box furnace, and kept at this temperature for 10–15 h. After this step, the reaction was cooled to room temperature by switching off the furnace.

$Cu_{2.93}Ba_6Sn_{4.4}S_{16}$ crystallizes as a cubic structure with the lattice parameter $a = 1453.3 \pm 0.3$ pm, a calculated density of 4.427 $g \cdot cm^{-3}$, and an energy gap of 2.3 eV (Ji et al. 2022b). For preparing this compound, the reactant elements were loaded in a molar ratio of Ba/Cu/Sn/S = 6:4:4:16 and mixed with KI salt flux in a carbonized silica ampoule. A 0.4 g of element mixtures were mixed with 0.4 g of KI and loaded into a carbonized silica tube. Then, the silica tube was flamed-sealed under a vacuum of 1 Pa and heated within a programmable box furnace. The sample was heated to 800°C in 10 h, maintained for 48 h at this temperature, and then slowly cooled to 650°C within 100 h. The sample was kept for 10 h at 650°C before the furnace was shut off. The KI flux was removed with deionized water and dark red compound was obtained. This compound is stable in air and water.

B.7.16 Tin–Copper–Lead–Sulfur

SnS_2–Cu_2S–PbS. It was confirmed that there are Cu_2SnS_3–PbS and Cu_4SnS_4–PbS quasibinary sections in this system (Olekseyuk et al. 2021). No quaternary compounds were found in the system.

B.7.17 Tin–Copper–Titanium–Sulfur

The quaternary compound $Cu_{3.1}Ti_{0.1875}Sn_{0.9}S_4$, which crystallizes as a cubic structure with the lattice parameter $a = 543.2 \pm 0.2$ pm, is formed in the Sn–Cu–Ti–S system (Hirayama et al. 2022). To synthesize this phase, Cu wire, Ti powder, Sn grains, and S powder were sealed in an evacuated quartz tube and heated at 1100°C for 24 h. The obtained sample was then pulverized with an agate mortar and put into a carbon die with a 10-mm inner diameter. In a hot-press sintering furnace, the sample was sintered at 750°C for 40 min in a flowing N_2 environment at a uniaxial pressure of 50 MPa.

B.7.18 Tin–Copper–Arsenic–Sulfur

SnS_2–Cu_2S–As_2S_3. The region of the glass formation in this system was determined through XRD (Figure B.7.1) (Bereznyuk et al. 2021). Glassy samples were synthesized from elemental substances and As_2S_3. The mixtures were heated in evacuated quartz ampoules up to 830°C with exposure for 24 h at 400°C and 600°C. The samples were kept at 830°C for 10 h, followed by quenching in a 25% NaCl solution with crushed ice.

B.7.19 Tin–Copper–Antimony–Sulfur

Cu_2SnS_3–Sb_2S_3. The phase diagram of this system, constructed through DTA, XRD, and metallography, is a eutectic type (Figure B.7.2) (Mammadov 2020a; Bereznyuk et al. 2022a). The eutectic contains 13 mol% Cu_2SnS_3 and crystallizes at 492°C (Bereznyuk et al. 2022a) [30 mol% Cu_2SnS_3 and 477°C (Mammadov 2020a)]. At 230°C, Sb_2S_3 dissolves 5 mol% Cu_2SnS_3 and the solubility of Cu_2SnS_3 in Sb_2S_3 is 8 mol%. The ingots for the investigation were annealed at 230°C for 500 h.

Cu_2SnS_3–$CuSbS_2$. The phase diagram of this system, constructed through DTA, XRD and metallography, is a eutectic type (Figure B.7.3) (Bereznyuk et al. 2022a). The eutectic contains ~7 mol% Cu_2SnS_3 and crystallizes at 523°C. The solubility based on Cu_2SnS_3 is 35 mol% $CuSbS_2$ and decreases with the temperature decreasing to ~12 mol% at 230°C. The solubility based on $CuSbS_2$ is negligible. The ingots for the investigation were annealed at 230°C for 500 h.

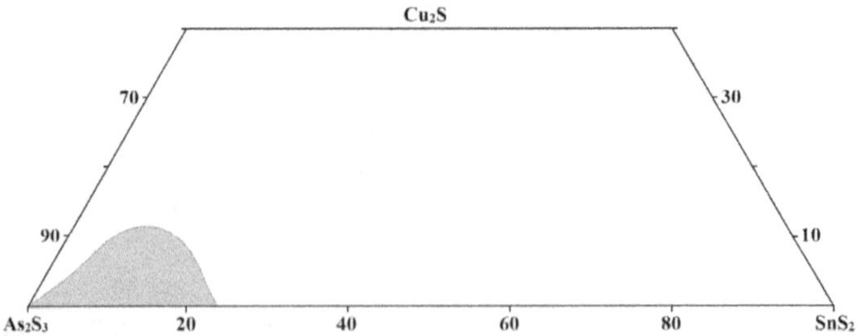

FIGURE B.7.1 Glass-forming region in the SnS_2–Cu_2S–As_2S_3 quasiternary system. (From Bereznyuk, O., et al., *Probl. khimii ta staloho rozvytku*, (4), 3, 2021.) Open access.

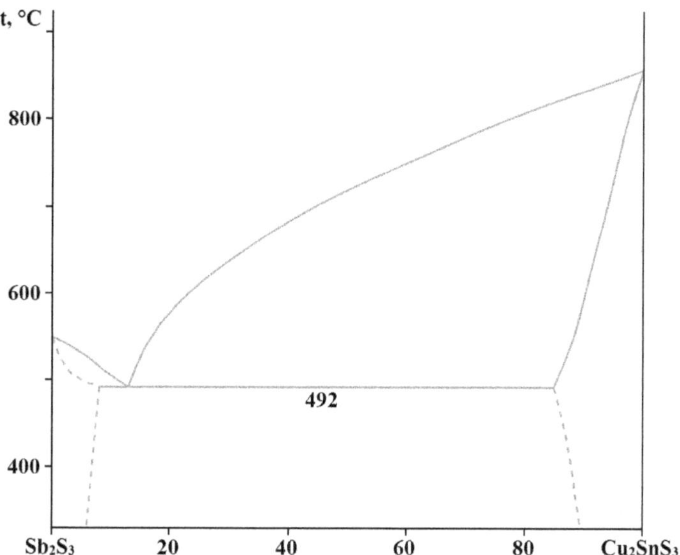

FIGURE B.7.2 Phase diagram of the Cu_2SnS_3–Sb_2S_3 system. (From Bereznyuk, O., et al., *Probl. khimii ta staloho rozvytku*, (4), 17, 2022.) Open access.

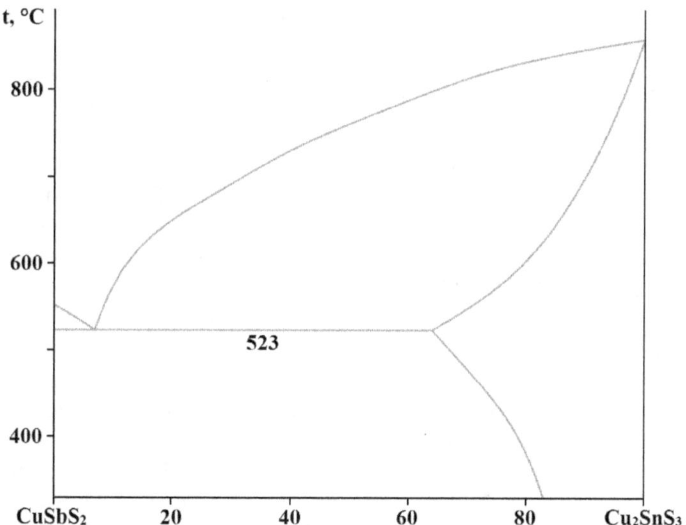

FIGURE B.7.3 Phase diagram of the Cu_2SnS_3–$CuSbS_2$ system. (From Bereznyuk, O., et al., *Probl. khimii ta staloho rozvytku*, (4), 17, 2022.) Open access.

Cu_2SnS_3–Cu_3SbS_3. The phase diagram of this system, constructed through DTA, XRD and metallography, is a eutectic type (Figure B.7.4) (Mammadov 2019; Bereznyuk et al. 2022a). The eutectic contains 20 mol% Cu_2SnS_3 and crystallizes at 593°C (Bereznyuk et al. 2022a) [25 mol% Cu_2SnS_3 and 507°C (Mammadov 2019)]. Phase transition based on Cu_3SbS_3 is eutectoid and takes place with decreasing temperature from 360°C to 319°C (Bereznyuk et al. 2022a). At 230°C, the solubility based on Cu_3SbS_3 reaches 10 mol%, Cu_2SnSe_3 and the solubility based on Cu_2SnS_3 is 7–8 mol% Cu_3SbS_3. The ingots for the investigation were annealed at 230°C for 500 h.

According to Mammadov (2019), the solubility of Cu_3SbS_3 in Cu_2SnS_3 at room temperature is 9 mol% and increases to 17 mol% at 507°C and Cu_3SbS_3 dissolves 7 and 19 mol% of Cu_2SnS_3 at room temperature and 507°C, respectively.

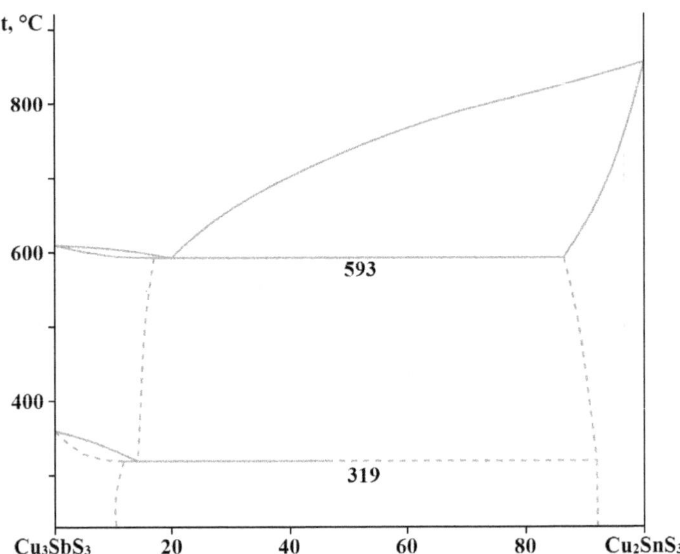

FIGURE B.7.4 Phase diagram of the Cu_2SnS_3–Cu_3SbS_3 system. (From Bereznyuk, O., et al., *Probl. khimii ta staloho rozvytku*, (4), 17, 2022.) Open access.

Cu_3SnS_4–Cu_3SbS_4. A continuous series of the solid solution can be obtained in this system (Chen et al. 2020b). With increasing level of Sn in the $Cu_3Sb_{1-x}Sn_xS_4$ solid solutions, the system shows a gradual evolution from semiconducting to metallic behavior. Polycrystalline samples of these solid solutions were prepared by mechanical alloying combined with spark plasma sintering. Starting powders of Cu, Sb, Sn, and S were loaded into a stainless steel jar with stainless steel balls and sealed in an inert atmosphere of argon. The powder mixture was milled at a rotation speed of 420 rpm for 20 h in a planetary ball mill. The powders were sintered at 450°C for 3 min, at a pressure of 50 MPa, using a spark plasma sintering furnace.

Cu_2S–SnS_2–Sb_2S_3. There are six two-phase equilibria and seven three-phase fields between binary and ternary compounds of this system at 230°C (Figure B.7.5) (Bereznyuk et al. 2022a). The Cu_4SnS_4–Cu_3SbS_3, $Cu_4Sn_7S_{16}$–Sb_2S_3, and $Cu_4Sn_7S_{16}$–Sb_2SnS_5 systems are the nonquasibinary section of this quasiternary system as all ternary compounds melt incongruently.

The region of the glass formation in this system was determined through XRD (Figure B.7.6) (Bereznyuk et al. 2021). Glassy samples were synthesized from elemental substances. The mixtures were heated in evacuated quartz ampoules up to 830°C with exposure for 24 h at 400°C and 600°C. The samples were kept at 830°C for 10 h, followed by quenching in a 25% NaCl solution with crushed ice.

According to the data of Pavan Kumar et al. (2021b), the $Cu_{22}Sn_{10-x}Sb_xS_{32}$ solid solution, which crystallizes as a cubic structure with the lattice parameter a = 1083.50 ± 0.01 pm for x = 0.75 and a = 1083.73 ± 0.01 pm for x = 1.0, is formed in the Sn–Cu–Sb–S system. The samples of this solid solution were synthesized by mechanical alloying. The stoichiometric amounts of Cu, Sn, Sb, and S were loaded into a 25 mL tungsten carbide jar containing seven balls of 10 mm under an Ar atmosphere. High-energy ball milling was performed in a planetary ball mill operating at room temperature at a disk rotation speed of 600 rpm for 6 h. The resulting powders were then ground and sieved down to 150 μm. Powders were densified at 550°C for 30 min under a pressure of 64 MPa. All sample preparations and handling of powders were performed in an Ar-filled glove box with oxygen contents of <1 ppm.

B.7.20 Tin–Copper–Vanadium–Sulfur

The quaternary compound $Cu_{26}V_2Sn_6S_{32}$, which crystallizes as a cubic structure with the lattice parameter a = 1082.89 ± 0.02 pm for $Cu_{27.3}V_{1.9}Sn_6S_{31.4}$ sintered at 750°C and a = 1077.01 ± 0.02 pm for $Cu_{25}V_{1.9}Sn_6S_{32.6}$ sintered at 600°C, is formed in the Sn–Cu–V–S system (Bourgès et al. 2016, 2018).

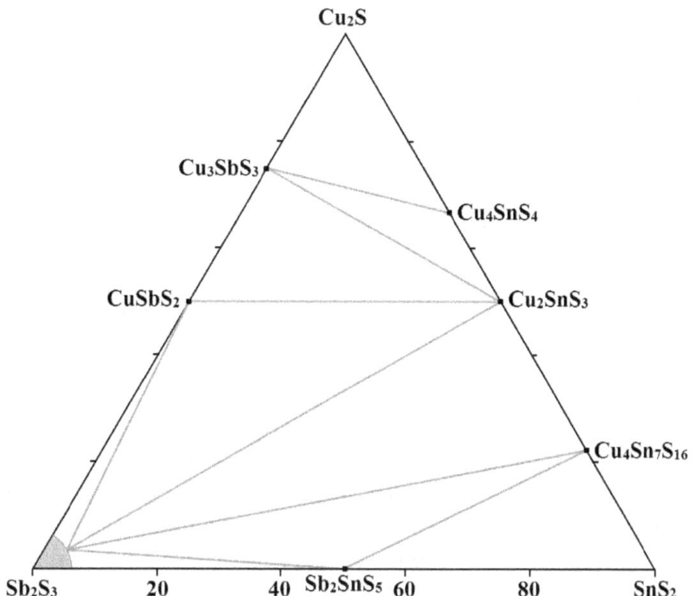

FIGURE B.7.5 Isothermal section of the SnS_2–Cu_2S–Sb_2S_3 quasiternary system at 230°C. (From Bereznyuk, O., et al., *Phys. Chem. Solid State*, **23**(1), 57, 2022.)

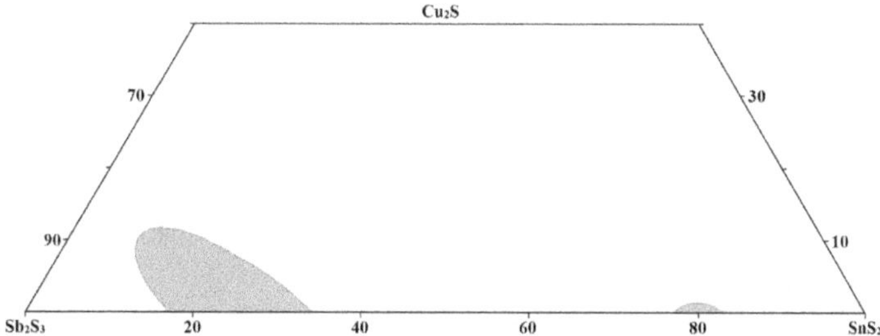

FIGURE B.7.6 Glass-forming region in the SnS_2–Cu_2S–Sb_2S_3 quasiternary system. (From Bereznyuk, O., et al., *Probl. khimii ta staloho rozvytku*, (4), 3, 2021.) Open access.

Mechanically alloyed powders of $Cu_{26}V_2Sn_6S_{32}$ were prepared by milling 5 g of Cu, V, S, and Sn in stoichiometric proportions in a planetary ball mill. The milling was performed for 12 h at a speed of 600 rpm under Ar atmosphere in a 45 mL tungsten carbide jar containing 10 balls of 10 mm diameter. It produces a fine black powder. The obtained powder was then densified using two different consolidation techniques: (1) spark plasma sintering during 45 min at 600°C under a uniaxial pressure of 64 MPa with cooling and heating rates of 50°C·min^{-1} with a slight over pressure of +50 hPa (Ar); (2) hot pressing at 750°C for 1 h under a uniaxial pressure of 70 MPa in Ar gas flow of 300 mL·min^{-1} with heating and cooling rates of 10°C·min^{-1} and 20°C·min^{-1}, respectively.

B.7.21 Tin–Copper–Niobium–Sulfur

The quaternary compound $Cu_{3.1}Nb_{0.25}Sn_{0.9}S_4$, which crystallizes as a cubic structure with the lattice parameter $a = 543.2 \pm 0.2$ pm, is formed in the Sn–Cu–Nb–S system (Hirayama et al. 2022). To synthesize this phase, Cu wire, Nb grains, Sn grains, and S powder were sealed in an evacuated quartz tube and heated at 1100°C for 24 h. The obtained sample was then pulverized with an agate mortar and put into a carbon die with a 10-mm inner diameter. In a hot-press sintering furnace, the sample was sintered at 750°C for 40 min in a flowing N_2 environment at a uniaxial pressure of 50 MPa.

B.7.22 Tin–Silver–Lead–Sulfur

SnS_2–Ag_2S–PbS. It was confirmed that in this system, the ternary compounds Ag_2SnS_3, $Ag_4Sn_3S_8$, and Ag_8SnS_6 form quasibinary sections with $PbSnS_3$ (Olekseyuk et al. 2021). There is also a two-phase equilibrium between Ag_8SnS_6 and PbS. No quaternary compounds were found in the system. The solubility based on Ag_8SnS_6 does not exceed 2 mol% PbS.

B.7.23 Tin–Silver–Phosphorus–Sulfur

Ag_8SnS_6–Ag_7PS_6. This system is characterized by the formation of a continuous series of the solid solutions between high-temperature modifications of Ag_8SnS_6 and Ag_7PS_6 (Figure B.7.7) (Bereznyuk et al. 2022d). The liquidus and solidus curves have a minimum at ~65 mol% Ag_7PS_6. At room temperature, there are two-phase regions within the interval of ~25–42 and ~65–73 mol% Ag_7PS_6 in the subsolidus part of the phase diagram.

The quaternary compound $Ag_7SnP_3S_{12}$, which is stable up to 465°C and crystallizes as a monoclinic structure with the lattice parameters $a = 1059.0 \pm 0.6$, $b = 1020.6 \pm 0.5$, $c = 2116.5 \pm 0.3$ pm, $\beta = 119.873 \pm 0.005°$, a calculated density of 3.879 g·cm^{-3}, and an energy gap of 2.25 eV, is formed in the Si–Ag–In–S system (Fan et al. 2017). Yellow single crystals of this compound were obtained by solid state reactions. Sn (0.2 mM), Ag (1.4 mM), S (2.4 mM), and red phosphorus (0.6 mM) were ground into a fine powder

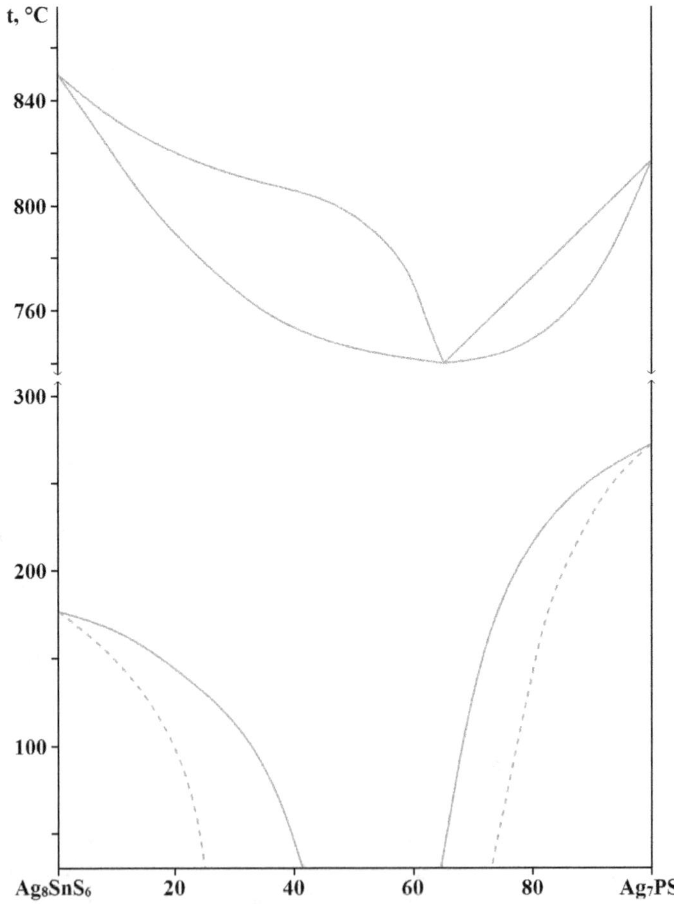

FIGURE B.7.7 Phase diagram of the Ag_8SnS_6–Ag_7PS_6 system. (From Bereznyuk, O., et al., *Probl. khimii ta staloho rozvytku*, (4), 3, 2022.)

in an agate mortar, pressed into a pellet, loaded into Pyrex tube, evacuated to 0.01 Pa, and flame-sealed. The tube was then placed into a computer-controlled furnace, heated from room temperature to 250°C at a rate of 50°C·h^{-1}, maintained at that temperature for 1 day, then heated to 600°C at a rate of 50°C·h^{-1}, maintained at that temperature for 4 days, and then slowly cooled to 250°C at a rate of 2.5°C·h^{-1}. It was finally cooled to room temperature in 12 h. Pure crystals of $Ag_7SnP_3S_{12}$ were hand-picked under a microscope.

B.7.24 Tin–Silver–Arsenic–Sulfur

SnS_2–Ag_2S–As_2S_3. The isothermal section of this quasiternary system at 230°C is given in Figure B.7.8 (Bereznyuk et al. 2022c). No quaternary compounds were found in the system. The extent of solid solutions based on binary and ternary compounds at this temperature is negligible. The alloys for the investigations were annealed at 230°C for 500 h. The glass-forming region in this system is presented in Figure B.7.9 (Bereznyuk et al. 2021, 2022c).

B.7.25 Tin–Silver–Antimony–Sulfur,

SnS_2–$AgSbS_2$. The phase diagram of this system, constructed through DTA, XRD, and metallography, is a eutectic type (Figure B.7.10) (Bereznyuk et al. 2022a). The eutectic contains 25 mol% SnS_2 and crystallizes at 468°C. At 230°C, the solubility regions based on $AgSbS_2$ and SnS_2 are 3 and 5 mol%, respectively. At the same time, the solubility based on SnS_2 increases to 30 mol% at the eutectic temperature. The ingots for the investigation were annealed at 230°C for 500 h.

Ag_2SnS_3–Sb_2S_3. The phase diagram of this system, constructed through DTA, XRD, metallography, and measurement of microhardness and density, is a eutectic type (Mammadov 2020b). The eutectic contains 60 mol% Ag_2SnS_3 and crystallizes at 477°C. At room temperature, Sb_2S_3 dissolves 10 mol% of Ag_2SnS_3 and Ag_2SnS_3 dissolves 3 mol% of Sb_2S_3. The ingots for the investigation were annealed at 230°C for 270 h. Subsequently, the quasibinary nature of this section was not confirmed by Bereznyuk et al. (2022a).

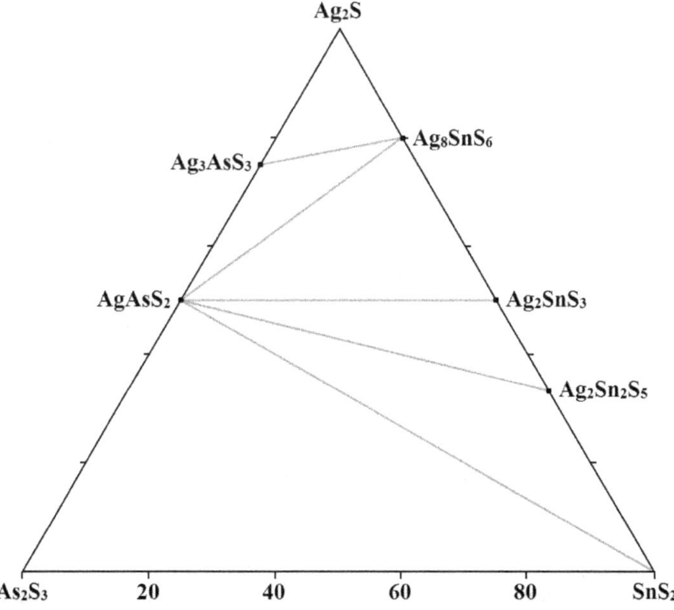

FIGURE B.7.8 Isothermal section of the SnS_2–Ag_2S–As_2S_3 quasiternary system at 230°C. (From Bereznyuk, O., et al., *Phys. Chem. Solid State*, **23**(1), 57, 2022.)

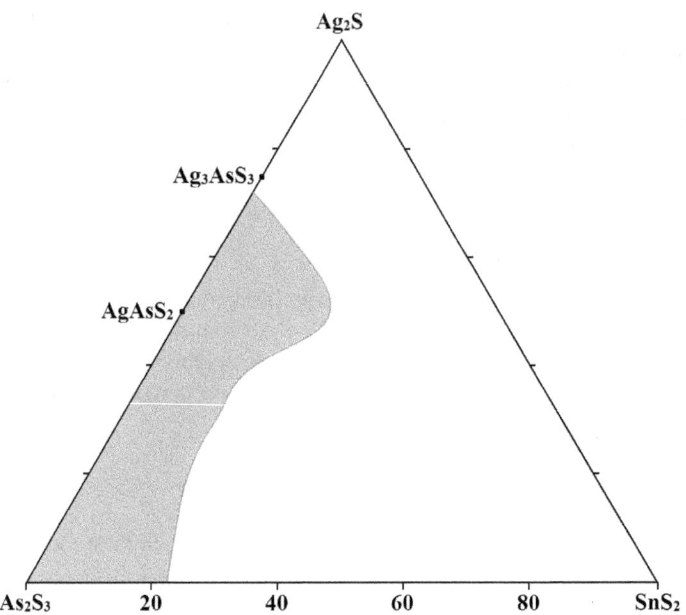

FIGURE B.7.9 Glass-forming region in the SnS_2–Ag_2S–As_2S_3 quasiternary system. (From Bereznyuk, O., et al., *Probl. khimii ta staloho rozvytku*, (4), 3, 2021.) Open access.

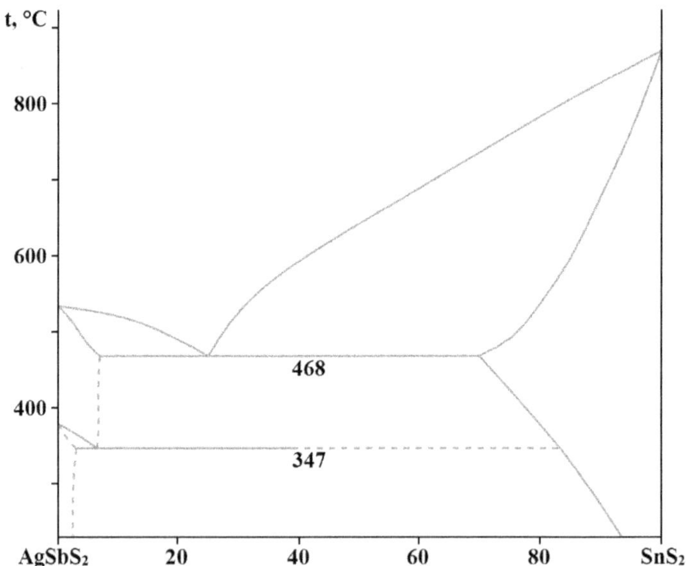

FIGURE B.7.10 Phase diagram of the SnS_2–$AgSbS_2$ system. (From Bereznyuk, O., et al., *Probl. khimii ta staloho rozvytku*, (4), 17, 2022.) Open access.

Ag_2SnS_3–$AgSbS_2$. The phase diagram of this system, constructed through DTA, XRD, and metallography, is a eutectic type (Figure B.7.11) (Bereznyuk et al. 2022a). The eutectic contains 30 mol% of Ag_2SnS_3 and crystallizes at 477°C. At 230°C, the solubility regions based on both compounds are 5 mol%, and at 477°C, $AgSbS_2$ dissolves 15 mol% of Ag_2SnS_3 and Ag_2SnS_3 dissolves 11 mol% of $AgSbS_2$. With an increase in the Ag_2SnS_3 content, a eutectoid process takes place and the temperature of $AgSbS_2$ phase transformation decreases by 25°C. The ingots for the investigation were annealed at 230°C for 500 h.

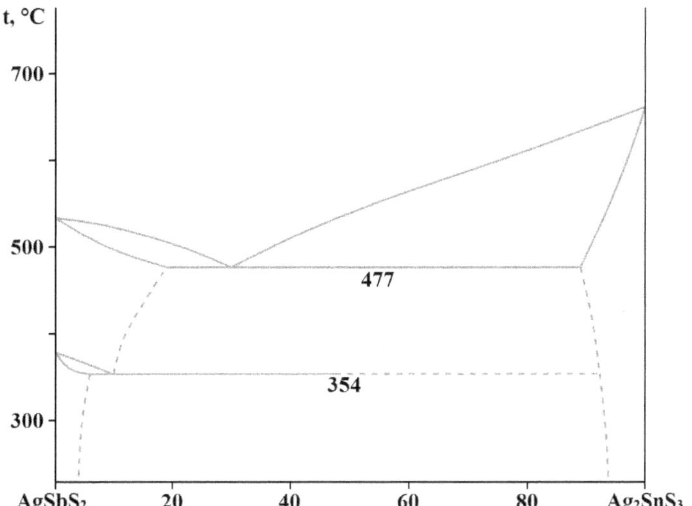

FIGURE B.7.11 Phase diagram of the Ag_2SnS_3–$AgSbS_2$ system. (From Bereznyuk, O., et al., *Probl. khimii ta staloho rozvytku*, (4), 17, 2022.) Open access.

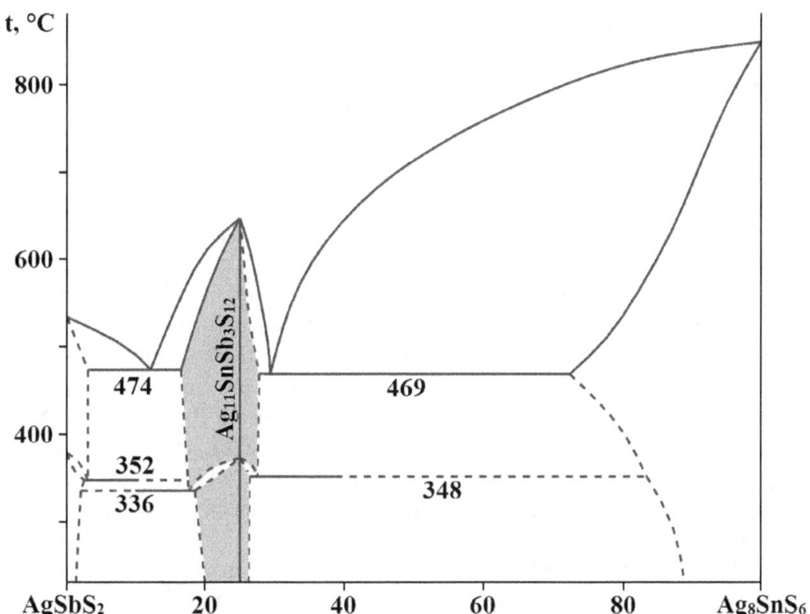

FIGURE B.7.12 Phase diagram of the Ag_8SnS_6–$AgSbS_2$ system. (From Bereznyuk, O., et al., *Probl. khimii ta staloho rozvytku*, (4), 17, 2022.) Open access.

Ag_2SnS_3–Ag_3SbS_3. This system is quasibinary at 230°C (Bereznyuk et al. 2022a). The $Ag_{11}SnSb_3S_{12}$ quaternary compound whose structure has not been established is formed in the system. The solubility based on the starting compounds does not exceed 5 mol%. The ingots for the investigation were annealed at 230°C for 500 h.

$Ag_4Sn_3S_8$–$AgSbS_2$. This system is nonquasibinary as $Ag_4Sn_3S_8$ melts incongruently (Bereznyuk et al. 2022a). At 230°C, the solubility based on the starting compounds does not exceed 5 mol%. The ingots for the investigation were annealed at 230°C for 500 h.

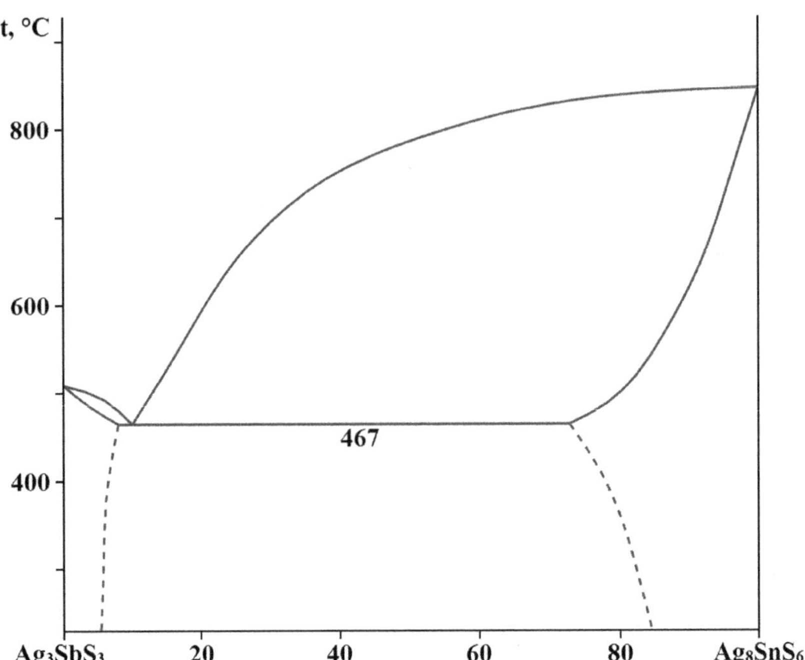

FIGURE B.7.13 Phase diagram of the Ag_8SnS_6–Ag_3SbS_3 system. (From Bereznyuk, O., et al., *Probl. khimii ta staloho rozvytku*, (4), 17, 2022.) Open access.

Ag_8SnS_6–$AgSbS_2$. The phase diagram of this system, constructed through DTA, XRD, and metallography, is presented in Figure B.7.12 (Bereznyuk et al. 2022a). The eutectics contain 12 and 30 mol% Ag_8SnS_6 and crystallize at 474°C and 469°C, respectively. The quaternary compound $Ag_{11}SnSb_3S_{12}$, which melts congruently at 647°C, undergoes polymorphic transformation at 376°C, and is a phase of changing composition, is formed in the system. The homogeneity region of this compound is within the interval from 16 to 27 mol% Ag_8SnS_6 at the eutectic temperatures and from 20 to 25 mol% Ag_8SnS_6 at 230°C. The solubility region based on $AgSbS_2$ is negligible and the solubility based on Ag_8SnS_6 is 5 mol% at 230°C and 25 mol% at 469°C. The ingots for the investigation were annealed at 230°C for 500 h.

Ag_8SnS_6–Ag_3SbS_3. The phase diagram of this system, constructed through DTA, XRD, and metallography, is a eutectic type (Figure B.7.13) (Bereznyuk et al. 2022a). The eutectic contains 10 mol% Ag_8SnS_6 and crystallizes at 467°C. The solubility based on Ag_3SbS_3 at 230°C reaches up to 5 mol% and based on Ag_8SbS_6 is 15 mol% and increases to 8 and 22 mol%, respectively, at the eutectic temperature. The ingots for the investigation were annealed at 230°C for 500 h.

SnS_2–Ag_2S–Sb_2S_3. The isothermal section of this quasiternary system at 230°C is displayed in Figure B.7.14 (Bereznyuk et al. 2022a,c). The $Ag_{11}SnSb_3S_{12}$ quaternary compound is formed in the system. Most of the compounds in the system have small solid-phase solubility. The alloys for the investigations were annealed at 230°C for 500 h. The glass-forming region in this system is presented in Figure B.7.15 (Bereznyuk et al. 2021, 2022c).

B.7.26 Tin–Silver–Oxygen–Sulfur

The quaternary compound $Ag_2Sn(S_2O_7)_3$, which crystallizes as a trigonal structure with the lattice parameters $a = 1655.60 \pm 0.03$ and $c = 1026.22 \pm 0.02$ pm at 153 K, is formed in the Sn–Ag–O–S system (Logemann et al. 2012). The reaction for obtaining this compound was performed in a thick-walled glass ampoule. The tube was loaded with $SnCl_4$ (1 mM), oleum (1 mL, 65% SO_3), and Ag_2SO_4 (1 mM), torch-sealed under vacuum, and placed in a resistance furnace. The ampoule was maintained at 250°C for 24 h and cooled down to room temperature at a rate of 1.8°C·h^{-1}. The colorless crystals are very moisture-sensitive and were separated from oleum in a glove box.

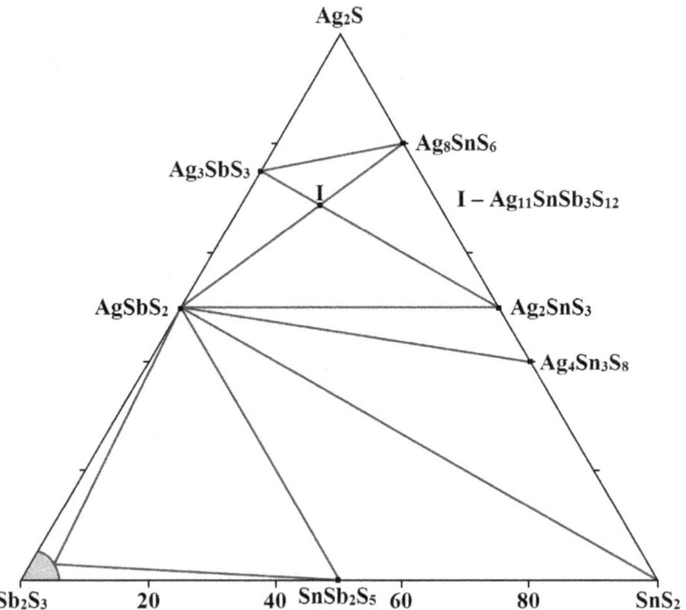

FIGURE B.7.14 Isothermal section of the SnS_2–Ag_2S–Sb_2S_3 quasiternary system at 230°C. (From Bereznyuk, O., et al., *Probl. khimii ta staloho rozvytku*, (4), 17, 2022.) Open access.

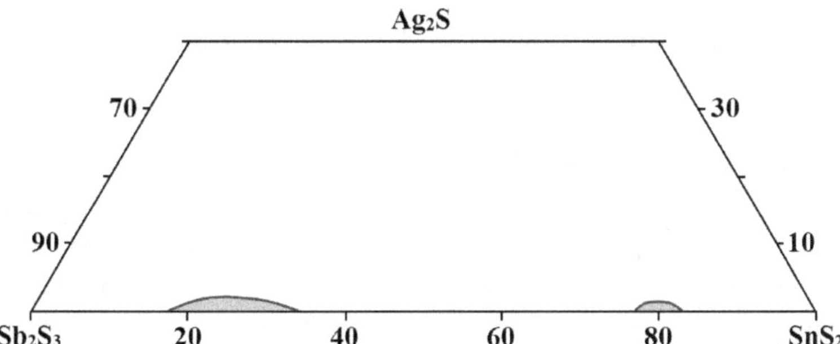

FIGURE B.7.15 Glass-forming region in the SnS_2–Ag_2S–Sb_2S_3 quasiternary system. (From Bereznyuk, O., et al., *Probl. khimii ta staloho rozvytku*, (4), 3, 2021.) Open access.

B.7.27 Tin–Silver–Bromine–Sulfur

The quaternary compound $Ag_6SnS_4Br_2$, which melts incongruently at 439°C and crystallizes as an orthorhombic structure with the lattice parameters $a = 667.050 \pm 0.010$, $b = 782.095 \pm 0.009$, and $c = 2314.04 \pm 0.03$ pm, is formed in the Sn–Ag–Br–S system (Mikolaĭchuk et al. 2010). The solid-phase synthesis of this compound was performed from the mixture $Ag_8SnS_6 + SnS_2 + 4AgBr$ in the quartz container preliminarily evacuated to ~1 Pa at 430°C followed by quenching in air. The alloy dissociation was suppressed by additional introduction of sulfur and $SnBr_2$ in amounts of 12 and 14 mg·cm^3 of the ampoule-free volume, respectively. The quaternary compound Ag_7SnS_5Br is not formed in the system.

B.7.28 Tin–Gold–Barium–Sulfur

The quaternary compound Au_2BaSnS_4, which is formed in this system, crystallizes as an orthorhombic crystal system ($a = 663.87 \pm 0.03$, $b = 1106.05 \pm 0.07$, and $c = 1096.76 \pm 0.06$ pm, and a calculated density of 6.418 g·cm^{-3}) in space group $C222_1$ instead of $P2_12_12$ as reported earlier (Teske et al. 2020).

B.7.29 Tin–Gold–Lanthanum–Sulfur

The quaternary compound $Au_{1.94}La_6Sn_2S_{14}$, which crystallizes as a hexagonal structure with the lattice parameters $a = 1040.01 \pm 0.19$, $c = 601.78 \pm 0.14$ pm, and a calculated density of 5.597 g·cm^{-3} at 100 K, is formed in the Sn–Au–La–S system (Akopov et al. 2021). The powder of this compound was synthesized using pre-arc-melted precursor and sulfur without flux, while the crystals were grown using a salt flux synthesis from a pre-arc-melted precursor and sulfur. The precursor was prepared using pieces of La, Au, and Sn. The sample with a total mass of 0.6–1 g was weighed out in a molar ratio of La/Au/Sn = 6:1:2.05. A slight excess of Sn was used to account for its evaporation during arc melting. The sample was then placed in an arc-melter on a copper hearth along with an oxygen getter material (zirconium metal). The arc-melter chamber was later sealed and evacuated for 20 min followed by purging with argon; this process was repeated three times to ensure no oxygen was present in the chamber. During arc melting, the getter was melted first to ensure the absorbance of any trace oxygen, and then the samples were heated for 1000°C for ~ 1 min at a current of ~ 80 A until molten, allowed to solidify, flipped, and re-arced two more times to ensure homogeneity under a current of ~ 100 A. The precursor ingots were then crushed into a powder using a steel Plattner-style diamond crusher.

The single crystals of $Au_{1.94}La_6Sn_2S_{14}$ were grown using pre-arc-melted precursor, S powder, and KI as a flux. The ratio of (precursor + sulfur) to KI flux was kept at a 1: 30 ratio by mass (for 0.1 g of reactants 3 g of KI was used). The reactants and KI were added to a silica ampoule, which was sealed under vacuum. The samples were heated to 950°C-1050°C over 12 h, allowed to stand over 72–120 h, and then cooled down to room temperature over 8 h. The KI flux was washed with deionized water, and the sample was then filtered under vacuum and dried. The crystals have a transparent yellow-gold appearance.

The bulk powder of the compound was synthesized using pre-arc melted precursor and sulfur powder. The sample was prepared by loading the La/Au/Sn precursor and sulfur in a fused silica ampoule in a 1:14 precursor to sulfur ratio by mass. The ampoule was then flame-sealed under vacuum. The sample was heated over 12 h to 950°C–1050°C, maintained for 72–120 h and then allowed to cool to room temperature over 8 h.

B.7.30 Tin–Magnesium–Oxygen–Sulfur

The quaternary compound $MgSn(SO_4)_3$, which crystallizes as a rhombohedral structure with the lattice parameters $a = 892.2 \pm 0.3$ pm and $\alpha = 55°17' \pm 0.06'$ or $a = 827.9$ and $c = 2260.0$ pm in the hexagonal setting and the calculated and experimental densities of 3.1 and 3.20 g·cm^{-3}, is formed in the Sn–Mg–O–S system (Perret and Couchot 1973; Perret 1975). This compound was obtained by oxidizing $SnCl_2{\cdot}2H_2O$ in H_2SO_4 and heated to 150°C (Patel 1953; Perret and Couchot 1973; Perret 1975). After cooling, $MgSO_4{\cdot}7H_2O$ was added. The solution, subjected to strong agitation, is then quickly brought to 250°C–270°C. Double salt begins to precipitate at 230°C. By maintaining heating for approximately 2 h, the final product appears to have better crystallinity. The precipitate was filtered through sintered glass, washed thoroughly with absolute alcohol, then with anhydrous ether, and finally carefully dried under vacuum at 250°C. A white powder of $MgSn(SO_4)_3$ was obtained. It is very hygroscopic and hydrolyzed by water.

B.7.31 Tin–Barium–Zinc–Sulfur

The quaternary compound $BaZnSnS_4$, which melts congruently at 836°C and crystallizes as an orthorhombic structure with the lattice parameters $a = 2139.8 \pm 0.3$, $b = 2186.5 \pm 0.3$, $c = 1270.7 \pm 1.7$ pm, a calculated density of 4.019 g·cm^{-3}, and energy gap of 3.25 eV, is formed in the Sn–Ba–Zn–S system (Li et al. 2021b). Single crystals of this compound were synthesized using Ba, ZnS, Sn, and S (molar ratio 1:1:1:3) with a total mass of 500 mg as the starting materials, and additional 400 mg KI as the flux. After being ground to fine powder and pressed into pellets, the mixture was loaded into the quartz tube and further evacuated to 0.01 Pa, and then sealed by flame. The tube was placed into a muffle furnace, heated from room temperature to 750°C, maintained at this temperature for 5 days, and finally cooled down to 400°C at a rate of 5°C·h^{-1}. The yellow crystals of the title compound were obtained, which are stable in water and air.

B.7.32 Tin–Zinc–Oxygen–Sulfur

The quaternary compound $ZnSn(SO_4)_3$, which crystallizes as a rhombohedral structure with the lattice parameters a = 894.1 pm and α = 55°03' or a = 826.2 and c = 2268.4 pm in the hexagonal setting, is formed in the Sn–Zn–O–S system (Perret and Couchot 1973). This compound was prepared in the same way as $MgSn(SO_4)_3$ but using $ZnSO_4{\cdot}7H_2O$ instead of $MgSO_4{\cdot}7H_2O$ (Patel 1953; Perret and Couchot 1973). A white powder of this compound was obtained. It is also very hygroscopic and hydrolyzed by water.

B.7.33 Tin–Cadmium–Oxygen–Sulfur

The quaternary compound $CdSn(SO_4)_3$, which crystallizes as a rhombohedral structure with the lattice parameters a = 923.1 pm and α = 53°24' or a = 829.6 and c = 2367.5 pm in the hexagonal setting, is formed in the Sn–Cd–O–S system (Perret and Couchot 1973). This compound was prepared in the same way as $MgSn(SO_4)_3$ but using anhydrous $CdSO_4$ instead of $MgSO_4{\cdot}7H_2O$ (Patel 1953; Perret and Couchot 1973). A white powder of this compound was obtained. It is also very hygroscopic and hydrolyzed by water.

B.7.34 Tin–Cadmium–Chlorine–Sulfur

The quaternary compound $CdSnSCl_2$, which crystallizes as an orthorhombic structure with the lattice parameters $a = 401.4 \pm 0.4$, $b = 1299.6 \pm 0.2$, $c = 947.1 \pm 0.9$ pm, a calculated density of 4.491 $g{\cdot}cm^{-3}$, and energy gap of 2.87 eV, is formed in the Sn–Cd–Cl–S system (Ran et al. 2021b). This compound was synthesized by mixing of SnS and $CdCl_2$ in a molar ratio of 1:1. The starting materials were sealed in evacuated silica tube, slowly heated to 550°C for 130 h, and maintained at that temperature for 200 h. Then, the samples were slowly cooled to 150°C at a rate of 2°$C{\cdot}h^{-1}$ (200 h) and then allowed to cool naturally to room temperature. Yellow irregular plate crystals of the title compound were finally obtained.

B.7.35 Tin–Cadmium–Bromine–Sulfur

The quaternary compound $CdSnSBr_2$, which crystallizes as an orthorhombic structure with the lattice parameters $a = 406.4 \pm 0.2$, $b = 1374.6 \pm 0.3$, $c = 962.1 \pm 0.2$ pm, a calculated density of 5.227 $g{\cdot}cm^{-3}$, and energy gap of 2.72 eV, is formed in the Sn–Cd–Br–S system (Ran et al. 2021b). This compound was prepared in the same way as $CdSnSCl_2$ but using $CdBr_2$ instead of $CdCl_2$.

B.7.36 Tin–Oxygen–Manganese–Sulfur

The quaternary compound $MnSn(SO_4)_3$, which crystallizes as a rhombohedral structure with the lattice parameters a = 908.9 pm and α = 54°09' or a = 827.4 and c = 2319.7 pm in the hexagonal setting, is formed in the Sn–O–Mn–S system (Perret and Couchot 1973). This compound was prepared in the same way as $MgSn(SO_4)_3$ was synthesized using anhydrous $MnSO_4$ instead of $MgSO_4{\cdot}7H_2O$ (Patel 1953; Perret and Couchot 1973). A white powder of this compound was obtained. It is also very hygroscopic and hydrolyzed by water.

B.7.37 Tin–Oxygen–Cobalt–Sulfur

The quaternary compound $CoSn(SO_4)_3$, which crystallizes as a rhombohedral structure with the lattice parameters a = 897.3 pm and α = 54°35' or a = 823.1 and c = 2283.4 pm in the hexagonal setting, is formed in the Sn–O–Co–S system (Perret and Couchot 1973). This compound was prepared in the same way as $MgSn(SO_4)_3$ was synthesized using anhydrous $CoSO_4{\cdot}7H_2O$ instead of $MgSO_4{\cdot}7H_2O$ (Patel 1953; Perret and Couchot 1973). A purplish pink powder of this compound was obtained. It is also very hygroscopic and hydrolyzed by water.

B.7.38 Tin–Oxygen–Nickel–Sulfur

The quaternary compound $NiSn(SO_4)_3$, which crystallizes as a rhombohedral structure with the lattice parameters a = 889.7 pm and α = 55°05' or a = 822.7 and c = 2256.9 pm in the hexagonal setting, is formed in the Sn–O–Ni–S system (Perret and Couchot 1973). This compound was prepared in the same way as $MgSn(SO_4)_3$ but using anhydrous $NiSO_4 \cdot 7H_2O$ instead of $MgSO_4 \cdot 7H_2O$ (Patel 1953; Perret and Couchot 1973). An yellow powder of this compound was obtained. It is also very hygroscopic and hydrolyzed by water.

B.8 Systems Based on Tin Selenides

B.8.1 Tin–Hydrogen–Nitrogen–Selenium

The quaternary compound $Sn_2Se_6(N_2H_4)_3(N_2H_5)_4$, which crystallizes as a triclinic structure with the lattice parameters a = 664.75 ± 0.06, b = 954.74 ± 0.09, c = 988.30 ± 0.10 pm, α = 94.110 ± 0.002°, β = 99.429 ± 0.002°, γ = 104.141 ± 0.002° and a calculated density of 2.618 g·cm^{-3}, is formed in the Sn–H–N–Se system (Mitzi 2005; Yuan and Mitzi 2009). Yellow crystals of this compound were formed by dissolving (in a nitrogen atmosphere, over several minutes, with stirring) of $SnSe_2$ (1 mM) in 1 mL of hydrazine and Se (1 mM), yielding, ultimately, a light-yellow solution. Evaporating the solution under flowing nitrogen gas over a period of several hours leads to the formation of a yellow crystalline powder. The powder loses weight while sitting on a balance at room temperature, reflecting a loss of solvent (i.e., hydrazine) from within the structure.

B.8.2 Tin–Lithium–Magnesium–Selenium

The quaternary compound $Li_2MgSnSe_4$, which crystallizes as an orthorhombic structure with the lattice parameters a = 840.2 ± 1.4, b = 718.1 ± 1.2, c = 672.8 ± 1.1 pm, a calculated density of 3.867 g·cm^{-3}, and an energy gap of 2.62 eV, is formed in the Sn–Li–Mg–Se system (Gao et al. 2021a). Single crystals of this compound were prepared using a melting method in a sealed quartz tube. The starting mixture (molar ratio of Li/Mg/Sn/Se = 2:1:1:4) was packaged in a graphite crucible in a glove box. After that, the graphite crucible was moved into a quartz tube, which was flame-sealed under a vacuum about 10^{-3} Pa. Then, the sample was heated to 880°C in 46 h, kept at 880°C for 50 h, and then cooled to room temperature in 48 h. Breaking the tube, the yellow $Li_2MgSnSe_4$ single crystals were harvested in the graphite crucible. It is worth mentioning that these crystals show strong moisture absorptions in air.

The syntheses of $Li_2MgSnSe_4$ powder sample was tried at a higher temperature. The mixtures of Li, Mg, Sn, and Se with a molar stoichiometric ratio were first weighed, ground, and sealed in quartz tube. The sealed sample was slowly heated to 900°C in 60 h in a muffle furnace, kept at this temperature for 100 h, and then cooled to room temperature in 100 h.

B.8.3 Tin–Lithium–Mercury–Selenium

The quaternary compound $Li_2HgSnSe_4$, which melts congruently at 665°C and crystallizes as an orthorhombic structure with the lattice parameters a = 1445.5 ± 0.2, b = 837.72 ± 0.10, c = 682.06 ± 0.08 pm, a calculated density of 5.219 g·cm^{-3}, and an energy gap of 1.93 eV, is formed in the Sn–Li–Hg–Se system (Gao et al. 2022b). To prepare this compound, a mixture of Li (2 mM), HgSe (1 mM), $SnSe_2$ (1 mM), and Se (1 mM) was loaded into a graphite crucible that avoids the harmful reaction between Li metal and tube wall and then put the graphite crucible into a vacuum-sealed silica tube. The silica tube was firstly heated to 700°C in 50 h, kept at 700°C for about 100 h, and slowly cooled down to room temperature within 5 days. DMF solvent was used to wash the final products. Many red crystals of the title compound were found in the tube. As for air-unstable Li, an Ar-filled glove box was used to complete the whole preparation process.

The crystal growth process of this compound was completed with a two-zone Bridgman–Stockbarger furnace. A tapered vacuum-sealed silica tube containing polycrystalline $Li_2HgSnSe_4$ (~ 6 g) was placed in the top zone of the furnace. The upper and bottom zones were heated to 680°C and 620°C within the furnace with a temperature gradient of~7–10°C·cm^{-1} at the crystallization point of $Li_2HgSnSe_4$. The crucible was lowered at a rate of ~2–3 mm per day. Flake-like single crystals could be picked from the final polycrystalline ingot.

B.8.4 Tin–Sodium–Copper–Selenium

Two quaternary compounds, $Na_3CuSnSe_4$ and $Na_6Cu_8Sn_3Se_{13}$, are formed in the Sn–Na–Cu–Se system. The first compound crystallizes as a monoclinic structure with the lattice parameters $a = 680.83 \pm 0.16$, $b = 1023.3 \pm 0.2$, $c = 1303.3 \pm 0.3$ pm, $\beta = 90.419 \pm 0.005°$, a calculated density of 4.148 g·cm^{-3}, and an energy gap of 2.07 eV (Yang et al. 2020b) [$a = 691$, $b = 1032$, $c = 1480$ pm, $\beta = 117.33°$, a calculated density of 4.02 g·cm^{-3}, and an energy gap of 0.58 eV (Klepp and Fabian 1997)]. To synthesize this compound, Na and Cu_2Se, Sn, and S powders were used. In order to ensure the reliability of raw material ratio, vacuum glove box was chosen to complete the weighing process and eliminate the effect of air oxidation. The raw mixture with a stoichiometric proportion of elements was firstly loaded into the graphite crucible and then the crucible was put into the silica tube. Using the flame gun and air extractor, silica tube was carefully vacuum-sealed with the internal vacuum of ~ 10^{-3} Pa in ampoule. Muffle furnace was used to complete the crystallization reaction and the setting temperature process was carries out as follows: firstly heated up to 850°C in 30 h to ensure the mixture melt completely and kept at this temperature within 4 days, then slowly cooled down to 200°C in 100 h, and finally quickly down to the room temperature in one day. Finally, many sub-millimeter-level red crystals were found under the optical microscope after washing with DMF solution. They easily absorb moisture in the air within several minutes.

$Na_6Cu_8Sn_3Se_{13}$ is stable up to approximately 633°C and crystallizes as a cubic structure with the lattice parameter $a = 1847.54 \pm 0.08$, a calculated density of 4.274 g·cm^{-3}, and an energy gap of 1.76 eV (Teri et al. 2021). This compound was synthesized at 160°C for 7 days in a thick Pyrex tube from the batch of Na_2CO_3 (0.10 mM), $CuCl_2{\cdot}2H_2O$ (0.10 mM), Sn powder (0.13 mM), Se powder (0.42 mM), and a mixed solvent of 1,4-diaminobutane (7.52 mM) and deionized water (6.88 mM). The products were washed with ethanol and deionized water to obtain black cubic crystals.

B.8.5 Tin–Sodium–Silver–Selenium

The quaternary compound $Na_3AgSnSe_4$, which crystallizes as a monoclinic structure with the lattice parameters $a = 853.4 \pm 0.7$, $b = 675.8 \pm 0.6$, $c = 1657.4 \pm 1.4$ pm, $\beta = 104.106 \pm 0.011°$, a calculated density of 4.380 g·cm^{-3}, and an energy gap of 1.98 eV, is formed in the Sn–Na–Ag–Se system (Yang et al. 2020b). This compound was prepared in the same way as $Na_3CuSnSe_4$ but using Ag block instead of Cu_2Se powder.

B.8.6 Tin–Sodium–Zinc–Selenium

The quaternary compound $Na_2ZnSn_2Se_6$, which crystallizes as an orthorhombic structure with the lattice parameters $a = 2460.51 \pm 0.07$, $b = 1351.64 \pm 0.04$, $c = 760.31 \pm 0.02$ pm, a calculated density of 4.321 g·cm^{-3}, and an energy gap of 2.05 eV, is formed in the Sn–Na–Zn–Se system (Ye et al. 2020). For the synthesis of the target compound, a mixture of the starting materials Na_2Se, Zn, Sn, and Se in a molar ratio of 1:1:2:6 was loaded into a graphite crucible and placed in a quartz tube. The tube was flame-sealed under vacuum (~0.01 Pa) and then placed in a muffle furnace, heated from room temperature to 600°C in 40 h, kept at that temperature for 96 h, and then cooled to room temperature with a rate of 4°C·h^{-1}. The product was washed with water and dried with ethanol. The pure red crystals of $Na_2ZnSn_2Se_6$ were manually picked out.

B.8.7 Tin–Sodium–Cadmium–Selenium

The quaternary compound $Na_2CdSn_2Se_6$, which crystallizes as an orthorhombic structure with the lattice parameters $a = 2519.84 \pm 0.14$, $b = 1349.33 \pm 0.09$, $c = 767.33 \pm 0.05$ pm, a calculated density of 4.427 g·cm^{-3}, and an energy gap of 2.16 eV, is formed in the Sn–Na–Cd–Se system (Ye et al. 2020). This compound was synthesized in the same way as $Na_2ZnSn_2Se_6$ but using Cd instead of Zn. The pure red crystals of $Na_2CdSn_2Se_6$ were manually picked out.

B.8.8 Tin–Sodium–Mercury–Selenium

The quaternary compound $Na_2Hg_3Sn_2Se_8$, which crystallizes as a tetragonal structure with the lattice parameters $a = 937.49 \pm 0.07$, $c = 979.03 \pm 0.15$ pm, a calculated density of 5.854 g·cm^{-3}, and an energy gap of 1.88 eV, is formed in the Sn–Na–Hg–Se system (Gao et al. 2022b). To synthesize this compound, a mixture of Na (2 mM), HgSe (3 mM), $SnSe_2$ (2 mM), and Se (1 mM) was loaded into a graphite crucible that avoids the harmful reaction between Na metal and tube wall and then the graphite crucible was put into vacuum-sealed silica tube. The silica tube was firstly heated to 600°C in 50 h, kept at 600°C for about 100 h, and slowly cooled down to room temperature within 5 days. The final products were carefully washed with DMF and red microcrystals appeared in the tube. As for air-unstable Na block, an Ar-filled glove box was used to complete the whole preparation process.

B.8.9 Tin–Potassium–Mercury–Selenium

The quaternary compound $K_2HgSnSe_4$, which is stable up to approximately 393°C and crystallizes as a tetragonal structure with the lattice parameters $a = 805.31 \pm 0.05$, $c = 695.01 \pm 0.08$ pm, a calculated density of 5.256 g·cm^{-3}, and an energy gap of 2.01 eV, is formed in the Sn–K–Hg–Se system (Teri et al. 2021). This compound was synthesized at 100°C for 7 days in a thick Pyrex tube from the batch of K_2CO_3 (0.049 mM), $Hg(CH_3COO)_2$ (0.049 mM), Sn powder (0.045 mM), and Se powder (0.20 mM) in a mixed solvent of ethylenediaminee (0.43 mM) and ethanol (0.47 mM). The products were washed with ethanol and deionized water to obtain orange block crystals.

B.8.10 Tin–Copper–Barium–Selenium

The quaternary compound $Cu_2BaSnSe_4$, which crystallizes as an orthorhombic structure with the lattice parameters $a = 1111.05 \pm 0.02$, $b = 1122.75 \pm 0.02$, $c = 674.36 \pm 0.01$ pm, and an energy gap of 1.72 eV, is formed in the Sn–Cu–Ba–Se system (Shin et al. 2016). To prepare this compound, a stoichiometric mixture of BaSe, SnSe, and CuSe were carefully ground/homogenized and cold-pressed into pellet inside a nitrogen-filled glove box. The pellet was placed inside a quartz tube, which was flame-sealed under dynamic vacuum (~7×10^{-5} Pa). The reaction mixture was then heated to 650°C during 5–6 h inside a box furnace, and kept at this temperature for 10–15 h. After this step, the reaction was cooled to room temperature by switching off the furnace.

B.8.11 Tin–Copper–Zinc–Selenium

The quaternary compound $Cu_2ZnSnSe_4$, which crystallizes as a tetragonal structure with the lattice parameters $a = 568.84 \pm 0.03$, $c = 1134.72 \pm 0.13$ pm, and a calculated density of 5.671 g·cm^{-3} for the single crystal obtaining by the Bridgman method and $a = 569.54 \pm 0.17$, $c = 1134.75 \pm 0.06$ pm, and a calculated density of 5.657 g·cm^{-3} for the single crystal grown by the chemical transport reaction method, is formed in the Sn–Cu–Zn–Se system (Nateprov et al. 2013).

B.8.12 Tin–Copper–Lead–Selenium

$SnSe_2$–Cu_2Se–PbSe. It was confirmed that in this system, the Cu_2SnSe_3–PbSe section is quasibinary and the alloys directly adjacent to $SnSe_2$ contain not three but four phases, which means that the tetrahedration of the Sn–Cu–Pb–Se system does not pass through $SnSe_2$, but through SnSe (Olekseyuk et al. 2021).

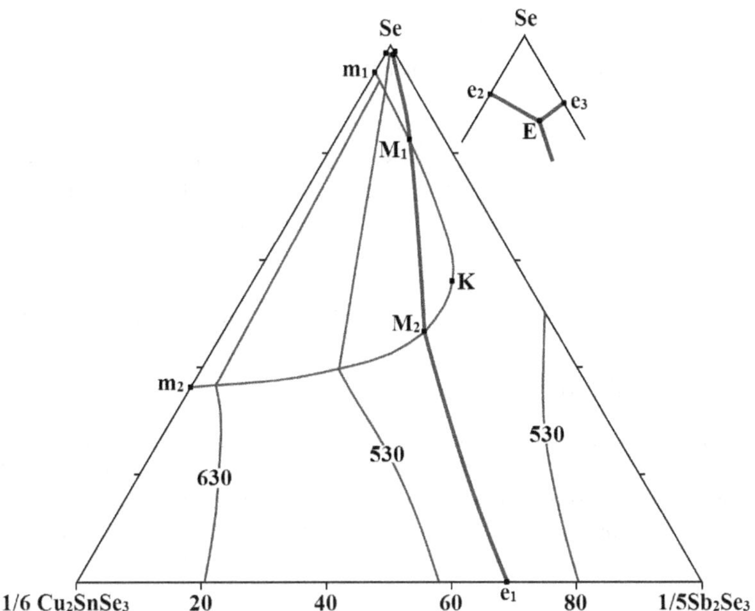

FIGURE B.8.1 Liquidus surface of the Cu_2SnSe_3–Sb_2Se_3–Se quasiternary system. (From Ismayilova, E.N., et al., *Condens. Matter Interphases*, **25**(1), 47, 2023.)

B.8.13 Tin–Copper–Antimony–Selenium

Cu_2SnSe_3–Sb_2Se_3–Se. The liquidus surface of this system (Figure B.8.1) consists of three fields of primary crystallization of solid solutions based on Cu_2SnSe_3 and Sb_2Se_3 as well as of elemental selenium (Ismayilova et al. 2023). The region of primary crystallization of Se is degenerated. A wide region of immiscibility exists in the system from the side of the Cu_2SnSe_3–Se system. Ternary eutectic *E* crystallizes at 217°C. Three vertical sections of this system were also constructed. The alloys for the investigations were first annealed at 380°C for 200 h and then at 180°C for 300 h.

B.8.14 Tin–Silver–Lead–Selenium

$SnSe_2$–Ag_2Se–PbSe. It was confirmed that in this system the Ag_8SnS_6–PbSe section is quasibinary and the alloys directly adjacent to $SnSe_2$ contain not three but four phases, which means that the tetrahedration of the Sn–Cu–Pb–Se system does not pass through $SnSe_2$, but through SnSe.

B.8.15 Tin–Strontium–Zinc–Selenium

The quaternary compound $SrZnSnSe_4$, which crystallizes as an orthorhombic structure with the lattice parameters $a = 2171.58 \pm 0.18$, $b = 2186.25 \pm 0.18$, $c = 1312.70 \pm 0.11$ pm, a calculated density of 5.009 g·cm^{-3}, and an energy gap of 1.82 eV, is formed in the Sn–Sr–Zn–Se system (Hou et al. 2022). To synthesize this compound, SrSe, Zn, Sn, and Se were stored in an Ar-filled glove box without oxygen and moisture, were ground, and then loaded into a silica tube coated with carbon in the molar radio of 1:1:1:3 to obtain the powder of $SrZnSnSe_4$ through high-temperature solid state reaction. The prepared powder was mixed with flux KI in the mass ratio of 1:1 and then loaded into the graphitic crucible placed in the silica tube. Next, the tube was sealed under vacuum (< 10^{-3} Pa) and put into a programmable muffle furnace with the following controlled heating ramp: heated to 800°C within 30 h, kept for two days at this temperature, and slowly cooled to room temperature at a rate of 3°C·h^{-1}. In the end, the product was cleaned with distilled water to get rid of the redundant KI and the red crystals that are stable in air were obtained.

B.8.16 Tin–Barium–Zinc–Selenium

The quaternary compound $BaZnSnSe_4$, which melts congruently at 762°C and crystallizes as an orthorhombic structure with the lattice parameters $a = 2224.0 \pm 0.7$, $b = 2270.5 \pm 0.7$, $c = 1319.7 \pm 0.4$ pm, a calculated density of 5.081 g·cm^{-3}, and energy gap of 1.88 eV, is formed in the Sn–Ba–Zn–Se system (Li et al. 2021b). Single crystals of this compound was synthesized using Ba, ZnSe, Sn, and Se (molar ratio 1:1:1:3) with a total mass of 500 mg as the starting materials and an additional 400 mg of KI as the flux. After being ground to fine powder and pressed into pellets, the mixture was loaded into a quartz tube and further evacuated to 0.01 Pa, and then sealed by flame. The tube was placed in a muffle furnace, heated from room temperature to 950°C, maintained at this temperature for 5 days, and finally cooled down to 400°C at a rate of 5°C·h^{-1}. The red crystals of the title compound were obtained, which are stable in water and air.

B.8.17 Tin–Barium–Gallium–Selenium

Two mare quaternary compounds, $Ba_8Ga_2Sn_7Se_{18}$ and $Ba_{10}Ga_2Sn_9Se_{22}$, are formed in the Sn–Ba–Ga–Se system (Li et al. 2017). Both compounds are stable up to 800°C under a nitrogen atmosphere. They crystallize in the orthorhombic structure with the lattice parameters $a = 1246.6 \pm 0.9$, $b = 935.8 \pm 0.6$, $c = 1814 \pm 2$ pm, a calculated density of 5.5 g·cm^{-3}, and energy gap of 1.65 eV for the first compound and $a = 938.4 \pm 0.2$, $b = 4483.4 \pm 0.6$, $c = 1241.6 \pm 0.2$ pm, a calculated density of 5.1 g·cm^{-3}, and energy gap of 1.67 eV for the second one.

The black block-shaped crystals of $Ba_8Ga_2Sn_7Se_{18}$ were first discovered by a reaction of Ba, Ga, Sn, and Se (molar ratio 7:2:3:14) with a total mass of 400 mg. The reactants were loaded into a graphite crucible, sealed in an evacuated silica tube under a vacuum of 0.01 Pa, then heated to 980°C within 50 h, kept there for 48 h, and then cooled to 400°C at a rate of 5°·C.h^{-1}. Black crystals of the title compound and $Ba_6Sn_6Se_{13}$ as a secondary phase together with a few red crystals of $Ba_5Ga_2Se_8$ were produced. Therefore, binary BaSe, Ga_2Se_3, and SnSe were used instead of chemical elements. The mixture of BaSe, Ga_2Se_3, and SnSe in a molar ratio of 8:1:7 with a total mass of about 500 mg was ground carefully and pressed into a pellet. Then the pellet was put into a graphite crucible in an evacuated silica jacket and subsequently heated to 750°C in 34 h, kept for 80 h, and then cooled to 300°C within 100 h before the furnace was turned off. The homogenous $Ba_8Ga_2Sn_7Se_{18}$ was produced.

The black block-shaped crystals of $Ba_{10}Ga_2Sn_9Se_{22}$ were initially obtained from a reaction of Ba, Ga, Sn, and Se in a 13:2:10:26 molar ratio with a total weight of 300 mg. And the pure phase was obtained by the reaction of cold-pressed pellet of the well-ground mixture of binary BaSe, Ga_2Se_3, and SnSe in a 10:1:9 molar ratio with a total weight of about 500 mg. The heating profile was heating to 750°C in 50 h, dwelling for 80 h, and then cooling to 300°C at a rate of 4°C·h^{-1}. The homogeneous crystals of $Ba_{10}Ga_2Sn_9Se_{22}$ were obtained. They are relatively moisture-sensitive.

B.8.18 Tin–Barium–Manganese–Selenium

The quaternary compound $BaMnSnSe_4$, which crystallizes as an orthorhombic structure with the lattice parameters $a = 2231.43 \pm 0.10$, $b = 2270.57 \pm 0.11$, $c = 1345.23 \pm 0.06$ pm, and a calculated density of 4.887 g·cm^{-3}, is formed in the Sn–Ba–Mn–Se system (Assoud et al. 2011). To synthesize this compound, the elements in the stoichiometric ratio were loaded in a silica tube, which was evacuated and sealed under dynamic vacuum, and then placed into a temperature-controlled furnace. The silica tube was heated to 850°C within 24 h, kept at 850°C for a period of 4 days, and then cooled to 200°C within 8 days. Thereafter, the furnace was turned off. The sample looked homogeneous, comprising mostly microcrystalline red powder. Because of the air-sensitive nature of Ba, all elements were handled in an argon-filled glove box.

B.8.19 Tin–Tellurium–Nickel–Selenium

The Sn–Te–Ni–Se phase diagram at 400°C was proposed by Musa and Chen (2021). The interfacial reactions in the $Ni/SnTe_{0.9}Se_{0.1}$ couple were systematically investigated at 400°C. It was shown that the reaction rate decreases with lower reaction temperatures and the reaction zone grows thicker with longer reaction time.

B.9 Systems Based on Tin Telluride

B.9.1 Tin–Lithium–Oxygen–Tellurium

The quaternary compound Li_2SnTeO_6, which crystallizes as an orthorhombic structure with the lattice parameters $a = 519.2 \pm 0.2$, $b = 492.7 \pm 0.2$, and $c = 851.3 \pm 0.5$ pm, is formed in the Sn–Li–O–Te system (Choisnet et al. 1989). The thermal stability of this compound is moderate: after 1 day at 800°C, it is transformed into other phases, whose nature has not been investigated. Pure Li_2SnTeO_6 may be obtained by a two-step procedure: Li_2SnO_3 was first prepared by a solid state reaction between Li_2CO_3 and SnO_2 (2 days at 1000°C); it was then mixed with a small excess (5–10 mol%) of TeO_2, the mixture was progressively heated up to 600°C, and then annealed for 2 days at 700°C.

B.9.2 Tin–Sodium–Antimony–Tellurium

The quaternary compound $NaSn_{18}SbTe_{20}$, which melts at 801°C and crystallizes as a cubic with the lattice parameter $a = 632.5 \pm 0.1$ pm and an energy gap of 0.1–0.15 eV, is formed in the Sn–Na–Sb–Te system (Guéguen et al. 2009). This compound was prepared as polycrystalline ingots in silica tube by mixing high-purity Na, Sn, Sb, and Te in the appropriate stoichiometric ratio. To prevent reaction between the sodium metal and silica, the tube was carbon-coated prior to use. All components (except Na) were loaded into silica tube under ambient atmosphere and the corresponding amount of Na was later added under an inert atmosphere in a dry glove box. The silica tube was then flame-sealed under a residual pressure of ~0.01 Pa, placed into a tube furnace (mounted on a rocking table), and heated at 980°C for 4 h to allow complete melting of all the components. While molten, the furnace was allowed to rock for 2 h to facilitate complete mixing and homogeneity of the liquid phase. The furnace was finally immobilized at the vertical position and was cooled from 980°C to 550°C over 43 h followed by a faster cool (6–8 h) to room temperature. The resulting ingot generally was silvery-metallic in color with a smooth surface.

B.9.3 Tin–Sodium–Bismuth–Tellurium

The quaternary compound $NaSn_{18}BiTe_{20}$, which melts at 811°C and crystallizes as a cubic structure with the lattice parameter $a = 633.7 \pm 0.1$ pm and an energy gap of 0.19 eV, is formed in the Sn–Na–Bi–Te system (Guéguen et al. 2009). It was prepared in the same way as $NaSn_{18}SbTe_{20}$ but using Bi instead of Sb.

B.9.4 Tin–Copper–Cadmium–Tellurium

The quaternary compound $Cu_2CdSnTe_4$, which crystallizes as a tetragonal structure with the lattice parameters $a = 619.8 \pm 0.1$ and $c = 1225.6 \pm 0.3$ pm, is formed in the Sn–Cu–Cd–Te system (Dong et al. 2014). This compound begins to decompose at 350°C and completely decomposes to CdTe, CuTe, SnO_2, and Te at 475°C. Hot pressing at 300°C and 150 MPa for 3 h under a N_2 flow resulted in a polycrystalline specimen with a density of 98% of the theoretical density. $Cu_2CdSnTe_4$ was synthesized by direct reaction of the elements. Cd shot and Cu, Sn, and Te powders were loaded into a silica ampoule, in a stoichiometric ratio of 2:1:1:4 and then sealed in a quartz tube under vacuum and heated to 700°C for 4 days before cooling to room temperature. The product was then ground into fine powder, cold-pressed into a pellet, and annealed at 350°C for 1 week.

B.9.5 Tin–Lead–Bismuth–Tellurium

SnTe–PbTe–Bi_2Te_3. 3D modeling and visualization of the liquidus and solidus surfaces of different phases of this quasiternary system were performed by Aghazade et al. (2023). The obtained analytical dependences of the liquidus temperatures on the composition made it possible to visualize the crystallization surfaces of all phases separately and all on one graph.

B.9.6 Tin–Antimony–Manganese–Tellurium

$SnSb_2Te_4$–$MnSb_2Te_4$. This system is nonquasibinary section of the SnTe–Sb_2Te_3–MnTe quasiternary system due to the incongruent melting of both compounds, but is stable in the subsolidus area where the continuous $Sn_xMn_{1-x}Sb_2Te_4$ solid solutions exist (Orujlu et al. 2022). The lattice parameters of these solid solutions increase linearly with increasing Sn content following Vegard's law.

$SnSb_4Te_7$–$MnSb_4Te_7$. This system is also the nonquasibinary section of the SnTe–Sb_2Te_3–MnTe quasiternary system as both compounds melt incongruently (Orujlu et al. 2022). The continuous $Sn_xMn_{1-x}Sb_4Te_7$ solid solutions are formed in the subsolidus area. The concentration dependence of the lattice parameters of these solid solutions is in good agreement with Vegard's law.

The SnTe–$MnSb_2Te_4$ and $SnSb_2Te_4$–MnTe systems are also nonquasibinary (Orujlu et al. 2022).

SnTe–Sb_2Te_3–MnTe. The isothermal section of this quasiternary system at 500°C is presented in Figure B.9.1 (Orujlu et al. 2022). Four regions of the solid solutions exist in the system at this temperature. There are also five two-phase and one three-phase regions in the system.

Six fields of the primary crystallization exist on the liquidus surface of this system (Figure B.9.2) (Orujlu et al. 2022). The primary crystallization of the low-temperature modification of MnTe occurs in a quite large area. The second-largest primary crystallization area belongs to the solid solution based on SnTe. There is one transition point on the liquidus surface: U (610°C) – L + α-MnTe ⇔(SnTe) $Sn_xMn_{1-x}Sb_2Te_4$.

All these systems were studied through DTA, XRD, and SEM. At the first stage, the MnTe-rich alloys were annealed at 700°C, while the SnTe-rich alloys were annealed at 600°C for one week. Then, the samples were subjected to the thermal treatment at 500°C for one month.

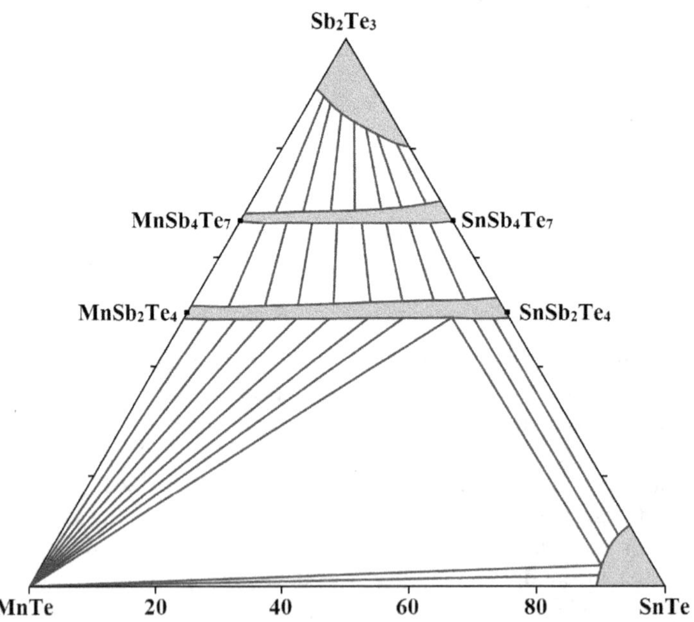

FIGURE B.9.1 Isothermal section of the SnTe–Sb_2Te_3–MnTe quasiternary system at 500°C. (From Orujlu, E.N., et al., *Calphad*, **76**, 102398, 2022.)

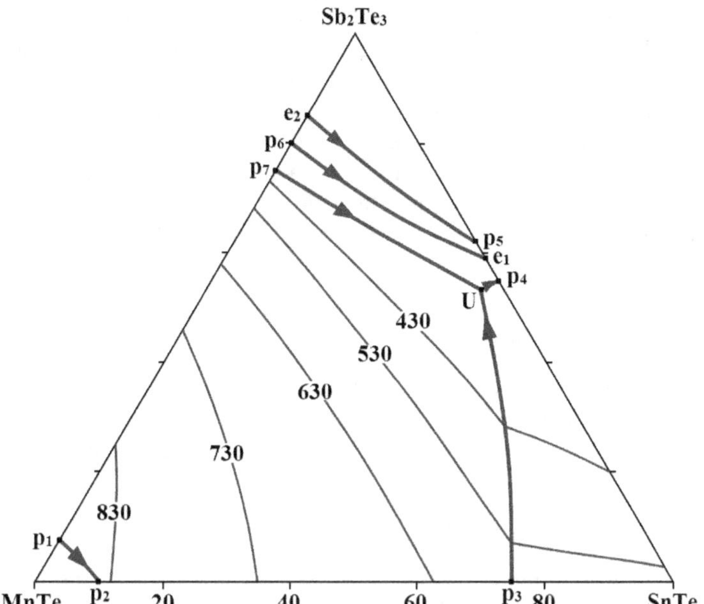

FIGURE B.9.2 Liquidus surface of the SnTe–Sb_2Te_3–MnTe quasiternary system. (From Orujlu, E.N., et al., *Calphad*, **76**, 102398, 2022.)

B.10 Systems Based on Lead Sulfide

B.10.1 Lead–Hydrogen–Oxygen–Sulfur

The quaternary compound $Pb_4(OH)_4][Pb(S_2O_3)_3]$, which has two polymorphic modifications, is formed in the Pb–H–O–S system (Kampf et al. 2023d). The first modification (mineral cubothioplumbite), crystallizes as a cubic structure with the lattice parameter $a = 1491.79 \pm 0.13$ pm and a calculated density of 5.748 g·cm^{-3} and the second one (mineral hexathioplumbite) crystallizes as a hexagonal structure with the lattice parameters $a = 1072.13 \pm 0.10$, $c = 865.41 \pm 0.06$ pm, and a calculated density of 5.531 g·cm^{-3}.

B.10.2 Lead–Copper–Dysprosium–Sulfur

The quaternary compound $CuDyPbS_3$, which crystallizes as an orthorhombic structure, is formed in the Pb–Cu–Dy–S system. According to Aliyev et al. (2023), it melts incongruently and its lattice parameters are as follows: $a = 1030$, $b = 394$, $c = 1290$ pm, and a calculated density 6.84 g·cm^{-3}.

B.10.3 Lead–Copper–Erbium–Sulfur

The quaternary compound $CuErPbS_3$, which crystallizes as an orthorhombic structure, is formed in the Pb–Cu–Er–S system. According to Aliyev et al. (2023), it melts incongruently and its lattice parameters are as follows: $a = 1026$, $b = 390$, $c = 1286$ pm, and a calculated density 6.90 g·cm^{-3}.

B.10.4 Lead–Silver–Phosphorus–Sulfur

The quaternary compound $Ag_7PbP_3S_{12}$, which is stable up to 392°C, has a phase transition at ~204°C, and crystallizes as a monoclinic structure with the lattice parameters $a = 1074.0 \pm 0.3$, $b = 989.9 \pm 0.3$, $c = 2107.8 \pm 0.5$ pm, $\beta = 118.678 \pm 0.011°$, a calculated density of 4.865 g·cm^{-3} at 25°C, and an energy gap of 2.09 eV and in hexagonal structure with the lattice parameters $a = 1072.1 \pm 0.3$, $c = 1008.3 \pm 0.5$

pm, and a calculated density of 4.766 g·cm^{-3} at 212°C, is formed in the Pb–Ag–P–S system (Fan et al. 2017). Orange-colored single crystals of this compound were obtained by solid state reactions. Pb (0.2 mM), Ag (1.4 mM), S (2.4 mM), and red phosphorus (0.6 mM) were ground into a fine powder in an agate mortar, pressed into a pellet, loaded into Pyrex tube, evacuated to 0.01 Pa, and flame-sealed. The tube was then placed into a computer-controlled furnace, heated from room temperature to 250°C at a rate of 50°C·h^{-1}, maintained at that temperature for 1 day, then heated to 600°C at a rate of 50°C·h^{-1}, maintained at that temperature for 4 days, and then slowly cooled to 250°C at a rate of 2.5°C·h^{-1}. It was finally cooled to room temperature in 12 h. Pure crystals of $Ag_7PbP_3S_{12}$ were hand-picked under a microscope.

B.10.5 Lead–Silver–Arsenic–Sulfur

The quaternary compound $AgPb_{18}As_{25}S_{56}$ (mineral interliveingite), which crystallizes as a monoclinic structure with the lattice parameters a = 840.90 ± 0.04, b = 791.14 ± 0.04, c = 7001.6 ± 0.3 pm, and β = 93.287 ± 0.002°, is formed in the Pb–Ag–As–S system (Topa et al. 2023c,d).

B.10.6 Lead–Silver–Bismuth–Sulfur

The quaternary compound $Ag_{2.25}Pb_{2.5}Bi_{4.25}S_{10}$, which crystallizes as an orthorhombic structure with the lattice parameters a = 408.4 ± 0.1, b = 1345.3 ± 0.4, c = 3393.2 ± 0.9 pm, and a calculated density of 7.03 g·cm^{-3}, is formed in the Pb–Ag–Bi–S system (Topa et al. 2010b). The crystals of this compound were grown using a Bridgman method from a sulfide melt produced by melting of a charge with approximate composition $AgPbBi_3S_6$ at 850°C.

B.10.7 Lead–Boron–Oxygen–Sulfur

Another quaternary compound, $Pb_6O_2(BO_3)SO_4$, which crystallizes as an orthorhombic structure with the lattice parameters a = 649.2 ± 0.1, b = 1164.5 ± 0.3, and c = 1790.4 ± 0.5 pm, is formed in the Pb–B–O–S system (Aurivillius 1983).

It was also shown that $Pb_4(BO_3)_2SO_4$ is a semiconductor with an energy gap of 3.65 eV (Xie et al. 2022a). This compound was synthesized by a high-temperature solid-phase method. PbO (4.4639 g), H_3BO_3 (0.6183 g), and $(NH_4)_2SO_4$ (0.6607 g) were fully mixed by ball milling. The mixture was poured into the crucibles and then put in the muffle furnace for calcination. To remove the carbonate and water, the mixture above was heated to 450°C at a heating rate of 5.3°C·min^{-1} and kept at that temperature for 2 h in air. Then, the temperature of the mixture was maintained up to 630°C for 6 h in the muffle furnace. After that, the product was cooled down and ground after calcination.

B.10.8 Lead–Gallium–Lanthanum–Sulfur

The quaternary compound $Ga_{1.6}La_3Pb_{0.1}S_7$, which crystallizes as a hexagonal structure with the lattice parameters a = 1019.02 ± 0.03, c = 606.61 ± 0.02 pm, and a calculated density of 4.6326 ± 0.0004 g·cm^{-3}, is formed in the Pb–Ga–La–S system (Blashko et al. 2022b). This compound was synthesized from the elements in a quartz container using a muffle furnace. The container was evacuated to 0.01 Pa and sealed. The step-by-step synthesis is carried out as follows: (1) heating the mixture to 600°C at a rate of 30°C·h^{-1}; (2) annealing for 100 h; (3) heating to 1100°C at a rate of 12°C·h^{-1}; (4) annealing for 2 h; (5) cooling to 500°C at a rate of 12°C·h^{-1}; and (6) homogenization annealing for 500 h. After reaching the equilibrium, the synthesized alloy was quenched in room-temperature water.

B.10.9 Lead–Gallium–Praseodymium–Sulfur

The quaternary compound $Ga_{1.6}Pr_3Pb_{0.1}S_7$, which crystallizes as a hexagonal structure with the lattice parameters a = 1000.34 ± 0.03, c = 605.87 ± 0.03 pm, and a calculated density of 4.8666 ± 0.0005 g·cm^{-3}, is formed in the Pb–Ga–Pr–S system (Blashko et al. 2022b). It was synthesized in the same way as $Ga_{1.6}La_3Pb_{0.1}S_7$ was prepared.

B.10.10 Lead–Indium–Bismuth–Sulfur

One more quaternary phase, $In_{8.38}Pb_5Bi_{1.62}S_{20}$, which also crystallizes as a monoclinic structure with the lattice parameters $a = 2706.5 \pm 0.4$, $b = 388.25 \pm 0.03$, $c = 1568.3 \pm 0.2$ pm, and $\beta = 103.59 \pm 0.01°$, is formed in the Pb–In–Bi–S system (Reis et al. 2012). To prepare this phase, a mixture of PbS (3.6 mM), In_2S_3 (1.8 mM), Bi_2S_3 (0.6 mM), and I_2 (40 mg, as a transport agent) was placed in a quartz ampoule of length ~20 cm and internal diameter 1.3 cm. The ampoule was sealed under vacuum and placed in a two-zone tube furnace. The end containing the reaction mixture (zone A) was heated to 670°C and the other end (zone C) to 600°C. After eight weeks, a carpet of comparatively large black needle-like crystals of the title phase had grown on bulk, which had formed in zone A. In the center of the ampoule (zone B), more of these crystals had been deposited. At the cool end of the ampoule (zone C), a conglomerate of crystals of a different phase had grown. The structural investigation and EPMA showed them to be $In_{3.7}Pb_4Bi_{2.3}S_{13}$, a compositional variant of $In_2Pb_4Bi_4S_{13}$.

B.10.11 Lead–Thallium–Arsenic–Sulfur

The quaternary compound $TlPb_{14}As_{17}S_{40}$ (mineral buynite), which crystallizes as a monoclinic structure with the lattice parameters $a = 791.8 \pm 0.1$, $b = 2571.1 \pm 0.4$, $c = 835.4 \pm 0.1$ pm, and $\beta = 90.626 \pm 0.002°$, is formed in the Pb–Tl–As–S system (Topa et al. 2023g,h).

B.10.12 Lead–Thallium–Oxygen–Sulfur

The quaternary compound $Tl_2Pb(SO_4)_2$, which crystallizes as a rhombohedral structure with the lattice parameters $a = 805.2$ pm, $\alpha = 40°41'$ or $a = 559.7$, $c = 2212.4$ pm in hexagonal setting and the calculated and experimental densities of 6.707 and 6.68 $g{\cdot}cm^{-3}$, respectively, is formed in the Pb–Tl–O–S system (Schwarz 1966). $Tl_2Pb(SO_4)_2$ was synthesized by annealing $PbSO_4$ and Tl_2SO_4 at 500°C (two times, 30 h + 31 h, air) and from the melt at 800°C-850°C by quenching.

B.10.13 Lead–Lanthanum–Yttrium–Sulfur

$PbS–Y_2S_3–La_2S_3$. The isothermal section of this system at 500°C is presented in Figure B.10.1 (Marchuk and Smitiukh 2021). The existence of the ternary phases was confirmed and it was found that quasibinary equilibria $La_2PbS_4–Y_2S_3$ and $Y_2PbS_4–La_2PbS_4$ occur in the quasiternary system. The wide two-phase region in the $Y_2S_3–La_2PbS_4$ system is caused by the existence of a $La_{2+2/3x}Pb_{1-x}S_4$ solid solution ($x = 0$-0.69).

B.10.14 Lead–Antimony–Tellurium–Sulfur

The quaternary phases $Pb_{3-x}Sb_{1+x}S_4Te_{2-\delta}$, which is stable at room temperature, have a narrow indirect energy gap, and crystallizes as a triclinic structure with the lattice parameters $a = 589.25 \pm 0.05$, $b = 589.99 \pm 0.05$, $c = 1519.06 \pm 0.14$ pm, $\alpha = 93.918 \pm 0.007°$, $\beta = 97.070 \pm 0.007°$, $\gamma = 90.255 \pm 0.007°$, and a calculated density of 6.6966 $g{\cdot}cm^{-3}$ for $Pb_{2.70}Sb_{1.29}S_4Te_{1.62}$ and $a = 592.01 \pm 0.05$, $b = 589.79 \pm 0.05$, $c = 1515.72 \pm 0.15$ pm, $\alpha = 93.928 \pm 0.007°$, $\beta = 97.054 \pm 0.008°$, and $\gamma = 90.074 \pm 0.007°$ for $Pb_{2.94}Sb_{1.05}S_4Te_{1.76}$, is formed in the Pb–Sb–Te–S system (Chen et al. 2017a).

Single crystals of these phases were grown using the self-flux method. Pb nuggets, Sb lumps, S powders, and Te granules were weighed according to a molar ratio of 3:1:4:2. The total mass of the starting materials was ~0.8 g. These starting materials were sealed in an evacuated fused silica tube in high vacuum (0.01 Pa) and were transferred into a tube furnace. The furnace was heated to 850°C in 15 h, maintained at this temperature for 2 h, slowly cooled to 600°C in 80 h, and then turned off. Single

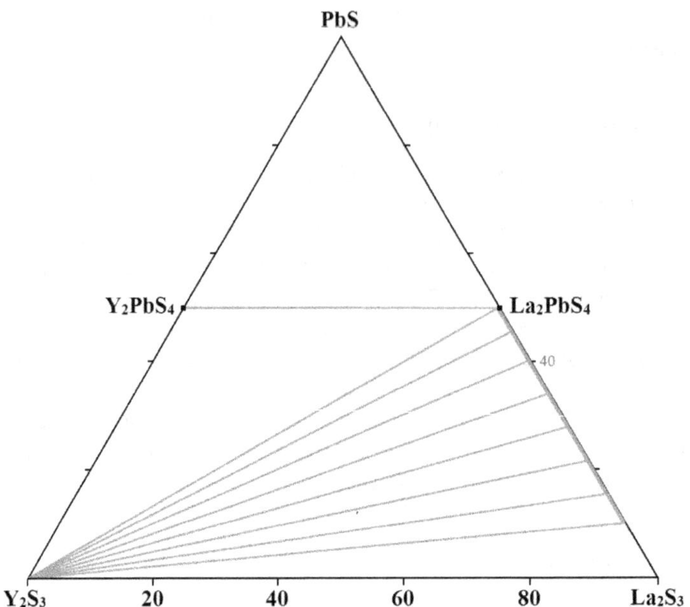

FIGURE B.10.1 Isothermal section of the PbS–Y_2S_3–La_2S_3 quasiternary system at 500°C. (From Marchuk, O., and Smitiukh, O., *Probl. khimii ta staloho rozvytku*, (1), 20, 2021.) Open access.

crystals were found on the surface of the final ingot. The crystals are planar shaped with metallic dark and mirror-like surfaces.

A large quantity of single crystals could be also obtained via the Te flux approach. The same starting materials as above were weighed with a molar ratio of 3:1:4:20, followed by the same sample heating process. After maintaining at 600°C for 5 h, the tube was taken out and centrifuged to get rid of the extra molten Te. The shiny single crystals were obtained in the bottom of the tube. The obtained single crystals are stable in air.

B.10.15 Lead–Bismuth–Tellurium–Sulfur

The quaternary compound $PbBi_4Te_4S_3$ (mineral clogauite), which crystallizes as a trigonal structure with the lattice parameters a = 427.7 ± 0.4, c = 2350 ± 10 pm (clogauite-12*H*), a = 427.8 ± 0.4, c = 4690 ± 30 pm (clogauite-24*H*), and a = 427.8 ± 0.4, c = 7040 ± 30 pm (clogauite-36*H*), is formed in the Pb–Bi–Te–S system (Cook et al. 2023, 2024).

B.10.16 Lead–Selenium–Tellurium–Sulfur

Samples of the solid solution of $PbS_xSe_yTe_{1-x-y}$ were investigated in the 4–200 K temperature range using electrical and X-ray methods by Lebedev and Sluchinskaya (1994). The region where low-temperature phase transition takes place was established (Figure B.10.2).

B.11 Systems Based on Lead Selenide

B.11.1 Lead–Copper–Bismuth–Selenium

The $Cu_2Pb_3Bi_8Se_{16}$ quaternary compound (mineral selenojunoite), which crystallizes as a monoclinic structure with the lattice parameters a = 2698 ± 1, b = 409.5 ± 0.1, c = 1735.3 ± 0.7 pm, and β = 127.74 ± 0.06°, is formed in the Pb–Cu–Bi–Se system (Gekimyants et al. 2023a,b).

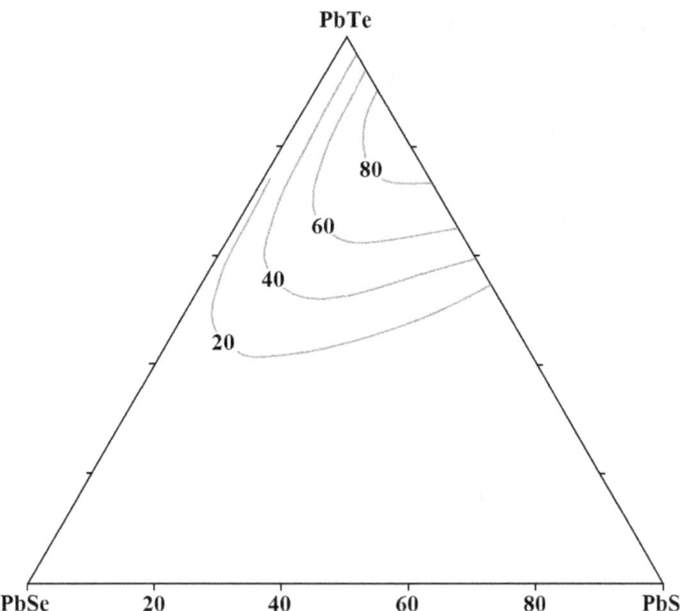

FIGURE B.10.2 The dependence of low-temperature phase transition (in K) on composition for $PbS_xSy_yTe_{1-x-y}$ solid solution. (From Lebedev, A.I. and Sluchinskaya I.A., *J. Alloys Compd.*, **203**(1–2), 51, 1994.)

B.11.2 Lead–Silver–Bismuth–Selenium

The quaternary compound $Ag_3Pb_4Bi_{11}Se_{22}$, which crystallizes as a monoclinic structure with the lattice parameters $a = 1385.3 \pm 0.3$, $b = 421.08 \pm 0.08$, $c = 1972.1 \pm 0.4$ pm, $\beta = 104.57 \pm 0.03°$, and a calculated density of 7.738 g·cm^{-3}, is formed in the Pb–Ag–Bi–Se system (Heinke et al. 2015). To synthesize this compound, a mixture of Pb, Ag, Bi, and Se was filled into silica glass ampoule in an argon atmosphere according to a nominal composition of $Ag_{0.68}Pb_{0.61}Bi_{2.28}Se_5$. The sample was heated to 600°C for 3 days and subsequently quenched in water. The product was thoroughly ground in an agate mortar, filled in a silica glass ampoule under argon atmosphere again, heated up to 600°C for 7 days, and afterward quenched in water. The material consisted of sintered crystallites of the title compound.

B.11.3 Lead–Gallium–Lanthanum–Selenium

The quaternary compound $La_3Pb_{0.1}Ga_{1.6}Se_7$, which crystallizes as a hexagonal structure with the lattice parameters $a = 1055.95 \pm 0.04$, $c = 638.68 \pm 0.03$ pm, and a calculated density of 5.9200 ± 0.0007 g·cm^{-3}, is formed in the Pb–Ga–La–Se system (Blashko 2022a). The synthesis of this compound was carried out from the elements in evacuated quartz ampoule according to the following scheme: heating to 700°C at a rate of 4°C·h^{-1}, annealing for 10 h, heating to 1100°C at a rate of 12°C·h^{-1}, annealing for 2 h, cooling to 500°C at a rate of 6°C·h^{-1}, annealing for 200 h, and quenching the ampoule with the product in water.

B.11.4 Lead–Gallium–Praseodymium–Selenium

The quaternary compound $Pr_3Pb_{0.1}Ga_{1.6}Se_7$, which crystallizes as a hexagonal structure with the lattice parameters $a = 1037.27 \pm 0.06$, $c = 637.72 \pm 0.05$ pm, and a calculated density of 6.192 ± 0.001 g·cm^{-3}, is formed in the Pb–Ga–Pr–Se system (Blashko 2022a). This compound was obtained in the same way as $La_3Pb_{0.1}Ga_{1.6}Se_7$ was prepared.

B.11.5 Lead–Bismuth–Tellurium–Selenium

$PbSe + Bi_2Te_3 \Leftrightarrow PbTe + Bi_2Se_3$. The "composition–structure" diagrams of the alloys in the PbSe–Bi_2Se_3 and PbTe–Bi_2Te_3 quasibinary systems and also diagonals of the cross-section PbTe–Bi_2Se_3 and PbSe–Bi_2Te_3 of this ternary mutual system were studied through XRD (Shelimova et al. 2015). There are a variety of laminated compounds in the system, and they belong to the following homologous series: $[PbTe]_m[Bi_2Te_3]_n$ and $[PbTe]_m[Bi_2(Te_{1-x}Se_x)_3]_n$ and also $[(PbSe)_5]_m[(Bi_2Se_3)_3]_n$. The compounds have a mixed laminated structure and are linked by a common structural pattern, particularly, by alternation of seven- and five-laminated packages in the direction of the hexagonal *c* axis. On the basis of the obtained data and the previous results, the scheme of the isothermal section of the ternary mutual system is plotted at 500°C, and the boundaries of the regions of quaternary solid solutions based on the $[PbTe]_m[Bi_2Te_3]_n$ homologous series were determined. The performed structural studies have shown that the PbTe–Bi_2Se_3 and PbSe–Bi_2Te_3 systems are not quasibinary.

B.11.6 Lead–Oxygen–Chlorine–Selenium

The quaternary compound $Pb_3Cl_2(SeO_3)_2$ (mineral wangkuirenite), which crystallizes as a monoclinic structure with the lattice parameters $a = 1342.66 \pm 0.04$, $b = 558.26 \pm 0.02$, $c = 1300.24 \pm 0.04$ pm, and $\beta = 94.287 \pm 0.002°$, is formed in the Pb–O–Cl–Se system (Yang 2023d,e).

B.12 Systems Based on Lead Telluride

B.12.1 Lead–Sodium–Antimony–Tellurium

The quaternary compound $NaPb_{18}SbTe_{20}$, which melts at 915°C and crystallizes as a cubic with the lattice parameter $a = 651.0 \pm 0.1$ pm and an energy gap of 0.37 eV, is formed in the Pb–Na–Sb–Te system (Guéguen et al. 2009). This compound was prepared as polycrystalline ingots in silica tube by mixing high-purity Na, Pb, Sb, and Te in the appropriate stoichiometric ratio. To prevent reaction between the sodium metal and silica, the tube was carbon-coated prior to use. All components (except Na) were loaded into silica tube under an ambient atmosphere and the corresponding amount of Na was later added under inert atmosphere in a dry glove box. The silica tube was then flame-sealed under a residual pressure of ~0.01 Pa, placed into a tube furnace (mounted on a rocking table), and heated at 980°C for 4 h to allow complete melting of all components. While molten, the furnace was allowed to rock for 2 h to facilitate complete mixing and homogeneity of the liquid phase. The furnace was finally immobilized at the vertical position and was cooled from 980°C to 550°C over 43 h followed by a faster cool (6–8 h) to room temperature. The resulting ingot generally was silvery-metallic in color with a smooth surface.

B.12.2 Lead–Sodium–Bismuth–Tellurium

The quaternary compound $NaPb_{18}BiTe_{20}$, which melts at 925°C and crystallizes as a cubic structure with the lattice parameter $a = 648.3 \pm 0.1$ pm and an energy gap of 0.25 eV, is formed in the Pb–Na–Bi–Te system (Guéguen et al. 2009). It was prepared in the same way as $NaPb_{18}SbTe_{20}$ but using Bi instead of Sb.

References

Acharya S., Pandey K., Basnet R., Acharya G., Nabi M.R.U., Wang J., Hu J. Single crystal growth and characterization of topological semimetal ZrSnTe, *J. Alloys Compd.*, **968**, 171903 (2023).

Aghazade A.I., Babanly V.I., Gojayeva I.M., Orujlu E.N., Mammadov A.N. 3D modeling of phase diagram of the ternary SnTe–PbTe–Bi_2Te_3 system, *Azerb. Chem. J.*, (2), 62–68 (2023).

Aguilera J., Varela M.T.B., Vázquez T. Procedure of synthesis of thaumasite, *Cement Concr. Res.*, **31**(8), 1163–1168 (2001).

Akopov G., Hewage N.W., Yox P., Viswanathan G., Lee S.J., Hulsebosch L.P., Cady S.D., Paterson A.L., Perras F.A., Xu W., Wu K., Mudryk Y., Kovnir K. Synthesis-enabled exploration of chiral and polar multivalent quaternary sulfides, *Chem. Sci.*, **12**(44), 14718–14730 (2021).

Akopov G., Viswanathan G., Hewage N.W., Yox P., Wu K., Kovnir K. Pd and octahedra do not get along: Square planar [PdS_4] units in non-centrosymmetric $La_6PdSi_2S_{14}$, *J. Alloys Compd.*, **902**, 163756 (2022).

Alakbarova T.M., Meyer H-J., Orujlu E.N., Amiraslanov I.R., Babanly M.B. Phase equilibria of the GeTe–Bi_2Te_3 quasi-binary system in the range 0–50 mol% Bi_2Te_3, *Phase Trans.*, **94**(5), 366–375 (2021).

Alakbarova T.M., Meyer H.-J., Orujlu E.N., Babanly M.B. A refined phase diagram of the GeTe–Bi_2Te_3 system, *Condens. Matter Interphases*, **24**(1), 11–18 (2022a).

Alakbarova T.M., Orujlu E.N., Babanly D.M. Imamaliyeva S.Z., Babanly M.B. Solid-phase equilibria in the $GeBi_2Te_4$–Bi_2Te_3–.Te– system and thermodynamic properties of compounds of the GeTe mBi_2Te_3–homologous series, *Phys. Chem. Solid State*, **23**(1), 25–33 (2022b).

Aleksandrov V.V. Glass formation and ionic conductivity in the glasses of the GeS_2–Ga_2S_3–MCl systems (M = Tl, Ag) [in Russian], *Fiz. khim. stekla*, **16**(3), 728–731 (1990).

Aliyev F.R. Synthesis and study of a novel 9P-type mixed layered tetradymite-like $GeBi_4Te_4$ compound in the Ge–Te–Bi system, *Phys. Chem. Solid State*, **22**(3), 401–406 (2021).

Aliyev F.R., Orujlu E.N., Babanly D.M. Synthesis and study of a new mixed-layered compound $GeBi_3Te_4$ belonging to the nBi_2–$m$$GeBi_2Te_4$ homologous series, *Bull. Univ. Karaganda, Chem.*, [1(105)], 92–98 (2022).

Aliyev O.M., Azhdarova D.S., Maksudova T.F., Ragimova V.M., Bayramova S.T. Synthesis, growth of monocrystals and properties of the compounds of $PbLnCuS_3$ (Ln–La, Nd, Sm, Gd, Dy, Er) type, *Azerb. Chem. J.*, (1), 183–190 (2023).

Aliyev O.M., Gasymova S.A. Phase equilibria in the YbS–SnS_2 system and physical-chemical properties of thiostannates [in Azerbaijanian], *Chem. Probl.*, (2), 387–391 (2009).

Aliyeva Z.M., Bagheri S.M., Aliev Z.S., Alverdiyev I.J., Yusibov Y.A., Babanly M.B. The phase equilibria in the Ag_2S–Ag_8GeS_6–Ag_8SnS_6 system, *J. Alloys Compd.*, **611**, 395–400 (2014).

Almoussawi B., Yao W.-D., Guo S.-P., Whangbo M.-H., Dupray V., Clevers S., Deng S., Kabbour H. Negative second harmonic response of Sn^{4+} in the fresnoite oxysulfide $Ba_2SnSSi_2O_7$, *Chem. Mater.*, **34**(10), 4375–4383 (2022).

Alverdiev I.Dzh., Imamalieva S.Z., Babanly D.M., Yusibov Yu.A., Tagiev D.B., Babanly M.B. Thermodynamic study of siver-tin selenides by the EMF method with Ag_4RbI_5 solid electrolyte, *Russ. J. Electrochem.*, **55**(5), 467–474 (2019). DIO: 10.1134/s1023193519050021

Alverdiev I.Dzh., Yusibov Y.A., Bagkheri S.M., Imamalieva S.Z., Babanly M.B. Thermodynamic study of Ag_8GeSe_6 by EMF with an Ag_4RbI_5 solid electrolyte, *Russ. J. Electrochem.*, **53**(5), 551–554 (2017).

Alverdiyev I.J. Refinement of phase diagram in the Cu_2S–GeS_2 system, *Chem. Probl.*, (3), 423–428 (2019).

Amiraslanova A.J., Babanly K.N., Imamaliyeva S.Z., Alverdiyev I.J., Yusibov Yu.A. Phase relations in the Ag_8SiS_6–Ag_8SiTe_6 system and characterization of solid solutions, *Azerb. Chem. J.*, (2), 169–177 (2023a).

Amiraslanova A.J., Babanly K.N., Imamaliyeva S.Z., Yusibov Yu.A., Babanly M.B. Phase equilibria in the Ag_8SiSe_6–Ag_8SiTe_6 system and characterization of the solid solutions $Ag_8SiSe_{6-x}Te_x$, *Appl. Chem. Eng.*, **6**(2), 11–19 (2023b).

Amiraslanova A.J., Mammadova A.T., Alverdiyev I.J., Yusibov Yu.A., Babanly M.B. $Ag_8GeS_6(Se_6)$–Ag_8GeTe_6 systems: phase relations, synthesis, and characterization of solid solutions, *Azerb. Chem. J.*, (1), 22–29 (2023c).

Amova A., Hristova-Vasileva T., Aljihmani L., Bineva I., Vassilev V. Region of glass formation and main physicochemical properties of glasses from the As_2Se_3–Ag_4SSe–PbTe system, *J. Alloys Compd.*, **573**, 32–36 (2013).

Amova A., Hristova-Vasileva T., Vassilev V. Phase equilibria in the Ag_4SSe–PbTe system, *Thermochim. Acta*, **531**, 42–45 (2012).

An Y., Ji M., Baiyin M., Liu X., Jia C., Wang D. A solvothermal synthesis and the structure of $K_4Ag_2Sn_3S_9{\cdot}2KOH$, *Inorg. Chem.*, **42**(14), 4248–4249 (2003).

An Y., Ye L., Ji M., Liu X., Menghe B., Jia C. A solvothermal synthesis and characterization of a new open-framework $K_4Ag_2Ge_3S_9{\cdot}H_2O$, *J. Solid State Chem.*, **177**(7), 2506–2510 (2004).

Angelova B., Dimitrov R., Khekimova A. Investigation of $CuSeO_3$ interaction with ZnS, PbS and FeS [in Bulgarian], *Nauchn. tr. Plovdiv. Univ. Ser. Khimiya*, **13**(3), 313–322 [1975(1976)].

Antao S.M., Hassan I., Parise J.B. The structure of danalite at high temperature obtained from synchrotron radiation and rietveld refinements, *Canad. Mineralog.*, **41**(6), 1413–1422 (2003).

Araki T. The crystal structure of linarite, reexamined, *Mineralog. J.*, **3**(5–6), 282–295 (1962).

Ardit M., Cruciani G., Dondi M., Garbarino G.L., Nestola F. Phase transitions during compression of thaumasite, $Ca_3Si(OH)_6(CO_3)(SO_4){\cdot}12H_2O$: A high-pressure synchrotron powder X-ray diffraction study, *Mineralg. Mag.*, **78**(5), 1193–1208 (2014).

Armbruster T., Hummel W. (Sb,Bi,Pb) ordering in sulfosalts: Crystal-structure refinement of a Bi-rich izoklakeite, *Am. Mineralog.*, **72**(7–8), 821–831 (1987).

Armbruster T., Stalder H.A., Oberhänsli R Antimon-reicher Giessenit vom Zervreilasee (Vals, Graubünder), *Schweiz. Mineral. Petrogr. Mitt.*, **64**(1–2), 21–26 (1984).

Armstrong J.A., Dann S.E., Neumann K., Marco J.F. Synthesis, structure and magnetic behaviour of the danalite family of minerals, $Fe_8[BeSiO_4]_6X_2$ (X = S, Se, Te), *J. Mater. Chem.*, **13**(5), 1229–1233 (2003).

Ashirov G.M., Babanly K.N., Mashadiyeva L.F., Yusibov Y.A., Babanly M.B. Phase equilibra in the Ag_2S–Ag_8GeS_6–Ag_8SiS_6 system and some properties of solid solutions, *Condens. Matter Interphases*, **25**(2), 292–301 (2023a).

Ashirov G.M., Babanly K.N., Mashadiyeva L.F., Yusibov Y.A., Babanly M.B. Phase equilibra in the Ag_2Se–Ag_8GeSe_6–Ag_8SiSe_6 system and characterization of the $Ag_8Si_{1-x}Ge_xSe_6$ solid solutions, *Chem. Probl.*, (3), 229–241 (2023b).

Assoud A., Kleinke H. A new polyselenide with a novel Se_7^{8-} unit: the structure of $Sr_{19-x}Pb_xGe_{11}Se_{44}$ with $x = 5.0$ and 6.4, *Eur. J. Inorg. Chem.*, **2017**(46), 5515–5520 (2017).

Assoud A., Kuropatwa B.A., Kleinke H. Barium manganese(II) selenostannate(IV), $BaMnSnSe_4$, *Acta Crystallogr.*, **E67**(12), i72 (2011).

Assoud A., Soheilnia N., Kleinke H. Crystal and electronic structure of the red semiconductor $Ba_4LaSbGe_3Se_{13}$ comprising the complex anion $[Ge_2Se_7–Sb_2Se_4–Ge_2Se_7]^{14-}$, *J. Solid State Chem.*, **177**(7), 2249–2254 (2004).

Aurivillius B. The crystal structure of a basic lead borate sulfate, $Pb_6O_2(BO_3)_2SO_4$, *Chem. Scr.*, **22**(4), 168–170 (1983).

Avdontceva M.S., Zolotarev A.A. Jr, Krivovichev S.V., Krzhizhanovskaya M.G., Sokol E.V., Kokh S.N, Bocharov V.N., Rassomakhin M.A, Zolotarev A.A. Fluorellestadite from burned coal dumps: crystal structure refinement, vibrational spectroscopy data and thermal behavior, *Mineral. Petrol.*, **115**(3), 271–281 (2021).

Averkieva G.K., Berdichevskiy G.V., Vaypolin A.A., Goryunova N.A., Prochukhan V.D. On the chemical interaction in the Cu–Ga–Ge–As–Se quinary system [in Russian], *Izv. AN SSSR. Neorgan. Mater.*, **4**(7), 1064–1066 (1968).

Averkieva G.K., Vaipolin A.A., Goriunova N.A. About some ternary compounds of the $A^I_2B^{IV}C^{VI}_3$-type and based on its solid solutions [in Russian], In: *Issled. po poluprovodnikam.* Kishinev: Kartia Moldoveniaske Publishers, 44–56 (1964).

Babanly N.B., Bulanova M.V., Mustafaeva A.N., Mammadov A.N. Determination and modeling of the liquidus surface, vapor pressure and immiscibility boundaries in the Cu–Pb–S system, *Azerb. Chem. J.*, (4), 35–42 (2021).

Bachechi F. Crystal structure of montbrayite, *Nature, Phys. Sci.*, **231**(20), 67–68 (1971).

Bachechi F. Synthesis and stability of montbrayite, Au_2Te_3, *Am. Mineralog.*, **57**(1–2), 146–154 (1972).

Bachmann H.-G., Zemann J. Die Kristallstruktur von Linarit, $PbCuSO_4(OH)_2$, *Acta Crystallogr.*, **14**(7), 747–753 (1961).

Badikov V.V., Badikov D.V., Shevyrdyaeva G.S., Laptev V.B., Melnikov A.A., Chekalin S.V. Optical and generation characteristics of new nonlinear $Ba_2Ga_8GeS_{16}$ and $Ba_2Ga_8(GeSe_2)S_{14}$ crystals for the mid-IR range, *Quant. Electron.*, **52**(3), 296–300 (2022b).

Bagheri S.M., Alverdiyev I.J., Aliev Z.S., Yusibov Y.A., Babanly M.B. Phase relationships in the $1.5GeS_2 + Cu_2GeSe_3 \Leftrightarrow 1.5GeSe_2 + Cu_2GeS_3$ reciprocal system, *J. Alloys Compd.*, **625**, 131–137 (2015).

Bagheri S.M., Alverdiyev I.J., Imamaliyeva S.Z., Babanly M.B. The phase equilibria in the Cu_8GeS_6–Cu_8GeSe_6 system and thermodynamic properties of solid solutions, *Chem. J.*, **4**(2), 26–31 (2014).

Bai C., Chu Y., Zhou J., Wang L., Luo L., Pan S., Li J. Two new tellurite halides with cationic layers: syntheses, structures, and characterizations of $CdPb_2Te_3O_8Cl_2$ and $Cd_{13}Pb_8Te_{14}O_{42}Cl_{14}$, *Inorg.Chem. Front.*, **9**(5), 1023–1030 (2022).

Baiyin M., Ye L., An Y., Liu X., Jia C., Ning G. A solvothermal synthesis and the structure of $(NH_4)_2Ag_6Sn_3S_{10}$, *Bull. Chem. Soc. Jpn.*, **78**(7), 1283–1284 (2005).

Baiyin M.-H., Ji M., Liu X., An Y.-L., Jia C.-Y., Ning G.-L. Solvothermal synthesis and investigation of $K_4Ag_2Sn_3S_9{\cdot}2H_2O$ with layered structure, *Chem. J. Chin. Univ.*,**25**(8), 1391–1394 (2004).

Baiyin M.-H., Liu F., Da L.-M., Bao Y.-S. Solvothermal synthesis and characterization of $Cs_4Sn_5S_{12}$ [in Chinese], *J. Synth. Cryst.*, **44**(3), 734–739 (2015).

Baker W.E. Hinsdalite pseudomorphous after pyromorphite from Dundas, Tasmania, *Pap. Proc. Royal Soc. Tasmania*, **97**, 129–132 (1963).

Ballirano P. The thermal behaviour of sacrofanite, *Eur. J. Mineral.*, **30**(3), 507–514 (2018).

Ballirano P., Bonaccorsi E. The crystal structure of sacrofanite, the 74 Å phase of the cancrinite group, *Acta Crystallogr.*, **A61**, Suppl., C382 (2005).

Ballirano P., Bonaccorsi E., Maras A., Merlino S. Crystal structure of afghanite, the eight-layer member of the cancrinite-group: evidence for long-range Si, Al ordering, *Eur. J. Mineral.*, **9**(1), 21–30 (1997).

Ballirano P., Bonaccorsi E., Maras A., Merlino S. The crystal structure of franzinite, the ten-layer mineral of the cancrinite group, *Canad. Mineralog.*, **38**(3), 657–668 (2000).

Ballirano P., Bonaccorsi E., Merlino S., Maras A. Carbonate groups in davyne: structural and crystal-chemical considerations, *Canad. Mineralog.*, **36**(5), 1285–1292 (1998).

Ballirano P., Maras A., Buseck P.R., Wang S., Yates A.M. Improved powder X-ray data for cancrinites. I: Afghanite, *Powder Diffr.*, **9**(1), 68–73 (1994).

Ballirano P., Maras A., Buseck P.R., Wang S., Yates A.M. Improved powder X-ray data for cancrinites II: Liottite and sacrofanite, *Powder Diffr.*, **10**(1), 13–19 (1995).

Ballirano P., Merlino S., Bonaccorsi E., Maras A. The crystal structure of liottite, a six-layer member of the cancrinite group, *Canad. Mineralog.*, **34**(5), 1021–1030 (1996).

Barde A., Joubert D.P. Investigation of structural, mechanical, dynamic, electronic and optical properties of seleno-germanates A_2GeSe_4 (A = Mg, Ca and γ-Sr) from first principles, *Mater. Today Commun.*, **22**, 100785 (2020).

Bardelli S., Ye Z., Wang F., Zhang B., Wang J. Synthesis, crystal and electronic structures, and nonlinear optical properties of $Y_4Si_3S_{12}$, *Z. anorg. und. allg. Chem.*, **648**(15), e202100388 (2022).

Bariand P., Cesbron F., Giraud R. Une nouvelle espèce minérale: l'afghanite de Sar-e-Sang, Badakhshan, Afghanistan. Comparaison avec les minéraux du groupe de la cancrinite, *Bull. Soc. fr. minéral. cristallogr.*, **91**(1), 34–42 (1968).

Barkov A.Y., Fleet M.E., Martin R.F., Tarkian M. Compositional variations in oulankaite and a new series of argentoan oulankaite from the Lukkulaisvaara layered intrusion, northern Russian Karelia, *Canad. Mineralog.*, **42**(2), 439–453 (2004).

Barkov A.Y., Men'shikov Y.P., Begizov V.D., Lednev A.I. Oulankaite, a new platinum-group mineral from the Lukkulaisvaara layered intrusion, Northern Karelia, Russia, *Eur. J. Mineral.*, **8**(2), 311–316 (1996).

Basu K., Bortnikov N.S., Moorkherjee A., Mozgova N.N., Tsepin A.I., Vyalsov L.N. Rare minerals from Rajpura-Dariba, Rajasthan, India. IV: A new Pb–Ag–Tl–Sb sulfosalt, rayite, *N. Jb. Miner., Mh.*, (7), 296–304 (1983).

Bayer G. New perovskite-type compounds A_2BTeO_6, *J. Am. Ceram. Soc.*, **46**(12), 604–605 (1963).

Bayliss P. Revised unit-cell dimensions, space group, and chemical formula of some metallic minerals, *Canad. Mineralog.*, **28**(4), 751–755 (1990).

Bayramova U.R. Determination of thermodynamic functions of phase transition of Cu_8SiSe_6 compound by the DSC method, *Chem. Probl.*, (2), 116–121 (2022a).

Bayramova U.R., Babanly K.N., Ahmadov E.I., Mashadiyeva L.F., Babanly M.B. Phase equilibria in the Cu_2S–Cu_8SiS_6–Cu_8GeS_6 system and thermodynamic functions of phase transitions of the $Cu_8Si_{(1-x)}Ge_xS_6$ argyrodite phases, *J. Phase Equilib. Diffus.*,**44**(5), 509–519 (2023).

Bayramova U.R., Poladova A., Mashadiyeva L. Synthesis and X-ray study of the $Cu_8Ge_{(1-x)}Si_xS_6$ solid solutions. *New Mater., Compd. Appl.*, **6**(3), 276–281 (2022b).

Bayramova U.R., Poladova A.N., Mashadiyeva L.F., Babanly M.B. Calorimetric determination of phase transitions of Ag_8BX_6 (B = Ge, Sn; X = S, Se) compounds, *Condens. Matter Interphases*, **24**(2), 187–195 (2022c).

Becker R., Brockner W., Schäfer H. Kristallstruktur und Schwingungsspektren des Di-Blei-Hexathiohypodiphosphates $Pb_2P_2S_6$, *Z. Naturforsch.*, **38**(8), 874–879 (1983).

Behrens H., Ostermeier J. Zur Kenntnis des Verhaltens von Nichtmetallchalkogeniden gegenüber flüssigem Ammoniak, VIII. Über Reaktionen der Sulfide des Siliciums, Germaniums und Zinns im Ammonosystem, *Chem. Ber.*, **95**(2), 487–499 (1962).

Belakovskiy D.I., Cámara F. New mineral names, *Am. Mineralog.*, **101**(2), 487–492 (2016).

Belakovskiy D.I., Cámara F. New mineral names, *Am. Mineralog.*, **104**(4), 625–629 (2019).

Belakovskiy D.I., Cámara F., Gagne O.C. New mineral names, *Am. Mineralog.*, **102**(3), 694–700 (2017a).

Belakovskiy D.I., Cámara F., Gagne O.C. New mineral names, *Am. Mineralog.*, **102**(11), 2341–2347 (2017b).

Belakovskiy D.I., Cámara F., Gagne O.C., Uvarova Y. New mineral names, *Am. Minetalog.*, **100**(10), 2352–2362 (2015).

Belakovskiy D.I., Cámara F., Gagne O.C., Uvarova Y. New mineral names, *Am. Mineralog.*, **102**(9), 1961–1968 (2017c).

Belakovskiy D.I., Cámara F., Gagne O.C., Uvarova Y. New mineral names, *Am. Mineralog.*, **101**(6), 1489–1497 (2016a).

Belakovskiy D.I., Cámara F., Gagne O.C., Uvarova Y. New mineral names, *Am. Mineralog.*, **101**(7), 1709–1716 (2016b).

Belakovskiy D.I., Cámara F., Gatta D. New mineral names, *Am. Mineralog.*, **99**(4), 870–875 (2014).

Belakovskiy D.I., Cámara F., Uvarova Y. New mineral names, *Am. Mineralog.*, **102**(7), 1565–1571 (2017d).

Belakovskiy D.I., Cámara F., Uvarova Y. New mineral names, *Am. Mineralog.*, **104**(7), 1108–1119 (2019).

Belakovskiy D.I., Gagne O.C. New mineral names, *Am. Mineralog.*, **100**(7), 1649–1654 (2015).

Belakovskiy D.I., Smith D.G.W., Welch M.D. New mineral names, *Am. Mineralog.*, **98**(4), 811–815 (2013).

Belakovskiy D.I., Uvarova Y., Gagne O.C. New mineral names, *Am. Mineralog.*, **101**(4), 1012–1019 (2016c).

Belakovskiy D.I., Welch M.D., Cámara F., Gatta D., Tait K.T. New mineral names, *Am. Mineralog.*, **97**(8–9), 1523–1530 (2012).

Bellagra H.K., Kogut Y.M., Piskach L.V. Component interaction in the quasi-ternary system PbSe–Ga_2Se_3–$GeSe_2$, *J. Phase Equilib. Diffus.*, **44**(1), 3–16 (2023).

Berdonosov P.S., Janson O., Olenev A.V., Krivovichev S.V., Rosner H., Dolgikh V.A., Tsirlin A.A. Crystal structures and variable magnetism of $PbCu_2(XO_3)_2Cl_2$ with X = Se, Te, *Dalton Trans.*, **42**(26), 9547–9554 (2013).

Bereznyuk O., Alriqiq M., Kogut Yu., Piskach L. Phase equilibria in the $Cu(Ag)_2S$–Sb_2S_3–SnS_2 systems [in Ukrainian], *Probl. khimii ta staloho rozvytku*, (4), 17–30 (2022a).

Berezniuk O., Petrus' I., Olekseyuk I., Smitiukh O., Zamuruyeva O., Nakhod V. The Ag_2S–GeS_2–P_2S_5 at 500 K, *J. Solid State Chem.*, **313**, 123340 (2022).

Bereznyuk O., Petrus' I., Olekseyuk I., Zamuruyeva O.V., Skipalskiy M.I. Phase equilibria, glass formation and optical properties of glasses in the Ag_2S–$B^{IV}S_2$–$C^{V}{}_2S_3$ systems (B^{IV} – Ge, Sn; C^{V} – As, Sb), *Phys. Chem. Solid State*, **23**(1), 57–61 (2022b).

Bereznyuk O., Petrus' I., Smitiukh O., Olekseyuk I. Glass formation in the quasiternary systems AI_2S–$B^{IV}S_2$–$C^{V}{}_2S_3$ (A^{I} – Cu, Ag; B^{IV} – Ge, Sn, C^{V}– As, Sb) [in Ukrainian], *Probl. khimii ta staloho rozvytku*, (4), 3–10 (2021).

Bereznyuk O., Petrus' I., Zamuruyeva O.V., Piskach L.V. Properties of the glasses in the Ag_2S–GeS_2–$As(Sb)_2S_3$ systems [in Ukrainian], *Nauk. Visn. Uzhgorod. Univ. Ser. Khimiya*, [2(48)], 29–37 (2022c).

Bereznyuk O., Piskach L. Interaction in the Cu_2S–Sb_2S_3–GeS_2 quasiternary system [in Ukrainian], *Probl. khimii ta staloho rozvytku*, (3), 3–12 (2023).

Bereznyuk O., Smitiukh O., Piskach L. Interaction along the $Cu(Ag)_7PS_6$–$Cu(Ag)_8Ge(Sn)S_6$ sections [in Ukrainian], *Probl. khimii ta staloho rozvytku*, (4), 3–16 (2022d).

Berlepsch P. Crystal structure and crystal chemistry of the homeotypes edenharterite ($TlPbAs_3S_6$) and jentschite ($TlPbAs_2SbS_6$) from Lengenbach, Binntal (Switzerland), *Schweiz. Mineral. Petrogr. Mitt.*, **76**(2), 147–157 (1996).

Berlepsch P., Armbruster T., Topa D. Structural and chemical variations in rathite, $Pb_8Pb4–_x(Tl_2As_2)_x(Ag_2As_2)As_{16}S_{40}$: modulations of a parent structure, *Z. Kristallogr.*, **217**(11), 581–590 (2002).

Berman H., Gonyer F.A. Re-examination of colusite, *Am. Mineralog.*, **24**(6), 377–381 (1939).

Bernardini G.P., Bonazzi P., Corazza M., Corsini F., Mazzetti G., Poggi L., Tanelli G. New data on the Cu_2FeSnS_4–Cu_2ZnSnS_4 pseudobinary system at 750° and 550°C, *Eur. J. Mineral.*, **2**(2), 219–225 (1990).

Bernert Th., Pfitzner A. Characterization of mixed crystals in the system $Cu_2Mn_xCo_{1-x}GeS_4$ and investigations of the tetrahedra volumes, *Z. anorg. und allg. Chem.*, **632**(7), 1213–1218 (2006).

Bernstein L.R. Renierite, $Cu_{10}ZnGe_2Fe_4S_{16}$–$Cu_{11}GeAsFe_4S_{16}$: a coupled solid solution series, *Am. Mineralog.*, **71**(1–2), 210–221 (1986).

Bernstein L.R., Reichel D.G., Merlino S. Renierite crystal structure refined from Rietveld analysis of powder neutron-diffraction data, *Am. Mineralog.*, **74**(9–10), 1177–1181 (1989).

Berry L.G. The unit cell of linarite, *Am. Mineralog.*, **36**(5–6), 511–512 (1951).

Biagioni C., Moëlo Y. Lead-antimony sulfosalts from Tuscany (Italy). XIX. Crystal chemistry of chovanite from two new occurrences in the Apuan Alps and its 8 Å crystal structure, *Mineralog. Mag.*, **81**(4), 811–831 (2017a).

Biagioni C., Moëlo Y. Lead-antimony sulfosalts from Tuscany (Italy). XVIII. New data on the crystal-chemistry of boscardinite, *Mineralog. Mag.*, **81**(1), 47–60 (2017b).

Biagioni C., Moëlo Y., Orlandi P., Stanley C.J., Evain M. Meerschautite, IMA 2013-061. CNMNC Newsletter No. 17, October 2013, page 3004, *Mineralog. Mag.*, **77**(7), 2997–3005 (2013a).

Biagioni C., Moëlo Y., Favreau G., Bourgoin V., Boulliard J.-C. Structure of Pb-rich chabournéite from Jas Roux, France, *Acta Crystallogr.*, **B71**(1), 81–88 (2015).

Biagioni C., Moëlo Y., Orlandi P. Lead-antimony sulfosalts from Tuscany (Italy). XV. (Tl–Ag)-bearing rouxelite from Monte Arsiccio mine: occurrence and crystal chemistry, *Mineralog. Mag.*, **78**(3), 651–661 (2014a).

Biagioni C., Moëlo Y., Orlandi P., Paar W.H. Andreadiniite, IMA 2014-049. CNMNC Newsletter No. 22, October 2014, page 1244, *Mineralog. Mag.*, **78**(5), 1241–1248 (2014b).

Biagioni C., Moëlo Y., Orlandi P., Paar W.H. Lead-antimony sulfosalts from Tuscany (Italy). XXIII. Andreadiniite, $CuAg_7HgPb_7Sb_{24}S_{48}$, a new oversubstituted (Cu,Hg)-rich member of the andorite homeotypic series from the Monte Arsiccio mine, Apuan Alps, *Eur. J. Mineral.*, **30**(5), 1021–1035 (2018a).

Biagioni C., Moëlo Y., Orlandi P., Stanley C.J. Lead-antimony sulfosalts from Tuscany (Italy). XVII. Meerschautite, $(Ag,Cu)_{5.5}Pb_{42.4}(Sb,As)_{45.1}S_{112}O_{0.8}$, a new expanded derivative of owyheeite from the Pollone mine, Valdicastello Carducci: occurrence and crystal structure, *Mineralog. Mag.*, **80**(4), 675–690 (2016a).

Biagioni C., Moëlo Y., Perchiazzi N., Demitri N., Lepore G.O. Lead-antimony sulfosalts from Tuscany (Italy). XXIV. Crystal structure of thallium-bearing chovanite, $TlPb_{26}(Sb,As)_{31}S_{72}O$, from the Monte Arsiccio Mine, Apuan Alps, *Minerals*, **8**(11), 535 (2018b).

Biagioni C., Orlandi P., Moëlo Y. Carducciite, IMA 2013-006. CNMNC Newsletter No. 16, August 2013, page 2702, *Mineralog. Mag.*, **77**(6), 2695–2709 (2013b).

Biagioni C., Orlandi P., Moëlo Y., Bindi L. Lead-antimony sulfosalts from Tuscany (Italy). Carducciite, $(AgSb)Pb_6(As,Sb)_8S_{20}$, a new Sb-rich derivative of rathite from the Pollone mine, Valdicastello Carducci: occurrence and crystal structure, *Mineralog. Mag.*, **78**(7), 1775–1793 (2014c).

Biagioni C., Pasero M., Moëlo Y., Zaccarini F., Paar W.H. Lead-antimony sulfosalts from Tuscany (Italy). XXII. Marcobaldiite, $\sim Pb_{12}(Sb_3As_2Bi)_{\Sigma 6}S_{21}$, a new member of the jordanite homologous series from the Pollone mine, Valdicastello Carducci, *Eur. J. Mineral.*, **30**(3), 581–592 (2018c).

Biagioni C., Pasero M., Moëlo Y., Zaccarini F., Paar W.H., Merlino S. Marcobaldiite, IMA 2015-109. CNMNC Newsletter No. 30, April 2016, page 410, *Mineralog. Mag.*, **80**(2), 407–413 (2016b).

Biagioni C., Pasero M., Zaccarini F. Tiberiobardiite, $Cu_9Al(SiO_3OH)_2(OH)_{12}(H_2O)_6(SO_4)_{1.5}\cdot 10H_2O$, a new mineral related to chalcophyllite from the Cretaio Cu prospect, Massa Marittima, Grosseto (Tuscany, Italy): Occurrence and crystal structure, *Minerals*, **8**(4), 152 (2018d).

Biagioni C., Pasero M., Zaccarini F. Tiberiobardiite, IMA 2016-096. CNMNC Newsletter No. 36, April 2017, page 404, *Mineralog. Mag.*, **81**(2), 403–409 (2017b).

Biagioni C., Sejkora J., Moëlo Y., Favreau G., Bourgoin V., Boulliard J.-C., Bonaccorsi E., Mauro D., Musetti S., Pasero M., Perchiazzi N., Ulmanová J. Ginelfite, IMA 2022-110. CNMNC Newsletter No 71, *Eur. J. Mineral.*, **35**(1), 79 (2023a).

Biagioni C., Sejkora J., Moëlo Y., Favreau G., Bourgoin V., Boulliard J.-C., Bonaccorsi E., Mauro D., Musetti S., Pasero M., Perchiazzi N., Ulmanová J. Ginelfite, IMA 2022-110. CNMNC Newsletter No 71, *Mineralog. Mag.*, **87**(2), 335 (2023b).

Bichler D., Zinth V., Johrendt D., Heyer O., Forthaus M.K., Lorenz T., Abd-Elmeguid M.M., Structural and magnetic phase transitions of the V_4-cluster compound GeV_4S_8, *Phys. Rev. B*, **77**(21), 212102 (2008).

Bideaux R.A., Nichols M.C., Williams S.A. The arsenate analogue of tsumebite, a new mineral, *Am. Mineralog.*, **51**(1–2), 258–259 (1966).

Bindi L., Biagioni C., Keuysch F.N. Oyonite, $Ag_3Mn_2Pb_4Sb_7As_4S_{24}$, a new member of the lillianite homologous series from the Uchucchacua base-metal deposit, Oyon District, Peru, *Minerals*, **8**(5), 192 (2018a).

Bindi L., Biagioni C., Keuysch F.N. Oyonite, IMA 2018-002. CNMNC Newsletter No. 43, June 2018, page 648, *Eur. J. Mineral.*, **30**(3), 647–652 (2018b).

Bindi L., Biagioni C., Keuysch F.N. Oyonite, IMA 2018-002. CNMNC Newsletter No. 43, June 2018, page 781, *Mineralog. Mag.*, **82**(3), 779–785 (2018c).

Bindi L., Biagioni C., Martini B., Salvetti A. Ciriottiite, $Cu(Cu,Ag)_3Pb_{19}(Sb,As)_{22}(As_2)S_{56}$, the Cu-analogue of sterryite from the Tavagnasco mining district, Piedmont, Italy, *Minerals*, **6**(1), 8 (2016a).

Bindi L., Biagioni C., Martini B., Salvetti A. Ciriottiite, IMA 2015-027. CNMNC Newsletter No. 26, August 2015, page 943, *Mineralog. Mag.*, **79**(4), 941–947 (2015).

Bindi L., Bonazzi P., Keutsch F.N. Menchettiite, IMA 2011-009. CNMNC Newsletter No. 10, October 2011, page 2550, *Mineralog. Mag.*, **75**(5), 2549–2561 (2011).

Bindi L., Cipriani C. Mazzettiite, $Ag_3HgPbSbTe_5$, a new mineral species from Findley Gulch, Saguache county, Colorado, USA, *Canad. Mineralog.*, **42**(6), 1739–1743 (2004a).

Bindi L., Cipriani C. Museumite, $Pb_5AuSbTe_2S_{12}$, a new mineral from the gold-telluride deposit of Sacarîmb, Metaliferi Mountains, western Romania, *Eur. J. Mineral.*, **16**(5), 835–838 (2004b).

Bindi L., Garavelli A., Pinto D., Pratesi G., Vurro F. Ordered distribution of I and Cl in the low-temperature crystal structure of mutnovskite, $Pb_4As_2S_6ICl$: An X-ray single-crystal study, *J. Solid State. Chem.*, **181**(2), 306–312 (2008).

Bindi L., Keutsch F.N., Bonazzi P. Menchettiite, $AgPb_{2.40}Mn_{1.60}Sb_3As_2S_{12}$, a new sulfosalt belonging to the lillianite series from the Uchucchacua polymetallic deposit, Lima Department, Peru, *Am. Mineralog.*, **97**(2–3), 440–446 (2012).

Bindi L., Menchetti S. Structural changes accompanying the phase transformation between leadhillite and susannite: A structural study by means of in situ high-temperature single-crystals X-ray diffraction, *Am. Mineralog.*, **90**(10), 1641–1647 (2005).

Bindi L., Nestola F., A. Guastoni A., Secco L. The crystal structure of dalnegroite, $Tl_{5-x}Pb_{2x}(As,Sb)_{21-x}S_{34}$: a masterpiece of structural complexity, *Mineralog. Mag.*, **74**(6), 999–1012 (2010).

Bindi L., Paar W.H., Lepore G.O. Montbrayite, $(Au,Ag,Sb,Pb,Bi)_{23}(Te,Sb,Pb,Bi)_{38}$, from the Robb-montbray Mine, Montbray, Québec: Crystal structure and revision of the chemical formula, *Canad. Mineralog.*, **56**(2), 129–142 (2018d).

Bindi L., Putz H., Paar W.H., Stanley C.J. Omariniite, $Cu_8Fe_2ZnGe_2S_{12}$, the germanium analogue of stannoidite, a new mineral species from Capillitas, Argentina, *Mineralog. Mag.*, **81**(5), 1151–1159 (2017).

Bindi L., Putz H., Paar W.H., Stanley C.J. Omariniite, IMA 2016-050. CNMNC Newsletter No. 33, October 2016, page 1140, *Mineralog. Mag.*, **80**(6), 1135–1144 (2016b).

Bindi L., Spry P.G., Bonazzi P., Makovicky E., Balić-Žunić T. Quadratite, $AgCdAsS_3$: Chemical composition, crystal structure, and OD character, *Am. Mineralog.*, **98**(1), 242–247 (2013).

Blashko N.M., Marchuk O.V. Smitiukh O.V., Fedorchuk A.O. The crystal structure of $La_3Pb_{0.1}Ga_{1.6}Se_7$ and $Pr_3Pb_{0.1}Ga_{1.6}Se_7$ [in Ukrainian[*Nauk. Visn. Uzhgorod. Univ., Ser. Khim.*, [2(48)], 10–15 (2022a).

Blashko N.M., Smitiukh O.V., Marchuk O.V. The crystal structure of $La_3Pb_{0.1}Ga_{1.6}S_7$ and $Pr_3Pb_{0.1}Ga_{1.6}S_7$ compounds, *Phys. Chem. Solid State*, **23**(1), 96–100 (2022b).

Bonaccorsi E. The crystal structure of giuseppettite, the 16-layer member of the cancrinite–sodalite group, *Microporous Mesoporous Mater.*, **73**(3), 129–136 (2004).

Bonaccorsi E., Ballirano P., Cámara F. The crystal structure of sacrofanite, the 74 Å phase of the cancrinite–sodalite supergroup, *Microporous Mesoporous Mater.*, **147**(1), 318–326 (2012).

Bonaccorsi E., Della Ventura G., Bellatrecia F., Merlino S. The thermal behaviour and dehydration of pitiglianoite, a mineral of the cancrinite-group, *Microporous Mesoporous Mater.*, **99**(3), 225–235 (2007).

Bonaccorsi E., Merlino S., Pasero M. Davyne: its structural relationships with cancrinite and vishnevite, *N. Jb. Miner. Mh.*, (3), 97–112 (1990).

Bonaccorsi E., Merlino S., Pasero M., Macedonio G. Microsommite: crystal chemistry, phase transitions, Ising model and Monte Carlo simulations, *Phys. Chem. Minerals*, **28**(8), 509–522 (2001).

Bonaccorsi E ., Orlandi P. Marinellite, a new feldspathoid of the cancrinite-sodalite group, *Eur. J. Mineral.*, **15**(6), 1019–1027 (2003).

Bonaccorsi E., Orlandi P. Second occurrence of pitiglianoite, a mineral of the cancrinite-group, *Atti. Soc. Tosc. Nat., Mem., Ser. A*, **103**, 193–195 (1996).

Bor F.Y., Tarassoff P. Solubility of oxygen in copper mattes, *Canad. Metallurg. Quart.*, **10**(4), 267–271 (1971).

Borisov S.V., Pervukhina N.V., Kurat'eva N.V., Magarill S.A., Kuchumov B.M. Crystal structure of natural Ag–Cu–Pb–Bi sulfide, *J. Struct. Chem.*, **58**(1), 78–83 (2017).

Borodaev Yu.S., Garavelli A., Kuzmina O.V., Mozgova N.N., Organova N.I., Trubkin N.V., Vurro F. Rare sulfosalts from Vulcano, Aeolian Islands, Italy. I. Se-bearing kirkiite, $Pb_{10}(Bi,As)_6(S,Se)_{19}$, *Canad. Mineralog.*, **36**(4), 1105–1114 (1998).

Bosi F., Hatert F., Pasero M., Mills S.J. IMA Commission on New Minerals, Nomenclature and Classification (CNMNC). Newsletter No. 76, *Eur. J. Mineral.*, **35**(6), 1073–1078 (2023).

Bosi F., Hatert F., Pasero M., Mills S.J. IMA Commission on New Minerals, Nomenclature and Classification (CNMNC). Newsletter No 76, *Mineralog. Mag.*, **88**(1), 105–109 (2024).

Bourgès C., Bouyrie Y., Supka A.R., Al Rahal Al Orabi R., Lemoine P., Lebedev O.I., Ohta M., Suekuni K., Nassif V., Hardy V., Daou R., Miyazaki Y., Fornari M., Guilmeau E. High-performance thermoelectric bulk colusite by process controlled structural disordering, *J. Am. Chem.Soc.*, **140**(6), 2186–2195 (2018).

Bourgès C., Gilmas M., Lemoine P., Mordvinova N.E., Lebedev O.I., Hug E., Nassif V., Malaman B., Daou R., Guilmeau E. Structural analysis and thermoelectric properties of mechanically alloyed colusites, *J. Mater. Chem. C*, **4**(31), 7455–7463 (2016).

Bourgès C., Lemoine P., Lebedev O.I., Daou R., Hardy V., Malaman B., Guilmeau E. Low thermal conductivity in ternary $Cu_4Sn_7S_{16}$ compound, *Acta Mater.*, **97**, 180–190 (2015).

Bousquet C., Vicente C.P., Krämer A., Tirado J.L., Olivier-Fourcade J., Jumas J.C. Electrochemical lithium insertion in a cation deficient thiospinel $Cu_{3.31}GeFe_4Sn_{12}S_{32}$, *J. Mater. Chem.*, **8**(6), 1399–1404 (1998).

Bowes C.L., Lough A.J., Malek A., Ozin G.A., Petrov S., Young D. Thermally stable self-assembling open-frameworks: Isostructural Cs^+ and $(CH_3)_4N^+$ iron germanium sulfides, *Chem. Ber.*, **129**(3), 283–287 (1996).

Boycheva S.V., Vassilev V.S., Ivanova Z.G. On the glass formation in the semiconducting $GeSe_2$–Sb_2Se_3–ZnTe, As_2Se_3–Sb_2Se_3–ZnSe and GeSe2–ZnTe–ZnSe systems, *J. Phys. D: Appl. Phys.*, **32**(4), 529–532 (1999).

Brodersen K., Procher H., Hummel H.-U. Zur Darstellung und Kristallstruktur von $PbZn(NCS)_4$, *Z. Naturforsch.*, **42B**(6), 679–681 (1987).

Bujnowski T.J., Guggenheim S., Kato T. Crystal structure determination of anandite-2*M* mica, *Am. Mineralog.*, **94**(8–9), 1144–1152 (2009).

Buller S., Koch C., Bensch W., Zalden P., Sittner R., Kremers S., Wuttig M., Schürmann U., Kienle L., Leichtweiss T., Janek J., Schönborn B. Influence of partial substitution of Te by Se and Ge by Sn on the properties of the Blu-ray phase-change material $Ge_8Sb_2Te_{11}$. *Chem. Mater.*, **24**(18), 3582–3590 (2012).

Burns P.C., Cooper M.A., Hawthorne F.C. Parakhinite, $Cu_2^{+3}PbTe^{6+}O_6(OH)_2$: crystal structure and revision of chemical formula, *Canad. Mineralog.*, **33**(1), 33–40 (1995).

Burns P.C., Pluth J.J., Smith J.V., Eng P., Steele I., Houseley R.M. Quetzalcoatlite: A new octahedral-tetrahedral structure from a $2 \times 2 \times 40$ μm³ crystal at the Advanced Photon Source-GSE-CARS Facility, *Am. Mineralog.*, **85**(3–4), 604–607 (2000).

Cabri L.J., Fleischer M., Pabst A. New mineral names, *Am. Mineralog.*, **66**(9–10), 1099–1103 (1981).

Caldera D., Quintero M., Morocoima M., Moreno E., Quintero E., Grima-Gallardo P., Bocaranda P., Henao J.A., Macías M.A., Briceño J.M., Mora A.E. Lattice parameters values and phase diagram for the $Cu_2Zn_{1-z}Mn_zGeSe_4$ alloy system, *J. Alloys Compd.*, **614**, 253–257 (2014).

Caldera D., Quintero M., Morocoima M., Quintero E., Grima P., Marchan N., Morenoa E., Bocaranda P., Delgado G.E., Mora A.E., Briceño J.M., Fernandez J.L. Lattice parameters values and phase diagram for the $Cu_2Zn_{1-z}Fe_zGeSe_4$ alloy system, *J. Alloys Compd.*, **457**(1–2), 221–224 (2008).

Calvez L., Ma H.-L., Lucas J., Zhang X.-H. Selenium-based glasses and glass ceramics transmitting light from the visible to the far-IR, *Adv. Mater.*, **19**(1), 129–132 (2007).

Cámara F., Bellatreccia F., Della Ventura G., Gunter M.E., Sebastiani M., Cavallo A. Kircherite, a new mineral of the cancrinite–sodalite group with a 36-layer stacking sequence: Occurrence and crystal structure, *Am. Mineralog.*, **97**(8–9), 1494–1504 (2012).

Cámara F., Bellatreccia F., Della Ventura G., Mottana A. Farneseite, a new mineral of the cancrinite–sodalite group with a 14-layer stacking sequence: occurrence and crystal structure, *Eur. J. Mineral.*, **17**(6), 839–846 (2005).

Cámara F., Bellatreccia F., Della Ventura G., Mottana A., Bindi L., Gunter M.E., Sebastiani M. Fantappièite, a new mineral of the cancrinite–sodalite group with a 33-layer stacking sequence: Occurrence and crystal structure, *Am. Mineralog.*, **95**(4), 472–480 (2010).

Cámara F., Gagné O.C., Belakovskiy D.I., Uvarova Y. New mineral names, *Am. Mineralog.*, **100**(5–6), 1319–1332 (2015).

Cámara F., Gagne O.C., Uvarova Y., Belakovskiy D.I. New mineral names, *Am. Mineralog.*, **99**(11–12), 2437–2444 (2014a).

Cámara F., Gatta G.D., Uvarova Y., Gagne O.C., Belakovskiy D.I. New mineral names, *Am. Mineralog.*, **99**(7), 1511–1518 (2014b).

Cámara F., Gatta G.D., Uvarova Y., Gagne O.C., Belakovskiy D.I. New mineral names, *Am. Mineralog.*, **99**(10), 2150–2158 (2014c).

Cámara F., Holtstam D., Jansson N., Jonsson E., Karlsson A., Langhof J., Majka J., Zetterqvist A. Zinkgruvanite, $Ba_4Mn^{2+}{}_4Fe^{3+}{}_2(Si_2O_7)_2(SO_4)_2O_2(OH)_2$, a new ericssonite-group mineral from Zinkgruvan Zn–Pb–Ag–Cu deposit, Askersund, Örebro County, Sweden, *Eur. J. Mineral.*, **33**(6), 659–673 (2021).

Cámara F., Holtstam D., Jansson N., Jonsson E., Karlsson A., Langhof J., Majka J., Zetterqvist A. Zinkgruvanite, IMA 2020-031. CNMNC Newsletter No. 56, *Eur. J. Mineral.*, **32**(4), 443–448 (2020a).

Cámara F., Holtstam D., Jansson N., Jonsson E., Karlsson A., Langhof J., Majka J., Zetterqvist A. Zinkgruvanite, IMA 2020-031. CNMNC Newsletter No. 56, *Mineralog. Mag.*, **84**(4), 623–627 (2020b).

Campostrini I., Gramaccioli C.M., Demartin F. Orlandiite, $Pb_3Cl_4(SeO_3){\cdot}H_2O$, a new mineral species, and an associated lead–copper selenite chloride from the Baccu Locci mine, Sardinia, Italy, *Canad. Mineralog.*, **37**(6), 1493–1498 (1999).

Cannillo E., Dal Negro A., Rossi G. The crystal structure of latiumite, a new type of sheet silicate, *Am. Mineralog.*, **58**(5–6), 466–470 (1973).

Cannillo E., Rossi G., Ungaretti L. The crystal structure of delhayelite, *Rend. Soc. Ital. Mineral. Petrol.*, **26**, 63–75 (1970).

Cao W., Mei D., Yang Y., Wu Yuanwang., Zhang L., Wu Yuandong, He X., Lin Z., Huang F. From $CuFeS_2$ to $Ba_6Cu_2FeGe_4S_{16}$: rational band gap engineering achieves large second-harmonic-generation together with high laser damage threshold, *Chem. Commun.*, **55**(96), 14510–14513 (2019).

Cao X.-L., Hu C.-L., Xu X., Kong F., Mao J.-G. $Pb_2TiOF(SeO_3)_2Cl$ and $Pb_2NbO_2(SeO_3)_2Cl$: small changes in structure induced a very large SHG enhancement, *Chem. Commun.*, **49**(85), 9965–9967 (2013).

Cao X.-L., Kong F., Hu C.-L., Xu X., Mao J.-G. $Pb_4V_6O_{16}(SeO_3)_3(H_2O)$, $Pb_2VO_2(SeO_3)_2Cl$, and $PbVO_2(SeO_3)F$: new lead(II)–vanadium(V) mixed-metal selenites featuring novel anionic skeletons, *Inorg. Chem.*, **23**(16), 8816–8824 (2014).

Casamento J., Lopez J.S., Moroz N.A., Olvera A., Djieutedjeu H., Page A., Uher C., Poudeu P.F.P. Crystal structure and thermoelectric properties of the $^{7,7}L$ lillianite homologue $Pb_6Bi_2Se_9$, *Inorg. Chem.*, **56**(1), 261–268 (2017).

Caye R., Laurent Y., Picot P., Pierrot R., Lévy C. La hocartite, Ag_2SnFeS_4, une nouvelle espиce minérale, *Bull. Soc. fr. minéral. cristallogr.*, **91**(4), 383–387 (1968).

Cesbron F., Bachet B., Oosterbosch R. La demesmaekerite, sélénite hydraté d'uranium, cuivre, et plomb, *Bull. Soc. fr. minéral. cristallogr.*, **88**, 422–425 (1965).

Cesbron F., Giraud R., Picot P., Pillard F. La vinciennite $Cu_{10}Fe_4Sn(As,Sb)S_{16}$, une nouvelle espèce minérale. Etude paragenétique du gîte type de Chizeuil, Saône-et-Loire, *Bull. Minéralog.*, **108**, 447–456 (1985).

Chai X., Chen W., Yan Q., Liu B., Jiang X., Guo G. $Rb_2MGe_3S_8$ (M = Zn, Cd): Non-centrosymmetry transformation led by structure change of $[MGe_3S_8]^{2-}$ unit [in Chinese], *Acta Chim. Sin.*, **80**(5), 633–639 (2022).

Chang J., Chen S., Chiu K., Wu H., Chen J. Liquidus projection of the Ag–Sn–Te ternary system, *Metall. Mater. Trans.*, **A45**(9), 3728–3740 (2014). 10.1007/s11661-014-2318-x

Chang J.-S., Chen S-W. Liquidus projection and isothermal section of the Sb–Se–Sn system, *Metall. Mater. Trans.*, **E4**(2–4), 89–100 (2017).

Chang L.L.Y. $Ag_{1.2}Sn_{0.9}Sb_3S_6$, a tin-bearing andorite phase, *Mineralog. Mag.*, **51**(363), 741–743 (1987).

Chang L.L.Y., Li X., Zheng C. The jamesonite — benavidesite series, *Canad. Mineralog.*, **25**(4), 667–672 (1987).

Chang L.L.Y, Walia D.S., Knowles C.R. Phase relations in the systems $PbS–Sb_2S_3–Bi_2S_3$ and $PbS–FeS–Sb_2S_3–Bi_2S_3$, *Econ. Geol.*, *Econ. Geol.*, **75**(2), 317–328 (1980).

Chang L.L.Y., Wu D., Knowles Ch.R. Phase relations in the system $Ag_2S–Cu_2S–PbS–Bi_2S_3$, *Econ. Geol.*, **83**(2), 405–418 (1988).

Chen B., Yang J.-H., Wang H.-D., Imai M., Ohta H., Michioka C., Yoshimura K., Fang M.-H. Magnetic properties of layered itinerant electron ferromagnet Fe_3GeTe_2, *J. Phys. Soc. Jpn.*, **82**(12), 124711_1–124711_7 (2013).

Chen G.-R., Wang M.-F., Lee C.-S. Synthesis and characterization of new multinary selenides $Sn_4In_5Sb_9Se_{25}$ and $Sn_{6.13}Pb_{1.87}In_{5.00}Sb_{10.12}Bi_{2.88}Se_{35}$, *J. Solid.State Chem.*, **307**, 122855 (2022a).

Chen H., Malliakas C.D., Narayan A., Fang L., Chung D.Y., Wagner L.K., Kwok W.-K., Kanatzidis M.G. Charge density wave and narrow energy gap at room temperature in 2D $Pb_{3-x}Sb_{1+x}S_4Te_{2-\delta}$ with square Te sheets, *J. Am. Chem. Soc.*, **139**(32), 11271–11276 (2017a).

Chen J., Lin C., Yang S., Jiang X., Shi S., Sun Y., Li B., Fang S., Ye N. Cd_4SiQ_6 (Q = S, Se): Ternary infrared nonlinear optical materials with mixed functional building motifs, *Cryst. Growth Des.*, **20**(4), 2489–2496 (2020a).

Chen K., Di Paola C., Laricchia S., Reece M.J., Weber C., McCabe E., Abrahams I., Bonini N. Structural and electronic evolution in the $Cu_3SbS_4–Cu_3SnS_4$ solid solution, *J. Mater. Chem. C.*, **8**(33), 11508–11516 (2020b).

Chen K., Lin C., Zhang S., Peng G., Chen Y., Zhao D., Fan H., Yang S., Wen X., Luo M., Lin Z., Ye N. $A_3Te(Zn_2Ge)Ge_2O_{14}$ (A = Sr, Ba, and Pb): New langasite mid-infrared nonlinear optical materials by rational chemical substitution, *Chem. Mater.*, **33**(15), 6012–6017 (2021a).

Chen M., Fang Y. The chemical composition and crystal parameters of calcium chlorosulfatosilicate, *Cem. Concr. Res.*, **19**(2), 184–188 (1989).

Chen M., Sun Y., Balladares E., Pizarro C., Zhao B., Experimental studies of liquid/spinel/matte/gas equilibria in the Si–Fe–O–Cu–S system at controlled $P(SO_2)$ 0.3 and 0.6 atm, *Calphad*, **66**, 101642 (2019).

Chen M.-M., Ma Z., Li B.-X., Wei W.-B., Wu X.-T., Lin H., Zhu Q.-L. $M_2As_2Q_5$ (M = Ba, Pb; Q = S, Se): a source of infrared nonlinear optical materials with excellent overall performance activated by multiple discrete arsenate anions, *J. Mater. Chem. C*, **9**(4), 1156–1163 (2021b).

Chen S.-W., Huang T.-Y., Hsu Y.-H., Kroupa A. Phase diagram of the Pb–Se–Sn system, *J. Electron. Mater.*, **49**(8), 4714–4729 (2020c).

Chen S.-W., Huang T.-Y., Hsu Y.-H., Liu J.X, Zemanova A., Kroupa A. Phase diagram of Pb–Se–Te system I: Experimental study, *Calphad*, **74**, 102310 (2021c).

Chen S.-W., Kroupa A., Du J.-Y., Zemanová A., Hutabalian Y., Vřešťál J., Chiu K.-C. Experimental and theoretical study of the Ag–Sn–Te phase diagram, *J. Phase Equilib. Diffus.*, **43**(2), 139–163 (2022b).

Chen W.-F., Liu B.-.W., Jiang X.-M., Guo G.-C. Infrared nonlinear optical performances of a new sulfide *β*-$PbGa_2S_4$, *J. Alloys Compd.*, **905**, 164090 (2022c).

Chen W.-F., Liu B.-W., Pei S.-M., Jiang X.-M., Guo G.-C. $[K_2PbX][Ga_7S_{12}]$ (X = Cl, Br, I): The first lead-containing cationic moieties with ultrahigh second-harmonic generation and band gaps exceeding the criterion of 2.33 eV, *Adv. Sci.*, **10**(13), 2207630 (2023).

Chen Y.G., Yang N., Jiang X.X., Guo Y., Zhang X.M. Pb@Pb8 basket-like-cluster-based lead tellurate-nitrate Kleinman-forbidden nonlinear-optical crystal: $Pb_9Te_2O_{13}(OH)(NO_3)_3$, *Inorg.Chem.*, **56**(14), 7900–7906 (2017b).

Cheng H., Tudi A., Wang P., Zhang K., Yang Z., Pan S. Design and synthesis of Ba_3SiSe_5 with suitable birefringence modulated via M^{IV} atoms in the Ba–M^{IV}–Q (M^{IV} = Si, Ge; Q = S, Se) system, *Dalton Trans.*, **50**(34), 11999–12005 (2021).

Cheng Y., Wu H., Yu H., Hu Z., Wang J., Wu Y. Rational design of a promising oxychalcogenide infrared nonlinear optical crystal, *Chem. Sci.*, **13**(18), 5305–5310 (2022).

Chernov A.N., Ilyukhin V.V., Maksimov B.A., Belov N.V. Crystal structure of innelite – $Na_2Ba_3(Ba,K,Mn)(Ca,Na)Ti(TiO_2)_2[Si_2O_7]_2(SO_4)_2$ [in Russian], *Kristallografiya*, **16**(1), 87–92 (1971).

Chesnokov B.V., Bazhenova L.F., Bushmakin A.F. Fluorellestadite $Ca_{10}[(SO_4),(SiO_4)]_6F_2$ — a new mineral [in Russian], *Zap. Vses. Mineralog. Obshch.*, **116**(6), 743–746 (1987).

Chiu C.-N., Hsu C.-M., Chen S.-W., Wu H.-J. Phase equilibria of the Sn–Bi–Te ternary system, *J. Electron. Mater.*, **41**(1), 22–31 (2012).

Choisnet J., Rulmont A., Tarte P. Ordering phenomena in the $LiSbO_3$ type structure: the new mixed tellurates Li_2TiTeO_6 and Li_2SnTeO_6, *J. Solid State Chem.*, **82**(2), 272–278 (1989).

Chondroudis K., Kanatzidis M.G. $Rb_4Sn_2Ag_4(P_2Se_6)_3$: First example of a quinary selenophosphate and an unusual Sn–Ag s^2-d^{10} interaction, *Inorg. Chem.*, **37**(12), 2848–2849 (1998).

Chorba O.Y., Filep M.Y., Pogodin A.I., Malakhovska T.O., Sabov M.Yu. Triangulation of the Cu–Sn–Se system [in Ukrainian], *Nauk. Visn. Uzhgorod. Univ., Ser. Khim.*, [2(46)], 22–27 (2021).

Choudhury A., Dorhout P.K. Alkali-metal thiogermanates: sodium channels and variations on the La_3CuSiS_7 structure type, *Inorg. Chem.*, **54**(3), 1055–1065 (2015).

Chu Y., Wu K., Su X., Han J., Yang Z., Pan S. Intriguing structural transition inducing variable birefringences in ABa_2MS_4Cl (A = Rb, Cs; M = Ge, Sn), *Inorg. Chem.*, **57**(18), 11310–11313 (2018).

Chudo H., Micioka C., Nakamura H., Yoshimura K. Magnetic and structural transitions of GeV_4S_8, *Physica B*, **378–380**, 1150–1151 (2006).

Chukanov N.V., Britvin S.N., Van, K.V., Möckel S., Zadov A.E. Kottenheimite, $Ca_3Si(OH)_6(SO_4)_2{\cdot}12H_2O$, a new member of the ettringite group from the Eifel area, Germany, *Canad. Mineralog.*, **50**(1), 55–63 (2012).

Chukanov N.V., Britvin S.N., Zadov A.E., Van, K.V. Kottenheimite, IMA 2011-038. CNMNC Newsletter No. 10, October 2011, page 2557, *Mineralog. Mag.*, **75**(5), 2549–2561 (2011).

Chukanov N.V., Rastsvetaeva R.K., Pekov I.V., Zadov A.E. Alloriite, $Na_5K_{1.5}Ca(Si_6Al_6O_{24})(SO_4)(OH)_{0.5}{\cdot}H_2O$, a new mineral of the cancrinite group [in Russian], *Zap. Ros. Mineralog. Obshch.*, **136**(1), 82–89 (2007).

Chukanov N.V., Rastsvetaeva R.K., Pekov I.V., Zadov A.E., Allori R., Zubkova N.V., Gister G., Pushcharovskiy D.Yu., Van K.V. Biachellaite $(Na,Ca,K)_8(Si_6Al_6O_{24})(SO_4)_2(OH)_{0.5}{\cdot}H_2O$, a new mineral of the cancrinite group [in Russian], *Zap. Ros. Mineralog. Obshch.*, **137**(3), 57–66 (2008).

Chukanov N.V., Rastsvetaeva R.K., Pekov I.V., Zadov A.E., Allori R., Zubkova N.V., Giester G., Puscharovsky D.Yu., Van K.V. Biachellaite, $(Na,Ca,K)_8(Si_6Al_6O_{24})(SO_4)_2(OH)_{0.5}{\cdot}H_2O$, a new mineral species of the cancrinite group, *Geol. Ore Deposits*, **51**(7), 588–594 (2009).

Chukanov N.V., Sapozhnikov A.N., Kaneva E.V., Varlamov D.A., Vigasina M.F. Bystrite, $Na_7Ca(Al_6Si_6O_{24})S_5^{2-}Cl^-$: formula redefinition and relationships with other four-layer cancrinite-group minerals, *Mineralog. Mag.*, **87**(3), 455–464 (2023).

Chukanov N.V., Zubkova N.V., Pekov I.V., Pushcharovsky D.Y. Sulfite analogue of alloriite from Sacrofano, Latium, Italy: Crystal chemistry and specific features of genesis, *Geol. Ore Deposits*, **63**(8), 793–804 (2021a).

Chukanov N.V., Zubkova N.V., Pekov I.V., Shendrik R.Y., Varlamov D.A., Vigasina M.F., Belakovskiy D.I., Britvin S.N., Yapaskurt V.O., Pushcharovsky D.Y. Sapozhnikovite, IMA 2021-030. CNMNC Newsletter No. 62, *Eur. J. Mineral.*, **33**(4), 482 (2021b).

Chukanov N.V., Zubkova N.V., Pekov I.V., Shendrik R.Y., Varlamov D.A., Vigasina M.F., Belakovskiy D.I., Britvin S.N., Yapaskurt V.O., Pushcharovsky D.Y. Sapozhnikovite, IMA 2021-030. CNMNC Newsletter No. 62, *Mineralog. Mag.*, **85**(4), 636–637 (2021c).

Chukanov N.V., Zubkova N.V., Pekov I.V., Shendrik R.Y., Varlamov D.A., Vigasina M.F., Belakovskiy D.I., Britvin S.N., Yapaskurt V.O., Pushcharovsky D.Y. Sapozhnikovite, $Na_8(Al_6Si_6O_{24})(HS)_2$, a new sodalite-group mineral from the Lovozero alkaline massif, Kola Peninsula, *Mineralog. Mag.*, **86**(1), 49–59 (2022).

Chukanov N.V., Zubkova N.V., Varlamov D.A., Pekov I.V., Belakovskiy D.I., Britvin S.N., Van K.V., Ermolaeva V.N., Vozchikova S.A., Pushcharovsky D.Y. Steudelite, IMA 2021-007. CNMNC Newsletter No. 61, *Eur. J. Mineral.*, **33**(3), 299–304 (2021d).

Chukanov N.V., Zubkova N.V., Varlamov D.A., Pekov I.V., Belakovskiy D.I., Britvin S.N., Van K.V., Ermolaeva V.N., Vozchikova S.A., Pushcharovsky D.Y. Steudelite, IMA 2021-007. CNMNC Newsletter No. 61, *Mineralog. Mag.*, **85**(3), 459–463 (2021e).

Chung M.-Y., Lee C.-S. Multinary selenides with unusual coordination environment of bismuth, *Inorg. Chem.*, **51**(24), 13328–13333 (2012).

Chung M.-Y., Lee C.-S. New quinternary selenides: Syntheses, characterizations, and electronic structure calculations, *J. Solid State Chem.*, **202**, 154–160 (2013).

Chutas N.I., Kress V.C., Ghiorso M.S., Sack R.O. A solution model for high-temperature PbS–$AgSbS_2$–$AgBiS_2$ galena, *Am. Mineralog.*, **93**(10), 1630–1640 (2008).

Cicirello G., Wu K., Wang J. Synthesis, crystal structure, linear and nonlinear optical properties of quaternary sulfides $Ba_6(Cu_2X)Ge_4S_{16}$ (X = Mg, Mn, Cd). *J. Solid State Chem.*, **300**, 122226 (2021a).

Cicirello G., Wu K., Zhang B.B., Wang J. Applying band gap engineering to tune the linear optical and non-linear optical properties of noncentrosymmetric chalcogenides $La_4Ge_3Se_xS_{12-x}$ (x = 0, 2, 4, 6, 8, 10), *Inorg. Chem.Front.*, **8**(22), 4914–4923 (2021b).

Cipriani C., Corazza M., Pratesi G. A new lead silicate sulfate hydroxide from Val Fucinaia, Tuscany, Italy, *Per. Mineral.*, **64**, 309–313 (1995).

Clark A.M., Criddle A.J., Roberts A.C., Bonardi M., Moffatt E.A. Feinglosite, a new mineral related to brackebuschite, from Tsumeb, Namibia, *Mineralog. Mag.*, **61**(405), 285–289 (1997).

Climent-Pascual E., de Paz J.R., Rodríguez-Carvajal J., Suard E., Sáez-Puche R. Synthesis and characterization of the ultramarine-type analog $Na_{8-x}[Si_6Al_6O_{24}]\cdot(S_2,S_3,CO_3)_{1-2}$, *Inorg. Chem.*, **48**(14), 6526–6533 (2009).

Cook N.J., Ciobanu C.L., Yao J., Stanley C.J., Liu W., Slattery A., Wade B. Clogauite, IMA 2023-062. CNMNC Newsletter No 76, *Eur. J. Mineral.*, **35**(6), 1074–1075 (2023).

Cook N.J., Ciobanu C.L., Yao J., Stanley C.J., Liu W., Slattery A., Wade B. Clogauite, IMA 2023-062. CNMNC Newsletter No 76, *Mineralog. Mag.*, **88**(1), 106 (2024).

Cook N.J., Wood S.A., Gebert W., Bernhardt H.J., Medenbach O. Crerarite, a new Pt–Bi–Pb–S mineral from the Cu–Bi–Ni–PGE deposit at Lac Sheen, Abitibi-Témiscamingue, Québec, Canada. *N. Jb. Miner. Mh.*, (12), 567–575 (1994).

Cooper M.A., Hawthorne F.C. Refinement of the crystal structure of zoned philipsbornite–hidalgoite from the Tsumeb mine, Namibia, and hydrogen bonding in the $D^{2+}G^{3+}{}_3(T^{5+}O)(TO_3OH)(OH)_6$ alunite structures, *Mineralog. Mag.*, **76**(4), 839–849 (2012).

Cooper M., Hawthorne F.C. The crystal structure of wherryite, $Pb_7Cu_2(SO_4)_4(SiO_4)_2(OH)_2$, a mixed sulfate-silicate with $[^{[6]}M(TO_4)_2\Phi]$ chains, *Canad. Mineralog.*, **32**(2), 373–380 (1994).

Cooper M.A., Hawthorne F.C., Back M.E. The crystal structure of khinite and polytypism in khinite and parakhinite, *Mineralog. Mag.*, **72**(3), 763–770 (2008).

Cooper M.A., Hawthorne F.C., Moffatt E. Steverustite, $Pb^{2+}{}_5(OH)_5[Cu^+(S^{6+}O_3S^{2-})_3](H_2O)_2$, a new thiosulphate mineral from the Frongoch Mine Dump, Devils Bridge, Ceredigion, Wales: description and crystal structure, *Mineralog. Mag.*, **73**(2), 235–250 (2009).

Coulon M., Heitz F., Le Bihan M.-Th. Contribution a l'étude structurale d'un sulfure de plomb, d'antimoine et d'étain: la Frankeite, *Bull. Soc. fr. minéral. cristallogr.*, **84**(4), 350–353 (1961).

Craig A.J., Shin S.H., Cho J.B., Balijapelly S., Kelly J.C., Stoyko S.S., Choudhury A., Jang J.I., Aitken J.A. Crystal structure, electronic structure, and optical properties of the novel $Li_4CdGe_2S_7$, a wide-bandgap quaternary sulfide with a polar structure derived from lonsdaleite, *Acta Crystallogr.*, **C78**(9), 470–480 (2022).

Cui S., Wu H., Hu Z., Wang J., Wu Y., Yu H. The antiperovskite-type oxychalcogenides $Ae_3Q[GeOQ_3]$ (Ae = Ba, Sr; Q = S, Se) with large second harmonic generation responses and wide band gaps, *Adv. Sci.*, **10**(4), 2204755 (2023).

Damidot D., Barnett S.J., Glasser F.P., Macphee D.E. Investigation of the $CaO–Al_2O_3–SiO_2–CaSO_4–CaCO_3–H_2O$ system at 25°C by thermodynamic calculation, *Adv. Cem. Res.*, **16**(2), 69–76 (2004).

Dang J., Wang N., Yao J., Wu Y., Lin Z., Mei D. $AgGaGeSe_4$: An infrared nonlinear quaternary selenide with good performance, *Symmetry*, **14**(7), 1426 (2022).

Dangel P.N., Wuensch B.J. The crystallography of colusite, *Am. Mineralog.*, **55**(9–10), 1787–1791 (1970).

Davaasuren B., Emwas A.-H., Rothenberger A. MAu_2GeS_4-chalcogel (M = Co, Ni): heterogeneous intra- and intermolecular hydroamination catalysts, *Inorg. Chem.*, **56**(16), 9609–9616 (2017).

Delgado G.E., Quintero E., Tovar R., Quintero M. X-ray powder diffraction study of the semiconducting alloy $Cu_2Cd_{0.5}Mn_{0.5}GeSe_4$, *Cryst. Res. Technol.*, **39**(9), 807–810 (2004).

Della Ventura G., Bellatreccia F., Parodi G.C., Cámara F., Piccinini M. Single-crystal FTIR and X-ray study of vishnevite, ideally $[Na_6(SO_4)][Na_2(H_2O)_2](Si_6Al_6O_{24})$, *Am. Mineralog.*, **92**(5–6), 713–721 (2007).

Demartin F., Gramaccioli C.M., Pilati T. The crystal structure of orlandiite, $Pb_3Cl_4(SeO_3)\cdot H_2O$, a complex case of twinning and disorder, *Canad. Mineralog.*, **41**(5), 1147–1153 (2003).

Deng T., Xing T., Brod M., Sheng Y., Qiu P., Veremchuk I., Song Q., Wei T.-R., Yang J., Snyder G.J., Grin Y., Chen L., Shi X. Discovery of high-performance thermoelectric copper chalcogenide using modified diffusion-couple high-throughput synthesis and automated histogram analysis technique, *Energy Environ. Sci.*, **13**(9), 3041–3053 (2020).

Dihn P., Klement R. Isomorphe Apatitarten, *Z. Elektrochem. Angew. Phys. Chem.*, **48**(6), 331–333 (1942).

Ding N., Kanatzidis M.G. Acid-induced conversions in open-framework semiconductors: From $[Cd_4Sn_3Se_{13}]^{6-}$ to $[Cd_{15}Sn_{12}Se_{46}]^{14-}$, a remarkable disassembly/reassembly process, *Angew. Chem.*, **118**(9), 1425–1429 (2006a).

Ding N., Kanatzidis M.G. Acid-induced conversions in open-framework semiconductors: From $[Cd_4Sn_3Se_{13}]^{6-}$ to $[Cd_{15}Sn_{12}Se_{46}]^{14-}$, a remarkable disassembly/reassembly process, *Angew. Chem. Int. Ed.*, **45**(9), 1397–1401 (2006b).

Dong Y., Eckert B., Wang H., Zeng X., Tritt T.M., Nolas G.S. Synthesis, crystal structure, and transport properties of $Cu_{2.2}Zn_{0.8}SnSe_{4-x}Te_x$ ($0.1 \leq x \leq 0.4$), *Dalton Trans.*, **44**(19), 9014–9019 (2015a).

Dong Y., Khabibullin A.R., Wei K., Ge Z.-H., Martin J., Salvador J.R., Woods L.M., Nolas G.S. Synthesis, transport properties, and electronic structure of $Cu_2CdSnTe_4$, *Appl. Phys. Lett.*, **104**(25), 252107_1–252107_4 (2014).

Dong Y., Wojtas L., Martin J., Nolas G.S. Synthesis, crystal structure, and transport properties of quaternary tetrahedral chalcogenides, *J. Mater. Chem. C*, **3**(40), 10436–10441 (2015b).

Dost J., Steinike U., Paudert R. Über die Komplexbildung zwischen Zink- und Blei(II)-thiocyanat, *Krist. und Techn.*, **3**(2), 247–254 (1968).

Doussier C., Moëlo Y., Léone P., Meerschaut A. $(Mn_{1-x}Pb_x)Pb_{10+y}Sb_{12-y}S_{26-y}Cl_{4+y}O$, a new oxy-chloro-sulfide with ~2 nm-spaced $(Mn,Pb)Cl_4$ single chains within a waffle-type crystal structure, *J. Solid State Chem.*, **180**(8), 2323–2334 (2007).

Doverspike K., Dwight K., Wold A. Preparation and characterization of $Cu_2ZnGeS_{4-y}Se_y$, *Chem. Mater.*, **2**(2), 194–197 (1990).

Doverspike K., Kershaw R., Dwight K., Wold A. Preparation and properties of the system $Cu_2Zn_{1-x}Fe_xGeS_4$, *Mater. Res. Bull.*, **23**(7), 959–964 (1988).

Drummond A.D., Trotter J., Thompson R.M., Gower J.A. Neyite; a new sulphosalt from Alice Arm, British Columbia, *Canad. Mineralog.*, **10**(1), 90–96 (1969).

Du K.-Z., Qi X.-H., Feng M.-L., Li J.-R., Wang X.-Z., Du C.-F., Zou G.-D., Wang M., Huang X.-Y. Synthesis, structure, band gap, and near-infrared photosensitivity of a new chalcogenide crystal, $(NH_4)_4Ag_{12}Sn_7Se_{22}$, *Inorg. Chem.*, **55**(11), 5110–5112 (2016). 10.1021/acs.inorgchem.6b00803

Duan R., Lin H., Wang Y., Zhou Y., Wu L. Non-centrosymmetric sulfides $A_2Ba_6MnSn_4S_{16}$ (A = Li, Ag): syntheses, structures and properties, *Dalton Trans.*, **49**(18), 5914–5920 (2020).

Duan R.-H., Li R.-A., Liu P.-F., Lin H., Wang Y., Wu L.-M. Modifying disordered sites with rational cations to regulate band-gaps and second-harmonic-generation responses markedly: $Ba_6Li_2ZnSn_4S_{16}$ vs. $Ba_6Ag_2ZnSn_4S_{16}$ vs. $Ba_6Li_{2.67}Sn_{4.33}S_{16}$, *Cryst. Growth Des.*, **18**(9), 5609–5616 (2018).

Duan R.-H., Liu P.-F., Lin H., Zheng Y.J., Yu J.-S., Wu X.-T., Huang-Fu S.-X., Chen L. $Ba_6Li_2CdSn_4S_{16}$: lithium substitution simultaneously enhances band gap and SHG intensity, *J. Mater. Chem.*, **5**(28), 7067–7074 (2017).

Duchardt M., Ruschewitz U., Adams S., Dehnen S., Roling B. Vacancy-controlled Na^+ superion conduction in $Na_{11}Sn_2PS_{12}$, *Angew. Chem.*, **130**(5), 1351–1355 (2018a).

Duchardt M., Ruschewitz U., Adams S., Dehnen S., Roling B. Vacancy-controlled Na^+ superion conduction in $Na_{11}Sn_2PS_{12}$, *Angew. Chem. Int. Ed.*, **57**(5), 1365–1369 (2018b).

Dunn P.J., Braithwaite R.S.W., Roberts A.C., Ramik R.A. Kegelite from Tsumeb, Namibia: a redefinition, *Am. Mineralog.*, **75**(5–6), 702–704 (1990).

Dunn P.J., Cabri L.J., Chao G.Y., Fleischer M., Francis C.A., Grice J.D., Jambor J.L., Pabst A. New mineral names, *Am. Mineralog.*, **69**(3–4), 406–412 (1984a).

Dunn P.J., Chao G.Y., Fitzpatrick J.J., Langley R.H., Fleischer M., Zilczer J.A. New mineral names, *Am. Mineralog.*, **71**(1–2), 227–232 (1986).

Dunn P.J., Chao G.Y., Fleischer M., Ferraiolo J.A., Langley R.H., Pabst A., Zilczer J.A. New mineral names, *Am. Mineralog.*, **70**(1–2), 214–221 (1985a).

Dunn P.J., Ferraiolo J.A., Fleischer M., Gobel V., Grice J.D., Langley R.H., Shigley J.E., Vanko D.A., Zilczer J. New mineral names, *Am. Mineralog.*, **70**(11–12), Pt. II, 1329–1335 (1985b).

Dunn P.J., Fleischer M., Burns R.G., Pabst A. New mineral names, *Am. Mineralog.*, **68**(3–4), 471–475 (1983a).

Dunn P.J., Fleischer M., Chao G.Y., Cabri L.J., Mandarino J.A. New mineral names, *Am. Mineralog.*, **68**(11–12), 1248–1252 (1983b).

Dunn P.J., Fleischer M., Langley R.H., Shigley J.E., Zilczer J.A. New mineral names, *Am. Mineralog.*, **70**(7–8), 871–881 (1985c).

Dunn P.J., Gobel V., Grice J.D., Puziewicz J., Shigley J.E., Vanko D.A., Zilczer J. New mineral names, *Am. Mineralog.*, **70**(3–4), 436–441 (1985d).

Dunn P.J., Grice J.D., Fleischer M., Pabst A. New mineral names, *Am. Mineralog.*, **69**(1–2), 210–215 (1984b).

Dunn P.J., Peacor D.R., Leavens P.B., Baum J.L. Charlesite, a new mineral of the ettringite group, from Franklin, New Jersey, *Am. Mineralog.*, **68**(9-10), 1033–1037 (1983c).

Edenharter A., Steins M., Bente K. Stukturuntersuchung an Montbrayit, Au_2Te_3, *Z. Kristallogr.*, Suppl., 63–64 (1991).

Edge R.A., Taylor H.F.W. Crystal structure of thaumasite, $[Ca_3Si(OH)_6 \cdot 12H_2O](SO_4)(CO_3)$, *Acta Crystallogr.*, **B27**(3), 594–601 (1971).

Effenberger H. Crystal structure and chemical formula of schmiederite, $Pb_2Cu_2(OH)_4(SeO_3)(SeO_4)$, with a comparison to linarite, $PbCu(OH)_2(SO_4)$, *Mineral. Petrol.*, **36**(1), 3–12 (1987).

Effenberger H. $PbCu_3(OH)(NO_3)(SeO_3)_3 \cdot 1/2H_2O$ und $Pb_2Cu_3O_2(NO_3)_2(SeO_3)_2$: Synthese und Kristallstrukturuntersuchung, *Monatsh. Chem.*, **117**(10), 1099–1106 (1986).

Effenberger H. The crystal structure of mammothite, $Pb_6Cu_4AlSbO_2(OH)_{16}Cl_4(SO_4)_2$, *Tschermaks Miner. Petr. Mitt.*, **34**(3-4), 279–288 (1985).

Effenberger H., Culetto F.J., Topa D., Paar W.H. The crystal structure of synthetic buckhornite, $[Pb_2BiS_3]$ $[AuTe_2]$, *Z. Kristollagr.*, **215**(1), 10–16 (2000).

Effenberger H., Paar W.H., Topa D., Culetto F.J.. Giester G. Towards the crystal structure of nagyagite, $[Pb(Pb,Sb)S_2][(Au,Te)]$, *Am. Mineralog.*, **84**(4), 669–676 (1999).

Egorov N.B. Synthesis and properties of $Na_6[Pb(S_2O_3)_4]{\cdot}6H_2O$, *Russ. J. Inorg. Chem.*, **55**(2), 179–185 (2010).

Endo T., Doi Y., Wakeshima M., Suzuki K., Matsuo Y., Tezuka K., Ohtsuki T., Shan Y.J., Hinatsu Y. Magnetic properties of the melilite-type oxysulfide $Sr_2MnGe_2S_6O$: magnetic interactions enhanced by anion substitution, *Inorg. Chem.*, **56**(5), 2459–2466 (2017).

Enikeev M.R. A new mineral, nasledovite, from the Altyn-Topkansk ore field [in Russian], *Dokl. AN Uzb. SSR*, (5), 13–16 (1958).

Ercit T.C., Piilonen P.C., Locock A.J., Kolitsch U., Rowe R. New mineral names, *Am. Mineralog.*, **91**(7), 1201–1209 (2006).

Ercit T.C., Piilonen P.C., Poirier G., Tait K.T. New mineral names, *Am. Mineralog.*, **94**(7), 1075–1083 (2009).

Ersundu A.E., Sayyed M.I., Kıbrıslı O., Akıllı V., Ersundu M.Ç. (2020). A thorough investigation of the Bi_2O_3–$PbCl_2$–TeO_2 system: glass forming region, thermal, physical, optical, structural, mechanical and radiation shielding properties, *J. Alloys Compd.*, **857**, 158279 (2021).

Evain M., Petricek V., Moëlo Y., Maurel C. First (3 + 2)-dimensional superspace approach to the structure of levyclaudite-(Sb), a member of the cylindrite-type minerals, *Acta Crystallogr.*, **B62**(5), 775–789 (2006).

Evstigneeva T.L., Genkin A.D., Troneva N.V., Filimonova A.A., Tsepin A.I. Shadlunite, a new sulfide of copper, iron, lead, manganese, and cadmium from copper–nickel ores [in Russian], *Zap. Vses. Mineralog. Obsh.*, **102**(1), 63–74 (1973).

Evsyunin V.G., Rastsvetaeva R.K., Sapozhnikov A.N., Kashaev A.A. Modulated structure of orthorhombic lazurite, *Crystallogr. Rep.*, **43**(6), 999–1002 (1998).

Evsyunin V.G., Sapozhnikov A.N., Kashaev A.A., Rastsvetaeva R.K. Crystal structure of triclinic lazurite, *Crystallogr. Rep.*, **42**(6), 938–945 (1997).

Evsyunin V.G., Sapozhnikov A.N., Rastsvetaeva R.K., Kashaev A.A. Crystal structure of the potassium-rich hayüne from Arissia (Italy) [in Russian], *Kristallografiya*, **41**(4), 659–662 (1996).

Fahey J.J., Daggett E.B., Gordon S.G. Wherryite, a new mineral from the Mammoth mine, Arizona, *Am. Mineralog.*, **35**(1-2), 93–98 (1950).

Fallah-Mehrjardi A., Hidayat T., Hayes P.C., Jak E. Experimental investigation of gas/slag/matte/tridymite equilibria in the Cu–Fe–O–S–Si system in controlled atmospheres: experimental results at 1473 K (1200°C) and $P(SO_2) = 0.25$ atm, *Metall. Mater. Trans.*, **B48**(6), 3017–3026 (2017).

Fallah-Mehrjardi A., Hidayat T., Hayes P.C., Jak E. Experimental investigation of gas/slag/matte/tridymite equilibria in the Cu–Fe–O–S–Si system in controlled gas atmospheres:experimental results at 1523 K (1250°C) and $P(SO_2) = 0.25$ atm, *Metall. Mater. Trans.*, **B49**(4), 1732–1739 (2018).

Fallah-Mehrjardi A., Hidayat T., Hayes P.C., Jak E. The effect of sulfur dioxide partial pressure on gas-slag-matte-tridymite equilibria in the Cu–Fe–O–S–Si system at 1200°C, *Miner. Process. Extr. Metall.*, **131**(1), 61–68 (2020a).

Fallah-Mehrjardi A., Hidayat T., Hayes P.C., Jak E. The influence of temperature and matte grade on gas-slag-matte-tridymite equilibria in the Cu–Fe–O–S–Si system at $p(SO_2) = 0.25$ atm, *Miner. Process. Extr. Metall.*, **131**(1), 53–60 (2020b).

Falmbigl M., Putzky D., Ditto J., Esters M., Bauers S.R., Ronning F., Johnson D.C. Influence of defects on the charge density wave of $([SnSe]_{1+\delta})_1(VSe_2)_1$ ferecrystals, *ACS Nano*, **9**(8), 8440–8448 (2015).

Fan Y.-H., Zeng H.-Y., Jiang X.-M., Zhang M.-J., Liu B.-W., Guo G.-C., Huang J.-S. Thiophosphates containing Ag^+ and lone-pair cations with interchiral double helix show both ionic conductivity and phase transition, *Inorg. Chem.*, **56**(2), 962–973 (2017).

Fang Y.N., Ritter C., White T.J. Crystal chemical characteristics of ellestadite-type apatite: implications for toxic metal immobilization, *Dalton Trans.*, **43**(42), 16031–16043 (2014).

Fang Y.N., Ritter C., White T. The crystal chemistry of $Ca_{10-y}(SiO_4)_3(SO_4)_3Cl_{2-x-2y}F_x$ ellestadite, *Inorg. Chem.*, **50**(24), 12641–12650 (2011).
Fayos J., Glasser F.P., Howie R.A., Lachowski E., Perez-Mendez M. Structure of dodecacalcium potassium fluoride dioxide tetrasilicate bis(sulphate), $KF{\cdot}2[Ca_6(SO_4)(SiO_4)_2O]$: a fluorine-containing phase encountered in cement clinker production process, *Acta Crystallogr.*, **C41**(6), 814–816 (1985).
Fedorchuk A.O., Gorgut G.P., Parasyuk O.V., Lakshminarayana G., Kityk I.V., Piasecki M. IR operated novel $Ag_{0.98}Cu_{0.02}GaGe_3Se_8$ single crystals, *J. Phys. Chem. Solids*, **72**(11), 1354–1357 (2011).
Fedorchuk A.O., Zhbankov O.Ye., Lakshminarayana G., Kityk I.V., Tokaichuk Y., Myronchuk G.L., Davydyuk G.Ye., Yakymchuk O.V., Parasyuk O.V. Synthesis and spectral features of Ag_2SnS_3 crystals, *Mater. Chem. Phys.*, **135**(2-3), 249–253 (2012).
Feng K., Kang l., Lin Z., Yao J., Wu J. Noncentrosymmetric chalcohalide $NaBa_4Ge_3S_{10}Cl$ with large band gap and IR NLO response, *J. Mater. Chem.*, **2**(23), 4590–4596 (2014).
Ferhat A., Ollitrault-Fichet R., Rivet J. Description du système ternaire Ag–Ge–Te, *J. Alloys Compd.*, **177**(2), 337–355 (1991).
Ferrenti A.M., Meschke V., Ghosh S., Davis J., Drichko N., Toberer E.S., McQueen T.M. Hydrothermal synthesis of ordered corkite, $PbFe_3(PO_4)(SO_4)(OH)_6$, a S = 5/2 kagomé antiferromagnet, *J. Solid State Chem.*, **317**, Pt. A, 123620 (2023).
Fikar O., Vřešťál J., Kroupa A., Chen S.-W. The study of the Pb–Se–Te phase diagram: Part 2 — The thermodynamic assessment of the Se–Te and Pb–Se–Te systems, *Calphad*, **74**, 102309 (2021).
Filut M.A., Rule A.C., Bailey S.W. Crystal structure refinement of anandite-2*Or*, a barium- and sulfur-bearing trioctahedral mica, *Am. Mineralog.*, **70**(11-12), Pt. II. 1298–1308 (1985).
Fleischer M. New mineral names, *Am. Mineralog.*, **39**(3-4), 402–408 (1954).
Fleischer M. New mineral names, *Am. Mineralog.*, **44**(11-12), 1321–1329 (1959).
Fleischer M. New mineral names, *Am. Mineralog.*, **45**(11-12), 1313–1317 (1960).
Fleischer M. New mineral names, *Am. Mineralog.*, **47**(5-6), 805–812 (1962).
Fleischer M. New mineral names, *Am. Mineralog.*, **48**(1-2), 209–217 (1963).
Fleischer M. New mineral names, *Am. Mineralog.*, **49**(11-12), 1774–1778 (1964).
Fleischer M. New mineral names, *Am. Mineralog.*, **50**(1-2), 261–268 (1965).
Fleischer M. New mineral names, *Am. Mineralog.*, **51**(3-4), Pt. 1, 529–534 (1966a).
Fleischer M. New mineral names, *Am. Mineralog.*, **51**(11-12), 1815–1820 (1966b).
Fleischer M. New mineral names, *Am. Mineralog.*, **52**(5-6), 925–929 (1967a).
Fleischer M. New mineral names, *Am. Mineralog.*, **52**(9-10), 1579–1589 (1967b).
Fleischer M. New mineral names, *Am. Mineralog.*, **53**(3-4), 507–511 (1968a).
Fleischer M. New mineral names, *Am. Mineralog.*, **53**(11-12), 2103–2106 (1968b).
Fleischer M. New mineral names, *Am. Mineralog.*, **54**(3-4), Pt. 1, 573–580 (1969a).
Fleischer M. New mineral names, *Am. Mineralog.*, **54**(9-10), 1495–1499 (1969b).
Fleischer M. New mineral names, *Am. Mineralog.*, **55**(7-8), 1444–1449 (1970).
Fleischer M. New mineral names, *Am. Mineralog.*, **58**(11-12), 1111–1115 (1973).
Fleischer M. New mineral names, *Am. Mineralog.*, **59**(7-8), 873–875 (1974).
Fleischer M. New mineral names, *Am. Mineralog.*, **63**(11-12), 1289–1291 (1978).
Fleischer M., Burns R.G., Cabri L.J., Chao G.Y., Hogarth D.D., Pabst A. New mineral names, *Am. Mineralog.*, **64**(11-12), 1329–1334 (1979a).
Fleischer M., Burns R.G., Cabri L.J., Francis C.A., Pabst A. New mineral names, *Am. Mineralog.*, **67**(3-4), 413–418 (1982a).
Fleischer M., Cabri L.J., Chao G.Y., Mandarino J.A., Pabst A. New mineral names, *Am. Mineralog.*, **67**(5-6), 621–624 (1982b).
Fleischer M., Cabri L.J., Nickel E.H., Pabst A. New mineral names, *Am. Mineralog.*, **62**(5-6), 593–600 (1977a).
Fleischer M., Cabri L.J., Pabst A. New mineral names, *Am. Mineralog.*, **62**(11-12), 1259–1262 (1977b).
Fleischer M., Cabri L.J., Pabst A. New mineral names, *Am. Mineralog.*, **65**(3-4), 406–408 (1980a).
Fleischer M., Chao G.Y., Cabri L.J. New mineral names, *Am. Mineralog.*, **60**(7-8), 736–739 (1975a).
Fleischer M., Chao G.Y., Francis C.A., Pabst A. New mineral names, *Am. Mineralog.*, **66**(1-2), 217–220 (1981).
Fleischer M., Chao G.Y., Kato A. New mineral names, *Am. Mineralog.*, **60**(5-6), Pt. 1, 485 489 (1975b).
Fleischer M., Chao G.Y., Pabst A. New mineral names, *Am. Mineralog.*, **64**(1-2), 241–245 (1979b).
Fleischer M., Chao G.Y., Pabst A. New mineral names, *Am. Mineralog.*, **64**(3-4), 464–467 (1979c).
Fleischer M., Mandarino J.A. New mineral names, *Am. Mineralog.*, **59**(3-4), 381–384 (1974).

Fleischer M., Mandarino J.A., Kato A. New mineral names, *Am. Mineralog.*, **53**(7-8), 1421–1427 (1968).

Fleischer M., Mandarino J.A., Pabst A. New mineral names, *Am. Mineralog.*, **65**(1-2), 205–210 (1980b).

Fleischer M., Mandarino J.A., Servos K., Toulmin P. New mineral names, *Am. Mineralog.*, **47**(9-10), 1216–1223 (1962).

Fleischer M., Pabst A. New mineral names, *Am. Mineralog.*, **68**(1-2), Pt. II, 280–283 (1983).

Fleischer M., Pabst A., Mandarino J.A., Chao G.Y., Cabri L.J. New mineral names, *Am. Mineralog.*, **61**(1-2), 174–186 (1976).

Foit F.F., Jr., New data on roeblingite, *Am. Mineralog.*, **51**(3-4), Pt. 1, 504–508 (1966).

Font-Altaba M. A thermal study of thaumasite, *Mineralog. Mag.*, **32**(250), 567–572 (1960).

Förster H.-J., Bindi L., Grundmann G., Stanley C.J. Quijarroite, $Cu_6HgPb_2Bi_4Se_{12}$, a new selenide from the El Dragón Mine, Bolivia, *Minerals*, **6**(4), 123 (2016a).

Förster H.-J., Bindi L., Stanley C.J., Grundmann G. Hansblockite, $(Cu,Hg)(Bi,Pb)Se_2$, the monoclinic polymorph of grundmannite: a new mineral from the Se mineralization at El Dragón (Bolivia), *Mineralog. Mag.*, **81**(3), 629–640 (2017).

Förster H.-J., Bindi L., Stanley C.J., Grundmann G. Hansblockite, IMA 2015-103. CNMNC Newsletter No. 30, April 2016, page 408, *Mineralog. Mag.*, **80**(2), 407–413 (2016b).

Förster H.-J., Bindi L., Stanley C.J., Grundmann G. Quijarroite, IMA 2016-052. CNMNC Newsletter No. 33, October 2016, page 1140, *Mineralog. Mag.*, **80**(6), 1135–1144 (2016c).

Francis C.A., Criddle A.J., Stanley C.J., Lange D.E., Shieh S., Francis J.C. Buckhornite, $AuPb_2BiTe_2S_3$, a new mineral species from Boulder County, Colorado, and new data for aikinite, tetradymite and calaverite, *Canad. Mineralog.*, **30**(4), 1039–1047 (1992).

Francotte J., Moreau J., Ottenburgs R., Lévy C. La briartite, $Cu_2(Fe,Zn)GeS_4$, une nouvelle espèce minérale, *Bull. Soc. fr. minéral. cristallogr.*, **88**, 432–437 (1965).

Frank-Kamenetskaya O.V., Rozhdestvenskaya I.V., Yanulova L.A. New data on the crystal structures of colusites and arsenosulvanites, *J. Struct. Chem.*, **43**(1), 89–100 (2002).

Friedrich D., Quintero M.A., Hao S., Laing C.C., Wolverton C., Kanatzidis M.G. $AInSn_2S_6$ (A = K, Rb, Cs) — layered semiconductors based on the SnS_2 structure, *Inorg. Chem.*, **61**(34), 13525–13531 (2022).

Frydrych R. Darstellung und Eigenschaften von Blei(IV)-selenit, $Pb(SeO_3)_2$, und der Alkali-triselenitoplumbate(IV), $M_2[Pb(SeO_3)_3]$, *Z. anorg. und allg. Chem.*, **427**(3), 260–264 (1976).

Gacemi A., Benbertal D., Gautier-Luneau I., Mosset A. Crystal structure of lead chloride thiocyanate, PbCl(SCN), *Z. Kristallogr., New Cryst. Str.*, **220**(3), 309–310 (2005).

Gagne O.C., Belakovskiy D.I., Cámara F., Uvarova Y. New mineral names, *Am. Mineralog.*, **102**(2), 466–470 (2017).

Galuskin E.V., Gfeller F., Armbruster T., Galuskina I.O., Vapnik Y., Murashko M., Włodyka R., Dzierżanowski P. New minerals with a modular structure derived from hatrurite from the pyrometamorphic Hatrurim Complex. Part I. Nabimusaite, $KCa_{12}(SiO_4)_4(SO_4)_2O_2F$, from larnite rocks of Jabel Harmun, Palestinian Autonomy, Israel, *Mineralog. Mag.*, **79**(5), 1061–1072 (2015a).

Galuskin E.V., Gfeller F., Armbruster T., Galuskina I.O., Vapnik Y., Murashko M., Włodyka R., Dzierżanowski P. Nabimusaite, IMA 2012-057. CNMNC Newsletter No. 15, February 2013, page 5, *Mineralog. Mag.*, **77**(1), 1–12 (2013).

Galuskin E.V., Gfeller F., Galuskina I.O., Armbruster T., Krzątała A., Vapnik Y., Kusz J., Dulski M., Gardocki M., Gurbanov A.G., Dzierżanowski P. New minerals with a modular structure derived from hatrurite from the pyrometamorphic rocks. Part III. Gazeevite, $BaCa_6(SiO_4)_2(SO_4)_2O$, from Israel and the Palestine Autonomy, South Levant, and from South Ossetia, Greater Caucasus, *Mineralog. Mag.*, **81**(3), 499–513 (2017).

Galuskin E.V., Gfeller F., Galuskina I.O., Armbruster T., Vapnik Y., Kusz J., Dulski M., Gardocki M., Dzierżanowski P. Gazeevite, IMA 2015-037. CNMNC Newsletter No. 26, August 2015, page 946, *Mineralog. Mag.*, **79**(4), 941–947 (2015b).

Galuskina I.O., Gfeller F., Galuskin E.V., Armbruster T., Vapnik Y., Dulski M., Gardocki M., Jeżak L., Murashko M. New minerals with modular structure derived from hatrurite from the pyrometamorphic rocks, part IV: Dargaite, $BaCa_{12}(SiO_4)_4(SO_4)_2O_3$, from Nahal Darga, Palestinian Autonomy, *Mineralog. Mag.*, **83**(1), 81–88 (2019).

Gao H., Zhang K., Abudurusuli A., Bai C., Yang Z., Lai K., Li J., Pan S. Syntheses, structures and properties of alkali and alkaline earth metal diamond-like compounds Li_2MgMSe_4 (M = Ge, Sn), *Materials*, **14**(20), 6166 (2021a).

Gao L., Bian G., Yang Y., Zhang B., Wu X., Wu K. Na_4SnS_4 and Na_4SnSe_4 exhibiting multifunctional physicochemical performances as potential infrared nonlinear optical crystals and sodium ion conductors, *New J. Chem.*, **45**(28), 12362–12366 (2021b).

Gao L., Wu X., Xu J., Tian X., Zhang B., Wu K. Rational combination of multiple structural groups on regulating nonlinear optical property in hexagonal Ln_3MGeS_7 polar crystals, *J. Alloys Compd.*, **900**, 163535 (2022a).

Gao L., Xu J., Tian X., Zhang B., Wu X., Wu K. $AgGaSe_2$-inspired nonlinear optical materials: tetrel selenides of alkali metals and mercury, *Chem. Mater.*, **34**(13), 5991–5998 (2022b).

Gao L., Yang Y., Zhang B., Wu X., Wu K. Triclinic layered $A_2ZnSi_3S_8$ (A = Rb and Cs) with large optical anisotropy and systematic research on the inherent structure-performance relationship in the $A_2M^{IIB}M^{IV}{}_3Q_8$ family, *Inorg. Chem.*, **60**(16), 12573–12579 (2021c).

Garavelli A., Mozgova N.N., Orlandi P., Bonaccorsi E., Pinto D., Moëlo Y., Borodaev Y.S. Rare sulfosalts from Vulcano, Aeolian Islands, Italy. VI. Vurroite, $Pb_{20}Sn_2(Bi,As)_{22}S_{54}\ Cl_6$, a new mineral species, *Canad. Mineralog.*, **43**(2), 703–711 (2005).

Garbev K., Ullrich A., Beuchle G., Bergfeldt B., Stemmermann P. Chlorellestadite (synth): Formation, structure, and carbonate substitution during synthesis of belite clinker from wastes in the presence of $CaCl_2$ and CO_2, *Minerals*, **12**(9), 1179 (2022).

Garg G., Bobev S., Roy A., Ghose J., Das D., Ganguli A.K. A new silicon-doped cation-deficient thiospinel, $Cu_{5.52(8)}Si_{1.04(8)}\square_{1.44}Fe_4Sn_{12}S_{32}$: crystal structure, Mössbauer studies, and electrical properties, *J. Solid State Chem.*, **161**(2), 327–331 (2001).

Gasymov V.A., Hasanov G.F. Synthesis and X-ray diffraction study of $Pb_4Yb_2S_7$ compound, *Chem. Probl.*, (3), 382–387 (2020).

Gatta G.D., McIntyre G.J., Swanson J.G., Jacobsen S.D. Minerals in cement chemistry: A single-crystal neutron diffraction and Raman spectroscopic study of thaumasite, $Ca_3Si(OH)_6(CO_3)(SO_4)\cdot 12H_2O$, *Am. Mineralog.*, **97**(7), 1060–1069 (2012).

Gekimyants V.M., Zubkova N.V., Pekov I.V., Agakhanov A.A., Ksenofontov D.A., Lavrov O.B., Yapaskurt V.O., Pautov L.A., Britvin S.N., Platonova N.V., Plechov P.Y., Pushcharovsky D.Y. Selenojunoite, IMA 2023-038. CNMNC Newsletter No 75, *Eur. J. Mineral.*, **35**(5), 892 (2023a).

Gekimyants V.M., Zubkova N.V., Pekov I.V., Agakhanov A.A., Ksenofontov D.A., Lavrov O.B., Yapaskurt V.O., Pautov L.A., Britvin S.N., Platonova N.V., Plechov P.Y., Pushcharovsky D.Y. Selenojunoite, IMA 2023-038. CNMNC Newsletter No 75, *Mineralog. Mag.*, **87**(6), 956 (2023b).

Gemmi M., Campostrini I., Demartin F., Gorelik T.E., Gramaccioli C.M. Structure of the new mineral sarrabusite, $Pb_5CuCl_4(SeO_3)_4$, solved by manual electrondiffraction tomography, *Acta Crystallogr.*, **B68**(1), 15–23 (2012).

Genkin A.D., Evstigneeva T.L., Vyal'sov L.N., Laputina I.P. Kharaelakhite $(Pt,Cu,Pb,Fe,Ni)_9S_8$ — a new sulfide of platinum, copper and lead [in Russian], *Mineralog. zhurn.*, **7**(1), 78–83 (1985).

Gfeller F., Galuskina I.O., Galuskin E.V., Armbruster T., Vapnik Y., Dulski M., Gardocki M., Jeżak, L., Murashko M. Dargaite, IMA2015-068. CNMNC Newsletter No. 28, December 2015, page 1860, *Mineralog.l Mag.*, **79**(7), 1859–1864 (2015).

Giacovazzo C., Menchetti S., Scordari F. The crystal structure of caledonite, $Cu_2Pb_5(SO_4)_3CO_3(OH)_6$, *Acta Crystallogr.*, **B29**(9), 1986–1990 (1973).

Giacovazzo C., Menchetti S., Scordari F. X-ray powder data for caledonite, *Mineralog. Mag.*, **40**(313), 536 (1976).

Giester G., Rieck B. Bechererite, $(Zn,Cu)_6Zn_2(OH)_{13}[(S,Si)(O,OH)_4]_2$, a novel mineral species from the Tonopah-Belmont mine, Arizona, *Am. Mineralog.*, **81**(1-2), 244–248 (1996).

Ginderow D., Cesbron F. Structure de la demesmaekerite, $Pb_2Cu_5(SeO_3)_6(UO_2)_2(OH)_6\cdot 2H_2O$, *Acta Crystallogr.*, **C39**(7), 824–827 (1983).

Gitzendanner R.L., DiSalvo F.J. Synthesis and structure of a new quinary sulfide halide: $LaCa_2GeS_4Cl_3$, *Inorg. Chem.*, **35**(9), 2623–2626 (1996).

Giuseppetti G., Tadini C. Corkite, $PbFe_3(SO_4)(PO_4)(OH)_6$, its crystal structure and ordered arrangement of the tetrahedral cations, *N. Jb. Miner., Mh.*, **1987**(1), 71–81 (1987).

Giuseppetti G., Tadini C. The crystal structure of 2*O* brittle mica: anandite, *Tschermaks Miner. Petr. Mitt.*, **18**(3), 169–184 (1972).

Gojayeva I.M., Babanly V.I., Aghazade A.I., Orujlu E.N.Experimental reinvestigation of the $PbTe–Bi_2Te_3$ pseudo-binary system, *Azerb. Chem. J.*, (2), 47–53 (2022).

Goryunova N.A., Averkieva G.K., Vaypolin A.A. On the possibility of obtaining single crystals of multicomponent alloys [in Russian], In: *Fizika*, Leningrad, Publish. House of Leningr. Inst. Civil Eng., 52–53 (1965).

Gottfried C. XXXVI. Die Raumgruppe des Helvins, *Z. Kristallogr.*, **65**(1-6), 425–427 (1927).

Graeser S. Giessenit — ein neues Pb–Bi–Sulfosalz aus dem Dolomit des Binnatals, *Schweiz. Mineral. Petrogr. Mitt.*, **43**(2), 471–478 (1963).

Graeser S., Edenharter A. Jentschite ($TlPbAs_2SbS_6$) — a new sulphosalt mineral from Lengenbach, Binntal (Switzerland), *Mineralog. Mag.*, **61**(404), 131–137 (1997).

Graeser S., Guggenheim R. Zur Unterscheidung von Wallisit und Hatchit. Eine Kombination rцntgenographischer und rasterelektronmikroskopischer Merhoden, *Schweiz. Mineral. Petrogr. Mitt.*, **58**(3), 215–222 (1978).

Graeser S., Harris D.C. Giessenite from Giessen near Binn, Switzerland: new data, *Canad. Mineralog.*, **24**(1), 19–20 (1986).

Graeser S., Lustenhouwer W., Berlepsch P. Quadratite $Ag(Cd,Pb)(As,Sb)S_3$ – a new sulfide mineral from Lengenbach, Binntal (Switzerland), *Schweiz. Mineral. Petrogr. Mitt.*, **78**(3), 489–494 (1998).

Grice J.D., Cooper M.A. Mammothite: a Pb–Sb–Cu–Al oxy-hydroxide-sulfate – hydrogen atom determination lowers space group symmetry, *Canad. Mineralog.*, **52**(4), 687–698 (2014).

Grubessi O., Mottana A., Paris E. Thaumasite from the Tschwinning mine, South Africa, *Tschermaks Miner. Petr. Mitt.*, **35**(3), 149–156 (1986).

Guéguen A., Poudeu P.F.P., Li C.-P., Moses S., Uher C., He J., Dravid V., Paraskevopoulos K.M., Kanatzidis M.G. Thermoelectric properties and nanostructuring in the p-type materials $NaPb_{18-x}Sn_xMTe_{20}$ (M = Sb, Bi), *Chem. Mater.*, **21**(8), 1683–1694 (2009).

Guittard M., Julien-Pouzol M. Les composés haxagonaux de type La_3CuSiS_7, *Bull. Soc. chim. France*, (7), 2467–2469 (1970).

Gulay L.D., Daszkiewicz M., Shemet V.Ya. Crystal structure of the RE_2PbS_4 (*RE* = Y, Dy, Ho, Er, and Tm) compounds and a comparison with the crystal structures of other rare earth lead chalcogenides, *Z. anorg. und allg. Chem.*, **634**(11), 1887–1895 (2008).

Gunatilleke W.D.C.B., May A.F., Hight Walker A.R., Biacchi A.J., Nolas G.S. Synthesis, crystal structure, and physical properties of $BaSnS_2$, *Phys. Stat. Sol, (RRL)*, **16**(5), 2100624 (2022).

Guo W.-H., Jiang X.-M., Liu B.-W., Wu J.-W., Li S.-F., Zeng H.-Y., Guo G.-C., Huang J.-S. Face-shared octahedral dimer $In_2O_7S_2$ in the non-centrosymmetric barium indiumsilicate oxysulfide $Ba_2In_2Si_3O_{10}S$, *Eur. J. Inorg. Chem.*, **2016**(12), 1846–1850 (2016).

Guseynov F.N. Thermodynamic properties of the $PbBi_6Te_{10}$ and $PbBi_4Te_7$ compounds, *Kimiya Probl.*, (3), 589–593 (2010).

Hahn S.-I., Kim P., Yun H. Synthesis and crystal structure of the new selenogermanate, $Cs_2CdGe_3Se_8$, *Bull. Korean Chem. Soc.*, **34**(4), 1250–1252 (2013).

Hålenius U., Hatert F., Pasero M., Mills S.J. IMA Commission on New Minerals, Nomenclature and Classification (CNMNC). Newsletter No. 41. New minerals and nomenclature modifications approved in 2017 and 2018, *Eur. J. Mineral.*, **30**(1), 183–186 (2018a).

Hålenius U., Hatert F., Pasero M., Mills S.J. IMA Commission on New Minerals, Nomenclature and Classification (CNMNC). Newsletter No. 41. New minerals and nomenclature modifications approved in 2017 and 2018, *Mineralog. Mag.*, **82**(1), 229–233 (2018b).

Hålenius U., Hatert F., Pasero M., Mills S.J. IMA Commission on New Minerals, Nomenclature and Classification (CNMNC). Newsletter No. 42. New minerals and nomenclature modifications approved in 2018, *Eur. J. Mineral.*, **30**(2), 403–408 (2018c).

Hålenius U., Hatert F., Pasero M., Mills S.J. IMA Commission on New Minerals, Nomenclature and Classification (CNMNC). Newsletter No. 42. New minerals and nomenclature modifications approved in 2018, *Mineralog. Mag.*, **82**(2), 445–451 (2018d).

Hall S.R., Szymański J.T., Stewart J.M. Kesterite, $Cu_2(Zn,Fe)SnS_4$, and stannite, $Cu_2(Fe,Zn)SnS_4$, structurally similar but distinct minerals, *Canad. Mineralog.*, **16**(2), 131–137 (1978).

Halyan V.V., Shevchuk M.V., Davydyuk G.Ye., Voronyuk S.V., Kevshyn A.H., Bulatetsky V.V. Glass formation region and X-ray analysis of the glassy alloys in $AgGaSe_2+GeS_2 \Leftrightarrow AgGaS_2+GeSe_2$ system, *Semicond. Phys., Quantum Electronics and Optoelectronics*, **12**(2), 138–142 (2009).

Hamdi M., Lafond A., Guillot-Deudon C., Hlel F., Gargouri M., Jobic S. Crystal chemistry and optical investigations of the $Cu_2Zn(Sn,Si)S_4$ series for photovoltaic applications, *J. Solid State Chem.*, **220**, 232–237 (2014).

Harada K., Nagashima K., Nakao K., Kato A. Hydroxylellestadite, a new apatite from Chichibu Mine, Saitama Prefecture, Japan, *Am. Mineralog.*, **56**(9-10), 1507–1518 (1971).

Harm S., Hatz A.-K., Moudrakovski I., Eger R., Kuhn A., Hoch C., Lotsch B.V. Lesson learned from NMR: characterization and ionic conductivity of LGPS-like Li_7SiPS_8, *Chem. Mater.*, **31**(4), 1280–1288 (2019).

Harm S., Hatz A.-K., Schneider C., Hoefer C., Hoch C., Lotsch B.V. Finding the right blend: interplay between structure and sodium ion conductivity in the system Na_5AlS_4–Na_4SiS_4, *Front. Chem.*, **8**, 90 (2020).

Harris D.C., Chen T.T. Benjaminite, reinstated as a valid species, *Canad. Mineralog.*, **13**(4), 402–407 (1975).

Harris D.C., Jambor J.L., Lachance G.R., Thorpe R.I. Tintinaite, the antimony analogue of kobellite, *Canad. Mineralog.*, **9**(3), 371–382 (1968).

Harris D.C., Roberts A.C., Criddle A.J. Izoklakeite, a new mineral species from Izok Lake, Northwest Territories, *Canad. Mineralog.*, **24**(1), 1–5 (1986).

Harris D.C., Roberts A.C., Criddle A.J. Jaskólskiite from Izok Lake, northwest Territories, *Canad. Mineralog.*, **22**(3), 487–491 (1984).

Hartenbach I., Nilges T., Schleid T. Thiosilicate der Selten-Erd-Elemente: IV. Die *quasi*-isostrukturellen Verbindungen $NaSm_3S_3[SiS_4]$, $CuCe_3S_3[SiS_4]$, $Ag_{0,63}Ce_3S_{2,63}Cl_{0,37}[SiS_4]$ und $Sm_3S_2Cl[SiS_4]$ – Synthese, Kristallstruktur und Untersuchungen zur Silberionendynamik, *Z. anorg. und allg. Chem.*, **633**(13-14), 2445–2452 (2007).

Hartmann F., Benkada A., Indris S., Poschmann M., Lühmann H., Duchstein P., Zahn D., Bensch W. Directed dehydration as synthetic tool for generation of a new Na_4SnS_4 polymorph: crystal structure, Na^+ conductivity, and influence of Sb-substitution, *Angew. Chem. Int. Ed.*, **61**(36), e202202182 (2022a).

Hartmann F., Benkada A., Indris S., Poschmann M., Lühmann H., Duchstein P., Zahn D., Bensch W. Gezielte Dehydratisierung als Synthesewerkzeug zur Herstellung eines neuen Na_4SnS_4-Polymorphs: Kristallstruktur, Na^+-Leitfähigkeit und Einfluss der Sb-Substitution, *Angew. Chem.*, **134**(36), e202202182 (2022b).

Hassan I., Antao S.M., Parise J.B. Haüyne: phase transition and high-temperature structures obtained from synchrotron radiation and Rietveld refinements, *Mineralog. Mag.*, **68**(3), 499–513 (2004).

Hassan I., Buseck P.R. Cluster ordering and antiphase domain boundaries in hauyne, *Canad. Mineralog.*, **27**(2), 173–180 (1989).

Hassan I., Grundy H.D. Structure of davyne and implications for stacking faults, *Canad. Mineralog.*, **28**(2), 341–349 (1990).

Hassan I., Grundy H.D. The crystal structure of hauyne at 293 and 153 K, *Canad. Mineralog.*, **29**(1), 123–130 (1991).

Hassan I., Grundy H.D. The crystal structures of helvite group minerals, $(Mn,Fe,Zn)_8(Be_6Si_6O_{24})S_2$, *Am. Mineralog.*, **70**(1-2), 186–192 (1985).

Hassan I., Grundy H.D. The structure of nosean, ideally $Na_8[Al_6Si_6O_{24}]SO_4{\cdot}H_2O$, *Canad. Mineralog.*, **27**(2), 165–172 (1989).

Hassan I., Peterson R.C., Grundy H.D. The structure of lazurite, ideally $Na_6Ca_2(Al_6Si_6O_{24})S_2$, a member of the sodalite group, *Acta Crystallogr.*, **C41**(6), 827–832 (1985).

Hawthorne F.C., Bladh K.W., Burke E.A.J., Grew E.S., Langley R.H., Puziewicz J., Roberts A.C., Schedler R.A., Shigley J.E., Vanko D.A. New mineral names, *Am. Mimeralog.*, **72**(1-2), 222–230 (1987).

Hawthorne F.C., Burke E.A.J., Ercit T.S., Grew E.S., Grice J.D., Jambor J.L., Puziewicz J., Roberts A.C., Vanko D.A. New mineral names, *Am. Mineralg.*, **73**(1-2), 189–199 (1988).

Hawthorne F.C., Cooper M.A. Khinite-4*O* [= khinite] and khinite-3*T* [= parakhinite], *Canad. Mineralog.*, **47**(2), 473–476 (2009).

Hawthorne F.C., Fleischer M., Grew E.S., Grice J.D., Jambor J.L., Puziewicz J., Roberts A.C., Vanko D.A., Zilczer J.A. New mineral names, *Am. Mimeralog.*, **71**(9-10), 1277–1282 (1986).

Hayek E., Hinterauer K. Sulfate von Metallen hoher Wertigkeitsstufe. *Monatsh. Chem,*, **82**(2), 205–209 (1951).

Haynes A.S., Liu T.-K., Frazer L., Lin J.-F., Wang S.-Y., Ketterson J.B., Kanatzidis M.G., Hsu K.-F. Second harmonic generation response of the cubic chalcogenide $Ba_{(6-x)}Sr_x[Ag_{(4-y)}Sn_{(y/4)}](SnS_4)_4$, *J. Solid State Chem.*, **248**, 119–125 (2017).

He J., Guéguen A., Sootsman J.R., Zheng J.C., Wu L., Zhu Y., Kanatzidis M.G., Dravid V.P. Role of self-organization, nanostructuring, and lattice strain on phonon transport in $NaPb_{18-x}Sn_xBiTe_{20}$ thermoelectric materials, *J. Am. Chem. Soc.*, **131**(49), 17828–17835 (2009).

He J., Zhang X., Guo P., Cheng Y., Zheng C., Huang F. Synthesis, structure, and optical properties of $K_{2.4}Ga_{2.4}M_{1.6}Q_8$ (M = Si, Ge; Q = S, Se) crystals and glasses, *RSC Adv.*, **6**(80), 76789–76794 (2016).

Heinke F., Meyer R., Wagner G., Oeckler O. Crystal structure determination of $Ag_3Pb_4Bi_{11}Se_{22}$ by microfocussed synchrotron radiation, *Z. anorg. und allg. Chem.*, **641**(2), 192–196 (2015).

Heinke F., Urban P., Werwein A., Fraunhofer C., Rosenthal T., Schwarzmüller S., Souchay D., Fahrnbauer F., Dyadkin V., Wagner G., Oeckler O. Cornucopia of structures in the pseudobinary system $(SnSe)_xBi_2Se_3$: a crystal-chemical copycat, *Inorg. Chem.*, **57**(8), 4427–4440 (2018).

Hendricks S.B. The crystal structure of alunite and the jarosites, *Am. Mineralog.*, **22**(6), 773–784 (1937).

Heo J.W., Banerjee A., Park K.H., Jung Y.S., Hong S.-T. New Na-ion solid electrolytes $Na_{4-x}Sn_{1-x}Sb_xS_4$ ($0.02 \leq x \leq 0.33$) for all-solid-state Na-ion batteries, *Adv. Energy Mater.*, **8**(11), 1702716 (2018).

Hess H., Keller P. Die Kristallstruktur von Queitit, $Pb_4Zn_2[SO_4|SiO_4|Si_2O_7]$, *Z. Kristallogr.*, **151**(1-4), 287–299 (1980).

Hidayat T., Fallah-Mehrjardi A., Hayes P.C., Jak E. Experimental investigation of gas/slag/matte/spinel equilibria in the Cu–Fe–O–S–Si system at 1473 K (1200°C) and $P(SO_2) = 0.25$ atm, *Metall. Mater. Trans.*, **B49**(4), 1750–1765 (2018).

Hidayat T., Fallah-Mehrjardi A.T., Hayes P.C., Jak E. Experimental study of gas/slag/matte/spinel equilibria and minor elements partitioning in the Cu–Fe–O–S–Si system. In: *Adv. in Molten Slags, Fluxes, and Salts: Proc. 10th Intern. Conf. Molten Slags, Fluxes and Salts (MOLTEN16)* / Eds. Reddy R.G. et al. The Minerals, Metals and Materials Society, 1207–1220 (2016).

Highcock R.W., Smith G.W., Vickers M.E. On the structural relationship between two polymorphs of $Pb_4SO_4(CO_3)_2(OH)_2$, *Acta Crystallogr.*, **A40**, Suppl., C-249 (1984).

Hîncu I. Asupra sistemului oxidant-reducător sulfură de metal greu (Cu, Zn, Pb) – azotat de amoniu [in Romanian], *Bull. Inst. politehn. Iaşi*, Sec.2, **17**(3-4), 35–47 (1971).

Hîncu I., Golgoţiu T. Contribution a l'étude des réaction en phase solide des systèmes $PbS–NH_4NO_3$ et $PbS–NH_4NO_3–NaCl$. Note III, *Bull. Inst. politehn. Iaşi*, Sec.2, **19**(1-2), 25–33 (1973).

Hirai T., Kurata K., Takeda Y. Derivation of new semiconducting compounds by cross substitution for group IV semiconductors and their semiconducting and thermal properties, *Solid-State Electronics*, **10**(10), 975–981 (1967).

Hirayama S., Suekuni K., Sauerschnig P., Ohta M., Ohtaki M. Cu–S-based thermoelectric compounds with a sphalerite-derived disordered crystal structure, *J. Solid State, Chem.*, **309**, 122960 (2022).

Hoffmann C., Armbruster, T., Giester G. Acentric structure (*P*3) of bechererite, $Zn_7Cu(OH)_{13}[SiO(OH)_3SO_4]$, *Am. Mineralog.*, **82**(9-10), 1014–1018 (1997).

Hogarth D.D. Afghanite: new occurrences and chemical composition, *Canad. Mineralog.*, **17**(1), 47–52 (1979).

Hogarth D.D., Griffin W.L. New data on lazurite, *Lithos*, **9**(1), 39–54 (1976).

Holloway, Jr W.M., Giordano T.J. Refinement of the crystal structure of helvite, $Mn_4(BeSiO_4)_3S$, *Acta Crystallogr.*, **B28**(1), 114–117 (1972).

Honig E., Shen H.-S., Yao G.-Q., Doverspike K., Kershaw R., Dwight K., Wold A. Preparation and characterization of $Cu_2Zn_{1-x}Mn_xGeS_4$, *Mater. Res. Bull.*, **23**(3), 307–312 (1988).

Hou F., Mei D., Zhang Y., Liang F., Wang J., Lu J., Lin Z., Wu Y. $SrZnSnSe_4$: A quaternary selenide with large second harmonic generation and birefringence, *J. Alloys Compd.*, **904**, 163944 (2022).

Huang-Fu S.-X., Shen J.-N., Lin H., Chen L., Wu L.-M. Supercubooctahedron $(Cs_6Cl)_2Cs_5[Ga_{15}Ge_9Se_{48}]$ exhibiting both cation and anion exchange, *Chem. Eur. J.*, **21**(27), 9809–9815 (2015).

Hudson-Edwards K.A., Smith A.M.L., Dubbin W.E., Bennett A.J., Murphy P.J., Wright K. Comparison of the structures of natural and synthetic Pb-Cu-jarosite-type compounds, *Eur. J. Mineral.*, **20**(2), 241–252 (2008).

Huseynov Dzh.I., Bayramov R.B., Mammadov A.M., Valiev V.K. Physical-chemical properties of the alloys of the SnSe–DySe system [in Azerbaijanian], *Chem. Probl.*, (1), 110–113 (2009).

Ibrahimova F.S. Ag–Sn–Se System: phase diagram, thermodynamics and modeling, *Azerb. Chem. J.*, (4), 84–93 (2019a).

Ibrahimova F.S. Calculation of standard thermodynamic functions of argyrodit Ag_8GeSe_6, *Chem. Probl.*, (3), 358–365 (2019b).

Imamaliyeva S.Z., Alakbarzade G.I., Gasymov V.A., Babanly M.B. Experimental study of the $Tl_4PbTe_3–Tl_9TbTe_6–Tl_9BiTe_6$ section of the Tl–Pb–Bi–Tb–Te system, *Mater. Res.*, **21**(4), e20180189 (2018a).

Imamaliyeva S.Z., Alakbarzade G.I., Mahmudova M.A., Amiraslanov I.R., Babanly M.B. Phase equilibria in the $Tl_4PbTe_3–Tl_9SmTe_6–Tl_9BiTe_6$ section of the Tl–Pb–Bi–Sm–Te system, *Acta Chim. Slov.*, **65**(2), 365–371 (2018b).

Imamaliyeva S.Z., Alakbarzade G.I., Mamedov A.N., Babanly M.B. Modeling the phase diagram of the $Tl_9TbTe_6–Tl_4PbTe_3–Tl_9BiTe_6$ system, *Azerb. Chem. J.*, (2), 6–12 (2021a).

Imamaliyeva S.Z., Alakbarzade G.I., Salimov Z.E., Izzatli S.B., Jafarov Ya.I., Babanly M.B. The Tl_4PbTe_3–Tl_9GdTe_6–Tl_9BiTe_6 isopleth section of the Tl–Pb–Bi–Gd–Te system, *Chem. Probl.*, (4), 496–504 (2018c).

Imamaliyeva S.Z., Guseynov F.N., Babanly M.B. Phase equilibria and properties of the solid solutions in the Tl_9NdTe_6–Tl_9BiTe_6–Tl_4PbTe_3 system [in Russian], *Azerb. khim. zhurn.*, (1), 49–54 (2009).

Imamaliyeva S.Z., Mamedov A.N., Babanly M.B. Modeling the phase diagram of the Tl_9GdTe_6–Tl_4PbTe_3–Tl_9BiTe_6 system, *New Mater., Compd. Appl.*, **5**(2), 79–86 (2021b).

Ismailova E.N., Bakhtiyarly I.B., Babanly M.B. Refinement of the phase diagram of the SnSe–Sb_2Se_3 system, *Chem. Probl.*, (2), 250–256 (2020).

Ismailova E.N., Mashadieva L.F., Babanly D.M., Shevel'kov A.V., Babanly M.B. Diagram of solid-phase equilibria in the SnSe–Sb_2Se_3–Se system and thermodynamic properties of tin antimony selenides, *Russ. J. Inorg. Chem.*, **66**(1), 96–103 (2021).

Ismayilova E.N., Mashadiyeva L.F., Bakhtiyarly I.B., Babanly M.B. Phase equilibria in the Cu_2SnSe_3–Sb_2Se_3–Se system, *Condens. Matter Interphases*, **25**(1), 47–54 (2023).

Israël R., de Gelder R., Smits J.M.M., Beurskens P.T., Eijt S.W.H., Rasing Th., van Kempen H., Maior M.M., Motrija S.F. Crystal structures of di-tin-hexa(seleno)hypodiphosphate, $Sn_2P_2Se_6$, in the ferroelectric and paraelectric phase. *Z. Kristallogr.*, **213**(1), 34–41 (1998).

Ito T-I., Miraoka H. Nakaséite, an andorite-like new mineral, *Z. Kristallogr.*, **113**(1-6), 93–98 (1960).

Ivanov V.G., Sapozhnikov A.N. The first finding of afghanite in the USSR [in Russian], *Zap. Vses. Mineralog. Obsh.*, **104**(3), 328–331 (1975).

Ivanov V.G., Sapozhnikov A.N., Piskunova L.F., Kashaev A.A. Tounkite $(Na,Ca,K)_8(Al_6Si_6O_{24})(SO_4)_2Cl{\cdot}H_2O$: A new cancrinite-like mineral [in Russian], *Zap. Ros. mineralog. obshch.*, **121**(2), 92–95 (1992).

Ivanov V.V., Pyatenko Yu.A. On the so-called kesterite [in Russian], *Zap. Vses. Mineralog. Obshch.*, **88**(2), 165–168 (1959).

Ivanov-Emin B.N., Kostrikin V.M. Contribution to the chemistry of germanium. IV. Selenogermanates of alkaline metals [in Russian], *Zhurn. obshch. khimii*, **17**(7), 1253–1256 (1947).

Jacobsen S.D., Smyth J.R., Swope R.J. Thermal expansion of hydrated six-coordinate silicon in thaumasite, $Ca_3Si(OH)_6(CO_3)(SO_4){\cdot}12H_2O$, *Phys. Chem. Minerals*, **30**(6), 321–329 (2003).

Jafarzadeh P., Menezes L.T., Cui M., Assoud A., Zhang W., Halasyamani P.S., Kleinke H. $BaCuSiTe_3$: A noncentrosymmetric semiconductor with $CuTe_4$ tetrahedra and ethane-like Si_2Te_6 units, *Inorg. Chem.*, **58**(17), 11656–11663 (2019).

Jambor J.L. New lead sulfantimonides from Madoc, Ontario. Part 2 – mineral descriptions, *Canad. Mineralog.*, **9**(2), 191–213 (1967).

Jambor J.L. New mineral names, *Am. Mineralog.*, **74**(9-10), 1215–1220 (1989).

Jambor J.L., Bladh K.W., Ercit T.S., Grice J.D., Grew E.S. New mineral names, *Am. Mineralog.*, **73**(7-8), 927–935 (1988).

Jambor J.L., Burke E.A.J. New mineral names, *Am. Mineralog.*, **75**(11-12), 1431–1437 (1990).

Jambor J.L., Burke E.A.J. New mineral names, *Am. Mineralog.*, **76**(11-12), 2020–2026 (1991).

Jambor J.L., Burke E.A.J. New mineral names, *Am. Mineralog.*, **78**(7-8), 845–849 (1993).

Jambor J.L., Dutrizac J.E. Beaverite–plumbojarosite solid solutions, *Canad. Mineralog.*, **21**(1), 101–113 (1983).

Jambor J.L., Dutrizac J.E. The synthesis of beaverite, *Canad. Mineralog.*, **23**(1), 47–51 (1985).

Jambor J.L., Grew E.S. New mineral names, *Am. Mineralog.*, **75**(7-8), 931–937 (1990).

Jambor J.L., Grew E.S. New mineral names, *Am. Mineralog.*, **79**(1-2), 185–189 (1994a).

Jambor J.L., Grew E.S. New mineral names, *Am. Mineralog.*, **79**(3-4), 387–391 (1994b).

Jambor J.L., Grew E.S., Roberts A.C. New mineral names, *Am. Mineralog.*, **81**(11-12), 1513–1518 (1996a).

Jambor J.L., Grew E.S., Roberts A.C. New mineral names, *Am. Mineralog.*, **82**(7-8), 820–823 (1997a).

Jambor J.L., Grew E.S., Roberts A.C. New mineral names, *Am. Mineralog.*, **85**(10), 1561–1565 (2000).

Jambor J.L., Grew E.S., Roberts A.C. New mineral names, *Am. Mineralog.*, **87**(10), 1509–1513 (2002).

Jambor J.L., Kovalenker V.A., Roberts A.C. New mineral names. *Am. Mineralog*, **80**(5-6), 630–635 (1995a).

Jambor J.L., Owens de A.R., Grice J.D., Feinglos M.N. Gallobeudantite, $PbGa_3[(AsO_4),(SO_4)]_2(OH)_6$, a new mineral species from Tsumeb, Namibia, and associated new gallium analogues of the alunite–jarosite family, *Canad. Mineralog.*, **34**(6), 1305–1315 (1996b).

Jambor J.L., Pertsev N.N., Roberts A.C. New mineral names, *Am. Mineralog.*, **80**(7-8), 845–850 (1995b).

Jambor J.L., Pertsev N.N., Roberts A.C. New mineral names, *Am. Mineralog.*, **82**(3-4), 430–433 (1997b).

Jambor J.L., Pertsev N.N., Roberts A.C. New mineral names, *Am. Mineralog.*, **82**(9-10), 1038–1041 (1997c).

Jambor J.L., Puziewicz J. New mineral names, *Am. Mineralog.*, **74**(3-4), 500–505 (1989).
Jambor J.L., Puziewicz J. New mineral names, *Am. Mineralog.*, **76**(9-10), 1728–1735 (1991).
Jambor J.L., Puziewicz J. New mineral names, *Am. Mineralog.*, **78**(3-4), 450–455 (1993a).
Jambor J.L., Puziewicz J. New mineral names, *Am. Mineralog.*, **78**(9-10), 1108–1112 (1993b).
Jambor J.L., Puziewicz J., Roberts A.C. New mineral names, *Am. Mineralog.*, **79**(7-8), 763–767 (1994a).
Jambor J.L., Puziewicz J., Roberts A.C. New mineral names, *Am. Mineralog.*, **80**(3-4), 404–409 (1995c).
Jambor J.L., Puziewicz J., Roberts A.C. New mineral names, *Am. Mineralog.*, **80**(11-12), 1328–1333 (1995d).
Jambor J.L., Puziewicz J., Roberts A.C. New mineral names, *Am. Mineralog.*, **83**(5-6), 652–656 (1998).
Jambor J.L., Puziewicz J., Roberts A.C. New mineral names, Am. Mineralog., **84**(4), 685–688 (1999).
Jambor J.L., Roberts A.C. New mineral names, *Am. Mineralog.*, **80**(1-2), 184–188 (1995a).
Jambor J.L., Roberts A.C. New mineral names, *Am. Mineralog.*, **80**(9-10), 1073–1077 (1995b).
Jambor J.L., Roberts A.C. New mineral names, *Am. Mineralog.*, **82**(11-12), 1261–1264 (1997).
Jambor J.L., Roberts A.C. New mineral names, *Am. Mineralog.*, **83**(3-4), 400–403 (1998).
Jambor J.L., Roberts A.C. New mineral names, *Am. Mineralog.*, **84**(9), 1464–1468 (1999).
Jambor J.L., Roberts A.C. New mineral names, *Am. Mineralog.*, **85**(1), 263–266 (2000).
Jambor J.L., Roberts A.C. New mineral names, *Am. Mineralog.*, **86**(1-2), 197–200 (2001).
Jambor J.L., Roberts A.C. New mineral names, *Am. Mineralog.*, **87**(7), 996–999 (2002).
Jambor J.L., Roberts A.C. New mineral names, *Am. Mineralog.*, **88**(2-3), 475–479 (2003).
Jambor J.L., Roberts A.C. New mineral names, *Am. Mineralog.*, **89**(11-12), 1826–1834 (2004).
Jambor J.L., Roberts A.C. New mineral names, *Am. Mineralog.*, **90**(1), 271–275 (2005).
Jambor J.L., Roberts A.C., Puziewicz J. New mineral names, *Am. Mineralog.*, **79**(5-6), 570–574 (1994b).
Jambor J.L., Vanko D.A. New mineral names, *Am. Mineralog.*, **74**(7-8), 946–951 (1989).
Jambor J.L., Vanko D.A. New mineral names, *Am. Mineralog.*, **75**(5-6), 706–713 (1990).
Jambor J.L., Vanko D.A. New mineral names, *Am. Mineralog.*, **76**(7-8), 1434–1440 (1991).
Jambor J.L., Vanko D.A. New mineral names, *Am. Mineralog.*, **77**(3-4), 446–452 (1992).
Jana S., Ishtiyak M., Govindaraj L., Arumugam S., Tripathy B., Malladi S.K., Niranjan M.K., Prakash J. Metal to insulator transition in $Ba_2Ge_2Te_5$: Synthesis, crystal structure, resistivity, thermal conductivity, and electronic structure, *Mater. Res. Bull.*, **147**, 111641 (2022).
Jana S., Yadav S., Swati, Niranjan M.K., Prakash J $Ba_{14}Si_4Sb_8Te_{32}(Te_3)$: hypervalent Te in a new structure type with low thermal conductivity, *Dalton Trans.*, **52**(42), 15426–15439 (2023).
Jarosch D., Zemann J. Yafsoanite: a garnet type calcium–tellurium(VI)–zinc oxide, *Mineral. Petrol.*, **40**(2), 111–116 (1989).
Jaszczak J.A., Rumsey M.S., Bindi L., Hackney S.A.,Wise M.A., Stanley C.J., Spratt J. Merelaniite, $Mo_4Pb_4VSbS_{15}$, a new molybdenum-essential member of the cylindrite GroupG, from the Merelani tanzanite deposit, Lelatema Mountains, Manyara Region, Tanzania, *Minerals*, **6**(4), 115 (2016).
Jelley E.E. The preparation and constitution of the thiostannates. Part I. Sodium ortho- and meta-thiostannate, *J. Chem. Soc.*, 1580–1582 (1933).
Ji B., Guderjahn E., Wu K., Syed T.H., Wei W., Zhang B., Wang J. Revisiting fhiophosphate $Pb_3P_2S_8$: A multifunctional material combining a nonlinear optical response and photocurrent response, *Phys. Chem. Chem. Phys.*, **23**(41), 23696–23702 (2021).
Ji B., Sarkar A., Wu K., Swindle A., Wang J. $A_2P_2S_6$ (A = Ba, Pb): a good platform to study polymorph effect and lone pairs effect to form acentric structure, *Dalton Trans.*, **51**(11), 4522–4531(2022a).
Ji B., Wang F., Wu K., Zhang B., Wang J. d^6 *versus* d^{10}, which is better for second harmonic generation susceptibility? A case study of $K_2TGe_3Ch_8$ (T = Fe, Cd; Ch = S, Se), *Inorg. Chem.*, **62**(1), 574–582 (2023).
Ji B., Wu K., Chen Y., Wang F., Rossini A.J., Zhang B., Wang. J. $Ba_6(Cu_xZ_y)Sn_4S_{16}$ (Z = Mg, Mn, Zn, Cd, In, Bi, Sn): High chemical flexibility resulting in good nonlinear-optical properties, *Inorg. Chem.*, **61**(5), 2640–2651 (2022b).
Jia Y.-J., Zhang X., Chen Y.-G., Jiang X., Song J.-N., Lin Z., Zhang X.-M. $PbBi(SeO_3)_2F$ and $Pb_2Bi(SeO_3)_2Cl_3$: coexistence of three kinds of stereochemically active lone-pair cations exhibiting excellent nonlinear optical properties, *Inorg. Chem.*, **61**(39), 15368–15376 (2022).
Jiang B., Qiu P., Eikeland E., Chen H., Song Q., Ren D., Zhang T., Yang J., Iversen B.B., Shi X., Chen L. Cu_8GeSe_6-based thermoelectric materials with an argyrodite structure, *J. Mater. Chem. C*, **5**(4), 943–952 (2017).
Jiang H.-L., Mao J.-G. Synthesis, crystal structure and characterization of the barium zinc tellurate disilicate: $Ba_3Zn_6[TeO_6][Si_2O_7]_2$, *Z. anorg. und. allg. Chem.*, **632**(12-13), 2053–2057 (2006).

Jiang Q.-Q., Ling L., Baiyin M. New quaternary selenogermanates of $Rb_2Ag_2GeSe_4$ and $Cs_2Cu_2GeSe_4$, *Inorg. Chem. Commun.*, **113**, 107794 (2020).

Johan Z., Dódony I., Morávek P., Pašava J. La buckhornite, $Pb_2AuBiTe_2S_3$, du gisement d'or de Jílové, République Tchèque, *C. r. Acad. sci.*, **318**(9), 1225–1231 (1994).

Johan Z., Kvaček M., Picot P. La petrovicite, $Cu_3HgPbBiSe_5$, un nouveau minéral, *Bull. Soc. fr. minéral. cristallogr.*, **99**(5), 310–313 (1976).

Johan Z., Mantienne J. Thallium-rich mineralization at Jas Roux, Hautes-Alpes, France: a complex epithermal, sediment-hosted, ore-forming system, *J. Czech Geol. Soc.*, **45**(1–2), 63–77 (2000).

Johan Z., Mantienne J., Picot P. La chabournéite, un nouveau minéral thallifère, *Bull. Minéralog.*, **104**(1), 10–15 (1981).

Johan Z., Picot P. La pirquitasite, Ag_2ZnSnS_4, un nouveau membre du groupe de la stannite, *Bull. Minéral.*, **105**(3), 229–235 (1982).

Johan Z., Picot P., Ruhlmann F. The ore mineralogy of the Otish Mountains uranium deposit, Quebec: skippenite, Bi_2Se_2Te, and watkinsonite, $Cu_2PbBi_4(Se,S)_8$, two new mineral species, *Canad. Mineralog.*, **25**(4), 625–638 (1987).

Jothi P.R., Scheifers J.P., Zhang Y., Alghamdi M., Stekovic D., Itkis M.E., Shi J., Fokwa B.P.T. $Fe_{5-x}Ge_2Te_2$ – a new exfoliable itinerant ferromagnet with high Curie temperature and large perpendicular magnetic anisotropy, *Phys. status solidi (RRL)*, **14**(3), 1900666 (2020).

Kaczorowski D., Melnychuk Kh.O., Marchuk O.V., Gulay L.D., Daszkiewicz M. Crystal structure and magnetic properties of novel La(Ce,Pr)R'$PbSi_2S_8$ (R' = Ce, Pr, Sm, Tb, Dy, Y, Ho and Er) compounds, *J. Solid State Chem.*, **290**, 121565 (2020).

Kaib T., Bron P., Haddadpour S., Mayrhofer L., Pastewka L., Järvi T.T., Moseler M., Roling B., Dehnen S. Lithium chalcogenidotetrelates: LiChT – synthesis and characterization of new Li^+ ion conducting Li/Sn/Se compounds, *Chem. Mater.*, **25**(15), 2961–2969 (2013).

Kaib T., Haddadpour S., Kapitein M., Bron P., Schröder C., Eckert H., Roling B., Dehnen S. New lithium chalcogenidotetrelates, LiChT: Synthesis and characterization of the Li^+-conducting tetralithium *ortho*-sulfidostannate Li_4SnS_4, *Chem. Mater.*, **24**(11), 2211–2219 (2012).

Kaib T., Kapitein M., Dehnen S. Synthesis and crystal structure of $[Li_8(H_2O)_{29}][Sn_{10}O_4S_{20}]\cdot 2H_2O$, *Z. anorg. und allg. Chem.*, **637**(12), 1683–1686 (2011).

Kampf A.R., Cooper M.A., Mills S.J., Housley R.M., Rossman G.R. Andychristyite, IMA 2015-024. CNMNC Newsletter No. 26, August 2015, page 943, *Mineralog. Mag.*, **79**(4), 941–947 (2015).

Kampf A.R., Cooper M.A., Mills S.J., Housley R.M., Rossman G.R. Lead-tellurium oxysalts from Otto Mountain near Baker, California: XII. Andychristyite, $PbCu^{2+}Te^{6+}O_5(H_2O)$, a new mineral with hcp stair-step layers, *Mineralog. Mag.*, **80**(6), 1055–1065 (2016).

Kampf A.R., Housley R.M., Marty J. Lead-tellurium oxysalts from Otto Mountain near Baker, California: III. Thorneite, $Pb_6(Te^{6+}{}_2O_{10})(CO_3)Cl_2(H_2O)$, the first mineral with edge-sharing octahedral tellurate dimers, *Am. Mineralog.*, **95**(10), 1548–1553 (2010a).

Kampf A.R., Housley R.M., Mills S.J., Rossman G.R., Marty J. Hagstromite, IMA 2019-093. CNMNC Newsletter No. 53, *Eur. J. Mineral.*, **32**(1), 212 (2020a).

Kampf A.R., Housley R.M., Mills S.J., Rossman G.R., Marty J. Hagstromite, IMA 2019-093. CNMNC Newsletter No. 53, *Mineralog. Mag.*, **84**(1), 161 (2020b).

Kampf A.R., Housley R.M., Mills S.J., Rossman G.R., Marty J. Hagstromite, $Pb_8Cu^{2+}(Te^{6+}O_6)_2(CO_3)Cl_4$, a new lead–tellurium oxysalt mineral from Otto Mountain, California, USA, *Mineralog. Mag.*, **84**(4), 517–523 (2020c).

Kampf A.R., Marty J., Brent T. Lead-tellurium oxysalts from Otto Mountain near Baker, California: II. Housleyite, $Pb_6CuTe_4O_{18}(OH)_2$, a new mineral with Cu–Te octahedral sheets, *Am. Mineralog.*, **95**(8-9), 1337–1342 (2010b).

Kampf A.R., Mills S.J., Celestian A.J., Ma C., Yang H., Thorne B. Flaggite, IMA 2021-044. CNMNC Newsletter No. 63, *Eur. J. Mineral.*, **33**(5), 640 (2021a).

Kampf A.R., Mills S.J., Celestian A.J., Ma C., Yang H., Thorne B. Flaggite, IMA 2021-044. CNMNC Newsletter No. 63, *Mineralog. Mag.*, **85**(6), 911 (2021b).

Kampf A.R., Mills S.J., Celestian A.J., Ma C., Yang H., Thorne B. Flaggite, $Pb_4Cu^{2+}{}_4Te^{6+}{}_2(SO_4)_2O_{11}(OH)_2(H_2O)$, a new mineral with stair-step-like HCP layers from Tombstone, Arizona, USA, *Mineralog. Mag.*, **86**(3), 397–404 (2022a).

Kampf A.R., Mills S.J., Housley R.M. The crystal structure of munakataite, $Pb_2Cu_2(Se^{4+}O_3)(SO_4)(OH)_4$, from Otto Mountain, San Bernardino County, California, USA, *Mineralog Mag.*, **74**(6), 991–998 (2010d).

Kampf A.R., Mills S.J., Housley R.M., Ma C., Thorne B. Tombstoneite, IMA 2021-053. CNMNC Newsletter No. 63, *Eur. J. Mineral.*, **33**(5), 642 (2021c).

Kampf A.R., Mills S.J., Housley R.M., Ma C., Thorne B. Tombstoneite, IMA 2021-053. CNMNC Newsletter No. 63, *Mineralog. Mag.*, **85**(6), 913 (2021d).

Kampf A.R., Mills S.J., Housley R.M., Marty J. Agaite, IMA 2011-115. CNMNC Newsletter No. 13, June 2012, page 812, *Mineralog. Mag.*, **76**(3), 807–817 (2012a).

Kampf A.R., Mills S.J., Housley R.M., Marty J. Fuettererite, IMA 2011-111. CNMNC Newsletter No. 13, June 2012, page 811, *Mineralog. Mag.*, **76**(3), 807–817 and **76**(5), 1281–1288 (2012b).

Kampf A.R., Mills S.J., Housley R.M., Marty J. Lead-tellurium oxysalts from Otto Mountain near Baker, California: IX. Agaite, $Pb_3Cu^{2+}Te^{6+}O_5(OH)_2(CO_3)$, a new mineral with CuO_5–TeO_6 polyhedral sheets, *Am. Mineralog.*, **98**(2-3), 512–517 (2013a).

Kampf A.R., Mills S.J., Housley R.M., Marty J. Lead-tellurium oxysalts from Otto Mountain near Baker, California: VIII. Fuettererite, $Pb_3Cu^{2+}{}_6Te^{6+}O_6(OH)_7C_{15}$, a new mineral with double spangolite-type sheets, *Am. Mineralog.*, **98**(2-3), 506–511 (2013b).

Kampf A.R., Mills S.J., Housley R.M., Marty J., Thorne B. Lead-tellurium oxysalts from Otto Mountain near Baker, California: V. Timroseite, $Pb_2Cu^{2+}{}_5(Te^{6+}O_6)_2(OH)_2$, and paratimroseite, $Pb_2Cu^{2+}{}_4(Te^{6+}O_6)_2(H_2O)_2$, two new tellurates with Te-Cu polyhedral sheets, *Am. Mineralog.*, **95**(10), 1560–1568 (2010c).

Kampf A.R., Mills S.J., Housley R.M., Rossman G.R., Marty J., Thorne B. Bairdite, IMA 2012-061. CNMNC Newsletter No. 15, February 2013, page 6, *Mineralog. Mag.*, **77**(1), 1–12 (2013c).

Kampf A.R., Mills S.J., Housley R.M., Rossman G.R., Marty J., Thorne B. Eckhardite, IMA 2012-085. CNMNC Newsletter No. 16, August 2013, page 2696, *Mineralog. Mag.*, **77**(6), 2695–2709 (2013d).

Kampf A.R., Mills S.J., Housley R.M., Rossman G.R., Marty J., Thorne B. Lead-tellurium oxysalts from Otto Mountain near Baker, California: X. Bairdite, $Pb_2Cu^{2+}{}_4Te^{6+}{}_2O_{10}(OH)_2(SO_4)(H_2O)$, a new mineral with thick HCP layers, *Am. Mineralog.*, **98**(7), 1315–1321 (2013e).

Kampf A.R., Mills S.J., Housley R.M., Rossman G.R., Marty J., Thorne B. Lead-tellurium oxysalts from Otto Mountain near Baker, California: XI. Eckhardite, $(Ca,Pb)Cu^{2+}Te\text{-}O_5(H_2O)$, a new mineral with HCP stair-step layers, *Am. Mineralog.*, **98**(8-9), 1617–1623 (2013f).

Kampf A.R., Mills S.J., Housley R.M., Rumsey M.S., Spratt J. Lead-tellurium oxysalts from Otto Mountain near Baker, California: VII. Chromschieffelinite, $Pb_{10}Te_6O_{20}(OH)_{14}(CrO_4)(H_2O)_5$, the chromate analog of schieffelinite, *Am. Mineralog.*, **97**(1), 212–219 (2012c).

Kampf A.R., Missen O.P., Mills S.J., Ma C., Housley R.M., Chorazewicz M., Marty J., Coolbaugh M. Matthiasweilite, IMA 2021-069. CNMNC Newsletter No 64, *Eur. J. Mineral.*, **34**(1), 3 (2022b).

Kampf A.R., Missen O.P., Mills S.J., Ma C., Housley R.M., Chorazewicz M., Marty J., Coolbaugh M. Matthiasweilite, IMA 2021-069. CNMNC Newsletter No 64, *Mineralog. Mag.*, **86**(1), 179 (2022c).

Kampf A.R., Missen O.P., Mills S.J., Ma C., Housley R.M., Chorazewicz M., Marty J., Coolbaugh M., Momma K. Matthiasweilite, $PbTe^{4+}O_3$, a new tellurite mineral from the Delamar mine, Lincoln County, Nevada, USA, *Canad. Mineralog.*, **60**(5), 805–814 (2022d).

Kampf A.R., Moore P.B., Jonsson E.J., Swihart G.H. Philolithite, a new mineral from Lengban, Värmland, Sweden, *Mineralog. Rec.*, **29**(3), 201–206 (1998).

Kampf A.R., Pluth J.J., Chen Y.-S., Roberts A.C., Housley R.M. Bobmeyerite, a new mineral from Tiger, Arizona, USA, structurally related to cerchiaraite and ashburtonite, *Mineralog. Mag.*, **77**(1), 81–91 (2013g).

Kampf A.R., Pluth J.J., Chen Y.-S., Roberts A.C., Housley R.M. Bobmeyerite, IMA 2012-019. CNMNC Newsletter No. 14, October 2012, page 1282, *Mineralog. Mag.*, **76**(5), 1281–1288 (2012d).

Kampf A.R., Smith J.B., Hughes J.M., Ma C., Emproto C. Cherokeeite, IMA 2022-016. CNMNC Newsletter No 68, *Eur. J. Mineral.*, **34**(5), 387 (2022e).

Kampf A.R., Smith J.B., Hughes J.M., Ma C., Emproto C. Cherokeeite, IMA 2022-016. CNMNC Newsletter No 68, *Mineralog. Mag.*, **86**(5), 855 (2022f).

Kampf A.R., Smith J.B., Hughes J.M., Ma C., Emproto C. Cuprocherokeeite, IMA 2022-086. CNMNC Newsletter No 70, *Eur. J. Mineral.*, **34**(6), 599 (2022g).

Kampf A.R., Smith J.B., Hughes J.M., Ma C., Emproto C. Cuprocherokeeite, IMA 2022-086. CNMNC Newsletter No 70, *Mineralog. Mag.*, **87**(1), 166–167 (2023a).

Kampf A.R., Smith J.B., Hughes J.M., Ma C., Emproto C. Dinilawiite, IMA 2023-061. CNMNC Newsletter No 76, *Eur. J. Mineral.*, **35**(6), 1074 (2023b).

Kampf A.R., Smith J.B., Hughes J.M., Ma C., Emproto C. Dinilawiite, IMA 2023-061. CNMNC Newsletter No 76, *Mineralog. Mag.*, **88**(1), 106 (2024).

Kampf A.R., Smith J.B., Hughes J.M., Ma C., Emproto C. Hayelasdiite, IMA 2022-021. CNMNC Newsletter No 68, *Eur. J. Mineral.*, **34**(5), 388 (2022h).

Kampf A.R., Smith J.B., Hughes J.M., Ma C., Emproto C. Hayelasdiite, IMA 2022-021. CNMNC Newsletter No 68, *Mineralog. Mag.*, **86**(5), 856 (2022i).

Kampf A.R., Smith J.B., Hughes J.M., Ma C., Emproto C. Haywoodite, IMA 2021-115. CNMNC Newsletter No 67, *Eur. J. Mineral.*, **34**(3), 360 (2022j).

Kampf A.R., Smith J.B., Hughes J.M., Ma C., Emproto C. Haywoodite, IMA 2021-115. CNMNC Newsletter No 67, *Mineralog. Mag.*, **86**(5), 850 (2022k).

Kampf A.R., Smith J.B., Hughes J.M., Ma C., Emproto C. Hydroredmondite, IMA 2021-073. CNMNC Newsletter No 64, *Eur. J. Mineral.*, **34**(1), 4 (2022l).

Kampf A.R., Smith J.B., Hughes J.M., Ma C., Emproto C. Hydroredmondite, IMA 2021-073. CNMNC Newsletter No 64, *Mineralog. Mag*, **86**(1), 180 (2022m).

Kampf A.R., Smith J.B., Hughes J.M., Ma C., Emproto C. New minerals from the Redmond Mine, North Carolina, USA: III. Cherokeeite and cuprocherokeeite, two new minerals containing $[Pb_2(Zn,Cu^{2+})(OH)_4]^{2+}$ chains, *Can. J. Mineral. Petrol.*, **61**(3), 635–647 (2023c).

Kampf A.R., Smith J.B., Hughes J.M., Ma C., Emproto C. New minerals from the Redmond Mine, North Carolina, USA: II. Cubothioplumbite and hexathioplumbite, two new minerals containing the cubane-like $[Pb_4(OH)_4]^{4+}$ structural unit, *Can. J. Mineral. Petrol.*, **61**(3), 623–633 (2023d).

Kampf A.R., Smith J.B., Hughes J.M., Ma C., Emproto C. New minerals from the Redmond Mine, North Carolina, USA: IV. Haywoodite and hanahanite, two new minerals containing gordaite-like sheets, *Can. J. Mineral. Petrol.*, **61**(6), 1137–1149 (2023e).

Kampf A.R., Smith J.B., Hughes J.M., Ma C., Emproto C. New minerals from the Redmond Mine, North Carolina, USA: I. Redmondite, hydroredmondite, and sulfatoredmondite, three minerals containing the novel $[Pb_8O_2Zn(OH)_6]^{8+}$ structural unit *Can. J. Mineral. Petrol.*, **61**(1), 189–202 (2023f).

Kampf A.R., Smith J.B., Hughes J.M., Ma C., Emproto C. Redmondite, IMA 2021-072. CNMNC Newsletter No 64, *Eur. J. Mineral.*, **34**(1), 4 (2022n).

Kampf A.R., Smith J.B., Hughes J.M., Ma C., Emproto C. Redmondite, IMA 2021-072. CNMNC Newsletter No 64, *Mineralog. Mag*, **86**(1), 180 (2022o).

Kampf A.R., Smith J.B., Hughes J.M., Ma C., Emproto C. Sulfatoredmondite, IMA 2021-089. CNMNC Newsletter No 65, *Eur. J. Mineral.*, **34**(1), 145 (2022p).

Kampf A.R., Smith J.B., Hughes J.M., Ma C., Emproto C. Sulfatoredmondite, IMA 2021-089. CNMNC Newsletter No 65, *Mineralog. Mag*, **86**(2), 356 (2022q).

Kampf A.R., Smith J.B., Hughes J.M., Ma C., Emproto C. Zincochenite, IMA 2022-025. CNMNC Newsletter No 68, *Eur. J. Mineral.*, **34**(5), 388 (2022r).

Kampf A.R., Smith J.B., Hughes J.M., Ma C., Emproto C. Zincochenite, IMA 2022-025. CNMNC Newsletter No 68, *Mineralog. Mag.*, **86**(5), 857 (2022s).

Kanatzidis M.G., Liao J.H., Marking G.A. Alkali metal quaternary chalcogenides and process for the preparation of thereof, USA Patent 5 614 128. Appl. № 606565. Filed 26.02.96; Data of Patent 25.03.1997 (1997a).

Kanatzidis M.G., Liao J.H., Marking G.A. Alkali metal quaternary chalcogenides and process for the preparation of thereof, USA Patent 5 618 471. Appl. № 606886. Filed 26.02.96; Data of Patent 08.04.1997 (1997b).

Karup-Møller S. Berryite from Greenland, *Canad. Mineralog.*, **8**(4), 414–423 (1966).

Karup-Møller S., Makovicky E. On pavonite, cupropavonite, benjaminite, and "oversubstituted" gustavite, *Bull. Minéralog.*, **102**(4), 351–367 (1979).

Kasatkin A.V., Makovicky E., Plášil J., Škoda R., Agakhanov A.A., Karpenko V.Y., Nestola F. Tsygankoite, $Mn_8Tl_8Hg_2(Sb_{21}Pb_2Tl)_{\Sigma 24}S_{48}$, a new sulfosalt from the Vorontsovskoe gold deposit, Northern Urals, Russia, *Minerals*, **8**(5), 218 (2018a).

Kasatkin A.V., Makovicky E., Plášil J., Škoda R., Agakhanov A.A., Karpenko V.Y. Tsygankoite, IMA 2017-088. CNMNC Newsletter No. 41, February 2018, page 184, *Eur. J. Mineral.*, **30**(1), 183–186 (2018b).

Kasatkin A.V., Makovicky E., Plášil J., Škoda R., Agakhanov A.A., Karpenko V.Y. Tsygankoite, IMA 2017-088. CNMNC Newsletter No. 41, February 2018, page 230, *Mineralog. Mag.*, **82**(1), 229–233 (2018c).

Kato A. Stannoidite, $Cu_5(Fe,Zn)_2SnS_8$, a new stannite-like mineral from the Konjo mine, Okayama Prefecture, Japan, *Bull. Nat. Sci. Mus. Tokyo*, **12**(1), 165–172 (1969).

Kato Y., Saito R., Sakano M., Mitsui A., Hirayama M., Kanno R. Synthesis, structure and lithium ionic conductivity of solid solutions of $Li_{10}(Ge_{1-x}M_x)P_2S_{12}$ (M = Si, Sn), *J. Power Sources*, **271**, 60–64 (2014).

Keutsch F.N., Zheng S.-L., Topa D., Stanley C. Ramosite, IMA 2019-099. CNMNC Newsletter No. 53, *Eur. J. Mineral.*, **32**(1), 213 (2020a).

Keutsch F.N., Zheng S.-L., Topa D., Stanley C. Ramosite, IMA 2019-099. CNMNC Newsletter No. 53, *Mineralog. Mag.*, **84**(1), 162 (2020b).

Kim A.A., Zayakina N.V., Lavrent'yev Yu.G. Yafsoanite, $(Zn_{1.38}Ca_{1.36}Pb_{0.26})_3TeO_6$, a new tellurium mineral, *Intern. Geol. Rev.*, **24**(11), 1295–1298 (1982a).

Kim A.A., Zayakina N.V., Lavrent'yev Y.G. Yafsoanite $(Zn_{1.38}Ca_{1.36}Pb_{0.26})_3TeO_6$ – a new tellurium mineral [in Russian], *Zap. Vses. Mineralog. Obshch.*, **111**(1), 118–121 (1982b).

Kim A.A., Zayakina N.V., Lavrent'ev Yu.G., Makhotko V.F. V,Si-variety of dugganite – the first find in the USSR, *Mineralog. Zhurn.*, **10**(6), 85–89 (1988).

Kim A.A., Zayakina N.V., Makhotko V.F. Kuksite $Pb_3Zn_3TeO_6(PO_4)_2$ and cheremnykhite $Pb_3Zn_3TeO_6(VO_4)_2$, new tellurates from the Kuranakh gold deposit (Central Aldan, Southern Yakutia) [in Russian], *Zap. Vses. Mineralog. Obshch.*, **119**(5), 50–57 (1990).

Kingston P.W. Studies of mineral sulphosalts. XXI – Nuffieldite, a new species, *Canad. Mineralog.*, **9**(4), 439–452 (1968).

Kissin S.A. A reinvestigation of the stannite (Cu_2FeSnS_4)–kesterite (Cu_2ZnSnS_4) pseudobinary system, *Canad. Mineralog.*, **27**(4), 689–697 (1989).

Kissin S.A., Owens de A.R. The crystallography of sakuraiite, *Canad. Mineralog.*, **24**(6), 679–683 (1986a).

Kissin S.A., Owens de A.R. The properties and modulated structure of potosiite from the Cassiar District, British Columbia, *Canad. Mineralog.*, **24**(1), 45–50 (1986b).

Kissin S.A., Owens de A.R. The relatives of stannite in the light of new data, *Canad. Mineralog.*, **27**(4), 673–688 (1989).

Klaska R., Jarchow O. Synthetischer Sulfat-Hydrocancrinit vom Mikrosommit-Typ, *Naturwissenschaften*, **64**(2), 93 (1977).

Klement R. Isomorpher Ersatz des Phosphors in Apatiten durch Silicium und Schwefel, *Naturwissenschaften*, **27**(4), 57–58 (1939).

Klement R., Dihn P. Isomorphe Apatitarten, *Naturwissenschaften*, **29**(20), 301 (1941).

Klepov V.V., Smith M.D., zur Loye H.-C. Targeted synthesis of uranium(IV) thiosilicates, *Inorg. Chem.*, **58**(13), 8275–8278 (2019).

Klepp K.O., Fabian F. Preparation and crystal structure of $Na_3CuSnSe_4$. A novel selenostannate(IV) with a layered structure, *Eur. J. Solid State Inorg. Chem.*, **34**(10), 1155–1163 (1997).

Klymovych O., Zmiy O., Olekseyuk I. Phase equlibria and the glass formation in the Cu_2Se–$SnSe_2$–As_2Se_3 system [in Ukrainian], *Nauk. Visnyk Skhidnoyevrop. Nats. Univ. im. Lesi Ukrainky. Ser. Khim. nauky*, [23(272)], 89–94 (2013).

Knill D.C., Young B.R. Thaumasite from Co. Down, Northern Ireland, *Mineralog. Mag.*, **32**(248), 416–418 (1960).

Knyazev A.V., Bulanov E.N., Smirnova N.N., Korokin V.Zh., Shushunov A.N., Blokhina A.G. Low-temperature heat capacity and thermal expansion of synthetic caracolite $Na_3Pb_2(SO_4)_3Cl$, *Thermochim. Acta*, **596**, 1–5 (2014).

Ko Y., Cahill C.L., Parise J.B. Novel layered sulfides of tin: Synthesis and structural characterization of $Cs_4Sn_5S_{12}{\cdot}2H_2O$ and $Sn_5S_{12}(N_2C_4H_{11})(N_4C_{10}H_{24})$, *J. Chem. Soc., Chem. Commun.*, (1), 69–70 (1994).

Kohatsu I., Wuensch B.J. Crystal chemistry of a new sulfosalt mineral, nuffieldite, $(Pb_{10}Bi_{10}Cu_4S_{28})$, *Am. Mineralog.*, **57**(1-2), 319 (1972).

Kohatsu I., Wuensch B.J. The crystal structure of nuffieldite, $Pb_2Cu(Pb,Bi)Bi_2S_7$, *Z. Kristallogr.*, **138**(1-6), 343–365 (1973).

Koike K., Yazawa A. Thermodynamic studies of the molten Cu_2S–FeS–SnS systems [in Japanese], *Shigen-to-Sozai*, **110**(1), 43–47 (1994).

Kolitsch U., Giester G. Elyite, $Pb_4Cu(SO_4)O_2(OH)_4{\cdot}H_2O$: crystal structure and new data, *Am. Mineralog.*, **85**(11-12), 1816–1821 (2000).

Kolitsch U., Tiekink E.R.T., Slade P.G., Taylor M.R., Pring P. Hinsdalite and plumbogummite, their atomic arrangements and disordered lead sites, *Eur. J. Mineral.*, **11**(6), 949–954 (1999).

Kondo R. The synthesis and crystallography of a group of new compounds belonging to the hauyne type structure [in Japanese], *J. Ceram. Assoc. Jpn*, **73**(832), 101–108 (1965).

Kong F., Jiang H.-L., Mao J.-G. $La_4(Si_{5.2}Ge_{2.8}O_{18})(TeO_3)_4$ and $La_2(Si_6O_{13})(TeO_3)_2$: Intergrowth of the lanthanum(III) tellurite layer with the XO_4 (X = Si/Ge) tetrahedral layer, *J. Solid State Chem.*, **181**(2), 263–268 (2008).

Kopylov N.I. The $Cu_3As–Cu_{2-x}S–PbS–Na_2S$ system [in Russian], *Zhurn. neorgan. khimii*, **22**(8), 2269–2273 (1977).

Kopylov N.I. The $PbS–Cu_2S–FeS–Na_2S$ system [in Russian], *Zhurn. neorgan. khimii*, **14**(6), 1702–1704 (1969).

Kopylov N.I., Kostenetskiy V.P., Toguzov M.Z., Rozhin A.V., Letyagin Yu.I. Interfacial equilibria in complex systems determining the modes of electric melting of semi-products of lead production [in Russian], In: *Sulfid. rasplavy tyazh. met.* M., 128–135 (1982).

Kopylov N.I., Smailov S.D., Smirnov M.P. Equilibrium in the $Pb–(Cu_3As–Fe_2As–PbS–Na_2S)$ system [in Russian], *Tsvet. met.*, (7), 12–14 (1996).

Kopylov N.I., Smailov S.D., Toguzov M.Z. Immissibility in the $Cu_3As–Fe_2As–PbS–Na_2S$ system [in Russian], *Zhurn. neorgan. khimii*, **33**(11), 2918–2922 (1988).

Kopylov N.I., Toguzov M.Z. On the issue of the melt immiscibility in the $Cu_3As–Fe_2As–PbS–Na_2S$ system [in Russian], *Zhurn. neorgan. khimii*, **47**(2), 326–330 (2002).

Kovalenker V.A., Evstigneeva T.L., Malov V.S., Trubkin N.V., Gorshkov A.I., Geinke V.R. Nekrasovite $Cu_{26}V_2Sn_6S_{32}$ – a new mineral of the colusite group [in Russian], *Mineralog. zhurn.*, **6**(2), 88–97 (1984a).

Kovalenker V.A., Troneva N.V., Bortnikov N.S., Basova G.V., Vyal'sov L.N. First occurrence of miharaite in the USSR [in Russian], *Dokl. AN SSSR*, **277**(3), 689–692 (1984b).

Kovrugin V.M., Colmont M., Terryn C., Colis S., Siidra O.I., Krivovichev S.V., Mentré O. pH controlled pathway and systematic hydrothermal phase diagram for elaboration of synthetic lead nickel selenites, *Inorg. Chem.*, **54**(5), 2425–2434 (2015).

Kravchenko S.M., Vlasova E.V., Kazakova M.E., Ilyukhin V.V., Abrashev K.K. Innelite – a new barium silicate [in Russian], *Dokl. AN SSSR*, **141**(5), 1198–1199 (1961).

Krebs B., Hürter H.-U. Selenostannate aus wäßriger Lösung: Darstellung und Struktur von $Na_4SnSe_4{\cdot}16H_2O$, *Z. anorg. allg. Chem.*, **462**(1), 143–151 (1980).

Krebs B., Jacobsen H.-J. Darstellung und Struktur von $Na_4GeSe_4{\cdot}14H_2O$, *Z. anorg. und allg. Chem.*, **421**(2), 97–104 (1976).

Krebs B., Müller H. Selenogemanat aus wäßriger Lözung: Darstellung und Structur von $Na_4Ge_2Se_6{\cdot}16H_2O$, *Z. anorg. und allg. Chem.*, **496**(1), 47–57 (1983).

Krebs B., Pohl S. Thiogermanate mit adamantanartigen $Ge_4S_{10}^{4-}$ Ionen, *Z. Naturforsch.*, **26B**(8), 853–854 (1971).

Krebs B., Pohl S., Schiwy W. Darstellung und Struktur von $Na_4Ge_2S_6{\cdot}14H_2O$ und $Na_4Sn_2S_6{\cdot}14H_2O$, *Z. anorg. allg. Chem.*, **393**(3), 241–252 (1972).

Krebs B., Pohl S., Schiwy W. Hexathio-digermanates and -distannates: A new type of dimeric tetrahedral ion, *Angew. Chem. Int. Ed. Engl.*, **9**(11), 897–898 (1970a).

Krebs B., Pohl S., Schiwy W. Hexathiodigermanate und -distannate: Ein neuer Typ dimerer Tetraeder-Ionen, *Angew. Chem.*, **82**(21), 884 (1970b).

Krebs B., Uhlen H. Selenostannate aus wäßriger Lösung: Darstellung und Struktur von $Na_4Sn_2Se_6{\cdot}13H_2O$, *Z. anorg. allg. Chem.*, **549**(6), 35–45 (1987).

Krebs B., Wallstab H.-J. Thio-hydroxoanionen des Germaniums: Darstellung, Structur und Eigenschaften von $Na_2GeS_2(OH)_2{\cdot}5H_2O$, *Z. Naturforsch.*, **36B**(11), 1400–1406 (1981a).

Krebs B., Wallstab H.-J. Thio-hydroxogermanates: a novel type of mixed tetrahedral anions, *Inorg. Chim. Acta*, **54**, L123–L124 (1981b).

Krivovichev S.V., Armbruster T., Yakovenchuk V.N. The crystal structure of krivovichevite, $Pb_3[Al(OH)_6](SO_4)(OH)$, *Canad. Mineralog.*, **47**(1), 153–158 (2009).

Krivovichev S.V., Filatov S.K., Burns P.C., Vergasova L.P. The crystal structure of allochalcoselite, $Cu^{+}Cu^{2+}{}_5PbO_2(SeO_3)_2Cl_5$, a mineral with well-defined Cu^+ and Cu^{2+} positions, *Canad. Mineralog.*, **44**(2), 507–514 (2006).

Krizan J.W., de la Cruz C., Andersen N.H., Cava R.J. Crystal structure and magnetic properties of the $Ba_3TeCo_3P_2O_{14}$, $Pb_3TeCo_3P_2O_{14}$, and $Pb_3TeCo_3V_2O_{14}$ langasites, *J. Solid State Chem.*, **203**, 310–320 (2013).

Krol' O.F., Chernov V.I., Shipovalov Yu.V., Khan G.A. Saryarkite: a new mineral [in Russian], *Zap. Vses. mineralog. obshch.*, **93**(2), 147–155 (1964).

Krüger G., Büttner M., Oliew G., Thilo E. Über Anhydride von Kieselsäure und Methyltrisilanol mit Schwefelsäure und deren Polymerisationsvermögen, *Z. anorg. und allg. Chem.*, **360**(1-2), 70–76 (1968).

Krymus A.S., Kityk I.V., Demchenko P., Parasyuk O.V., Myronchuk G.L., Khyzhun O.Y., Piasecki M. $AgGaSiSe_4$: growth, crystal and band electronic structure, optoelectronic and piezoelectric properties, *Mater. Res. Bull.*, **95**, 177–184 (2017).

Kryukova G.N., Heuer H., Wagner G, Doering T., Bente K. Synthetic $Cu_{0.507(5)}Pb_{8.73(9)}Sb_{8.15(8)}I_{1.6}S_{20.0(2)}$ nanowires, *J. Solid State Chem.*, **178**(1), 376–381 (2005).

Kudoh Y., Takéuchi Y. The superstructure of stannoidite, *Z. Kristallogr.*, **144**(1-6), 145–160 (1976).

Kuliev A.Z., Kakhramanov K.Sh., Babaev G.M., Repen'ko S.I. Investigation of the physico-chemical compatibility of the PbS–PbSe–PbTe alloys with Fe, Co, Ni and NiSb [in Russian], *Izv. AN SSSR. Neorgan. mater.*, **10**(11), 2078 (1974).

Kumar V.P., Guilmeau E., Raveau B., Caignaert V., Varadaraju U.V. A new wide band gap thermoelectric quaternary selenide $Cu_2MgSnSe_4$, *J. Appl. Phys.*, **118**(15), 155101_1–155101_8 (2015).

Kuo D.-H., Tsega M. Hole mobility enhancement of Cu-deficient $Cu_{1.75}Zn(Sn_{1-x}Al_x)Se_4$ bulks, *J. Solid State Chem.*, **206**, 134–138 (2013).

Kuo D.-H., Wubet W. Mg dopant in $Cu_2ZnSnSe_4$: An *n*-type former and a promoter of electrical mobility up to 120 $cm^2 \cdot V^{-1} \cdot s^{-1}$, *J. Solid State Chem.*, **215**, 122–127 (2014).

Kupčik V. Die Kristallstruktur des Minerals Eclarit, $(Cu,Fe)Pb_9Bi_{12}S_{28}$, *Tschermaks Miner. Petr. Mitt.*, **32**(4), 259–269 (1984).

Kusachi I., Shiraishi N., Shimada K., Ohnishi M., Kobayashi S. CO_3-rich charlesite from the Fuka mine, Okayama Prefecture, Japan, *J. Mineral. Petrol. Sci.*, **103**(1), 47–51 (2008).

Laflamme J.H.G., Roberts A.C., Criddle A.J., Cabri L.J. Owensite, $(Ba,Pb)_6(Cu,Fe,Ni)_{25}S_{27}$, a new mineral species from the Wellgreen Cu–Ni–Pt–Pd deposit, Yukon, *Canad. Mineralog.*, **33**(3), 665–670 (1995).

Lai W.-H., Haynes A.S., Frazer L., Chang Y.-M., Liu T.-K., Lin J.-F., Liang I.-C., Sheu H.-S., Ketterson J.B., Kanatzidis M.G., Hsu K.-F. Second harmonic generation response optimized at various optical wavelength ranges through a series of cubic chalcogenides $Ba_6Ag_{2.67+4\delta}Sn_{4.33-\delta}S_{16-x}Se_x$, *Chem. Mater.*, **27**(4), 1316–1326 (2015).

Lam A.E., Groat L.A., Ercit T.S. The crystal structure of dugganite, $Pb_3Zn_3Te^{6+}As_2O_{14}$, *Canad. Mineralog.*, **36**(3), 823–830 (1998).

Lam A.E., Groat L.A., Grice J.D., Ercit T.S. The crystal structure of choloalite, *Canad. Mineralog.*, **37**(3), 721–729 (1999).

Lampe von F., Grimmer A.-R., Wallis B. Preparation and characterization of the calcium silicate sulfate chloride $Ca_4(SiO_4)(SO_4)Cl_2$, *Cem. Concr. Res.*, **19**(4), 595–602 (1989).

Landon R.E., Mogilnor A.H. Colusite, a new mineral of the sphalerite group, *Am. Mineralog.*, **18**(12), 528–533 (1933).

Large R.R., Mumme W.G. Junoite, "wittite", and related seleniferous bismuth sulfosalts from Juno Mine, Northern Territory, Australia, *Econ. Geol.*, **70**(2), 369–383 (1975).

Launay S., Mahé P., Quarton M. Crystal data for thirteen tellurates $M^IBaNbTe_2O_9$ (M^I = Li, Na, K, Rb, Ag) and $M^{II}_{0.5}BaNbTe_2O_9$ (M^{II} = Mg, Ca, Sr, Ba, Cu, Zn, Cd, Pb), *Powder Diffr.*, **4**(1), 31–33 (1989a).

Launay S., Mahé P., Quarton M. Synthese et caracterisation de nouveaux tellurates à valence mixte, *J. Less Common Metals*, **147**(2), 161–166 (1989b).

Lavela P., Tirado J.L., Morales J., Olivier-Fourcade J., Jumas J.-C. Lithium intercalation and copper extraction in spinel sulfides of general formula $Cu_2MSn_3S_8$ (M = Mn, Fe, Co, Ni), *J. Mater. Chem.*, **6**(1), 41–47 (1996).

Lebedev A.I., Sluchinskaya I.A. Low-temperature phase transitions in some quaternary solid solution of IV-VI semiconductors, *J. Alloys Compd.*, **203**(1-2), 51–54 (1994).

Lee F. The submicroscopic structure of wenkite, *Am. Mineralog.*, **59**(9-10), 1135–1136 (1974).

Lee S.-P., Huang C.-H., Chan T.-S., Chen T.-M. New Ce^{3+}-activated thiosilicate phosphor for LED lighting – synthesis, luminescence studies, and applications, *ACS Appl. Mater. Interf.*, **6**(10), 7260–7267 (2014).

Leoni L., Mellini M., Merlino S., Orlandi P. Cancrinite-like minerals: new data and crystal chemical considerations, *Rend. Soc. Ital. Mineral. Petrol.*, **35**(2), 713–719 (1979).

Leoni L., Mellini M., Merlino S., Orlandi P. Crystal chemistry of latiumite, *Rend. Soc. Ital. Mineral. Petrol.*, **33**(2), 531–537 (1977).

Leonov V.V., Popkov A.G. Investigation of the properties of the Bi_2Te_3–Sb_2Te_3–SnTe–GeTe alloys [in Russian], *Izv. AN SSSR. Neorgan. mater.*, **16**(8), 1488–1489 (1980).

Levcenko S., Nateprov A., Kravtsov V., Guc M., Peréz-Rodríguez A., Izquierdo-Roca V., Fontané X., Arushanov E. Structural study and Raman scattering analysis of $Cu_2ZnSiTe_4$ bulk crystals, *Opt. Express*,, **22**(S7), A1936–A1943 (2014).

Li C., Feng K., Tu H., Yao J., Wu Y. Four new chalcohalides, $NaBa_2SnS_4Cl$, KBa_2SnS_4Cl, KBa_2SnS_4Br and $CsBa_2SnS_4Cl$: Syntheses, crystal structures and optical properties, *J. Solid State Chem.*, **227**, 104–109 (2015a).

Li C., Lin Z., Kang L., Lin Zh., Huang H., Yao J., Wu Y. Sn_2SiS_4, synthesis, structure, optical and electronic properties. *Opt. Mater.*, **47**, 379–385 (2015b).

Li G.-M., Wu K., Pan S.-L. $Li_2BaSiSe_4$: A new metal-mixed selenide with outstanding second-harmonic generation [in Chinese], *Chin. Sci. Bull.*, **64**(16), 1671–1678 (2019a).

Li G.-M., Wu K., Huang Y., Yang Z., Pan S. $[Ge_2S_5(S_2)]^{4-}$, a new NLO-active unit leading to asymmetric structure discovered in $Li_2Cs_4Ge_2S_5(S_2)Cl_2$: an experimental and theoretical study, *Chem. Eur. J.*, **25**(21), 5440–5444 (2019b).

Li J., Li X.-H., Yao W.-D., Liu W., Guo S.-P. Three-in-one strategy constructing the first high-performance nonlinear optical sulfide crystallizing with the $P4_3$ space group, *Small*, **19**(38), 2303090 (2023a).

Li J., Zhou W., Yao W.-D., Liu W., Guo S.-P. Three-in-one tetrahedral functional units constructing diamondlike $Ag_2In_2SiS_{3.06}Se_{2.94}$ with high-performance nonlinear-optical activity, *Inorg. Chem.*, **62**(33), 13179–13183 (2023b).

Li M.-Y., Li B.-X., Lin H., Ma Z., Wu L.-M., Wu X.-T., Zhu Q.-L. $Sn_2Ga_2S_5$: A polar semiconductor with exceptional infrared nonlinear optical properties originating from the combined effect of mixed asymmetric building motifs, *Chem. Mater.*, **31**(16), 6268–6275 (2019c).

Li M.-Y., Li B.-X., Lin H., Shi Y.-F., Ma Z., Wu L.-M., Wu X.-T., Zhu Q.-L. Ternary mixed-metal Cd_4GeS_6: Remarkable nonlinear-optical properties based on a tetrahedral-stacking framework, *Inorg. Chem.*, **57**(15), 8730–8734 (2018a).

Li P.-F., Kong F., Mao J.-G. $M^{II}{}_2M_3{}^{III}F_3(Te_6F_2O_{16})$ (M^{II} = Pb, Ba; M^{III} = Al, Ga): New mixed anionic tellurites with isolated Te_6 coplanar rings, *J. Solid State Chem.*, **286**, 121288 (2020a).

Li R.-A., Zhou Z., Lian Y.-K., Jia F., Jiang X., Tang M.-C., Wu L.-M., Sun J., Chen L. A_2SnS_5: Structural incommensurate modulation exhibiting strong second-harmonic generation and high laser induced damage threshold (A = Ba, Sr), *Angew. Chem. Int. Ed.*, **59**(29), 11861–11865 (2020b).

Li S.-F., Jiang X.-M., Fan Y.-H., Liu B.-W., Zeng H.-Y., Guo G.-C. New strategy for designing promising mid-infrared nonlinear optical materials: narrowing band gap for large nonlinear optical efficiency and reducing thermal effect for high laser-induced damage threshold, *Chem. Sci.*, **9**(26), 5700–5708 (2018b).

Li X.-H., Chen Y.-T., Xue H., Guo S.-P., $KInSi_{1.32}Sn_{0.68}Se_6$: an infrared nonlinear optical material containing three types of tetrahedral units, *Inorg. Chem.*, **59**(9), 5823–5827 (2020c).

Li X.-H, Shi Z.-H, Yang M, Liu W, Guo S.-P. $Sn_7Br_{10}S_2$: The first ternary halogen-rich chalcohalide exhibiting a chiral structure and pronounced nonlinear optical properties, *Angew. Chem. Int. Ed.*, **61**(9), e202115871 (2022).

Li X.-H., Suen N.-T., Chi Y., Sun Y., Gong A., Xue H., Guo S.-P. Partial congener substitution induced centrosymmetric to noncentrosymmetric transformation witnessed by $K_3Ga_3(Ge_{7-x}M_x)Se_{20}$ (M = Si, Sn) and their nonlinear optical properties, *Inorg. Chem.*, **58**(19), 13250–13257 (2019d).

Li X.-H., Yao W.-D., Wei Y.-L., Guo S.-P. Three-in-one strategy constructing a series of hybrid tetrahedral motif-based selenides with balanced second-order nonlinear optical performance, *Inorg. Chem.*, **60**(9), 6641–6648 (2021a).

Li Y., Song X., Liu Y., Guo Y., Sun Y., Ji M., You Z., An Y. Syntheses, structures, and photocatalytic properties of open-framework Ag–Sn–S compounds. *Dalton Trans.*, **49**(33), 11708–11714 (2020d).

Li Y.-H., Liu Y., Guo Y.-K., Sun Y., Ji M., You Z.-L., An Y.-L. Cotemplating assembly and structural variation of three-dimensional open-framework sulfides, *Inorg. Chem.*, **58**(21), 14289–14293 (2019e).

Li Y.-N., Chen Z.-X., Yao W.-D., Tang R.-L., Guo S.-P. Heterovalent cations substitution to design asymmetric chalcogenides with promising nonlinear optical performances, *J. Mater. Chem. C*, **9**(27), 8659–8665 (2021b).

Li Y.-Y., Wang J.-Q., Liu P.-F., Lin H., Chen L., Wu L.-M. Centrosymmetric to noncentrosymmetric structural transformation of new quaternary selenides induced by isolated dimeric $[Sn_2Se_4]$ units: from $Ba_8Ga_2Sn_7Se_{18}$ to $Ba_{10}Ga_2Sn_9Se_{22}$, *RSC Adv.*, **7**(14), 8082–8089 (2017).

Lian Y.-K., Li R.-A., Liu X., Wu L.-M., Chen L. $Sr_6(Li_2Cd)A_4S_{16}$ (A = Ge, Sn): How to go beyond the band gap limitation via site-specific modification, *Cryst. Growth Des.*, **20**(12), 8084–8089 (2020).

Liao J.H., Marking G.M., Hsu K.F., Matsushita Y., Ewbank M.D., Borwick R., Cunningham P., Rosker M.J., Kanatzidis M.G. α- and β-$A_2Hg_3M_2S_8$ (A = K, Rb; M = Ge, Sn): polar quaternary chalcogenides with strong nonlinear optical response, *J. Am. Chem. Soc.*, **125**(31), 9484–9493 (2003).

Lin Z., Feng K., Tu H., Kang L., Lin Z., Yao Y., Wu Y. Three new chalcohalides, $Ba_4Ge_2PbS_8Br_2$, $Ba_4Ge_2PbSe_8Br_2$ and $Ba_4Ge_2SnS_8Br_2$: Syntheses, crystal structures, band gaps, and electronic structures, *J. Alloys Compd.*, **611**, 422–426 (2014).

Liu B.-W., Hu C.-L., Zeng H.-Y., Jiang X.-M., Guo, G.-C. Strong SHG response via high orientation of tetrahedral functional motifs in polyselenide $A_2Ge_4Se_{10}$ (A = Rb, Cs), *Adv. Opt. Mater.*, **6**(13), 1800156 (2018).

Liu H., Chang L.L.Y. Knowles C.R. The Mn isotype of andorite and uchucchacuaite, *Canad. Mineralog.*, **32**(1), 185–188 (1994).

Liu H., Song Z., Wu H., Hu Z., Wang J., Wu Y., Yu H. $[Ba_2F_2][Ge_2O_3S_2]$: An unprecedented heteroanionic infrared nonlinear optical material containing three typical anions, *ACS Mater. Lett.*, **4**(9), 1593–1598 (2022a).

Liu Q.-Q., Liu X., Wu L.-M., Chen L. $SrZnGeS_4$: A dual-waveband nonlinear optical material with a transparency spanning UV/Vis and far-IR spectral regions, *Angew. Chem.*, **134**(28), e202205587 (2022b).

Liu Q.-Q., Liu X., Wu L.-M., Chen L. $SrZnGeS_4$: A dual-waveband nonlinear optical material with a transparency spanning UV/Vis and far-IR spectral regions, *Angew. Chem. Int. Ed.*, **62**(28), e2022055872022 (2022c).

Liu X., Peng J., Xiao X., Xiong Z., Huang G., Chen B., He Z., Huang W. Crystal growth, characterization, and thermal annealing of nonlinear optical crystals $AgGaGe_nSe_{2(n+1)}$ (n = 1.5, 1.75, 2, 3, 4, 5, and 9) for mid-infrared application, *Inorg. Chem.*, **61**(17), 6562–6573 (2022d).

Liu Y., Song X.-D., Zhang R.-C., Zhou F.-Y., Zhang J.-W., Jiang X.-M., Ji M., An Y.-L. Solvothermal syntheses and characterizations of four quaternary copper sulfides $BaCu_3MS_4$ (M = In, Ga) and $BaCu_2MS_4$ (M = Sn, Ge). *Inorg. Chem.*, **58**(22), 15101–15109 (2019).

Livingstone A., Russel J.D. X-ray powder data for susannite and its distinction from leadhillite, *Mineralog. Mag.*, **49**(354) 759–761 (1985).

Livingstone A., Ryback G., Fejer E.E., Stanley C.J. Mattheddleite, a new mineral of the apatite group from Leadhills, Strathclyde Region, *Scot. J. Geol.*, **23**(1), 1–8 (1987).

Livingstone A., Sarp H. Macphersonite, a new mineral from Leadhills, Scotland, and Saint-Prix, France – a polymorph of leadhillite and susannite, Mineralog. Mag., **48**(347), 277–282 (1984).

Locock A.J., Ercit T.S., Kjellman J., Piilonen P.C. New mineral names, *Am. Mineralog.*, **91**(11-12), 1945–1954 (2006a).

Locock A.J., Piilonen P.C., Ercit T.S., Rowe R. New mineral names, *Am. Mineralog.*, **91**(1), 216–224 (2006b).

Logemann C., Gunzelmann D., Klüner T., Senker J., Wickleder M.S. Reactions with oleum under harsh conditions: characterization of the unique $[M(S_2O_7)_3]^{2-}$ ions (M = Si, Ge, Sn) in $A_2[M(S_2O_7)_3]$ (A = NH_4, Ag), *Chem. Eur. J.*, **18**(48), 15495–15503 (2012).

Logemann C., Klüner T., Wickleder M.S. The $[Si(S_2O_7)_3]^{2-}$ anion: a first example of octahedral silicon coordination by three chelating inorganic ligands, *Chem. Eur. J.*, **17**(3), 758–760 (2011).

Loose A., Sheldrick W.S. Solventothermal synthesis and structure of lamellar alkali metal selenidostannates(IV) $A_4Sn_4Se_{10}{\cdot}xH_2O$ with 20- (A = Cs; x = 3.2) and 36-membered pores (A = K, Rb; x = 4.5, 1.5). *J. Solid State Chem.*, **147**(1), 146–153 (1999a).

Loose A., Sheldrick W.S. Solventothermal synthesis of the lamellar selenidostanates(IV) $A_2Sn_4Se_9{\cdot}H_2O$ (A = Rb, Cs) and $Cs_2Sn_2Se_6$, *Z. Naturforsch.*, **53B**(3), 349–354 (1998).

Loose A., Sheldrick W.S. Synthesis and structure of the bicycle $[Sn_4Se_{11}]^{6-}$ anion in $Na_6Sn_4Se_{11}{\cdot}22H_2O$, *Z. anorg. und allg. Chem.*, **627**(9), 2051–2052 (2001).

Loose A., Sheldrick W.S. Synthesis and structure of $[Sn_4Se_{11}]^{6-}$, $[Sn_2Se_5{}^{2-}]$, and $[Sn_3Se_7{}^{2-}]$-model anions for the solventothermal reaction pathway to lamellar selenidostanates(IV), *Z. anorg. und allg. Chem.*, **625**(2), 233–240 (1999b).

Loose A., Sheldrick W.S. Quaternary selenidogermanates(IV) $A_3[AgGe_4Se_{10}]{\cdot}2H_2O$ and $A_2[MnGe_4Se_{10}]{\cdot}3H_2O$ (A = Rb, Cs) with open framework structures, *Z. Naturforsch.*, **52B**(6), 687–692 (1997).

López-Vergara F., Galdámez A., Manríquez V., Barahona P., Peña O. $Cu_2Mn_{1-x}Co_xSnS_4$: Novel kësterite type solid solutions, *J. Solid State. Chem.*, **198**, 386–391 (2013).

López-Vergara F., Galdámez A., Manríquez V., Barahona P., Peña O. Magnetic properties and crystal structure of solid-solution $Cu_2Mn_xFe_{1-x}SnS_4$ chalcogenides with stannite-type structure, *Phys. stat. sol. (b)*, **251**(5), 958–964 (2014).

Lu R., Olvera A., Bailey T.P., Fu J., Su X., Veremchuk I., Yin Z., Buchanan B., Uher C., Tang X., Grin Y., Poudeu P.F.P. High carrier mobility and ultralow thermal conductivity in the synthetic layered superlattice $Sn_4Bi_{10}Se_{19}$, *Mater. Adv.*, **2**(7), 2382–2390 (2021).

Ly T.T., Park J., Kim K., Ahn H.-B., Lee N.J., Kim K., Park T.-E., Duvjir G., Lam N.H., Jang K., You C.-Y., Jo Y., Kim S.K., Lee C., Kim S., Kim J. Direct observation of Fe–Ge ordering in $Fe_{5-x}GeTe_2$ crystals and resultant helimagnetism, *Adv. Funct. Mater.*, **31**(17), 2009758 (2021).

Lychmanyuk O.S., Gulay L.D., Olekseyuk I.D. Isothermal section of the Y_2Se_3–Cu_2Se–$GeSe_2$ system at 870 K and crystal structure of the $Y_3CuGeSe_7$ compound, *Pol. J. Chem.*, **80**(3), 463–469 (2006).

Lunts A. Mineralogy and genesis of genthelvine from rare-earth pegmatites of alkaline granites of the Europian part of the USSR [in Russian], *Izv AN LatvSSR*, [6(191)], 75–84 (1963).

Ma M., Dang J., Wu Y., Jiang X., Mei D. Optimal design of mid-infrared nonlinear-optical crystals: from $SrZnSnSe_4$ to $SrZnSiSe_4$, *Inorg. Chem.*, **62**(17), 6549–6553 (2023).

Ma Y.-X., Hu C.-L., Li B.-X., Kong F., Mao J.-G. $PbCdF(SeO_3)(NO_3)$: A nonlinear optical material produced by synergistic effect of four functional units, *Inorg. Chem.*, **57**(18), 11839–11846 (2018).

Makovicky E., Balić-Žunić T., Topa D. The crystal structure of neyite, $Ag_2Cu_6Pb_{25}Bi_{26}S_{68}$, *Canad. Mineralog.*, **39**(5), 1365–1376 (2001).

Makovicky E., Karup-Møller S. New data on giessenite from the Bjørkesen sulfide deposit at Otoften, northern Norway, *Canad. Mineralog.*, **24**(1), 21–25 (1986).

Makovicky E., Mumme W.G. The crystal structure of benjaminite $Cu_{0.50}Pb_{0.40}Ag_{2.30}Bi_{6.80}S_{12}$, *Canad. Mineralog.*, **17**(3), 607–618 (1979).

Makovicky E., Mumme W.G., Nørrestam R. The crystal structures of izoklakeite, dadsonite and jaskolskiite, *Acta Crystallogr.*, **A40**(1), Suppl., C-246 (1984).

Makovicky E., Nørrestam R. The crystal structure of jaskolskiite, $Cu_xPb_{2+x}(Sb,Bi)_{2-x}S_5$ ($x \approx 0.2$), a member of the meneghinite homologous series, *Z. Kristallog.*, **171**(1-4), 179–194 (1985).

Makovicky E., Petříček V., Dušek M., Topa D. The crystal structure of franckeite, $Pb_{21.7}Sn_{9.3}Fe_{4.0}Sb_{8.1}S_{56.9}$, *Am. Mineralog.*, **96**(11-12), 1686–1702 (2011).

Makovicky E., Stöger B., Topa D. The incommensurately modulated crystal structure of roshchinite, $Cu_{0.09}Ag_{1.04}Pb_{0.65}Sb_{2.82}As_{0.37}S_{6.08}$, *Z. Kristallogr.*, **233**(3-4), 255–267 (2018a).

Makovicky E., Topa D. The crystal structure of barikaite. *Mineralog. Mag.*, **77**(8), 3093–3104 (2013).

Makovicky E., Topa D. The crystal structure of jasrouxite, a Pb–Ag–As–Sb member of the lillianite homologous series, *Eur. J. Mineral.*, **26**(1), 145–155 (2014).

Makovicky E., Topa D., Paar W.H. The definition and crystal structure of clino-oscarkempffite, $Ag_{15}Pb_6Sb_{21}Bi_{18}S_{72}$, *Eur. J. Mineral.*, **30**(3), 569–579 (2018b).

Malinko S.V., Chukanov N.V., Dubinchuk V.T., Zadov A.E., Koporulina E.V. Buryatite, $Ca_3(Si,Fe^{3+},Al)(SO_4)[B(OH)_4](OH)_5O{\cdot}12H_2O$, a new mineral [in Russian], *Zap. Vseros. Mineralog. Obshch.*, **130**(2), 72–78 (2001).

Mamedov A.N., Akhmedova N.Ya., Asadov C.M., Babanly N.B., Mamedov E.I. Thermodynamic analysis and defection formation in alloys on the basis of lead selenide containing copper, *Chem. Probl.*, (1), 16–25 (2019).

Mamedov A.N., Kulieva S.A. Thermodynamic functions of formation of Ln_2SnSe_4 (Ln – La, Ce, Pr, Nd) compounds [in Azerbaijanian], *Kimya Problem.*, (4), 525–527 (2012).

Mammadov Sh.G. Phase formation in the Cu_2SnS_3–Sb_2S_3 system [in Russian], *Tomsk State Univ. J. Chem.*, (18), 18–26 (2020a).

Mammadov Sh.G. The Ag_2SnS_3–Sb_2S_3 quasibinary section [in Russian], *Izv. Sarat. Univ. Nov. ser. Ser. Khimiya. Biologiya. Ekologiya*, **20**(1), 49–54 (2020b).

Mammadov Sh.H. Phase equilibrium in Cu_2SnS_3–Cu_3SbS_3 system [in Russian], *Tomsk State Univ. J. Chem.*, (15), 26–35 (2019).

Mandarino J.A. Barquillite $Cu_2(Cd,Fe)GeSe_4$, *Mineral. Rec.*, **30**(5), 399–405 (1999).

Mandarino J.A. New minerals, *Canad. Mineralog.*, **39**(6), 1751–1760 (2001).

Mandarino J.A. New minerals, *Canad. Mineralog.*, **41**(3), 803–828 (2003).

Manos M.J., Chrissafis K., Kanatzidis M.G. Unique pore selectivity for Cs^+ and exceptionally high NH^{4+} exchange capacity of the chalcogenide material $K_6Sn[Zn_4Sn_4S_{17}]$, *J. Am. Chem. Soc.*, **128**(27), 8875–8883 (2006).

Manos M.J., Ding N., Kanatzidis M.G. Layered metal sulfides: Exceptionally selective agents for radioactive strontium removal, *Proc. Nat. Acad. Sci. USA*, **105**(10), 3696–3699 (2008).

Manos M.J., Iyer R.G., Quarez E., Liao J.H., Kanatzidis M.G. $\{Sn[Zn_4Sn_4S_{17}]\}^{6-}$: A robust open framework based on metal-linked penta-supertetrahedral $[Zn_4Sn_4S_{17}]^{10-}$ clusters with ion-exchange properties, *Angew. Chem. Int. Ed.*, **44**(23), 3552–3555 (2005).

Manos M.J., Kanatzidis M.G. Use of hydrazine in the hydrothermal synthesis of chalcogenides: the neutral framework material $[Mn_2SnS_4(N_2H_4)_2]$, *Inorg. Chem.*, **48**(11), 4658–4660 (2009).

Marchuk O., Smitiukh O. Analysis of crystal structures of the initial phases of the quasiternary Y_2S_3–La_2S_3–PbS system [in Ukrainian], *Probl. khimii ta staloho rozvytku*, (1), 20–25 (2021).

Martan H., Weiss J. Metall-Schwefelstickstoff-Verbindungen. 16 Reaktionsprodukte von Blei- und Zinnsalzen mit S_4N_4. Die Strukturen von $PbN_2S_2{\cdot}NH_3$, PbN_2S_2 und $SnCl_4{\cdot}2S_4N_4$, *Z. anorg. und allg. Chem.*, **514**(7), 107–114 (1984).

Martin S.M., Sills J.A. ^{29}Si and ^{27}Al MASS-NMR studies of $Li_2S + Al_2S_3 + SiS_2$ glasses, *J. Non-Cryst. Solids*, **135**(2-3), 171–181 (1991).

Martucci A., Cruciani G. *In situ* time resolved synchrotron powder diffraction study of thaumasite, *Phys. Chem. Minerals*, **33**(10), 723–731 (2006).

Marumo F., Nowacki W. The crystal structure of hatchite, $PbTlAgAs_2S_5$, *Z. Kristallogr.*, **125**(1-6), 249–265 (1967).

Marushko L.P. Phase equilibria in the Ag_2CdGeS_4–$AgCd_2GaS_4$ system at 820 K [in Ukrainian], *Nauk. Visnyk Volyns'k. Nats. Univ. im. Lesi Ukrainky. Ser. Khim. nauky*, (17), 59–62 (2012a).

Marushko L.P. Phase equilibria in the Cu_2CdGeS_4–$AgCdGeS_4$ system at 820 K [in Ukrainian], *Nauk. Visnyk Volyns'k. Nats. Univ. im. Lesi Ukrainky. Ser. Khim. nauky*, (17), 110–114 (2012b).

Marushko L.P., Piskach L.V., Parasyuk O.V., Ivashchenko I.A., Olekseyuk I.D. Quasi-ternary system Cu_2GeS_3–Cu_2SnS_3–CdS, *J. Alloys Compd.*, **484**(1-2), 147–153 (2009a).

Marushko L.P., Piskach L.V., Parasyuk O.V., Olekseyuk I.D., Volkov S.V., Pekhnyo V.I. The reciprocal system $Cu_2GeS_3 + 3CdSe \Leftrightarrow Cu_2GeSe_3 + 3CdS$, *J. Alloys Compd.*, **473**(1-2), 94–99 (2009b).

Matsubara S., Mouri T., Miyawaki R., Yokoyama K., Nakahara M. Munakataite, a new mineral from the Kato mine, Fukuoka, Japan, *J. Mineral. Petrol. Sci.*, **103**(5), 327–332 (2008).

Matsunaga T., Yamada N. A study of highly symmetrical crystal structures, commonly seen in high-speed phase-change materials, using synchrotron radiation, *Jpn. J. Appl. Phys.*, **41**(3B), 1674–1678 (2002).

Maurel C., Moëlo Y. Synthèse de la nuffieldite dans le système Bi–Pb–Sb–Cu–S, *Canad. Mineralog.*, **28**(4), 745–749 (1990).

May A.F., Bridges C.A., McGuire M.A., Physical properties and thermal stability of $Fe_{5-x}GeTe_2$ single crystals, *Phys. Rev. Mater.*, **3**(10), 104401_1–104401_104401_15 (2019a).

May A.F., Du M.-H., Cooper V.R., McGuire M.A. Tuning magnetic order in the van der Waals metal Fe_5GeTe_2 by cobalt substitution, *Phys. Rev. Mater.*, **4**(7), 074008 (2020).

May A.F., Ovchinnikov D., Zheng Q., Hermann R., Calder S., Huang B., Fei Z., Liu Y., Xu X., McGuire M.A. Ferromagnetism near room temperature in the cleavable van der Waals crystal Fe_5GeTe_2, *ACS Nano*, **13**(4), 4436–4442 (2019b).

McConnell D. The substitution of SiO_4- and SO_4-groups for PO_4-groups in the apatite structure; ellestadite, the end-member, *Am. Mineralog.*, **22**(9), 977–986 (1937).

McDonald A.M., Chao G.Y. Haineaultite, a new hydrated sodium calcium titanosilicate from Mont Saint-Hilaire, Quebec: description, structure determination and genetic implications, *Canad. Mineralog.*, **42**(3), 769–780 (2004).

McDonald A.M., Stanley C.J., Ross K.C., Nestola F. Zincobriartite, IMA 2015-094. CNMNC Newsletter No. 29, February 2016, page 203, *Mineralog. Mag.*, **80**(1), 199–205 (2016).

McLean W.J. Confirmation of the mineral species wherryite, *Am. Mineralog.*, **55**(3-4), Pt. 1, 505–508 (1970).

Meerschaut A., Palvadeau P., Moëlo Y., Orlandi P. Lead-antimony sulfosalts from Tuscany (Italy). IV. Crystal structure of pillaite, $Pb_9Sb_{10}S_{23}ClO_{0.5}$, an expanded monoclinic derivative of hexagonal $Bi(Bi_2S_3)_9I_3$, from the zinkenite group, *Eur. J. Mineral.*, **13**(4), 779–790 (2001).

Mei D., Cao W., Wang N., Jiang X., Zhao J., Wang W., Dang J., Zhang S., Wu Y., Rao P., Lin Z. Breaking through the "3.0 eV wall" of energy band gap in mid-infrared nonlinear optical rare earth chalcogenides by charge-transfer engineering, *Mater. Horiz.*, **8**(8), 2330–2334 (2021).

Meisser N., Schenk K., Berlepsch P., Brugger J., Bonin M., Criddle A.J., Thélin P., Bussy F. Pizgrischite, $(Cu,Fe)Cu_{14}PbBi_{17}S_{35}$, a new sulfosalt from the Swiss Alps: description, crystal structure and occurrence, *Canad. Mineralog.*, **45**(5), 1229–1245 (2007).

Mellini M., Merlino S., Rossi G. The crystal structure of tuscanite, *Am. Mineralog.*, **62**(11-12), 1114–1120 (1977).

Melullis M., Clérac R., Dehnen S. Ternary Mn/Ge/Se anions from reactions of $[Ba_2(H_2O)_9][GeSe_4]$: Synthesis and characterization of compounds containing discrete or polymeric $[Mn_6Ge_4Se_{17}]^{6-}$ units, *Chem. Commun.*, (48), 6008–6010 (2005).

Melullis M., Dehnen S. Syntheses, crystal structures, UV-vis spectra and first NMR spectra of new potassium salts of chalcogenogermanates, *Z. anorg. und allg. Chem.*, **633**(13-14), 2159–2167 (2007).

Mel'nikov O.K., Litvin B.N., Fedosova S.P. Obtaining the group of helvin compounds [in Russian], In: *Gidrotermalnyi sintez kristallov*. M.: Nauka Publish., 167–174 (1968).

Melnychuk Kh.O., Smitiukh O.V., Marchuk O.V., Mazur N.V., Yukhymchuk V.O. Structure investigation of $Ce_{0.5}R_{1.5}PbSi_2S_8$ and $Pr_{1.5}R_{0.5}PbSi_2S_8$ (R – Tb, Y, Er) chalcogenides [in Ukrainian], *Nauk. Visn. Uzhgorod. Univ., Ser. Khim.*, [1(43)], 6–15 (2020).

Menezes L.T., Assoud A., Zhang W., Halasyamani P.S., Kleinke H. Effect of Pb iubstitution in $Sr_{2-x}Pb_xGeSe_4$ on crystal structures and nonlinear optical properties predicted by DFT calculations, *Inorg. Chem.*, **59**(20), 15028–15035 (2020).

Merlino S. The crystal structure of wenkite, *Acta Crystallogr.*, **30**(5), 1262–1266 (1974).

Merlino S., Mellini M., Bonaccorsi E., Pasero M., Leoni L., Orlandi P. Pitiglianoite, a new feldspathoid from southern Tuscany, Italy: Chemical composition and crystal structure, *Am. Mineralog.*, **76**(11–12), 2003–2008 (1991).

Merlino S., Orlandi P. Carraraite and zaccagnaite, two new minerals from the Carrara marble quarries: their chemical compositions, physical properties, and structural features, *Am. Mineralog.*, **86**(10), 1293–1301 (2001).

Merlino S., Orlandi P. Liottite, a new mineral in the cancrinite-davyne group, *Am. Mineralog.*, **62**(3-4), 321–326 (1977).

Mesbah A., Prakash J., Lebègue S., Stojko W., Ibers J.A. Syntheses, crystal structures, and electronic properties of $Ba_8Si_2US_{14}$ and $Ba_8SiFeUS_{14}$, *Solid State Sci.*, **48**, 120–124 (2015).

Miehe G. Crystal structure of kobellite, *Nature, Phys. Sci.*, **231**(23), 133–134 (1971).

Mikolaĭchuk A.G., Moroz N.V., Demchenko P.Y. Synthesis and electrical conductivity of a new $Ag_6SnS_4Br_2$ superionic compound, *Phys. Solid State*, **52**(2), 237–240 (2010).

Mill B.V. Formation of phases with the $Ca_3Ga_2Ge_4O_{14}$ structure in the $AO–TeO_3–Ga_2O_3–XO_2$ (A = Pb, Ba, Sr; X = Si, Ge) and $PbO–TeO_3–MO–GeO_2$ (M = Zn, Co) systems, *Russ. J. Inorg. Chem.*, **55**(10), 1611–1616 (2010).

Mill B.V., Synthesis of dugganite $Pb_3TeZn_3As_2O_{14}$ and its analogues, *Russ. J. Inorg. Chem.*, **54**(8), 1205–1209 (2009).

Mills S.J., Kampf A.R., Christy A.C., Housley R.M., Thorne B., Chen Y.-S., Steele I.M. Favreauite, a new selenite mineral from the El Dragón mine, Bolivia, *Eur. J. Mineral.*, **26**(6), 771–781 (2014a).

Mills S.J., Kampf A.R., Housley R.M., Christy A.C., Thorne B., Chen Y.-S., Steele I.M. Favreauite, IMA 2014-013. CNMNC Newsletter No. 20, June 2014, page 557, *Mineralog. Mag.*, **78**(3), 549–558 (2014b).

Mills S.J., Kampf A.R., Kolitsch U., Housley R.M., Raudsepp M. The crystal chemistry and crystal structure of kuksite, $Pb_3Zn_3Te^{6+}P_2O_{14}$, and a note on the crystal structure of yafsoanite, $(Ca,Pb)_3Zn(TeO_6)_2$, *Am. Mineralog.,* **95**(7), 933–938 (2010).

Mills S.J., Kampf A.R., Momma K., Housley R.M., Marty J. Müllerite, IMA 2019-060. CNMNC Newsletter No. 52, *Eur. J. Mineral.*, **32**(1), 4 (2020a).

Mills S.J., Kampf A.R., Momma K., Housley R.M., Marty J. Müllerite, IMA 2019-060. CNMNC Newsletter No. 52, *Mineralog. Mag.*, **83**(6), 889 (2019).

Mills S.J., Kampf A.R., Momma K., Housley R.M., Marty J. Müllerite, the Fe-analogue of backite from Otto Mountain, California, USA, *Canad. Mineralog.*, **58**(4), 413–419 (2020b).

Mills S.J., Kolitsch U., Miyawaki R., Groat L.A., Poirier G. Joëlbruggerite, $Pb_3Zn_3(Sb^{5+},Te^{6+})As_2O_{13}(OH,O)$, the Sb^{5+} analog of dugganite, from the Black Pine mine, Montana, *Am. Mineralog.*, **94**(7), 1012–1017 (2009a).

Mills S.J., Madsen I.C., Grey I.E., Birch W.D. *In situ* XRD study of the thermal decomposition of natural arsenian plumbojarosite, *Canad. Mineralog.*, **47**(3), 683–696 (2009b).

Milodowski A.E., Morgan D.J. Thermal reactions of leadhillite $Pb_4SO_4(CO_3)_2(OH)_2$, *Clay Minerals*, **19**(5), 825–841 (1984).

Mintser E.F. Benjaminite – $(Cu,Ag)_2Pb_2Bi_4S_9$ [in Russian], *Dokl. AN SSSR*, **174**(3), 675–678 (1967).

Missen O.P., Kampf A.R., Mills S.J., Housley R.M., Spratt J., Welch M.D., Coolbaugh M.F., Marty J., Chorazewicz M., Ferraris C. The crystal structures of the mixed-valence tellurium oxysalts tlapallite,

$(Ca,Pb)_3CaCu_6[Te^{4+}_3Te^{6+}O_{12}]_2(Te^{4+}O_3)_2(SO_4)_2{\cdot}3H_2O$, and carlfriesite, $CaTe^{4+}_2Te^{6+}O_8$, *Mineralog. Mag.*, **83**(4), 539–549 (2019a).

Missen O.P., Mills S.J., Spratt J., Welch M.D., Birch W.D., Rumsey M.S., Vylita J. The crystal-structure determination and redefinition of eztlite, $Pb^{2+}_2Fe^{3+}_3(Te^{4+}O_3)_3(SO_4)O_2Cl$, *Mineralog. Mag.*, **82**(6), 1355–1367 (2018).

Missen O.P., Rumsey M.S., Kampf A.R., Mills S.J., Back M.E., Spratt J. The discreditation of oboyerite and a note on the crystal structure of plumbotellurite, *Mineralog. Mag.*, **83**(6), 791–797 (2019b).

Mitzi D.B. Synthesis, structure, and thermal properties of soluble hydrazinium germanium(IV) and tin(IV) selenide salts, *Inorg. Chem.*, **44**(10), 3755–3761 (2005).

Mitzi D.B., Kosbar L.L., Murray C.E., Copel M., Afzali A. High-mobility ultrathin semiconducting films prepared by spin coating, *Nature*, **428**(6980), 299–303 (2004).

Miyawaki R., Hatert F., Pasero M., Mills S.J. IMA Commission on New Minerals, Nomenclature and Classification (CNMNC). Newsletter No. 51, *Eur. J. Mineral.*, **31**(5-6), 1099–1104 (2019a).

Miyawaki R., Hatert F., Pasero M., Mills S.J. IMA Commission on New Minerals, Nomenclature and Classification (CNMNC). Newsletter No. 51, *Mineralog. Mag.*, **83**(5), 757–761 (2019b).

Miyawaki R., Hatert F., Pasero M., Mills S.J. IMA Commission on New Minerals, Nomenclature and Classification (CNMNC). Newsletter No. 60, *Eur. J. Mineral.*, **33**(2), 203–208 (2021a).

Miyawaki R., Hatert F., Pasero M., Mills S.J. IMA Commission on New Minerals, Nomenclature and Classification (CNMNC). Newsletter No. 60, *Mineralog. Mag.*, **85**(3), 454–458 (2021b).

Miyawaki R., Hatert F., Pasero M., Mills S.J. IMA Commission on New Minerals, Nomenclature and Classification (CNMNC). Newsletter No. 63, *Eur. J. Mineral.*, **33**(5), 639–646 (2021c).

Miyawaki R., Hatert F., Pasero M., Mills S.J. IMA Commission on New Minerals, Nomenclature and Classification (CNMNC). Newsletter No. 63, *Mineralog. Mag.*, **85**(6), 910–915 (2021d).

Miyawaki R., Matsubara S., Hashimoto E. Elyite from the Mizuhiki mine, Fukushima Prefecture, Japan, *Bull. Nat. Sci. Mus., Ser. C*, **23**(1-2), 27–33 (1997).

Mizuguchi Y., Hijikata Y., Abe T., Moriyoshi C., Kuroiwa Y., Goto Y., Miura A., Lee S., Torii S., Kamiyama T., Lee C.H.,. Ochi M., Kuroki K. Crystal structure, site selectivity, and electronic structure of layered chalcogenide $LaOBiPbS_3$, *Europhys. Lett.*, **119**(2), 26002_1–26002_5 (2017).

Moëlo Y., Balitskaya O., Mozgova N., Sivtsov A. Chloro-sulfosels de l'indice plombo-antimonifère des Cougnasses (Hautes-Alpes), *Eur. J. Mineral.*, **1**(3), 381–390 (1989).

Moëlo Y., Guillot-Deudon C., Evain M., Orlandi P., Biagioni C. Comparative modular analysis of two complex sulfosalt structures: sterryite, $Cu(Ag,Cu)_3Pb_{19}(Sb,As)_{22}(As–As)S_{56}$, and parasterryite, $Ag_4Pb_{20}(Sb,As)_{24}S_{58}$, *Acta Crystallogr.*, **B68**(5), 480–492 (2012).

Moëlo Y., Makovicky E., Karup-Møller S., Cervelle B., Maurel C. La lévyclaudite, $Pb_8Sn_7Cu_3(Bi,Sb)_3S_{28}$, une nouvelle espèce à structure incommensurable, de la série de la cylindrite, *Eur. J. Mineral.*, **2**(5), 711–723 (1990).

Moëlo Y., Meerschaut A. Makovicky E. Refinement of the crystal structure of nuffieldite, $Pb_2Cu_{1.4}(Pb_{0.4}Bi_{0.4}Sb_{0.2})Bi_2S_7$: structural relationships and genesis of complex lead sulfosalt structures, *Canad. Mineralog.*, **35**(6), 1497–1508 (1997).

Moëlo Y., Orlandi P., Guillot-Deudon C., Biagioni C., Paar W.H., Evain M. Lead–antimony sulfosalts from Tuscany (Italy). XI. The new mineral species parasterryite, $Ag_4Pb_{20}(Sb_{14.5}As_{9.5})_{\Sigma 24}S_{58}$, and associated sterryite, $Cu(Ag,Cu)_3Pb_{19}(Sb,As)_{22}(As–As)S_{56}$, from the Pollone mine, Tuscany, Italy, *Canad. Mineralog.*, **49**(2), 623–638 (2011).

Moëlo Y., Orlandi P., Guillot-Deudon C., Biagioni C., Paar W.H., Evain M. Parasterryite, IMA 2010-033. CNMNC Newsletter No 5, October 2010, page 860, *Mineralog. Mag.*, **74**(5), 859–862 (2010).

Moëlo Y., Oudin E., Picot P., Caye R. L'uchucchacuante, $AgMnPb_3Sb_5S_{12}$, une nouvelle espèce minérale de la série de l'andorite, *Bull. Minéralog.*, **107**(5), 597–604 (1984).

Moëlo Y., Palvadeau P., Meisser N., Meerschaut A. Structure cristalline d'une ménéghinite naturelle pauvre en cuivre, $Cu_{0,58}Pb_{12,72}(Sb_{7,04}Bi_{0,24})S_{24}$, *C. r. Acad. sci., Geosci.*, **334**(8), 529–536 (2002).

Moh G.H. Tin-containing mineral systems. Part II: Phase relations and mineral assemblages in Cu–Fe–Zn–Sn–S system, *Chem. Erde*, **34**(1), 1–61 (1975).

Møller C.K. The structure of $Pb(NH_4)_2(SO_4)_2$ and related compounds, *Acta Chem. Scand.*, **8**(1), 81–87 (1954).

Moore P.B., Kampf A.R., Sen Gupta P.K. The crystal structure of philolithite, a trellis-like open framework based on cubic closest-packing of anions, *Am. Mineralog.*, **85**(5-6), 810–816 (2000).

Moore P.B., Shen J. Roeblingite, $Pb_2Ca_6(SO_4)_2(OH)_2(H_2O)_4[Mn(Si_3O_9)_2]$: its crystal structure and comments on the lone pair effect, *Am. Mineralog.*, **69**(11-12), 1173–1179 (1984).

Moreno E., Quintero M., Morocoima M., Quintero E., Grima P., Tovar R., Bocaranda P., Delgado G.E., Contreras J.E., Mora A.E., Briceño J.M., Godoy R.A., Fernandez J.L., Henao J.A., Macías M.A. Lattice parameter values and phase transitions for the $Cu_2Cd_{1-z}Mn_zSnSe_4$ and $Cu_2Cd_{1-z}Fe_zSnSe_4$ alloys, *J. Alloys Compd.*, **486**(1-2), 212–218 (2009).

Morgan W.C Genthelvite and bertrandite from the Cairngorm Mountains, Scotland, *Mineralog. Mag.*, **36**(277), 60–63 (1967).

Moroz M., Demchenko P., Prokhorenko M., Prokhorenko S., Pereviznyk O., Rudyk B., Solyak L., Reshetnyak O. Thermodynamic properties of the $AgSnTe_2$ compound [in Ukrainian], *Visnyk L 'viv. Univ. Ser. khim.*, (61), Pt. 2, 83–93 (2020).

Moroz M.V., Prokhorenko M.V. Phase equilibria and thermodynamic properties of phases in the Ag–Cd–Sn–Se system, *Inorg. Mater.*, **51**(8), 799–805 (2015).

Moroz M.V., Prokhorenko M.V., Rudyk B.P. Thermodynamic properties of phases of the Ag–Ge–Te system, *Russ. J. Electrochem.*, **50**(12), 1177–1181 (2014).

Morris R.C. Osarizawaite from Western Australia, *Am. Mineralog.*, **47**(9-10), 1079–1093 (1962).

Morris R.C. Osarizawaite from Western Australia – a correction, *Am. Mineralog.*, **48**(7-8), 947 (1963).

Mottana A., Fiori S., Parodi G.C. Improved X-ray powder diffraction data for franckeite, *Powder Diffr.*, **7**(2), 112–114 (1992).

Mozgova N.N., Nenasheva S.N., Borodaev Y.S., Yudovskaya M.A. Nuffieldite from the Maleevskoe massive sulfide deposit, Russia, *Canad. Mineralog.*, **32**(2), 359–364 (1994).

Müller H., Kockelmann W., Johrendt D. The magnetic structure and electronic ground states of Mott insulators GeV_4S_8 and GaV_4S_8, *Chem. Mater.*, **18**(8), 2174–2180 (2006).

Mumme W.G. Junoite, $Cu_2Pb_3Bi_8(S,Se)_{16}$, a new sulfosalt from Tennant Creek, Australia: Its crystal structure and relationship with other bismuth sulfosalts, *Am. Mineralog.*, **60**(7-8), 548–558 (1975).

Mumme W.G. Proudite from Tennant Creek, Northern Territory, Australia: its crystal structure and relationship with weibullite and wittite, *Am. Mineralog.*, **61**(9-10), 839–852 (1976).

Mumme W.G. Seleniferous lead-bismuth sulphosalts from Falun, Sweden: weibullite, wittite, and nordstriimite, *Am. Mineralog.*, **65**(7-8), 789–796 (1980a).

Mumme W.G. The crystal structure of nordströmite, $CuPb_3Bi_7(S,Se)_{14}$ from Falun, Sweden: a member of the junoite homologous series, *Canad. Mineralog.*, **18**(3), 343–352 (1980b).

Mumme W.G. The crystal structure of paděraite, a mineral of the cuprobismutite series, *Canad. Mineralog.*, **24**(3), 513–521 (1986).

Mumme W.G., Gable R.W., Wilson N. A crystal structure determination of (selenian) kobellite from the Boliden Mine, Sweden, *N. Jb. Miner. Abh. (J. Min. Geochem.)*, **191**(1), 109–115 (2013).

Mumme W.G., Scott T.R. The relationships between basic ferric sulfate and plumbojarosite, *Am. Mineralog.*, **51**(3-4), 443–453 (1966).

Mumme W.G., Topa D., Makovicky E. Proudite: a redetermination of its crystal structure and the proudite–felbertalite homologous series, *Canad. Mineralog.*, **47**(1), 25–38 (2009).

Mumme W.G., Watts J.A. Additional physical, optical and X-ray data for pekoite, *Canad. Mineralog.*, **14**(4), 578 (1976a).

Mumme W.G. Watts J.A. Pekoite, $CuPbBi_{11}S_{18}$, a new member of the bismuthinite-aikinite mineral series: its crystal structure and relationship with naturally- and synthetically-formed members, *Canad. Mineralog.*, **14**(3), 322–333 (1976b).

Murciego A., Pascua M.I., Babkine J., Dusausoy Y., Medenbach O., Bernhardt H.-J. Barquillite, $Cu_2(Cd,Fe)GeS_4$, a new mineral from the Barquilla deposit, Salamanca, Spain, *Eur. J. Mineral.*, **11**(1), 111–118 (1999).

Murdoch J. X-ray investigation of colusite, germanite and renierite, *Am. Mineralog.*, **38**(9-10), 794–801 (1953).

Musa A.F., Chen S.-W. Interfacial reactions in Ni/Se–Sn, Ni/Se–Te, Ni/Sn–Te and Ni/Se–Sn–Te couples, *J. Electron. Mater.*, **50**(8), 4346–4357 (2021).

Nagaoka A., Yoshino K., Kakimoto K., Nishioka K. Phase diagram of the Ag_2SnS_3–ZnS pseudobinary system for Ag_2ZnSnS_4 crystal growth, *J. Cryst. Growth*, **555**, 125967 (2021).

Nagel A. The crystal structure of a thallium sulfosalt, $Tl_8Pb_4Sb_{21}As_{19}S_{68}$, *Z. Kristallogr.*, **150**(1-4), 85–106 (1979).

Nagel A., Nowacki W. Die Kristallstruktur eines Tl-Sulfosaltez, $Tl_8Pb_4Sb_{21}As_{19}S_{68}$, *Z. Kristallogr.*, **149**(1-2), 147 (1979).

Nateprov A., Kravtsov V.Ch., Gurieva, G., Schorr S. Single crystal X-ray structure investigation of $Cu_2ZnSnSe_4$, *Surf. Engin. Appl. Electrochem.*, **49**(5), 423–426 (2013).

Navarro J.A., Glasser F.P. Crystal chemistry of calcium silicates: Germanate and chromate analogues of $KF{\cdot}2[Ca_6(SO_4)(SiO_4)_2O]$, *Cem. Concr. Res.*, **15**(6), 1051–1054 (1985).

Naydych T.L., Zhbankov O.Ye., Mazurets' I.I. Glass forming region in the $AgGaS_2$–$Cd_{0.5}GaS_2$–GeS_2 system [in Ukrainian], *Nauk. Visnyk Volyns'k. Nats. Univ. im. Lesi Ukrainky. Ser. Khim. nauky*, (4), 80–83 (2006).

Nedoshovenko E.G., Turkina E.Yu., Tver'yanovich Yu.S., Borisova Z.U. Glassforming and interaction of the components in the GeS_2–Ga_2S_3–NaCl system [in Russian], *Vestn. Leningr. Univ. Ser. 4*, (2), 52–57 (1986).

Nenasheva S.N., Borodaev Yu.S., Mozgova N.N., Sivtsov A.V., Ryabeva E.G. First discovery of copper-free benjaminite [in Russian], *New Data on Minerals*, Moscow, **34**, 152–156 (1987).

Nenasheva S.N., Efimov A.V., Sivtsov A.V., Mozgova N.N. Borodaevite $[Ag_5(Fe,Pb)_1Bi_7]_{13}(Sb,Bi)_2S_{17}$ – a new mineral [in Russian], *Zap. Vseros. mineralog. obsh.*, **121**(4), 113–120 (1992).

Nestola F., Guastoni A., Bindi L., Secco L. Dalnegroite, $Tl_{5-x}Pb_{2x}(As,Sb)_{21-x}S_{34}$, a new thallium sulphosalt from Lengenbach quarry, Binntal, Switzerland, *Mineralog. Mag.*, **73**(6), 1027–1032 (2009).

Newton R.C., Goldsmith J.R. Stability of the end-member scapolites: $3NaAlSi_3O_8{\cdot}NaCl$, $3CaAl_2Si_2O_8{\cdot}CaCO_3$, $3CaAl_2Si_2O_8{\cdot}CaSO_4$, *Z. Kristallogr.*, **143**(1-6), 333–353 (1976).

Ni Y., Wu H., Xiao R., Huang C., Wang Z., Mao M., Qi M., Cheng G. Growth and optical properties of single $AgGaGe_5Se_{12}$ (AGGSe) crystal, *Opt. Mater.*, **42**, 458–461 (2015).

Nichols M.C. The structure of tsumebite, *Am. Mineralog.*, **51**(1-2), 267 (1966).

Nimis P., Molin G., Visona D. Crystal chemistry of danalite from Daba Shabeli Complex (N Somalia), *Mineralog Mag.*, **60**(2), 375–379 (1996).

Norako M.E., Greaney M.G., Brutchey R.L. Synthesis and characterization of wurtzite-phase copper tin selenide nanocrystals, *J. Am. Chem. Soc.*, **134**(1), 23–26 (2012).

Nørby P., Eikeland E., Overgaard J., Johnsen S., Iversen B.B. Expanding the structural versatility of thiostannate (IV) complexes, *CrystEngComm*, **17**(11), 2413–2420 (2015).

Nørby P., Overgaard J., Christensen P.S., Richter B, Song X., Dong M., Han A., Skibsted J., Iversen B.B., Johnsen S. $(NH_4)_4Sn_2S_6{\cdot}3H_2O$: crystal structure, thermal decomposition, and precursor for textured thin film, *Chem. Mater.*, **26**(15), 4494–4504 (2014).

Nouadji M., Ivanova Z.G., Poulain M., Zavadil J., Attaf A. Glass formation, physicochemical characterization and phototoluminescence properties of new Sb_2O_3–PbO–ZnO and Sb_2O_3–PbO–ZnS systems, *J. Alloys Compd.*, **549**, 158–162 (2013).

Novikova E.M., Vasil'ev M.G., Evseev V.A., Ershova S.A. Phase diagram of the Sn–GaAs–ZnSe quasiternary system from the Sn-rich side [in Russian], *Deposited in VINITI*, № 3039–74Dep (1974).

Nowacki W., Iitaka Y., Bürki H. Structural investigations on sulfosalts from the Lengenbach, Binn Valley, Switzerland, *Acta Crystallogr.*, **13**(12), 1006–1007 (1960).

Nowacki W., Iitaka Y., Bürki H. Structural investigations on sulfosalts from the Lengenbach, Binn Valley (Ct. Wallis). Part 2, *Schweiz. Mineral. Petrogr. Mitt.*, **41**(1), 103–116 (1961).

Nuffield E.W. Benjaminite, *Am. Mineralog.*, **38**(5-6), 550–552 (1953).

Nuffield E.W. Benjaminite – a re-examination of the type material, *Canad. Mineralog.*, **13**(4), 394–401 (1975).

Nuffield E.W. Cupropavonite from Hall's Valley, Park County, Colorado, *Canad. Mineralog.*, **18**(2), 181–184 (1980).

Nuffield E.W. Observations on kobellite, *Univ. Toronto Studies: VI. Geol. Ser.*, **52**, 86–89 (1948).

Nuffield E.W., Harris D.C. Studies of mineral sulpho-salts: XX. Berryite, a new species, *Canad. Mineralog.*, **8**(4), 407–413 (1966).

Ohmasa M., Mariolacos K. The crystal structure of $(Pb_{1-x},Bi_x)Bi_2Cu_2Cu_{2-x}S_5I_2$ (x = 0.88), *Acta Crystallogr.*, **B30**(11), 2640–2643 (1974).

Ohta M., Chung D.Y., Kunii M., Kanatzidis M.G. Low lattice thermal conductivity in $Pb_5Bi_6Se_{14}$, $Pb_3Bi_2S_6$, and $PbBi_2S_4$: promising thermoelectric materials in the cannizzarite, lillianite, and galenobismuthite homologous series, *J. Mater. Chem. A*, **2**(47), 20048–20058 (2014).

Ohtsuki T., Kitakaze A., Sugaki A. Synthetic minerals with quaternary components in the system Cu–Fe–Sn–S – Synthetic sulfide minerals (X), *Sci. Repts. Tohoku Univ. Ser. III*, **14**(3), 269–282 (1980).

Olekseyuk I., Kogut Yu., Piskach L. Phase equilibria at the isothermal sections of the $Ag(Cu)_2X$–PbX–SnX_2 (X = S, Se) quasiternary systems at room temperature [in Ukrainian], *Probl. khimii ta staloho rozvytku*, (1), 14–30 (2021).

Olekseyuk I.D., Marushko L.P., Ivashchenko I.A., Piskach L.V., Parasyuk O.V. Phase equilibria between the quaternary semiconductors $A^I_2B^{II}C^{IV}X_4$ (A^I – Cu; B^{II} – Zn, Cd; C^{IV} – Si, Ge, Sn, X – S, Se), *Chem. Met. Alloys*, **12**(3-4), 51–60 (2019).

Olekseyuk I.D., Ostap'yuk T.A., Yukhymuk T.V., Zmiy O.F. Phase interactions on isothermal sections of the Ag_2Se–Ge(Sn)Se_2–Sb_2Se_3 systems 570 K [in Ukrainian], *Nauk. Visnyk Volyns'k. Nats. Univ. im. Lesi Ukrainky. Ser. Khim. nauky*, (29), 35–40 (2009).

Olvera A., Shi G., Djieutedjeu H., Page A., Uher C., Kioupakis E., Poudeu P.F.P. $Pb_7Bi_4Se_{13}$: A lillianite homologue with promising thermoelectric properties, *Inorg. Chem.*, **54**(3), 746–755 (2015).

Omar M.S. Crystal growth and investigation of the solid solutions of the system $CuGe_2P_3$–I_2–IV–VI_3, *Mater. Res. Bull.*, **25**(6), 691–698 (1990).

Onac B.P., Effenberger H., Ettinger K., Panzaru S.C., Hydroxylellestadite from Cioclovina Cave (Romania): Microanalytical, structural, and vibrational spectroscopy data, *Am. Mineralog.*, **91**(11-12), 1927–1931 (2006).

Origlieri M.J., Downs R.T. Schaurteite, $Ca_3Ge(SO_4)_2(OH)_6{\cdot}3H_2O$. *Acta Crystallogr.*, **E69**(2), i6 (2013).

Orlandi P., Biagioni C., Bonaccorsi E., Moëlo Y., Paar W.H. Lead-antimony sulfosalts from Tuscany (Italy). XII. Boscardinite, $TlPb_4(Sb_7As_2)_{\Sigma 9}S_{18}$, a new mineral species from the Monte Arsiccio mine: Occurrence and crystal structure, *Canad. Mineralog.*, **50**(2), 235–251 (2012).

Orlandi P., Biagioni C., Bonaccorsi E., Moëlo Y., Paar W.H. Protochabournéite, IMA 2011-054. CNMNC Newsletter No. 11, December 2011, page 2888, *Mineralog. Mag.*, **75**(6), 2887–2893 (2011).

Orlandi P., Biagioni C., Moëlo Y., Bonaccorsi E., Paar W.H. Lead-antimony sulfosalts from Tuscany (Italy). XIII. Protochabournéite, ~ $Tl_2Pb(Sb_{9-8}As_{1-2})_{\Sigma 10}S_{17}$, from the Monte Arsiccio mine: Occurrence, crystal structure and relationship with chabournéite, *Canad. Mineralog.*, **51**(3), 475–494 (2013).

Orlandi P., Leoni L., Mellini M., Merlino S. Tuscanite, a new mineral related to latiumite, *Am. Mineralog.*, **62**(11-12), 1110–1113 (1977).

Orlandi P., Meerschaut A., Moëlo Y., Palvadeau P., Leone Ph. Lead-antimony sulfosalts from Tuscany (Italy). VIII. Rouxelite, $Cu_2HgPb_{22}Sb_{28}S_{64}(O,S)_2$, a new sulfosalt from Buca della Vena mine, Apuan Alps: definition and crystal structure, *Canad. Mineralog.*, **43**(3), 919–933 (2005).

Orlandi P., Meerschaut A., Palvadeau P., Merlino S. Lead-antimony sulfosalts from Tuscany (Italy). V. Definition and crystal structure of moëloite, $Pb_6Sb_6S_{14}(S_3)$, a new mineral from the Ceragiola marble quarry. *Eur. J. Mineral.*, **14**(3), 599–606 (2002).

Orlandi P., Merlino S., Duchi G., Vezzalini G. Colusite: a new occurrence and crystal chemistry, *Canad. Mineralog.*, **19**(3), 423–427 (1981).

Orlandi P., Moëlo Y., Biagioni C. Lead-antimony sulfosalts from Tuscany (Italy). X. Dadsonite from the Buca della Vena mine and Bi-rich izoklakeite from the Seravezza marble quarries, *Per. Mineral.*, **79**(1), 113–121 (2010).

Orlandi P., Moëlo Y., Meerschaut A., Palvadeau P. Lead-antimony sulfosalts from Tuscany (Italy). III. Pillaite, $Pb_9Sb_{10}S_{23}ClO_{0.5}$, a new Pb–Sb oxy-chloro-sulfosalt, from Buca della Vena mine,, *Eur. J. Mineral.*, **13**(3), 605–610 (2001).

Orlandi P., Moëlo Y., Meerschaut A., Palvadeau P., Leone Ph. Lead-antimony sulfosalts from Tuscany (Italy). VI. Pellouxite, ~ $(Cu,Ag)_2Pb_{21}Sb_{23}S_{55}ClO$, a new oxy-chloro-sulfosalt from Buca della Vena mine, Apuan Alps, *Eur. J. Mineral.*, **16**(5), 839–844 (2004).

Orujlu E.N., Aliev Z.S., Babanly M.B. The phase diagram of the MnTe–SnTe–Sb_2Te_3 ternary system and synthesis of the iso- and aliovalent cation-substituted solid solutions, *Calphad*, **76**, 102398 (2022).

Orujlu E.N., Seidzade A.E., Aliev Z.S., Amiraslanov I.R., Babanly M.B. Synthesis and crystal structure of a new 9P-type layered van der Waals compound $SnBi_4Te_4$, *Chem. Probl.*, (1), 40–48 (2020).

Ostapyuk T.A., Zmiy O.F., Ivashchenko I.A., Olekseyuk I.D. The Cu_2Se–PbSe–As_2Se_3 system, *Chem. Met. Alloys*, **7**(1-2), 164–169 (2014).

Ottenburgs R., Goethals H. Synthèse et polymorphisme de la briartite, *Bull. Soc. fr. minéral. cristallogr.*, **95**(4), 458–463 (1972).

Oudin E., Picot P., Pillard F., Moëlo Y., Burke E.A.J., Zakrzewski M.A. La bénavidésite, $Pb_4(Mn,Fe)Sb_6S_{14}$, un noveau minéral de la série de la jamesonite, *Bull. Minéralog.*, **105**(2), 166–169 (1982).

Ozawa T., Saitow A., Hori H. Chemistry and crystallography of Bi-rich izoklakeite from the Otome mine, Yamanashi Prefecture, Japan and discussion of the izoklakeite-giessenite series, *Mineralog. J.*, **20**(4), 179–187 (1998).

Paar W.H., Chen T.T., Kupcik V., Hanke K. Eclarit, $(Cu,Fe)Pb_9Bi_{12}S_{28}$, ein neues Sulfosalz von Bärenbad, Hollersbachtal, Salzburg, Österreich, *Tschermaks Miner. Petr. Mitt.*, **32**(2-3), 103–110 (1983).

Paar W.H., Mereiter K., Braithwaite R.S.W., Keller P., Dunn P.J. Chenite, $Pb_4Cu(SO_4)_2(OH)_6$, a new mineral, from Leadhills, Scotland, *Mineralog. Mag.*, **50**(355), 129–135 (1986).

Paar W.H., Moëlo Y., Mozgova N.N., Organova N.I., Stanley C.J., Roberts A.C., Culetto F.J., Effenberger H.S., Topa D., Putz H., Sureda R.J., de Brodtkorb M.K. Coiraite, $(Pb,Sn^{2+})_{12.5}As_3Fe^{2+}Sn^{4+}{}_5S_{28}$: a franckeite-type new mineral species from Jujuy Province, NW Argentina, *Mineralog. Mag.*, **72**(5), 1083–1101 (2008).

Pajares I., De la Torre Á.G., Martínez-Ramírez S., Puertas F., Blanco-Varela M.-T. Quantitative analysis of mineralized white Portland clinkers: The structure of fluorellestadite, *Powder Diffr.*, **17**(4), 281–286 (2002).

Pal V., Kumar B., Paliwal M. Phase equilibria study in Ga–Sn–Te system using thermodynamic modeling and experimental validation, *J. Phase Equilib. Diffus.*, **44**(5), 642–653 (2023).

Palache C., Richmond W.E. Caledonite, *Am. Mineralog.*, **24**(7), 441–445 (1939).

Palvadeau P., Meerschaut A., Orlandi P., Moëlo Y. Lead-antimony sulfosalts from Tuscany (Italy). VII. Crystal structure of pellouxite, ~ $(Cu,Ag)_2Pb_{21}Sb_{23}S_{55}ClO$, an expanded monoclinic derivative of $Ba_{12}Bi_{24}S_{48}$ hexagonal sub-type (zinkenite group), *Eur. J. Mineral.*, **16**(5), 845–855 (2004).

Pamplin B.R., Hasoon F.S. The system InAs–$ZnGeAs_2$–$CuInSe_2$–Cu_2GeSe_3, *Progr. Cryst. Growth Character.*, **10**, 213–215 (1985).

Panigrahi G., Yadav S., Jana S., Sundaramoorthy M., Arumugam S., Niranjan M.K., Prakash J. $Y_3Fe_{0.5}SiSe_7$: A new cation-deficient quaternary mixed transition metal chalcogenide with extremely low thermal conductivity, *Solid State Sci.*, **138** 107133 (2023).

Papageorgakis J. Wenkit, ein neues Mineral von Candoglia, *Schweiz. Mineral. Petrogr. Mitt.*, **42**(1), 269–274 (1962).

Paradis-Fortin L., Lemoine P., Guilmeau E., Malaman B., Elkaïm E., Zitolo A., Cordier S., Guélou G., Raveau B., Prestipino C. Resolution of the cationic distribution in synthetic germanite $Cu_{22}Fe_8Ge_4S_{32}$ by an experimental combinatorial approach based on synchrotron resonant powder diffraction data: A case study and guidelines for analogous compounds, *Chem. Mater.*, **34**(16), 7434–7445 (2022).

Parmentier A.B., Smet P.F., Bertram F., Christen J., Poelman D. Structure and luminescence of $(Ca,Sr)_2SiS_4{:}Eu^{2+}$ phosphors, *J. Phys. D: Appl. Phys.*, **43**(8), 085401–085407 (2010).

Patel S.R. The preparation of the double sulphates of tin and certain bivalent metals, *Proc. Ind. Acad. Sci.*, **37A**(1), 38–40 (1953).

Pattiaratchi D.B., Saari E., Sahama Th.G. Anandite, a new barium iron silicate from Wilagedera, North Western Province, Ceylon, *Mineralog. Mag.*, **36**(277), 1–4 (1967).

Pauling L. The structure of sodalite and helvite, *Z. Kristallogr.*, **74**(1-6), 213–225 (1930).

Pavan Kumar V., Lemoine P., Carnevali V., Guélou G., Lebedev O.I., Boullay P. Raveau B., Al Rahal Al Orabi R., Fornari M., Prestipino C., Menut D., Candolfi C., Malaman B., Juraszek J., Guilmeau E. Ordered sphalerite derivative $Cu_5Sn_2S_7$: A degenerate semiconductor with high carrier mobility in the Cu–Sn–S diagram, *J. Mater. Chem. A*, **9**(17), 10812–10826 (2021a).

Pavan Kumar V., Lemoine P., Carnevali V., Guélou G., Lebedev O.I., Raveau B., Al Rahal Al Orabi R., Fornari M., Candolfi C., Prestipino C., Menut D., Malaman B., Juraszek J., Suekuni K., Guilmeau E. Local-disorder-induced low thermal conductivity in degenerate semiconductor $Cu_{22}Sn_{10}S_{32}$, *Inorg. Chem.*, **60**(21), 16273–16285 (2021b).

Pažout R., Dušek M. Crystal structure of natural orthorhombic $Ag_{0.71}Pb_{1.52}Bi_{1.32}Sb_{1.45}S_6$, a lillianite homologue with N = 4; comparison with gustavite, *Eur. J. Mineral.*, **22**(5), 741–750 (2010).

Pažout R., Dušek M. Natural monoclinic $AgPb(Bi_2Sb)_3S_6$, an Sb-rich gustavite, *Acta Crystallogr.*, **C65**(11), i77–i80 (2009).

Pažout R., Plášil J., Dušek M., Sejkora J., Dolníček Z. Holubite, $Ag_3Pb_6(Sb_8Bi_3)_{\Sigma 11}S_{24}$, from Kutná Hora, Czech Republic, a new member of the andorite branch of the lillianite homologous series, *Mineralog. Mag.*, **87**(4), 582–590 (2023a).

Pažout R., Plášil J., Dušek M., Sejkora J., Dolníček Z. Holubite, IMA 2022-112. CNMNC Newsletter No 72, *Eur. J. Mineral.*, **35**(2), 286 (2023b).

Pažout R., Plášil J., Dušek M., Sejkora J., Dolníček Z. Holubite, IMA 2022-112. CNMNC Newsletter No 72, *Mineralog. Mag.*, **87**(3), 513 (2023c).

Pažout R., Plášil J., Dušek M., Sejkora J., Ilinca G. Lazerckerite, IMA 2022-113. CNMNC Newsletter No 72, *Eur. J. Mineral.*, **35**(2), 286–287 (2023d).

Pažout R., Plášil J., Dušek M., Sejkora J., Ilinca G. Lazerckerite, IMA 2022-113. CNMNC Newsletter No 72, *Mineralog. Mag.*, **87**(3), 513 (2023e).

Pažout R., Sejkora J. Staročeskéite, $Ag_{0.70}Pb_{1.60}(Bi_{1.35}Sb_{1.35})_{\Sigma 2.70}S_6$, from Kutná Hora, Czech Republic, a new member of the lillianite homologous series, *Mineralog. Mag.*, **82**(4), 993–1005 (2018).

Pažout R., Sejkora J. Staročeskéite, IMA 2016-101. CNMNC Newsletter No. 36, April 2017, page 405, *Mineralog. Mag.*, **81**(2), 403–409 (2017).

Peacock M.A., Thompson R.M. Montbrayite, a new gold telluride, *Am. Mineralog.*, **31**(11-12), 515–526 (1946).

Peacor D.R., Dunn P.J., Schnorrer-Köhler G., Bideaux R.A. Mammothite, a new mineral from Tiger, Arizona and Laurium, Greece, *Mineralog. Rec.*, **16**(2), 117–120 (1985).

Pekov I.V., Agakhanov A.A., Zubkova N.V., Belakovskiy D.I., Vigasina M.F., Britvin S.N., Turchkova A.G., Sidorov E.G. Philoxenite, $(K,Na,Pb)_4(Na,Ca)_2(Mg,Cu)_3(Fe^{3+}{}_{0.5}Al_{0.5})(SO_4)_8$, a new mineral from fumarole exhalations of the Tolbachik volcano, Kamchatka, Russia, *Zap. Ros. mineralog. obshch.*, **149**(4), 67–77 (2020a).

Pekov I.V., Britvin S.N., Agakhanov A.A., Turchkova A.G., Zhegunov P.S., IMA 2023-025. Viskontite, CNMNC Newsletter No 74, *Eur. J. Mineral.*, **35**(4), 663 (2023a).

Pekov I.V., Britvin S.N., Agakhanov A.A., Turchkova A.G., Zhegunov P.S., IMA 2023-025. Viskontite, CNMNC Newsletter No 74, *Mineralog. Mag.*, **87**(5), 786 (2023b).

Pekov I.V., Britvin S.N., Agakhanov A.A., Vigasina M.F., Turchkova A.G., Zhegunov P.S. Paramolybdomenite, IMA 2023-025. CNMNC Newsletter No 74, *Eur. J. Mineral.*, **35**(4), 662 (2023c).

Pekov I.V., Britvin S.N., Agakhanov A.A., Vigasina M.F., Turchkova A.G., Zhegunov P.S. Paramolybdomenite, IMA 2023-025. CNMNC Newsletter No 74, *Mineralog. Mag.*, **87**(5), 785 (2023d).

Pekov I.V., Chukanov N.V., Britvin S.N., Kabalov Y.K., Göttlicher J., Yapaskurt V.O., Zadov A.E., Krivovichev S.V., Schüller W., Ternes B. The sulfite anion in ettringite-group minerals: a new mineral species hielscherite, $Ca_3Si(OH)_6(SO_4)(SO_3)\cdot 11H_2O$, and the thaumasite–hielscherite solid-solution series, *Mineralog. Mag.*, **76**(5), 1133–1152 (2012).

Pekov I.V., Chukanov N.V., Kabalov Y.K., Britvin S.N., Göttlicher J., Yapaskurt V.O., Zadov A.E., Krivovichev S.V., Schüller W., Ternes B. Hielscherite, IMA 2011-037. CNMNC Newsletter No. 10, October 2011, page 2556, *Mineralog. Mag.*, **75**(5), 2549–2561 (2011).

Pekov I.V., Chukanov N.V., Kulikova I.M., Belakovsky D.I. Phosphoinnelite, $Ba_4Na_3Ti_3Si_4O_{14}(PO_4,SO_4)_2(O,F)_3$, a new mineral species from peralkaline pegmatite of the Kovdor pluton, Kola Peninsula, *Geol. Ore Deposits*, **49**(7), 530–536 (2007).

Pekov I.V., Sereda E.V., Zubkova N.V., Yapaskurt V.O., Chukanov N.V., Britvin S.N., Lykova I.S. Pushcharovsky D.Y. Genplesite, $Ca_3Sn(SO_4)_2(OH)_6\cdot 3H_2O$, a new mineral of the fleischerite group: first occurrence of a tin sulfate in nature, *Eur. J. Mineral.*, **30**(2), 375–382 (2018).

Pekov I.V., Sereda E.V., Zubkova N.V., Yapaskurt V.O., Chukanov N.V., Britvin S.N., Lykova I.S. Pushcharovsky D.Y. Genplesite, IMA 2014-034. CNMNC Newsletter No. 21, August 2014, page 803, *Mineralog. Mag.*, **78**(4), 797–804 (2014).

Pekov I.V., Zubkova N.V., Agakhanov A.A., Belakovskiy D.I., Vigasina M.F., Britvin S.N., Turchkova A.G., Sidorov E.G., Pushcharovsky D.Y. Philoxenite, IMA 2015-108. CNMNC Newsletter No. 30, April 2016, page 410, *Mineralog. Mag.*, **80**(2), 407–413 (2016a).

Pekov I.V., Zubkova N.V., Agakhanov A.A., Chukanov N.V., Belakovskiy D.I., Sidorov E.G., Britvin S.N., Pushcharovsky D.Y. Eleomelanite, IMA 2015-118. CNMNC Newsletter No. 30, April 2016, page 412, *Mineralog. Mag.*, **80**(2), 407–413 (2016b).

Pekov I.V., Zubkova N.V., Agakhanov A.A., Chukanov N.V., Belakovskiy D.I., Sidorov E.G., Britvin S.N., Turchkova A.G., Pushcharovsky D.Yu. Eleomelanite, $(K_2Pb)Cu_4O_2(SO_4)_4$, a new mineral species from the Tolbachik Volcano, Kamchatka, Russia, *Canad. Mineralog.*, **58**(5), 625–636 (2020b).

Perret R. Crystal data for magnesium–tin (IV) double sulphate, *J. Appl. Cryst.*, **8**(2), 336 (1975).

Perret R., Bouillet A.-M. Les apatites–sulfates $Na_3Cd_2(SO_4)_3Cl$ et $Na_3Pb_2(SO_4)_3Cl$, *cminéral. cristallogr.*, **98**(4), 254–255 (1975).

Perret R., Couchet P. Sur une famille rhomboédrique de sulfates doubles d'étain IV $M^{II}Sn(SO_4)_3$ (M^{II} = Mg, Mn, Co, Ni, Zn, Cd), *C. r. Acad. sci. Sér. C*, **276**(1), 85–87 (1973).

Perret R., Couchot P., Bouteiller B., Thrierr-Sorel A. Sulfates triples d'étain (IV), *C. r. Acad. sci. Sér. C*, **279**, 465–467 (1974).

Petrova I.V., Pobedimskaya E.A., Bryzgalov I.A. Crystal structure of the mikharaite $Cu_4FePbBiS_6$ [in Russian], *Dokl. AN SSSR*, **299**(1), 123–127 (1988).

Petruk W. Larosite, a new copper–lead–bismuth sulphide, *Canad. Mineralog.*, **11**(4), 886–891 (1972).

Philippo S., Hatert F., Bruni Y. Vignola P. Luxembourgite, IMA 2018-154. CNMNC Newsletter No. 49, *Eur. J. Mineral.*, **31**(3), 654 (2019a).

Philippo S., Hatert F., Bruni Y. Vignola P. Luxembourgite, IMA 2018-154. CNMNC Newsletter No. 49, *Mineralog. Mag.*, **83**(3), 480 (2019b).

Philippo S., Hatert F., Bruni Y. Vignola P., Sejkora J. Luxembourgite, $AgCuPbBi_4Se_8$, a new mineral species from Bivels, Grand Duchy of Luxembourg, *Eur. J. Mineral.*, **32**(4), 449–455 (2020).

Pierrot R. Revue bibliographique des modifications apportées à la nomenclature minéralogique, *Bull. Soc. fr. minéral. cristallogr.*, **91**(3), 300–307 (1968).

Piilonen P.C., Ercit T.S., Roberts A.C. New mineral names, *Am. Mineralog.*, **90**(4), 768–773 (2005a).

Piilonen P.C., Grew E.S., Ercit T.S., Roberts A.C. New mineral names, *Am. Mineralog.*, **90**(7), 1227–1233 (2005b).

Piilonen P.C., Locock A., Grew E.S. New mineral names, *Am. Mineralog.*, **90**(11-12), 1945–1952 (2005c).

Piilonen P.C., Locock A.J., Rowe R., Ercit T.S. New mineral names, *Am. Mineralog.*, **92**(4), 703–707 (2007).

Piilonen P.C., Poirier G. New mineral names, *Am. Mineralog.*, **95**(8-9), 1357–1361 (2010).

Piilonen P.C., Poirier G., Tait K.T. New mineral names, *Am. Mineralog.*, **93**(11-12), 1941–1946 (2008).

Piilonen P.C., Rowe R., Ercit T.S., Locock A.J. New mineral names, *Am. Mineralog.*, **91**(8-9), 1452–1457 (2006).

Pinto D., Balić-Žunić T., Bonaccorsi E., Borodaev Y.S., Garavelli A., Garbarino C., Makovicky E., Mozgova N.N., Vurro F. Rare sulfosalts from Vulcano, Aeolian Islands, Italy. VII. Cl-bearing galenobismutite, *Canad. Mineralog.*, **44**(2), 442–457 (2006).

Pinto D., Bonaccorsi E., Balić-Žunić T., Makovicky E. The crystal structure of vurroite, $Pb_{20}Sn_2(Bi,As)_{22}S_{54}Cl_6$: OD-character, polytypism, twinning, and modular description, *Am. Mineralog.*, **93**(5-6), 713–727 (2008).

Pliego-Cuervo Y.B., Glasser F.P. The role of sulphates in cement clinkering reactions: Phase formation and melting in the system $CaO–Ca_2SiO_4–CaSO_4–K_2SO_4$, *Cem. Concr. Res.*, **7**(5), 477–481 (1977).

Pobedimskaya E.A., Rastsvetaeva R.K., Terentieva L.E., Sapozhnikov A.N. Crystal structure of afghanite [in Russian], *Dokl. AN SSSR*, **320**(4), 882–886 (1991a).

Pobedimskaya E.A., Terentieva L.E., Sapozhnikov A.N., Kashaev A.A., Dorokhova G.I. Crystal structure of bystrite [in Russian], *Dokl. AN SSSR*, **319**(4), 873–878 (1991).

Poduska K.M., Cario L., DiSalvo F.J., Min K., Halasyamani P.S. Structural studies of a cubic, high-temperature (α) polymorph of Pb_2GeS_4 and the isostructural $Pb_{2-x}Sn_xGeS_{4-y}Se_y$ solid solution, *J. Alloys Compd.*, **335**(1-2), 105–110 (2002).

Pogodin A.I., Filep M.Y., Shender I.O., Kokhan O.P., Studenyak I.P. Interaction in the $Ag_6PS_5I–Ag_7GeS_5I$ and $Ag_7GeS_5I–Ag_7SiS_5I$ systems [in Ukrainian], *Nauk. Visn. Uzhgorod. Univ., Ser. Khim.*, [1(45)], 42–46 (2021).

Pogodin A.I., Kokhan O.P., Solomon A.M., Izay V.Yu., Stasyuk Yu.M., Studenyak I.P., Tsimbota M.Yu. Growth of Cu_7GeS_5I, Ag_7GeS_5I and $(Cu_{1-x}Ag_x)_7GeS_5I$ solid solutions single crystals [in Ukrainian], *Nauk. Visn. Uzhgorod. Univ., Ser. Khim.*, [2(36)], 7–9 (2016).

Pogodin A.I., Luchynets' M.M., Filep M.Yo., Kokhan O.P., Studenyak I.P., Kush P. Synthesis, growth and structural properties of the $(Cu_{1-x}Ag_x)_7GeSe_5I$ solid solutions [in Ukrainian], *Nauk. Visn. Uzhgorod. Univ., Ser. Fiz.*, (45), 7–13 (2019).

Pogu A., Jaschin P.W., Varma K.B.R., Vidyasagar K. Syntheses, structural variants and characterization of $A_2CdSn_2S_6$ (A = Cs, Rb and K) compounds. *J. Solid State Chem.*, **277**, 713–720 (2019).

Pohl S., Krebs B. Darstellung und Struktur von $Cs_4Ge_4S_{10}{\cdot}3H_2O$, *Z. anorg. und. allg. Chem.*, **424**(3), 265–272 (1976).

Poirier G., Ercit T.S., Tait K.T., Piilonen P.C., Rowe R. New mineral names, *Am. Mineralog.*, **94**(2-3), 399–408 (2009)

Poirier G., Piilonen P.C. New mineral names, *Am. Mineralog.*, **92**(10), 1776–1779 (2007).

Porter Y., Halasyamani P.S. A low temperature method for the synthesis of new lead selenite chlorides: $Pb_3(SeO_3)(SeO_2OH)Cl_3$ and $Pb_3(SeO_3)_2Cl_2$, *Inorg. Chem.*, **40**(12), 2640–2641 (2001).

Powell D.W., Thomas R.G., Williams P.A., Birch W.D., Plimer I.R. Choloalite: synthesis and revised chemical formula, *Mineralog. Mag.*, **58**(392), 505–508 (1994).

Prakash J., Mesbah A., Lebègue S., Ibers J.A. Synthesis, crystal structure, and electronic structure of $Ba_2GeTe_3(Te_2)$. *Solid State Sci.*, **97**, 105974 (2019).

Pring A. The crystal chemistry of the sartorite group minerals from Lengenbach, Binntal, Switzerland – a HRTEM study, *Schweiz. Mineral. Petrogr. Mitt.*, **81**(1), 69–87 (2001).

Pringle G.J., Thorpe R.I. Bohdanowiczite, junoite and laitakarite from the Kidd Creek mine, Timmins, Ontario, *Canad. Mineralog.*, **18**(3), 353–360 (1980).

Pršek J., Ozdín D., Sejkora J. Eclarite and associated Bi sulfosalts from the Brezno-Hviezda occurrence (Nízke Tatry Mts, Slovak Republic). *N. Jb. Miner. Abh.*, **185**(2), 117–130 (2008).

Pushcharovkiy D.Yu., Yamnova N.A., Khomyakov A.P. Crystal structure of high-potassium vishnevite [in Russian], *Kristallografiya*, **34**(1), 67–70 (1989).

Quintero E., Tovar R., Quintero M., Delgado G.E., Morocoima M., Caldera D., Ruiz J., Mora A.E., Briceño M., Fernandez J.L. Lattice parameter values and phase transitions for the $Cu_2Cd_{1-z}Mn_zGeSe_4$ and $Cu_2Cd_{1-z}Fe_zGeSe_4$ alloys, J. *Alloys Compd.*, **432**(1-2), 142–148 (2007).

Quintero E., Tovar R., Quintero M., Morocoima M., Ruiz J., Delgado G., Broto J.M., Rakoto H. Structural characterization and magnetic properties for the semiconducting semimagnetic system $Cu_2Cd_{1-z}Mn_zGeSe_4$ alloys, *Physica B: Condens. Matter*, **320**(1-4), 384–387 (2002).

Radosavljevic S.A., Rakic S.M., Stojanovic J.N., Radosavljevic-Mihajlovic A.S. Occurrence of petrukite in Srebrenica orefield, Bosnia and Herzegovina, *N. Jb. Miner. Abh.*, **181**(1), 21–26 (2005).

Ramos E.P., Zhang Z., Assoud A., Kaup K., Lalère F., Nazar L.F. Correlating ion mobility and single crystal structure in sodium-ion chalcogenide-based solid state fast ion conductors: $Na_{11}Sn_2PnS_{12}$ (Pn = Sb, P), *Chem. Mater.*, **30**(21), 7413–7417 (2018).

Ran M.-Y., Ma Z., Chen H., Li B., Wu X.-T., Lin H., Zhu Q.-L. Partial isovalent anion substitution to access remarkable second-harmonic generation response: a generic and effective strategy for design of IR NLO materials, *Chem. Mater.*, **32**(13), 5890–5896 (2020).

Ran M.-Y., Ma Z., Wu X.-T., Lin H., Zhu Q.-L. $Ba_2Ge_2Te_5$: a ternary NLO-active telluride with unusual one-dimensional helical chains and giant second harmonic-generation tensors, *Inorg. Chem. Front.*, **8**(22), 4838–4845 (2021a).

Ran M.-Y., Zhou S.-H., Li B., Wei W., Wu X.-T., Lin H., Zhu Q.-L. Enhanced second-harmonic-generation efficiency and birefringence in melillite oxychalcogenides $Sr_2MGe_2OS_6$ (M = Mn, Zn, and Cd), *Chem. Mater.*, **34**(24), 3853–3861 (2022).

Ran M.-Y., Zhou S.-H., Wei W., Li B.-X., Wu X.-T., Lin H., Zhu Q.-L. Rational design of a rare-earth oxychalcogenide $Nd_3[Ga_3O_3S_3][Ge_2O_7]$ with superior infrared nonlinear optical performance, *Small*, **19**(19), 2300248 (2023).

Ran M.-Y., Zhou S.-H., Wei W., Song B.-J., Shi Y.-F., Wu X.-T., Lin H., Zhu Q.-L. Quaternary chalcohalides $CdSnSX_2$ (X = Cl or Br) with neutral layers: syntheses, structures, and photocatalytic properties, *Inorg. Chem.*, **60**(5), 3431–3438 (2021b).

Rastsvetaeva R.K., Bolotina N.B., Sapozhnikov A.N., Kashaev A.A., Schoenleber A., Chapuis G. Average structure of cubic lazurite with a three-dimensional incommensurate modulation, *Crystallogr. Rep.*, **47**(3), 404–407 (2002).

Rastsvetaeva R.K., Chukanov N.V. Model of the crystal structure of biachellaite as a new 30-layer member of the cancrinite group, *Crystallogr. Rep.*, **53**(6), 981–988 (2008).

Rastsvetaeva R.K., Ivanova A.G., Chukanov N.V., Verin I.A. Crystal structure of alloriite [in Russian]. *Dokl. Akad. Nauk*, **415**(2), 242–246 (2007).

Reis I., Krämer V., Seiler A., Topa D., Keller E. $Pb_{5.0(1)}In_{8.4(1)}Bi_{1.6(1)}S_{20}$, a new quaternary lead indium bismuth sulfide, *Acta Crystallogr.*, **C68**(3), i12–i16 (2012).

Reshak A.H., Kogut Y.M., Fedorchuk A.O., Zamuruyeva O.V., Myronchuk G.L., Parasyuk O.V., Kamarudin H., Auluck S., Plucinski K.J., Bila J. Electronic and optical features of the mixed crystals $Ag_{0.5}Pb_{1.75}Ge(S_{1-x}Se_x)_4$, *J. Mater. Chem. C*, **1**(31), 4667–4675 (2013).

Riccardi R., Gout D., Gauthier G., Guillen F., Jobic S., Garcia A., Huguenin D., Macaudière P., Fouassier C., Brec R. Structural investigations and luminescence properties of the $Ce_3(SiS_4)_2X$ (X = Cl, Br, I) family and the $La_{3-x}Ce_x(SiS_4)_2I$ ($0 \leq x \leq 1$) solid solution, *J. Solid State Chem.*, **147**(1), 259–268 (1999).

Richards W.D., Tsujimura T., Miara L.J., Wang Y., Kim J.C., Ong S.P., Uechi I., Suzuki N., Ceder G. Design and synthesis of the superionic conductor $Na_{10}SnP_2S_{12}$, *Nat. Commun.*, **7**(1), 11009 (2016).

Riedel E ., Morlock W. Spinelle mit substituierten Nichtmetallteilgittern. VI. Röntgenographische und elektronische Eigenschaften, Mößbauer- und IR-Spektren des Spinellsystems $CuCrSn(S_{1-x}Se_x)_4$, *Z. anorg. und allg. Chem.*, **438**(1), 233–241 (1978).

Roberts A.C. A triclinic unit cell for oboyerite. *Geol. Surv. Can. Pap.*, **80**(1B) 295 (1980).

Rodot H. Solutions solides entre composés semiconducteurs binaires et ternaires, In: *Proc. Intern. Conf. Semicond. Phys.*, Prague, 1010–1014 (1961).

Rothballer J., Bachhuber F., Rommel S.M., Söhnel T., Weihrich R. Origin and effect of In–Sn ordering in $InSnCo_3S_2$: a neutron diffraction and DFT study, *RSC Adv.*, **4**(79), 42183–42189 (2014).

Rouse R.C., Dunn P.J. A contribution to the crystal chemistry of ellestadite and silicate sulfate apatites, *Am. Mineralog.*, **67**(1-2), 90–96 (1982).

Rozenberg K.A., Rastsvetaeva R.K., Chukanov N.V. Crystal structures of oxalate-bearing cancrinite with an unusual arrangement of CO_3 groups and sulfate-rich davyne, *Crystallogr. Rep.*, **54**(5), 793–799 (2009).

Rozenberg K.A., Sapozhnikov A.N., Rastsvetaeva R.K., Bolotina N.B., Kashaev A.A. Crystal structure of a new representative of the cancrinite group with a 12-layer stacking sequence of tetrahedral rings, *Crystallogr. Rep.*, **49**(4), 635–642 (2004).

Rozhdestvenskaya I.V., Zayakina N.V., Kim A.A. Crystalline structure of Zn-, Ca-tellurate – yafsoanite [in Russian], *Mineralog. zhurn.*, **6**(2), 75–79 (1984).

Ruan T.-T, Wang W.-W., Hu C.-L., Xu X., Mao J.-G. $Pb_4(BO_3)_2(SO_4)$ and $Pb_2[(BO_2)(OH)](SO_4)$: New lead(II) borate-sulfate mixed-anion compounds with two types of 3D network structures, *J. Solid State Chem.*, **260**, 39–45 (2018).

Rudashevskiy N.S., Mochalov A.G., Begizov V.D., Men'shikov Yu.P., Shumskaya N.I. Inaglyite, $Cu_3Pb(Ir,Pt)_8S_{16}$, a new mineral [in Russian], *Zap. Vses. Mineralog. Obshch.*, **113**(6), 712–717 (1984a).

Rudashevskiy N.S., Mochalov A.G., Trubkin N.V., Gorshkov A.I., Men'shikov Yu.P., Shumskaya N.I. Konderite, $Cu_3Pb(Rh,Pt,Ir)_8S_{14}$, a new mineral [in Russian], *Zap. Vses. Mineralog. Obshch.*, **113**(6), 703–712 (1984b).

Rumsey M.S., Jaszczak J.A., Bindi L., Hackney S.A., Wise M.A., Stanley C., Spratt J. Merelaniite, IMA 2016-042. CNMNC Newsletter No. 33, page 1137, *Mineralog. Mag.*, **80**(6), 1135–1144 (2016).

Russell J.D., Milodowski A.E., Fraser A.R., Clark D.R. New IR and XRD data for leadhillite of ideal composition, *Mineralog. Mag.*, **47**(344), 371–375 (1983).

Ruzin E., Dehnen S. Influence of the counterions on the structures of ternary Zn/Sn/Se anions: synthesis and properties of $[Rb_{10}(H_2O)_{14.5}][Zn_4(\mu_4\text{-}Se)_2(SnSe_4)_4]$ and $[Ba_5(H_2O)_{32}][Zn_5Sn(\mu_3\text{-}Se)_4(SnSe_4)_4]$, *Z. anorg. und allg. Chem.*, **632**(5), 749–755 (2006).

Ruzin E., Fuchs A., Dehnen S. Fine-tuning of optical properties with salts of discrete or polymeric, heterobimetallic telluride anions $[M_4(\mu_4\text{-}Te)(SnTe_4)_4]^{10-}$ (M = Mn, Zn, Cd, Hg) and ${}^3_\infty\{[Hg_4(\mu_4\text{-}Te)(SnTe_4)_3]^{6-}\}$. *Chem. Commun.*, (46), 4796–4798 (2006a).

Ruzin E., Jakobi S., Dehnen S. Syntheses, structures and reactivity of novel hydrates of *ortho*-sulfidostannte salts, *Z. anorg. und allg. Chem.*, **634**(6-7), 995–1001 (2008).

Ruzin E., Kracke A., Dehnen S. Efficient synthesis and properties of single-crystalline $[SnTe_4]^{4-}$ salts, *Z. anorg. und allg. Chem.*, **632**(6), 1018–1026 (2006b).

Sachanyuk V.P., Olekseyuk I.D., Parasyuk O.V. X-ray powder diffraction study of the $Cu_2Cd_{1-x}Mn_xSnSe_4$ alloys, *Phys. stat. sol. (a)*, **203**(3), 459–465 (2006).

Sahama Th.G., Hytönen K. Delhayelite, a new silicate from the Belgian Congo, *Mineralog. Mag.*, **32**(244), 6–9 (1959).

Saidov A.S., Usmonov Sn.N., Rakhmonov U.Kh., Kurmantayev A.N., Bahtybayev A.N. Multicomponent solid solutions $(ZnSe)_{1-x-y}(Si_2)_x(GaP)_y$, *J. Mater. Sci. Res.*, **1**(2), 150-156 (2012).

Saint-Jean S.J., Hansen S. Nonstoichiometry in chlorellestadite, *Solid State Sci.*, **7**(1), 97–102 (2005).

Saint-Jean S.J., Jøns E., Lundgaard N., Hansen S. Chlorellestadite in the preheater system of cement kilns as an indicator of HCl formation, *Cem. Concr. Res.*, **35**(3), 431–437 (2005).

Sakharova M.S., Bryzgalov I.A., Eremin N.I., Petrova I.V. About the find of miharaite in silver-bearing veins in the North-East of the USSR [in Russian], *Dokl. AN SSSR*, **284**(2), 456–459 (1985).

Sapozhnikov A.N. Indexing of additional reflections on the X-ray Debye diffraction patterns of lazurite concerning the study of modulation of its structure [in Russian], *Zap. Vses. Mineralog. Obshch.*, **119**(1), 110–116 (1990).

Sapozhnikov A.N. On the modulated structure of lazurite from the ore deposits of the Southwestern Pamirs [in Russian], *Kristallografiya*, **37**(4), 889–893 (1992).

Sapozhnikov A.N., Bolotina N.B., Chukanov N.V., Kaneva E.V., Shendrik R.Y., Vigasina M.F., Ivanova L.A. Slyudyankaite, IMA 2021-062a. CNMNC Newsletter No 65, *Eur. J. Mineral.*, **34**(1), 143–144 (2022a).

Sapozhnikov A.N., Bolotina N.B., Chukanov N.V., Kaneva E.V., Shendrik R.Y., Vigasina M.F., Ivanova L.A. Slyudyankaite, IMA 2021-062a. CNMNC Newsletter No 65, *Mineralog. Mag*, **86**(2), 354 (2022b).

Sapozhnikov A.N., Ivanov V.G., Piskunova L.F., Kashaev A.A., Terent'eva L.E., Pobedimskaya E.A. Bystrite $Ca(Na,K)_7(Si_6Al_6O_{24})(S_3)_{1.5}\cdot H_2O$ – a new cancrinite-like mineral [in Russian], *Zap. Vses. Mineralog. Obsh.*, **120**(3), 97–100 (1991).

Sapozhnikov A.N., Kaneva E.V., Cherepanov D.I., Suvorova L.F., Levitsky V.I., Ivanova L.A., Reznitsky L.Z. Vladimirivanovite, IMA 2010-070. CNMNC Newsletter No. 8, April 2011, page 291, *Mineralog. Mag.*, **75**(2), 289–294 (2011a).

Sapozhnikov A.N., Kaneva E.V., Cherepanov D.I., Suvorova L.F., Levitsky V.I., Ivanova L.A., Reznitsky L.Z. Vladimirivanovite $Na_6Ca_2[Al_6Si_6O_{24}](SO_4,S_3,S_2,Cl)_2\cdot H_2O$ – a new mineral from sodalite group [in Russian], *Zap. Ros. Mineralog. Obsh.*, **140**(5), 36–45 (2011b).

Sapozhnikov A.N., Kaneva E.V., Cherepanov D.I., Suvorova L.F., Levitsky V.I., Ivanova L.A., Reznitsky L.Z. Vladimirivanovite, $Na_6Ca_2[Al_6Si_6O_{24}](SO_4,S_3,S_2,Cl)_2\cdot H_2O$, a new mineral of sodalite group. *Geol. Ore Deposits,* **54**(7), 557–564 (2012).

Sapozhnikov A.N., Kaneva E.V., Suvorova L.F., Levitsky V.I., Ivanova L.A. Sulfhydrylbystrite, $Na_5K_2Ca(Al_6Si_6O_{24})(S_5)(SH)$, a new mineral with the LOS framework, and re-interpretation of bystrite: cancrinite-group minerals with novel extra-framework anions, *Mineralog. Mag.*, **81**(2), 383–402 (2017).

Sapozhnikov A.N., Kaneva E.V., Suvorova L.F., Levitsky V.I., Ivanova L.A., Mitichkin M.A., Barash I.G. Sulfhydrylbystrite, IMA 2015-010. CNMNC Newsletter No. 25, June 2015, page 534, *Mineralog. Mag.*, **79**(3), 529–535 (2015).

Sapozhnikov A.N., Tauson V.L., Lipko S.V., Shendrik R.Yu., Levitskii V.I., Suvorova L.F., Chukanov N.V., Vigasina M.F. On the crystal chemistry of sulfur-rich lazurite, ideally $Na_7Ca(Al_6Si_6O_{24})(SO_4)(S_3)^{-}\cdot nH_2O$, *Am. Mineralog.*, **106**(2), 226–234 (2021).

Sarma D., Malliakas C.D., Subrahmanyam K.S., Islam S.M., Kanatzidis M.G. $K_{2x}Sn_{4-x}S_{8-x}$ (x = 0.65-1): a new metal sulfide for rapid and selective removal of Cs^+, Sr^{2+} and UO_2^{2+} ions, *Chem. Sci.*, **7**(2), 1121–1132 (2016).

Sarp H., Burri G. Étude de la schmiederite de la mine Condor, La Rioja (Sierra de Cacheuta) Argentine, un séléniate et sélénite hydraté de plomb et de cuivre, *Schweiz. Mineral. Petrogr. Mitt.*, **67**(3), 219–223 (1987).

Sarp H., Černy R. Barrotite, IMA 2011-061a. CNMNC Newsletter No. 16, August 2013, page 2698, *Mineralog. Mag.*, **77**(6), 2695–2709 (2013).

Sarp H., Černy R., Pushcharovsky D.Yu., Schouwink P., Teyssier J., Williams P.A., Babalik H., Mari G. La barrotite, $Cu_9Al(HSiO_4)_2[(SO_4)(HAsO_4)_{0.5}](OH)_{12}\cdot 8H_2O$, un nouveau minéral de la mine de Roua (Alpes-Maritimes, France), *Riviéra Sci.*, **98**, 3–22 (2014).

Sarp H., Deferne J. La chessexite, un nouveau minéral, *Schweiz. Mineral. Petrogr. Mitt.*, **62**(3), 337–341 (1982).

Sassi S., Candolfi C., Delaizir G., Migot S., Ghanbaja J., Gendarme C., Dauscher A., Malaman B., Lenoir B. Crystal structure and transport properties of the homologous compounds $(PbSe)_5(Bi_2Se_3)_{3m}$ (m = 2, 3), *Inorg. Chem.*, **57**(1), 422–434 (2018).

Sato E., Nakai I., Miyawaki R., Matsubara S. Crystal structures of alunite family minerals: beaverite, corkite, alunite, natroalunite, jarosite, svanbergite, and woodhouseite, *N. Jb. Miner. Abh.*, **185**(3), 313–322 (2009).

Sato E., Nakai I., Terada Y., Tsutsumi Y., Yokoyama K., Miyawaki R., Matsubara S. Beaverite-(Zn),$Pb(Fe_2Zn)(SO_4)_2(OH)$, a new member of the alunite group, from Mikawa Mine, Niigata Prefecture, Japan, *Mineralog. Mag.*, **75**(2), 375–377 (2011).

Sato E., Nakai I., Terada Y., Tsutsumi Y., Yokoyama K., Miyawaki R., Matsubara S. Study of Zn-bearing beaverite $Pb(Fe_2Zn)(SO_4)_2(OH)_6$ obtained from Mikawa mine, Niigata Prefecture, Japan, *J. Mineral. Petrol. Sci.*, **130**(2), 141–144 (2008).

Schiwy W., Blutau C., Gäthje D., Krebs B. Darstellung und Struktur von $K_2SnS_3\cdot 2H_2O$, *Z. anorg. und allg. Chem.*, **412**(1), 1–10 (1975).

Schiwy W., Krebs B. $Sn_{10}O_4S_{20}^{8-}$: Ein neuer Typ eines Polyanions, *Angew. Chem.*, **87**(12), 451–452 (1975a).

Schiwy W., Krebs B. $Sn_{10}O_4S_{20}^{8-}$: a new type of polyanion, *Angew. Chem. Int. Ed. Engl.*, **14**(6), 436 (1975b).

Schiwy W., Pohl S., Krebs B. Darstellung und Struktur von $Na_4SnS_4\cdot 14H_2O$, *Z. anorg. und allg. Chem.*, **402**(1), 77–86 (1973).

Schofield P.F., Wilson C.C., Knight K.S., Kirk C.A. Proton location and hydrogen bonding in the hydrous lead copper sulfates linarite, $PbCu(SO_4)(OH)_2$, and caledonite, $Pb_5Cu_2(SO_4)_3CO_3(OH)_6$, *Canad. Mineralog.*, **47**(3), 649–662 (2009).

Schwarz H. Doppelverbindungen vom Typ $Me^{I}_2Me^{II}(X^{VI}O_4)_2$ mit der Struktur von $Sr_3(PO_4)_2$. I. Sulfate, *Z. anorg. allg. Chem.*, **344**(1-2), 41–55 (1966).

Schwarz H. Verbindungen mit Apatitstruktur. II. Apatite des Typs $M^{II}_{10}(X^{VI}O_4)_3(X^{IV}O_4)_3F_2$ (M^{II} = Sr, Pb; X^{VI} = S, Cr; X^{IV} = Si, Ge), *Z. anorg. allg. Chem.*, **356**(1-2), 36–45 (1967a).

Schwarz H. Verbindungen mit Apatitstruktur. III. Apatite des Typs $Pb_6K_4(X^{V}O_4)_4(X^{VI}O_4)_2$ (X^{V} = P, As; X^{VI} = S, Se), *Z. anorg. allg. Chem.*, **356**(1-2), 29–35 (1967b).

Schwarz R., Giese H. Beiträge zur Chemie des Germaniums, III. Mitteil.: Sulfo- und Pergermanate, *Ber. Deutsch. Chem. Ges. (A and B Series)*, **63**(4), 778–782 (1930).

Schwarzmüller S., Hoai-Thuong Tran V., Yang F., Oeckler O. The sodium antimony telluridogermanate(III) $Na_9Sb[Ge_2Te_6]_2$, *Z. anorg. und allg. Chem.*, **645**(16), 1037–1042 (2019).

Schwarzmüller S., Souchay D., Günther D., Gocke A., Dovgaliuk I., Miller S.A., Snyder G.J., Oeckler O. Argyrodite-type $Cu_8GeSe_{6-x}Te$ ($0 \leq x \leq 2$): temperature-dependent crystal structure and thermo-electric properties. *Z. anorg. und allg. Chem.*, **644**(24), 1915–1922 (2018).

Seidzade A.E., Orujlu E.N., Babanly D.M., Imamaliyeva S.Z., Babanly M.B. Solid-phase equilibria in the $SnTe–Sb_2Te_3$–Te system and the thermodynamic properties of the tin–antimony tellurides, *Russ. J. Inorg. Chem.*, **67**(5), 683–690 (2022).

Seidzade A.E., Orujlu E.N., Doert T., Amiraslanov I.R., Aliev Z.S., Babanly M.B. An updated phase diagram of the $SnTe–Sb_2Te_3$ system and the crystal structure of the new compound $SnSb_4Te_7$, *J. Phase Equilib. Diffus.*, **42**(3), 373–378 (2021).

Semenov E.I., Khomyakov A.P., Cherepivskaya G.E., Ugryumova N.G. Sodian cancrinites of the Lovozero alkaline massif [in Russian], *Mineralog. Zhurn.*, **6**(2), 50–54 (1984).

Sevryukov N.N., Salikova G.E. The $GeS_2–Na_2S–H_2O$ system at 25°C [in Russian], *Zhurn. neorgan. khimii*, **15**(6), 1634–1639 (1970).

Sevryukov N.N., Salikova G.E., Dolganev V.P. On the composition and some properties of the thiogermanates of the alkali metals [in Russian], *Zhurn. neorgan. khimii*, **14**(1), 26–31 (1969).

Seyidzade A.E., Aghayeva A.A., Orujlu E.N., Imamaliyeva S.Z. Thermodynamic properties of Sb_2Te_3-based solid solutions in the $SnTe–Sb_2Te_3$ system, *Azerb. Chem. J.*, (1), 83–88 (2022).

Shafagatova G.G., Alikhanov R.A., Valiyev V.G., Dzhafarova S.Z. Phase equilibria in the PrSe–PbSe system [in Azerbaijanian], *Kimiya Probl.*, (3), 412–414 (2012).

Shang M., Halasyamani P.S. Mixed lone-pair and mixed anion compounds: $Pb_3(SeO_3)(HSeO_3)Br_3$, $Pb_3(SeO_3)(OH)Br_3$, $CdPb_8(SeO_3)_4Cl_4Br_6$ and $RbBi(SeO_3)F_2$, *J. Solid State Chem.*, **282**, 121121 (2020).

Shchipalkina N.V., Vereshchagin O.S., Chukanov N.V., Gorelova L.A., Bocharov V.N., Pekov I.V. High-temperature behavior of the UV-luminescent sodalite-type natural compound $Na_8(Al_6Si_6O_{24})(HS)_2$: Comparative study by *in situ* single-crystal X-ray diffraction and Raman spectroscopy, *J. Solid State Chem.*, **323**, 124067 (2023).

Sheldrick W.S. Zur Kenntnis von $Cs_4Sn_5S_{12}{\cdot}2H_2O$, ein Cäsium(I)-Thiostannat(IV) mit fünf- und sechsfach koordiniertem Zinn, *Z. anorg. und allg. Chem.*, **562**(1), 23–30 (1988).

Sheldrick W.S., Schaaf B. Darstellung und Kristallstruktur von $Rb_2Sn_3S_7{\cdot}2H_2O$ und $Rb_4Sn_2Se_6$, *Z. anorg. und allg.Chem.*, **620**(6), 1041–1045 (1994).

Shelimova L.E., Karpinskii O.G., Konstantinov P.P., Avilov E.S., Kretova M.A., Lubman G.U., Nikhezina I.Yu., Zemskov V.S. Composition and properties of compounds in the $PbSe–Bi_2Se_3$ system, *Inorg. Mater.*, **46**(2), 120–126 (2010).

Shelimova L.E., Karpinskii O.G., Zemskov V.S. X-ray diffraction study of ternary layered compounds in the $PbSe–Bi_2Se_3$ system, *Inorg. Mater.*, **44**(9), 927–931 (2008).

Shelimova L.E., Zemskov V.S., Avilov E.S., Kretova M.A., Nikhezina I.Yu., Mikhailova A.B. Solid solutions based on laminated chalcogenides of bismuth and lead in ternary reciprocal Pb,Bi||Se,Te system, *Inorg. Mater.: Appl. Res.*, **6**(4), 298–304 (2015).

Shemet V.Ya., Gulay L.D., Olekseyuk I.D. Isothermal sections of the $Y_2Se_3–Cu_2Se$–Sn(Pb)Se systems at 870 K and crystal structure of the $Y_{4.2}Pb_{0.7}Se_7$ compound, *Pol. J. Chem.*, **79**(8), 1315–1326 (2005).

Shevchuk M. Phase equilibria in the section $AgGaSe_2–GeS_2$ [in Ukrainian], *Nauk. Visnyk Volyns'k. Nats. Univ. im. Lesi Ukrainky. Ser. Khim. nauky*, [23(272)], 83–86 (2013a).

Shevchuk M. Phase equilibria in the section $AgGaS_2–GeSe_2$ [in Ukrainian], *Nauk. Visnyk Volyns'k. Nats. Univ. im. Lesi Ukrainky. Ser. Khim. nauky*, [24(273)], 8–12 (2013b).

Shevchuk M. The $AgGaS_2 + GeSe_2 \Leftrightarrow AgGaSe_2 + GeS_2$ system [in Ukrainian], *Nauk. Visnyk Volyns'k. Nats. Univ. im. Lesi Ukrainky. Ser. Khim. nauky*, [24(273)], 15–21 (2013c).

Shevchuk M.V., Atuchin V.V., Kityk A.V., Fedorchuk A.O., Romanyuk Y. E., Całus S., Yurchenko O.M., Parasyuk O.V. Single crystal preparation and properties of the $AgGaGeS_4$–$AgGaGe_3Se_8$ solid solution, *J, Cryst. Growth*, **318**(1), 708–712 (2011).

Shevchuk M.V., Olekseyuk I.D. The $AgGaGeS_4$–$AgGaGe_3Se_8$ system [in Ukrainian], *Nauk. Visnyk Volyns'k. Derzh. Univ. im. Lesi Ukrainky. Ser. Khim. nauky*, (4), 87–89 (2006).

Shevchuk M.V., Olekseyuk I.D. The $AgGaSe_2$ + SnS_2 ⇔ $AgGaS_2$ + $SnSe_2$ system, *Fiz. khim. tv. tila*, **11**(2), 386–390 (2010).

Shi Y.-F., Li X.-F., Zhang Y.-X., Lin H., Ma Z., Wu L.-M., Wu X.-T., Zhu Q.-L. $[(Ba_{19}Cl_4)(Ga_6Si_{12}O_{42}S_8)]$: a two-dimensional wide-band-gap layered oxysulfide with mixed-anion chemical bonding and photocurrent response, *Inorg. Chem.*, **58**(10), 6588–6592 (2019a).

Shi Y.-F., Ma Z., Li B.-X., Wu X.-T., Lin H., Zhu Q.-L. Phase matching achieved by isomorphous substitution in IR nonlinear optical material $Ba_2SnSSi_2O_7$ with an undiscovered $[SnO_4S]$ functional motif, *Mater. Chem. Front.*, **6**(20), 3054–3061 (2022).

Shi Y.-F., Zhou S.-H., Li B., Liu Y., Wu X.-T., Lin H., Zhu Q.-L. $Ba_5Ga_2SiO_4S_6$: a phase-matching nonlinear optical oxychalcogenide design via structural regulation originated from heteroanion introduction, *Inorg. Chem.*, **62**(1), 464–473 (2023).

Shi Z.-H., Chi Y., Sun Z.-D., Liu W., Guo S.-P. $Sn_2Ga_2S_5$: A type of IR nonlinear-optical material, *Inorg. Chem.*, **58**(18), 12002–12006 (2019b).

Shi Z.-H., Yang M., Yao W.-D., Liu W., Guo S.-P. $SnPQ_3$ (Q = S, Se, S/Se): A series of lone-pair cationic chalcogenophosphates exhibiting balanced NLO activity originating from SnQ_8 units, *Inorg. Chem.*, **60**(18), 14390–14398 (2021).

Shimizu M., Moh G.H., Kato A. Potosiite and incaite from the Hoei mine, Japan, *Mineral. Petrol.*, **46**(2), 155–161 (1992).

Shin D., Saparov B., Zhu T., Huhn W.P., Blum V., Mitzi D.B. $BaCu_2Sn(S,Se)_4$ – earth-abundant chalcogenides for thin-film photovoltaics, *Chem. Mater.*, **28**(13), 4771–4780 (2016).

Shkodin V.G., Malyshev V.P. Thermodynamics of the lead selenide and lead telluride interaction with soda at the carbon presence [in Russian], *Tr. Ural'skogo n.-i. i proectn. in-ta medn. prom-sti*, (11), 307–310 (1969).

Shuvalov R.R., Vergasova L.P., Semenova T.F., Filatov S.K., Krivovichev S.V., Siidra O.I., Rudashevsky N.S. Prewittite, $KPb_{1.5}Cu_6Zn(SeO_3)_2O_2Cl_{10}$, a new mineral from Tolbachik fumaroles, Kamchatka peninsula, Russia: Description and crystal structure, *Am. Mineralog.*, **98**(2-3), 463–469 (2013).

Shvedov G.I., Barkov A.Y. Oulankaite, a Pd-rich stannosulphotelluride, and associated platinum-group minerals from the Dzhaltul microgabbroic intrusion, Krasnoyarskiy kray, Russia, *N. Jb. Miner. Abh. (J. Min. Geochem.)*, **192**(3), 229–237 (2015).

Sieke C., Schleid T. $NaPr_9S_2[SiO_4]_6$: Ein Sulfidsilicat des Praseodyms mit Bromapatitstruktur, *Z. anorg. und allg. Chem.*, **623**(9), 1345–1346 (1997).

Sieke C., Schleid T. $Sm_4S_3[Si_2O_7]$ und $NaSm_9S_2[SiO_4]_6$: Zwei Sulfidsilicate mit dreiwertigem Samarium, *Z. anorg. und allg. Chem.*, **625**(1), 131–136 (1999).

Silverstein H.J., Cruz-Kan K., Hallas A.M., Zhou H.D., Donaberger R.L., Hernden B.C., Bieringer M., Choi E.S., Hwang J.M., Wills A.S., Wiebe C.R. $Pb_3TeCo_3V_2O_{14}$: a potential multiferroic Co bearing member of the dugganite series, *Chem. Mater.*, **24**(4), 664–670 (2012).

Silverstein H.J., Sharma A.Z., Cruz-Kan K., Zhou H.D., Huq A., Flacau R., Wiebe C.R. Complex long-range magnetic ordering in the Mn-bearing dugganite $Pb_3TeMn_3P_2O_{14}$, *J. Solid State Chem.*, **204**, 102–107 (2013a).

Silverstein H.J., Sharma A.Z., Stoller A.J., Cruz-Kan K., Flacau R., Donaberger R.L., Zhou H.D., Manuel P., Huq A., Kolesnikov A.I., Wiebe C.R. Phase diagram and magnetic structures of the Co-bearing dugganites $Pb_3TeCo_3A_2O_{14}$ (A = V, P), *J. Phys.: Condens. Matter*, **25**(24), 246004 (2013b).

Sineva S., Fallah-Mehrjardi A., Hidayat T., Shevchenko M., Shishin D., Hayes P.C., Jak E. Experimental investigation of gas/slag/matte/tridymite equilibria in the Cu–Fe–O–S–Si–Al–Ca–Mg system in controlled gas atmosphere: experimental results at 1473 K (1200°C), 1573 K (1300 C) and $p(SO_2)$ = 0.25 atm, *J. Phase Equilib. Diffus.*, **41**(3), 243–256 (2020).

Siuda R., Kruszewski Ł., Olds T.A. Borzęckiite, IMA 2018-146a. CNMNC Newsletter No 70, *Eur. J. Mineral.*, **34**(6), 591–592 (2022).

Siuda R., Kruszewski Ł., Olds T.A. Borzęckiite, IMA 2018-146a. CNMNC Newsletter No 70, *Mineralog. Mag.*, **87**(1), 160 (2023).

Slade T., Gvozdetskyi V., Wilde J.M., Kreyssig A., Gati E., Wang L.-L., Mudryk Y., Ribeiro R.A., Pecharsky V.K., Zaikina J.V., Bud'ko S.L., Canfield P.C. A low temperature structural transition in canfieldite, Ag_8SnS_6, single crystals, *Inorg. Chem.*, **60**(24), 19345–19355 (2021).

Smith R.L., Simons F.S., Vlisidis A.C. Hidalgoite, a new mineral, *Am. Mineralog.*, **38**(11–12), 1218–1224 (1953).

Sokolova M.N., Smol'yaninova N.N., Golovanova T.I., Chukanov N.V., Dmitrieva M.T. Delhayelite crystals from ristschorrites of the Rasvumchorr plateau (Khibiny massif), *New Data on Minerals*, Moscow, **40**, 115–118 (2005).

Sosedko T.A., Kasatov B.K., Furmakova L.N., Lipatova E.A. New data on cancrinite–vishnevite group minerals [in Russian], *Zap. Vses. Mineralog. Obshch.*, **118**(5), 78–84 (1989).

Spiridonov E.M. Maikainite $Cu_{20}(Fe,Cu)_6Mo_2Ge_6S_{32}$ and ovamboite $Cu_{20}(Fe,Cu,Zn)_6W_2Ge_6S_{32}$: New minerals in massive sulfide base meral ores, *Dokl. Earth Sci.*, **393**(9), 1329–1332 (2003).

Spiridonov E.M., Badalov A.S., Kovachev V.V. Stibiocolusite $Cu_{26}V_2(Sb,Sn,As)_6S_{32}$: a new mineral [in Russian], *Dokl. RAN*, **324**(2), 411–414 (1992). **324**(2), 411–414 (1992a).

Spiridonov E.M., Chvileva T.N. Bogdanovite, $Au_5(Cu,Fe)_3(Te,Pb)_2$, a new mineral of the group of intermetallic compounds of gold [in Russian], *Vestn. Mosk. Gos. Univ. Geol.*, (1), 44–52 (1979).

Spiridonov E.M., Kachalovskaya V.M., Kovachev V.V., Krapiva L.Ya. Germanocolusite $Cu_{26}V_2(Ge,As)_6S_{32}$ – a new mineral [in Russian], *Vestn. Mosk. Univ., Ser. 4, Geol.*, (6), 50–54 (1992b).

Spirovski F., Reiner C., Deiseroth H.-J., Kienle L., Mikus H.. M_3GeTe_2 and M_5GeTe_2 – new layered tellurides (M = Fe, Ni), *Z. anorg. und allg. Chem.*, **632**(12-13), 2103 (2006).

Sportouch S., Bastea M., Brazis P., Ireland J., Kannewurf C.R., Uher C., Kanatzidis M.G. Thermoelectric properties of the cubic family of compounds $AgPbBiQ_3$ (Q = S, Se, Te). Very low thermal conductivity materials, *Mat. Res. Soc. Symp. Proc.*, **545,** 123–130 (1998).

Springer G. The pseudobinary system Cu_2FeSnS_4–Cu_2ZnSnS_4 and its mineralogical significance, Canad. Mineralog., 11(2), 535–541 (1972).

Spry P.G., Merlino S., Wang S., Zhang X., Busek P.R. New occurrences and refined crystal chemistry of colusite, with comparisons to arsenosulvanite, *Am. Mineralog.*, **79**(7–8), 750–762 (1994).

Środek D., Galuskina I.O., Galuskin E., Dulski M., Książek M., Kusz J., Gazeev V. Chlorellestadite, $Ca_5(SiO_4)_{1.5}(SO_4)_{1.5}Cl$, a new ellestadite-group mineral from the Shadil-Khokh volcano, South Ossetia, *Mineral. Petrol.*, **112**(5), 743–752 (2018).

Środek D., Książek M., Galuskina I.O., Kusz J., Dulski M., Gazeev V., Galuskin E. Chlorellestadite, IMA 2017-013. CNMNC Newsletter No. 37, June 2017, page 532, *Eur. J. Mineral.*, **29**(3), 529–533 (2017a).

Środek D., Książek M., Galuskina I.O., Kusz J., Dulski M., Gazeev V., Galuskin E. Chlorellestadite, IMA 2017-013. CNMNC Newsletter No. 37, June 2017, page 532, *Mineralog. Mag.*, **81**(3), 737–742 (2017b).

Stahl J., Shlaen E., Johrendt D. The van der Waals ferromagnets $Fe_{5-\delta}GeTe_2$ and $Fe_{5-\delta-x}Ni_xGeTe_2$ – crystal structure, stacking Faults, and magnetic properties, *Z. anorg. und allg. Chem.*, **644**(24), 1923–1929 (2018).

Stanley C.J., Roberts A.C., Harris D.C. New data for nagyagite, *Mineralog. Mag.*, **58**(3), 479–482 (1994).

Stanley C.R. Hinsdalite and other products of oxidation at the Daisy Creek stratabound copper-silver prospect, northwestern Montana, *Canad. Mineralog.*, **25**(2), 213–220 (1987).

Steele I.M., Pluth J.J., Livingstone A. Crystal structure of macphersonite ($Pb_4SO_4(CO_3)_2(OH)_2$): comparison with leadhillite, *Mineralog. Mag.*, **62**(4), 451–459 (1998).

Steele I.M., Pluth J.J., Livingstone A. Crystal structure of mattheddleite: a Pb, S, Si phase with the apatite structure, *Mineralog. Mag.*, **64**(5), 915–921 (2000).

Steele I.M., Pluth J.J., Livingstone A. Crystal structure of susannite, $Pb_4SO_4(CO_3)_2(OH)_2$ a trimorph wirh macphersonite and leadhillite, *Eur. J. Mineral.*, **11**(3), 493–499 (1999).

Stöwe K. Zur Struktur und Dotierung von Selenosilikaten: die Kristallstruktur von Er_2SeSiO_4 und $Er_{3.75}Ca_{0.25}Se_{2.75}Cl_{0.25}Si_2O_7$, *Z. Naturforsch.*, **49B**(6), 733–740 (1994).

Strunz H. Isotypie der Verbindungen $PbK_2[SO_4]_2$ und $Ca_3[PO_4]_2$, *Naturwissenschaften*, **30**(16), 242 (1942).

Studenyak I.P., Kokhan O.P., Kranjčec M., Bilanchuk V.V., Panko V.V. Influence of S-Se substitution on chemical and physical properties of $Cu_7Ge(S_{1-x}Se_x)_5I$ superionic solid solutions, *J. Phys. Chem. Solids*, **68**(10), 1881–1884 (2007a).

Studenyak I.P., Kokhan O.P., Kranjčec M., Hrechyn M.I., Panko V.V. Crystal growth and phase interaction studies in the Cu_7GeS_5I–Cu_7SiS_5I superionic system, *J. Cryst. Growth*, **306**(2), 326–329 (2007b).

Studenyak I.P., Pogodin A.I., Kokhan O.P., Kavaliukė V., Šalkus T., Kežionis A., Orliukas A.F. Crystal growth, structural and electrical properties of $(Cu_{1-x}Ag_x)_7GeS_5I$ superionic solid solutions, *Solid State Ionics*, **329**, 119–123 (2019).

Studenyak I.P., Pogodin A.I., Shender I.A., Bereznyuk S.M., Filep M.J., Kokhan O.P., Kúš P. Preparation and electrical conductivity of $(Cu_{0.5}Ag_{0.5})_7SiS_5I$-based superionic ceramics, *J. Alloys Compd.*, **854**, 157131 (2021).

Sudarsanan K. Structure of hydroxyellestadite, *Acta Crystallogr.*, **B36**(7), 1636–1639 (1980).

Sugaki A., Shima H., Kitakaze A. Miharaite, $Cu_4FePbBiS_6$, a new mineral from the Mihara mine, Okayama, Japan, *Am. Mineralog.*, **65**(7-8), 784–788 (1980).

Sun M., Zhang X., Li C., Liu W., Lin Z., Yao J. Highly polarized $[GeOTe_3]$ motif-driven structural order promotion and an enhanced second harmonic generation response in the new nonlinear optical oxytelluride $Ba_3Ge_2O_4Te_3$, *J. Mater. Chem. C*, **10**(1), 150–159 (2022a).

Sun M., Zhang X., Xing W., Li Z., Liu W., Lin Z., Yin W., Yao J. Synthesis and characterizations of two tellurides β-$BaGa_2Te_4$ and $Ba_5Ga_2Ge_3Te_{12}$ with flexible chain structure, *Inorg. Chem.*, **60**(19), 14793–14802 (2021).

Sun M., Zhang X., Xing W., Uykur E., Yin W., Lin Z., Yao J. $Ba_6In_2Ge_2Te_{15}$: a THz birefringent material with an intriguing quasi-$[Te_5]^{4-}$ chain possessing large optical anisotropy and an ultrawide transmission range, *Inorg. Chem. Front.*, **9**(14), 3421–3427 (2022b).

Sun Y., Chen M., Balladares E., Pizarro C., Contreras C., Zhao B. Effect of CaO on the liquid/spinel/matte/gas equilibria in the Si–Fe–O–Cu–S system at controlled $P(SO_2)$ 0.3 and 0.6 atm, *Calphad*, **69**, 101751 (2020a).

Sun Y., Chen M., Balladares E., Pizarro C., Contreras C., Zhao B. Effect of MgO on the liquid/spinel/matte/gas equilibria in the Si–Fe–Mg–O–Cu–S system at controlled $P(SO_2)$ 0.3 and 0.6 atm, *Calphad*, **70**, 101803 (2020b).

Sun Y., Suzuki K., Hori S., Hirayama M., Kanno R. Superionic conductors: $Li_{10+\delta}[Sn_ySi_{1-y}]_{1+\delta}P_{2-\delta}S_{12}$ with a $Li_{10}GeP_2S_{12}$-type structure in the Li_3PS_4–Li_4SnS_4–Li_4SiS_4 quasi-ternary system, *Chem. Mater.*, **29**(14), 5858–5864 (2017).

Sun Y-L., Ablimit A., Zhai H.-F., Bao J.-K., Tang Z.T., Wang X.-B., Wang N.-L., Feng C.-M., Cao G.H. Design and synthesis of a new layered thermoelectric material $LaPbBiS_3O$, *Inorg. Chem.*, **53**(20), 11125–11129 (2014).

Szymański J.T. The crystal structure of beudantite, $Pb(Fe,Al)_3[(As,S)O_4]_2(OH)_6$, *Canad. Mineralog.*, **26**(4), 923–932 (1988).

Szymański J.T. The crystal structure of owensite, $(Ba,Pb)_6(Cu,Fe,Ni)_{25}S_{27}$, a new member of the djerfisherite group, *Canad. Mineralog.*, **33**(3), 671–678 (1995).

Szymański J.T. The crystal structure of plumbojarosite, $Pb[Fe_3(SO_4)_2(OH)_6]_2$, *Canad. Mineralog.*, **23**(4), 659–668 (1985).

Taguchi Y. On osarizawaite, a new mineral of the alunite group, from the Osarizawa mine, Japan, *Mineralog. J.*, **3**(4), 181–194 (1961).

Tait K.T., Belakovskiy D., Welch M.D., Gatta G.D., Cámara F. New mineral names, *Am. Mineralog.*, **97**(7), 1260–1265 (2012).

Tait K.T., DiCecco V., Ball N.A., Hawthorne F.C., Kampf A.K. Backite, $Pb_2Al(TeO_6)Cl$, a new tellurate mineral from the Grand Central mine, Tombstone Hills, Cochise County, Arizona: description and crystal structure, *Canad. Mineralog.*, **52**(6), 935–942 (2014b).

Tait K.T., DiCecco V., Cooper M.A., Ball N.A., Hawthorne F.C. Backite, IMA 2013-113. CNMNC Newsletter No. 19, February 2014, page 169, *Mineralog. Mag.*, **78**(1), 165–170 (2014a).

Takéuchi Y., Ohmasa M., Nowacki W. The crystal structure of wallisite, $PbTlCuAs_2S_5$, the Cu analogue of hatchite, $PbTlAgAs_2S_5$, *Z. Kristallog.*, **127**(5-6), 349–365 (1968).

Takéuchi Y., Takagi J., Yamanaka T. The crystal structure of $PbS{\cdot}2Bi_2S_3$, *Proc. Jpn. Acad.*, **50**(4), 317–321 (1974).

Tang C., Xing W.-H., Liang F., Sun M., Tang J., Lin Z., Yao J., Chen K., Wu J., Yin W., Bin K. Structural modification from centrosymmetric $Rb_4Hg_2Ge_2S_8$ to noncentrosymmetric $(Na_3Rb)Hg_2Ge_2S_8$: Mixed alkali-metals strategy for infrared nonlinear optical material design, *J. Mater. Chem. C*, **10**(9), 3300–3306 (2022).

Tang G., Yang Z., Luo L., Chen W. Preparation and properties of $GeSe_2$–Ga_2Se_3–KBr new chalcohalide glasses, *J. Alloys Compd.*, **459**(1-2), 472–476 (2008).

Tanibata N., Noi K., Hayashi A., Tatsumisago M. Preparation and characterization of Na_3PS_4–Na_4GeS_4 glass and glass-ceramic electrolytes, *Solid State Ionics*, **320**, 193–198 (2018).

Tarasenko M.S., Berezin A.S., Kiryakov A.S., Piryazev D.A., Filatova I.Yu., Naumov N.G. Synthesis, crystal structure and photoluminescence of Eu^{3+} or Tb^{3+} doped solid solutions $(Y_{1-x}RE_x)_4S_3(Si_2O_7)$, *J. Solid State Chem.*, **265**, 36–41 (2018).

Tarasenko M.S., Duritsyn R.V., Potapov D.A., Kuratieva N.V., Ryadun A.A., Naumov N.G. Double yttrium thiosilicates $AYSiS_4$ (A = Rb, Cs): synthesis, structure, optical properties, *J. Struct. Chem.,* **63**(12), 1988–1996 (2022).

Tatsumisago M., Hirai K., Hirata T., Takahashi M., Minami T. Structure and properties of lithium ion conducting oxysulfide glasses prepared by rapid quenching, *Solid State Ionics*, **86-88**, Pt. 1, 487–490 (1996).

Tatsumisago M., Hirai K., Minami T., Takada K., Kondo S. Superionic conduction in rapidly quenched Li_2S–SiS_2–Li_3PO_4 glasses, *J Ceram. Soc. Jpn.*, **101**(11), 1315–1317 (1993).

Teertstra D.K., Schindler M., Sherriff B.L., Hawthorne F.C. Silvialite, a new sulfate-dominant member of the scapolite group with an Al-Si composition near the *I4/m–P42/n* phase transition, *Mineralog. Mag.*, **63**(3), 321–329 (1999).

Teri G., Li N., Bai S., Namila E., Baiyin M. Synthesis, crystal structure, photocatalysis, photocurrent response: one-dimensional $K_2HgSnSe_4$ and three-dimensional $Na_6Cu_8Sn_3Se_{13}$, *CrystEngComm*, **23**(35), 6079–6085 (2021).

Teske C.L. $Ba_2ZnGe_2S_6O$: Ein neues Oxidsulfid mit Tetraedergerüststruktur. *Z. Naturforsch.*, **35B**(6), 672–675 (1980).

Teske C.L. Darstellung und Kristallstruktur von $Ba_3CdSn_2S_8$ mit einer Anmerkung über $Ba_6CdAg_2Sn_4S_{16}$, *Z. anorg. und allg. Chem.*, **522**(3), 122–130 (1985a).

Teske C.L. Über Oxidsulfide mit Åkermanitstruktur $CaLaGa_3S_6O$, $SrLaGa_3S_6O$, $La_2ZnGa_2S_6O$ und $Sr_2ZnGe_2S_6O$, *Z. anorg. und allg. Chem.*, **531**(12), 52–60 (1985b).

Teske C.L., Terraschke H., Mangelsen S., Bensch W. Re-investigation of barium-gold(I)-tetra-thiostannate(IV), $Ba[Au_2SnS_4]$, with short $Au^I \cdots Au^I$ separation showing luminescence properties, *Z. anorg. und allg. Chem.*, **646**(21), 1716–1721 (2020).

Teske C.L., Vetter O. Präparative und röntgenographische Untersuchung am System $Cu_{2-x}Ag_xBaSnS_4$, *Z. anorg. und allg. Chem.* , **426**(3), 281–287 (1976).

Thilo E., Winkler A. Über $K_2Si(S_2O_7)_3$, eine weitere SiO_2–SO_3-Verbindung, *Z. anorg. und allg. Chem.*, **365**(3-4), 180–184 (1969).

Thompson R.M. Danalite from British Columbia, *Canad. Mineralog.*, *6*(1), 68–71 (1957).

Tian X., Zhang X., Xiao Y., Wu X., Zhang B., Yang D., Wu K. From oxides to oxysulfides: the mixed-anion GeS_3O unit induces huge improvement in the nonlinear optical effect and optical anisotropy for potential nonlinear optical materials, *RSC Adv.*, **12**(25), 16296–16300 (2022).

Tilley C.E., Henry N.F.M. Latiumite (sulphatic potassium–calcium–aluminium silicate), a new mineral from Albano, Latium, Italy, *Mineralog. Mag.*, **30**(220), 39–45 (1953).

Toguzov M.Z., Kopylov N.I., Sychev A.P. The $Cu_{2-x}S$–PbS–FeS–ZnS system [in Russian], *Zhurn. neorgan. khimii*, **25**(8), 2237–2240 (1980).

Tomashyk V. Multinary alloys based on II–VI semiconductors, *London: CRC Press*, (2015).

Topa D., Keutsch F.N., Makovicky E., Kolitsch U., Paar W. Polloneite, a new complex Pb(-Ag)-As-Sb sulfosalt from the Pollone mine, Apuan Alps, Tuscany, Italy, *Mineralog. Mag.*, **81**(6), 1303–1322 (2017a).

Topa D., Keutsch F.N., Makovicky E., Kolitsch U., Paar W. Polloneite, IMA 2014-093. CNMNC Newsletter No. 24, April 2015, page 249, *Mineralog. Mag.*, **79**(2), 247–251 (2015).

Topa D., Kolitsch U. The crystal chemistry of rathite based on new electron-microprobe data and single-crystal structure refinements: the role of thallium, *Minerals*, **8**(10), 466 (2018).

Topa D., Kolitsch U., Makovicky E., Favreau G., Stanley C., Bourgoin V., Boulliard J.-C . Écrinsite, IMA 2015-099. CNMNC Newsletter No. 29, February 2016, page 204, *Mineralog. Mag.*, **80**(1), 199–205 (2016a).

Topa D., Kolitsch U., Makovicky E., Stanley C. Écrinsite, $AgTl_3Pb_4As_{11}Sb_9S_{36}$, a new thallium-rich homeotype of baumhauerite from the Jas Roux sulphosalt deposit, Parc National des Écrins, Hautes-Alpes, France, *Eur. J. Mineral.*, **29**(4), 689–700 (2017b).

Topa D., Kolitsch U., Stoeger B., Keutsch F., Stanley C. Dewitite, IMA 2019-098. CNMNC Newsletter No. 63, *Eur. J. Mineral.*, **33**(5), 645 (2021a).

Topa D., Kolitsch U., Stoeger B., Keutsch F., Stanley C. Dewitite, IMA 2019-098. CNMNC Newsletter No. 63, *Mineralog. Mag.*, **85**(6), 915 (2021b).

Topa D., Makovicky E. Argentobaumhauerite: name, chemistry, crystal structure, comparison with baumhauerite, and position in the Lengenbach mineralization sequence, *Mineralog. Mag.*, **80**(5), 819–840 (2016).

Topa D., Makovicky E. Eclarite: new data and interpretations, *Canad. Mineralog.*, **50**(2), 371–386 (2012).

Topa D., Makovicky E. The crystal structure of padĕraite, $Cu_7(X_{0.33}Pb_{1.33}Bi_{11.33})_{\Sigma13}S_{22}$, with X = Cu or Ag: new data and interpretation, *Canad. Mineralog.*, **44**(2), 481–495 (2006).

Topa D., Makovicky E., Balić-Žunić T. What is the reason for the doubled unit-cell volumes of copper–lead-rich pavonite homologues? The crystal structures of cupromakovickyite and makovickyite, *Canad. Mineralog.*, **46**(2), 515–523 (2008).

Topa D., Makovicky E., Favreau G., Bourgoin V., Boulliard J.-C., Zagler G., Putz H. Jasrouxite, a new Pb–Ag–As–Sb member of the lillianite homologous series from Jas Roux, Hautes-Alpes, France, *Eur. J. Mineral.*, **25**(6), 1031–1038 (2013a).

Topa D., Makovicky E., Favreau G., Bourgoin V., Boulliard J.-C., Zagler G., Putz H. Jasrouxite, IMA 2012-058. CNMNC Newsletter No. 15, February 2013, page 5, *Mineralog. Mag.*, **77**(1), 1–12 (2013b).

Topa D., Makovicky E., Ilinca G., Dittrich H. Cupromakopavonite, $Cu_8Ag_3Pb_4Bi_{19}S_{38}$, a new mineral species, its crystal structure and the cupropavonite homologous series, *Canad. Mineralog.*, **50**(2), 295–312 (2012).

Topa D., Makovicky E., Paar W.H. Clino-oscarkempffite, IMA 2012-086. CNMNC Newsletter No. 16, August 2013, page 2696, *Mineralog. Mag.*, **77**(6), 2695–2709 (2013c).

Topa D., Makovicky E., Paar W.H., Stanley C.J., Roberts A.C. Oscarkempffite, IMA 2011-029. CNMNC Newsletter No. 10, October 2011, page 2555, *Mineralog. Mag.*, **75**(5), 2549–2561 (2011).

Topa D., Makovicky E., Putz H. The crystal structure of angelaite, $Cu_2AgPbBiS_4$, *Canad. Mineralog.*, **48**(1), 145–153 (2010a).

Topa D., Makovicky E., Putz H., Mumme W.G. The crystal structure of berryite $Cu_3Ag_2Pb_3Bi_7S_{16}$, *Canad. Mineralog.*, **44**(2), 465–480 (2006).

Topa D., Makovicky E., Putz H., Zagler G., Tajjedin H. Arsenquatrandorite, IMA 2012-087. CNMNC Newsletter No. 16, August 2013, page 2696, *Mineralog. Mag.*, **77**(6), 2695–2709 (2013d).

Topa D., Makovicky E., Putz H., Zagler G., Tajjedin H. Barikaite, IMA 2012-055. CNMNC Newsletter No. 15, February 2013, page 4, *Mineralog. Mag.*, **77**(1), 1–12 (2013e).

Topa D., Makovicky E., Schimper H.J., Dittrich H. The crystal structure of a synthetic orthorhombic N = 8 member of the lillianite homologous series, *Canad. Mineral.*, **48**(5), 1127–1135 (2010b).

Topa D, Makovicky E., Sejkora J., Dittrich H. The crystal structure of watkinsonite, $Cu_2PbBi_4Se_8$, from the Zálesí uranium deposit, Czech Republic, *Canad. Mineralog.*, **48**(5), 1109–1118 (2010c).

Topa D., Makovicky E., Tajjedin H., Putz H., Zagler G. Barikaite, $Pb_{10}Ag_3(Sb_8As_{11})S_{19}S_{40}$, a new member of the sartorite homologous series, *Mineralog. Mag.*, **77**(7), 3039–3046 (2013f).

Topa D., Paar W.H. Cupromakovickyite, $Cu_8Pb_4Ag_2Bi_{18}S_{36}$, a new mineral species of the pavonite homologous series, *Canad. Mineralog.*, **46**(2), 503–514 (2008).

Topa D., Paar W.H., Makovicky E., Stanley C.J., Roberts A.C. Oscarkempffite, $Ag_{10}Pb_4(Sb_{17}Bi_9)_{\Sigma26}S_{48}$, a new Sb-Bi member of the lillianite homologous series, *Mineralog. Mag.*, *Mineralog. Mag.*, **80**(5), 809–817 (2016b).

Topa D., Paar W.H., Putz H., Zagler G., Brodtkorb de M.K., Stanley C.J., Roberts A.C., Makovicky E. Mineralogical data on angelaite, $Cu_2AgPbBiS_4$, from the Los Manantiales district, Chubut, Argentina, *Canad. Mineralog.*, **48**(1), 139–144 (2010d).

Topa D., Sicher P., Keutsch F., Kolitsch U., Stanley C. Baiamareite, IMA 2023-044. CNMNC Newsletter No 75, *Eur. J. Mineral.*, **35**(5), 892–893 (2023a).

Topa D., Sicher P., Keutsch F., Kolitsch U., Stanley C. Baiamareite, IMA 2023-044. CNMNC Newsletter No 75, *Mineralog. Mag.*, **87**(6), 956 (2023b).

Topa D., Stoeger B., Keutsch F., Kolitsch U., Stanley C. Interliveingite, IMA 2022-144. CNMNC Newsletter No 72, *Eur. J. Mineral.*, **35**(2), 292–293 (2023c).

Topa D., Stoeger B., Keutsch F., Kolitsch U., Stanley C. Interliveingite, IMA 2022-144. CNMNC Newsletter No 72, *Mineralog. Mag.*, **87**(3), 518 (2023d).

Topa D., Stoeger B., Keutsch F., Kolitsch U., Stanley C. Lasmanisite, IMA 2022-128. CNMNC Newsletter No 72, *Eur. J. Mineral.*, **35**(2), 289 (2023e).

Topa D., Stoeger B., Keutsch F., Kolitsch U., Stanley C. Lasmanisite, IMA 2022-128. CNMNC Newsletter No 72, *Mineralog. Mag.*, **87**(3), 515 (2023f).

Topa D., Stoeger B., Kolitsch U., Keutsch F., Raber T., Stanley C. Buynite, IMA 2023-049. CNMNC Newsletter No 75, *Eur. J. Mineral.*, **35**(5), 893 (2023g).

Topa D., Stoeger B., Kolitsch U., Keutsch F., Raber T., Stanley C. Buynite, IMA 2023-049. CNMNC Newsletter No 75, *Mineralog. Mag.*, **87**(6), 957 (2023h).

Topa D., Stoeger B., Kolitsch U., Keutsch F., Stanley C. Hayyanite, IMA 2023-048. CNMNC Newsletter No 75, *Eur. J. Mineral.*, **35**(5), 893 (2023i).

Topa D., Stoeger B., Kolitsch U., Keutsch F., Stanley C. Hayyanite, IMA 2023-048. CNMNC Newsletter No 75, *Mineralog. Mag.*, **87**(6), 956–957 (2023j).

Topa D., Stoeger B., Kolitsch U., Keutsch F., Stanley C., Raber T. Vallouiseite, IMA 2023-051. CNMNC Newsletter No 75, *Eur. J. Mineral.*, **35**(5), 894 (2023k).

Topa D., Stoeger B., Kolitsch U., Keutsch F., Stanley C., Raber T. Vallouiseite, IMA 2023-051 CNMNC Newsletter No 75, *Mineralog. Mag.*, **87**(6), 957 (2023l).

Topa D., Stoeger B., Kolitsch U., Keutsch F., Stanley C. Sardashtite, IMA 2022-140. CNMNC Newsletter No 72, *Eur. J. Mineral.*, **35**(2), 291–292 (2023m).

Topa D., Stoeger B., Kolitsch U., Keutsch F., Stanley C. Sardashtite, IMA 2022-140. CNMNC Newsletter No 72, *Mineralog. Mag.*, **87**(3), 517 (2023n).

Topa D., Stoeger B., Kolitsch U., Stanley C. Montpelvouxite, IMA 2022-137. CNMNC Newsletter No 72, *Eur. J. Mineral.*, **35**(2), 291 (2023o).

Topa D., Stoeger B., Kolitsch U., Stanley C. Montpelvouxite, IMA 2022-137. CNMNC Newsletter No 72,*Mineralog. Mag.*, **87**(3), 516–517 (2023p).

Triviño Vázquez F. New facts concerning the identification of coattings in heat recovery cyclones in dopol kiln, *Cem. Concr. Res.*, **12**(4), 485–496 (1982).

Tsuchiya N., Takéuchi Y. Fine texture of hauyne having a modulated structure, *Z. Kristallogr.*, **173**(3–4), 273–281 (1985).

Tver'yanovich Yu.S., Nedoshovenko E.G., Aleksandrov V.V., Turkina E.Yu., Tver'yanovich A.S., Sokolov I.A. Chalcogenide glasses containing metal chlorides [in Russian], *Zhurn. neorgan. khimii*, **22**(1), 13–19 (1996).

Urban P., Schneider M.N., Seemann M., Wright J.P., Oeckler O. Information on real-structure phenomena in metastable GeTe-rich germanium antimony tellurides $(GeTe)_nSb_2Te_3$ ($n \geq 3$) by semiquantitative analysis of diffuse X-ray scattering, *Z. Kristallogr.*, **230**(6), 369–384 (2015).

Usman M., Smith M.D., Morrison G., Klepov V.V., Zhang W., Halasyamani P.S., zur Loyer H.-C. Molten alkali halide flux growth of an extensive family of noncentrosymmetric rare earth sulfides: structure and magnetic and optical (SHG) properties, *Inorg. Chem.*, **58**(13), 8541–8550 (2019).

Vassilev V., Aljihmani L., Parvanova V. Phase equilibria in the Ag_4SSe–SnTe system, *J. Therm. Anal. Calorim.*, **75**(1), 63–71 (2004).

Vassilev V., Ivanova Z.G., Aljihmani L., Ctrnoskova E., Cernosek Z. Glass-forming regions, properties, and structure of the chalcogenide As_2Se_3–$GeSe_2$–SnTe (Ag_4SSe) systems, *Mater. Lett.*, **59**(1), 85–87 (2005).

Vassilev V.S., Boycheva S.V., Petkov P. Glass formation in the $GeSe_2(As_2Se_3)$–Sb_2Se_3–CdTe, *Mater. Lett.*, **52**(1-2), 126–129 (2002a).

Vassilev V.S., Ivanova Z.G., Dospeiska E.S., Boycheva S.V. Multicomponent $GeSe_2$–CdI_2–TeO_2 (Bi_2O_3) systems: glass formation and properties, *J. Phys. Chem. Solids*, **63**(5), 815–819 (2002b).

Venevtsev Y.N., Politova E.D., Zhdanov G.S. Tellurium containing ferroelectrics: X-ray data, dielectric properties, phase transitions, *Ferroelectrics*, **8**(1), 489–490 (1974).

Verchenko V.Yu., Tsirlin A.A., Sobolev A.V., Presniakov I.A., Shevelkov A.V., Ferromagnetic order, strong magnetocrystalline anisotropy, and magnetocaloric effect in the layered telluride $Fe_{3-\delta}GeTe_2$, *Inorg. Chem.*, **54**(17), 8598–8607 (2015).

Vergasova L.P., Krivovichev S.N., Britvin S.N., Filatov S.K., Berns P.K., Anan'ev V.V. Allochalcoselite, $Cu^+Cu_5^{2+}PbO_2(SeO_3)_2Cl_5$, a new mineral from volcanic exhalations (Kamchatka, Russia) [in Russian], *Zap. Ros. mineralog. obshch.*, **134**(3), 70–74 (2005).

Vicente C.P., Tirado J.L., Bousquet C., Olivier Fourcade J., Jumas J.C. Cation deficient $Cu_{4-x}GeCo_4Sn_{12}S_{32}$ thiospinels: electrochemical behaviour and induced structural modifications, *J. Mater. Chem.*, **9**(10), 2567–2572 (1999).

Vymazalová A., Kozlov V.V., Laufek F., Stanley C.J., Shkilev I.A. Okruginite, IMA 2022-096. CNMNC Newsletter No 71, *Eur. J. Mineral.*, **35**(1), 77 (2023a).

Vymazalová A., Kozlov V.V., Laufek F., Stanley C.J., Shkilev I.A. Okruginite, IMA 2022-096. CNMNC Newsletter No 71, *Mineralog. Mag.*, **87**(2), 332 (2023b).

Wang J., Cheng Y., Wu H., Hu Z., Wang J., Wu Y., Yu H. $Sr_3[SnOSe_3][CO_3]$: A heteroanionic nonlinear optical material containing planar π-conjugated $[CO_3]$ and heteroleptic $[SnOSe_3]$ anionic groups, *Angew. Chem. Int. Ed.*, **61**(21), e202201616 (2022a).

Wang J., Li P., Cai T., Yang D.-D., Xiong W.-W. Four two-dimensional ternary selenides based on group 13 and 14 metals: Syntheses, crystal structures, and electrochemical properties, *J. Solid State Chem.*, **263**, 88–93 (2018).

Wang P., Abudoureheman M., Chen Z. Experimental and theoretical studies of the ternary thiophosphate $PbPS_3$ featuring ethane-like $[P_2S_6]^{4-}$ unit, *Dalton Trans.*, **49**(47), 17221–17229 (2020a).

Wang P., Abudoureheman M., Zhang K., Zheng J., Chen Z., Wu Q. $Ag_4SnGe_2S_7$: A noncentrosymmetric chalcogenide in I_4–II–IV_2–VI_7 system with non-diamond-like structure featuring 1D $_{\infty}[SnGe_2S_8]^{6-}$ infinite chain, *Inorg. Chem.*, **61**(39), 15303–15309 (2022b).

Wang R., Bu K., Zhang X., Gu Y., Xiao Y., Zhan Z., Huang F. A novel two-dimensional oxysulfide $Sr_{3.5}Pb_{2.5}Sb_6O_5S_{10}$: synthesis, crystal structure, and photoelectric properties, *J. Mater. Chem C*, **8**(32), 11018–11021 (2020b).

Wang R., Liang F., Liu X., Xiao Y., Liu Q., Zhang X., Wu L.-M., Chen L., Huang F. Heteroanionic melilite oxysulfide: a promising infrared nonlinear optical candidate with a strong second-harmonic generation response, sufficient birefringence, and wide bandgap, *ACS Appl. Mater. Interf.*, **14**(20), 23645–23652 (2022c).

Wang S., Buseck P.R. Cylindrite: The relation between its cylindrical shape and modulated structure, *Am. Mineralog.*, **77**(7-8), 758–764 (1992).

Wang X., Wang H.H., Makarenko B., Jacobson A.J. Molecular polyselenide anions sandwiched between cationic extended metal hydroxides: Synthesis and structures of $[Sr_2Sn(OH)_6(H_2O)_6]Se_4$ and $[Sr_2Sn(OH)_6(H_2O)_5]Se_3{\cdot}H_2O$, *Z. anorg. und allg. Chem.*, **638**(15), 2538–2541 (2012).

Wang Y., Luo M., Zhao P., Che X., Cao Y., Huang F. $Sr_4Pb_{1.5}Sb_5O_5Se_8$: a new mid-infrared nonlinear optical material with a moderate SHG response, *CrystEngComm*, **22**(20), 3526–3530 (2020c).

Wang Y., Lü X., Zheng C., Liu X., Chen Z., Yang W., Lin J., Huang F. Chemistry design towards a stable sulfide-based superionic conductor $Li_4Cu_8Ge_3S_{12}$, *Angew. Chem.*, **131**(23), 7755–7759 (2019).

Wang Y., Lü X., Zheng C., Liu X., Chen Z., Yang W., Lin J., Huang F. Chemistry design towards a stable sulfide-based superionic conductor $Li_4Cu_8Ge_3S_{12}$, *Angew. Chem. Int. Ed.*, **58**(23), 7673–7677 (2019).

Warner T.E., Bancells M.M., Lund P.B., Lund F.W., Ravnsbæk D.B. On the thermal stability of manganese(II) sulfate and its reaction with zeolite A to form the sodalite $Na_6Mn_2[Al_6Si_6O_{24}](SO_4)_2$, *J. Solid State Chem.*, **277**, 434–440 (2019).

Wedel B., Sugiyama K., Hiraga K., Itagaki K. Zur Kristallchemie des ersten Blei-Zink-Silicium-Telluroxids: $PbZn_4SiTeO_{10}$, *Z. Naturforsch.*, **54B**(4), 469–472 (1999).

Weihrich R., Anusca I. Half antiperovskites. III. Crystallographic and electronic structure effects in $Sn_{2-x}In_xCo_3S_2$, *Z. anorg. und allg. Chem.*, **632**(2), 1531–1537 (2006).

Weil M., Shirkhanlou M. Incorporation of sulfate or selenate groups into oxotellurates(IV): II. Compounds with divalent lead, *Z. anorg. und allg. Chem.*, **643**(12), 757–765 (2017).

Weil M., Shirkhanlou M., Stürzer T. Phase formation studies of lead(II) copper(II) oxotellurates: The crystal structures of dimorphic $PbCuTeO_5$, $PbCuTe_2O_6$, and $[Pb_2Cu_2(Te_4O_{11})](NO_3)_2$, *Z. anorg. und allg. Chem.*, **645**(3), 347–353 (2019).

Welch M.D., Cooper M.A., Hawthorne F.C., Criddle A.J. Symesite, $Pb_{10}(SO_4)O_7Cl_4(H_2O)$, a new PbO-related sheet mineral: Description and crystal structure, *Am. Mineralog.*, **85**(10), 1526–1533 (2000).

Wenk H.-R. New X-ray data for wenkite, *Schweiz. Mineral. Petrogr. Mitt.*, **46**(1), 85–88 (1966).

Wenk H.-R. The structure of wenkite, *Z. Kristallogr.*, **137**(2-3), 113–126 (1973).

Willett R.D., Vij A., Imhof J.M., Cleary D.A. Redetermination of the crystal structure of $[Na_4(H_2O)_{14}]SnS_4$: a compound containing a novel chain of aquated sodium ions, *J. Chem. Crystallogr.*, **30**(6), 405–410 (2000).

Williams P.A., Hatert F., Pasero M., Mills S.J. IMA Commission on New Minerals, Nomenclature and Classification (CNMNC). Newsletter No. 14. New minerals and nomenclature modifications approved in 2012, *Mineralog. Mag.*, **76**(5), 1281–1288 (2012).

Williams S.A. Choloalite, $CuPb(TeO_3)_2{\cdot}H_2O$, a new mineral, *Mineralog. Mag.*, **44**(1), 55–57 (1981).

Williams S.A. Cuzticite and eztlite, two new tellurium minerals from Moctezuma, Mexico, *Mineralog. Mag.*, **46**(339), 257–259 (1982).

Williams S.A. Elyite, basic lead-copper sulfate, a new mineral from Nevada, *Am. Mineralog.*, **57**(3-4), 364–367 (1972).

Williams S.A. Khinite, parakhinite, and dugganite, three new tellurates from Tombstone, Arizona, *Am. Mineralog.*, **63**(9-10), 1016–1019 (1978).

Williams S.A. Quetzalcoatlite, $Cu_4Zn_8(TeO_3)_3(OH)_{18}$, a new mineral from Moctezuma, Sonora, *Mineralog. Mag.*, **39**(303), 261–263 (1973).

Williams S.A. Schieffelinite, a new lead tellurate–sulphate from Tombstone, Arizona, *Mineralog. Mag.*, **43**(330), 771–773 (1980).

Williams S.A., Duggan M. Tlapallite, a new mineral from Moctezuma, Sonora, Mexico, *Mineralog. Mag.*, **42**(322), 183–186 (1978).

Wu J., Huang W., Liu H.-G., He Z., Chen B., Zhu S., Zhao B., Lei Y., Zhou X. Investigation of the thermal properties and crystal growth of the nonlinear optical crystal $AgGaS_2$ and $AgGaGeS_4$, *Cryst. Growth Des.*, **20**(5), 3140–3153 (2020).

Wu M., Emge T.J., Huang X., Li J., Zhang Y. Designing and tuning properties of a three-dimensional porous quaternary chalcogenide built on a bimetallic tetrahedral cluster $[M_4Sn_3S_{13}]^{5-}$ (M = Zn/Sn), *J. Solid State Chem.*, **181**(3), 415–422 (2008).

Wu M., Su W., Jasutkar N., Huang X., Li J. An open-framework bimetallic chalcogenide structure $K_3Rb_3Zn_4Sn_3Se_{13}$ built on a unique $[Zn_4Sn_3Se_{16}]^{12-}$ cluster: synthesis, crystal structure, ion exchange and optical properties, *Mater. Res. Bull.*, **40**(1), 21–27 (2005).

Wu Z.-X., Chen W.-F., Liu B.-W., Jiang X.-M., Guo G.-C. $K_2CdGe_3S_8$: A new infrared nonlinear optical sulfide, *Symmetry*, **15**(1), 236 (2023).

Xia M., Li R.K. Structural variety in zinc telluro-phosphates: syntheses, crystal structures and characterizations of $Sr_2Zn_3Te_2P_2O_{14}$, $Pb_2Zn_3Te_2P_2O_{14}$ and $Ba_2Zn_2TeP_2O_{11}$, *Dalton Trans.*, **45**(17), 7492–7499 (2016).

Xiang-Ping G., Watanabe M., Ohkawa M., Hoshino K., Shibata Y., Desong C. Felbertalite and related bismuth sulfosalts from the Funiushan copper skarn deposit, Nanjing, China, *Canad. Mineralog.*, **39**(6), 1641–1652 (2001).

Xiao C., Fard Z.H., Sarma D., Song T.-B., Xu C., Kanatzidis M.G. Highly efficient separation of trivalent minor actinides by a layered metal sulfide ($KInSn_2S_6$) from acidic radioactive waste, *J. Am. Chem. Soc.*, **139**(46), 16494–16497 (2017).

Xie H., Wang F., Liao B., Liao X., Chen J., Yu Y., Hou S., Fan X. A novel double anion layered photocatalyst $Pb_4(BO_3)_2SO_4$ with enhanced photocatalytic performance for antibiotic degradation, *Chem. Eng. J.*, **429**, 132344 (2022a).

Xie W., Yun Y., Deng L., Li G., Pan S. Second-harmonic generation-positive $Na_2Ga_2SiS_6$ with a broad band gap and a high laser damage threshold, *Inorg. Chem.*, **61**(19), 7546–7552 (2022b).

Xing W., Fang P., Wang N., Li Z., Lin Z., Yao J., Yin W., Kang B. Two mixed-anion units of $[GeOSe_3]$ and $[GeO_3S]$ originating from partial isovalent anion substitution and inducing moderate second harmonic generation response and large birefringence, *Inorg. Chem.*, **59**(22), 16716–16724 (2020).

Xiong S., Liu Z., Yang L., Ma Y., Xu W., Bai J., Chen H. Anion and cation co-doping of Na_4SnS_4 as sodium superionic conductors, *Mater. Today Phys.*, **15**, 100281 (2020).

Xiong W.-W., Miao J., Li P.-Z., Zhao Y., Liu B., Zhang Q. $\{[M(NH_3)_6][Ag_4M_4Sn_3Se_{13}]\}_\infty$ (*M* = Zn, Mn): Three-dimensional chalcogenide frameworks constructed from quaternary metal selenide clusters with two different transition metals, *J. Solid State Chem.*, **218**, 146–150 (2014).

Yadav S., Jana S., Panigrahi G., Malladi S.K., Niranjan M.K., Prakash J. Five coordinated Mn in $Ba_4Mn_2Si_2Te_9$: synthesis, crystal structure, physical properties, and electronic structure, *Dalton Trans.*, **51**(24), 9265–9277 (2022).

Yadav S., Panigrahi G., Niranjan M.K., Prakash J. Ba_3GeTeS_4: A new quaternary heteroanionic chalcogenide semiconductor, *J. Solid State Chem.*, **323**, 124028 (2023).

Yahia H.B., Motohashi K., Mori S., Sakuda A., Hayashi A. Synthesis, structure and properties of Na_4GeS_4, *J. Alloys Compd.*, **960**, 170600 (2023).

Yakovenchuk V.N., Pakhomovsky Y.A., Men'shikov Y.P., Mikhailova J.A., Ivanyuk G.Yu. Krivovichevite, $Pb_3[Al(OH)_6](SO_4)(OH)$, a new mineral species from the Lovozero alkaline massif, Kola Peninsula, Russia, *Canad. Mineralog.*, **45**(3), 451–456 (2007).

Yan H., Kuwabara A., Smith M.D., Yamaura K., Tsujimoto Y., zur Loye H.-C. Flux crystal growth, structure, and optical properties of the new germanium oxysulfide $La_4(GeS_2O_2)_3$, *Cryst. Growth. Des.*, **20**(6), 4054–4061 (2020).

Yan H., Matsushita Y., Yamaura K., Tsujomoto Y. $La_3Ga_3Ge_2S_3O_{10}$: An ultraviolet nonlinear optical oxysulfide designed by anion-directed band gap engineering, *Angew. Chem. Int. Ed.*, **60**(51), 26561–26565 (2021).

Yang H., Andrade M.B., Downs R.T., Gibbs R.B., Jenkins R.A. Raygrantite, IMA 2013-001. CNMNC Newsletter No. 16, August 2013, page 2700, *Mineralog. Mag.*, **77**(6), 2695–2709 (2013a).

Yang H., Andrade M.B., Downs R.T., Gibbs R.B., Jenkins R.A. Raygrantite, $Pb_{10}Zn(SO_4)_6(SiO_4)_2(OH)_2$, a new mineral isostructural with iranite, from the Big Horn Mountains, Maricopa County, Arizona, USA, *Canad. Mineralog.*, **54**(3), 625–634 (2016).

Yang H., Downs R.T., Evans S.H., Feinglos M.N., Tait K.T. Crystal structure of uchucchacuaite, $AgMnPb_3Sb_5S_{12}$, and its relationship with ramdohrite and fizélyite, *Am. Mineralog.*, **96**(7), 1186–1189 (2011a).

Yang H., Downs R.T., Evans S.H., Pinch W.W. Terrywallaceite, $AgPb(Sb,Bi)_3S_6$, isotypic with gustavite, a new mineral from Mina Herminia, Julcani Mining District, Huancavelica, Peru, *Am. Mineralog.*, **98**(7), 1310–1314 (2013b).

Yang H., Ran M.-Y., Zhou S.-H., Wu X.-T., Lin H., Zhu Q.-L. Rational design *via* dual-site aliovalent substitution leads to an outstanding IR nonlinear optical material with well-balanced comprehensive properties, *Chem. Sci.*, **13**(36), 10725–10733 (2022a).

Yang H., Zhou S.-H., Ran M.-Y., Wu X.-T., Lin H., Zhu Q.-L. Melilite oxychalcogenide $Sr_2FeGe_2OS_6$: a phase-matching IR nonlinear optical material realized by isomorphous substitution, *Inorg. Chem. Front.*, **10**(7), 2030–2038 (2023a).

Yang H., Zhou S.-H., Ran M.-Y., Wu X.-T., Lin H., Zhu Q.-L. Oxychalcogenides as promising ultraviolet nonlinear optical candidates: experimental and theoretical studies of $AEGeOS_2$ (AE = Sr and Ba), *Inorg. Chem.*, **61**(39), 15711–15720 (2022b).

Yang H., Gibbs R.B., Sousa F.X., Downs R.T. Evanichite, IMA 2022-033. CNMNC Newsletter No 68, *Eur. J. Mineral.*, **34**(5), 390 (2022c).

Yang H., Gibbs R.B., Sousa F.X., Downs R.T. Evanichite, IMA 2022-033. CNMNC Newsletter No 68, *Mineralog. Mag.*, **86**(5), 858 (2022d).

Yang H., Gibbs R.B., Sousa F.X., Downs R.T. Evanichite, $Pb_6Cr^{3+}(Cr^{6+}O_4)_2(SO_4)(OH)_7FCl$, from Tiger, Arizona, USA, the first mineral containing both Cr^{3+} and Cr^{6+}, *Can. J. Mineral. Petrol.,* 61(2), 419–429 (2023b).

Yang H., Gu X., Gibbs R.B., Downs R.T. Murphyite, $Pb(TeO_4)$, the Te-analogue of raspite, a new mineral from tombstone, Arizona, USA, *Can. J. Mineral. Petrol.*, **61**(2), 401–409 (2023c).

Yang H., Gu X., Gibbs R.B., Downs R.T. Wangkuirenite, IMA 2023-030. CNMNC Newsletter No 74, *Eur. J. Mineral.*, **35**(4), 663–664 (2023d).

Yang H., Gu X., Gibbs R.B., Downs R.T. Wangkuirenite, IMA 2023-030. CNMNC Newsletter No 74, *Mineralog. Mag.*, **87**(5), 786–787 (2023e).

Yang H., Gu X., Gibbs R.B., Scott M.M. Murphyite, IMA 2021-107. CNMNC Newsletter No. 66, *Eur. J. Mineral.*, **34**(2), 256 (2022e).

Yang H., Gu X., Gibbs R.B., Scott M.M. Murphyite, IMA 2021-107. CNMNC Newsletter No. 66, *Mineralog. Mag.*, **86**(2), 361 (2022f).

Yang H., Gu X., Jenkins R.A., Gibbs R.B., McGlasson J.A., Scott, M.M. Petermegawite, IMA 2021-079. CNMNC Newsletter No 64, *Eur. J. Mineral.*, **34**(1), 5 (2022g).

Yang H., Gu X., Jenkins R.A., Gibbs R.B., McGlasson J.A., Scott, M.M. Petermegawite, IMA 2021-079. CNMNC Newsletter No 64, *Mineralog. Mag.*, **86**(1), 181 (2022h).

Yang H., Gu X., McGlasson J.A., Gibbs R.B. Guangyuanite, IMA 2022-124. CNMNC Newsletter No 72, *Eur. J. Mineral.*, **35**(2), 289 (2023f).

Yang H., Gu X., McGlasson J.A., Gibbs R.B. Guangyuanite, IMA 2022-124. CNMNC Newsletter No 72, *Mineralog. Mag.*, **87**(3), 515 (2023g).

Yang H., Gu X., Scott M.M., Jenkins R.A., Gibbs R.B., McGlasson J.A., Downs R.T. Petermegawite, $Al_6(Se^{4+}O_3)_3[SiO_3(OH)](OH)_9 \cdot 10H_2O$, a new Al-bearing selenite mineral, from the El Dragón Mine, Potosí, Bolivia, *Can. J. Mineral. Petrol.*, **61**(5), 987–998 (2023h).

Yang H., McGlasson J.A., Gibbs R.B., Downs R.T. Franksousaite, IMA 2021-096. CNMNC Newsletter No 65, *Eur. J. Mineral.*, **34**(1), 146–147 (2022i).

Yang H., McGlasson J.A., Gibbs R.B., Downs R.T. Franksousaite, IMA 2021-096. CNMNC Newsletter No 65, *Mineralog. Mag.*, **86**(2), 357 (2022j).

Yang H., McGlasson J.A., Gibbs R.B., Downs R.T. Franksousaite, $PbCu(SeO_4)(OH)_2$, the Se^{6+} analogue of linarite, a new mineral from the El Dragón mine, Potosí, Bolivia. *Mineralog. Mag.*, **86**(5), 792–798 (2022k).

Yang H., Pinch W.W., Downs R.T., Evans S.H. Terrywallaceite, IMA 2011-017. CNMNC Newsletter No. 10, October 2011, page 2552, *Mineralog. Mag.*, **75**(5), 2549–2561 (2011b).

Yang R., Wang J., Wu F., Wei Q., Xue M. First-principles calculations to investigate structural, electronic and optical properties of *Cmc*2_1-Ge_2As_2X (X = S, Se, Te and Po) under pressure effect, *J. Phys. Chem. Solids*, **176**(12), 111231 (2023i).

Yang Y., Chu Y., Zhang B., Wu K., Pan S. Unique unilateral-chelated mode-induced d–p–π interaction enhances second-harmonic generation response in new Ln_3LiMS_7 family, *Chem. Mater.*, **33**(11), 4225–4230 (2021).

Yang Y., Guo Y., Zhang B., Wang T., Chen Y.-G., Hao X., Yu X., Zhang X.-M. Lead tellurite crystals $BaPbTe_2O_6$ and $PbVTeO_5F$ with large nonlinear-/linear-optical responses due to active lone pairs and distorted octahedral, *Inorg.Chem.*, **61**(3), 1538–1545 (2022l).

Yang Y., Ibers J.A. Accidental silicon-containing compounds: crystal structures of $La_3Al_{0.44}Si_{0.93}S_7$, $BaSm_4(SiO_4)_3Se$, and monoclinic and orthorhombic $Ln_2(SiO_4)Te$ (Ln = Nd and Sm), *J. Solid State Chem.*, **155**(2), 433–440 (2000).

Yang Y., Song M., Zhang J., Gao L., Wu X., Wu K. Coordinated regulation on critical physiochemical performances activated from mixed tetrahedral anionic ligands in new series of $Sr_6A_4M_4S_{16}$ (A = Ag, Cu; M = Ge, Sn) nonlinear optical materials, *Dalton Trans.*, **49**(11), 3388–3392 (2020a).

Yang Y., Wu K., Zhang B., Wu X., Lee M.-H. One-dimensional double chains in sodium-based quaternary chalcogenides displaying intriguing red emission and large optical anisotropy, *Inorg. Chem.*, 59(4), 2519–2526 (2020b).

Yanulova M.K., Baigulov E.M., Zaitseva R.P. Wenkite in ores from the Karagayla deposit [in Russian], *Zap. Vses. Mineralog. Obshch.*, **100**(4), 492–498 (1971).

Yao H.-G., Zhang R.-C., Ji S.-H., Ji M., An Y.-L., Ning G.-L. Mineralizer effect on the synthesis of two types of one-dimensional chains silver-selenogermanate and selenostannate, *Inorg. Chem. Commun.*, **13**(11), 1296–1298 (2010).

Ye Z., Bardelli S., Wu K., Sarkar A., Swindle A., Wang J. Synthesis, crystal growth, electronic properties and optical properties of $Y_6IV_{2.5}S_{14}$ (IV = Si, Ge) *Z. anorg. und. allg. Chem.*, **648**(2), e202100271 (2021).

Ye R., Cheng X., Liu B.-W., Jiang X.-M., Yang L.-Q., Deng S., Guo G.-C. Strong nonlinear optical effect attained by atom-response-theory aided design in the $Na_2M^{II}M^{IV}{}_2Q_6$ (M^{II} = Zn, Cd; M^{IV} = Ge, Sn; Q = S, Se) chalcogenide system, *J. Mater. Chem. C*, **8**(4), 1244–1247 (2020).

You F., Liang F., Huang Q., Hu Z., Wu Y., Lin Z. $Pb_2GaF_2(SeO_3)_2Cl$: band engineering strategy by aliovalent substitution for enlarging bandgap while keeping strong second harmonic generation response, *J. Am. Chem. Soc.*, **141**(2), 748–752 (2019).

Yu H., Young J., Wu H., Zhang W., Rondinelli J.M., Halasyamani P.S. Electronic, crystal chemistry, and nonlinear optical property relationships in the dugganite $A_3B_3CD_2O_{14}$ family (A = Sr, Ba or Pb; B = Mg or Zn; C = Te or W, and D = P or V), *J. Am. Chem. Soc.*, **138**(14), 4984–4989 (2016a).

Yu H., Zhang W., Young J., Rondinelli J.M., Halasyamani P.S. Bidenticity-enhanced second harmonic generation from Pb chelation in $Pb_3Mg_3TeP_2O_{14}$, *J. Am. Chem. Soc.*, **138**(1), 88–91 (2016b).

Yu Z., Shang S.-L., Gao Y., Wang D., Li X., Liu Z.-K., Wang D. A quaternary sodium superionic conductor – $Na_{10.8}Sn_{1.9}PS_{11.8}$, *Nano Energy*, **47**, 325–330 (2018).

Yuan D., Jin S., Liu N., Shen S., Lin Z., Li K., Chen X. Tuning magnetic properties in quasi-two-dimensional ferromagnetic $Fe_{3-y}Ge_{1-x}As_xTe_2$ ($0 \leq x \leq 0.85$), *Mater. Res. Express*, **4**(3), 036103 (2017).

Yuan F.-Y., Huang Y.-Z., Zhang H., Lin C.-S., Chai G., Cheng W.-D. $Ba_4GeSb_2Se_{11}$: An infrared nonlinear optical crystal with a V-shaped $Se_3{}^{2-}$ group possessing a large contribution to the SHG response, *Inorg. Chem.*, **60**(20), 15593–15598 (2021a).

Yuan F.-Y., Huang Y.-Z., Zhang H., Zhou A.-Y., Cheng W.-D., Lin C.-S. Synthesis and characterization of a new quaternary selenide $Ba_4Sn_3GeSe_9$ containing $[SnGeSe_5]^{4-}$ and $[Sn_2Se_4]^{4-}$ units, *Chin. J. Struct. Chem.*, **40**(7), 962–966 (2021b).

Yuan M., Dirmyer M., Badding J., Sen A., Dahlberg M., Schiffer P. Controlled assembly of zero-, one-, two-, and three-dimensional metal chalcogenide structures, *Inorg. Chem.*, **46**(18), 7238–7240 (2007).

Yuan M., Mitzi D.B. Solvent properties of hydrazine in the preparation of metal chalcogenide bulk materials and films, *Dalton Trans.*, (31), 6078–6088 (2009).

Yudovskaya M.A., Trubkin N.V., Koporulina E.V., Belakovsky D.I., Mokhov A.V., Kuznetsova M.V., Golovanova T.I. Abramovite, $Pb_2SnInBiS_7$, a new mineral species from fumaroles of the Kudryavy volcano, Kurile Islands, Russia, *Geol. Ore Deposits*, **50**(7), 551–555 (2008).

Yusibov Yu.A., Aliyeva Z.M., Babanly M.B. Thermodynamic properties of the Cu_2GeSe_3, *Azerb. Chem. J.*, (1), 108–114 (2023).

Zachariasen W.H. X-ray examination of colusite, $(Cu,Fe,Mo,Sn)_4(S,As,Te)_{3\text{-}4}$, *Am. Mineralog.*, **18**(12), 534–537 (1933).

Zakrzewski M.A. Jaskólskiite, a new Pb–Cu–Sb–Bi sulfosalt from the Vena deposit, Sweden, *Canad. Mineralog.*, **22**(3), 481–485 (1984).

Zakrzewski M.A., Makovicky E. Izoklakeite from Vena, Sweden, and kobellite homologous series, *Canad. Mineralog.*, **24**(1), 7–18 (1986).

Zambonini F. Sur la palmiérite du Vésuve et les minéraux qui l'accompagnent, *C. r. Acad. sci.*, **172**, 1419–1422 (1921).

Zelenski M., Balić-Žunić T., Bindi L., Garavelli A., Makovicky E., Pinto D., Vurro F. First occurrence of iodine in natural sulfosalts: The case of mutnovskite, $Pb_2AsS_3(I,Cl,Br)$, a new mineral from the Mutnovsky volcano, Kamchatka Peninsula, Russian Federation, *Am. Mineralog.*, **91**(1), 21–28 (2006).

Zelenski M., Garavelli A., Pinto D., Vurro F., Moëlo Y., Bindi L., Makovicky E., Bonaccorsi E. Tazieffite, $Pb_{20}Cd_2(As,Bi)_{22}S_{50}Cl_{10}$, a new chloro-sulfosalt from Mutnovsky volcano, Kamchatka Peninsula, Russian Federation, *Am. Mineralog.*, **94**(10), 1312–1324 (2009).

Zemskov V.S., Shelimova L.E., Konstantinov P.P., Avilov E.S., Kretova M.A., Nikhezina I.Y. Thermoelectric materials with low heat conductivity based on $PbSe–Bi_2Se_3$ compounds. *Inorg. Mater.: Appl. Res.*, **2**(5), 405–413 (2011).

Zeng H.-Y., Zheng F.-K., Guo G.-C., Huang J.-S. Syntheses and single-crystal structures of La_3AgSnS_7, $Ln_3M_xMS_7$ (Ln = La, Ho, Er; M=Ge, Sn; $1/4 \leq x \leq 1/2$), *J. Alloys Compd.*, **458**(1-2), 123–129 (2008).

Zhang B., Feng M.-L., Li J., Hu Q., Qi X.-H., Huang X.-Y., Syntheses, crystal structures, and optical and photocatalytic properties of four small-amine-molecule-directed M–Sn–Q (M = Zn, Ag; Q = S, Se) compounds, *Cryst. Growth Des.*, **17**(3) 1235–1244 (2017a).

Zhang D., Li G., Peng Y., Li L. Synthesis of a ternary thiostannate with 3D channel decorated by hydronium for high proton conductivity. *Inorg. Chem.*, **56**(1), 208–212 (2017b).

Zhang G., Li P., Ding J., Liu Y., Xiong W.-W., Nie L., Wu T., Zhao Y., Tok A.I.Y., Zhang Q. Surfactant-thermal syntheses, structures, and magnetic properties of Mn–Ge–sulfides/selenides, *Inorg. Chem.*, **53**(19), 10248–10256 (2014).

Zhang G.-Q., Yao Z.-Y., Zhang J., Luo H.-B., Kong Y.-R., Zou Y., Tian Z.-F., Ren X.-M. Open-framework chalcogenide $(H_3O)KCu_6Ge_2S_8{\cdot}nH_2O$ exhibiting high mixed proton-electron conduction, *J. Phys. Chem. C*, **125**(13), 7034-7043 (2021).

Zhang L., Mei D., Wu Y., Shen C., Hu W., Zhang L., Li J., Wu Y., He X. Syntheses, structures, optical properties, and electronic structures of $Ba_6Cu_2GSn_4S_{16}$ (G = Fe, Ni) and $Sr_6D_2FeSn_4S_{16}$ (D = Cu, Ag), *J. Solid State Chem.*, 272 69–77 (2019a).

Zhang N., Xu Q.-T., Shi Z.-H., Yang M., Guo S.-P. Characterizations and nonlinear-optical properties of pentanary transition-metal oxysulfide $Sr_2CoGe_2OS_6$, *Inorg. Chem.*, **61**(43), 17002–17006 (2022).

Zhang R.-C., Zhang J.-C., Cao Z., Wang J.-J., Liang S.-S., Cong H.-J., Wang H.-J., Zhang D.-J., An Y.-L. Unusual flexibility of microporous sulfides during ion exchange, *Inorg. Chem.*, **57**(21), 13128–13136 (2018a).

Zhang S., Liang F., Gong P., Yang Y., Lin Z. $Na_4CdGe_2S_7$: A sodium-rich quaternary wide-band-gap chalcogenide with two-dimensional $[Ge_2CdS_7]_\infty$ layers, *Inorg. Chem.*, **59**(22), 16132–16136 (2020).

Zhang S.-Y., Hu C.-L., Li P.-X., Jiang H.-L. Mao J.-G. Syntheses, crystal structures and properties of new lead(II) or bismuth(III) selenites and tellurite, *Dalton Trans.*, **41**(31), 9532–9542 (2012a).

Zhang S.-Y., Hu C.-L., Mao J.-G. New mixed metal selenites and tellurites containing Pd^{2+} ions in a square planar geometry, *Dalton Trans.*, **41**(7), 2011–2017 (2012b).

Zhang X., Wang Q., Ma Z., He J., Wang Z., Zheng C., Lin J., Huang F. Synthesis, structure, multiband optical, and electrical conductive properties of a 3D open cubic framework based on $[Cu_8Sn_6S_{24}]^{z-}$ clusters, *Inorg. Chem.*, **54**(11), 5301–5308 (2015).

Zhang X., Xiao Y., Wang R., Fu P., Zheng C., Huang F. Synthesis, crystal structures and optical properties of noncentrosymmetric oxysulfides $AeGeS_2O$ (Ae = Sr, Ba), *Dalton Trans.*, **48**(39), 14662–14668 (2019b).

Zhang Z., Ramos E., Lalère F., Assoud A., Kaup K., Hartman P., Nazar L.F. $Na_{11}Sn_2PS_{12}$: a new solid state sodium superionic conductor, *Energy Environ. Sci.*, **11**(1), 87–93 (2018b).

Zhao C., Zhang B., Tian X., Zhou G., Xu J., Wu K. $Na_6Sn_3P_4S_{16}$: Sn(II)-chelated PS_4 groups inspired an ultrastrong SHG response, *Inorg. Chem. Front.*, **10**(19), 5726–5733 (2023).

Zhao Z.-Y., Liu Q.-L., Zhao X. DFT calculations study of structural, electronic, and optical properties of $Cu_2ZnSn(S_{1-x}Se_x)_4$ alloys, *J. Alloys Compd.*, **618**, 248–253 (2015).

Zhou J., Fan Z., Zhang K., Yang Z., Pan S., Li J. $Rb_2CdSi_4S_{10}$: novel $[Si_4S_{10}]$ T2-supertetrahedra-contained infrared nonlinear optical material with large band gap, *Mater. Horiz.*, **10**(2), 619–624 (2023).

Zhou M., Ruan B., Dong Q., Yang Q., Gu Y., Chen L., Yi J., Shi Y., Chen G., Ren Z. A low-temperature face-centered-cubic polymorph of Cu_6GeWSe_8 with a semiconductor-like conductivity, *J. Alloys Compd.*, **968**, 171903 (2023).

Zhou W., Geng M., Yan M., Suen N.-T., Liu W., Guo S.-P. Alkali metal partial substitution-induced improved second-harmonic generation and enhanced laser-induced damage threshold for Ag-based sulfides, *Inorg. Chem. Front.*, **9**(15), 3779–3787 (2022a).

Zhou W., Liu W., Guo S.-P. $(Na_{0.74}Ag_{1.26})BaSnS_4$: A new $AgGaS_2$-type nonlinear optical sulfide with a wide band gap and high laser induced damage threshold, *Chem. Eur. J.*, **28**(61), e202202063 (2022b).

Zhou W., Shi Z.-H., Liu W., Guo S.-P. Noncentrosymmetric chalcohalide $K_2Ba_3Ge_3S_9Cl_2$: A new nonlinear optical material with remarkable laser-induced damage threshold, *J. Alloys Compd.*, **895**, Pt. 2, 162602 (2022c).

Zhou W., Yao W.-D., Zhang Q., Xue H., Guo S.-P. Introduction of Li into Ag-based noncentrosymmetric sulfides for high-performance infrared nonlinear optical materials, *Inorg. Chem.*, **60**(7), 5198–5205 (2021a).

Zhou W., Zhang Q., Yao W.-D., Xue H., Guo S.-P. Stepwise Li substitution induced structure evolution and improved nonlinear optical performance for diamond-like sulfides, *Inorg. Chem.*, **60**(16), 12536–12544 (2021b).

Zidarov N., Shivachev B., Kalvachev Y., Stanchev H. Caledonite from Madan ore district, Central Rhodopes, South Bulgaria, *C. r. Acad. Bulg. sci.*, **60**(12), 1311–1316 (2007).

Zimmermann C., Anson C.E., Weigend F., Clérac R., Dehnen S. Unusual syntheses, structures, and electronic properties of compounds containing ternary, T3-type supertetrahedral M/Sn/S anions $[M_5Sn(\mu_3\text{-}S)_4(SnS_4)_4]^{10-}$ (M = Zn, Co), *Inorg. Chem.*, **44**(16), 5686–5695 (2005).

Zobac O., Richter K.W., Kroupa A. Experimental phase diagram of the Ag–Se–Sn system at 250, 400 and 550°C, *J. Phase Equilib. Diffus.*, **43**(1), 32–42 (2022a).

Zobac O., Zemanova A., Chen S.-W., Kroupa A. CALPHAD-type assessment of the Pb–Se–Sn system, *J. Phase Equilib. Diffus.*, **43**(2), 243–255 (2022b).

Zobac O., Zizka R., Roupcová P., Kroupa A. Experimental study of the Ni–Se–Sn phase diagram isothermal sections at 800 K, 1000 K and 1100 K, *J. Phase Equilib. Diffus.*, **44**(4), 594–605 (2023).

Zubkova N.V., Pushcharovsky D.Yu, Giester G., Tillmanns E., Pekov I.V., Kleimenov D.A. The crystal structure of arsentsumebite, $Pb_2Cu[(As,S)O_4]_2(OH)$, *Mineral. Petrol.*, **75**(1-2), 79–88 (2002).

Index

www.ingramcontent.com/pod-product-compliance
Lightning Source LLC
LaVergne TN
LVHW081255100826
845148LV00005B/888

* 9 7 8 0 3 6 7 6 4 2 1 7 4 *